D. Demus, J. Goodby, G. W. Gray,
H.-W. Spiess, V. Vill

Physical Properties
of Liquid Crystals

Related Reading from WILEY-VCH

D. Demus, J. Goodby, G. W. Gray,
H.-W. Spiess, V. Vill (Eds.)

Handbook of Liquid Crystals
ISBN 3-527-29502-X (4-Volume Set)

J. L. Serrano (Ed.)
Metallomesogens
ISBN 3-527-29296-9

D. Demus, J. Goodby, G. W. Gray,
H.-W. Spiess, V. Vill

Physical Properties
of Liquid Crystals

 WILEY-VCH

Weinheim · New York · Chichester
Brisbane · Singapore · Toronto

Prof. Dietrich Demus
Veilchenweg 23
06118 Halle
Germany

Prof. John W. Goodby
School of Chemistry
University of Hull
Hull, HU6 7RX
U.K.

Prof. George W. Gray
Merck Ltd.
Liquid Crystals
Merck House
Poole BH15 1TD
U.K.

Prof. Hans-Wolfgang Spiess
Max-Planck-Institut für
Polymerforschung
Ackermannweg 10
55128 Mainz
Germany

Dr. Volkmar Vill
Institut für Organische Chemie
Universität Hamburg
Martin-Luther-King-Platz 6
20146 Hamburg
Germany

This book was carefully produced. Nevertheless, authors, editors and publisher do not warrant the information contained therein to be free of errors. Readers are advised to keep in mind that statements, data, illustrations, procedural details or other items may inadvertently be inaccurate.

Library of Congress Card No. applied for.

A catalogue record for this book is available from the British Library.

Deutsche Bibliothek Cataloguing-in-Publication Data:

Physical properties of liquid crystals / ed. by Dietrich Demus ... – 1.
Aufl. – Weinheim ; New York ; Chichester ; Brisbane ; Singapore ;
Toronto : Wiley-VCH, 1999
   ISBN 3-527-29747-2

The Editors

D. Demus
studied chemistry at the Martin-Luther-University, Halle, Germany, where he was also awarded his Ph. D. In 1981 he became Professor, and in 1991 Deputy Vice-Chancellor of Halle University. From 1992–1994 he worked as a Special Technical Advisor for the Chisso Petrochemical Corporation in Japan. Throughout the period 1984–1991 he was a member of the International Planning and Steering Committee of the International Liquid Crystal Conferences, and was a non-executive director of the International Liquid Crystal Society. Since 1994 he is active as a Scientific Consultant in Halle. He has published over 310 scientific papers and 7 books and he holds 170 patents.

J. W. Goodby
studied for his Ph. D. in chemistry under the guidance of G. W. Gray at the University of Hull, UK. After his post-doctoral research he became supervisor of the Liquid Crystal Device Materials Research Group at AT&T Bell Laboratories. In 1988 he returned to the UK to become the Thorn-EMI/STC Reader in Industrial Chemistry and in 1990 he was appointed Professor of Organic Chemistry and Head of the Liquid Crystal Group at the University of Hull. In 1996 he was the first winner of the G. W. Gray Medal of the British Liquid Crystal Society.

G. W. Gray
studied chemistry at the University of Glasgow, UK, and received his Ph. D. from the University of London before moving to the University of Hull. His contributions have been recognised by many awards and distinctions, including the Leverhulme Gold Medal of the Royal Society (1987), Commander of the Most Excellent Order of the British Empire (1991), and Gold Medallist and Kyoto Prize Laureate in Advanced Technology (1995). His work on structure/property relationships has had far reaching influences on the understanding of liquid crystals and on their commercial applications in the field of electro-optical displays. In 1990 he became Research Coordinator for Merck (UK) Ltd, the company which, as BDH Ltd, did so much to commercialise and market the electro-optic materials which he invented at Hull University. He is now active as a Consultant, as Editor of the journal "Liquid Crystals" and as author/editor for a number of texts on Liquid Crystals.

H. W. Spiess
studied chemistry at the University of Frankfurt/Main, Germany, and obtained his Ph. D. in physical chemistry for work on transition metal complexes in the group of H. Hartmann. After professorships at the University of Mainz, Münster and Bayreuth he was appointed Director of the newly founded Max-Planck-Institute for Polymer Research in Mainz in 1984. His main research focuses on the structure and dynamics of synthetic polymers and liquid crystalline polymers by advanced NMR and other spectroscopic techniques.

V. Vill
studied chemistry and physics at the University of Münster, Germany, and acquired his Ph. D. in carbohydrate chemistry in the group of J. Thiem in 1990. He is currently appointed at the University of Hamburg, where he focuses his research on the synthesis of chiral liquid crystals from carbohydrates and the phase behavior of glycolipids. He is the founder of the LiqCryst database and author of the Landolt-Börnstein series *Liquid Crystals.*

List of Contributors

Barois, P. (**IV**:6.1)
Centre de Recherche P. Pascal
Ave. A. Schweitzer
33600 Pessac Cedex
France

Blinov, L. M. (**IV**:9)
Institute of Crystallography
Russian Academy of Sciences
Leninsky prosp. 59
117333 Moscow
Russia

Bouligand, Y. (**IV**:7)
Histophysique et Cytophysique
Ecole Pratique des Hautes Etudes
10, rue A. Bocquel
49100 Angers
France

Cladis, P. E. (**IV**:6.3 and **IV**:6.4)
Advanced Liquid Crystal Technologies, Inc.
P.O. Box 1314
Summit, NJ 07902
USA

Dunmur, D. A.; Toriyama, K. (**IV**:1–5)
Department of Chemistry
University of Southampton
Southampton SO17 1BJ
U.K.

Goodby, J. W. (**II**)
School of Chemistry
University of Hull
Hull, HU6 7RX
U.K.

Gray, G. W. (**I** and **II**)
Merck Ltd.
Liquid Crystals
Merck House
Poole BH15 1TD
U.K.

Jérôme, B. (**IV**:10)
FOM-Institute for Atomic and Molecular
Physics
Kruislaan 407
1098 SJ Amsterdam
The Netherlands

Kapustina, O. (**IV**:11)
Shvernik St. 4
117036 Moscow
Russia

Leslie, F. M. (**III**:1)
University of Strathclyde
Dept. of Mathematics
Livingstone Tower
26 Richmond Street
Glasgow G1 1XH
U.K.

Noack, F.[†] (**IV**:13)
IV. Physikalisches Institut
Universität Stuttgart
Pfaffenwaldring 57
70550 Stuttgart
Germany

Osipov, M. (**III**:2)
Institute of Crystallography
Academy of Science of Russia
Leninski pr. 59
Moscow 117333
Russia

Palffy-Muhoray, P. (**IV**:12)
Kent State University
Liquid Crystal Institute
Kent, OH 44242
USA

Pollmann, P. (**IV**:6.2.4)
Universität GH Paderborn
FB 13, Chemie u. Chemietechnik
FG Physikal. Chemie
33095 Paderborn
Germany

Schneider, F.; Kneppe, H. (**IV**:8)
Universität GH Siegen
Physikal. Chemie
57068 Siegen
Germany

Thoen, J. (**IV**:6.2.1)
Katholieke Universiteit Leuven
Departement Natuurkunde
Laboratorium voor Akoestiek en
Thermische Fysica
Celstijnenlaan 200 D
3001 Leuven (Heverlee)
Belgium

Wedler, W. (**IV**:6.2.2 and **IV**:6.2.3)
Tektronix, Inc.; CPID
M/S 63-424
26600 SW Parkway
Wilsonville, OR 97070-1000
USA

Wilson, M. R. (**III**:3)
Department of Chemistry
University of Durham
South Road
Durham, DH1 3LE
U.K.

Contents

Chapter I
Introduction and Historical Development

George W. Gray

1 Introduction

It is with a sense of responsibility that I begin this summary of the historical development of liquid crystals, because one of the two authors of the original *Handbook of Liquid Crystals* of 1980 [1] was Professor Hans Kelker, a friend and a very well informed authority on the history of the subject and the personalities involved in the earlier stages of its emergence. Those who attended the Twelfth International Liquid Crystal Conference in Freiburg in 1988, which marked the centenary of the discovery of liquid crystals, and heard Professor Kelker's plenary lecture – *Some Pictures of the History of Liquid Crystals* [2] – which was part of a conference session devoted to a historical review of the field, will know this. Here he demonstrated that he was in possession of a very wonderful collection of manuscripts and photographs relating to the scientists who, in the latter part of the nineteenth century and the early part of the twentieth century, laid the foundations of our present-day knowledge of liquid crystals.

I am not in that privileged situation, but I have worked in the field for 50 years, beginning my first experiments on aromatic carboxylic acids in October 1947. I have therefore worked through approaching half of the historical span of the subject, including the most recent years during which the subject has expanded and deepened so markedly. I hope this first-hand experience will counterbalance my lack of detailed historical knowledge of the earlier years, as possessed by Professor Kelker. Were he alive today, I hope he would not disapprove of what I write in this chapter.

The history of the development of liquid crystals may be divided into three phases:

1. The period from their discovery in the latter part of the nineteenth century through to about 1925, the years during which the initial scepticism by some that a state of matter was possible in which the properties of anisotropy and fluidity were combined, through to a general acceptance that this was indeed true, and publication of a first classification of liquid crystals into different types.

2. The period from 1925 to about 1960, during which general interest in liquid crystals was at a fairly low level. It was a niche area of academic research, and only relatively few, but very active, scientists were devoted to extending knowledge of liquid crystals. Two world wars and their aftermaths of course contributed greatly to the retardation of this field during this period. Taking the aftermaths of the wars into account, probably at least 15 years were effectively lost to progress during this second phase.

3. The period from 1960 until today is by contrast marked by a very rapid development in activity in the field, triggered of course by the first indications that technological applications could be found for liquid crystals. These early indications

were justified, and led to today's strong electro-optical display industry. The quest for new applications stimulated research and the flow of financial support into the involved areas of chemistry, physics, electrical and electronic engineering, biology, etc. As a marker of this activity, the numbers of papers and patents published in 1968 was about 2000 and this had risen to 6500 in 1995.

2   The Early Years up to About 1925

The question as to when liquid crystals were discovered must now be addressed. In pinpointing a discovery, it is necessary to distinguish simple observations of an unusual phenomenon or effect from observations that develop into an understanding of the meaning and significance of that phenomenon or effect. If we accept that the latter criteria must be met to justify the word discovery, then the credit for the discovery of liquid crystals must go to Friederich Reinitzer, a botanist of the Institute for Plant Physiology of the German University of Prague, who in a paper submitted on May 3, 1888 [3], described his observations of the colored phenomena occurring in melts of cholesteryl acetate and cholesteryl benzoate. In addition, he noted the "double melting" behavior in the case of cholesteryl benzoate, whereby the crystals transformed at 145.5 °C into a cloudy fluid, which suddenly clarified only on heating to 178.5 °C. Subsequent cooling gave similar color effects (but see later) to those observed on cooling the melt of cholesteryl acetate. Today of course we know that the colored phenomena reported by Reinitzer are characteristic of many cholesteric or chiral nematic (N*) liquid crystal phases.

In his article [3], Reinitzer acknowledges that other workers before him had observed curious color behavior in melts of cholesteryl systems. He mentions that Planar in Russia and Raymann in Paris had noted violet colors reflected from cholesteryl chloride and that Lobisch in Germany had observed a bluish-violet flourescence in the case of cholesteryl amine and cholesteryl chloride. Two things distinguish these earlier observations from those of Reinitzer. These are Reinitzer's recording of the "double melting" property of cholesteryl benzoate, and the fact that Reinitzer carried out preliminary studies on thin films of cholesteryl benzoate and noted the range of spectral colors reflected as the temperature decreased until crystallization occurred and the complementary nature of the colored light when the sample was viewed in transmission. Moreover, Reinitzer knew of the excellent work of the German physicist Professor Otto Lehmann, then at the Polytechnical School at Aachen, in designing and developing polarization microscopes, and recognized that Lehmann could advise on the optical behavior of his cholesteryl esters.

The approach to Lehmann was made in March 1888 and the correspondence is excellently documented in Kelker and Knoll's article [2]. This interaction led to agreement that Reinitzer's materials were homogeneous systems of which Lehmann wrote in August 1889: "It is of high interest for the physicist that crystals can exist with a softness, being so considerable that one could call them nearly liquid." This led quickly to the submission by Lehmann, by then at the University of Karlsruhe, of his paper *Über fliessende Kristalle* to the *Zeitschrift für Physikalische Chemie* [4].

Significantly, this uses for the first time the term liquid crystal. As a consequence of the above events and the development of our understanding of liquid crystals which stemmed from them, we must clearly acknowledge Reinitzer as the true discoverer of liquid crystals and the date of the event as March 14, 1888.

It should be noted that the discovery related exclusively to materials we now class as thermotropic liquid crystals, wherein the liquid crystal phases form either on heating crystals or on cooling isotropic liquids, that is, as a consequence of thermal effects. In addition to thermotropic liquid crystals, a second class of fluid anisotropic materials is known, namely, lyotropic liquid crystals where the disruptive effect on the crystal lattice involves a solvent (often water), coupled where necessary with thermal change. Here, the order of the crystal is broken down by the solvent and the molecules form micelles which then arrange themselves in an ordered way, while allowing fluidity. Excess of solvent completes the decrease in order and an isotropic solution is formed. Observations of anisotropy and optical birefringence in such systems were indeed made well before Reinitzer's discovery, but like the observations of Planar, Raymann, and Lobisch, there was no followthrough to a realization of the full significance of what was being seen. These observations were made by Mettenheimer [5], Valentin [6], and Virchow [7] in the period 1834–1861, and involved studies of biological samples derived from nerve tissue, for example, myelin, a complex lipoprotein which can be separated into fractions and which forms a sheath round nerve cells. In water-containing sodium oleate, these sheaths develop what have been called myelinic forms visible microscopically, especially in polarized light, as fluid, birefringent entities. Progress on these anisotropic systems was however impeded by the complexity and lack of reproducibility of the biological systems involved, and whilst predating the studies of Reinitzer and Lehmann are not generally regarded as marking the discovery of liquid crystals.

Following publication of his paper in 1889 [4], Lehmann continued work with liquid crystals and indeed dominated the scene in the late 1800s and the early part of the twentieth century, continuing to publish on liquid crystals until the year of his death in 1922.

Turning to purely synthetic materials, unlike the cholesteryl esters which were of natural origin, examples of liquid crystal behavior were found in these by Lehmann in 1890. The materials were azoxy ethers prepared by Gattermann and Ritschke [8]. The next ten years or so saw studies of p-methoxycinnamic acid and in 1902 the synthesis by Meyer and Dahlem [9] of the first smectogen, ethyl p-azoxybenzoate, although not recognized structurally for what it was at that time. Through studying such materials, Lehmann did however recognize that all liquid crystals are not the same, and indeed in 1907 he examined the first liquid crystal material exhibiting two liquid crystal phases. This material had what was later shown to be a smectic A (SmA) and a cholesteric (N*) phase. Significantly in the context of much later work in the field of applications, he also reported on the aligning effects of surfaces on liquid crystals.

Despite the growing number of compounds shown to exhibit liquid crystal phases (and in a short number of years Vorländer contributed about 250), the acceptance of liquid crystals as a novel state of matter was not universal. Tammann in particular [10] persisted in the view that liquid crystals were colloidal suspensions, and was in bitter argument with Lehmann and Schenk who upheld the view that they were homogeneous systems existing in a new state distinct from the crystalline solid and isotropic liquid states. Nernst [11] too did not subscribe to the latter view and believed that liquid crystals were mixtures of tautomers.

There was however a steadily growing body of evidence supporting the view that liquid crystals represent a true state of matter and acceptance of this slowly grew, aided by the excellent reviews of 1905 by Schenk (*Kristalline Flüssigkeiten und flüssige Kristalle*) [12] and Vorländer (*Kristallinisch-flüssige Substanzen*) [13]. There then followed the important review of optical effects by Stumpf [14] and, much later, an important paper was that by Oseen [15] on a kinetic theory of liquid crystals. The real seal of acceptance of liquid crystals for what they are, i.e., a fascinating and distinct state of matter, was however given in 1922 in the famous publication by G. Friedel [16] in the *Annales de Physique*, entitled *Les États Mesomorphes de la Matière*.

Here, in connection with Friedel's article and on a personal note, I well remember my research supervisor, Professor and later Sir Brynmor Jones, sending me to the library to find the appropriate journal, requiring that I produce a complete translation from French of all 273 pages in order to be "fully familiar with all that had been written". This I dutifully did in the fullness of time, and on taking my translation to show my supervisor, he then reached up to a shelf and withdrew a black notebook saying "now you can compare the quality of your translation with mine!" I learned much from that exercise, as will anyone who repeats it today.

In addition to containing a wealth of information on microscopic techniques and materials, Friedel's article represented in 1922 the first classification of liquid crystals into types, i.e., nematic, smectic and cholesteric. Today, of course, cholesterics are known simply as chiral nematics with no need that they be derived from cholesterol, and we recognize the existence of several polymorphic smectic forms, whereas Friedel allowed for only one (today's smectic A; SmA).

Friedel did however understand the layered nature of smectics, firstly through the stepped edges possessed by smectic droplets with a free surface, and secondly through his detailed studies of the optical microscopic textures of thin films of smectic phases. He understood the optical discontinuities, i.e., the defects, of the smectic focal-conic texture and saw the relationship of the black lines delineating ellipses of different eccentricities and their associated hyperbolae in terms of focal-conic "domains" which may be divided into a series of parallel, curved surfaces known as Dupin cyclides. He also understood that the optically extinct homeotropic textures of smectics of the type he studied gave positive uniaxial interference figures consistent with systems of layers lying flat to the surface. His microscopic studies demonstrated the immense value of the optical microscope as a precise scientific instrument in studies of all types of liquid crystal phases.

Friedel's article, coupled with the publications on synthesis and studies of new liquid crystal materials by organic chemists in Germany, notably Vorländer (see, for example, his monograph *Chemische Kristallographie der Flüssigkeiten* of 1924 [17]), firmly cemented in place all the earlier observations, providing a firm basis on which to build the future structure of the subject.

Before moving on to phase two of the history, we might just return to Reinitzer, the discoverer of liquid crystals, and recognize the quality of his powers of observation, for not only did he focus on the color effects and double melting, but also he noted the blue color appearing in the isotropic melt just before the sample turned into the cloudy cholesteric phase. About this, he said the following: "there appeared (in the clear melt) at a certain point a deep blue colour which spread rapidly through the whole mass and almost as quickly disappeared, again leaving in its place a uniform turbidity. On further cooling, a similar colour effect appeared for the second time to be followed by crystallisation of the mass and a simultaneous disappearance of the colour effect." The turbid state and the second color effect were of course due to the cholesteric phase, but the first transient blue color we now know was associated with the optically isotropic 'blue phases' we are familiar with today. Although Lehmann believed that this transient effect represented a different state, the full significance of Reinitzer's observations had to wait until the 1980s when these isotropic cubic phases became a focus of attention in condensed matter physics.

A further point concerning the first phase of our history of liquid crystals is about nomenclature, a matter about which scientists of today still love to argue. In the early years, however, the debate was sparked by Friedel who strongly objected to Lehmann's term liquid crystal, on the basis that liquid crystals were neither true liquids nor true crystals. The term does of course remain in widespread use today, simply because the juxtaposition of two contradictory terms carries an element of mystery and attraction. Friedel preferred the term mesomorphic to describe the liquid crystal state, and the associated term mesophase, reflecting the intermediate nature of these phases between the crystalline and isotropic liquid states. These terms are again widely used today and coexist happily with the Lehmann terminology. A useful term springing from Friedel's nomenclature is the word mesogen (and also nematogen and smectogen), used to describe a material that is able to produce mesophases. The associated term mesogenic is used by some to indicate that a material does form liquid crystal phases and by others to indicate that a compound is structurally suited to give mesophases, but may not, if, for

example, the melting point of the crystalline state is too high. Then the isotropic liquid is produced directly from the crystal, and, on cooling, crystallization may occur too quickly for even a monotropic liquid crystal phase to form. Yet this compound may show strong tendencies to be mesomorphic if binary phase diagrams of state are examined using a standard material as the second component. My view is that the term mesogenic should be used to describe a structural compatibility with mesophase formation, without the requirement that a phase is actually formed. After all, if the compound does really form a mesophase, the description of it as mesomorphic is perfectly adequate.

Finally, on the subject of nomenclature, Friedel of course gave us today's terms smectic and nematic with their well-known Greek derivations.

3   The Second Phase from 1925 to 1959

In the first part of this period, Vorländer and his group in Halle contributed strongly to the growing number of compounds known to form liquid crystal phases, some showing up to three different mesophases. Based upon his work came the recognition that elongated molecular structures (lath- or rod-like molecules) were particularly suited to mesophase formation. His work also showed that if the major axis of a molecule were long enough, protrusions could be tolerated without sacrifice of the liquid crystal properties. Thus 1,4-disubstituted naphthalenes with a strong extension of the major axis through the 1,4-substituents were liquid crystalline, despite the protruding second ring of the naphthalene core. It is interesting that Vorländer records that the materials behaved as liquid crystallline resins or lacquers (an early thought perhaps about the potential of liquid crystals for applications).

In his book *Nature's Delicate Phase of Matter*, Collings [18] remarks that over 80 doctoral theses stemmed from Vorländer's group in the period 1901–1934. Further evidence of Vorländer's productivity is found in the fact that five of the 24 papers presented at the very important and first ever symposium on liquid crystals held in 1933 under the auspices of the Faraday Society in London, Liquid Crystals and Anisotropic Fluids – A General Discussion [19], were his. Perhaps the most important consequence of Vorländer's studies was that in laying down the foundations of the relationship between molecular structure and liquid crystal properties, attention was focused upon the molecules as the fundamental structural units of the partially ordered phases. Up to then, even Lehmann had been uncertain about the units involved in the ordering and what occurred at the actual transitions.

The Faraday Meeting of 1933 was of great importance in bringing together the small number of active, but often isolated, scientists involved at that time in liquid crystal research. This propagated knowledge and understanding, but, as we shall see, it also generated some dispute.

As early as 1923, de Broglie and E. Friedel (the son of G. Friedel) had shown [20] that X-ray reflections could be obtained from a system of sodium oleate containing water, and that the results were consistent with a lamellar or layered structure. This X-ray work was extended [21] in 1925 to Vorländer's thermotropic ethyl *p*-azoxybenzoate, confirming G. Friedel's conclusions of a layered structure stemming from his microscopic studies of smectic defect structures. Further, in the period 1932–1935, Herrmann [22], who also contributed to the 1933 Faraday Discussion, was deci-

sive in confirming the lamellar nature of smectics by X-ray studies which included Vorländer's material exhibiting more than one smectic phase. The latter work substantiated a change from a statistical order in the layers of one smectic to a hexagonal ordering in the lower temperature phase. A tilted lamellar structure was also found by Herrmann for some thallium soaps [23].

Amongst other names of historical interest featured on the Faraday Discussion program were, for example, Fréedericksz and Zolina (forces causing orientation of an anisotropic liquid), Zocher (magnetic field effects on nematics), Ostwald, Lawrence (lyotropic liquid crystals), Bernal, Sir W. H. Bragg (developing the concept of Dupin cyclides in relation to Friedel's earlier studies of focal-conics), and also Ornstein and Kast who presented new arguments in favor of the swarm theory, which was first put forward in 1907–1909 by Bose [24]. This theory had proposed that the nematic phase consisted of elongated swarms of some 10^6 molecules, and in the 1930s much effort was given to proving the existence of these swarms, which were used to explain some, but not all, of the physical properties of nematics. However, at the 1933 Faraday Meeting, the presentation of Oseen [25] and the strong reservations expressed by Zocher during the discussions were already casting shadows of doubt on the swarm theory. Today of course we accept that definitive proof of the existence of swarms was never obtained, and by 1938 Zocher was expressing further strong reservations about the theory [26], proposing alternatively that the nematic phase is a continuum, such that the molecular orientation changes in a continuous manner throughout the bulk of the mesophase. This was called the distortion hypothesis and together with Oseen's work marked the beginning of the modern continuum theory of liquid crystals. However, developments here had to wait until after the second world war when Frank [27], at a further Faraday discussion in 1958, and consequent upon his re-examination of Oseen's treatment, presented it as a theory of curvature elasticity, to be advanced in the next historical phase by names such as Ericksen, Leslie, de Gennes, and the Orsay Group in France.

The period following the war up until the 1950s is also significant for the work of Chatelain [28], in collaboration with Falgueirettes. Using surface alignment techniques, they measured the refractive indices of different nematics, and Chatelain produced his theoretical treatment of the values of the ordinary and extraordinary indices of refraction of an oriented nematic melt.

Following Vorländer, other chemists were now becoming interested in new liquid crystal materials, and in the early 1940s we find publications on structure/property relations by Weygand and Gabler [29]. Later, in the 1950s, Wiegand [30] in Germany and the author in the UK were also making systematic changes in the structures of mesogens to establish the effects on liquid crystal behavior. The author's work included not only systematic modifications to aromatic core structures, but also studies of many homologous series, establishing clearly that within series systematic changes in transition temperature always occur, within the framework of a limited number of patterns of behavior. In the period 1951–1959, the author published some 20 papers on structure/property relations in liquid crystals. These are rather numerous to reference here, but in the account of the third historical phase from 1960 until today, reference to relevant reviews and books is given.

Lyotropic liquid crystals also progressed during this second phase. Lawrence's paper at the 1933 Faraday meeting discussed the phase diagrams for different compositions of fatty acid salts, recognizing the different phase types involved and the transitions undergone with change of temperature and/or water content. Examples of aromatic materials, including dyes, giving lyotropic phases were also found, and solvents other than water as the lyophase were explored.

The early work of Robinson et al. [31] was also done in this period. This involved solutions of poly-γ-benzyl-L-glutamate in organic solvents. These solutions exhibited the selective light reflecting properties of thermotropic cholesteric liquid crystals.

This period of history also saw the publication of work by Eaborn and Hartshorne [32] on di-isobutylsilandiol, which generated a mesophase. This was a puzzling result at the time, as the molecular shape was inconsistent with views of the time that liquid crystal formation required rod-shaped molecules. Light would be shed on this only after the discovery of liquid crystal phases formed by disc-shaped molecules in the early 1970s.

Finally, it should be noted that in this period, in 1957, a very important review on liquid crystals was published by Brown and Shaw [33]. This did much to focus the attention of other scientists on the subject and certainly contributed to the increase in liquid crystal research, which was to herald the strong developments in the early 1970s.

The period 1925–1959 may be usefully summarized now. Although the level of activity in the field was limited, important developments did occur in relation to:

– the influence of external fields (electric and magnetic) on liquid crystals;
– the orienting influences of surfaces;
– measurements of the anisotropic physical properties of aligned liquid crystals;
– the range of new liquid crystal materials and structure/property relationships;

- the development of theories of the liquid crystal state ranging from the swarm theory to the emerging continuum theory;
- increased awareness of the value of polarizing optical microscopy for the identification of mesophases, the determination of transition temperatures, and reaching a fuller understanding of defect textures.

4 The Third Phase from 1960 to the Present Time

The first ten years of this period saw several important developments which escalated interest and research in liquid crystals. Among these, there was the publication by Maier and Saupe [34] of their papers on a mean field theory of the nematic state, focusing attention on London dispersion forces as the attractive interaction amongst molecules and upon the order parameter. This theory must be regarded as the essential starting point for the advances in theoretical treatments of the liquid crystal state which followed over the years.

There was also much activity in the field of new liquid crystal materials, notably by Demus et al. [35] in Germany and by the author who, in 1962, produced his monograph *Molecular Structure and the Properties of Liquid Crystals* [36], published by Academic Press.

Also, further X-ray studies began to advance knowledge of the structure of liquid crystal phases, particularly smectics. The work of de Vries and Diele should be mentioned, and later on, notably that by Levelut and co-workers in France and Leadbetter in England (see, for example, [37, 38]), work which culminated in the 1980s in a clear structural elucidation and classification of smetic liquid crystals. This distin-

guished the true lamellar smectics with little or no correlation between layers from lamellar systems, previously regarded as smectics, which possess three-dimensional order and are really soft crystals. Today, the true smectics are labeled SmA, for example, and the crystal phases are referred to simply by a letter such as K, or by CrK. The phase once known as SmD and first observed by the author and co-workers in laterally nitro substituted biphenyl carboxylic acids [39] is now recognized [40] as a cubic liquid crystal phase. Several other examples of cubic thermotropic liquid crystal phases are now known [41, 42].

Such studies focused attention on the microscopic textures of liquid crystal phases. The defects characterizing these textures are now well understood through rather beautiful studies by workers such as Kléman [43], and from textures it is now possible to go a long way towards characterizing the phase behavior of new materials. A great deal of work on phase characterization has been done, and two reference sources are important [44, 45]. Through such detailed studies of phase behavior, new phenomena were often recognized and explained, for example, the re-entrance phenomenon through the work of Cladis [46], and the existence of the blue phases (BPI, BPII, and BPIII) through the work of several groups [47]. We should remember of course that Reinitzer did observe blue phases many years earlier and knew, without understanding the situation, that something occurred between the isotropic liquid and the N^* phase on cooling many chiral materials. Reflectance microscopy played a big part in the eventual elucidation of these phases as cubic phases involving double twist cylinders.

The widened interest in liquid crystals exemplified above had its origins in a number of events, such as the publication of the Brown and Shaw review [33] and the author's monograph [36], but a very big part was played by the launch of the now regular International Liquid Crystal Conferences (ILCCs). We owe these now biennial conferences to Glenn Brown, the first one being organised by him in 1965 in Kent State University in Ohio. Attended by some 90 delegates from different countries, this meeting was most important in providing that small group of research workers with an identity and the opportunity to meet and discuss problems and new ideas. Glenn Brown was indeed successful in obtaining funding for the second ILCC, held again in Kent State University, in 1968. The liquid crystal community owes a great deal to Glenn Brown for his vision in making these meetings possible. He created for liquid crystals a community of scientists within which the vigorous development of new research results and the technological innovation of the 1970s was to generate and continue unabated to the present time.

At the 1965 ILCC Meeting, the focus on applications was on Fergason's presentations on thermography using cholesteric liquid crystals, but in 1968, the meeting was attended by a group of researchers from Radio Corporation of America (RCA) in Princeton, where work under Heilmeier, Castellano, Goldmacher, and Williams was being done on display devices based on liquid crystals. The liquid crystal community was having its eyes opened to the potential of liquid crystals for application in electro-optical displays. The seminal work of the RCA group, initially on dynamic scattering displays, cannot be overstressed for its importance to the field. Two years later, at the 1970 ILCC in Berlin, display applications of liquid crystals were being discussed freely, and later when the patents of Schadt and Helfrich and of Fergason on the twisted nematic liquid crystal electro-optical display mode came into the public domain, activity

intensified. At the fourth ILCC in 1972, again at Kent State, display applications dominated the meeting, and two years later at the 1974 ILCC in Stockholm, the author's presentation was on materials for use in *five* different display types, i.e., dynamic scattering, Fréedericksz, twisted nematic, cholesteric memory, and cholesteric-nematic phase change display devices.

Electro-optical liquid crystal display devices were now well established, and the twisted nematic device was obviously the superior one, based as it was upon a field effect in a pure nematic of positive dielectric anisotropy rather than upon the conductivity anisotropy, generated by ionic dopants in nematics of negative dielectric anisotropy, as in dynamic scattering displays.

In 1970, the author and co-workers obtained a research grant from the UK Ministry of Defence for work on room temperature liquid crystal materials that would function well in electro-optic displays. When our attention was directed to liquid crystal materials of positive dielectric anisotropy rather than negative dielectric anisotropy, we were able to make rapid progress, drawing on the store of fundamental knowledge relating molecular structure to liquid crystal properties, and, as a consequence, following patenting, the synthesis and behavior of the 4-alkyl- and 4-alkoxy-4'-cyanobiphenyls, designed for twisted nematic displays, were published in 1973 [48]. The history of the events leading up to the discovery of the cyanobiphenyls has been nicely documented by Hilsum [49], the originator of the research program under the author at the University of Hull and the coordinator of the associated programs on physics/devices at DRA, Malvern, and eventually on commercial production at BDH Ltd (now Merck UK Ltd).

The advent of the cyanobiphenyls made available the materials for the manufacture of high quality, reliable liquid crystal displays, and in so doing provided the secure basis upon which today's burgeoning electro-optical liquid crystal display device industry rests. The availability of these materials also spawned intense interest in other related families of materials, and during the later 1970s, the cyclohexane analogs of the biphenyls [50] and the pyrimidine analogs [51] became available, widening the choice of physical properties available to device engineers.

The market place welcomed the first simple, direct drive twisted nematic liquid crystal (LC) displays, but in so doing, created a demand for devices capable of portraying more complex data, particularly important for the display of Chinese/Japanese characters. Multiplex driven liquid crystal displays with some capability in this direction required the exploration of more complex mixtures incorporating ester components. However, the limitations of multiplex addressing were quickly exposed, encouraging interest in new device forms and addressing techniques. One development was the discovery of the supertwisted nematic display [52, 53] and of addressing methods for twisted nematic displays using thin film transistors. These two possibilities have progressed forwards successfully, each having to overcome its own particular problems. For a survey, see the review by Schadt [54]. As a result, in the late 1980s and early 1990s really excellent full color LC displays for direct view and for projection, involving where necessary high definition resolution, have come to the fore and dominate today's marketplace. The devices have steadily improved in viewing angle, brightness, definition, and color quality, and the most up to date displays (supertwisted nematic and active matrix twisted nematic) are technological products of great quality which lead the liquid crystal display device

industry into the new millennium in a most confident mood.

Work in other display areas has of course occurred. Through the seminal work of R. B. Meyer and the research of Clark and Lagerwall [55] on surface stabilized ferroelectric liquid crystal devices based on chiral smectic C liquid crystal materials, the potential for ferroelectric devices has been fully explored in recent years. With their faster switching capability, they are attractive, and the difficulties over addressing schemes and the manufacture of ferroelectric displays will perhaps soon be overcome to give the marketplace a further liquid crystal device.

In the search for novel materials, particularly new ferroelectric materials, new phase types were also discovered, notably the antiferroelectric phase [56] which, with tristable switching characteristics, also has potential for display use, possibly overcoming some of the difficulties with ferroelectric systems and providing a further display device of high quality.

A further development in the display area concerns liquid crystal devices using in-plane-switching techniques, giving much improved viewing angle of the display [57, 58]. Here the molecules switch across the surface of the display cell, and this technology is now being adopted by three companies. Plasma switching of other types of liquid crystal display is another interesting technology awaiting further development [59].

Going back to the earlier 1970s, with the advent of the cyanobiphenyls and later the cyclohexane and pyrimidine analogs, not only was the display device industry provided with a wealth of novel, useful materials, but also those in fundamental research were given a range of stable, room temperature liquid crystal materials for study by a growing panoply of experimental and theoretical techniques. Fundamental research therefore moved forward rapidly with very good readily available materials and the funding for such work that was released by the potential for technological applications. Publications of research papers and patents escalated, as mentioned earlier, and in the journals *Molecular Crystals and Liquid Crystals* and *Liquid Crystals* the field now has its own dedicated literature shop windows.

The knowledge base in fundamental science was also extended greatly by the discovery in Chandrasekhar's group [60] of liquid crystals formed by disc-shaped molecules. Capable of forming discotic nematic phases and a range of columnar phases, these materials currently attract much interest and technological applications for them are possible.

This period also saw the growth of work on liquid crystal polymers both of the main chain and the side group varieties, and amongst others, the names of Blumstein and Finkelmann are associated with the first advances in this field which attracts many workers today (see [61, 62]). Main chain liquid crystal polymers became known for their ability to form high tensile strength fibers and moldings, and side group liquid crystal polymers also have applications in optical components such as brightness enhancement films for display devices. Anisotropic networks, elastomers, and gels are also intriguing systems that have been reviewed by Kelly [63].

Fundamental studies of liquid crystals have also opened up much interest in metallomesogens, i.e., materials incorporating metal centers and capable of giving liquid crystal phases. Many materials ranging from the calamitic (rod-like) type to the disc-like metallophthalocyanines are now known and with appropriate metal centers have interesting magnetic characteristics. Indeed, such have been the developments in

this area that a text devoted to the subject has been published [64].

If this section of this chapter hardly reads as a chronological history, this is because the last 15 years have seen so many developments in different directions that a simple pattern of evolution does not exist. Instead, developments have occurred in an explosive way, emanating outward from the core of fundamental knowledge acquired up to the end of the 1970s. We have already looked briefly at display applications, liquid crystals from disc-shaped molecules, liquid crystal polymers, and metallomesogens, but to follow all the developments of recent years radiating out from the central core is hardly possible in a short chapter like this.

Only a few topics can be selected for brief mention, and if some areas of development are excluded through shortness of space, the author can at least feel confident that justice is done to them in the four volume *Handbook of Liquid Crystals*.

4.1 Lyotropic Liquid Crystals

Activity here may have been less intense than in the thermotropic liquid crystal area. Important developments began, however, through the work of the Swedish group under Ekwall, Fontell, Lawson, and Flautt, and the 1970s saw the publication by Winsor of his R-theory of fused micellar phases [65] and of Friberg's valuable book *Lyotropic Liquid Crystals* [66].

The importance of lyotropic liquid crystals in the oil industry, the food industry, and the detergent industry is high because of the need to know the exact behavior of amphiphile/water/oil systems and the role played by the micellar phases in the context of efficient extraction of oil from natural rocks and whether a system will flow, the processes of baking and the uses of emulsifiers, and the general efficiency of soaps and detergents. Especial mention should be made of the work of Tiddy and his studies of the complex phase relations involving lamellar, hexagonal, and viscous isotropic (cubic) micellar phases and his extension of the field to include zwitterionic systems (see, for example, [67]), as well as cationic and anionic amphiphiles. The importance of lyotropic liquid crystal concepts in relation to biological systems and the role played therein by liquid crystals must also be noted. A recent review of lyotropic liquid crystals by Hiltrop is to be found in [68].

4.2   Theory

Theoretical treatments of liquid crystals such as nematics have proved a great challenge since the early models by Onsager and the influential theory of Maier and Saupe [34] mentioned before. Many people have worked on the problems involved and on the development of the continuum theory, the statistical mechanical approaches of the mean field theory and the role of repulsive, as well as attractive forces. The contributions of many theoreticians, physical scientists, and mathematicians over the years has been great – notably of de Gennes (for example, the Landau–de Gennes theory of phase transitions), McMillan (the nematic–smectic A transition), Leslie (viscosity coefficients, flow, and elasticity), Cotter (hard rod models), Luckhurst (extensions of the Maier–Saupe theory and the role of flexibility in real molecules), and Chandrasekhar, Madhusudana, and Shashidhar (pretransitional effects and near-neighbor correlations), to mention but some. The devel-

opment of these theories and their significance are fully documented in the second edition of Chandrasekhar's excellent monograph [69] entitled simply *Liquid Crystals*, and supported by earlier reviews in [70].

In many of the above studies, the Gay–Berne potential describing the interactions between anisotropic particles has been used, and this can be separated into repulsive and attractive parts, enabling studies of the roles played by each in mesophase formation. Computer simulation has been used to investigate Gay–Berne fluids, and phase diagrams giving isotropic, nematic, smectic, and crystalline phases have been produced (see, for example, Hashim et al. [71]). Simulations aimed ultimately at the prediction of the phase behavior of compounds of given molecular structure (the molecular dynamics approach), avoiding synthesis, is another area of growth, supported by the increasing power of computers, and results from studies of molecular mechanics, used in the determination of molecular structure and lowest energy conformations, are proving to be increasingly useful (see Chap. III, Sec. 3 of this volume [72]).

4.3   Polymer Dispersed Liquid Crystals (PDLCs) and Anchoring

The behavior of liquid crystals at surfaces is of course of great importance in normal flat panel electro-optical displays, and the subject of anchoring is an area of strong research activity, where the work of Barbero and Durand is particularly noteworthy [73].

In recent years, PDLCs have attracted much attention because sheets consisting of droplets of nematic liquid crystal in an amorphous polymer matrix can be made by a number of techniques. The orientation in the nematic droplets can be manipulated by electric fields, changing the appearance of the sheet from cloudy to clear, and opening up possibilities for electrically switchable windows and panels, and for large area signs and advertising boards. The high voltages initially required for the operation of such systems have been reduced considerably and problems of off-axis haze in the viewing panels have been diminished. The full impact of such switchable panels has not yet been realized, and some of the problems have focused attention on the fact that in such systems the liquid crystal is constrained in a confined geometry, and the ratio of surface contact area to bulk volume is high. This has led to important research on liquid crystals in confined geometries and a text on this subject by Crawford and Zumer was published in 1996 [74]. This valuable book embraces other important aspects of confinement than that in PDLC systems, and includes porous polymer network assemblies in nematic liquid crystals, polymer stabilized cholesterics with their implications for reflective cholesteric displays, liquid crystal gel dispersions, filled nematics, and anisotropic gels.

4.4   Materials and New Phases

The applications of liquid crystals have unquestionably added incentive to the quest for new liquid crystal materials with superior properties such as viscosity, elastic constants, transition temperatures, and stability. In recent years this has catalyzed work on chiral materials as dopants for ferroelectric displays and for antiferroelectric materials with structures avoiding the number of potentially labile ester groups that were present in the original materials in which antiferroelectric properties were discovered.

The quest for new materials, whether driven by their potential for applications or simply by natural scientific curiosity about structure/property relations (i.e., as part of fundamental research programs) has always been and is today a vital part of the liquid crystal scenario. Indeed, it is often the case that the free-thinking fundamental research on new materials opens the door to new applications or improvements in existing device performance. The fascination of the liquid crystal field in fact derives from this continuing materials–knowledge–applications knock-on effect. Importantly, it can occur in both directions.

Thus studies of the significance of the position of the location of the double bond in a terminal alkyl chain led to the alkenylbicyclohexane systems [75], which today provide excellent materials for supertwisted nematic devices, and the work on ferroelectric materials led to the discovery of the antiferroelectric phase.

A notable example of phase discovery was that of the twist grain boundary smectic A* phase (TGBA*) by Goodby et al. in 1989 [76]. This new liquid crystal phase is a frustrated smectic in which the opposing tendencies to twist and be lamellar are accommodated through a regular array of dislocations, the possibility of which was predicted by de Gennes [77] in 1972 when he saw the analogous roles played by the director in an SmA phase and the magnetic vector potential in a superconductor. The analogous TGBC* phase is also now known.

In the field of natural products, there is growing interest in the increasing range of carbohydrates that are being found to be liquid crystalline [78–80], swinging interest back to the role of liquid crystals in biological systems, where recent studies of lyotropic mesomorphism in deoxyguanosine cyclic monophosphates [81] and of the significance of hydrogen bonding [82] and charge transfer [83] in low molar mass liquid crystals may be of further relevance.

Studies of the liquid crystal properties of terminally [84] and laterally [85] connected di-, tri- and tetra-mesogens are probing the behavior in the area between monomeric liquid crystals and liquid crystal polymers. The intercalated smectic phase (SmA_{cal}) has been found [84] and the somewhat related organosiloxanes [86] are proving to be most interesting materials with properties intermediate between low molar mass and polymeric liquid crystals, developed apparently through a microsegregation of the siloxane parts of the molecules into layers, leaving the organic moieties as appendages. Materials with both ferroelectric and antiferroelectric properties are provided and, significantly for applications, have almost temperature insensitive tilt angles and polarizations.

Finally, a breakdown of the division between calamitic liquid crystals and the columnar phases formed by disc-shaped molecules has occurred through the discovery of materials that give both types of mesophase in single compounds. For example, some six-ring, double swallow-tailed mesogens reported by Weissflog et al. [87] exhibit the very interesting phase sequence SmC–oblique columnar–SmC–nematic, combining not only calamitic and columnar phases, but also a re-entrant SmC phase.

5 Conclusions

Thus in the above selected areas and in many others that cannot be mentioned specifically here, the development of knowledge and understanding of liquid crystal systems goes on in a manner that is quite prolific.

It remains to be said that much has happened in the short history of liquid crystals and that the field is vigorous in research today, both in the areas of fundamental science and application driven investigations. Saying that the position is healthy today may lead to the question: "What of the future?" Fortunately, the author is acting here in the capacity of historian, and it is not in the historian's job description to predict the future. It will only be said that the future prospects for liquid crystals look healthy, but they will only be maintained so if fundamental research by scientists of imagination is adequately funded to enable the exploration of new ideas, new aspects, and new possibilities, because history does demonstrate that many of the discoveries significant for applications and technology derive from sound basic science or a sound knowledge of established basic science.

6 References

[1] H. Kelker, W. Hatz, *Handbook of Liquid Crystals*, VCH, Deerfield Beach, FL **1980**.
[2] H. Kelker, P. M. Knoll, *Liq. Cryst.* **1989**, *5*, 19.
[3] F. Reinitzer, *Monatsh. Chem.* **1888**, *9*, 421; for an English translation see *Liq. Cryst.* **1989**, *5*, 7.
[4] O. Lehmann, *Z. Phys. Chem.* **1889**, *4*, 462.
[5] C. Mettenheimer, *Corr. Blatt Verein gem. Arb. Förderung wissenschaftl. Heilkunde* **1857**, *24*, 331.
[6] G. Valentin, *Die Untersuchung der Pflanzen- und der Tiergewebe im polarisierten Licht*, Engelmann, Leipzig **1861**.
[7] R. Virchow, *Virchows Arch. path. Anat. Physiol.* **1853**, *6*, 571.
[8] L. Gattermann, A. Ritschke, *Ber. Dtsch. Chem. Ges.* **1890**, *23*, 1738.
[9] F. Meyer, K. Dahlem, *Liebigs Ann.* **1902**, *320*, 122.
[10] G. Tammann, *Ann. Physik.* **1901**, *4*, 524; **1902**, *8*, 103; **1906**, *19*, 421.
[11] W. Nernst, *Z. Elektrochem.* **1906**, *12*, 431.
[12] R. Schenk, *Kristalline Flüssigkeiten und flüssige Kristalle*, Engelmann, Leipzig **1905**.
[13] D. Vorländer, *Kristallinisch-flüssige Substanzen*, Enke, Stuttgart **1905**.

[14] F. Stumpf, *Radioakt. Elektron.* **1918**, *XV*, 1.

[15] C. W. Oseen, *Die anisotropen Flüssigkeiten*, Borntrager, Berlin **1929**.

[16] G. Friedel, *Ann. Physique* **1922**, *18*, 273.

[17] D. Vorländer, *Chemische Kristallographie der Flüssigkeiten*, Akadem. Verlagsanstalt, Leipzig **1924**.

[18] P. J. Collings, *Nature's Delicate Phase of Matter*, Princeton U. P., **1990**.

[19] General Discussion Meeting held by the Faraday Society, April **1933**.

[20] M. de Broglie, E. Friedel, *C. R.* **1923**, *176*, 738.

[21] E. Friedel, *C. R.* **1925**, *180*, 269.

[22] K. Herrmann, A. H. Krummacher, *Z. Kristallogr.* **1932**, *81*, 31; **1935**, *92*, 49.

[23] K. Herrmann, *Trans. Faraday Soc.* **1933**, *29*, 27.

[24] E. Bose, *Phys. Z.* **1907**, *8*, 513; **1908**, *9*, 708; **1909**, *10*, 230; E. Bose, F. Conrat, *Phys. Z.* **1908**, *9*, 916.

[25] C. W. Oseen, *Trans. Faraday Soc.* **1933**, *29*, 883.

[26] H. Zocher, *Trans. Faraday Soc.* **1933**, *29*, 1062; H. Zocher, G. Ungar, *Z. Phys. Chem.* **1938**, *110*, 529; see also H. Zocher in *Liquid Crystals and Plastic Crystals*, Vol. 1 (Eds.: G. W. Gray, P. A. Winsor), Ellis Horwood, Chichester **1974**, Chap. 3.1, p. 64.

[27] F. C. Frank, *Discuss. Faraday Soc.* **1958**, *59*, 958.

[28] P. Chatelain, *C. R.* **1948**, *227*, 136; *Bull. Soc. Franc. Mineral. Crist.* **1954**, *77*, 353; **1955**, *78*, 262; P. Chatelain, O. Pellet, *Bull. Soc. Franc. Mineral. Crist.* **1950**, *73*, 154; J. Falgueirettes, *C. R.* **1952**, *234*, 2619.

[29] C. Weygand, R. Gabler, *J. Prakt. Chem.* **1940**, *155*, 332; C. Weygand, R. Gabler, J. Hoffmann, *Z. Phys. Chem.* **1941**, *50B*, 124; C. Weygand, R. Gabler, N. Bircon, *J. Prakt. Chem.* **1941**, *158*, 26.

[30] C. Wiegand, *Z. Naturforsch.* **1954**, *3b*, 313; **1955**, *9b*, 516.

[31] C. Robinson, *Trans. Faraday Soc.* **1956**, *52*, 571; C. Robinson, J. C. Ward, R. B. Beevers, *Discuss. Faraday Soc.* **1958**, *25*, 29.

[32] C. Eaborn, N. H. Hartshorne, *J. Chem. Soc.* **1955**, 549.

[33] G. H. Brown, W. G. Shaw, *Chem. Rev.* **1957**, *57*, 1097.

[34] W. Maier, A. Saupe, *Z. Naturforsch.* **1958**, *13a*, 564; **1959**, *14a*, 882; **1960**, *15a*, 287.

[35] D. Demus, L. Richter, C. E. Rurup, H. Sackmann, H. Schubert, *J. Phys. (Paris)* **1975**, *36*, C1-349.

[36] G. W. Gray, *Molecular Structure and the Properties of Liquid Crystals*, Academic, London **1962**.

[37] J. Doucet, A. M. Levelut, M. Lambert, *Phys. Rev. Lett.* **1974**, *32*, 301; J. Doucet, A. M. Levelut, M. Lambert, M. Liebert, L. Strzelecki, *J. Phys. (Paris)* **1975**, *36*, C1-13; J. J. Benattar, A. M. Levelut, L. Liebert, F. Moussa, *J. Phys. (Paris)* **1979**, *40*, C3-115.

[38] A. J. Leadbetter in *The Molecular Physics of Liquid Crystals* (Eds.: G. W. Gray, G. R. Luckhurst), Academic, London **1979**, Chap. 13 and references therein and later in *Thermotropic Liquid Crystals* (Ed.: G. W. Gray), Wiley and Sons, New York **1987**, Chap. 1 and references therein.

[39] G. W. Gray, B. Jones, F. Marson, *J. Chem. Soc.* **1957**, 393; D. Demus, G. Kunicke, J. Neelson, H. Sackmann, *Z. Naturforsch.* **1968**, *23a*, 84; S. Diele, P. Brand, H. Sackmann, *Mol. Cryst. Liq. Cryst.* **1972**, *17*, 84 and 163.

[40] G. E. Etherington, A. J. Leadbetter, X. J. Wang, G. W. Gray, A. R. Tajbakhsh, *Liq. Cryst.* **1986**, *1*, 209.

[41] D. Demus, A. Gloza, H. Hartung, I. Rapthel, A. Wiegeleben, *Cryst. Res. Technol.* **1981**, *16*, 1445.

[42] A. M. Levelut, B. Donnio, D. W. Bruce, *Liq. Cryst.* **1997**, *22*, 753.

[43] M. Kléman, *Liquid Crystals and Plastic Crystals*, Vol. 1 (Eds.: G. W. Gray, P. A. Winsor), Ellis Horwood, Chichester **1974**, Chap. 3.3, p. 76.

[44] D. Demus, L. Richter, *Textures of Liquid Crystals*, VEB Deutscher Verlag für Grundstoffindustrie, **1978**.

[45] G. W. Gray, J. W. Goodby, *Smectic Liquid Crystals, Textures and Structures*, Leonard Hill, Glasgow **1984**.

[46] P. E. Cladis, *Phys. Rev. Lett.* **1975**, *35*, 48; P. E. Cladis, R. K. Bogardus, W. B. Daniels, G. N. Taylor, *Phys. Rev. Lett.* **1977**, *39*, 720.

[47] D. Coates, G. W. Gray, *Phys. Lett.* **1973**, *45A*, 115; D. Armitage, F. P. Price, *J. Appl. Phys.* **1976**, *47*, 2735; H. Stegemeyer, K. Bergmann in *Liquid Crystals of One- and Two-Dimensional Order* (Eds.: W. Helfrich, A. Heppke), Springer, Berlin **1980**, p. 161; H. Stegemeyer, T. H. Blumel, K. Hiltrop, H. Onusseit, F. Porsch, *Liq. Cryst.* **1986**, *1*, 3; P. P. Crooker, *Liq. Cryst.* **1989**, *5*, 751.

[48] G. W. Gray, K. J. Harrison, J. A. Nash, *Electron. Lett.* **1973**, *9*, 130; G. W. Gray, K. J. Harrison, J. A. Nash, J. Constant, J. S. Hulme, J. Kirton, E. P. Raynes, *Liquid Crystals and Ordered Fluids*, Vol. 2 (Eds.: R. S. Porter, J. F. Johnson), Plenum, New York **1973**, p. 617.

[49] C. Hilsum in *Technology of Chemicals and Materials for Electronics* (Ed.: E. R. Howells), Ellis Horwood, Chichester **1991**, Chap. 3.

[50] R. Eidenschink, D. Erdmann, J. Krause, L. Pohl, *Angew. Chem., Int. Ed. Engl.* **1977**, *16*, 100.

[51] A. Boller, M. Cereghetti, M. Schadt, H. Scherrer, *Mol. Cryst. Liq. Cryst.* **1977**, *42*, 215.

[52] C. M. Waters, E. P. Raynes, V. Brimmel, *S. I. D. Proc. Jpn. Display* **1983**, 396.

[53] T. J. Scheffer, J. Nehring, *Appl. Phys. Lett.* **1984**, *45*, 1021.

[54] M. Schadt in *Liquid Crystals* (Eds.: H. Baumgartel, E. U. Frank, W. Grunbein; Guest Ed.: H. Stegemeyer), Springer, N. Y. **1994**, Chap. 6, p. 195.

[55] N. A. Clark, S. T. Lagerwall, *Appl. Phys. Lett.* **1980**, *36*, 899.

[56] A. D. L. Chandani, E. Gorecka, Y. Ouchi, H. Takezoe, A. Fukuda, *Jpn. J. Appl. Phys.* **1988**, *27*, L729.

[57] Society for Information Display, *Information Display* **1996**, *12* (12), 13.

[58] *Eurodisplay '96* **1997**, *16* (1), 16.

[59] S. Kataoka, *International Lecture*, Institute of Electrical and Electronic Engineers, Savoy Place, London, March **1997**.

[60] S. Chandrasekhar, B. K. Sadashiva, K. A. Suresh, *Pramana* **1977**, *9*, 471.

[61] *Side Chain Liquid Crystal Polymers* (Ed.: C. B. McArdle), Blackie, Glasgow **1989**.

[62] *Liquid Crystalline Polymers* (Ed.: C. Carfagna), Pergamon, Oxford **1994**.

[63] S. M. Kelly, *Liq. Cryst. Today* **1996**, *6* (4), 1.

[64] *Metallomesogens* (Ed.: J. L. Serrano), VCH, Weinheim **1996**.

[65] P. A. Winsor in *Liquid Crystals and Plastic Crystals*, Vol. 1 (Eds.: G. W. Gray, P. A. Winsor), Ellis Horwood, Chichester **1974**, Chap. 5, p. 199.

[66] S. Friberg, *Lyotropic Liquid Crystals* (Ed.: R. F. Gould), Advances in Chemistry Series, American Chemical Society, Washington, DC **1976**.

[67] H. Morgans, G. Williams, G. J. T. Tiddy, A. R. Katritzky, G. P. Savage, *Liq. Cryst.* **1993**, *15*, 899.

[68] K. Hiltrop in *Liquid Crystals* (Eds.: H. Baumgartel, E. U. Frank, W. Grunbein; Guest Ed.: H. Stegemeyer), Springer, N. Y. **1994**, Chap. 4, p. 143.

[69] S. Chandrasekhar, *Liquid Crystals*, 2nd ed., Cambridge U. P., Cambridge **1992**.

[70] *The Molecular Physics of Liquid Crystals* (Eds.: G. R. Luckhurst, G. W. Gray), Academic, London **1979**.

[71] R. Hashim, G. R. Luckhurst, S. Romano, *J. Chem. Soc., Faraday Trans.* **1995**, *21*, 2141; G. R. Luckhurst, R. A. Stephens, R. W. Phippen, *Liq. Cryst.* **1990**, *8*, 451.

[72] M. R. Wilson in *Handbook of Liquid Crystals*, Vol. 1 (Eds.: D. Demus, J. W. Goodby, G. W. Gray, H. W. Spiess and V. Vill), Wiley-VCH, Weinheim **1998**, p. 53.

[73] G. Barbero, G. Durand in *Liquid Crystals in Complex Geometries* (Eds.: G. P. Crawford, S. Zumer), Taylor and Francis, London **1996**, Chap. 2, p. 21.

[74] *Liquid Crystals in Complex Geometries* (Eds.: G. P. Crawford, S. Zumer), Taylor and Francis, London **1996**.

[75] M. Schadt, R. Buchecker, K. Muller, *Liq. Cryst.* **1989**, *5*, 293.

[76] J. W. Goodby, M. A. Waugh, S. M. Stein, E. Chin, R. Pindak, J. S. Patel, *Nature, Lond.* **1989**, *337*, 449; *J. Am. Chem. Soc.* **1989**, *111*, 8119; G. Strajer, R. Pindak, M. A. Waugh, J. W. Goodby, J. S. Patel, *Phys. Rev. Lett.* **1990**, *64*, 1545.

[77] P. G. de Gennes, *Solid State Commun.* **1972**, *10*, 753.

[78] J. W. Goodby, *Mol. Cryst. Liq. Cryst.* **1984**, *110*, 205; J. W. Goodby, J. A. Haley, G. Mackenzie, M. J. Watson, D. Plusquellec, V. Ferrieres, *J. Mater. Chem.* **1995**, *5*, 2209.

[79] H. Prade, R. Miethchen, V. Vill, *J. Prakt. Chem.* **1995**, *337*, 427; J. W. Goodby, J. A. Haley, M. J. Watson, G. Mackenzie, S. M. Kelly, P. Letellier, P. Gode, G. Goethals, S. Ronco, B. Harmouch, P. Martin, P. Villa, *Liq. Cryst.* **1997**, *22*, 497.

[80] M. A. Marcus, P. L. Finn, *Liq. Cryst.* **1988**, *3*, 38.

[81] G. P. Spada, S. Bonazzi, A. Gabriel, S. Zanella, F. Ciuchi and P. Mariani, *Liq. Cryst.* **1997**, *22*, 341.

[82] K. Willis, J. E. Luckhurst, D. J. Price, J. M. Frechet, H. Kihara, T. Kato, G. Ungar, D. W. Bruce, *Liq. Cryst.* **1996**, *21*, 585 and references therein.

[83] K. Praefcke, D. Singer, A. Eckert, *Liq. Cryst.* **1994**, *16*, 53; B. Neumann, D. Joachimi, C. Tschierske, *Liq. Cryst.* **1997**, *22*, 509.

[84] A. E. Blatch, I. D. Fletcher, G. R. Luckhurst, *Liq. Cryst.* **1995**, *18*, 801.

[85] J. Anderesch, C. Tschierske, *Liq. Cryst.* **1996**, *21*, 51; J. Anderesch, S. Diele, D. Lose, C. Tschierske, *Liq. Cryst.* **1996**, *21*, 103.

[86] M. Ibn-Elhaj, A. Skoulios, D. Guillon, J. Newton, P. Hodge, H. J. Coles, *Liq. Cryst.* **1995**, *19*, 373.

[87] W. Weissflog, M. Rogunova, I. Letkos, S. Diele, G. Pelzl, *Liq. Cryst.* **1995**, *19*, 541.

Chapter II
Guide to the Nomenclature and Classification of Liquid Crystals

John W. Goodby and George W. Gray

1 Introduction

Nomenclature in liquid crystal systems is a nonsystematic language that is still, like any modern language, very much alive. Thus, many changes to currently acceptable terms, introductions of new notations, and deletions of out-moded notation have been made since the conception of the currently used nomenclature system. As the nomenclature system is in somewhat of a fluxional state it is not wise to assume that all definitions and accompanying notations are sacrosanct. Nevertheless, in some areas the topic of nomenclature has settled down into an internationally accepted, but unrecognized (by Scientific Societies) notation system, while in other areas, where research is still very active, changes to notation are still common. Members of the International Liquid Crystal Society (ILCS) and the International Union of Pure and Applied Chemists (IUPAC) are, however, attempting to create the first widely accepted naming system for liquid crystals. The descriptions and notations that follow are in agreement with the current proposals of the ILCS and IUPAC.

Notation for liquid crystals really started with the naming of the nematic and smectic phases in the early 1920s by Friedel [1]. However, it was the discovery of the existence of a variety of smectic phases in the 1950s–60s which lead Sackmann and Demus to propose the current lettering scheme for smectic liquid crystals [2]. Originally only three smectic phases were defined, SmA, SmB and SmC, but more followed rapidly as new phases were discovered. The notation introduced by Sackmann and Demus was dependent on the thermodynamic properties of mesophases and their ability to mix with one another, thus the miscibility of a material of undefined phase type with a standard material of known/defined mesophase morphology became the criterion for phase classification. Immiscibility, on the other hand has no special significance. Consequently all materials should have become standardized with those labelled by Sackmann and Demus.

Shortly after the introduction of the notation system, confusion set in with the notations for the phases G and H becoming interchanged (which was later resolved by agreement between the Hull and Halle Research Groups [3]. In addition, the D phase had been introduced as a smectic phase but later it turned out to be cubic; the B phase was split into two, the tilted B and orthogonal B phases, which were later to be redefined as the B and G phases; two E phases were thought to exist, one being uniaxial and the other biaxial, that were later defined as all being biaxial; and of course there was the perennial problem as to whether or not a phase was a soft crystal or a real smectic phase. This latter debate finally gave rise to a change in notation for soft crystal phases, the Sm notation being dropped, but with the hangover of the B notation being used in both smectic and soft crystal phases.

As our understanding of smectic phases increased, and structural studies using X-ray diffraction became more prevalent, there was an attempt to use crystallographic notation to describe the structures of smectic phases and, in addition, subscripts and superscripts were introduced to describe certain structural features, for example the subscript 2 was introduced to describe bilayer structures. By and large, however, there has been a general resistance to moving over to a full blown crystallographic notation system simply because there is a general feeling that a small change in structure within a miscibility class would lead to an unnecessary change in notation, consequently leading to complications and confusion.

To some degree problems of notation did arise with the naming of columnar mesophases. Originally they were called discotic liquid crystals, and indeed they also acquired a crystallographic notation. Both of these notations have, however, fallen out of favor and the naming of the state has been redefined. As research in disc-like systems remains relatively active, it is to be expected that further phases will be discovered, and as our understanding of the structures of these phases increases changes may be made to our current notation.

Notation in chiral phases is in flux basically because many new phases have been recently discovered, for example, antiferroelectric phases, blue phases and twist grain boundary phases. Even the use of an asterisk to indicate the presence of chirality is a hotly debated topic because chiral systems have broken symmetries and sometimes helical structures. Thus the debate is an issue over which aspect of chirality does the asterisk represent – broken symmetry or helicity. Thus further developments and changes in notation may be expected to occur in this area in the future.

Another problem that exists in notation is the relationship between thermotropic and lyotropic nomenclature. In some cases continuous behavior has been seen between thermotropic and lyotropic mesophases suggesting that they should share the same notation; however, at this point in time this has not occurred.

The notation scheme given below will be used wherever possible in the *Physical Properties of Liquid Crystals*; however for the reasons given above, readers should take care in its implementation, and they should also remember that the literature has suffered many changes over the years and so nomenclature used years ago may not tally with today's notation.

2   General Definitions

Liquid crystal state – recommended symbol LC – a mesomorphic state having long-range orientational order and either partial positional order or complete positional disorder.

Mesomorphic state – a state of matter in which the degree of molecular order is intermediate between the perfect three-dimensional, long-range positional and orientational order found in solid crystals and the absence of long-range order found in isotropic liquids, gases and amorphous solids.

Liquid crystal – a substance in the liquid crystal state.

Crystal phase – phase with a long-range periodic positional/translational order.

Liquid phase – phase with no long-range periodic or orientational order.

Mesophase or *liquid crystal phase* – phase that does not possess long-range positional ordering, but does have long-range orientational order. A phase occurring over a defined range of temperature or pressure or concentration within the mesomorphic state.

Thermotropic mesophase – a mesophase formed by heating a solid or cooling an isotropic liquid, or by heating or cooling a thermodynamically stable mesophase.

Lyotropic mesophase – a mesophase formed by dissolving an amphiphilic mesogen in suitable solvents, under appropriate conditions of concentration and temperature.

Calamitic mesophase – a mesophase formed by molecules or macromolecules with rod or lath-like molecular structures.

Columnar phase – phase that is formed by stacking of molecules in columns. Note that sugars, etc. are not necessarily discotic; discotic reflects a disc-like molecular shape. Also phasmids are columnar, but not necessarily discotic.

Mesogen (mesomorphic compound) – a compound that under suitable conditions of temperature, pressure and concentration can exist as a mesophase.

Calamitic mesogen – a mesogen composed of molecules or macromolecules with rod or lath-like molecular structures.

Discotic mesogen – a mesogen composed of relatively flat, disc- or sheet-shaped molecules.

Pyramidal or bowlic mesogen – a mesogen composed of molecules derived from a semi-rigid conical core.

Polycatenary mesogen – a mesogen composed of molecules having an elongated rigid core with several flexible chains attached to the end(s).

Swallow-tailed mesogen – a mesogen composed of molecules with an elongated rigid core with a flexible chain attached at one end and a branched flexible chain, with branches of about the same length at the other.

Mesogenic dimers, trimers etc. – a mesogen consisting of molecules with two, three, or more linked mesogenic units usually of identical structure.

Sanidic mesogen – a mesogen composed of molecules with board-like shapes.

Amphiphilic mesogen – a compound composed of molecules consisting of two parts with contrasting character, which may be hydrophilic and hydrophobic, that is lipophobic and lipophilic.

Amphotropic material – a compound which can exhibit thermotropic as well as lyotropic mesophases.

3 Structural Features

Molecules of liquid crystalline compounds are subdivided into the *central core* (mesogenic group), the *linking groups*, and *lateral groups* as well as *terminal groups*, depending on whether or not the groups lie along the long axis of the molecule. In relation to disc-like molecules, the central rigid region is called the core and the outer region the periphery; linking groups have the same definition as for calamitic rod-like systems.

The term *mesogenic* means that the structure is generally compatible with mesophase formation. A compound that forms real mesophases is however *mesomorphic*.

4   Polymeric Liquid Crystals

Liquid crystalline polymers are classified either as *main-chain* liquid crystalline polymers (MCLCP), as polymers with mesogenic *side groups* (SGLCP), or as liquid crystalline elastomers.

5   Notation of Thermotropic Liquid Crystalline Properties

To denote one phase transition, the abbreviation T together with an index is used (for example, T_{N-I} for a nematic–isotropic transition; T_{C-A} for a smectic C–smectic A transition).

The complete transition sequence is characterized by the description of the solid state (Sec. 5.1), the liquid crystalline transitions (Sec. 5.2) and the clearing parameter (Sec. 5.3).

The phase symbol is followed by the upper temperature limit as measured during the heating and not during the cooling cycle.

Crystal types should be arranged in an ascending order of transition temperature.

If a transition temperature of a liquid crystalline phase is lower than the melting point, this phase only occurs monotropically.

Monotropic transitions appear in round brackets.

In this context, square brackets mean virtual transitions.

Examples:

Cr 34 N 56 I designates a compound melting at 34 °C into the nematic phase; at 56 °C

it changes into the isotropic phase; normal behavior.

Cr 56.5 (SmA 45) I designates a compound melting at 56.5 °C into the isotropic phase. Below 45 °C, a monotropic A phase exists.

Cr 120 B 134 I [N 56 I_e] designates a compound melting at 120 °C into the crystal B phase. At 134 °C the isotropic phase is formed. The virtual nematic clearing point of 56 °C is one obtained by extrapolation from mixtures.

Cr_1 78 Cr_2 212 N ? I_{decomp} designates a compound with a crystal to crystal transition at 78 °C and a melting point of 212 °C to a nematic phase. The clearing point is unknown because decomposition takes place.

Cr_1 112 (Cr_2 89) I designates a material with a metastable crystal phase formed on cooling the isotropic melt slowly and having a lower melting point than the stable crystal phase.

5.1   Description of the Solid State

Cr   crystalline phase
Cr_2 second crystalline phase
g   glassy state
T_g glass-transition temperature

5.1.1   Description of Soft Crystals

The following phases should just be called B, E, etc. (i.e. retaining their historic classification, but losing their smectic code letter Sm) because they are no longer regarded as true smectics as they have long range positional order. In fact they are soft or disordered crystals. The single letter notation (B, E, etc.) is preferred to B_{Cr}, E_{Cr} etc.

B   crystal B
E   crystal E
E_{mod} crystal E phase with modulated
       in-plane structure
G   crystal G
H   crystal H
J   crystal J
K   crystal K

Tilted Chiral Soft Crystal Phases

A superscript asterisk (*) is used throughout to denote the presence of chirality.

J*   tilted chiral crystal J
H*   tilted chiral crystal H
G*   tilted chiral crystal G
K*   tilted chiral crystal K

Chiral Orthogonal Soft Crystal Phases

A superscript asterisk (*) is used throughout to denote the presence of chirality.

B*   chiral orthogonal crystal B phase
E*   chiral orthogonal crystal E phase

5.2 Description of the Liquid Crystalline Phases

5.2.1 Nematic and Chiral Nematic Phases

n   director
N   nematic
N*   chiral nematic (cholesteric)
N_u uniaxial nematic phase
N_b biaxial nematic phase (sanidic phase)
also
N_∞^* infinite pitch cholesteric, i.e. at a helix
       inversion or compensation point

BP   blue phases. These are designated BP_I,
       BP_{II}, BP_{III} or BP_{fog}, and BPS.

5.2.2 Smectic Liquid Crystals

Use Sm for smectic instead of S (unless spelt out), this avoids subscripts and double subscripts

SmA   Smectic A
          SmA_1 monolayer
          SmA_2 bilayer
          SmA_d interdigitated bilayer
          $Sm\tilde{A}$ modulated bilayer
SmB or   Smectic B; hexatic B
SmB_{hex} (SmB preferred)
SmC   Smectic C
          SmC_1 monolayer
          SmC_2 bilayer
          SmC_d interdigitated bilayer
          $Sm\tilde{C}$ modulated bilayer
          SmC_{alt} alternating tilt phase –
          see smectic O
SmF   Smectic F
SmI   Smectic I
SmM   Smectic M – not found common
          use so far
SmO   Smectic O – not found common
          use so far. Use carefully in relation
          to smectic C_{alt} and antiferroelec-
          tric smectic C* phase
          SmO and SmM should be defined
          in the text

Intercalated mesophases take the subscript c, e.g., SmA_c.

Biaxial variants of uniaxial smectic phases take the subscript b, e.g., SmA_b.

5.2.3 Chiral Smectic Liquid Crystals

Usually given the notation * to indicate the presence of chirality, e.g., SmA*, SmC*.

Twist Grain Boundary Phases (TGB)

Orthogonal TGB Phases

TGBA* structure based on the smectic A phase. The phase can be either commensurate or incommensurate depending on the commensurability of the helical structure with respect to the rotation of the smectic A blocks in the phase.

TGBB* proposed structure based on a helical hexatic B phase.

Tilted TGB Phases

TGBC phase poses the following problems for nomenclature:

TGBC where the normal smectic C* helix is expelled to the screw dislocations.

TGBC* phase where the blocks have a local helix associated with the out of plane structure.

Possibly too early to define notation yet; therefore spell out notation in text.

Tilted Chiral Phases

SmC*   chiral C phase
SmC_∞^* infinite pitch smectic C* – ferroelectric
SmC_α^* unwound antiferroelectric phase occurring at higher temperature, and above a ferroelectric SmC* phase in a phase sequence
SmC_γ^* ferrielectric phases that occur on cooling ferroelectric C* phases
SmC_A^* antiferroelectric C* phase

Try not to use α, β and γ notations, simply spell out the phase type, e.g. ferrielectric SmC* phase, or SmC* (ferri).

The above notations may also be applied to SmI* and SmF* phases as and when required – for example, SmI_A^* is an antiferroelectric I* phase.

Orthogonal Phases

SmA*   chiral orthogonal smectic A
SmB*/   chiral orthogonal hexatic smectic
SmB_{hex}^* B phase (SmB* preferred)

Analogous Achiral Systems

Here, antiferroelectric-like structures are often observed, i.e. zigzag layer ordering. The following designation should be used and reference to antiferroelectric-like ordering should be suppressed:

SmC_{alt} alternating tilt smectic C phase – see also smectic O for cross-referencing

Unknown Phases

Label as SmX_1, SmX_2, etc.

5.2.4   Columnar Phases

N_D nematic discotic phase
Col_h hexagonal discotic
Col_{ho} ordered hexagonal columnar phase
Col_{hd} disordered hexagonal columnar phase
Col_{ro} ordered rectangular columnar phase
Col_{rd} disordered rectangular columnar phase
Col_t tilted columnar phase
ϕ phasmidic phase, but Col is preferred

For nematic phases – spell out positive and negative birefringent situations in the text. In addition spell out if the nematic phase is composed of single molecular entities or short columns which exist in a disordered nematic state.

Care should be taken when using the term discotic; columnar is preferred.

Unknown Discotic Phases

Col_1, Col_2, etc.

5.2.5 Plastic Crystals

Rotationally (3-D) disordered crystals that may be derived from globular molecules leading to isotropic phases. This classification does not apply to crystal smectic phases composed of elongated molecules, although these could be described as anisotropic plastic crystals.

5.2.6 Condis Crystals

Crystals in which the positional and conformational order in the packing of molecules arranged in parallel is lost to some degree.

5.2.7 Cubic

Cubic thermotropic liquid crystalline phases are designated Cub.

CubD   Cubic D phase

5.2.8 Re-entrants

Use the same notation as for nematic or other appropriate phase with subscript re-entrant, i.e., N_{re} or SmC_{re} or Col_{hre}.

5.3 Description of the Clearing Parameters

I   isotropic, standard case
I_{decomp} decomposition at clearing temperature
I_e extrapolated temperature

6 Stereochemistry

(#)   unknown chirality

One Chiral Centre (R or S)

(S)   chiral (S)
(R)   chiral (R)
(S)/(R) racemate

Two Chiral Centres

This situation is more complex and depends upon whether the two centres are the same or different. If they are the same the notation should be: chiral (*S,S* or *R,R*); or for racemic materials (*S,S* or *R,R*); meso-compounds (optically inactive by internal compensation should be denoted as (*S,R* or *R,S*).

7 References

[1] G. Friedel, *Ann. Physique* **1922**, *18*, 272.
[2] H. Sackmann, D. Demus, *Mol. Cryst. Liq. Cryst.* **1966**, *2*, 81.
[3] D. Demus, J. W. Goodby, G. W. Gray, H. Sackmann, *Mol. Cryst. Liq. Cryst. Lett.* **1980**, *56*, 311.

Chapter III
Theory of the Liquid Crystalline State

1 Continuum Theory for Liquid Crystals

Frank M. Leslie

1.1 Introduction

Continuum theory for liquid crystals has its origins in the work of Oseen [1] and Zocher [2] in the 1920s. The former essentially derived the static version of the theory for nematics that has been used extensively in device modelling, while the latter successfully applied it to Fréedericksz transitions [3]. Later Frank [4] gave a more direct formulation of the energy function employed in this theory, and stimulated interest in the subject after a period of relative dormancy. Soon thereafter Ericksen [5] set the theory within a mechanical framework, and generalized his interpretation of static theory to propose balance laws for dynamical behaviour [6]. Drawing upon Ericksen's work, Leslie [7] used ideas prevalent in continuum mechanics to formulate constitutive equations, thus completing dynamic theory.

This continuum theory models many static and dynamic phenomena in nematic liquid crystals rather well, and various accounts of both the theory and its applications are available in the books by de Gennes and Prost [8], Chandrasekhar [9], Blinov [10] and Virga [11], and also in the reviews by Stephen and Straley [12], Ericksen [13], Jenkins [14] and Leslie [15]. Given this success of continuum theory for nematics, much current interest in continuum modelling of liquid crystals now centres upon appropriate models for smectic liquid crystals, liquid crystalline polymers and (to a lesser extent) lyotropics. Otherwise, interest in nematics is largely confined to studies of behaviour at solid interfaces, this including discussions as to whether or not one should include an additional surface term in the Frank–Oseen energy as proposed by Nehring and Saupe [16], although there has also been some activity into the modelling of defects, particularly using a modified theory proposed recently for this purpose by Ericksen [17].

In this section I aim to describe in some detail continuum theory for nematics, and also to draw attention to some points of current interest, particularly surface conditions and surface terms. At this juncture it does seem premature to discuss new developments concerning smectics, polymers and lyotropics, although a brief discussion of an equilibrium theory for certain smectics seems appropriate, given that it relates to earlier work on this topic. Throughout, to encourage a wider readership, we endeavour to employ vector and matrix notation, avoiding use of Cartesian tensor notation.

1.2   Equilibrium Theory for Nematics

1.2.1   The Frank–Oseen Energy

Continuum theory generally employs a unit vector field $\mathbf{n}(\mathbf{x})$ to describe the alignment of the anisotropic axis in nematic liquid crystals, this essentially ignoring variations in degrees of alignment which appear to be unimportant in many macroscopic effects. This unit vector field is frequently referred to as a director. In addition, following Oseen [1] and Frank [4], it commonly assumes the existence of a stored energy density W such that at any point

$$W = W(\mathbf{n}, \nabla\mathbf{n}) \tag{1}$$

the energy is therefore a function of the director and its gradients at that point. Since nematic liquid crystals lack polarity, $\mathbf{n}$ and $-\mathbf{n}$ are physically indistinguishable and therefore one imposes the condition that

$$W(\mathbf{n}, \nabla\mathbf{n}) = W(-\mathbf{n}, -\nabla\mathbf{n}) \tag{2}$$

and invariance to rigid rotations requires that

$$W(\mathbf{n}, \nabla\mathbf{n}) = W(\underset{\sim}{P}\mathbf{n}, \underset{\sim}{P}\nabla\mathbf{n}\underset{\sim}{P}^T) \tag{3}$$

where $\underset{\sim}{P}$ is any proper orthogonal matrix, the superscript denoting the transpose of the matrix. While the above suffices for chiral nematics or cholesterics, for non-chiral nematics, invariably referred to simply as nematics, material symmetry requires that (Eq. 3) be extended to

$$W(\mathbf{n}, \nabla\mathbf{n}) = W(\underset{\sim}{Q}\mathbf{n}, \underset{\sim}{Q}\nabla\mathbf{n}\underset{\sim}{Q}^T) \tag{4}$$

where the matrix $\underset{\sim}{Q}$ belongs to the full orthogonal group, the function is therefore isotropic rather than hemitropic.

Oseen and Frank both consider an energy function that is quadratic in the gradients of the director $\mathbf{n}$, in which case the condi-

tions (Eq. 2) and (Eq. 3) lead to

$$\begin{aligned}
2W = {} & K_1(\text{div }\mathbf{n})^2 + K_2(\mathbf{n} \cdot \text{curl }\mathbf{n} + q)^2 \\
& + K_3|\mathbf{n} \times \text{curl }\mathbf{n}|^2 + (K_2 + K_4) \\
& \cdot \text{div}[(\mathbf{n} \cdot \text{grad})\mathbf{n} - (\text{div }\mathbf{n})\mathbf{n}]
\end{aligned} \tag{5}$$

where the Ks and q are constants. The above is the form appropriate to cholesterics or chiral nematics, q being related to the natural pitch of the helical configurations found in these materials, through

$$p = 2\pi/q. \tag{6}$$

For ordinary nematics, however, the conditions (Eq. 2) and (Eq. 4) yield

$$\begin{aligned}
2W = {} & K_1(\text{div }\mathbf{n})^2 + K_2(\mathbf{n} \cdot \text{curl }\mathbf{n})^2 \\
& + K_3|\mathbf{n} \times \text{curl }\mathbf{n}|^2 + (K_2 + K_4) \\
& \cdot \text{div}[(\mathbf{n} \cdot \text{grad})\mathbf{n} - (\text{div }\mathbf{n})\mathbf{n}]
\end{aligned} \tag{7}$$

the coefficient q necessarily zero for such materials. This latter energy can alternatively be expressed as

$$\begin{aligned}
2W = {} & (K_1 - K_2 - K_4)(\text{div }\mathbf{n})^2 \\
& + K_2 tr(\nabla\mathbf{n}\nabla\mathbf{n}^T) + K_4 tr(\nabla\mathbf{n})^2 \\
& + (K_3 - K_2)(\mathbf{n} \cdot \text{grad})\mathbf{n} \cdot (\mathbf{n} \cdot \text{grad})\mathbf{n}
\end{aligned} \tag{8}$$

$tr\underset{\sim}{A}$ and $\underset{\sim}{A}^T$ denoting the trace and transpose of the matrix $\underset{\sim}{A}$, respectively. This latter form can be more convenient for some purposes. It is common to refer to the constants K_1, K_2 and K_3 as the splay, twist and bend elastic constants, respectively, while the K_4 term is sometimes omitted since the last term in the form (Eq. 7) can clearly be expressed as a surface integral.

Given that nematics tend to align uniformly with the anisotropic axis everywhere parallel, Ericksen [18] argues that this must represent a state of minimum energy, and thus assumes that

$$W(\mathbf{n}, \nabla\mathbf{n}) > W(\mathbf{n}, 0), \quad \nabla\mathbf{n} \neq 0 \tag{9}$$

and as a consequence the coefficients in (Eq. 7) or (Eq. 8) must satisfy

$$\begin{aligned}
& K_1 > 0, \quad K_2 > 0, \quad K_3 > 0, \\
& 2K_1 > K_2 + K_4 > 0, \quad K_2 > K_4
\end{aligned} \tag{10}$$

Jenkins [19] discusses the corresponding restrictions placed upon the energy (Eq. 5) by assuming that the characteristic twisted helical configuration represents a minimum of energy.

To conclude this section we note an identity derived by Ericksen [6] from the condition (Eq. 3) by selecting

$$\underset{\sim}{P} = \underset{\sim}{I} + \varepsilon \underset{\sim}{R}, \quad \underset{\sim}{R}^T = -\underset{\sim}{R} \tag{11}$$

$\underset{\sim}{I}$ being the unit matrix and ε a small parameter. With this choice one can quickly show that

$$\mathbf{n} * \frac{\partial W}{\partial \mathbf{n}} + \nabla \mathbf{n} \left(\frac{\partial W}{\partial \nabla \mathbf{n}} \right)^T + \nabla \mathbf{n}^T \frac{\partial W}{\partial \nabla \mathbf{n}}$$

$$= \frac{\partial W}{\partial \mathbf{n}} * \mathbf{n} + \frac{\partial W}{\partial \nabla \mathbf{n}} \nabla \mathbf{n}^T + \left(\frac{\partial W}{\partial \nabla \mathbf{n}} \right)^T \nabla \mathbf{n} \tag{12}$$

a result required below. In the above, the notation $\mathbf{a} * \mathbf{b}$ represents the 3×3 matrix with $(i, j)^{\text{th}}$ element $a_i b_j$.

1.2.2 A Virtual Work Formulation

The approach adopted by Ericksen [5] to equilibrium theory for both nematic and cholesteric liquid crystals appeals to a principle of virtual work, which for any volume V of material bounded by surface S takes the form

$$\delta \int_V W \, dv = \int_V (\mathbf{F} \cdot \delta \mathbf{x} + \mathbf{G} \cdot \Delta \mathbf{n}) \, dv$$

$$+ \int_S (\mathbf{t} \cdot \delta \mathbf{x} + \mathbf{s} \cdot \Delta \mathbf{n}) \, ds \tag{13}$$

where

$$\Delta \mathbf{n} = \delta \mathbf{n} + (\delta \mathbf{x} \cdot \text{grad}) \mathbf{n} \tag{14}$$

$\mathbf{F}$ denotes body force per unit volume, $\mathbf{t}$ surface force per unit area, and $\mathbf{G}$ and $\mathbf{s}$ are generalized body and surface forces, respectively. With the common assumption of incompressibility the virtual displacement $\delta \mathbf{x}$ is not arbitrary, but is subject to the constraint

$$\text{div} \, \delta \mathbf{x} = 0 \tag{15}$$

and of course the variations $\delta \mathbf{n}$ and $\Delta \mathbf{n}$ are constrained by

$$\delta \mathbf{n} \cdot \mathbf{n} = \Delta \mathbf{n} \cdot \mathbf{n} = 0 \tag{16}$$

due to $\mathbf{n}$ being of fixed magnitude.

Through consideration of an arbitrary, infinitesimal, rigid displacement in which $\Delta \mathbf{n}$ is zero, it quickly follows from (Eq. 13) that

$$\int_V \mathbf{F} \, dv + \int_S \mathbf{t} \, ds = 0 \tag{17}$$

which of course expresses the fact that the resultant force is zero in equilibrium. Similarly, consideration of an arbitrary, infinitesimal, rigid rotation $\boldsymbol{\omega}$, in which

$$\delta \mathbf{x} = \boldsymbol{\omega} \times \mathbf{x}, \quad \Delta \mathbf{n} = \boldsymbol{\omega} \times \mathbf{n} \tag{18}$$

yields from (Eq. 13) following rearrangement of the triple scalar products

$$\int_V (\mathbf{x} \times \mathbf{F} + \mathbf{n} \times \mathbf{G}) \, dv + \int_S (\mathbf{x} \times \mathbf{t} + \mathbf{n} \times \mathbf{s}) \, ds = 0 \tag{19}$$

which one interprets as a statement that the resulting moment is zero in equilibrium. Hence (Eq. 19) relates the generalized forces $\mathbf{G}$ and $\mathbf{s}$ to the body and surface moments $\mathbf{K}$ and $\boldsymbol{\ell}$, respectively, through

$$\mathbf{K} = \mathbf{n} \times \mathbf{G}, \quad \boldsymbol{\ell} = \mathbf{n} \times \mathbf{s} \tag{20}$$

which allows the determination of the generalized body force.

By first expressing the left hand side of the statement of virtual work (Eq. 13) as using (Eq. 15)

$$\delta \int_V W \, dv = \int_V (\delta W + (\delta \mathbf{x} \cdot \text{grad}) W) \, dv \tag{21}$$

this taking account of the change of volume in the virtual displacement, and then reorganizing the resultant volume integral so that it has a similar format to the right hand side of (Eq. 13), it is possible to obtain expressions for the surface force $\mathbf{t}$ and the generalized surface force **s**, and also two equations required to hold in equilibrium [5]. Denoting by $\boldsymbol{\nu}$ the unit outwards normal to the surface S and bearing in mind the constraints (Eq. 15) and (Eq. 16), one finds from the surface integrals

$$\mathbf{t} = -p\boldsymbol{\nu} + \underset{\sim}{T^s}\boldsymbol{\nu}, \quad \underset{\sim}{T^s} = -\nabla \mathbf{n}^T \frac{\partial W}{\partial \nabla \mathbf{n}}$$

$$\mathbf{s} = \beta\mathbf{n} + \underset{\sim}{S^s}\boldsymbol{\nu}, \quad \underset{\sim}{S^s} = \frac{\partial W}{\partial \nabla \mathbf{n}} \tag{22}$$

and also from the volume integrals

$$\mathbf{F} - \operatorname{grad} p + \operatorname{div} \underset{\sim}{T^s} = 0,$$

$$\mathbf{G} - \frac{\partial W}{\partial \mathbf{n}} + \operatorname{div} \underset{\sim}{S^s} = \gamma\mathbf{n}, \tag{23}$$

with p an arbitrary pressure due to incompressibility, and β and γ arbitrary scalars due to the director having fixed magnitude. In equations (Eq. 23) the divergence applies to the second of the indices of the matrices $\underset{\sim}{T^s}$ and $\underset{\sim}{S^s}$. The former of (Eq. 23) clearly represents the point form of the balance of forces (Eq. 17), while the latter can be shown to be the point form of the balance of moments (Eq. 19), this requiring some manipulation involving the identity (Eq. 12).

1.2.3 Body Forces and Moments

While the action of gravity upon a liquid crystal is identical to that on other materials, external magnetic and electric fields have a rather different effect upon these anisotropic liquids than they do on isotropic materials. Both can give rise to body forces and moments as is to be expected from rather simple arguments common in magnetostatics and electrostatics. To fix ideas, consider a magnetic field **H** which induces a magnetization **M** in the material, and this in turn gives rise to a body force **F** and a body moment **K**, given by

$$\mathbf{F} = (\mathbf{M} \cdot \operatorname{grad})\mathbf{H}, \quad \mathbf{K} = \mathbf{M} \times \mathbf{H} \tag{24}$$

In an isotropic material the induced magnetization is necessarily parallel to the field and the couple is zero, but for a nematic or cholesteric liquid crystal the magnetization can have an anisotropic contribution of the form

$$\mathbf{M} = \chi_\perp \mathbf{H} + \chi_a \mathbf{n} \cdot \mathbf{H} \mathbf{n}, \quad \chi_a = \chi_\parallel - \chi_\perp \tag{25}$$

$\chi_\parallel$ and $\chi_\perp$ denoting the diamagnetic susceptibilities when **n** and **H** are parallel and perpendicular, respectively. As a consequence a body moment can occur given by

$$\mathbf{K} = \chi_a \mathbf{n} \cdot \mathbf{H} \mathbf{n} \times \mathbf{H} \tag{26}$$

and it immediately follows from the first of equations (Eq. 20) that

$$\mathbf{G} = \chi_a \mathbf{n} \cdot \mathbf{H} \mathbf{H} \tag{27}$$

any contribution parallel to the director being simply absorbed in the scalar γ in equations (Eq. 23). In general the anisotropy of the diamagnetic susceptibility χ_a is positive, but it is also rather small. Consequently one can ignore the influence of the liquid crystal upon the applied field.

Similar expressions arise for an electric field **E**, this creating an electric displacement **D** of the form

$$\mathbf{D} = \varepsilon_\perp \mathbf{E} + \varepsilon_a \mathbf{n} \cdot \mathbf{E} \mathbf{n}, \quad \varepsilon_a = \varepsilon_\parallel - \varepsilon_\perp \tag{28}$$

$\varepsilon_\parallel$ and $\varepsilon_\perp$ denoting the corresponding dielectric permittivities, which gives rise to similar body forces and moments. However, as Deuling [20] points out, there can be one important difference between the effects of magnetic and electric fields upon liquid crystals, in that an electric field can give rise to significant permittivities, and thus one must allow for the influence of the liquid crystal upon the applied field by employing the appropriate reduced version of Maxwell's equations.

Associated with a magnetic field there is an energy

$$\psi = \frac{1}{2}\mathbf{M} \cdot \mathbf{H} \tag{29}$$

and if one regards the energy ψ as simply a function of the director $\mathbf{n}$ and position $\mathbf{x}$, then one can write

$$\mathbf{F} = \frac{\partial \psi}{\partial \mathbf{x}}, \quad \mathbf{G} = \frac{\partial \psi}{\partial \mathbf{n}} \qquad (30)$$

the former using the fact that in equilibrium the magnetic field is irrotational. However, if the field applied is dependent upon the director, as can occur for an electric field, then (Eq. 30) are not valid.

1.2.4 The Equilibrium Equations

The equations (Eq. 23) representing balance of forces and moments constitute six equations for four unknowns, two components of the unit vector $\mathbf{n}$ and the scalars p and γ. However, this apparent overdeterminacy does not materialize if the external body force and moment meet a certain requirement [5]. To see this combine the two equations as follows

$$\mathbf{F} + \nabla \mathbf{n}^T \mathbf{G} - \operatorname{grad} p + \operatorname{div} \underset{\sim}{T}^s$$
$$+ \nabla \mathbf{n}^T \operatorname{div} \underset{\sim}{S}^s - \nabla \mathbf{n}^T \frac{\partial W}{\partial \mathbf{n}} = 0 \qquad (31)$$

and by appeal to equations (Eq. 22) this reduces to

$$\mathbf{F} + \nabla \mathbf{n}^T \mathbf{G} = \operatorname{grad}(p + W) \qquad (32)$$

clearly limiting the body force and moment. However, when (Eq. 30) applies, the above at once yields

$$p + W - \psi = p_o \qquad (33)$$

where p_o is an arbitrary constant pressure, and thus the equation for the balance of forces integrates to give the pressure, removing the potential overdeterminacy. Also the balance of moments becomes

$$\operatorname{div} \underset{\sim}{S}^s - \frac{\partial W}{\partial \mathbf{n}} + \frac{\partial \psi}{\partial \mathbf{n}} = \gamma \mathbf{n} \qquad (34)$$

which can be written in forms more convenient for particular problems as discussed by Ericksen [21].

Frequently one selects a particular form for the director $\mathbf{n}$ so that it is immediately a unit vector, this representation invariably involving two angles θ and ϕ so that

$$\mathbf{n} = \mathbf{f}(\theta, \phi), \quad \mathbf{n} \cdot \frac{\partial \mathbf{f}}{\partial \theta} = \mathbf{n} \cdot \frac{\partial \mathbf{f}}{\partial \phi} = 0 \qquad (35)$$

Initially for purposes of illustration it is convenient to restrict θ and ϕ to be functions of a single Cartesian coordinate, say z, and consider the application of an external magnetic field $\mathbf{H}$, so that (Eq. 30) holds. In this event

$$W = W(\theta, \phi, \theta', \phi'), \quad \psi = \chi(\mathbf{x}, \theta, \phi) \qquad (36)$$

where the prime denotes differentiations with respect to z. Employing the chain rule, one obtains

$$\mathbf{n}' = \frac{\partial \mathbf{f}}{\partial \theta} \theta' + \frac{\partial \mathbf{f}}{\partial \phi} \phi' \qquad (37)$$

and thus

$$\frac{\partial W}{\partial \theta} = \frac{\partial W}{\partial \mathbf{n}} \cdot \frac{\partial \mathbf{f}}{\partial \theta} + \frac{\partial W}{\partial \mathbf{n}'} \cdot \left(\frac{\partial^2 \mathbf{f}}{\partial \theta^2} \theta' + \frac{\partial^2 \mathbf{f}}{\partial \theta \partial \phi} \phi' \right)$$

$$\frac{\partial W}{\partial \theta'} = \frac{\partial W}{\partial \mathbf{n}'} \cdot \frac{\partial \mathbf{f}}{\partial \theta}, \quad \frac{\partial \chi}{\partial \theta} = \frac{\partial \psi}{\partial \mathbf{n}} \cdot \frac{\partial \mathbf{f}}{\partial \theta} \qquad (38)$$

with similar expressions for the derivatives with respect to ϕ and ϕ'. Combining the above it follows that

$$\left(\frac{\partial W}{\partial \theta'} \right)' - \frac{\partial W}{\partial \theta} + \frac{\partial \chi}{\partial \theta}$$
$$= \left[\left(\frac{\partial W}{\partial \mathbf{n}'} \right)' - \frac{\partial W}{\partial \mathbf{n}} + \frac{\partial \psi}{\partial \mathbf{n}} \right] \cdot \frac{\partial \mathbf{f}}{\partial \theta} \qquad (39)$$

and similarly

$$\left(\frac{\partial W}{\partial \phi'} \right)' - \frac{\partial W}{\partial \phi} + \frac{\partial \chi}{\partial \phi}$$
$$= \left[\left(\frac{\partial W}{\partial \mathbf{n}'} \right)' - \frac{\partial W}{\partial \mathbf{n}} + \frac{\partial \psi}{\partial \mathbf{n}} \right] \cdot \frac{\partial \mathbf{f}}{\partial \phi} \qquad (40)$$

Hence, employing Eqs. (34) and (35) the former can be rewritten as

$$\left(\frac{\partial W}{\partial \theta'}\right)' - \frac{\partial W}{\partial \theta} + \frac{\partial \chi}{\partial \theta} = 0,$$

$$\left(\frac{\partial W}{\partial \phi'}\right)' - \frac{\partial W}{\partial \phi} + \frac{\partial \chi}{\partial \phi} = 0 \tag{41}$$

For the general case when θ and ϕ are arbitrary functions of Cartesian coordinates, essentially a repetition of the above leads to

$$\text{div}\left(\frac{\partial W}{\partial \nabla \theta}\right) - \frac{\partial W}{\partial \theta} + \frac{\partial \chi}{\partial \theta} = 0,$$

$$\text{div}\left(\frac{\partial W}{\partial \nabla \phi}\right) - \frac{\partial W}{\partial \phi} + \frac{\partial \chi}{\partial \phi} = 0 \tag{42}$$

which are clearly more convenient to use than (Eq. 34).

In addition, as Ericksen [21] also shows, the above reformulation can be extended to include curvilinear coordinate systems (y_1, y_2, y_3) introduced by

$$\mathbf{x} = \mathbf{x}(y_1, y_2, y_3). \tag{43}$$

Denoting the Jacobian of this transformation by

$$J = \frac{\partial(x_1, x_2, x_3)}{\partial(y_1, y_2, y_3)} \tag{44}$$

and introducing the notation

$$\overline{W} = JW = \overline{W}\left(y_i, \theta, \phi, \frac{\partial \theta}{\partial y_i}, \frac{\partial \phi}{\partial y_i}\right)$$

$$\overline{\psi} = J\psi = \overline{\chi}(y_i, \theta, \phi) \tag{45}$$

where y_i is short for y_1, y_2, and y_3, and similarly for the partial derivatives, one can show that (Eq. 34) can be recast as

$$\left(\frac{\partial \overline{W}}{\partial \theta_{,i}}\right)_{,i} - \frac{\partial \overline{W}}{\partial \theta} + \frac{\partial \overline{\chi}}{\partial \theta} = 0,$$

$$\left(\frac{\partial \overline{W}}{\partial \phi_{,i}}\right)_{,i} - \frac{\partial \overline{W}}{\partial \phi} + \frac{\partial \overline{\chi}}{\partial \phi} = 0 \tag{46}$$

where

$$\theta_{,i} = \frac{\partial \theta}{\partial y_i}, \quad \phi_{,i} = \frac{\partial \phi}{\partial y_i} \tag{47}$$

and the repeated index is summed over the values 1, 2 and 3.

Rather clearly Eqs. (41), (42) and (46) are appropriate forms of the Euler–Lagrange equations for the integral

$$\int_V (W - \psi) \mathrm{d}v \tag{48}$$

this the formulation of equilibrium theory initially adopted by, for example, Oseen [1] and Frank [4].

For electric fields, certainly for the special cases generally considered, the outcome is rather similar in that the equation for the balance of forces integrates to give the pressure, and one can recast that for balance of moments in the same way as above. However, a general treatment does not appear to be available.

1.2.5 Boundary Conditions

In general the choice of boundary conditions for the alignment at a liquid crystal–solid interface is one of two options, either strong or weak anchoring [8]. Strong anchoring as the term suggests implies that the alignment is prescribed at the boundary by a suitable prior treatment of the solid surface and remains fixed in the presence of competing agencies to realign it. Most commonly this fixed direction is in the plane of the surface (planar) or it is perpendicular (homeotropic), but it need not be so. Weak anchoring, first proposed by Papoular and Rapini [22] assigns an energy to the liquid crystal–solid interface, and assumes a balance between the moment or torque in the liquid crystal from the Frank–Oseen energy and that arising from the interfacial energy. Denoting by w this latter energy per unit area, the simplest assumption is a dependence upon the director **n** and a fixed direction $\mathbf{n}_o$ at the interface, so that

$$w = w(\mathbf{n}, \mathbf{n}_o) \qquad (49)$$

but equally one can have

$$w = w(\mathbf{n}, \boldsymbol{\nu}, \boldsymbol{\tau}) \qquad (50)$$

where $\boldsymbol{\nu}$ is again the unit normal and $\boldsymbol{\tau}$ a fixed unit vector on the surface. By essentially a repetition of the approach of section 2, Jenkins and Barratt [23] show that this leads to the boundary condition

$$\frac{\partial W}{\partial \nabla \mathbf{n}} \boldsymbol{\nu} + \frac{\partial w}{\partial \mathbf{n}} = \lambda \mathbf{n} \qquad (51)$$

W again denoting the Frank–Oseen energy and λ an arbitrary scalar.

If one introduces the representation (Eq. 35), it follows that

$$W = W(\theta, \phi, \nabla\theta, \nabla\phi), \quad w = \omega(\theta, \phi) \qquad (52)$$

and using the methods of the previous section the boundary condition (Eq. 51) becomes

$$\frac{\partial W}{\partial \nabla \theta} \cdot \boldsymbol{\nu} + \frac{\partial \omega}{\partial \theta} = 0, \quad \frac{\partial W}{\partial \nabla \phi} \cdot \boldsymbol{\nu} + \frac{\partial \omega}{\partial \phi} = 0 \quad (53)$$

Frequently these boundary conditions can be reduced to equations for θ and ϕ at the interface by eliminating the derivatives of θ and ϕ, after they have been obtained from the equilibrium equations.

More recently, however, it has become apparent that the situation can be more complex with the realization that surface anchoring in nematics can be bistable, as found by Jerome, Pieranski and Boix [24] and Monkade, Boix and Durand [25]. Shortly thereafter, Barberi, Boix and Durand [26] showed that one can switch the surface alignment from one anchoring to the other using an electric field. As Nobili and Durand [27] discuss, one must consider rather more complex forms for the surface energy

than previously in order to model these effects adequately, and they measure some relevant parameters, Sergan and Durand [28] describing further measurements.

1.2.6 Proposed Extensions

Some 25 years ago, Nehring and Saupe [16] proposed that one should add terms linear in second gradients of the director to the Frank–Oseen energy, this ultimately entailing the inclusion of a single additional term, namely

$$K_{13} \, \mathrm{div}((\mathrm{div}\ \mathbf{n})\mathbf{n}) \qquad (54)$$

which clearly proves to be a surface term in the sense that the volume integral integrates to give a surface integral over the boundary. On the grounds of a microscopic calculation they argue that the coefficient in (Eq. 54) is comparable in magnitude to the other coefficients in the Frank–Oseen energy, and so the term should be added.

More recently Oldano and Barbero [29] include such a term and consider the variational problem for static solutions to conclude that in general there are no continuous solutions to this problem, since it is not possible to satisfy all of the boundary conditions that arise in the variational formulation. These are of two types, one corresponding to weak anchoring as described in the previous section, but the other lacking a physical interpretation. This has given rise to some controversy, initially with Hinov [30], but later with contributions from Barbero and Strigazzi [31], Barbero, Sparavigna and Strigazzi [32], Barbero [33], Pergamenshchik [34] and Faetti [35,36]. In general the arguments rest solely upon a resolution of the variational problem, and mostly favour the inclusion of higher order derivatives in the bulk elastic energy in different ways to overcome this difficulty. More recently Barbero and Durand [37] relate this additional term to temperature-induced surface transitions in the alignment, and Barberi, Barbero, Giocondo and Moldovan [38] attempt to measure the corresponding coefficient. Also Lavrentovich and Pergamenshchik [39] argue that such a term explains their observations of stripe domains in experiments with very thin films, and also provide a measurement of the coefficient. While these latter developments relating to experimental observations are of interest, it is rather early to draw any conclusions.

A further recent innovation is due to Ericksen [17] who proposes an extension to the Frank–Oseen theory in order to improve solutions modelling defects. To this end he incorporates some variation in the degree of alignment or the order parameter, and therefore proposes an energy of the form

$$W = W(s, \mathbf{n}, \nabla s, \nabla \mathbf{n}) \tag{55}$$

where $\mathbf{n}$ is again a unit vector describing alignment and s is a scalar representing the degree of order or alignment. By allowing this scalar to tend to zero as one approaches point or line defects, it is possible to avoid the infinite energies that can occur with the Frank–Oseen energy. Some account of analyses based on this development are to be found in the book by Virga [11].

1.3   Equilibrium Theory for Smectic Liquid Crystals

1.3.1   An Energy Function for SmC Liquid Crystals

The first to give serious thought to continuum theory for smectics appears to have been de Gennes' group at Orsay [8]. Amongst other things they present an ener-

gy function for SmC liquid crystals [40], albeit restricted to small perturbations of planar layers. More recently, however, Leslie, Stewart, and Nakagawa [41] derive an energy for such smectics which is not limited to small perturbations, but which is identical to the Orsay energy when so restricted [42]. Below our aim is to present this energy, and show its relationship to that proposed by the Orsay group.

One can conveniently describe the layering in smectic liquid crystals by employing a density wave vector $\mathbf{a}$, which following Oseen [43] and de Gennes [8] is subject to the constraint

$$\operatorname{curl} \mathbf{a} = 0 \tag{56}$$

provided that defects or dislocations in the layering are absent. To describe SmC configurations de Gennes adds a second vector $\mathbf{c}$ perpendicular to $\mathbf{a}$ and in the direction of inclination of the tilt of alignment with respect to $\mathbf{a}$. Leslie, Stewart and Nakagawa [41] invoke two simplifications to reduce mathematical complexity that clearly restrict the range of applicability of their energy, but equally appear very reasonable in many cases. They firstly assume that the layers, although deformable, remain of constant thickness, and also that the angle of tilt with respect to the layer normal remains constant, the latter excluding thermal and pretransitional effects. With these assumptions there is no loss of generality in choosing both $\mathbf{a}$ and $\mathbf{c}$ to be unit vectors, identifying $\mathbf{a}$ with the layer normal, the constraint (Eq. 56) still applying.

As in nematic theory one assumes the existence of an elastic energy W which here is a function of both $\mathbf{a}$ and $\mathbf{c}$ and their gradients, so that

$$W = W(\mathbf{a}, \mathbf{c}, \nabla \mathbf{a}, \nabla \mathbf{c}) \tag{57}$$

The absence of polarity in the alignment implies that the energy is independent of a simultaneous change of sign in **a** and **c**, and thus

$$W(\mathbf{a}, \mathbf{c}, \nabla\mathbf{a}, \nabla\mathbf{c}) = W(-\mathbf{a}, -\mathbf{c}, -\nabla\mathbf{a}, -\nabla\mathbf{c}) \tag{58}$$

Invariance to rigid rotations adds the requirement that

$$W(\mathbf{a}, \mathbf{c}, \nabla\mathbf{a}, \nabla\mathbf{c}) = W(\underline{P}\mathbf{a}, \underline{P}\mathbf{c}, \underline{P}\nabla\mathbf{a}\underline{P}^T, \underline{P}\nabla\mathbf{c}\underline{P}^T) \tag{59}$$

for all proper orthogonal matrices $\underline{P}$. While these conditions suffice for chiral materials, for non-chiral materials symmetry requires that (Eq. 59) is extended to include all orthogonal matrices.

Assuming a quadratic dependence upon gradients one finds for non-chiral materials that [44]

$$\begin{aligned}
2W = {}& K_1^a(tr\nabla\mathbf{a})^2 + K_2^a(\mathbf{c}\cdot\nabla a\mathbf{c})^2 \\
& + 2K_3^a(tr\nabla\mathbf{a})(\mathbf{c}\cdot\nabla a\mathbf{c}) \\
& + K_1^c(tr\nabla\mathbf{c})^2 + K_2^c tr(\nabla\mathbf{c}\nabla\mathbf{c}^T) \\
& + K_3^c(\nabla\mathbf{c}\cdot\nabla\mathbf{c}) \\
& + 2K_4^c(\nabla\mathbf{c}\cdot\nabla\mathbf{c}\mathbf{a}) \\
& + 2K_1^{ac}(tr\nabla\mathbf{c})(\mathbf{c}\cdot\nabla a\mathbf{c}) \\
& + 2K_2^{ac}(tr\nabla\mathbf{c})(tr\nabla\mathbf{a}) \tag{60}
\end{aligned}$$

the coefficients being constants. This expression omits three surface terms associated with

$$tr(\nabla\mathbf{c})^2 - (tr\nabla\mathbf{c})^2, \quad tr(\nabla\mathbf{a})^2 - (tr\nabla\mathbf{a})^2,$$
$$tr(\nabla\mathbf{c}\nabla\mathbf{a}) - (tr\nabla\mathbf{a})(tr\nabla\mathbf{c}), \tag{61}$$

partly on the grounds that the above suffices for present purposes.

For chiral materials Carlsson, Stewart and Leslie [45] show that one must add two terms

$$2K_4^a\mathbf{b}\cdot\nabla a\mathbf{c}, \quad 2K_5^c\mathbf{b}\cdot\nabla c\mathbf{a} \tag{62}$$

plus an additional surface term associated with div **b**, where **b** completes the orthonormal triad with the vectors **a** and **c**, so that

$$\mathbf{b} = \mathbf{a} \times \mathbf{c} \tag{63}$$

The latter of (Eq. 62) describes the characteristic twist of the **c** director about the layer

normal, while the former implies a non-planar equilibrium configuration, and consequently is generally omitted.

One can of course employ any two members of the orthonormal triad **a**, **b** and **c** to describe smectic configurations, and employing **b** and **c** the energy (Eq. 60) becomes [44]

$$\begin{aligned}
2W = {}& A_{12}(\mathbf{b}\cdot\text{curl }\mathbf{c})^2 + A_{21}(\mathbf{c}\cdot\text{curl }\mathbf{b})^2 \\
& + 2A_{11}(\mathbf{b}\cdot\text{curl }\mathbf{c})(\mathbf{c}\cdot\text{curl }\mathbf{b}) \\
& + B_1(\text{div }\mathbf{b})^2 + B_2(\text{div }\mathbf{c})^2 \\
& + B_3\left[\frac{1}{2}(\mathbf{b}\cdot\text{curl }\mathbf{b} + \mathbf{c}\cdot\text{curl }\mathbf{c})\right]^2 \\
& + 2B_{13}(\text{div }\mathbf{b})\left[\frac{1}{2}(\mathbf{b}\cdot\text{curl }\mathbf{b} + \mathbf{c}\cdot\text{curl }\mathbf{c})\right] \\
& + 2C_1(\text{div }\mathbf{c})(\mathbf{b}\cdot\text{curl }\mathbf{c}) \\
& + 2C_2(\text{div }\mathbf{c})(\mathbf{c}\cdot\text{curl }\mathbf{b}) \tag{64}
\end{aligned}$$

As Carlsson, Stewart and Leslie [42] discuss, the terms associated with the coefficients A_{12}, A_{21}, B_1, B_2 and B_3 represent independent deformations of the uniformly aligned planar layers, the remaining terms being coupling terms. Also, as the notation for the coefficients anticipates, this expression reduces to the Orsay energy [40] when one considers small perturbations of uniformly aligned planar layers. The coefficients in the two energies (Eq. 60) and (Eq. 64) are related by

$$\begin{aligned}
& K_1^a = A_{21}, \quad K_2^a = 2A_{11} + A_{12} + A_{21} + B_3 - B_1, \\
& \qquad 2K_3^a = -(2A_{11} + 2A_{21} + B_3) \\
& K_1^c = B_2 - B_3, \quad K_2^c = B_3, \quad K_3^c = B_1 - B_3, \\
& \qquad K_4^c = B_{13} \\
& K_1^{ac} = C_1 + C_2, \quad K_2^{ac} = -C_2 \tag{65}
\end{aligned}$$

as Leslie, Stewart, Carlsson and Nakagawa [44] demonstrate.

1.3.2   Equilibrium Equations

Adopting a similar approach to that by Ericksen for nematics, Leslie, Stewart and Nakagawa [41] assume a principle of virtual work for a volume V of smectic liquid crystal bounded by a surface S of the form

$$\delta \int_V W \, dv = \int_V (\mathbf{F} \cdot \delta\mathbf{x} + \mathbf{G}^a \cdot \Delta\mathbf{a} + \mathbf{G}^c \cdot \Delta\mathbf{c}) \, dv$$
$$+ \int_S (\mathbf{t} \cdot \delta\mathbf{x} + \mathbf{s}^a \cdot \Delta\mathbf{a} + \mathbf{s}^c \cdot \Delta\mathbf{c}) \, ds$$

(66)

where

$$\Delta\mathbf{a} = \delta\mathbf{a} + (\delta\mathbf{x} \cdot \text{grad})\mathbf{a},$$
$$\Delta\mathbf{c} = \delta\mathbf{c} + (\delta\mathbf{x} \cdot \text{grad})\mathbf{c}$$

(67)

$\mathbf{F}$ denotes external body force per unit volume, $\mathbf{G}^a$ and $\mathbf{G}^c$ generalized external body forces per unit volume, $\mathbf{t}$ surface force per unit area, and $\mathbf{s}^a$ and $\mathbf{s}^c$ generalized surface forces per unit area. As before, on account of the assumed incompressibility, the virtual displacement is not arbitrary, but is subject to

$$\text{div } \delta\mathbf{x} = 0$$

(68)

and the variation $\delta\mathbf{a}$ must satisfy

$$\text{curl } \delta\mathbf{a} = \text{curl}(\Delta\mathbf{a} - \nabla\mathbf{a}\delta\mathbf{x}) = 0$$

(69)

on account of (Eq. 56). In addition of course

$$\mathbf{a} \cdot \delta\mathbf{a} = \mathbf{c} \cdot \delta\mathbf{c} = \mathbf{a} \cdot \Delta\mathbf{a} = \mathbf{c} \cdot \Delta\mathbf{c} = 0$$
$$\mathbf{a} \cdot \delta\mathbf{c} + \mathbf{c} \cdot \delta\mathbf{a} = \mathbf{a} \cdot \Delta\mathbf{c} + \mathbf{c} \cdot \Delta\mathbf{a} = 0$$

(70)

given that $\mathbf{a}$ and $\mathbf{c}$ are mutually orthogonal unit vectors.

By a repetition of the arguments for a nematic regarding an infinitesimal, rigid displacement and rotation, one deduces that

$$\int_V \mathbf{F} \, dv + \int_S \mathbf{t} \, ds = 0$$
$$\int_V (\mathbf{x} \times \mathbf{F} + \mathbf{a} \times \mathbf{G}^a + \mathbf{c} \times \mathbf{G}^c) \, dv$$
$$= \int_S (\mathbf{x} \times \mathbf{t} + \mathbf{a} \times \mathbf{s}^a + \mathbf{c} \times \mathbf{s}^c) \, ds = 0$$

(71)

representing balance of forces and moments, respectively. Hence, as above, it is possible to relate the generalized forces to a body moment $\mathbf{K}$ and surface moment ℓ as follows

$$\mathbf{K} = \mathbf{a} \times \mathbf{G}^a + \mathbf{c} \times \mathbf{G}^c, \quad \ell = \mathbf{a} \times \mathbf{s}^a + \mathbf{c} \times \mathbf{s}^c$$

(72)

essentially allowing the determination of the generalized body forces.

By rewriting the left hand side of (Eq. 66) just as before, it follows that the surface force and generalized surface forces are given respectively by [41]

$$\mathbf{t} = -p\boldsymbol{v} + \nabla\mathbf{a}^T(\boldsymbol{\beta} \times \boldsymbol{v}) + \underset{\sim}{T}^s\boldsymbol{v},$$

$$\underset{\sim}{T}^s = -\nabla\mathbf{a}^T \frac{\partial W}{\partial \nabla\mathbf{a}} - \nabla\mathbf{c}^T \frac{\partial W}{\partial \nabla\mathbf{c}}$$

$$\mathbf{s}^a = \alpha\mathbf{a} + \mu\mathbf{c} + \boldsymbol{v} \times \boldsymbol{\beta} + \underset{\sim}{S}^a\boldsymbol{v}, \quad \underset{\sim}{S}^a = \frac{\partial W}{\partial \nabla\mathbf{a}},$$

$$\mathbf{s}^c = \lambda\mathbf{c} + \mu\mathbf{a} + \underset{\sim}{S}^c\boldsymbol{v}, \quad \underset{\sim}{S}^c = \frac{\partial W}{\partial \nabla\mathbf{c}}$$

(73)

where $\boldsymbol{v}$ denotes the unit surface normal, p an arbitrary pressure due to the assumed incompressibility, $\boldsymbol{\beta}$ an arbitrary vector arising from the constraint (Eq. 69), and α, λ and μ arbitrary scalars stemming from the constraints (Eq. 70). In addition one obtains the balance laws

$$\mathbf{F} - \text{grad } p - \nabla\mathbf{a}^T(\text{curl } \boldsymbol{\beta}) + \text{div } \underset{\sim}{T}^s = 0$$

$$\text{div } \underset{\sim}{S}^a - \frac{\partial W}{\partial\mathbf{a}} + \mathbf{G}^a + \gamma\mathbf{a} + \kappa\mathbf{c} + \text{curl } \boldsymbol{\beta} = 0$$

$$\text{div } \underset{\sim}{S}^c - \frac{\partial W}{\partial\mathbf{c}} + \mathbf{G}^c + \tau\mathbf{c} + \kappa\mathbf{a} = 0$$

(74)

γ, κ and τ being arbitrary scalars again arising from the constrains (Eq. 70). The first of equations (Eq. 74) is clearly the point form of the balance of forces, the first of equations (Eq. 71), and the remaining two are equivalent to the second of (Eq. 71) representing balance of moments, this requiring the generalization of the identity (Eq. 12). Also one can show that the surface moment ℓ is given by

$$\ell = (\boldsymbol{\beta} \cdot \mathbf{a})\boldsymbol{v} - (\mathbf{a} \cdot \boldsymbol{v})\boldsymbol{\beta} + \mathbf{a} \times \frac{\partial W}{\partial \nabla\mathbf{a}}\boldsymbol{v}$$
$$+ \mathbf{c} \times \frac{\partial W}{\partial \nabla\mathbf{c}}\boldsymbol{v}$$

(75)

this combining Eqs. (72) and (73).

If external forces and moments are absent, it is possible to combine equations (Eq. 74) to obtain the integral

$$p + W = p_o \qquad (76)$$

where p_o is an arbitrary constant pressure. With certain restrictions on the external body force and moment, a similar result follows when these terms are present. Hence as for a nematic, the balance of forces need not concern us except possibly to compute surface forces, and we therefore have two Euler–Lagrange type equations representing balance of moments.

Boundary conditions for the above theory tend to be very similar to those employed in nematics, with little so far emerging by way of genuine smectic boundary conditions.

1.4 Dynamic Theory for Nematics

1.4.1 Balance Laws

To derive a dynamic theory one can of course extend the above formulation of equilibrium theory employing generalized body and surface forces as in the initial derivation [7,15]. Here, however, we prefer a different approach [46], which, besides providing an alternative, is more direct in that it follows traditional continuum mechanics more closely, although introducing body and surface moments usually excluded, as well as a new kinematic variable to describe alignment of the anisotropic axis.

As above we assume that the liquid crystal is incompressible, and thus the velocity vector $\mathbf{v}$ is subject to the constraint

$$\text{div } \mathbf{v} = 0 \qquad (77)$$

with the result that conservation of mass reduces to the statement that the density ρ is

constant in a homogeneous material. As before the director $\mathbf{n}$ is constrained to be of unit magnitude. Since thermal effects are excluded, our two balance laws are those representing conservation of linear and angular momentum, which for a volume V of liquid crystal bounded by surface S take the forms

$$\frac{\mathrm{d}}{\mathrm{d}t} \int_V \rho \mathbf{v} \, \mathrm{d}v = \int_V \mathbf{F} \, \mathrm{d}v + \int_S \mathbf{t} \, \mathrm{d}s$$

$$\frac{\mathrm{d}}{\mathrm{d}t} \int_V \rho \mathbf{x} \times \mathbf{v} \, \mathrm{d}v$$

$$= \int_V (\mathbf{x} \times \mathbf{F} + \mathbf{K}) \, \mathrm{d}v + \int_S (\mathbf{x} \times \mathbf{t} + \boldsymbol{\ell}) \, \mathrm{d}s \qquad (78)$$

where $\mathbf{F}$ and $\mathbf{K}$ are body force and moment per unit volume, $\mathbf{t}$ and $\boldsymbol{\ell}$ surface force and moment per unit area, respectively, $\mathbf{x}$ the position vector, and the time derivative the material time derivative. The inertial term associated with the director in the latter of equations (Eq. 78) is omitted on the grounds that it is negligible.

The force and moment on a surface with unit normal $\boldsymbol{v}$ are given by

$$\mathbf{t} = -p\boldsymbol{v} + \underset{\sim}{T}\boldsymbol{v}, \quad \boldsymbol{\ell} = \underset{\sim}{L}\boldsymbol{v} \qquad (79)$$

where p is an arbitrary pressure arising from the assumed incompressibility, and $\underset{\sim}{T}$ and $\underset{\sim}{L}$ are the stress and couple stress matrices or tensors, respectively. As a consequence one can express (Eq. 78) in point form

$$\rho \frac{\mathrm{d}\mathbf{v}}{\mathrm{d}t} = \mathbf{F} - \text{grad } p + \text{div } \underset{\sim}{T},$$

$$0 = \mathbf{K} + \hat{\mathbf{T}} + \text{div } \underset{\sim}{L} \qquad (80)$$

where $\hat{\mathbf{T}}$ is the axial vector associated with the asymmetric matrix $\underset{\sim}{T}$, so that

$$\hat{T}_x = T_{zy} - T_{yz}, \quad \hat{T}_y = T_{xz} - T_{zx},$$

$$\hat{T}_z = T_{yx} - T_{xy}, \qquad (81)$$

and the divergence is with respect to the second index of both $\underset{\sim}{T}$ and $\underset{\sim}{L}$. Finally we remark that for isotropic liquids $\mathbf{K}$ and $\underset{\sim}{L}$ are generally assumed to be absent, so that angular momentum reduces to $\hat{\mathbf{T}}$ being zero, or equivalently that the stress matrix $\underset{\sim}{T}$ is symmetric.

1.4.2   A Rate of Work Hypothesis

Somewhat analogous to our earlier principle of virtual work, we here assume that the rate of working of external forces and moments either goes into increasing the kinetic and elastic energies, or is dissipated as viscous dissipation. Thus for a volume V of liquid crystal bounded by surface S

$$\int_V (\mathbf{F} \cdot \mathbf{v} + \mathbf{K} \cdot \mathbf{w}) \, dv + \int_S (\mathbf{t} \cdot \mathbf{v} + \boldsymbol{\ell} \, \mathbf{w}) \, ds$$

$$= \frac{d}{dt} \int_V \left(\frac{1}{2} \rho \mathbf{v} \cdot \mathbf{v} + W \right) dv + \int_V D \, dv \quad (82)$$

in which $\mathbf{w}$ denotes the local angular velocity, W is the Frank–Oseen energy, and D represents the rate of viscous dissipation. With the aid of equations (Eq. 79) the above can be expressed in point form, and following some simplification through use of equations (Eq. 80) one obtains

$$tr(\underset{\sim}{T} \underset{\sim}{V}^T) + tr(\underset{\sim}{L} \underset{\sim}{W}^T) - \mathbf{w} \cdot \hat{\mathbf{T}} = \frac{dW}{dt} + D \quad (83)$$

where the velocity and angular velocity gradient matrices $\underset{\sim}{V}$ and $\underset{\sim}{W}$ take the forms

$$\underset{\sim}{V} = \left[\frac{\partial v_i}{\partial x_j} \right], \quad \underset{\sim}{W} = \left[\frac{\partial w_i}{\partial x_j} \right] \quad (84)$$

respectively.

Noting that the material time derivative of the director $\mathbf{n}$ is given by

$$\frac{d}{dt} \mathbf{n} = \mathbf{w} \times \mathbf{n} \quad (85)$$

and also that one can show that

$$\frac{d}{dt} (\nabla \mathbf{n}) = \nabla \left(\frac{d\mathbf{n}}{dt} \right) - \nabla \mathbf{n} \underset{\sim}{V} \quad (86)$$

it is possible by using the chain rule to express the material derivative of the energy W in a form linear in the velocity gradients,

angular velocity gradients and the angular velocity. Also by appeal to the identity (Eq. 12), the resultant expression cancels the contributions from the static terms on the left hand side of (Eq. 83), given by Eqs. (20) and (22), and thus the rate of work postulate reduces to

$$tr(\underset{\sim}{T}^d \underset{\sim}{V}^T) + tr(\underset{\sim}{L}^d \underset{\sim}{W}^T) - \mathbf{w} \cdot \hat{\mathbf{T}}^d = D > 0 \quad (87)$$

the superscript d denoting dynamic contributions, and noting that the rate of viscous dissipation is necessarily positive. Alternatively of course one can argue from the linearity of these terms in the velocity and angular velocity gradients that the static contributions are indeed given by our earlier expressions.

To complete our dynamic theory, it is necessary to prescribe forms for the dynamic stress and couple stress. The simplest choice consistent with known effects appears to be that at a material point

$$\underset{\sim}{T}^d \text{ and } \underset{\sim}{L}^d \text{ are functions of } \underset{\sim}{V}, \mathbf{n} \text{ and } \mathbf{w} \quad (88)$$

all evaluated at that point at the given instant. However, since dependence upon the gradients of the angular velocity is excluded, it follows at once from the inequality (Eq. 87), given that these gradients occur linearly, that

$$\underset{\sim}{L}^d = 0 \quad (89)$$

in agreement with the earlier formulation [7]. Hence the rate of viscous dissipation inequality reduces to

$$D = tr(\underset{\sim}{T}^d \underset{\sim}{V}^T) - \mathbf{w} \cdot \hat{\mathbf{T}} > 0 \quad (90)$$

which we employ below to limit viscous coefficients.

1.4.3   The Viscous Stress

Invariance at once requires that the assumption (Eq. 88) for the dynamic part of the stress matrix be replaced by

$\underset{\sim}{T}^d$ is a function of $\underset{\sim}{A}$, **n** and $\boldsymbol{\omega}$ (91)

all evaluated at that point at that instant, where

$$2\underset{\sim}{A} = \underset{\sim}{V} + \underset{\sim}{V}^T, \quad \boldsymbol{\omega} = \mathbf{w} - \frac{1}{2}\,\mathrm{curl}\,\mathbf{v} \quad (92)$$

$\underset{\sim}{A}$ being the familiar rate of strain tensor or matrix, and $\boldsymbol{\omega}$ the angular velocity relative to the background rotation of the continuum. While it is possible to derive the viscous stress from the above assumption [46], we opt here for a more direct approach, and replace (Eq. 91) by

$\underset{\sim}{T}^d$ is a function of $\underset{\sim}{A}$, **n** and **N**   (93)

all evaluated at that point at that instant, where using (Eq. 85)

$$\mathbf{N} = \boldsymbol{\omega} \times \mathbf{n} = \frac{\mathrm{d}}{\mathrm{d}t}\mathbf{n} - \frac{1}{2}(\underset{\sim}{V} - \underset{\sim}{V}^T)\mathbf{n} \quad (94)$$

this restricting the dependence upon the relative rotation, essentially discounting the component parallel to the director **n**, which can be shown to be zero in any event [46].

Nematic symmetry requires that (Eq. 93) be an isotropic function and independent of a change of sign in the director **n**. Thus assuming a linear dependence upon $\underset{\sim}{A}$ and **N** one finds that

$$\underset{\sim}{T}^d = \alpha_1 \mathbf{n} \cdot \underset{\sim}{A}\mathbf{n}\,\mathbf{n} * \mathbf{n} + \alpha_2 \mathbf{N} * \mathbf{n} + \alpha_3 \mathbf{n} * \mathbf{N}$$
$$+ \alpha_4 \underset{\sim}{A} + \alpha_5 \underset{\sim}{A}\mathbf{n} * \mathbf{n} + \alpha_6 \mathbf{n} * \underset{\sim}{A}\mathbf{n}, \quad (95)$$

where again $\mathbf{a} * \mathbf{b}$ denotes the matrix with $(i,j)^{\text{th}}$ element $a_i b_j$, and the αs are constants. Ignoring thermal effects, the above is also the form that one obtains for cholesterics or chiral nematics.

Somewhat straightforwardly it follows from the above that

$$\hat{\mathbf{T}}^d = \mathbf{n} \times \mathbf{g} \quad (96)$$

where the vector **g** takes the form

$$\mathbf{g} = -\gamma_1 \mathbf{N} - \gamma_2 \underset{\sim}{A}\mathbf{n},$$
$$\gamma_1 = \alpha_3 - \alpha_2, \quad \gamma_2 = \alpha_6 - \alpha_5 \quad (97)$$

Also employing Eqs. (92), (94) and (96) the viscous dissipation inequality (Eq. 90) becomes

$$D = tr(\underset{\sim}{T}^d\underset{\sim}{A}) - \boldsymbol{\omega} \cdot \hat{\mathbf{T}}^d = tr(\underset{\sim}{T}^d\underset{\sim}{A}) - \mathbf{g} \cdot \mathbf{N} > 0$$
$$(98)$$

The above of course restricts possible values for the viscous coefficients and one can readily deduce from it that [47]

$$\alpha_4 > 0, \quad 2\alpha_4 + \alpha_5 + \alpha_6 > 0, \quad (99)$$
$$3\alpha_4 + 2\alpha_5 + 2\alpha_6 + 2\alpha_1 > 0$$
$$\gamma_1 > 0, \quad (\alpha_2 + \alpha_3 + \gamma_2)^2 < 4\gamma_1(2\alpha_4 + \alpha_5 + \alpha_6)$$

the calculation aided by choosing axes parallel to **n** and **N**. However, in many cases it proves simpler to deduce consequences of (Eq. 98) by writing down the dissipation function for the particular flow under consideration, rather than try to derive results from the above inequalities.

By invoking Onsager relations, Parodi [48] argues that one restrict the viscous coefficients to satisfy the relationship

$$\gamma_2 = \alpha_3 + \alpha_2 \quad (100)$$

but subsequently Currie [49] shows that this relationship also follows as a result of a stability argument. As a consequence this condition between the viscous coefficients is now generally accepted, and leads to some simplification in the use of the theory. For example, when (Eq. 100) holds, Ericksen [21] shows that the viscous stress and the vector **g** follow directly from the dissipation function D through

$$\underset{\sim}{T}^d = \frac{1}{2}\frac{\partial D}{\partial \underset{\sim}{V}}, \quad \mathbf{g} = -\frac{1}{2}\frac{\partial D}{\partial \dot{\mathbf{n}}}, \quad \dot{\mathbf{n}} = \frac{\mathrm{d}}{\mathrm{d}t}\mathbf{n} \quad (101)$$

results that we require below.

1.4.4   Equations of Motion

If one combines the Eqs. (20), (22), (79), (89) and (96), it is possible to express the balance law for angular moment, the second of equations (Eq. 80) in the form

$$\mathbf{n} \times \left(\operatorname{div} \underset{\sim}{S}^s - \frac{\partial W}{\partial \mathbf{n}} + \mathbf{G} + \mathbf{g} \right) = 0 \qquad (102)$$

this involving some manipulation and the use of the identity (Eq. 12), a result perhaps more easily derived employing Cartesian tensor notation. Equivalently (Eq. 102) becomes

$$\operatorname{div} \underset{\sim}{S}^s - \frac{\partial W}{\partial \mathbf{n}} + \mathbf{G} + \mathbf{g} = \gamma \mathbf{n} \qquad (103)$$

a rather natural extension of the second of equations (Eq. 23) given the result (Eq. 89). With angular momentum in this form one can simplify the balance law for linear moment, the first of equations (Eq. 80), in a manner rather similar to the derivation of (Eq. 32) to give

$$\rho \frac{d\mathbf{v}}{dt} = \mathbf{F} + \nabla \mathbf{n}^T \mathbf{G} - \operatorname{grad}(p + W)$$
$$+ \nabla \mathbf{n}^T \mathbf{g} + \operatorname{div} \underset{\sim}{T}^d \qquad (104)$$

Further, if the body force and moment satisfy (Eq. 30), this last equation reduces to

$$\rho \frac{d\mathbf{v}}{dt} = -\operatorname{grad} \tilde{p} + \nabla \mathbf{n}^T \mathbf{g} + \operatorname{div} \underset{\sim}{T}^d,$$
$$\tilde{p} = p + W - \psi, \qquad (105)$$

which is clearly simpler than the original.

As in Section 1.2.4, if one introduces a representation for the director $\mathbf{n}$ referred to Cartesian axes that trivially satisfies the constraint upon it, say

$$\mathbf{n} = \mathbf{f}(\theta, \phi), \quad \mathbf{n} \cdot \frac{\partial \mathbf{f}}{\partial \theta} = \mathbf{n} \cdot \frac{\partial \mathbf{f}}{\partial \phi} = 0 \qquad (106)$$

it also follows that

$$(107)$$
$$\nabla \mathbf{n} = \frac{\partial \mathbf{f}}{\partial \theta} * \nabla \theta + \frac{\partial \mathbf{f}}{\partial \phi} * \nabla \phi, \quad \dot{\mathbf{n}} = \frac{\partial \mathbf{f}}{\partial \theta} \dot{\theta} + \frac{\partial \mathbf{f}}{\partial \phi} \dot{\phi}$$

As before, if an external field satisfying (Eq. 30) is present, one has

$$(108)$$
$$W = W(\theta, \phi, \nabla\theta, \nabla\phi), \quad \psi = \chi(\mathbf{x}, \theta, \phi)$$
$$D = 2\Delta(\theta, \phi, \dot{\theta}, \dot{\phi}, \nabla\mathbf{v}), \quad \dot{\theta} = \frac{d\theta}{dt}, \quad \dot{\phi} = \frac{d\phi}{dt}$$

and proceeding as earlier and noting that

$$\frac{\partial \Delta}{\partial \dot{\theta}} = \frac{1}{2} \frac{\partial D}{\partial \dot{\mathbf{n}}} \cdot \frac{\partial \mathbf{f}}{\partial \theta}, \quad \frac{\partial \Delta}{\partial \dot{\phi}} = \frac{1}{2} \frac{\partial D}{\partial \dot{\mathbf{n}}} \cdot \frac{\partial \mathbf{f}}{\partial \phi} \qquad (109)$$

one can show that Eqs. (103) and (105) can be recast as

$$\operatorname{div}\left(\frac{\partial W}{\partial \nabla \theta} \right) - \frac{\partial W}{\partial \theta} + \frac{\partial \chi}{\partial \theta} - \frac{\partial \Delta}{\partial \dot{\theta}} = 0$$

$$\operatorname{div}\left(\frac{\partial W}{\partial \nabla \phi} \right) - \frac{\partial W}{\partial \phi} + \frac{\partial \chi}{\partial \phi} - \frac{\partial \Delta}{\partial \dot{\phi}} = 0 \qquad (110)$$

and

$$\rho \dot{\mathbf{v}} = -\operatorname{grad} \tilde{p} + \operatorname{div}\left(\frac{\partial \Delta}{\partial \underset{\sim}{V}} \right) - \frac{\partial \Delta}{\partial \dot{\theta}} \nabla\theta - \frac{\partial \Delta}{\partial \dot{\phi}} \nabla\phi$$

$$(111)$$

respectively, the divergence in the latter with respect to the second index. As Ericksen [21] shows, it is possible to present the above reformulation of the equations in terms of curvilinear coordinates. However, given our rather restrictive notation, an attempt to summarise this here is more likely to confuse than to enlighten, and therefore we refer the interested reader to the original paper.

Boundary conditions for dynamic theory simply add the customary non-slip hypothesis upon the velocity vector to the conditions described above for static theory, with only a very occasional reference to the inclusion of a surface viscosity for motions of the director at a solid interface.

1.5   References

[1] C. W. Oseen, *Ark. Mat. Astron. Fys.* **1925**, *19A*, 1.
[2] H. Zocher, *Phys. Z.* **1927**, *28*, 790.
[3] H. Zocher, *Trans. Faraday Soc.* **1933**, *29*, 945.
[4] F. C. Frank, *Disc. Faraday Soc.* **1958**, *25*, 19.
[5] J. L. Ericksen, *Arch. Rat. Mech. Anal.* **1962**, *9*, 371.
[6] J. L. Ericksen, *Trans. Soc. Rheol.* **1961**, *5*, 23.

[7] F. M. Leslie, *Arch. Rat. Mech. Anal.* **1968**, *28*, 265.

[8] P. G. de Gennes, J. Prost, *The Physics of Liquid Crystals*, 2nd edn, Oxford University Press **1993**.

[9] S. Chandrasekhar, *Liquid Crystals*, 2nd edn. Cambridge University Press **1992**.

[10] L. M. Blinov, *Electro-optical and Magneto-optical Properties of Liquid Crystals*, J. Wiley and Sons, New York **1983**.

[11] E. G. Virga, *Variational Theories for Liquid Crystals*, Chapman and Hall, London **1994**.

[12] M. J. Stephen, J. P. Straley, *Rev. Mod. Phys.* **1974**, *46*, 617.

[13] J. L. Ericksen, *Adv. Liq. Cryst.* **1976**, *2*, 233.

[14] J. T. Jenkins, *Ann. Rev. Fluid. Mech.* **1978**, *10*, 197.

[15] F. M. Leslie, *Adv. Liq. Cryst.* **1979**, *4*, 1.

[16] J. Nehring, A. Saupe, *J. Chem. Phys.* **1971**, *54*, 337.

[17] J. L. Ericksen, *Arch. Rat. Mech. Anal.* **1991**, *113*, 97.

[18] J. L. Ericksen, *Phys. Fluids* **1966**, *9*, 1205.

[19] J. T. Jenkins, *J. Fluid. Mech.* **1971**, *45*, 465.

[20] H. J. Deuling, *Mol. Cryst. Liq. Cryst.* **1972**, *19*, 123.

[21] J. L. Ericksen, *Q. J. Mech. Appl. Math.* **1976**, *29*, 203.

[22] A. Rapini, M. Papoular, *J. Phys. (Paris) Colloq.* **1969**, *30*, C4, 54.

[23] J. T. Jenkins, P. J. Barratt, *Q. J. Mech. Appl. Math.* **1974**, *27*, 111.

[24] B. Jerome, P. Pieranski, M. Boix, *Europhys. Lett.* **1988**, *5*, 693.

[25] M. Monkade, M. Boix, G. Durand, *Europhys. Lett.* **1988**, *5*, 697.

[26] R. Barberi, M. Boix, G. Durand, *Appl. Phys. Lett.* **1989**, *55*, 2506.

[27] M. Nobili, G. Durand, *Europhys. Lett.* **1994**, *25*, 527.

[28] V. Sergan, G. Durand, *Liq. Cryst.* **1995**, *18*, 171.

[29] C. Oldano, G. Barbero, *J. Phys. (Paris) Lett.* **1985**, *46*, 451.

[30] H. P. Hinov, *Mol. Cryst. Liq. Cryst.* **1987**, *148*, 157.

[31] G. Barbero, A. Strigazzi, *Liq. Cryst.* **1989**, *5*, 693.

[32] G. Barbero, A. Sparavigna, A. Strigazzi, *Nuovo Cim. D* **1990**, *12*, 1259.

[33] G. Barbero, *Mol. Cryst. Liq. Cryst.* **1991**, *195*, 199.

[34] V. M. Pergamenshchik, *Phys. Rev. E* **1993**, *48*, 1254.

[35] S. Faetti, *Phys. Rev. E* **1994**, *49*, 4192.

[36] S. Faetti, *Phys. Rev. E* **1994**, *49*, 5332.

[37] G. Barbero, G. Durand, *Phys. Rev. E* **1993**, *48*, 1942.

[38] R. Barberi, G. Barbero, M. Giocondo, R. Moldovan, *Phys. Rev. E* **1994**, *50*, 2093.

[39] O. D. Lavrentovich, V. M. Pergamenshchik, *Phys. Rev. Lett.* **1994**, *73*, 979.

[40] Orsay Group on Liquid Crystals, *Solid State Commun.* **1971**, *9*, 653.

[41] F. M. Leslie, I. W. Stewart, M. Nakagawa, *Mol. Cryst. Liq. Cryst.* **1991**, *198*, 443.

[42] T. Carlsson, I. W. Stewart, F. M. Leslie, *Liq. Cryst.* **1991**, *9*, 661.

[43] C. W. Oseen, *Trans. Faraday Soc.* **1933**, *29*, 883.

[44] F. M. Leslie, I. W. Stewart, T. Carlsson, M. Nakagawa, *Cont. Mech. Thermodyn.* **1991**, *3*, 237.

[45] T. Carlsson, I. W. Stewart, F. M. Leslie, *J. Phys. A* **1992**, *25*, 2371.

[46] F. M. Leslie, *Cont. Mech. Thermodyn.* **1992**, *4*, 167.

[47] F. M. Leslie, *Q. J. Mech. Appl. Math.* **1966**, *19*, 357.

[48] P. Parodi, *J. Phys. (Paris)* **1970**, *31*, 581.

[49] P. K. Currie, *Mol. Cryst. Liq. Cryst.* **1974**, *28*, 335.

2 Molecular Theories of Liquid Crystals

Mikhail A. Osipov

2.1 Introduction

Molecular theories can provide important additional information about the properties and structure of liquid crystals because they enable one to understand (at least partially and in a simplified way) the onset of liquid crystalline order at the microscopic level. However, the development of a realistic molecular theory for liquid crystals appears to be a challenging problem for two main reasons. First, the intermolecular interaction potentials are not known exactly. These potentials are generally expected to be rather complex reflecting the relatively complex structure of typical mesogenic molecules. Second, it is hardly possible to do good statistical mechanics with such complex potentials even if they are known. At present there exists no regular method to calculate even the pair correlation function for a simple anisotropic fluid and one has to rely on rather crude approximations. The latter difficulty can, in principle, be overcome by means of computer simulations. Such simulations, however, appear to be very time consuming if realistic molecular models are employed. Even in the case of simple potentials one sometimes needs very large systems (for example, up to 8000 particles to simulate the

polar smectic phase with long-range dipole–dipole interactions [1]). On the other hand, such simulations with realistic potentials usually require extensive interpretation, as in the case of a real experiment. In fact, one would have to make too many simulations to trace a relation between the features of the molecular structure and the macroscopic parameters of liquid crystal phases in an empirical way. Thus, in general, computer simulations are not a substitution for a molecular theory, but merely an independent source of information. In particular, computer simulations provide a unique tool for testing and generating various approximations that are inevitably used in any molecular theory.

Thus, the primary goal of a molecular theory is to obtain a qualitative insight into the molecular origin of various effects in liquid crystals. First of all, it is important to understand which properties of liquid crystals are actually determined by some basic and simple characteristics of molecular structure (like, for example, the elongated or disc-like molecular shape or the polarizability anisotropy), and which properties are particularly sensitive to the details of molecular structure, including flexibility, biaxiality or even the location of particular elements of structure. In this short chapter we will not be able to address all of these questions. However, we make an attempt to present the basics of the molecular theory of the nematic and simple smectic phases. We will also provide some additional references for a reader with more specific interests.

We start with the microscopic definitions and discussion of the nematic and smectic order parameters and then proceed with some elementary information about anisotropic intermolecular interactions in liquid crystals. Then we discuss in more detail the main molecular theories of the nematic–isotropic phase transition and conclude with a consideration of molecular models for smectic A and smectic C phases.

2.2 Microscopic Definition of the Order Parameters for Nematic and Smectic Phases

2.2.1 Uniaxial Nematic Phase

The uniaxial nematic phase possesses a quadrupole-type symmetry and is characterized by the order parameter $Q_{\alpha\beta}$ which is a symmetric traceless second-rank tensor:

$$Q_{\alpha\beta} = S\left(n_\alpha n_\beta - \frac{1}{3}\delta_{\alpha\beta}\right) \qquad (1)$$

where the unit vector n is the director that specifies the preferred orientation of the primary molecular axes. We note that the corresponding macroscopic axis is nonpolar and therefore it is better represented by the quadratic combination $n_\alpha n_\beta$. The quantity S is the scalar order parameter that characterizes the degree of nematic ordering.

From the microscopic point of view the orientation of a rigid molecule i can be spec-

ified by the unit vectors a_i and b_i in the direction of long and short molecular axes, respectively; $(a_i \cdot b_i) = 0$. It should be noted that the definition of the primary molecular axis is somewhat arbitrary for a molecule without symmetry elements. Sometimes the axis a_i is taken along the main axis of the molecular inertia tensor. However, in many cases it is quite difficult to know beforehand which molecular axis will actually be ordered along the director. This can be a particularly difficult problem for a guest molecule in a nematic host [2].

Here we assume for simplicity that the primary axis is well defined. Then the scalar order parameter is given by the average

$$S = \left\langle P_2\left((a_i \cdot b_i)\right)\right\rangle \qquad (2)$$

where $P_2(x)$ is the second Legendre polynomial and the brackets $\langle\ldots\rangle$ denote the statistical averaging.

We note that the averaging in Eq. (2) is performed with the one-particle distribution function $f_1((a \cdot n))$ that determines the probability of finding a molecule with a given orientation of the long axis at a given point in space. In the uniaxial nematic phase the distribution function depends only on the angle ω between the long axis and the director. Then Eq. (2) can be rewritten as

$$S = \frac{1}{2}\int P_2(\cos\omega) f_1(\cos\omega)\,\mathrm{d}\cos\omega \qquad (3)$$

In mixtures of liquid crystals, the molecules of different components may possess different degrees of nematic ordering. In this case the nematic order parameters, S_α, for different components α are calculated separately using different distribution functions $f_\alpha(\cos\omega)$.

The definition of the order parameter (Eq. 2) is not entirely complete because one has to know the orientation of the director. In the general case the director appears self-consistently as a result of the breakdown of symmetry during the phase transition and sometimes its orientation is unknown beforehand (as, for example, in simulations). Thus, from the theoretical point of view it is more consistent to define directly the tensor order parameter as a true thermodynamic average:

$$Q_{\alpha\beta} = \left\langle a_\alpha a_\beta - \frac{1}{3}\delta_{\alpha\beta}\right\rangle = S\left(n_\alpha n_\beta - \frac{1}{3}\delta_{\alpha\beta}\right) \quad (4)$$

where the order parameter S is given by Eq. (2).

The molecular statistical definition of the order parameter can be clarified if one considers, for example, the magnetic susceptibility of the nematic phase. The anisotropic part of the susceptibility tensor $\chi_{\alpha\beta}$ can be considered as an order parameter [3] because it vanishes in the isotropic phase and is nonzero in the nematic phase. In addition, the macroscopic magnetic susceptibility can be written as a sum of contributions from individual molecules:

$$\chi_{\alpha\beta} = \rho \sum_i \left\langle \chi^{\mathrm{M}}_{\alpha\beta} \right\rangle \quad (5)$$

where $\chi^{\mathrm{M}}_{\alpha\beta}$ is the molecular susceptibility tensor and ρ is the number density.

Equation (4) is valid with high accuracy because induced magnetic dipoles interact only weakly. By contrast, interaction between the induced electric dipoles is strong and produces substantial local field effects which do not allow one to express the dielectric permittivity of the nematic phase in terms of the molecular parameters in a simple way (see, for example, [4]).

The molecular susceptibility $\chi^{\mathrm{M}}_{\alpha\beta}$ can be diagonalized:

$$\chi^{\mathrm{M}}_{\alpha\beta} = \chi_{11} b_\alpha b_\beta + \chi_{22} c_\alpha c_\beta + \chi_{33} a_\alpha a_\beta \quad (6)$$

where the orthogonal unit vectors a, b and c are the principal axes.

Now we have to average Eq. (5) to get the expression for the order parameter. In the nematic phase the orientational distribution function depends only on the primary molecular axis a. Thus the two short axes b and c are completely equivalent in the statistical sense and one obtains

$$\ldots \left\langle b_\alpha b_\beta \right\rangle = \left\langle c_\alpha c_\beta \right\rangle = \frac{1}{2}\left(\delta_{\alpha\beta} - \left\langle a_\alpha a_\beta \right\rangle\right) \quad (7)$$

where we have used the general relation

$$\delta_{\alpha\beta} = a_\alpha a_\beta + b_\alpha b_\beta c_\alpha c_\beta$$

Finally the macroscopic magnetic susceptibility can be expressed as

$$\chi_{\alpha\beta} = \bar{\chi}\delta_{\alpha\beta} + \Delta\chi \left\langle a_\alpha a_\beta - \frac{1}{3}\delta_{\alpha\beta}\right\rangle \quad (8)$$

where $\bar{\chi} = \rho(\chi_{11} + \chi_{22} + \chi_{33})$ is the average susceptibility and $\Delta\chi = \rho(\chi_{33} - (\chi_{11} + \chi_{22})/2)$ is the anisotropy of the susceptibility.

One can readily see from Eq. (7) that the traceless part of the average magnetic susceptibility is proportional to the nematic tensor order parameter $Q_{\alpha\beta}$ given by Eq. (3).

2.2.2   Biaxial Nematic Phase

Biaxial nematic ordering has been observed so far only in lyotropic systems. At the same time tilted smectic phases are also biaxial and thus the expressions presented in this section can be of more general use.

In the biaxial nematic phase it is possible to define two orthogonal directors n and m. In this case the magnetic susceptibility tensor can be rewritten as

$$\chi_{\alpha\beta} = \bar{\chi}\,\delta_{\alpha\beta} + \Delta\chi\,Q_{\alpha\beta} + \Delta\chi_\perp\,B_{\alpha\beta} \quad (9)$$

where $\Delta\chi_\perp = \rho(\chi_{11} - \chi_{22})$ is the transverse anisotropy of the molecular susceptibility that represents biaxiality, and where

$$B_{\alpha\beta} = \left\langle\left(b_\alpha b_\beta - c_\alpha c_\beta\right)\right\rangle \quad (10)$$

Now the traceless part of the magnetic susceptibility is a sum of two terms proportional to two tensor order parameters of the biaxial nematic phase: $Q_{\alpha\beta}$ and $B_{\alpha\beta}$.

In the uniaxial nematic phase $B_{\alpha\beta}=0$ and the tensor order parameter $Q_{\alpha\beta}$ is uniaxial. By contrast, in the biaxial phase the order parameter $Q_{\alpha\beta}$ can be written as a sum of a uniaxial and a biaxial part:

$$Q_{\alpha\beta} = S\left(n_\alpha n_\beta - \frac{1}{3}\delta_{\alpha\beta}\right)$$

$$+\Delta Q\left(m_\alpha m_\beta - l_\alpha l_\beta\right) \tag{11}$$

where the unit vector $l=[n\times m]$ and where S is the largest eigenvalue of the tensor $Q_{\alpha\beta}$ (for prolate molecules).

The parameters S and ΔQ can be expressed as [8]

$$S = \langle P_2\left((a \cdot n)\right)\rangle \tag{12}$$

$$\Delta Q = \frac{1}{3}\left(\langle P_2\left((a \cdot m)\right)\rangle - \langle P_2\left((a \cdot l)\right)\rangle\right) \tag{13}$$

The tensor order parameter $B_{\alpha\beta}$ can be expressed in the same general form as $Q_{\alpha\beta}$:

$$B_{\alpha\beta} = S'\left(n_\alpha n_\beta - \frac{1}{3}\delta_{\alpha\beta}\right)$$

$$+D\left(m_\alpha m_\beta - l_\alpha l_\beta\right) \tag{14}$$

where

$$S' = \langle P_2\left((b \cdot n)\right)\rangle - \langle P_2\left((c \cdot n)\right)\rangle \tag{15}$$

$$D = \frac{1}{3}\left(\langle P_2\left((b \cdot m)\right)\rangle - \langle P_2\left((b \cdot l)\right)\rangle \right.$$
$$\left. + \langle P_2\left((c \cdot m)\right)\rangle - \langle P_2\left((c \cdot l)\right)\rangle\right) \tag{16}$$

Here the parameter S' characterizes the tendency of the molecular short axes to be ordered along the main director n and the parameter ΔQ describes the ordering of the primary axis a along the director m. In the case of perfect ordering of the primary molecular axes ($S=1$) the parameter

$S'=0$, $\Delta Q=0$ and the biaxial ordering in the phase is described by the parameter $B_{\alpha\beta}=D\left(m_\alpha m_\beta - l_\alpha l_\beta\right)$.

2.2.3 Smectic A and C Phases

Smectic ordering in liquid crystals is usually characterized by the complex order parameter $\rho_\alpha e^{i\psi}$ introduced by de Gennes [3]. Here $\rho_\alpha=\langle\cos(q \cdot r)\rangle$ is the amplitude of the density wave, ψ is the phase and q is the wave vector. This order parameter appears naturally in the Fourier expansion of the one-particle density $\rho(r)$.

The order parameter of the SmC phase appears to be more complex because in this phase the director is not parallel to the wave vector q. In a simple case, it is just possible to use the tilt angle Θ as an order parameter. However, this parameter does not specify the direction of the tilt and thus it is analogous to the scalar nematic order parameter S. The full tensor order parameter of the SmC phase can be constructed in several different ways. One is to define the pseudo-vector w [8, 9] that describes the rotation of the director with respect to the smectic plane normal:

$$w_\alpha = \delta_{\alpha\beta\gamma} Q_{\beta\nu} \hat{q}_\gamma \hat{q}_\nu \tag{17}$$

where $\delta_{\alpha\beta\gamma}$ is the Levi–Civita antisymmetric tensor, $\hat{q}$ is the unit vector along q and $Q_{\alpha\beta}$ is the (biaxial) nematic order parameter.

The pseudo vector order parameter w vanishes in the SmA phase. If we neglect the biaxiality of the SmC phase, the order parameter $Q_{\alpha\beta}$ is expressed in terms of the director n and the vector order parameter (Eq. 16) is simplified:

$$w = (n \cdot \hat{q})[n \times \hat{q}] \tag{18}$$

We note that the vector w is perpendicular to the tilt plane in the SmC phase and the absolute value of w is related to the tilt angle, $w = \sin 2\Theta/2 \approx \Theta$ at small $\Theta^2 \ll 1$. This pseudo vector order parameter can be used both in the theory of nonchiral and chiral SmC phases [8, 9].

2.3 Anisotropic Intermolecular Interactions in Liquid Crystals

2.3.1 Hard-core Repulsion

Liquid crystals are composed of relatively large molecules with strongly anisotropic shapes. In general, there exist a number of anisotropic interactions between such molecules that can be responsible for the nematic ordering. Historically the first consistent molecular theory of the nematic phase was proposed by Onsager [10]. Onsager showed that the nematic ordering can be stabilized by the hard-core repulsion between rigid rod-like molecules without any attraction forces. The steric repulsion between rigid particles is a limiting case of a strong short-range repulsion interaction that does not allow molecules to penetrate each other. The corresponding model interaction potential is discontinuous and can be written as

$$V_s (1,2) = \Omega\left(r_{12} - \xi_{12}\right) \tag{19}$$

where $\Omega(x)$ is a step function, $\Omega(r_{12} - \xi_{12}) = \infty$ if $r_{12} < \xi_{12}$ and thus the molecules penetrate each other; otherwise $\Omega(r_{12} - \xi_{12}) = 0$. Here ξ_{12} is the minimum distance of approach between the centers of the molecules 1 and 2 for a given relative orientation. The function ξ_{12} is determined by the molecular shape and depends on the relative orientation of the two molecules. For hard spheres $\xi_{12} = D$ where D is the diameter of the sphere. For any two prolate

molecules, ξ_{12} varies between the diameter D and the length L.

2.3.2 Electrostatic and Dispersion Interactions

Hard-core repulsion between anisotropic molecules, discussed in the previous subsection, can be the driving force of the I–N transition in lyotropic systems. In contrast, in thermotropic liquid crystals the transition occurs at some particular temperature and therefore some attraction interaction must be involved. The corresponding molecular theory, based on anisotropic dispersion interactions, was proposed by Maier and Saupe [11, 12].

The dispersion (or Van der Waals) interaction appears in the second-order perturbation theory. The initial interaction potential is the electrostatic one. For the two molecules i and j the electrostatic interaction energy can be written as

$$V_{el} (1,2) = \iint \frac{e(r_i)e(r_j)}{\left|R_i - R_j + r_i - r_j\right|}\, dr_i\, dr_j \tag{20}$$

where $e(r_i)$ is the charge distribution of the molecule i and R_i is the position of the center of mass.

The electrostatic interaction can be expanded in terms of molecular multipoles. For neutral molecules one obtains

$$V_{el}(i,j) = \frac{1}{R_{ij}^3} U_{dd}(i,j) + \frac{1}{R_{ij}^4} U_{dq}(i,j) + \ldots \tag{21}$$

where the first term is the dipole–dipole interaction potential

$$U_{dd} (i,j) = \left(\vec{\mu}_i \cdot \vec{\mu}_j\right) - 3\left(\vec{\mu}_i \cdot u_{ij}\right)\left(\vec{\mu}_j \cdot u_{ij}\right) \tag{22}$$

and the second term is the dipole–quadrupole interaction energy

$$U_{dq} (i,j) = U_{dq}^{ij} - U_{dq}^{ji} \tag{23}$$

where

$$U_{\text{dq}}^{ij} = \frac{3}{2} Tr \, \mathbf{Q}_j \left(\bar{\mu}_i \cdot \mathbf{u}_{ij} \right) + 3 \left(\bar{\mu}_i \cdot \mathbf{Q}_j \cdot \mathbf{u}_{ij} \right)$$
$$- \frac{15}{2} \left(\bar{\mu}_i \cdot \mathbf{Q}_j \cdot \bar{\mu}_i \right) \left(\bar{\mu}_i \cdot \mathbf{u}_{ij} \right) \qquad (24)$$

and where U_{dq}^{ij} is obtained by permutation of the indices i and j in Eq. (23). Here $\mathbf{R}_{ij} = \mathbf{R}_i - \mathbf{R}_j$ is the intermolecular vector, $\bar{\mu}_i$ is the molecular dipole and $\mathbf{Q}_i$ is the molecular quadrupole tensor, $\mathbf{u}_{ij} = \mathbf{R}_{ij}/|\mathbf{R}_{ij}|$.

The dispersion interaction energy is obtained in the second-order perturbation theory:

$$V_{\text{disp}}(i,j) = \sum_{ni,nj}^{\prime} \qquad (25)$$

$$\cdot \frac{\langle o_i \, o_j | V_{\text{el}}(i,j) | n_i \, n_j \rangle \langle n_i \, n_j | V_{\text{el}}(i,j) | o_i \, o_j \rangle}{E_{oioj} - E_{ninj}}$$

where $|o_i\rangle$ and $|n_i\rangle$ represent the ground state and the excited state of the molecule i, respectively and $E_{oioj} - E_{ninj}$ is the excitation energy of the system.

Substituting the multipole expansion (Eq. 20) into Eq. (24) and taking into account the leading term (that contains r_{ij}^{-6}) one obtains

$$V_{\text{disp}}(i,j) \approx \frac{1}{r_{ij}^6} \sum_{ni,nj}^{\prime} \qquad (26)$$

$$\cdot \frac{\langle o_i \, o_j | U_{\text{dd}}(i,j) | n_i \, n_j \rangle \langle n_i \, n_j | U_{\text{dd}}(i,j) | o_i \, o_j \rangle}{E_{oioj} - E_{ninj}}$$

where $U_{\text{dd}}(i,j)$ is given by Eq. (21).

We note that the full dispersion interaction energy (Eq. 24) can be approximated to by the dipole–dipole term (Eq. 25) only if the molecules are sufficiently far apart. However, Eq. (25) is often used in molecular theories of liquid crystals to draw qualitative conclusions. For example, the potential (Eq. 25) has been used in the Maier–Saupe theory.

The dipole–dipole dispersion interaction can be simplified if we assume that the

molecular short axes are oriented randomly around the long axes a_i and a_j. Then it is valid to average the potential (Eq. 25) over all b_i and b_j with the constraints $(b_i \cdot a_i) = 0$, $(b_j \cdot a_j) = 0$. It can be shown that the averaging results in the following expression for the effective uniaxial potential [13, 14]:

$$V_{\text{eff}} \left(a_i, u_{ij}; a_j \right)$$
$$= \text{const} - J_{ij}^2 \left(a_i \cdot u_{ij} \right)^2 - J_{ji}^2 \left(a_j \cdot u_{ij} \right)^2$$
$$- J_{ij} \left[\left(a_i \cdot a_j \right) - \left(a_i \cdot u_{ij} \right) \left(a_j \cdot u_{ij} \right) \right]^2 \qquad (27)$$

Here the coefficients can be expressed in terms of the electric dipole and quadrupole matrix elements [13, 14].

In the nematic phase there is no positional order and the molecular centers are distributed randomly. If one neglects the positional correlations, the interaction potential (Eq. 26) can be further simplified by averaging over all directions of the intermolecular vector. The resulting effective potential appears to be very simple:

$$V_{\text{eff}} \left(\left(a_i \cdot a_j \right) \right) = -\frac{1}{60} E r_{ij}^{-6} (\Delta \alpha)^2 P_2 \left(\left(a_i \cdot a_j \right) \right) \qquad (28)$$

where $\Delta \alpha$ is the anisotropy of the molecular polarizability and E is the average excitation energy.

Equation (27) presents a simple anisotropic attraction potential that favors nematic ordering. This potential has been used in the original Maier–Saupe theory [11, 12]. We note that the interaction energy (Eq. 27) is proportional to the anisotropy of the molecular polarizability $\Delta \alpha$. Thus, this anisotropic interaction is expected to be very weak for molecules with low dielectric anisotropy. Such molecules, therefore, are not supposed to form the nematic phase. This conclusion, however, is in conflict with experimental results. Indeed, there exist a number of materials (for example, cyclohexylcyclohexanes [15]) which form the nematic phase but exhibit very small anisotropy of the polarizability. This is an indication that anisotropic dispersion forces do not make the major contribution to the stabilization of the nematic phase. On the other hand, the well known success of the Maier–Saupe theory (as discussed below) is mainly determined by its mathematical form and not by the particular intermolecular interaction that has been taken into account.

As shown by Gelbart and Gelbart [16], the predominant orientational interaction in nematics must be the isotropic dispersion attraction modulated by the anisotropic molecular hard core. The isotropic part of the dispersion interaction is generally larger than the anisotropic part because it is proportional to the average molecular polarizability $\bar{\alpha}$. And the anisotropy of this effective potential comes from that of the asymmetric molecular shape. Thus this effective potential is a combination of intermolecular attraction and repulsion. It can be written as

$$V_{\text{eff}}(1,2) = J_{\text{att}}\left(r_{12}\right)\Theta\left(r_{12} - \xi_{12}\right) \tag{29}$$

where the step-function $\Theta(r_{12}-\xi_{12})$ determines the steric cut-off. $\Theta(r_{12}-\xi_{12})=0$ if the molecules penetrate each other (i.e. if $r_{12}<\xi_{12}$) and $\Theta(r_{12}-\xi_{12})=1$ otherwise.

We note that Eq. (24) contains also the induction interaction, that is the interaction between the permanent multipoles of the molecule i and the polarizability of the molecule j. This interaction corresponds to the terms with $n_i=0$ or $n_j=0$. The induction interaction can play an important role if the molecular hard core is strongly polar [18].

Electrostatic interactions between molecules with permanent electric multipoles are also strongly anisotropic. The corresponding interaction potentials, however, vanish after the integration over the intermolecular vector r_{ij} and thus they do not contribute in the mean-field approximation. The dipole–dipole and dipole–quadrupole potentials vanish also after an orientational averaging because they are polar. At the same time, the electrostatic interactions can be very important if the molecules possess large permanent dipoles. In this case the dipole–dipole interaction gives rise to strong short-range dipolar correlations including the formation of dimers with antiparallel dipoles. At present the statistical theory of strongly polar nematics is in its early stage (see, however, [6, 7]).

2.3.3   Model Potentials

Realistic intermolecular interaction potentials for mesogenic molecules can be very complex and are generally unknown. At the same time molecular theories are often based on simple model potentials. This is justified when the theory is used to describe some general properties of liquid crystal phases that are not sensitive to the details on the interaction. Model potentials are constructed in order to represent only the qualitative mathematical form of the actual interaction energy in the simplest possible way. It is interesting to note that most of the popular model potentials correspond to the first terms in various expansion series. For example, the well known Maier–Saupe potential $JP_2((a_i \cdot n))$ is just the first nonpolar term in the Legendre polynomial expansion of an arbitrary interaction potential between two uniaxial molecules, averaged over the intermolecular vector r_{ij}:

$$\bar{V}\left((a_i \cdot a_j)\right) = \int dr_{ij}\, V\left(a_i, r_{ij}, a_j\right)$$
$$= J_0 + J_2\, P_2\left((a_i \cdot a_j)\right) + .. \tag{30}$$

where we have taken into account only nonpolar terms. Here $V(a_i, r_{ij}, a_j)$ is an arbitrary interaction potential between two uniaxial rigid molecules. It depends only on the long axes a_i, a_j and on the intermolecular vector r_{ij}.

The partially averaged potential (Eq. 29) can be used in the molecular theory of the nematic–isotropic transition (also being supplemented by the P_4 term [19]). However, several other properties of nematics cannot be described in this way. For example, the full anisotropy of the Frank elastic constants can be accounted for only taking into account the explicit dependence of the interaction potential on the intermolecular vector [20]. In this case appropriate model potentials can be obtained using some more general expansion of the full potential $V(a_i, r_{ij}, a_j)$. This potential can be expanded in terms of the spherical invariants

$$V\left(a_i, r_{ij}, a_j\right) = \sum_{l,m,k} J_{lmk}\left(r_{ij}\right) T^{lmk}\left(a_i, u_{ij}, a_j\right) \tag{31}$$

The set $(T^{lmk}(a_i, u_{ij}, a_j))$ is a complete orthogonal set of basis functions [86] that contain the vector a_i to the power l, the vector u_{ij} to the power m and the vector a_j to the power k. The explicit expressions for the lower order invariants have been given, for example, by Van der Meer [14]. The invariants with one zero index are just Legendre polynomials. For example $T^{202}(a_i, u_{ij}, a_j) = P_2((a_i \cdot a_j))$.

The invariants with $l + m + k$ odd are pseudoscalars and therefore the corresponding coupling constants $J^{lmk}(r_{ij})$ are pseudoscalars as well. These terms can appear only in the interaction potential between chiral molecules. The first nonpolar chiral term of the general expansion (Eq. 30) reads:

$$V*\left(a_i, r_{ij}, a_j\right) = J*\left(r_{ij}\right)\left(\left[a_i \times a_j\right] \cdot r_{ij}\right) \tag{32}$$

The potential (Eq. 31) promotes the twist of the long axes of neighboring molecules and

is widely used in the statistical theory of cholesteric ordering [13].

Finally we note that there exist some special model potentials that combine an attraction at large separation and repulsion at short distances. The most popular potential of this kind is the Gay–Berne potential [22] which is a generalization of the Lennard–Jones potential for anisotropic particles. The Gay–Berne potential is very often used in computer simulations but not in the molecular theory because it is rather complex.

2.4 Molecular Theory of the Nematic Phase

2.4.1 Mean-field Approximation and the Maier–Saupe Theory

The simplest molecular theory of the nematic–isotropic (N–I) transition can be developed in the mean-field approximation. According to the general definition, in the mean-field approximation one neglects all correlations between different molecules. This is obviously a crude and unrealistic approximation but, on the other hand, it enables one to obtain very simple and useful expressions for the free energy. This approximation also appears to be sufficient for a qualitative description of the N–I transition. More precise and detailed theories of the nematic state are based on more elaborate statistical models that will be discussed briefly in Sec. 2.4.3.

In the language of statistical mechanics the mean-field approximation is equivalent to the assumption that the pair distribution function $f_2(1, 2)$ can be represented as a product of the two one-particle distributions, that is $f_2 = f_1(1) f_2(2)$. The same representation is applied also to all n-particle distribution functions. This definition can be used to derive the mean-field expression for the free energy.

We note that the general Gibbs expression for the free energy F can be written in terms of the N-particle distribution function $f_N = Z^{-1} \exp(-H/kT)$ in the following way:

$$F = \int H f_N \, d\Gamma - kT \int f_N \ln f_N \, d\Gamma \qquad (33)$$

where H is the Hamiltonian of the system and dT denotes the integration over all microscopic variables. On the level of pair interactions the Hamiltonian H can be represented as a sum of the kinetic energy plus the sum of interaction potentials for all molecular pairs

$$H = \sum_i E_k^i + \sum_{i,j} V(i,j) \qquad (34)$$

In the mean-field approximation the N-particle distribution is factorized as $f_N = \Pi_i f_1(i)$ and the general expression (Eq. 32) is reduced to the following mean-field free energy (in the absence of the external field):

$$F = \text{const} \qquad (35)$$
$$+ \frac{1}{2}\rho^2 \int V(\omega_i, \omega_j, r_{ij}) f_1(\omega_i, r_i)$$
$$\cdot f_1(\omega_j, r_j) \, d\omega_i \, d\omega_j \, dr_i \, dr_j$$
$$+ \rho kT \int f_1(\omega_i, r_i) \cdot \ln f_1(\omega_i, r_i) \, d\omega_i \, dr_i$$

where ω_i specifies the orientation of the molecule i. In Eq. (34) the free energy is represented as a functional of the one particle distribution function $f_1(i)$. The equilibrium distribution is determined by minimization of the free energy (Eq. 34):

$$f_i(\omega_i, r_i) = \frac{1}{Z} \exp\left[-\beta\rho \int V(\omega_i, r_{ij}, \omega_j)\right.$$
$$\left. \cdot f_1(\omega_j, r_j) \, d\omega_j \, dr_j\right] \qquad (36)$$

where

$$Z = \int \exp\left[-\beta\rho \int V(\omega_i, r_{ij}, \omega_j)\right.$$
$$\left. \cdot f_1(\omega_j, r_j) \, d\omega_j \, dr_j\right] d\omega_i \, dr_i \qquad (37)$$

It should be noted that Eqs. (35) and (36) are rather general and not restricted to the nematic phase. The distribution function $f_1(i)$ depends on the position r_i; therefore it can describe the molecular distribution in smectic phases as well.

In the nematic phase there is no positional order and the distribution function $f_1(i)$ depends only on the molecular orientation. In this case Eq. (35) can be simplified:

$$f_1(\omega_1) = \frac{1}{Z_0} \exp\left[-\beta U_{MF}(\omega_i)\right] \qquad (38)$$

where the mean-field potential $U_{MF}(\omega_1)$ is given by

$$U_{MF}(\omega_i) = \int \tilde{V}(\omega_1, \omega_2) f_1(\omega_2) \, d\omega_2 \qquad (39)$$

Here the effective pair potential $\tilde{V}(\omega_1, \omega_2)$ reads:

$$\tilde{V}(\omega_1, \omega_2) = \int U(\omega_1, r_{12}, \omega_2) \, dr_{12} \qquad (40)$$

One can readily see from Eq. (37) that in the mean field approximation each molecule feels some average mean-field potential produced by other molecules. This mean-field potential is just the pair interaction energy averaged over the position and orientation of the second molecule.

It is important to note that the mathematical form of Eqs. (37) and (38) appears to be rather general and goes far beyond the mean-field approximation. In fact, the one-particle distribution can always be written in the form of Eq. (37) with some unknown one-particle potential. In several advanced statistical theories this effective potential can be explicitly expressed in terms of the correlation functions. For example, such an expression can be obtained in the density functional theory discussed in Sec. 2.4.4.

The mean-field approximation was originally developed to describe lattice systems like ferromagnetics or ferroelectric crystals. In the case of liquids, however, some formal problems arise. In Eq. (39) the integral over r_{12} diverges because any attraction interaction diverges if the molecules are allowed to penetrate each other. The obvious solution to this problem is to take into account the hard-core repulsion that restricts the minimum distance between attraction centers. This can be done by introducing a steric cut-off into the integral in Eq. (39). Then the attraction potential is substituted by the effective potential $V_{\text{eff}}(1, 2) = V(1, 2)\, \Theta(r_{12} - \xi_{12})$ where $\Theta(r_{12} - \xi_{12})$ is a step-function (see Eq. 29). We note however, that the introduction of a steric cut-off is equivalent to taking account of simple short-range steric correlations. These correlations also give rise to the additional contribution to the free energy that is called packing entropy. This entropy is discussed in detail in Sec. 2.4.2.

If one neglects the asymmetry of the molecular shape (i.e. puts the function $\xi_{12} = D = \text{const}$ in the effective potential) and uses the dipole–dipole dispersion interaction potential (Eq. 27), one arrives at the Maier–Saupe theory. In this theory the interaction potential contains only the $P_2((\mathbf{a}_1 \cdot \mathbf{a}_2))$ term and as a result it is possible to obtain the closed equation for the nematic order parameter S. Substituting the potential $V(1, 2) = -J(r_{12}) P_2((\mathbf{a}_1 \cdot \mathbf{a}_2))\, \Theta(r_{12} - D)$ into Eqs. (27–29), multiplying both sides of Eq. (27) by $P_2((\mathbf{a}_1 \cdot \mathbf{a}_2))$ and integrating over $\mathbf{a}_2$, we obtain the equation

$$
S = \frac{1}{Z_0} \int P_2(\cos\Theta)
$$
$$
\cdot \exp\left[\frac{\rho}{\tilde{T}} S P_2(\cos\Theta)\right] \frac{d\cos\Theta}{2} \quad (41)
$$

where $\cos\Theta = (\mathbf{a} \cdot \mathbf{n})$ and the dimensionless temperature $\tilde{T} = kT/J_0$ where $J_0 = \int_D^\infty J(r) r^2 dr$.

The free energy of the nematic phase can be written in the form

$$
F = \frac{1}{2}\rho^2 J_0 S^2 - kT
$$
$$
\cdot \ln \int \exp\left[\frac{\rho}{\tilde{T}} S P_2(\cos\Theta)\right] \frac{d\cos\Theta}{2} \quad (42)
$$

Equations (40) and (41) describe the first order nematic–isotropic transition. At high temperatures Eq. (40) has only the isotropic solution $S = 0$. At $\tilde{T} \approx 0.223$ two other solutions appear. One of them is always unstable but the other one does correspond to the minimum of the free energy F and characterizes the nematic phase. The actual nematic–isotropic phase transition takes place when the free energy of the nematic phase becomes equal to that of the isotropic phase. This happens at $\tilde{T} = \tilde{T}_{N-I} \approx 0.220$. At the transition temperature the order parameter $S \approx 0.44$.

Finally, the isotropic solution loses its stability at $\tilde{T} = 0.2$ and below this temperature there exists only the nematic stable solution. Thus, $\tilde{T} \approx 0.223$ is the upper limit of metastability of the nematic phase and $\tilde{T} = 0.2$ is the lower limit of metastability of the isotropic phase.

The remarkable feature of the Maier–Saupe theory lies in its simplicity and universality. In particular, the temperature variation of the order parameter and its value at the transition point are predicted to be universal, that is independent of intermolecular forces and the molecular structure. This prediction appears to be supported by experiments. In Fig. 1 the results of the theory are compared with some experimental data for the parameter S presented by Luckhurst et al. [23]. One can see that the agreement is surprisingly good taking into account the number of approximations and

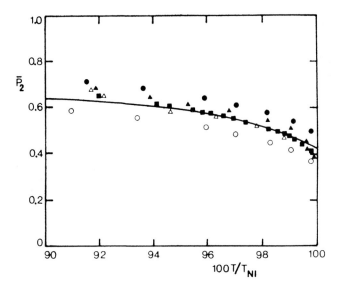

Figure 1. The temperature dependence of the order parameter S, at constant pressure, for 4,4′-dimethoxyazobenzene (●), 4,4′-diethoxyazobenzene (△), anisaldazine (▲), 2,4-nonadienoic acid (■) and 2,4-undecadienoic acid (○). The curve is predicted by the Maier–Saupe theory (after Luckhurst [23]).

simplifications involved in the Maier–Saupe theory.

It should be noted, however, that the agreement between the Maier–Saupe theory and experiment is not so good if one considers some other parameters of the N–I transition. For example, the discontinuity in entropy is overestimated several times. In particular, the difference between the transition temperature $T_{N–I}$ and the lower limit of stability of the isotropic phase T^* is strongly overestimated. In the Maier–Saupe theory the parameter $\gamma = (T_{N–I} - T^*)/T_{N–I}$ is about 0.1 while experimentally $T_{N–I} - T^* \approx 1 - 2\,C$ and therefore $\gamma \sim (3-6)10^{-3}$. This means that the mean-field theory overestimates the difference between the isotropic and the nematic phase. The discrepancy is clearly related to the neglect of short-range orientational correlations between anisotropic molecules that make the local structure rather similar in both phases.

The more serious problem in the original Maier–Saupe theory is related to the choice of the anisotropic dispersion interaction potential (Eq. 27) (see the detailed discussion

in [24]). An estimate of the transition temperature $T_{N–I}$ yields the value that is an order of magnitude too small [24, 25]. Moreover, there exist materials (for example, cyclohexylcyclohexanes [15] or alkylbicyclooctanes [24]) which possess very low anisotropy of the molecular polarizability but nevertheless form stable nematic phases. These examples enable us to conclude that anisotropic dispersion forces certainly cannot be a dominant general mechanism of the stabilization of the nematic phase.

It was first noticed by Gelbart and Gelbart [16] that the predominant anisotropic interaction in nematics results from a coupling between the isotropic attraction and the anisotropic hard-core repulsion. This coupling is represented by the effective potential $V_{\text{eff}}(1, 2) = V(1, 2) \Theta(r_{12} - \xi_{12})$. This potential can be averaged over all orientations of the intermolecular vector and then can be expanded in Legendre polynomials. The first term of the expansion has the same structure as the Maier–Saupe potential $J(r_{12})P_2((\boldsymbol{a}_1 \cdot \boldsymbol{a}_2))$ but with the coupling constant J determined by the anisotropy of the molecular shape and the *average* molecular polarizability rather than by the polarizability *anisotropy*. We note that the particular interpretation of the coupling constant J does not influence any of the results of the Maier–Saupe theory except for the absolute value of the transition temperature. Thus, the success of the Maier–Saupe theory is mainly determined by its general mathematical structure and by the form of the model potential.

2.4.2 Short-range Orientational Correlations

There are several ways to improve the Maier–Saupe theory. One way is to take into account the asymmetry of the molecular shape and to account for the excluded volume effects. This approach will be discussed in the next subsection. The second possible way is to improve the statistical part of the theory by taking into account some intermolecular correlations. Finally, the third way is to improve the model potential.

Strong short-range orientational correlations can be conveniently taken into account in the cluster approximation. The simplest version of the cluster approximation in the theory of liquid crystals was proposed by Ypma et al. [26] who used the general approach of Callen and Strieb developed in the theory of ferromagnetism.

In the two-particle cluster approximation the interaction between two neighboring molecules is taken into account exactly while the interaction with the rest of the nearest neighbors is treated in the mean-field approximation. The two-particle Hamiltonian is written as

$$H_2(1,2) = -J P_2\big((a_1 \cdot a_2)\big)$$
$$+\left(1 - \frac{1}{\sigma}\,\phi\Big[P_2\big((a_1 \cdot n)\big) + P_2\big((a_2 \cdot n)\big)\Big]\right) \quad (43)$$

where σ is the number of nearest neighbors and the parameter ϕ is the strength of the mean-field which is to be determined in a self-consistent way. The pair distribution function is assumed to have the Boltzmann form with the Hamiltonian (Eq. 42):

$$f_2(1,2) = \frac{1}{Z_2}\exp\big[-\beta H_2(1,2)\big] \quad (44)$$

The one-particle distribution function is given by the same type of expression as in the mean-field theory:

$$f_1(1,2) = \frac{1}{Z_1}\exp\big[-\beta\phi P_2\big((a_1 \cdot n)\big)\big] \quad (45)$$

The parameter ϕ is determined by the self-consistency relation that states that the order parameter S, calculated with the one-particle distribution function (Eq. 44), must be equal to the one calculated with the pair distribution function (Eq. 43):

$$\int P_2(a_1 \cdot n)f_1(1)\,da_1 =$$
$$\int P_2\big((a_1 \cdot n)\big)f_2(1,2)\,da_1\,da_2 \quad (46)$$

Finally from Eqs. (42)–(45) one can obtain the following simple expression for the free energy

$$\beta F = -\frac{1}{2}N\sigma\ln Z_2 + N(\sigma - 1)\ln Z_1 \quad (47)$$

where

$$Z_1 = \int \exp\big[-\beta\phi P_2\big((a_1 \cdot n)\big)\big]da_1$$
$$Z_2 = \int \exp\big[-\beta H_2(1,2)\big]da_1\,da_2 \quad (48)$$

The self-consistency relation (Eq. 45) is obtained by minimization of the free energy (Eq. 46) with respect to ϕ.

The use of a cluster approximation reduces the discrepancy between theory and experiment. For example, the difference $T_{N-I} - T^*$ is reduced several times compared to the Maier–Saupe theory. A detailed comparison of different versions of the cluster approximation with experimental data has been given by Chandrasekhar [61].

Finally we note that it is possible to include the $P_4(\cos\Theta)$ term in the model potential of the Maier–Saupe theory, as has been done by Luckhurst et al. [19] and Chandrasekhar and Madhusudana [28]. This procedure also leads to some quantitative improvement.

2.4.3   Excluded Volume Effects and the Onsager Theory

As discussed in the previous subsection, it is important to take into account the excluded volume effects even in a simple mean-field theory based on an anisotropic attraction interaction. The excluded volume effects are determined by hard-core repulsion that does not allow molecules to penetrate each other. It is interesting to note that by doing so we already go beyond the formal mean-field approximation. Indeed, with excluded volume effects the internal energy of the nematic phase can be written as

$$U = \frac{1}{2}\rho^2 \int V_s(1,2) h_2^0(1,2)$$
$$\cdot f_1(1) f_1(2) d(1) d(2) \qquad (49)$$

where $h_2^0(1, 2) = \exp[-\beta V_s(1, 2)]$ is, in fact, a simple correlation function between the rigid molecules 1 and 2. Here V_s (1, 2) is the steric repulsion potential (Eq. 19).

It is obvious that in this approximation we do take into account some short-range steric correlations between rigid molecules. We note that these correlations contribute not only to the internal energy but also to the entropy of the nematic. Excluded volume effects restrict the molecular mo-

tion and therefore the total entropy of the fluid is reduced. This additional contribution to the entropy is called the packing entropy.

A simple expression for the packing entropy at low densities was first derived by Onsager [10] who considered nematic ordering in a system of long rigid rods. In this system the rods interact only sterically and are supposed to be very long, $L/D \gg 1$, where L is the length and D is the diameter of the rod. At low densities it is possible to express the free energy of such a system in the form of the virial expansion:

$$\beta F = \rho \ln \rho + \rho$$
$$\cdot \int f_1(\omega_1)\big[\ln f_1(\omega_1) - 1\big]d\omega_1$$
$$+ \frac{1}{2}\rho^2 \int f_1(\omega_1) f_1(\omega_2) \qquad (50)$$
$$\cdot B(\omega_1, \omega_2)d\omega_1 d\omega_2 + \dots$$

where $B(\omega_1, \omega_2)$ is the excluded volume for the two rods:

$$B(\omega_1, \omega_2) = \int dr_{1\,2}\big(\exp[-\beta V_s(1, 2)] - 1\big)(51)$$

For two spherocylinders the excluded volume is expressed as

$$B(1, 2) = 2 L^2 D\big|\sin\gamma_{1\,2}\big| + 2\pi LD^2 + \frac{4}{3}\pi d^3$$
$$(52)$$

where $\gamma_{1\,2}$ is the angle between the long axes of the two spherocylinders.

All terms in Eq. (49) are purely entropical in nature because the system is athermal. The second term is the orientational entropy and the third one is the packing entropy that is related to the second virial coefficient for two rigid rods. The expansion in Eq. (49) is actually performed in powers of the packing fraction $\eta = \rho v_0 \approx pi\,\rho D^2 L \ll 1$ if $L/D \gg 1$, where v_0 is the volume of a spherocylinder. At a very low volume fraction of rods the higher order terms in the expansion can be neglected [29].

We note that the free energy (Eq. 49) has the same general mathematical form as the mean-field free energy (Eq. 34). In the gas consisting of long rods, the role of the effective anisotropic potential is played by the excluded volume $B(a_1, a_2)$ multiplied by kT. Thus it is not surprising that the minimization of the free energy (Eq. 49) yields practically the same equation for the orientational distribution function as in the Maier–Saupe theory:

$$f_1((a_1 \cdot n)) = \frac{1}{Z_0}$$
$$\cdot \exp\left[-I\int |\sin\gamma_{12}| f_1((a_2 \cdot n)) \frac{da_2}{4\pi}\right] \quad (53)$$

where $I = 2\rho L^2 D \approx \eta L/\pi D$.

From the mathematical point of view the only difference between Eq. (52) and the corresponding equation in the Maier–Saupe theory is in the form of the effective interaction potential. In the Maier–Saupe theory the potential is $-J P_2(\cos\gamma_{12})$ while in the Onsager theory the potential has a different form $kTI|\sin\gamma_{12}|$ that also contains a substantial contribution from higher-order Legendre polynomials. However, from the physical standpoint the most important difference between the Maier–Saupe and the Onsager theories is in the nature of the transition. Maier–Saupe theory describes the N–I transition in thermotropic liquid crystals where the ordering appears at some particular temperature. By contrast, in the Onsager theory the transition occurs when the volume fraction of rods is increased. In Eq. (52) the bifurcation point (pseudo-second-order transition) corresponds to $I = 1$ and thus the critical packing fraction appears to be of the order D/L, $\eta = \pi D/L \ll 1$ if $D/L \ll 1$. This is a crude estimate of the actual transition density. Therefore, in the case of very long rods the transition takes place at very low density. In the limiting case of $L/D \to \infty$ the Onsager theory, which is based

on the virial expansion, appears to be asymptotically exact [29].

The actual N–I transition for a gas consisting of long rods is more complex because the system separates into a dilute isotropic phase and the more concentrated nematic phase that already possesses a high degree of orientational order. This is related to the fact that the homogeneous system of long rods appears to be mechanically unstable (with respect to density fluctuations) within some density interval around $\eta_{cr} = \pi D/L$. The two coexisting densities ρ_1 and ρ_2 can be determined in the usual way by equating the chemical potentials and the pressures of the two phases:

$$\mu_1(\rho_1) = \mu_2(\rho_2)$$
$$P_1(\rho_1) = P_2(\rho_2) \quad (54)$$

where the pressure P and the chemical potential μ can be expressed in terms of the free energy in the following way:

$$P = -\left(\frac{\partial F}{\partial V}\right)_{T,N} = \rho \frac{\partial(F/V)}{\partial\rho} - \frac{F}{V} \quad (55)$$

$$\mu = -\left(\frac{\partial F}{\partial N}\right)_{T,V} = \rho \frac{\partial(F/V)}{\partial\rho} \quad (56)$$

Equations (53) and (54), supplemented by Eq. (52) for the orientational distribution function and Eq. (49) for the free energy, can be used to determine the coexisting densities and the value of the order parameter at the transition. Eq. (52) was first solved approximately by Onsager who used the trial function

$$f_1(\cos\Theta) = \left(\frac{\alpha}{4\pi\sinh\alpha}\right)\cosh(\alpha\cos\Theta) \quad (57)$$

and thus reduced the integral Eq. (52) to the equation for the single parameter α. The nematic order parameter S can be expressed in terms of α as

$$S = 1 - \frac{3}{\alpha} \cosh \alpha + \frac{3}{\alpha^2} \qquad (58)$$

Eq. (52) has been solved numerically by Lasher [30] and Lekkerkerker et al. [31] by employing the Legendre polynomial expansion and by Lee and Meyer [32] by a direct numerical method. The results do not differ significantly from those obtained with the Onsager trial function (Eq. 55) and therefore the approximation (Eq. 55) appears to be sufficient for most practical purposes. A more detailed description of the Onsager theory and its generalizations can be found in reviews [33–36]. Here we do not consider it any more because this chapter is focused on the theory of thermotropic liquid crystals. In this context the main consequence of the Onsager approach is the conclusion that the excluded volume effects can be very important in stabilizing the nematic phase. This is expected to be true also in the case of thermotropic nematics because they are also composed of molecules with relatively rigid cores. Therefore, the packing entropy must be taken into account in any consistent theory of the N–I transition. We discuss this contribution in more detail in the following subsection.

2.4.4　Packing Effects in Thermotropic Nematics

The Onsager expression for the packing entropy is valid at very low densities and the theory can be applied directly to dilute solutions of rigid particles like tobacco mosaic virus or helical synthetic polypeptides. At the same time typical thermotropic nematics are composed of molecules with an axial ratio of the order of 3 or 4 and the packing fraction is of the order of 1. Thus, the direct use of the Onsager theory for such systems can result in large errors. The Onsager expression for the packing entropy has been generalized to the case of condensed nematic phases by several authors [37–40] using different approximations. The corresponding expression for the packing entropy can be written in the following general way:

$$S_p = -\frac{1}{2} \lambda(\eta) \rho^2 k_B$$
$$\cdot \int \Theta\left(r_{12} - \xi_{12}\right) f_1\left((a_1 \cdot n)\right)$$
$$\cdot f_1\left((a_2 \cdot n)\right) da_1\, da_2\, dr_{12} \qquad (59)$$

Here the coefficient $\lambda(\eta)$ depends on the packing fraction η. In the Onsager theory, which corresponds to the limit of small η, the factor $\lambda = 1$. The frequently used approximations for the packing entropy have been derived from the scaled particle theory [37] and were proposed by Parsons and Lee [38, 41]. The equations of state derived from various molecular theories for a condensed solution of hard rods are given in the review of Sato and Teramoto [36]. One can readily see that even for long rods with $L/D = 50$ the Onsager equation of state deviates significantly from the results of the Parsons–Lee or scaled particle theory already at relatively small packing fractions $\eta \approx 0.3$. We note that recently the Parsons–Lee approximation has been tested by Jackson et al. [5] for the system of relatively short spherocylinders with axial ratios $L/D = 3$ and $L/D = 5$. It has been shown that the results agree very well with Parsons–Lee theory up to packing fractions of $\eta = 0.5$. In the Parsons–Lee approximation the function $\lambda(\eta)$ is written in a simple form

$$\lambda_{PL} = \frac{1}{4} \frac{4 - 3\eta}{(1 - \eta)^3} \qquad (60)$$

Taking into account the packing entropy (Eq. 57), one can write the model free energy for the nematic phase

$$F/V = kT\rho \int f_1((a_1 \cdot n))$$

$$\cdot \ln f_1((a_1 \cdot n)) da_1 + \frac{1}{2} kT\rho^2 \lambda(\eta)$$

$$\cdot \int f_1((a_1 \cdot n)) f_1((a_2 \cdot n))$$

$$\cdot V_{excl}((a_1 \cdot a_2)) da_1 da_2 + \frac{1}{2}\rho^2$$

$$\cdot \int f_1((a_1 \cdot n)) f_1((a_2 \cdot n)) V_{att}((a_1,a_2,r_{12}))$$

$$\cdot g_{HC}(a_1,a_2,r_{12}) da_1 da_2 dr_{12} \qquad (61)$$

where $V_{excl}(1, 2)$ is the excluded volume for the two rigid particles, $V_{att}(1, 2)$ is the attraction interaction potential and $g_{HC}(1, 2)$ is the pair correlation function for the reference hard-core fluid.

For simplicity one can approximate the pair correlation function by the steric cut-off $\Theta(r_{12}-\xi_{12})=\exp(-\beta V_s(1, 2))$. In this case the free energy (Eq. 59) corresponds to the so-called generalized Van der Waals theory considered in detail by Gelbart and Barboy [43, 44].

We note that in Eq. (59) the attraction interaction is taken into account as a perturbation. This means that one neglects a part of the orientational correlations determined by attraction. A simple way to improve the theory is to combine Eq. (59) with the two-particle cluster approach [40].

The first two terms in Eq. (59) represent the free energy of the reference hard-core fluid. The general structure of this reference free energy is similar in different approaches but the particular dependence on the packing fraction η can be quite different. This depends on the particular approach (used in the theory of hard-sphere fluids) that has been generalized to the nematic state. For example, the Parsons–Lee approximation is based on the Carnahan–Starling equation of state, the approach of Ypma and Vertogen is a generalization of the Percus–Yevick approximation and the approach of

Cotter is based on the scaled particle theory. The alternative way is to use the so-called y-expansion of the hard-core free energy proposed by Gelbart and Barboy [44]. This is an expansion in powers of $\eta/(1-\eta)$ which is much more reliable at high densities compared with the usual virial expansion in powers of η. The y-expansion has also been used by Mulder and Frenkel [51] in the interpretation of the results of computer simulations for a system of hard ellipsoids.

The free energy (Eq. 59) with some modifications has been used in the detailed description of the N–I transition by Gelbart et al. [42, 43], Cotter [37] and Ypma and Vertogen [40]. The work of Ypma and Vertogen also contains a critical comparison with other approaches. One interesting conclusion of this analysis is related to the effective axial ratio of a mesogenic molecule. It has been shown that a good agreement with the experiment for the majority of the parameters of the N–I transition can be achieved only if one assumes that the effective geometrical anisotropy of the mesogenic molecule is much smaller than its actual value. This result has been interpreted in terms of molecular clusters that are supposed to be the building units of the nematic phase. The anisotropy of such a cluster is assumed to be much smaller than that of a single molecule. On the other hand, this result can be attributed to the quantitative inaccuracy of the free energy (Eq. 59). It is not excluded that the theory based on Eq. (59) overestimates a contribution from the excluded volume effects in thermotropic nematics.

The same problem can be viewed in a different way. According to the results of computer simulations [45] the nematic ordering in an athermal system of elongated rigid particles is formed only if the axial ratio is more than three. At the same time, the effective value of L/D for typical mesogenic molecules is usually assumed to be more than three. Thus, any dense fluid, composed of such molecules, must be in the nematic phase at all temperatures. In reality, however, some additional contribution from attractive forces is required to stabilize the nematic phase. As a result, we conclude that there exists some very delicate balance between repulsion and attraction in real thermotropic liquid crystals. The discrepancy from experiment can also be related to molecular flexibility; the anisotropy of the hard core may not be sufficiently large to stabilize the nematic phase. Thus, a fully consistent molecular theory of nematic liquid crystals must take into account the molecular flexibility in some way. This is a particularly difficult problem and so far the flexibility has been accounted for only in the context of the generalized mean-field theory applied separately to each small molecular fragment (see the review of Luckhurst and references therein [46]).

2.4.5   The Role of Molecular Biaxiality

The majority of the existing molecular theories of nematic liquid crystals are based on simple uniaxial molecular models like spherocylinders. At the same time typical mesogenic molecules are obviously biaxial. (For example, the biaxiality of the phenyl ring is determined by its breadth-to-thickness ratio which is of the order of two.) If this biaxiality is important, even a very good statistical theory may result in a poor agreement with experiment when the biaxiality is ignored. Several authors have suggested that even a small deviation from uniaxial symmetry can account for important features of the N–I transition [29, 42, 47, 48].

In the uniaxial nematic phase composed of biaxial molecules the orientational distribution function depends on the orientation of both the molecular long axis a and the short axis b, i.e. $f_1(1) = f_1((a \cdot n), (b \cdot n))$. The influence of the biaxiality on the distribution function is suitably described by the order parameter D:

$$D = \langle P_2((b \cdot n)) \rangle - \langle P_2((c \cdot n)) \rangle \qquad (62)$$

which characterizes the difference in the tendencies of the two short axes b and c to orient along the director. The order parameter D appears to be rather small (roughly of the order 0.1 [47]) and it is often neglected in simple molecular theories. However, it has some influence on the N–I transition as discussed in ref. [47, 49]. At the same time, except for the parameter D, the molecular biaxiality does not directly manifest itself in the Maier–Saupe theory. If one neglects the parameter D, the equations of the theory will depend on some effective uniaxial potential which is equal to the true potential between biaxial molecules averaged over independent rotations of the two molecules about their long axes.

However, the contribution of the biaxiality is nontrivial if the hard-core repulsion between biaxial molecules is taken into account. The hard-core repulsion is described by the Maier function $\exp[-\beta V_s(1, 2)]$ that depends nonlinearly on the repulsive potential $V_s(1, 2)$. In this case the free energy depends on the effective uniaxial potential that is an average of the Maier function. This theory accounts for some biaxial steric correlations and the result of such averaging cannot be interpreted as a hard-core repulsion between some uniaxial particles. Thus, it is the biaxiality of the molecular shape that seems to be of primary importance.

The biaxiality of molecular shape can be directly taken into account in the context of the Onsager theory. The first attempt to do this has been made by Straley [49]. How-

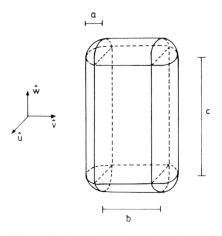

Figure 2. Spheroplatelet as a simple model for a biaxial particle [50].

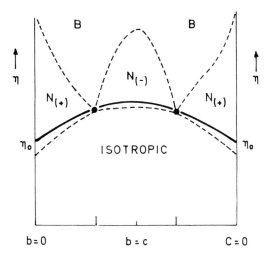

Figure 3. Phase diagram of the hard spheroplatelet fluid in the Onsager approximation for a family of particles with constant volume and different axial ratio. The points $b=0$ and $c=0$ correspond to the spherocylinders (after Mulder [50]).

ever, the full analyses can be performed only if the explicit expression for the excluded volume of the two biaxial particles is available. This expression has been derived by Mulder [50] for two spheroplatelets (see Fig. 2) with an arbitrary relative orientation. The excluded volume for such biaxial particles can be written in the analytical form

$$
V_{\text{excl}}(\omega_1, \omega_2)
$$

$$
= \frac{32\pi a^3}{3} + 8\pi\, a^2(b+c) + 8abc
$$

$$
+ 4abc\left[|v_1 \times w_2| + |v_2 \times w_1|\right]
$$

$$
+ 4ab^2 |v_1 \times v_2| + 4ac^2 |w_1 \times w_2|
$$

$$
+ b^2 c\left[|u_1 \times v_2| + |v_1 \times u_2|\right]
$$

$$
+ bc^2\left[|u_1 \times w_2| + |w_1 \times u_2|\right] \tag{63}
$$

where the unit vectors w, u and v are in the direction of the main axes of the spheroplatelet and the parameters a, b and c are shown on Fig. 2. The plausible constant volume section of the phase diagram of a fluid composed of such biaxial particles, obtained in the Onsager approximation, is shown in Fig. 3. The limiting points $c=0$ and $b=0$

correspond to the same spherocylinder with diameter α. In this limiting case the theory is reduced to the usual Onsager theory. The midpoint $b=c$ corresponds to the plate-like particle with the C_4 symmetry axis. Such particles form the uniaxial discotic nematic phase. For $0<c<b$ and for $0<b<c$ the particles are biaxial. However, for $c\gg b>c$ and for $b\gg c>1$ they are rod-like while for $b\approx c$ they are plate-like. Thus, somewhere between these two domains one should find the crossover shape that corresponds to a boundary between the transitions into two different nematic phases N_+ and N_-. In the N_- phase the molecular planes (and the long axes) are oriented approximately along the director, while in the N_+ phase they are oriented perpendicular to the director. The crossover shape corresponds also to the transition from the isotropic to the biaxial nematic phase. For very long spheroplatelets the crossover shape is characterized by the relation $b\approx c^{1/2}$ in dimensionless units.

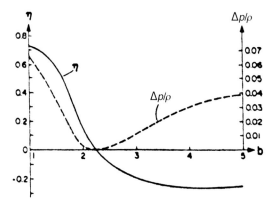

Figure 4. The dependence of the order parameter and the discontinuity in density at the N–I phase transition. The liquid is composed of hard rectangular parallelepipeds with dimensions $a = 1$, b and $c = 5$ (after Gelbart and Barboy [42]).

The phase diagram, shown in Fig. 3, was obtained from the Onsager approximation and, therefore, it is expected to be correct only at low densities. In the case of liquid densities the role of shape biaxiality has been analyzed by Gelbart [43]. Gelbart has considered the nematic ordering of rectangular parallelepipeds having dimensions $a < b < c$ (with a and c fixed, $c > a$) with the help of y-expansion. In Fig. 4, taken from [43], the nematic order parameter η and the discontinuity in density at the transition point are presented as a function of the breadth of the parallelepiped b that varies between $a = 1$ and $c = 5$. One can readily see that both the order parameter and the density gap decrease strongly with the increasing molecular biaxiality b/a. For the limiting case of $b = a = 1$, the rod limit, the order parameter at the transition is very large. However, for values of b/a slightly less than 2 both the order parameter and the density gap are rather close to the typical experimental data. Finally, the value $b/a \approx 2.25$ corresponds to the crossover shape and at this point the order parameter and the density gap vanish identically.

Thus we see that the effect of shape biaxiality can indeed be very strong. It seems that this can be the main reason why the molecular theories, based on rod-like molecular models, overestimate the first orderness of the nematic–isotropic transition.

2.4.6 Density Functional Approach to the Statistical Theory of Liquid Crystals

One can find in the literature a large number of molecular theories of liquid crystals proposed by different authors using different approximations. However, the majority of these theories and the corresponding expressions for the free energy can be derived in a systematic way with the help of some general approaches. A very fruitful approach of this kind is the density functional theory that has been first applied to liquid crystals by Sluckin and Shukla [52] and by Singh [53].

In the density functional approach the free energy of a liquid crystal is represented as a functional of the one-particle density $\rho(r, \omega) = \rho(r) f_1(\omega)$ where $\rho(r)$ is the number density and $f_1(r, \omega)$ is the one-particle distribution function. The equilibrium distribution function is determined, as usual, by minimization of the free energy functional $F[\rho]$. The general structure of this functional is unknown, of course, but the functional derivatives are known and can be expressed in terms of the correlation functions.

In the general case the free energy can be represented as a sum of two terms, $F = \Phi + H$ where Φ is the free energy of the system of noninteracting particles

$$\Phi = kT \int \rho(x) \left[\ln \rho(x) \Lambda - 1 \right] dx$$

$$+ \int \rho(x) U_e(x) dx \qquad (64)$$

where $x = (r, \omega)$, $U_e(x)$ is the external potential and Λ is a constant. The potential H is determined by the intermolecular interactions and its functional derivatives are related to the direct correlation functions. For example, the second derivative can be written as

$$\delta^2 H / \delta\rho(x_1)\delta\rho(x_2) = -kTC_2(x_1, x_2) \quad (65)$$

where $C_2(x_1, x_2)$ is the direct pair correlation function related to the full correlation function by the Ornstein–Zernike equation. Now Eq. (63) can be used to develop a theory of the N–I transition. For this purpose one can perform the functional Taylor expansion of the free energy of the nematic phase around its value in the isotropic phase. The expansion is performed in powers of $\Delta\rho$ where $\Delta\rho = \rho_N - \rho_I$ is the difference between the one-particle densities in the nematic and isotropic phases. Now the free energy of the nematic phase up to the second order in $\Delta\rho$ reads:

$$F_N = F_I + kT \quad (66)$$
$$\cdot \int \rho_N(x_1)[\ln\rho_N(x_1) - 1]dx_1$$
$$+ \int \rho_N(x_1)U_e/x_1\,dx_1$$
$$- \int C_1(x_1)\delta\rho(x_1)dx_1$$
$$- \frac{1}{2}\int C_2(x_1, x_2)\delta\rho(x_1)\delta\rho(x_2)dx_1\,dx_2$$

where $C_1 = \delta H/\delta\rho$ and where the direct correlation functions $C_1(1)$ and $C_2(1, 2)$ are calculated for the isotropic phase. The higher order terms in the expansion (Eq. 64) depend on the higher order direct correlation functions.

Taking into account the equilibrium condition for the isotropic distribution function ρ_I:

$$\ln\rho_I + C_1 - \rho_I U_e(1) = const$$

we arrive at the following self-consistent equation for the one-particle distribution function

$$\rho_N(r_1, \omega_1)$$
$$= \rho_I \exp[-\beta U_e(r_1, \omega_1)$$
$$+ \int C_2(r_{12}, \omega_1, \omega_2)\delta\rho(r_1, \omega_1)$$
$$\cdot \delta\rho(r_2, \omega_2)dr_2\,d\omega_2] \quad (67)$$

This general equation applies both to nematic and smectic phases because the one-particle density may depend on position. In the case of a uniform nematic phase without an external potential we obtain

$$f_1(\omega_1) =$$
$$Z^{-1}\exp\left[\int \tilde{C}_2(\omega_1, \omega_2)\delta\rho(\omega_2)d\omega_2\right] \quad (68)$$

We note that Eq. (66) has practically the same mathematical form as the mean-field equation (Eq. 39) for the orientational distribution function. In the density functional approach the role of the effective pair potential is played by the direct correlation function $\tilde{C}_2(\omega_1, \omega_2)$. In the case of uniaxial molecules the function $\tilde{C}_2(1, 2) = \tilde{C}_2((a_1 \cdot a_2))$. Expanding this function in Legendre polynomials, truncation after the P_2 term and substituting into Eq. (66), we arrive exactly at the Maier–Saupe equation for the orientational distribution function. This means that the general mathematical structure of the Maier–Saupe theory is not restricted to the mean-field approximation. The same equation has been derived in the context of a very general density functional approach. The only approximation has been related to the neglect of many-body direct correlation functions. Thus, the main equations of the Maier–Saupe theory remain valid generally on the level of pair correlations. This seems to be the main reason why this approach appears to be so successful in spite of its simplicity.

In the density functional approach the parameters of the N–I transition can be expressed in terms of the direct correlation function. Unfortunately, this function cannot be calculated exactly even for simple models and thus the general theory has to be accompanied by some practical approximations. We note that in this way it is possible to derive various approximate expressions for the free energy that have been obtained in various molecular theories. In other words, in the context of the density functional approach, various molecular theories usually correspond to some approximations for the direct correlation functions. We can illustrate this idea by the following examples.

The extended mean-field theory corresponds to the following approximate direct correlation function:

$$C_{\text{MF}}(1,2) \approx -V_{\text{att}}(1,2)\Theta(r_{12}-\xi_{12})/kT \quad (69)$$

The free energy of the Onsager theory can be obtained substituting the direct-correlation function by its first virial term

$$C_s(1,2) \approx \exp\left[-\beta V_s(1,2)\right] - 1 \quad (70)$$

Several more elaborate molecular theories correspond to the perturbative approximation

$$C_2(1,2) \approx C_{\text{HC}}(1,2) + g_{\text{HC}}(1,2)V_{\text{att}}(1,2) \quad (71)$$

where the functions $C_{\text{HC}}(1,2)$ and $g_{\text{HC}}(1,2)$ are calculated for the reference hard-core system.

Further approximations, including the ones discussed in Sec. 2.4.3, can be obtained by substituting $g_{\text{HC}}(1,2)$ with the steric cut-off function $\Theta(r_{12}-\xi_{12})$ and by using the Parsons approximation for the $C_{\text{HC}}(1,2)$:

$$C_{\text{HC}}(1,2) = C_{\text{hs}}(r_{12}/\xi_{12}) \quad (72)$$

where $C_{\text{hs}}(1,2)$ is the direct correlation function for a hard-sphere fluid. The more detailed discussion of various approxima-

tions for the direct correlation function can be found in the paper of Sluckin [54]. Recently some of these approximations have been tested against computer simulations [55].

The brief discussion of the density functional theory, presented above, enables one to conclude that this is a very powerful approach. It appears to be particularly helpful in the derivation of general expressions for various elasticity coefficients of the nematic phase. Indeed, it is also possible to expand the free energy of the distorted nematic state with respect to the homogeneous state in the same way as in Eq. (64). In this case the difference in the one-particle densities of the two states $\delta\rho$ is proportional to the gradients of the director and can be arbitrarily small. Then the functional expansion appears to be quantitatively correct. In this way it is possible to derive formally exact expressions for the Frank elastic constants [56], helical twisting power [57] and flexoelectric coefficients [58]. It should be noted, however, that so far the density functional theory has been formulated only for a system of rigid molecules. Thus, the corresponding general expressions are restricted to this simple class of molecular model.

2.5 Molecular Models for Simple Smectic Phases

2.5.1 Mean-field Theory of the Nematic–Smectic A Transition

Smectic phases are characterized by some positional order and therefore the one-particle distribution function $f_1(r, \omega)$ depends both on position r and the orientation ω. In the simplest smectic A phase there exists only one macroscopic direction that is parallel to the wave vector $\boldsymbol{k}$ of the periodic smectic structure. By contrast, in the smectic C phase there are two different macroscopic axes because the director is tilted with respect to $\boldsymbol{k}$ and thus the phase appears to be biaxial. As a result the distribution function of the smectic C phase generally depends on the orientation of both molecular long and short axes, if the molecules are biaxial.

For the smectic A phase the one-particle distribution function can be expanded in a complete set of basic functions:

$$f_1(\boldsymbol{r},\boldsymbol{a}) = f_1(\cos\omega,z)$$

$$= \sum_{l,m} f_{l,m}\, P_l(\cos\omega)\cos(kmz) \qquad (73)$$

where $\cos\omega = (\boldsymbol{a}\cdot\hat{\boldsymbol{k}})$ and where $\hat{\boldsymbol{k}} = \boldsymbol{k}/k$.

From Eq. (71) one can readily see that the dominant (i.e. lowest order) order parameters for the nematic–smectic A transition are:

$$S\langle P_2(\cos\omega)\rangle$$

$$\sigma = \langle P_2(\cos\omega)\cos(kz)\rangle$$

$$\tau = \langle\cos(kz)\rangle \qquad (74)$$

Here S is the usual nematic order parameter, τ is the purely translational order parameter and the parameter σ characterizes a coupling between orientational and translational ordering.

The simple theory of the nematic–smectic A transition has been proposed by McMillan [59] (and independently by Kobayashi [60]) by extending the Maier–Saupe approach to include the possibility of translational ordering. The McMillan theory is a classical mean-field theory and therefore the free energy is given by the general Eq. (34). For the smectic A phase it can be rewritten as

$$F = \mathrm{const} + \frac{1}{2}\rho^2 \int V(\boldsymbol{a}_i,\boldsymbol{a}_j,\boldsymbol{r}_{ij})$$

$$\cdot f_1(\boldsymbol{a}_i,\boldsymbol{r}_i)\, f_1(\boldsymbol{a}_j,\boldsymbol{r}_j)\,\mathrm{d}\boldsymbol{a}_i\,\mathrm{d}\boldsymbol{a}_j\,\mathrm{d}\boldsymbol{r}_i\,\mathrm{d}\boldsymbol{r}_j$$

$$+ \rho kT \int f_1(\boldsymbol{a}_i,\boldsymbol{r}_i)\ln f_i(\boldsymbol{a}_i,\boldsymbol{r}_i)\mathrm{d}\boldsymbol{a}_i\,\mathrm{d}\boldsymbol{r}_i \qquad (75)$$

where ρ is the mean density.

In the McMillan model the pair interaction potential is specified as

$$\qquad (76)$$

$$V(\boldsymbol{a}_i,\boldsymbol{a}_j,\boldsymbol{r}_{ij}) = -J_2(r_{12})\big(\delta + P_2\big((\boldsymbol{a}_1\cdot\boldsymbol{a}_2)\big)\big)$$

where δ is a dimensionless constant.

McMillan used a particular form for the coupling constant $J_2(r_{12})$:

$$J_2(r_{12}) = \frac{V}{Nr_0^3\pi^{3/2}}\exp\left[-(r_{12}/r_0)^2\right] \qquad (77)$$

where r_0 is some length of the order of the length of the rigid molecular core.

We note that the model potential (Eq. 74) is strongly simplified because in Eq. (74) the positional and orientational degrees of molecular freedom are decoupled. It has been pointed out by many authors [74–76] that in the general case the interaction potential must depend on the coupling between the intermolecular vector $\boldsymbol{r}_{12}$ and the molecular primary axes $\boldsymbol{a}_1$ and $\boldsymbol{a}_2$. We discuss the role of these terms in the next subsection.

The coupling constant (Eq. 75) can be expanded in Fourier series retaining only the leading term:

$$J_2(r_{12}) \approx -V\big[1 + \alpha\cos(kz)\big] \qquad (78)$$

where

$$\alpha = 2\exp\left[-(kr_0/2)\right] \qquad (79)$$

In the McMillan model the parameter α characterizes the strength of the interaction that induces the smectic ordering. The parameter α decreases with the increasing smectic period $d = 2\pi/k$ which is of the order of molecular length. Thus α is supposed to increase with increasing chain length.

The mean-field equilibrium distribution function is given by the general Eq. (35). Substituting Eq. (76) into Eq. (75) and then into Eq. (35) we obtain the distribution function of the smectic A phase

$$f_1(\cos\omega, z)$$

$$= Z^{-1}\exp\left[\frac{V_0}{kt}(\delta\alpha\tau\cos(kz)+\alpha\sigma\cos(kz)\right.$$

$$\left. \cdot P_2(\cos\omega)+SP_2(\cos\omega)\right] \qquad (80)$$

Multiplying both sides of Eq. (77) with the functions $P_2(\cos\omega)$, $\cos(kz)$ and $\cos(kz)$ $P_2(\cos\omega)$ and integrating over ω and z, one obtains the equations for the three order parameters

$$S=(d/Z)\int P_2(\cos\omega)$$

$$\cdot\exp\left[(V_0/kT)SP_2(\cos\omega)\right]$$

$$\cdot I_0(\kappa)\,d\cos\omega$$

$$\sigma=(d/Z)\int P_2(\cos\omega)$$

$$\cdot\exp\left[(V_0/kT)SP_2(\cos\omega)\right]$$

$$\cdot I_1(\kappa)\,d\cos\omega$$

$$\tau=(d/Z)\int\exp\left[(V_0/kT)SP_2(\cos\omega)\right]$$

$$\cdot I_0(\kappa)\,d\cos\omega \qquad (81)$$

where $\kappa=(V_0/kT)[\alpha\sigma P_2(\cos\omega)+\delta\alpha\tau]$ and the function $I_n(\kappa)$ is the n-th order modified Bessel function that appears after the integration over z.

Equation (78) together with the expressions (Eqs. 73 and 74) for the free energy can be used to calculate numerically the parameters of the nematic–smectic A phase transition. The corresponding phase diagram, taken from the original paper by McMillan [59], that includes the isotropic, nematic and smectic A phases is shown in Fig. 5. The inset in Fig. 5 also presents a typical phase diagram of a homologous series of compounds showing the transition temperature versus alkyl chain length.

In general the McMillan theory provides a good qualitative and sometimes even quantitative description of the nematic–smectic A phase transition. The theory accounts successfully for the decrease in the transition entropy with the breadth of the ne-

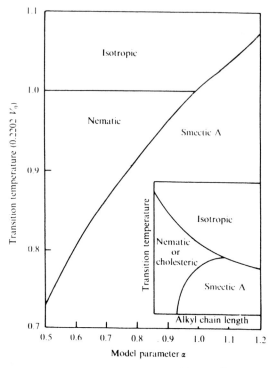

Figure 5. Phase diagram of a liquid crystal system according to the McMillan theory. Inset: typical phase diagram for a homologous series of compounds (after McMillan [59]).

matic phase and even enables one to locate the tricritical point in a reasonable way. A more detailed discussion of the McMillan theory can be found, for example, in the book of Chandrasekhar [61].

The McMillan theory has been further refined by several authors [62–64] to improve the quantitative agreement with experiment. However, the basic structure of the theory remains the same. This theory presents another example of a successful application of a simple mean-field approach. On the other hand, there are several limitations of the McMillan theory that cannot be ignored. Firstly, the theory is based on the semi phenomenological potential that does not allow determination of the smectic period in a self-consistent way. Secondly, the model potential (Eq. 74) is of the attractive type and therefore the McMillan theory does not consider the relative role of intermolecular attraction and hard-core repulsion in the stabilization of the smectic A phase. This problem appears to be of particular importance after it was shown by computer simulations that the smectic A phase can be formed in a system of hard spherocylinders without any attraction interaction. Finally, the potential (Eq. 74) does not depend on a coupling between orientational and translational degrees of freedom. This means that McMillan theory does not account for an interaction that forces the director to be normal to the smectic plane. The corresponding free energy is then unstable with respect to smectic C fluctuations.

Several authors have used different approaches to overcome these limitations in order to develop a more sophisticated statistical theory of the smectic A phase. We will discuss briefly some of the recent theories in the following two subsections.

2.5.2 Phase Diagram of a Hard-rod Fluid

The computer simulation studies of Frenkel et al. [45] indicate that the excluded volume effects for molecular hard cores must play an important role in the stabilization of the smectic phase. In particular, it has been shown that hard spherocylinders, interacting only via hard-core repulsion, can form nematic, smectic A, and columnar phases.

In the system of very long spherocylinders the nematic–isotropic transition can be quantitatively described by the Onsager theory discussed in Sec. 2.3.2. At the same time, this approximation is expected to provide only a qualitative description of the nematic–smectic A transition in the same

system. The reason is that long spherocylinders undergo a transition into the nematic phase at very small packing fractions $\eta \sim D/L \ll 1$. By contrast, the nematic–smectic A transition is expected to occur at large $\eta \sim 1$. The corresponding critical packing fractions can be estimated in the following way [65]. The Onsager theory is based on the virial expansion in powers of the number density $\rho = \eta/v_0$ where v_0 is the molecular volume. The transition to the nematic phase is determined by a balance between the orientational entropy $\rho \int f_1(1) \ln f_1(1) \, d(1)$ which is a maximum in the disordered state and the packing entropy $\sim \rho^2 \langle V_{excl}(1,2) \rangle$ which is a maximum in the orientationally ordered state. Thus the critical packing fraction is estimated as $\eta_{N-I} \sim v_0/\langle V_{excl} \rangle$. For long rods one finds $\langle V_{excl} \rangle \sim L^2 D$ and $v_0 \sim D^2 L$. Thus $\eta_{N-I} \sim D/L \ll 1$.

We note that in the system of long rods the nematic phase is strongly ordered. Then the transition into the smectic A phase is expected to take place in the nearly perfectly aligned system of rods. For parallel rods, however, the excluded volume $\langle V_{excl} \rangle \sim D^2 L$ and thus $\eta_{N-A} \sim 1$. This means that one cannot rely on the virial expansion even in the case of very long rods.

One possibility for improving the theory is to take into account higher order terms in the virial expansion. This has been done by Mulder for an aligned hard-rod fluid [66]. Mulder has taken into account the third- and fourth-order terms and has been able to obtain the numerical values of the transition density and the smectic period in very good agreement with the results of computer simulations [67]. The critical packing fraction and the dimensionless smectic wavelength observed are $\eta_{N-A} = 0.36$ and $\lambda \approx 1.27$ while the theoretical results are $\eta_{N-A} \approx 0.37$ and $\lambda \approx 1.34$ [66]. Recently Poniwierski performed an asymptotic analysis of the nematic–smectic A transition in the system of rods with orientational freedom and in the limit $L/D \to \infty$ [65]. He has shown that orientational fluctuations do not destroy the smectic A phase but the transition is shifted towards higher densities.

A different and more sophisticated approach to the theory of the nematic–smectic A transition is based on the nonlocal density functional theory developed for inhomogeneous hard-core fluids [68]. The nonlocal free energy functional is defined in the following way [69–71]:

$$F[\rho(r)] = F_{id}[\rho(r)] + H[\rho(r)] \qquad (82)$$

where $F_{id}[\rho]$ is the ideal gas contribution (see Sec. 2.4.4).

The excess of free energy $H[\rho]$ is assumed to have a form resembling the local density approximation:

$$H[\rho(r)] = \int \rho(r) \, \Delta\psi(\bar{\rho}(r)) dr \qquad (83)$$

where $\Delta\psi$ is the excess of free energy per particle and $\bar{\rho}(r)$ is some auxiliary density that depends on $\rho(r)$. For a homogeneous isotropic fluid $\bar{\rho} = \rho$. In the inhomogeneous state $\bar{\rho}(r)$ is related to $\rho(r)$ in a nonlocal way:

$$\bar{\rho}(r) = \int w_{eff}(r - r') \rho(r') dr' \qquad (84)$$

where $\omega_{eff}(r - r')$ is some weighting function.

The form of the function $\omega_{eff}(r)$ is different in different versions of the smoothed-density approximation proposed by Somoza and Tarazona [71, 72] and by Poniwierski and Sluckin [69, 73]. The density functional model of Somoza and Tarazona is based on the reference system of parallel hard ellipsoids that can be mapped into hard spheres. In the Poniwierski and Sluckin theory the effective weight function is determined by the Maier function for hard spherocylinders and the expression for $\Delta\psi(\rho)$ is obtained from the Carnahan–Starling ex-

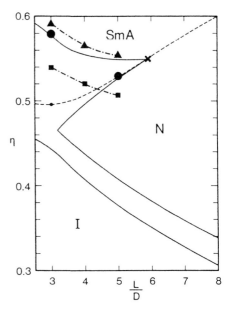

Figure 6. Phase diagram of a fluid of hard spherocylinders in the (axial ratio/order parameter) plane. The circles are the simulation results for the smectic A transition [45]. The N–SmA transition obtained in [45] is denoted by squares N and triangles SmA (after Poniwierski and Sluckin [69]).

cess of free energy for hard spheres with the same packing fraction.

Both groups have obtained phase diagrams for a fluid of hard spherocylinders that are in qualitative agreement with the results of computer simulations [45]. Figure 6 shows the phase diagram in the (L/D), η plane, obtained by Poniwierski and Sluckin [69], because it gives a more reasonable value for the tricritical point. One can see from Fig. 6 that the nematic–smectic A transition is first order for $L/D < 5.9$ and it is second order for $L/D > 5.9$. The location of the I–N–SmA triple point in Fig. 6 is in good agreement with simulations. The same good agreement has also been obtained by Somoza and Tarazona [71].

2.5.3 The Role of Intermolecular Attraction

The results of several molecular theories that describe the smectic ordering in a system of hard spherocylinders enable us to conclude that the contribution from hard-core repulsion can be described by the smoothed-density approximation. On the other hand, a realistic theory of thermotropic smectics can only be developed if the intermolecular attraction is taken into account. The interplay between hard-core repulsion and attraction in smectic A liquid crystals has been considered by Kloczkowski and Stecki [17] using a very simple model of hard spherocylinders with an additonal attractive r^{-6} potential. Using the Onsager approximation, the authors have obtained equations for the order parameters that are very similar to the ones found in the McMillan theory but with explicit expressions for the model parameters. The more general analysis has been performed by Mederos and Sullivan [76] who have treated the anisotropic attraction interaction by the mean-field approximation while the hard-core repulsion has been taken into account using the nonlocal density functional approach proposed by Somoza and Tarazona.

In [76] the intermolecular attraction potential has been taken in the form

$$(85)$$

$$V_{att}(1, 2) = V_1(r_{12}) + V_2(r_{12}) P_2((\boldsymbol{a}_1 \cdot \boldsymbol{a}_2))$$
$$+ V_3(r_{12}) \big[P_2((\boldsymbol{a}_1 \cdot \boldsymbol{u}_{12})) + P_2((\boldsymbol{a}_2 \cdot \boldsymbol{u}_{12})) \big]$$

where $\boldsymbol{u}_{12} = \boldsymbol{r}_{12}/r_{12}$.

We note that the last term in Eq. (82) has been omitted from the McMillan–Kobayashi theory. This term explicitly describes a coupling between the molecular long axis and the intermolecular vector. The effect of such coupling seems to be very important in smectic liquid crystals because this energy is obviously minimized when the molecules are packed in layers with their long axes parallel to the layer normal. (That is if $r_{12} \perp \boldsymbol{a}_1 \| \boldsymbol{a}_2$). The magnitude of the coupling constant V_3 in the last term is comparable to that of V_2 for rod-like molecules with an elongation typical of that for mesogens [77] and can even be predominant for weakly anisotropic molecules [78]. It should be noted also that the last term in Eq. (82) does not contribute to the free energy of the nematic phase in the mean-field approximation as it vanishes after averaging over all orientations of the intermolecular vector.

Using the specific expressions for the coupling constants in Eq. (82), Mederos and Sullivan obtained the temperature–density phase diagrams shown in Figs. 7a and b. These two diagrams have been obtained for the same value of the geometrical anisotropy $\sigma_\| / \sigma_\perp$ and for different values of the reduced strength of the symmetry breaking potential given by the last term in Eq. (82). We see that the smectic phase is stabilized with the increasing strength of the symmetry breaking potential. By contrast, the nematic phase tends to disappear. Thus the coupling between orientational and translational degrees of freedom is important indeed and it should also be taken into account in the description of the hard-core repulsion in smectic phases. It is not excluded, however, that the role of such interaction is overestimated in the Mederos–Sullivan theory because in this treatment the hard-core repulsion alone does not lead to the smectic A phase.

2.5.4 Smectic A–Smectic C Transition

The transition from the smectic A phase into the smectic C one is accompanied by the tilt of the molecular long axes with respect to

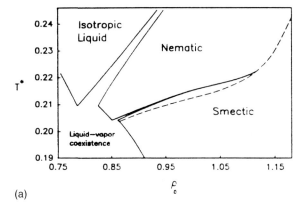

(a)

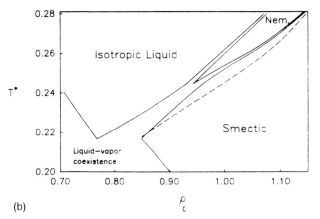

(b)

Figure 7. Temperature-density phase diagrams for $\sigma_{\parallel}\sigma_{\perp}=1.8$ and at $\varepsilon_3/\varepsilon_1=0.28$ (a) and $\varepsilon_3/\varepsilon_1=0.34$ (b) for a liquid crystal with both steric repulsion and attraction interactions between molecules. The parameter ε_3 measures the strength of the attraction potential that depends on the coupling between orientation and translation (after Mederos and Sullivan [76]).

the smectic plane normal e. We note that the resulting structure is unfavorable from the packing point of view [79]. Tilted molecules occupy more area in the smectic plane and therefore in the smectic C phase there is more excluded area in the layer than in the smectic A phase. As a result the packing entropy is decreased. This means that the smectic C phase is not expected to be formed in the system of hard rods, and, indeed, it has not been found in computer simulations of Veerman and Frenkel [45].

Thus, the smectic C phase can be formed only if there exists some specific intermolecular interaction that favors the tilt. Different interactions of this kind have been considered in the literature [79, 80–84] and some early theories have been analyzed in detail by Van der Meer [14].

Different molecular models for the SmC phase can be separated into two main classes that actually correspond to different molecular mechanisms of the smectic A–smectic C transition. Some models (for example, those of McMillan and Meyer [81] and Wulf [80]) imply that the molecular rotation about the long axis is frozen out in the smectic C phase. It seems to be even more important that in these models the smectic A–smectic C transition is governed by the ordering of molecular short axes while the tilt of the long axes occurs as a consequence. By contrast, in other models [82] the transition is directly related to the tilt of the long axes and the biaxiality of the smectic C phase is neglected.

We note that the more recent molecular models for the smectic C phase [79, 83, 84] also fall into one of these classes. For example, the model of Somoza and Tarazona [83] is based on steric interactions between molecules with biaxial shape. This interaction is assumed to be the driving force of the transition into the biaxial smectic C phase. At the same time, in the theory of Van der Meer and Vertogen [79] the molecular tilt is caused by the induction interaction between the off-center transverse dipoles and the polarizable core of neighboring molecules. This induction interaction is quadratic in dipole and therefore the free rotation around the molecular long axis does not destroy the smectic C phase. The recent theory of Poniwierski and Sluckin is based on the uniaxial molecular model in which hard cylinders carry axial quadrupoles.

It should be noted that the assumption of a strongly asymmetric orientational distribution of molecular short axes in the smectic C phase seems to be in contradiction with experiments [85]. Some other models also do not have any experimental support so far. Goodby et al. [87] and de Jeu [88] have studied the influence of electric dipole and molecular shape on the stability of the smectic C phase. The results do not support the models of Wulf [80] and Cabib and Benguigui [82] but reveal the importance of transverse dipoles. Thus there is some experimental evidence in favor of the model proposed by Van der Meer and Vertogen.

In this model the molecular tilt is determined by induction interaction between the off-center dipole and the polarizable core of the neighboring molecules. After averaging over the rotation around the molecular long axes the corresponding interaction potential reads [79]:

$$V_{ind}(1,2) = -J_{ind}\left(r_{12}(a_1 \cdot u_{12})^2\right) \quad (86)$$

where the coupling constant $J_{ind} \propto \bar{\alpha}\mu^2$ and

where μ is the molecular dipole and $\bar{\alpha}$ is the average molecular polarizability.

The interaction energy (Eq. 83) promotes the tilt of the director in the smectic C phase. This can be seen in the following way. Let us consider the case of perfect nematic ordering. Then the potential (Eq. 83) is reduced to $-J_{ind}(n \cdot u_{12})^2$. Now let us average this potential over all orientations of the intermolecular unit vector u_{12} within the smectic plane. For any two molecules within one plane the vector $u_{12} \perp e$ and one obtains

$$\langle V_{ind}(1,2)\rangle = -J_{ind}\cos^2\Theta \quad (87)$$

where Θ is the tilt angle, $\cos\Theta = (n \cdot e)$.

In the model of Van der Meer and Vertogen the induction interaction energy (Eq. 83) is counterbalanced by the hard-core repulsion coupled with isotropic attraction between molecular hard cores. The resulting interaction potential is presented in the form of an expansion:

$$
\begin{aligned}
V_{eff}(1,2) = (v_1 - J_{ind}) \\
\cdot [(a_1 \cdot u_{12})^2 + (a_2 \cdot u_{12})^2 \\
+ v_2(a_1 \cdot a_2)(a_1 \cdot u_{12})(a_2 \cdot u_{12}) \\
+ v_3(a_1 \cdot u_{12})^2(a_2 \cdot u_{12})^2]
\end{aligned}
\quad (88)
$$

where the coefficients v_1, v_2 and v_3 are expressed in terms of the shape anisotropy and the attraction interaction strength.

In Eq. (85) the constant v_1 is positive and thus, without the induction interaction, the potential (Eq. 85) stabilizes the smectic A phase. Taking into account the packing entropy, Van der Meer and Vertogen have obtained the following simple free energy in the case of perfect nematic ordering:

$$
\Delta F = \frac{1}{2}D_0\langle\cos\phi\rangle^2 - kT\ln\frac{1}{2\pi}
$$
$$
\cdot \int d\phi\,\exp[\beta D_0\langle\cos\phi\rangle\cos\phi] \quad (89)
$$

where

$$D_0 = 2\left[(1+T/T_p)B_2 - C_2\right]$$
$$\cdot P_2(\cos\Theta) - (1-T/T_p)$$
$$\cdot B_4 P_4(\cos\Theta) \qquad (90)$$

Here $\langle\cos\phi\rangle$ is the smectic order parameter and the coefficients B_2, C_2 and B_4 are presented in [79]. The induction interaction strength is adsorbed in C_2.

The simple free energy (Eq. 86) can be used to describe the transitions between nematic, smectic A and smectic C phases. The second order smectic A–smectic C transition temperature is given by

$$T_{AC} = T_p\left(\frac{3C_2}{3B_2 - 5B_4} - 1\right) \qquad (91)$$

In this model the temperature variation of the tilt angle is the reduced temperature scale and does not depend on any molecular parameters. The corresponding temperature variation for the smectic A–smectic C and nematic–smectic C transitions is shown

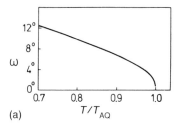

(a)

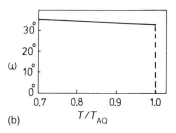

(b)

Figure 8. Temperature dependence of the tilt angle in the SmC phase near the second-order A–C transition (a) and near the first-order N–C transition (b) (after Van der Meer and Vertogen [79]).

in Fig. 8a and b, respectively, taken from [79].

The molecular theory of Van der Meer and Vertogen is based on a specific molecular model that is not in contradiction with experiment. At the same time Barbero and Durand [89] have shown that the molecular tilt is an intrinsic property of any layered quadrupolar structure. This idea has been used by Poniwierski and Sluckin in their model [84] that presents a rather general mechanism for the stabilization of the smectic C phase. It is interesting to note that the mathematical form of the interaction potential in the Poniwierski–Sluckin theory is similar to the potential (Eq. 85). The energy of electrostatic interaction between two axial quadrupoles, employed in [84], can be written as

$$U_{qq}(1,2) = \frac{3}{4}\frac{q^2}{r^5}$$
$$\cdot [1 + 2(\boldsymbol{a}_1 \cdot \boldsymbol{a}_2)^2 - 5(\boldsymbol{a}_1 \cdot \boldsymbol{u}_{12})^2 - 5(\boldsymbol{a}_2 \cdot \boldsymbol{u}_{12})^2$$
$$- 20(\boldsymbol{a}_1 \cdot \boldsymbol{a}_2)(\boldsymbol{a}_1 \cdot \boldsymbol{u}_{12})(\boldsymbol{a}_2 \cdot \boldsymbol{u}_{12})$$
$$+ 35(\boldsymbol{a}_1 \cdot \boldsymbol{u}_{12})^2 (\boldsymbol{a}_2 \cdot \boldsymbol{u}_{12})^2] \qquad (92)$$

The last three terms of the potential (Eq. 90) have the same mathematical form as the corresponding terms in Eq. (85). This means that the quadrupolar-type potential appears to be a good model potential for the theory of smectic A–smectic C transition.

The interaction potentials (Eqs. 85 and 90) essentially depend on a coupling between the molecular orientation and the intermolecular vector. We note that this coupling could be neglected in the first approximation in the theory of the nematic–smectic A transition, as it is done, for example, in the McMillan theory. At the same time this coupling just determines the effect in the theory of transition into the smectic C phase.

Finally we note that both theories, discussed above, neglect the biaxiality of the

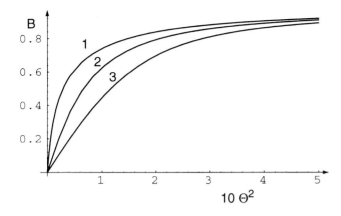

Figure 9. Biaxiality order parameter B as a function of the tilt angle in the SmC phase. The curves (3)–(1) correspond to the increasing strength of the biaxial part of the interaction potential [8].

smectic C phase. In real systems some biaxiality is always induced by the tilt and it is only a question of how large the corresponding contribution is to the free energy. In Fig. 9 we present the value of the biaxiality order parameter B (that determines the nonpolar ordering of molecular short axes in the smectic C phase) as a function of the tilt angle. This dependence has been calculated theoretically in [8] using a simple model. The value of the biaxiality order parameter seems to be overestimated. However, even if the actual value is several times smaller, the problem of interaction between biaxiality and the tilt in the smectic C phase deserves further attention.

2.6   Conclusions

The literature on the molecular theory of liquid crystals is enormous and in this chapter we have been able to cover only a small part of it. We have mainly been interested in the models for the nematic–isotropic, nematic–smectic A and smectic A–smectic C phase transitions. The existing theory includes also extensive calculations of the various parameters of the liquid crystal phases: Frank elastic constants, dielectric susceptibility, viscosity, flexoelectric coefficients and so on. This part of the molecular theory of liquid crystals remains completely beyond the scope of this review. We also did not consider the theory of more complex liquid crystalline phases including the cholesteric phase, the ferroelectric smectic C* phase, re-entrant phases, bilayer and incommensurate smectic phases and phases with hexatic ordering. The majority of these theories, however, employ the same general ideas and approximations that have been discussed above. Thus the present review presents some basic information necessary for the understanding of the existing molecular theory of liquid crystals.

Acknowledgements

The author would like to acknowledge interesting and stimulating discussions on the molecular theory of liquid crystals with T. J. Sluckin, D. A. Dunmur, G. R. Luckhurst, S. Hess, E. M. Terentjev, B. Mulder, A. Somoza and P. I. C. Teixeira.

2.7   References

[1] B. Berardi, S. Orlandi, C. Zannoni, *Chem. Phys. Lett.* **1996**, *261*, 363.
[2] A. Ferrarini, G. J. Moro, P. L. Nordio, G. R. Luckhurst, *Mol. Phys.* **1992**, *77*, 1.
[3] P. G. de Gennes, *The Physics of Liquid Crystals*, Oxford University Press, Oxford, **1974**.
[4] E. M. Averjanov, M. A. Osipov, *Sov. Phys. Uspihi* **1990**, *161*, 93.

[5] S. C. McGrother, D. C. Williamson, G. Jackson, *J. Chem. Phys.* **1996**, *104*, 6755.

[6] D. A. Dunmur, P. Palffy-Muhoray, *Mol. Phys.* **1992**, *76*, 1015; A. G. Vanakaras, D. J. Photinos, *Mol. Phys.* **1995**, *85*, 1089.

[7] A. Perera, G. N. Patey, *J. Chem. Phys.* **1989**, *91*, 3045.

[8] M. A. Osipov, S. A. Pikin, *J. Phys. (France)* **1995**, *II/5*, 1223.

[9] S. A. Pikin, M. A. Osipov in *Ferroelectric Liquid Crystals* (Ed.: G. W. Taylor), Gordon and Breach, New York, **1992**.

[10] L. Onsager, *Ann. N. Y. Acad. Sci.* **1949**, *51*, 627.

[11] W. Maier, A. Saupe, *Z. Naturforsch.* **1959**, *14a*, 882.

[12] W. Maier, A. Saupe, *Z. Naturforsch.* **1960**, *15a*, 287.

[13] B. W. Van der Meer, G. Vertogen in *Molecular Physics of Liquid Crystals* (Eds.: G. R. Luckhurst, G. W. Gray), Academic Press, New York, **1979**.

[14] B. W. Van der Meer, Thesis, Groningen, **1979**.

[15] L. Pohl, R. Eidenschink, J. Krause, G. Weber, *Phys. Lett. A* **1978**, *65*, 169.

[16] W. M. Gelbart, A. Gelbart, *Mol. Phys.* **1977**, *33*, 1387.

[17] A. Kloczkowski, J. Stecki, *Mol. Phys.* **1985**, *55*, 689.

[18] M. A. Osipov, H. Stegemeyer, A. Sprick, *Phys. Rev. E* **1996**, *54*, 6387.

[19] R. L. Humphries, P. G. James, G. R. Luckhurst, *J. Chem. Soc., Faraday Trans. II* **1972**, *68*, 1031.

[20] R. G. Priest, *J. Phys. (France)* **1975**, *36*, 437; *J. Chem. Phys.* **1976**, *65*, 408.

[21] B. W. Van der Meer, G. Vertogen, *Phys. Lett. A* **1979**, *71*, 488.

[22] J. G. Gray, B. J. Berne, *J. Chem. Phys.* **1981**, *74*, 3116.

[23] G. R. Luckhurst in *The Molecular Physics of Liquid Crystals* (Eds.: G. R. Luckhurst, G. W. Gray), Academic Press, London, **1979**.

[24] H. Toriumi, E. T. Samulski, *Mol. Cryst. Liq. Cryst.* **1983**, *101*, 163.

[25] W. H. de Jeu, J. van der Veen, *Mol. Cryst. Liq. Cryst.* **1977**, *40*, 1.

[26] J. G. T. Ypma, G. Vertogen, H. T. Koster, *Mol. Cryst. Liq. Cryst.* **1976**, *37*, 57.

[27] N. V. Madhusudana, S. Chandrasekhar, *Solid State Commun.* **1973**, *13*, 377.

[28] S. Chandrasekhar, N. V. Madhusudana, *Acta Crystallogr.* **1971**, *27*, 303.

[29] J. P. Straley, *Mol. Cryst. Liq. Cryst.* **1973**, *24*, 7.

[30] G. Lasher, *J. Chem. Phys.* **1970**, *53*, 4141.

[31] H. N. W. Lekkerkerker, Ph. Coulon, R. van der Haegen, R. Deblieck, *J. Chem. Phys.* **1984**, *80*, 3247.

[32] Sin-Doo Lee, R. B. Meyer, *J. Chem. Phys.* **1986**, *84*, 3443.

[33] G. J. Vroege, H. N. W. Lekkerkerker, *Rep. Progr. Phys.* **1992**, *55*, 1241.

[34] A. N. Semenov, A. R. Khokhlov, *Sov. Phys. Usp.* **1988**, *31*, 988.

[35] T. Odijk, *Macromolecules* **1986**, *19*, 2313.

[36] T. Sato, A. Teramoto in *Advances in Polymer Science*, Vol. 126, Springer-Verlag, Berlin, Heidelberg, **1996**.

[37] M. A. Cotter in *Molecular Physics of Liquid Crystals* (Eds.: G. R. Luckhurst, G. W. Gray), Academic Press, New York, **1979**.

[38] Sin-Doo Lee, *J. Chem. Phys.* **1987**, *87*, 4972.

[39] A. R. Khokhlov, A. N. Semenov, *J. Stat. Phys.* **1985**, *38*, 161.

[40] J. G. J. Ypma, G. Vertogen, *Phys. Rev. A* **1978**, *17*, 1490.

[41] J. D. Parsons, *Phys. Rev. A* **1979**, *19*, 1225.

[42] W. M. Gelbart, B. Barboy, *Acc. Chem. Res.* **1980**, *13*, 290.

[43] W. M. Gelbart, *J. Phys. Chem.* **1982**, *86*, 4298.

[44] B. Barboy, W. M. Gelbart, *J. Stat. Phys.* **1980**, *22*, 709.

[45] J. A. C. Veerman, D. Frenkel, *Phys. Rev. A* **1990**, *41*, 3237; D. Frenkel, *J. Phys. Chem.* **1988**, *92*, 3280.

[46] G. R. Luckhurst in *Recent Advances in Liquid Crystalline Polymers* (Ed.: L. L. Chapoy), Elsevier, London, **1986**, p. 105.

[47] G. R. Luckhurst, C. Zannoni, P. L. Nordio, U. Segre, *Mol. Phys.* **1975**, *330*, 1345.

[48] C. S. Shih, R. Alben, *J. Chem. Phys.* **1972**, *57*, 3055.

[49] J. P. Straley, *Phys. Rev. A* **1974**, *10*, 1881.

[50] B. Mulder, *Phys. Rev. A* **1989**, *39*, 360.

[51] B. M. Mulder, D. Frenkel, *Mol. Phys.* **1985**, *55*, 1171.

[52] T. J. Sluckin, P. Shukla, *J. Phys. A: Math. Gen.* **1983**, *16*, 1539.

[53] Y. Singh, *Phys. Rev. A* **1984**, *30*, 583.

[54] T. J. Sluckin, *Mol. Phys.* **1983**, *49*, 221.

[55] M. P. Allen et al., *Phys. Rev. E* **1995**, *52*, R95.

[56] A. Poniwierski, J. Stecki, *Mol. Phys.* **1979**, *38*, 1931.

[57] M. A. Osipov, Chap. 1 in *Liquid Crystalline and Mesomorphic Polymers* (Eds.: V. P. Shibaev, L. Lam), Springer-Verlag, New York, **1994**.

[58] M. A. Osipov, S. A. Pikin, E. M. Terentjev, *Sov. Scientific Rev.*, harwood acad. publ. **1989**, *11*, 191.

[59] W. L. McMillan, *Phys. Rev. A* **1971**, *4*, 1238; **1972**, *6*, 936.

[60] C. Kobayashi, *Mol. Cryst. Liq. Cryst.* **1971**, *13*, 137.

[61] S. Chandrasekhar, *Liquid Crystals*, 2nd edn., Cambridge University Press, **1992**.

[62] L. Shen, H. K. Sim, Y. M. Shin, C. W. Woo, *Mol. Cryst. Liq. Cryst.* **1977**, *39*, 299.

[63] H. T. Tan, *Phys. Rev. A* **1977**, *16*, 1715.

[64] P. J. Photinos, A. Saupe, *Phys. Rev. A* **1975**, *13*, 1926.

[65] A. Poniwierski, *Phys. Rev. A* **1992**, *45*, 5605.

[66] B. Mulder, *Phys. Rev. A* **1987**, *35*, 3095.

[67] A. Stroobants, H. N. W. Lekkerkerker, D. Frenkel, *Phys. Rev. Lett.* **1986**, *57*, 1452; *Phys. Rev. A* **1987**, *36*, 2929.

[68] P. Tarazona, *Phys. Rev. A* **1985**, *41*, 2672; W. A. Curtin, N. W. Ashcroft, *Phys. Rev. A* **1985**, *32*, 2909; T. F. Meister, D. M. Kroll, *Phys. Rev. A* **1985**, *31*, 4055.

[69] A. Poniwierski, T. J. Sluckin, *Phys. Rev. A* **1991**, *43*, 6837.

[70] A. Poniwierski, T. J. Sluckin, *Mol. Phys.* **1991**, *73*, 199.

[71] A. M. Somoza, P. Tarazona, *Phys. Rev. A* **1990**, *41*, 965.

[72] A. M. Somoza, P. Tarazona, *Phys. Rev. Lett.* **1988**, *61*, 2566; *J. Chem. Phys.* **1989**, *91*, 517.

[73] A. Poniwierski, R. Holyst, *Phys. Rev. Lett.* **1988**, *61*, 2461.

[74] L. Senbetu, C.-W. Woo, *Phys. Rev. A* **1978**, *17*, 1529.

[75] M. D. Lipkin, D. W. Oxtoby, *J. Chem. Phys.* **1983**, *79*, 1939.

[76] L. Mederos, D. E. Sullivan, *Phys. Rev. A* **1989**, *39*, 854.

[77] B. Tjipto-Margo, D. E. Sullivan, *J. Chem. Phys.* **1988**, *88*, 6620.

[78] C. G. Gray, K. E. Gubbins, *Theory of Molecular Fluids*, Clarendon Press, Oxford, **1984**.

[79] B. W. Van der Meer, G. Vertogen, *J. Phys. Colloq. (France)* **1979**, *40*, C3–222.

[80] A. Wulf, *Phys. Rev. A* **1975**, *11*, 365.

[81] A. Meyer, W. L. McMillan, *Phys. Rev. A* **1974**, *9*, 899.

[82] D. Cabib, L. Benguigui, *J. Phys. (France)* **1977**, *38*, 419.

[83] A. M. Somoza, P. Tarazona, *Phys. Rev. Lett.* **1988**, *61*, 2566.

[84] A. Poniwierski, T. J. Sluckin, *Mol. Phys.* **1991**, *73*, 199.

[85] H. Hervet, F. Volino et al., *J. Phys. (Lett.) France* **1974**, *35*, L-151; *Phys. Rev. (Lett.)* **1975**, *34*, 451; P.-J. Bos, J. Pirs et al., *Mol. Cryst. Liq. Cryst.* **1977**, *40*, 59.

[86] L. Blum, A. J. Torruella, *J. Chem. Phys.* **1972**, *56*, 303.

[87] J. W. Goodby, G. W. Gray, D. G. McDonnell, *Mol. Cryst. Liq. Cryst. (Lett.)* **1977**, *34*, 183.

[88] W. H. de Jeu, *J. Phys. (France)* **1977**, *38*, 1263.

[89] G. Barbero, G. Durand, *Mol. Cryst. Liq. Cryst.* **1990**, *179*, 57.

3 Molecular Modelling

Mark R. Wilson

Over the past 10 years, rapid progress has been made in the field of molecular modelling. Advances have been led by two important factors: the increase in speed and reduction in cost of modern computers, and an accompanying improvement in the accuracy and ease of use of molecular modelling software. Modelling packages are now commonly available in many laboratories. Their ability accurately to predict molecular structures of simple organic molecules makes them a useful tool in the study of liquid crystals. In this article developments in molecular modelling are discussed in the context of liquid crystal systems. The article is divided into two main sections: Sec. 3.1 covers the main molecular modelling techniques that are currently available; whilst Sec. 3.2 covers specific applications of these techniques in the study of liquid crystal molecules.

3.1 Techniques of Molecular Modelling

3.1.1 Molecular Mechanics

Molecular mechanics is the simplest and most commonly used molecular modelling

technique. It is concerned with the determination of molecular structure, and is of particular relevance to liquid crystal chemists concerned with the design of appropriate molecular structures, or to physicists interested in calculating molecular properties. The molecular mechanics approach has been reviewed in a number of places [1–4], and so here only the basics of the technique are described. The standard approximation employed in molecular mechanics is to consider a molecule as a collection of atoms held together by elastic restoring forces. These forces are described by simple functions that characterise the distortion of each structural feature within a molecule. Usually separate functions exist for each bond stretch, bond bend, and dihedral angle; as well as for each nonbonded interaction. Together these functions make up the molecular mechanics *force field* for a particular molecule. The steric energy E can then be defined with reference to the force field. E has no physical meaning in itself, but can be thought of as measuring how the energy of a particular molecular conformation varies from a hypothetical ideal geometry where all bonds, bond angles, etc. have their ideal (or natural) values.

$$E = \sum_{\substack{\text{bond} \\ \text{lengths}}} E_{\text{bond}} + \sum_{\substack{\text{bond} \\ \text{angles}}} E_{\text{angle}} \qquad (1)$$

$$+ \sum_{\substack{\text{dihedral} \\ \text{angles}}} E_{\text{torsion}} + \sum_{i=1}^{N} \sum_{j<i}^{N} (E_{\text{nb}ij} + E_{\text{el}ij})$$

The symbols in Eq. (1) have the following meanings: E_{bond} is the energy of a bond that is stretched or compressed from its natural value, E_{angle} is the energy of a bond angle that is distorted from its natural value, E_{torsion} is the energy of a dihedral angle that is distorted from its natural value, $E_{\text{nb}ij}$ and $E_{\text{el}ij}$ are respectively the Lennard–Jones and electrostatic nonbonded interactions between the pair of atoms i and j, and N is the number of atoms in the system.

Bond stretches are usually characterised by a harmonic potential of the form

$$E_{\text{bond}} = \frac{1}{2} K_{\text{b}} (l - l_0)^2 \qquad (2)$$

where l is the stretched or compressed bond length, l_0 is the natural bond length for an undistorted bond, and K_{b} is a force constant characterising the bond distortion. If a carbon–carbon bond is stretched from its ideal lowest energy value of $l_0 = 1.523$Å, this results in a contribution to the steric energy in Eq. (1). Similarly, E_{angle} and E_{torsion} characterise bond angle and dihedral angle perturbations through a harmonic potential and a truncated Fourier series respectively:

$$E_{\text{angle}} = \frac{1}{2} K_{\theta} (\theta - \theta_0)^2,$$

$$E_{\text{torsion}} = \sum_{m} \frac{1}{2} K_{\phi_m} [1 + \cos(m\phi - \delta)] \qquad (3)$$

where θ and θ_0 are distorted and natural bond angles, ϕ is a dihedral angle, δ is a phase angle, and K_{θ} and K_{ϕ_m} are force constants. Finally, 12–6 and Coulomb potentials are often used for nonbonded interactions:

$$E_{\text{nb}ij} = \frac{A_{ij}}{r_{ij}^{12}} - \frac{C_{ij}}{r_{ij}^{6}}, \quad E_{\text{el}ij} = \frac{q_i q_j}{r_{ij}} \qquad (4)$$

Here, r_{ij} is the distance between atoms i and j, $A_{ij} = (A_{ii} A_{jj})^{1/2}$, $C_{ij} = (C_{ii} C_{jj})^{1/2}$, and q_i and q_j are the partial electronic charges on atoms i and j. A_{ii} and C_{ii} can be expressed in terms of the Lennard–Jones parameters ε and σ: $A_{ii} = 4\, \varepsilon_{ii}\, \sigma_{ii}^{12}$ and $C_{ii} = 4\, \varepsilon_{ii}\, \sigma_{ii}^{6}$.

Each molecular conformation has a different value of the steric energy. So, although E has no direct physical significance by itself, the differences between steric energies of any two conformations is equivalent to the energy difference between them. The terms in Eq. (1)–(4) are not unique, and the exact form of the potential functions differs from one force field to another. However, all force fields rely on the fact that a specific interaction is similar in every molecule (i.e. a pure C–C bond stretch in ethane is similar to a C–C stretch in decane or in a large liquid crystal molecule). In parameterising the force field, all the force constants are carefully optimised to predict the structures and relative conformational energies of a control set of small molecules.

Currently, a number of excellent force fields exist in the literature. For low molecular weight organic liquid crystals, the MM3 force field [5] (and its predecessors MM2 and MM1 [6, 7]) generally produce excellent structures and good conformational energies. Molecules containing mainly alkyl chains and saturated rings are described well by most force fields. However, calculations involving some functional groups commonly used in liquid crystal molecules (e.g. –N=N–, F, and CN will yield less accurate structures and energies. Parameters associated with these groups are often marked *preliminary* within force fields. This simply reflects the lack of molecules with these functional groups in the force field control set. The range of valence states adopted by metals, the lack of metals in force field control sets, and polarization effects associated with metal ions (which are not handled well by traditional force fields) combine to make structural predictions for metal-containing liquid crystal structures much less reliable than predictions for organic liquid crystals.

Once an appropriate force field has been chosen, the aim of a molecular mechanics study is to optimise molecular geometry by minimising E. In an *energy-minimised* conformation the strain associated with the steric energy will be spread throughout the molecule. In practice, molecular mechanics packages consist of a force field combined with energy minimisation routines, and often a graphical user interface (GUI) to provide an easy mechanism for carrying out molecular calculations. A typical molecular mechanics calculation consists of the following steps:

– build a trial molecular structure by providing coordinates from a crystal structure or generating *drawn* structures via a GUI;
– minimise the energy of the trial structure to provide a *minimum energy conformation*;
– undertake a search for other energy minima by adjusting dihedral angles within a molecule and re-minimising the steric energy for each conformation.

The final result of this process is a series of potential energy minima, one of which will be the global minimum and represent the lowest energy conformation of the molecule. For many liquid crystal molecules, many conformations exist that are similar in energy. This makes the tasks of finding the global energy minimum and characterising molecular structure difficult. New techniques for conformational searching have recently made this process easier [8–10]. However, the problem of energy minimisation on a multidimensional surface is still a difficult one, and for complicated molecules it is not always possible to guarantee that all relevant conformations have been found.

A single molecule with N_{conf} energy minima can be thought of as a N_{conf} state system with the probability P_j of the molecule being in state j given by the Boltzmann distribution

$$P_j = \frac{\exp\left(-\Delta E_j / k_B T\right)}{\sum\limits_{i=1}^{N_{conf}} \exp\left(\Delta E_i / k_B T\right)} \tag{5}$$

where ΔE_i is the energy of conformation i relative to the lowest energy conformation and k_B is Boltzmann's constant. The use of Eq. (5) provides a simple weighting for each of the conformational states occupied at a particular temperature T.

3.1.2   Molecular Dynamics and Monte Carlo Simulation

One of the ways of circumventing the problem of finding multiple energy minima of complex molecules is to turn to more sophisticated techniques that are capable of sampling phase space efficiently without the need to home in on particular minimum energy conformations. The two most useful techniques are molecular dynamics (MD) and the Monte Carlo (MC) method. Both approaches make use of the same types of potential functions used in molecular mechanics, but are designed to sample conformation space such that a Boltzmann distribution of states is generated. MC and MD techniques for molecular systems have been widely reviewed [11–14], and only the basics of the two methods are described below.

In molecular dynamics Newton's equations of motion are solved for the system of atoms interacting via a potential such as that of Eq. (1). For each atom, the force $\mathbf{F}_i$ is given by

$$\mathbf{F}_i = -\nabla E_i \tag{6}$$

where E_i is the interaction energy of atom i in the force field. Typically, each atom is given a velocity sampled from a Maxwell–Boltzmann distribution, and the equations of motion are solved using finite difference techniques [13]. Simulations are broken down into a series of small time steps δt, and at each time step atomic forces are calculated and used to advance the velocities and atomic positions forward in time (see the schematic in Fig. 1). In the simplest form of MD the total energy of the system is conserved. However, it is usually more useful to employ a thermostat (such as the Nosé–Hoover thermostat [15]) in order to carry out MD calculations at constant temperature. When molecular mechanics force fields are used, the size of time step chosen depends on the fastest motion in the system, which is invariably a bond stretch. As a general rule, energy conservation improves dramatically as time steps are reduced, and δt should be at leat 25 times smaller than the period of the fastest motion in the system. For this reason, it is usual practice to constrain bond lengths in an atomic simulation using the SHAKE procedure [16]. This approximation works well because bond stretches are usually of sufficiently high frequency to be decoupled from bond bending motion and torsional angle rotations. With SHAKE, a typical MD time step is 2 fs; without bond length constraints, this must be reduced to at least 0.5 fs, with a consequential increase in computer time required for simulation.

Many molecular mechanics packages now include MD as an option. In a typical MD simulation of a single molecule, the molecule is slowly warmed from an energy-minimised (zero kelvin) structure to the required average temperature over a period of a few picoseconds. Simulations are then carried out for the desired length of time, with molecular conformations saved periodically for later analysis throughout the course of the simulation run. Later analysis of these conformations is then able to provide time-averaged information for a single molecule at the temperature of the simulation. This approach yields useful data on dihedral angle distributions, moment of inertia ellipsoids, average dipole moments, etc.

However, single molecule molecular dynamics for liquid crystal molecules can often be problematic. Many liquid crystal systems have torsional energy barriers in excess of 12 kJ mol^{-1} separating conformations of similar energy (see Sec. 3.2.1). Such barriers can be difficult to cross during the course of a short MD simulation, and this can result in molecules becoming periodically trapped in regions of phase space. This has led to the development of stochastic dynamics techniques where random noise added to the equations of motion is de-

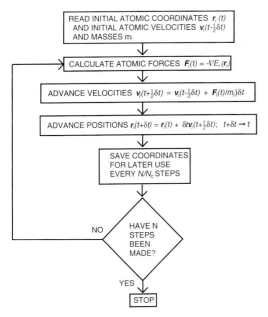

Figure 1. Schematic representation of a molecular dynamics simulation. The scheme for integrating the equations of motion is known as the leapfrog algorithm. The figure shows a flow diagram involving N_c samples of the coordinates for a simulation of N steps.

signed to mimic the effects of molecules colliding with a heat bath of surrounding molecules [17]. This process can lead to much faster barrier crossing rates than in standard MD. However, long simulations of 10–100 ns may still be needed if good dihedral angle distributions are required.

In Monte Carlo simulations the molecular mechanics interaction potential [Eq. (1)] can be used directly without the need to calculate atomic forces. Molecules must usually be represented by a set of internal coordinates (bond lengths, bond angles, and torsional angles), and the Metropolis approach is used to sample the configurational part of the partition function [11,12,14]. The Metropolis scheme involves making random changes to bond angles and torsional angles (bond lengths are usually held fixed) [11,18]. The energy of a new confirmation E_{new} is then calculated and compared with the previous energy E_{old}. If the trial energy is lower than E_{old}, the trial move is accepted. If the trial energy is higher than E_{old}, the move is accepted if $\exp[-(\Delta E)/k_B T]$ is greater than a random number between 0 and 1, where $\Delta E = E_{new} - E_{old}$. Consequently, over the course of a MC simulation, moves are accepted with a probability[1] of $\exp[-(\Delta E)/k_B T]$.

Single molecule MC simulations sample phase space much more efficiently than the corresponding MD calculations. Trial rotations about dihedral angles provide a mechanism to overcome the large energy barriers between liquid crystal conformers. The drawback of such calculations is complexity. They require a consistent representation of molecular structure, where Cartesian coordinates can be generated from a single reference point in terms of a set of specified

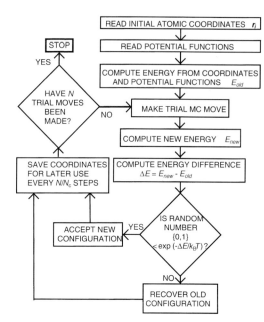

Figure 2. Schematic representation of a Metropolis Monte Carlo simulation. This scheme is suitable for most soft potentials, but constraints must be handled carefully (see footnote to text). N trial moves are attempted, with coordinate sampling carried out every N/N_c trial moves.

internal coordinates [18]; therefore it is harder to write general MC programs to handle any molecular structure. Despite this limitation, MC calculations are starting to become available in some modelling packages. Figure 2 shows a schematic diagram for the Metropolis method. The overall MC methodology is similar to that used in molecular dynamics, with conformations being periodically saved for later analysis after an initial equilibration period. A new technique has recently arisen that mixes stochastic dynamics with Monte Carlo sampling of dihedral angles [19]. This can greatly increase the barrier crossing rate for stochastic dynamics, and thereby reduce the length of runs required for efficient conformational sampling.

MD and MC are not restricted to single molecules, and in the last few years several

---

[1] When bond lengths are constrained, a small correction factor should be introduced in the configuration sampling to take account of the constraints [18].

simulations have appeared that have attempted to simulate mesogens in the liquid and liquid crystal phases [21–31]. These studies all make use of periodic boundary conditions, allowing a section of bulk fluid to be simulated without the need to worry about edge effects at the walls of the simulation box. However, care must always be taken to ensure that the simulation box is sufficiently large that atomic and centre of mass radial distribution functions are able to decay to a value of unity within the dimensions of the periodic box.

In contrast to single molecule calculations, in bulk simulations MD is preferable to Metropolis MC. A trial MC move involving a small rotation of a dihedral angle near the centre of a large rod-like molecule can lead to a large movement in the terminal parts of the molecule, resulting in collisions with neighbours in the bulk. This results in a large number of small rotations being rejected, and consequently a rather poor sampling of phase space occurs. Modern MC techniques such as configurational bias Monte Carlo are starting to tackle such problems successfully [32]; however, they have not as yet been applied to realistic simulations of liquid crystal systems. In contrast to MC methods, collisions with neighbours in the bulk fluid enable individual molecules within an MD simulation to sample phase space more efficiently than in the single molecule case.

The drawback with bulk simulations is their cost in terms of computer time. Computer time increases with the square of the number of atomic sites (truncation of short range interactions and the use of neighbour lists can improve this slightly), so a typical MD simulation of a few hundred molecules will involve the calculation of energies and forces for several thousand atomic sites at each time step. Because of this, it is usual in bulk simulations to employ the *united atom* approximation [21], in which carbons and attached hydrogens are replaced by single *extended* atomic sites. Use of this approximation requires a different force field to that employed in standard molecular mechanics studies. Internal molecular structures tend to be less accurately modelled by united atom force fields than by all-atom force fields. However, united-atom force fields have generally been designed with intermolecular forces in mind, and may well produce better intermolecular interaction energies than some all-atom force fields. In a series of papers [33–46], Jorgensen has carried out a large number of MC calculations for small molecules aimed at producing a set of transferable OPLS parameters (optimised parameters for liquid simulation) that model the thermodynamic properties of small molecules very well. These have recently been combined with the AMBER force field [47–50] to produce a combined AMBER/OPLS force field [51], which is ideal for the simulation of liquid crystal systems within the united-atom approximation [21]. Other united-atom force fields include the CHARMM force field [52, 53] and the AMBER force field itself.

In bulk MD simulations an initial equilibrium period of 200–300 ps is usually required to bring torsional angles into thermal equilibrium at the simulation temperature. However, molecular reorientation occurs on much longer timescales. Extrapolation from simpler models of liquid crystals suggest that the growth of a nematic liquid crystal from an isotropic liquid may require 1–10 ns of simulation time. This is currently at the limit of what can be achieved for atomic systems. However, the few bulk simulations that have appeared (see Sec. 3.2.4) suggest that this is a very exciting area of modelling that will develop strongly over the next few years.

3.1.3 Quantum Mechanical Techniques

The most natural way to determine molecular structure is by a direct quantum mechanical treatment of a molecule. This involves the solution of the Schrödinger equation for a given nuclear configuration, followed by the systematic adjustment of nuclear positions till the energy of a molecule is minimised. At the present time, quantum mechanical techniques are still extremely expensive, so that a full ab initio minimisation of molecular structure is only available for simple molecules such as methane [54]. Typically, computer time increases with the fourth power of the number of basis functions required in the calculation, and quantum energy minimisation is therefore extremely expensive for liquid crystal molecules. Despite this, quantum mechanical calculations are becoming useful in two guises. Firstly, accurate single point calculations can be carried out on energy-minimised structures produced by cheaper techniques such as molecular mechanics. Such calculations can provide reasonably accurate predictions for electric and magnetic properties of molecules [55]. However, it should be stressed that the molecular mechanics structure may not be the same as the molecular structure that would have been generated had the molecular geometry been allowed to relax in a quantum mechanical calculation. Secondly, semi-empirical quantum techniques have made rapid developments in the last few years [56]. These techniques are well suited to large liquid crystal molecules, and are starting to become useful in both the optimisation of molecular structures and in the determination of molecular properties [57].

3.2 Applications of Molecular Modelling

3.2.1 Determination of Molecular Structure

The determination of molecular structure is the simplest application of molecular modelling. In the first instance, a series of molecular mechanics calculations can produce a set of conformational energy minima for a liquid crystal that provide excellent information on molecular shape. For example, the mesogen 4-(*trans*-4-n-propylcyclohexyl]benzonitrile (PCH3) possesses a number of conformational energy minima corresponding to two possible chair conformations of the cyclohexane ring and rotations about torsional angles a, b, and c in Fig. 3. After an initial minimisation of a trial geometry, the next stage of conformational searching involves the driving of individual torsional angles [58]. The results of such calculations for PCH3 using the MM3 force field [5] are shown in Figs. 4a–c [59]. In its minimum-energy conformation the phenyl ring lies in the symmetry plane of the cyclohexyl ring, with the two torsional angles corresponding to the label c in Fig. 3 equal at 118° and −118°. However, the energy barrier corresponding to rotation about c is rather small, about 7.5 kJ mol^{-1} (Fig. 4c). The dihedral angle b involves the rotation of the propyl chain with respect to the cyclohexane ring. Figure 4b shows the standard

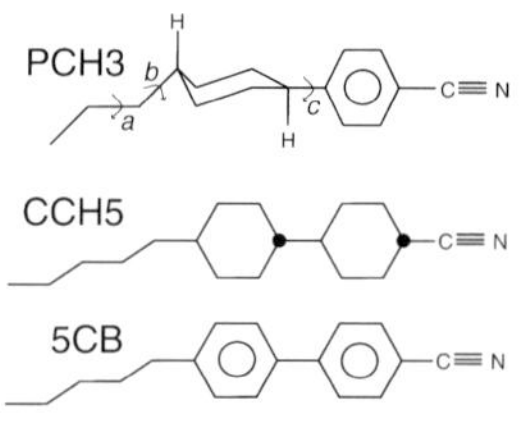

Figure 3. Structures of some common cyano-mesogens.

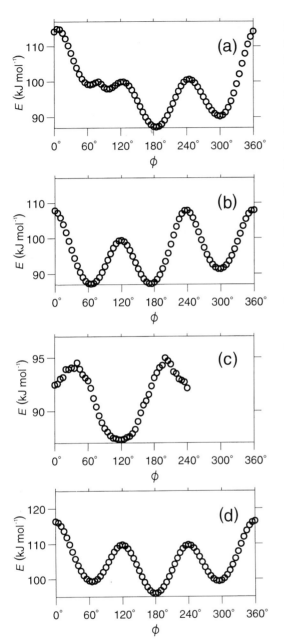

Figure 4. Typical torsional energy barriers for liquid crystal molecules. (a) Dihedral angle *a* for PCH3. (b) Dihedral angle *b* for PCH3. (c) Dihedral angle *c* for PCH3. (d) The terminal end-of-chain dihedral angle for PCH5. Torsional energy barriers are calculated using the MM3 force field [5] and the MacroModel molecular modelling package [60].

shape for this interaction potential, with two equal lowest energy conformations and a third energy minimum at 4.13 kJ mol^{-1} above the two low-energy states. The barriers to rotation occur at approximately 0° (20.69 kJ mol^{-1}), 120° (12.25 kJ mol^{-1}), and 240° (20.69 kJ mol^{-1}). Finally, the dihedral angle *a* is the end-of-chain dihedral for the propyl chain. Rotation about *a* produces two *gauche* conformations of unequal energy. The lowest energy *gauche* conformation is at 300° (3.05 kJ mol^{-1} above the ground state), whilst the second *gauche* conformation is split into two energy minima, the lowest of which occurs at 97.2° at an energy of 10.75 kJ mol^{-1} above the ground state. The energy barriers to rotation occur at 5° (27.8 kJ mol^{-1}), 80° (12.9 kJ mol^{-1}), 125° (12.8 kJ mol^{-1}), and 245° (13.5 kJ mol^{-1}). The assymmetry of the *gauche* conformations in Fig. 4a arises from strong repulsive interactions between the hydrogens on the terminal methyl group and axial cyclohexyl hydrogens as the alkyl chain rotates and collides with the core of the molecule. If the carbon chain of PCH3 is extended by two units (producing PCH5) then the terminal methyl group of the chain no longer interacts strongly with the cyclohexyl ring when rotation occurs about the end-of-chain dihedral angle (Fig. 4d). Consequently, the torsional angle energy profile of Fig. 4d is similar to that found in most liquid crystals with long alkyl chains: *gauche* conformations lie approximately 3.4 kJ mol^{-1} above the *trans* conformation, and the energy barriers to rotation are 20.5 kJ mol^{-1} and 13.9 kJ mol^{-1}.

From the dihedral angle energy profiles, a number of candidates for local potential energy minima can be identified and individually optimised. For PCH3, this results in the energy minima shown in Table 1, with the structure of the lowest energy conformation shown in Fig. 5. PCH3 exhibits a nematic phase between 36° and 46°C [61], and

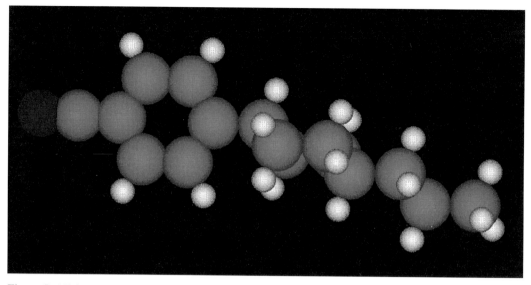

Figure 5. Minimum-energy conformation of PCH3.

Table 1. Local potential energy minima for PCH3 [59].

Energy (kJ mol^{-1})	Dihedral angle a	Dihedral angle b	Population at 46°C
87.34	−175.5°	173.2°	34.65
87.34	175.4°	64.9°	34.65
90.39	−60.6°	176.6°	10.98
90.39	60.6°	61.6°	10.98
91.47	180.0°	−62.2°	7.31
98.09	−97.2°	62.2°	0.60
98.09	97.2°	175.6°	0.60
102.41	−93.8°	−67.5°	0.12
102.41	93.9°	−57.1°	0.12

so application of Eq. (5) at $T=46°C$ for the nine conformations in Table 1 yields appropriate weights for conformer populations, suggesting that at least five conformers have significant populations at 46°C. Concerted rotation about torsional angles in the cyclohexane ring produces a chair conformation with both the phenyl and propyl groups in axial positions. MM3 calculations show that the lowest energy conformation for this ring is 28.5 kJ mol^{-1} above the ground state, meaning that this conformation is unlikely

to be significantly populated even at very high temperatures.

Recently, Dunmur et al. [57] have calculated the energy-minimised structures and rotational energy barriers for a number of chiral dopants using the MM2 force field and the techniques discussed above. In this work MM2 structures were further refined using the semi-empirical SCF quantum mechanical program MOPAC [62] (Sec. 3.1.3). Their results are shown in Fig. 6.

3.2.2   Determination of Molecular Properties

Energy-minimised structures generated by molecular mechanics or semi-empirical methods can be used as a starting point for the calculation of molecular properties. Dunmur et al. [57] have used MOPAC to calculate the dipole moments for the minimum-energy structures shown in Fig. 6. However, results have so far had mixed success. Calculated dipole moments can be compared with those measured in dilute so-

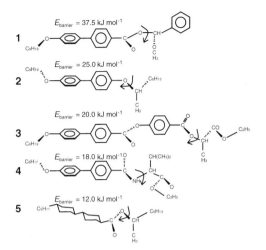

Figure 6. Energy-minimised structures for a series of chiral dopants adapted from the work of Dunmur et al. [57]. Structures were generated from MM2/MOPAC calculations, and $E_{barrier}$ is the (highest) rotational energy barrier calculated from the MM2 force field for the indicated bonds.

lutions of apolar solvents. For the materials in Fig. 6, good agreement was obtained with solution measurements for compounds **3** and **5**, but rather poor agreement was found for compounds **1**, **2**, and **4**. The errors have been attributed to the failure to take other molecular conformations into account in calculating the net dipole moment. In principle, this can be done using the analysis above (Sec. 3.2.1), but, in practice, the large number of possible conformations for compounds **1–5** makes this a rather lengthy process. Dunmur et al. [57] have also reported calculations of fragment dipoles for dipolar groups attached to the chiral centre in compounds **1–5**. They report reasonable correlation between transverse fragment dipole moments and measurements of spontaneous polarization (P_s) in SmC host solvents. In future, the single molecule Monte Carlo approach (described in Sec. 3.1.2) linked to a semi-empirical quantum method may provide a mechanism to generate useful conformationally averaged properties.

As yet, this combined technique has not been used for liquid crystal systems.

3.2.3 Determination of Intermolecular Potentials

Wilson and Dunmur have used molecular mechanics techniques to model the interaction energies of isolated pairs of liquid crystal molecules [58, 63, 64]. In this approach lowest energy conformations for liquid crystal molecules were first generated using MM2 [7], and these conformations were used to explore the potential energy surface for two molecules interacting via nonbonded terms of the form given in Eq. (4). For the mesogens 4-*n*-pentyl-4′-cyanobiphenyl (5CB) and 4-(*trans*-4-*n*-pentylcyclohexyl)cyclohexylcarbonitrile (CCH5, Fig. 3), distinct lowest energy *dimers* were found, corresponding to parallel and antiparallel arrangements of molecular dipoles, with the antiparallel configuration energetically favoured [63]. The removal of partial charges from the calculations was found to have a negligible effect on the spatial configuration of molecular pairs, but largely removed the energy difference between parallel and antiparallel arrangements. Wilson and Dunmur concluded that dispersive forces provide the dominant factor in causing molecular association in liquid crystals, but that dipolar effects are important in determining the balance between parallel and antiparallel molecular pairs. These conclusions fit well with results from dielectric [65] and light scattering [66, 67] studies of molecular association in dilute solutions of mesogens.

Luckhurst and Simmonds [68] have used Lennard–Jones pair potentials to characterise the molecular interaction potential $U_{av}(\boldsymbol{u}_1, \boldsymbol{u}_2, \boldsymbol{R})$ for two liquid crystal molecules with molecular long axes in the directions u_1 and u_2, where R is the vector between the respective centres of mass. In order to do this successfully, molecular biaxiality must be projected out by taking Boltzmann-weighted averages of the interaction potential over rotations about both molecular long axes:

$$U_{av}(u_1, u_2, R)$$

$$= \frac{\int_0^{2\pi}\int_0^{2\pi} U_{LJ}\exp(-U_{LJ}/k_BT)d\alpha_1\,d\alpha_2}{\int_0^{2\pi}\int_0^{2\pi}\exp(-U_{LJ}/k_BT)d\alpha_1\,d\alpha_2} \quad (7)$$

where α_1 and α_2 are the rotation angles about the molecular long axes of molecules 1 and 2, and $U_{LJ}=U_{LJ}(u_1, u_2, R)$ is the interaction energy of two molecules for given u_1, u_2, and R. As above, U_{LJ} is equal to a sum of all nonbonded pair interactions between molecules 1 and 2. Luckhurst and Simmonds went on to show that it is possible to take $U_{av}(u_1, u_2, R)$ and use it to parameterise a version of the Gay–Berne potential [68]. This *simplified* single site potential can then be used to carry out bulk simulations of liquid crystal mesophases.

3.2.4 Large-Scale Simulation of Liquid Crystals

Bulk atomistic simulations of liquid crystal mesophases are extremely time-consuming and currently represent the limit of what can be achieved with today's computers. However, in the past few years a number of (mainly) *united-atom* models of small mesogens have started to appear in the literature. These simulations are summarised in Table 2, and snapshots of molecules taken from a MD simulation of CCH5 are shown in Fig. 7. Many of the studies in Table 2 suffer from common drawbacks, namely small numbers of molecules and rather short simulation

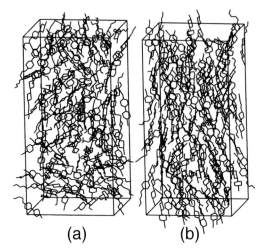

Figure 7. Snapshots showing the structure of CCH5 in liquid and liquid crystalline phases. (a) Isotropic phase at 390 K. (b) Nematic phase at 350 K, $S=0.62$. (Molecular coordinates are from the simulations carried out by Mark Wilson and Mike Allen [23], courtesy of the authors.)

times. However, the very fact that such simulations can now be attempted bodes well for future developments in this field.

In atomistic simulations, molecular order can be characterised in a number of ways. For rigid rodlike molecules, orientational ordering may be defined by reference to the molecular long axis vector a_j:

$$Q_{\alpha\beta} = \frac{1}{N_{mol}}\sum_{j=1}^{N_{mol}}\frac{3}{2}a_{j\alpha}a_{j\beta} - \frac{1}{2}\delta_{\alpha\beta} \quad (8)$$

where N_{mol} is the number of molecules, and $\alpha, \beta = x, y, z$. The director n is the eigenvector associated with the largest eigenvalue (λ_+) of Q, and λ_+ can be equated to the uniaxial order parameter S:

$$S = \lambda_+ = \langle P_2(n \cdot a)\rangle = \langle P_2(\cos\theta)\rangle \quad (9)$$

where P_2 is a second-rank Legendre polynomial, θ is the angle between a molecule and the director, and the angular brackets denote an ensemble average. For the flexible molecules often used in atomistic simulations, a_j can be assigned to the eigenvec-

Table 2. Atomistic simulations of bulk liquid crystal systems.

Mesogen	Ref.	N_{mol}	Study	Notes
5CB	[20]	64	NpT MD bond constraints	60 ps simulation for charged and uncharged systems
EBBA[a]	[24]	60	NpT MC (all-atom)	312 K, 10^3 moves per molecule (nematic $S=0.8$)
pHB[b]	[23]	8, 16	Energy minimisation and NVT MD (all-atom model)	300 K, 70 ps (crystal) 500 K, 116 ps (nematic)
CCH5	[21] [22]	128	NpT MD bond constraints	Up to 720 ps 390 K (isotropic) 370 K (nematic $S=0.38$) 350 K (nematic $S=0.62$)
THE5[c]	[25]	54	NVT MD bond and angle constraints	380 K, 100 ps (discotic columnar)
nOCB[d] $n=5, 6, 7, 8$	[26]	64	NpT MD bond and angle constraints	Up to 180 ps 5OCB, 331 K 6OCB, 330 K, 339 K, 359 K 7OCB, 337 K; 8OCB, 359 K
2MBCB[e]	[30]	32	NpT MD	230 K, 10 ps with/without twisted periodic boundaries
HBA[f]	[29]	125	NpT MD bond constraints (all-atom model)	Up to 590 ps, 475 K (nematic $S=0.86$)
PCH5	[31]	50, 100	NpT MD bond constraints	Up to 360 ps, 333 K (nematic, isotropic)
5CB	[28]	80	NpT MD bond constraints	300 K, 1600 ps (nematic $S=0.6$)

[a] 4-Ethoxybenzylidene-4'-*n*-butylaniline;
[b] truncated *tetramer* segment of the liquid crystal polyester of 4-hydroxybenzoic acid [phenyl-4-(4-benzoyloxy-) benzoyloxy benzoate];
[c] hexakispentyloxytriphenylene;
[d] 4-alkoxy-4'-cyanobiphenyl;
[e] (+)-4-(2-methylbutyl)-4'-cyanobiphenyl;
[f] *tetramer* segment of the liquid crystal polyester of 4-hydroxybenzoic acid.

tor associated with the smallest eigenvalue of the inertia tensor,

$$I_{j\alpha\beta} = \sum_i m_i (r_i^2 \delta_{\alpha\beta} - r_{i\alpha} r_{i\beta}) \qquad (10)$$

where m_i are atomic masses and the atomic distance vector r_i is measured relative to the molecular centre of mass. Alternatively, individual values of S can be calculated for different parts of the molecule. Wilson and Allen [22] (for CCH5) and Cross and Fung

[28] (for 5CB) have considered the ordering of individual segments within a molecule. Their simulations demonstrate a classic odd–even effect in the ordering of individual bonds in the alkyl chain: odd bonds have much higher order parameters than even bonds. For CCH5 [22], the orientational distribution function $f(\cos\theta)$ for odd bonds is strongly peaked along the director. In contrast, $f(\cos\theta)$ exhibits a broad distri-

bution for even bonds, and peaks at an angle to the director.

It is interesting to look at the dihedral angles for the alkyl chains in 5CB and CCH5. In the nematic phase at 350 K, Wilson and Allen found that almost 50% of molecules have the fully extended all-*trans* (ttt) conformation. However, dihedral angle distributions are temperature-dependent. The observed dihedral angle distribution $S(\phi)$ can be written in terms of an effective torsional potential $E_{eff}(\phi)$ (conformational free energy) [22, 28]:

$$S(\phi) = C\exp\left(-\frac{E_{eff}(\phi)}{k_B T}\right) \qquad (11)$$

where C is a normalization factor,

$$E_{eff}(\phi) = E_{ext}(\phi) + E_{torsion}(\phi) + E_{int}(\phi) \qquad (12)$$

where $E_{torsion}(\phi)$ is the torsional angle potential (see, e.g., Eq. (3)), $E_{int}(\phi)$ is due to internal nonbonded interactions (mainly 1–4 interactions), and $E_{ext}(\phi)$ depends on the local molecular environment of a molecule in a bulk fluid. $E_{eff}(\phi)$ can therefore be used to monitor the effect of the nematic field on molecular structure [28]. As the nematic phase is entered, the effect of $E_{ext}(\phi)$ is to favour conformations where bonds lie along the molecular long axis. In CCH5 this leads to a small change in shape, so that molecules become more elongated in the nematic phase.

For the five carbon chains of 5CB and CCH5, the three lowest energy chain conformations correspond to ttt, gtt, tgt, and ttg (Fig. 8). Whilst ttt is the lowest energy conformation for 5CB and CCH5, molecular mechanics predicts that the preferred ordering of *gauche* conformations is gtt, tgt, ttg owing to favourable chain ring nonbonded interactions, which reduce the value of $E_{int}(\phi)$ for *gauche* conformations close to the core [22]. However, in the nematic phase the effect of $E_{ext}(\phi)$ is such as to produce

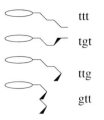

ttt

tgt

ttg

gtt

Figure 8. Chain conformations for the mesogen CCH5.

the preferred ordering tgt, ttg, gtt for *gauche* conformations in both 5CB and CCH5. This order arises because (unlike tgt) the *gauche* conformations ttg, gtt cause bonds to lie at an angle to the director, and so are strongly disfavoured by the local structure of the nematic fluid. The contributions of $E_{ext}(\phi)$ to $E_{eff}(\phi)$ significantly increase both the energy differences between *gauche* and *trans* conformers and the rotational energy barriers between conformers. This is something that is largely (and incorrectly) ignored in molecular mechanics studies of mesogens (Sec. 3.2.1).

A number of bulk simulations have attempted to study the dynamical properties of liquid crystals. Translational diffusion coefficients are available from the Einstein relation, which is valid for long times t:

$$D = \frac{1}{2t}\left\langle|r_i(t) - r_i(t_0)|^2\right\rangle \qquad (13)$$

where r_i is the centre of mass position of molecule i at time t and t_0 is an appropriate time origin. D may be resolved to monitor diffusion parallel ($D_\parallel$) and perpendicular ($D_\perp$) to the director separately. Table 3 lists values of $D_\parallel$ and $D_\perp$ from a variety of studies. In most cases a clear anisotropy in diffusion is seen, with diffusion along the nematic director favoured. The one exception in Table 3 is for the molecule THE5. In this case the simulations are for a discotic columnar phase, with diffusion favoured perpendicular to the direction of order. Both $D_\parallel$ and $D_\perp$ are very small for THE5.

The small number of molecules present in the simulations in Table 2 means that no

Table 3. Diffusion constants from atomistic simulations of liquid crystals.

Mesogen	Ref.	T (K)	$D_{\parallel}$ (10^{-9} m^2 s^{-1})	$D_{\perp}$ (10^{-9} m^2 s^{-1})	$D_{\parallel}/D_{\perp}$	S
5CB	[28]	300	1.12	0.30	3.73	0.60
CCH5	[21]	350	0.554	0.192	2.89	0.62
		370	1.076	0.517	2.08	0.38
PCH5	[31]	333	0.157	0.046	3.4	0.58
5OCB	[27]	331	0.36	0.188	1.91	0.53
6OCB		339	0.33	0.168	1.96	0.50
7OCB		337	0.316	0.176	1.80	0.50
8OCB		342	0.282	0.177	1.59	0.47
THE5	[26]	380	0.00984	0.0189	0.05	0.95

convincing atomistic simulations of smectic mesophases currently exist. However, a number of studies of bilayers and Langmuir–Blodgett films have been made in recent years [69–72], and these are starting to prove useful in studying molecular ordering in layered structures. One atomistic study has appeared that attempts to study a chiral system by molecular dynamics [30]. In this study twisted periodic boundary conditions were introduced to look at a (pseudo) chiral nematic phase. In this initial study only a very small system (32 2MBCB molecules) was used, so few definitive conclusions were available. However, this technique could, in principle, be extended to look at large chiral systems in the future.

Finally, it should be stressed that no one has yet proved the thermodynamic stability of atomistic model mesophases by growing a nematic phase directly from an isotropic liquid. This is a relatively easy process for single site models [68], but is extremely expensive for atom-based models, requiring long simulations (of the order of 10 ns) on systems of several hundred molecules. On account of this, data from the simulations in Table 2 should be treated as preliminary at this stage. However, the rapid increases in speed (and reduction in cost) of modern computers suggest that definitive atomistic simulations may be only a few years away.

3.3 References

[1] N. L. Allinger, *Rev. Phys. Org. Chem.* **1976**, *13*, 1–82.
[2] O. Burkert, N. L. Allinger, *Molecular Mechanics*, ACS Monograph 177, American Chemical Society, Washington, DC **1982**.
[3] D. B. Boyd, K. B. Lipkowitz, *J. Chem. Educ.* **1982**, *59*, 269–274.
[4] M. R. Wilson, Ph. D. Thesis, University of Sheffield **1988**, Chap. 1.
[5] N. L. Allinger, Y. H. Yuh, J. Lii, *J. Am. Chem. Soc.* **1989**, *111*, 8511–8582.
[6] N. L. Allinger, *J. Am. Chem. Soc.* **1977**, *99*, 8127–8134.
[7] N. L. Allinger, M. T. Tribble, M. A. Miller, D. H. Wertz, *J. Am. Chem. Soc.* **1971**, *93*, 1637–1647.
[8] A. J. Hopfinger, R. A. Pearlstein, *J. Comput. Chem.* **1984**, *5*, 486–499.
[9] D. M. Ferguson, D. J. Raber, *J. Am. Chem. Soc.* **1989**, *111*, 4371–4378.
[10] G. Chang, W. C. Guida, W. C. Still, *J. Am. Chem. Soc.* **1989**, *111*, 4379–4386.
[11] W. L. Jorgensen, *J. Phys. Chem.* **1983**, *87*, 5304–5314.
[12] D. Levesque, J. J. Weis, J. P. Hansen in *Applications of the Monte Carlo Method in Statistical Physics* (Ed.: K. Binder), Topics in Current Physics 36, Springer-Verlag, Berlin **1984**, Chap. 2.
[13] M. P. Allen, D. J. Tildesley, *Computer Simulation of Liquids*, Oxford University Press, Oxford **1987**, Chap. 3.
[14] M. P. Allen, D. J. Tildesley, *Computer Simulation of Liquids*, Oxford University Press, Oxford **1987**, Chap. 4.
[15] W. G. Hoover, *Phys. Rev.* **1985**, *A31*, 1695–1697.
[16] J. P. Ryckaert, G. Ciccotti, H. J. C. Berendsen, *J. Comput. Phys.* **1977**, *23*, 327–341.
[17] W. F. van Gunsteren, H. J. C. Berendsen, *Mol. Simul.* **1988**, *1*, 173–185.

[18] S. Leggetter, D. J. Tildesley, *Mol. Phys.* **1989**, *68*, 519–546.

[19] F. Guarnieri, W. C. Still, *J. Comput. Chem.* **1994**, *15*, 1302–1310.

[20] S. J. Picken, W. F. van Gunsteren, P. Th. van Duijnen, W. H. de Jeu, *Liq. Cryst.* **1989**, *6*, 357–371.

[21] M. R. Wilson, M. P. Allen, *Mol. Cryst. Liq. Cryst.* **1991**, *198*, 465–477.

[22] M. R. Wilson, M. P. Allen, *Liq. Cryst.* **1992**, *12*, 157–176.

[23] B. Jung, B. L. Schürmann, *Mol. Cryst. Liq. Cryst.* **1990**, *185*, 141–153.

[24] A. V. Komolkin, Yu. V. Molchanov, P. P. Yakutseni, *Liq. Cryst.* **1989**, *6*, 39–45.

[25] I. Ono, S. Kondo, *Mol. Cryst. Liq. Cryst.* **1991**, *8*, 69.

[26] I. Ono, S. Kondo, *Bull. Chem. Soc. Jpn.* **1992**, *65*, 1057–1061.

[27] I. Ono, S. Kondo, *Bull. Chem. Soc. Jpn.* **1993**, *66*, 633–638.

[28] C. W. Cross, B. M. Fung, *J. Chem. Phys.* **1994**, *101*, 6839–6848.

[29] J. Huth, T. Mosell, K. Nicklas, A. Sariban, J. Brickmann, *J. Phys. Chem.* **1994**, *98*, 7685–7691.

[30] M. Yoneya, H. J. C. Berendsen, *J. Phys. Soc. Jpn.* **1994** , *63*, 1025–1030.

[31] G. Krömer, D. Paschek, A. Geiger, *Ber. Bunsenges. Phys. Chem.* **1993**, *97*, 1188–1192.

[32] J. I. Siepmann, D. Frenkel, *Mol. Phys.* **1992**, *75*, 59–70.

[33] G. Kaminski, E. M. Duffy, T. Matsui, W. L. Jorgensen, *J. Phys. Chem.* **1994**, *98*, 13 077–13 082.

[34] E. M. Duffy, W. L. Jorgensen, *J. Am. Chem. Soc.* **1994**, *116*, 6337–6343.

[35] W. L. Jorgensen, E. R. Laird, T. B. Nguyen, J. Tiradorives, *J. Comput. Chem.* **1993**, *14*, 206–215.

[36] W. L. Jorgensen, T. B. Nguyen, *J. Comput. Chem.* **1993**, *14*, 195–205.

[37] J. M. Briggs, T. Matsui, W. L. Jorgensen, *J. Comput. Chem.* **1990**, *11*, 958–971.

[38] W. L. Jorgensen, J. M. Briggs, *Mol. Phys.* **1988**, *63*, 547–558.

[39] W. L. Jorgensen, J. Tiradorives, *J. Am. Chem. Soc.* **1988**, *110*, 1666–1671.

[40] W. L. Jorgensen, *J. Phys. Chem.* **1986**, *90*, 6379–6388.

[41] W. L. Jorgensen, J. Gao, *J. Phys. Chem.* **1986**, *90*, 2174–2182.

[42] W. L. Jorgensen, *J. Phys. Chem.* **1986**, *90*, 1276–1284.

[43] W. L. Jorgensen, J. D. Madura, *Mol. Phys.* **1985**, *56*, 1381–1392.

[44] W. L. Jorgensen, C. J. Swenson, *J. Am. Chem. Soc.* **1985**, *107*, 1489–1496.

[45] W. L. Jorgensen, C. J. Swenson, *J. Am. Chem. Soc.* **1985**, *107*, 569–578.

[46] W. L. Jorgensen, J. D. Madura, C. J. Swenson, *J. Am. Chem. Soc.* **1984**, *106*, 6638–6646.

[47] U. C. Singh, P. K. Weiner, J. Caldwell, P. A. Kollman, *AMBER 3.0*, University of California, San Francisco **1987**.

[48] P. K. Weiner, P. A. Kollmann, *J. Comput. Chem.* **1981**, *2*, 287–303.

[49] S. J. Weiner, P. A. Kollman, D. T. Nguyen, D. A. Case, *J. Comput. Chem.* **1986**, *7*, 230–252.

[50] S. J. Weiner, P. A. Kollman, D. A. Case, U. C. Singh, C. Ghio, G. Alagona, S. Profeta, P. Weiner, *J. Am. Chem. Soc.* **1984**, *106*, 765–784.

[51] W. L. Jorgensen, J. Tirado-Rives, *J. Am. Chem. Soc.* **1988**, *110*, 1666–1671.

[52] B. R. Gelin, M. Karplus, *Biochemistry* **1979**, *18*, 1256–1268.

[53] B. R. Brooks, R. E. Bruccoleri, B. D. Olafson, D. J. States, S. Swaminathan, M. Karplus, *J. Comput. Chem.* **1983**, *4*, 187–217.

[54] M. J. M. Pepper, I. Shavitt, P. V. Schleyer, M. N. Glukhovtsev, R. Janoschek, M. Quack, *J. Comput. Chem.* **1995**, *16*, 207–225.

[55] C. E. Dykstra, J. D. Augspurger, B. K. Kirtman, D. J. Malik in *Reviews in Computational Chemistry* (Eds.: K. B. Lipkowitz, D. B. Boyd), VCH, New York **1990**, Chap. 3.

[56] J. J. P. Stewart in *Reviews in Computational Chemistry* (Eds.: K. B. Lipkowitz, D. B. Boyd), VCH, New York **1990**, Chap. 2.

[57] D. A. Dunmur, M. Grayson, S. K. Roy, *Liq. Cryst.* **1994**, *16*, 95–104.

[58] M. R. Wilson, Ph. D. Thesis, University of Sheffield **1988**, Chap. 3.

[59] M. R. Wilson, unpublished work.

[60] *MacroModel V 3.5 X, Interactive Molecular Modelling System*, Department of Chemistry, Columbia University, New York **1992**.

[61] K. J. Toyne in *Thermotropic Liquid Crystals* (Ed.: G. W. Gray), Wiley, New York **1987**, Chap. 2.

[62] J. J. P. Stewart, *J. Comput. Aided Mol. Design* **1990**, *4*, 1–45.

[63] M. R. Wilson, D. A. Dunmur, *Liq. Cryst.* **1989**, *5*, 987–999.

[64] D. A. Dunmur, M. R. Wilson, *Mol. Simul.* **1989**, *4*, 37–59.

[65] K. Toriyama, D. A. Dunmur, *Mol. Phys.* **1985**, *56*, 479–484.

[66] K. Toriyama, D. A. Dunmur, *Mol. Cryst. Liq. Cryst.* **1986**, *139*, 123–142.

[67] K. Toriyama, D. A. Dunmur, *Liq. Cryst.* **1986**, *1*, 169–180.

[68] G. R. Luckhurst, P. S. J. Simmonds, *Mol. Phys.* **1993**, *80*, 233–252.

[69] P. van der Ploeg, H. J. C. Berendsen, *J. Chem. Phys.* **1982**, *76*, 3271–3276.

[70] P. van der Ploeg, H. J. C. Berendsen, *Mol. Phys.* **1983**, *49*, 233–248.

[71] A. Biswas, B. L. Schürmann, *J. Chem. Phys.* **1991**, *95*, 5377–5386.

[72] M. A. Moller, D. J. Tildesley, K. S. Kim, N. Quirke, *J. Chem. Phys.* **1991**, *94*, 8390–8401.

Chapter IV
Physical Properties

1 Tensor Properties of Anisotropic Materials

David Dunmur and Kazuhisa Toriyama

Liquid crystals are anisotropic, so like non-cubic crystals some of their properties depend on the direction along which they are measured. Such properties are known as tensor properties, and in order to provide a formal basis for the description of orientation-dependent physical properties of liquid crystals, we will give a brief introduction to tensors. An authoritative account of the tensor properties of crystals has been written by Nye [1], but liquid crystals are not explicitly dealt with. A convenient way of categorizing tensor properties is through their behavior on changing the orientation of a defining axis system. A scalar or zero rank tensor property is independent of direction, and examples are density, volume, energy or any orientationally averaged property such as the mean polarizability or mean electric permittivity (dielectric constant). The orientation dependence of a vector property such as dipole moment μ can be understood by considering how the components of the dipole moment change as the axis system is rotated. In Fig. 1 μ_β ($\beta=x, y, z$) are the components in the original coordinate frame, while $\mu'_\alpha(\alpha=X, Y, Z)$ are the components in the new axis system. If the quantities $a_{\alpha\beta}$ are the nine direction cosines between the axes of the two coordi-

nate frames, the transformation law for the vector property becomes:

$$\mu'_\alpha = a_{\alpha x}\,\mu_x + a_{\alpha y}\,\mu_y + a_{\alpha z}\,\mu_z = a_{\alpha\beta}\,\mu_\beta \quad (1)$$

where the repeated suffix β implies summation over all values of $\beta=x, y, z$. A consequence of the transformation law Eq. (1), is that under inversion all the components change sign, $\mu'_\alpha=-\mu_\alpha$: this is a polar vector. There are some quantities (axial or pseudo-

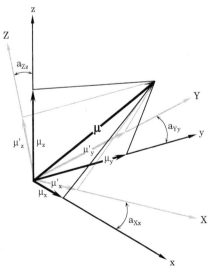

Figure 1. Change in vector components on rotating axes.

vectors) which transform according to:

$$\mu'_\alpha = -a_{\alpha\beta}\,\mu_\beta \qquad (2)$$

These are usually associated with magnetic phenomena (angular momentum, magnetic moments and magnetic fields are axial vectors) and related properties. The nine direction cosines can conveniently be represented as a 3×3 matrix, but the components of this matrix ($a_{\alpha\beta}$) are not independent, since for orthogonal axes the sums of squares of components in columns or rows are unity, while the sums of products of components in adjacent rows or columns are zero. It is clear that one set of orthogonal axes can be rigidly related to another by just three angles, and Euler angles (θ, ϕ, ψ) provide a consistent definition of three such angles which are frequently used. The Euler angles are defined as follows. Assuming that the axes (x, y, z) and (X, Y, Z) are initially coincident, rotation around $Z=z$, by an angle ψ gives $X\to x'$, $Y\to y'$, rotation about the new axis y' by θ gives $Z\to z$, $x'\to x''$, and finally rotation about z by an amount ϕ gives $x''\to x$ and $y'\to y$, see Fig. 2 [2]. The direction cosine matrix represented by $a_{\alpha\beta}$, can now be expressed in terms of Euler angles by:

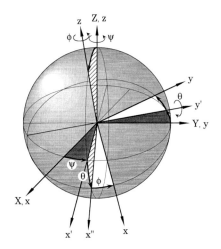

Figure 2. Euler angles.

sor quantity depends on another vector or tensor quantity. A simple example of importance to liquid crystals is the second rank tensor property electric susceptibility χ, which relates electric polarization (vector) to an applied electric field (vector). Neglecting nonlinear effects, the electric polarization P_α is proportional to the magnitude of the applied electric field E, but for anisotropic materials it may not be in the same direction. Consider a linear (one-dimensional) array of charges aligned at an angle

$$a_{\alpha\beta} = \mathbf{A} = \begin{bmatrix} a_{Xx} & a_{Xy} & a_{Xz} \\ a_{Yx} & a_{Yy} & a_{Yz} \\ a_{Zx} & a_{Yz} & a_{Zz} \end{bmatrix}$$

$$= \begin{bmatrix} \cos\theta\cos\phi\cos\psi - \sin\phi\sin\psi & -\cos\theta\sin\phi\cos\psi - \cos\phi\sin\psi & \sin\theta\cos\phi \\ \cos\theta\cos\phi\sin\psi + \sin\phi\cos\psi & -\cos\theta\sin\phi\sin\psi + \cos\phi\cos\psi & \sin\theta\sin\phi \\ -\sin\theta\cos\psi & \sin\theta\sin\psi & \cos\theta \end{bmatrix} \qquad (3)$$

Having established the transformation law Eq. (1), first rank tensor quantities (i.e. vectors) may be defined as those properties which transform according to Eq. (1). Higher order tensors can be defined in terms of different transformation laws, and they arise in a general sense when one vector or ten-

to an applied electric field (see Fig. 3). A separation of charge will be induced by the field resulting in a polarization necessarily along the direction of the array of charges: the electric field has induced a component of the polarization in a direction orthogonal

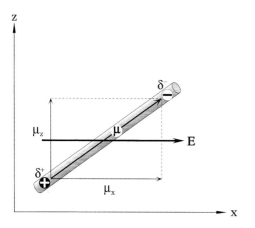

Figure 3. The polarization of a linear array of charges by a field.

to the field. Such a result cannot be described in terms of a simple proportionality between the polarization and the field ($P_\alpha \neq \chi E_\alpha$), because a field along the x-axis has induced a polarization along the z-axis. Obviously the susceptibility also depends on direction, and for the example given, the relationship between polarization and electric field can be written as (for the configuration depicted in Fig. 3, $E_z=0$):

$$P_x = \chi_{xx} E_x + \chi_{xz} E_z$$
$$P_z = \chi_{zx} E_x + \chi_{zz} E_z \qquad (4)$$

This is readily extended to three dimensions, and using the notation introduced above:

$$P_\alpha = \chi_{\alpha\beta} E_\beta \qquad (5)$$

where $\chi_{\alpha\beta}$ represents nine coefficients or a 3×3 matrix of nine quantities. The electric susceptibility of an anisotropic material can be represented by a matrix of nine components, and we can now consider how such a second rank tensor property changes as a result of rotation with respect to an orthogonal axis system. The transformation rule for second rank tensor properties is:

$$\chi'_{\alpha\beta} = a_{\alpha\gamma} a_{\beta\delta} \chi_{\gamma\delta} \qquad (6)$$

where, as before, the repeated suffixes γ and δ indicate a summation over all possible values (x, y, z). Thus the expression for each component of the transformed property, for example χ_{xy}, contains nine terms, although in practice symmetry usually reduces this number. It is possible to use an alternative transformation rule for second rank tensor properties, equivalent to Eq. (6), but which can be expressed in terms of the matrix of direction cosines, such that:

$$\chi' = \mathbf{A} \chi \mathbf{A}^t \qquad (7)$$

where $\mathbf{A}^t$ is the transpose of the direction cosine matrix.

Although the simplest direction-dependent property is a vector, most physical properties of liquid crystals are higher order tensors. All tensor properties can be categorized by their transformation properties under rotation of the coordinate frame, and the transformation law for a third rank tensor such as the piezoelectric tensor $P_{\alpha\beta\gamma}$ is:

$$P'_{\alpha\beta\gamma} = a_{\alpha\delta} a_{\beta\varepsilon} a_{\gamma\phi} P_{\delta\varepsilon\phi} \qquad (8)$$

The tensor rank of a property is also established by the number of components it has: first rank (vector) 3, second rank 9, third rank 27 and so on. This definition is correct for three dimensions, but in two dimensions the number of components for a tensor property of rank n becomes 2^n. It must be emphasized that not all the components are necessarily independent, and symmetry can provide relationships between them, thereby reducing the number that have to be separately measured.

Many physical properties can be described in terms of the response of a system to an external force or perturbation. This force might be an electric field or a magnetic field, a mechanical force (stress), a torque or a combination of these. The effect of an external perturbation may be described in terms of a new tensor property, or a modification to an existing one; polarization by an electric field results in an induced polarization which adds to any permanent polarization already present. To lowest order, the induced polarization is linear in the applied field, but nonlinear terms can be important for strong fields, as in nonlinear optics. The nonlinear contributions represented in Eq. (9) below are described in terms of third $\chi_{\alpha\beta\gamma}$ and fourth rank tensors $\chi_{\alpha\beta\gamma\delta}$:

$$p_\alpha = p_\alpha^{(0)} + \sum_{\beta=x,y,z} \chi_{\alpha\beta} E_\beta$$
$$+ \sum_{\beta,\gamma=x,y,z} \chi_{\alpha\beta\gamma} E_\beta E_\gamma$$
$$+ \sum_{\beta,\gamma,\delta=x,y,z} \chi_{\alpha\beta\gamma\delta} E_\beta E_\gamma E_\delta \quad (9)$$

This equation includes a number of pairs of repeated suffixes, and each pair indicates a summation over all possible values. Having understood this convention (the Einstein summation convention) it is no longer necessary to write the summation signs explicitly in an equation. It is sometimes useful to represent the physical properties in terms of their contribution to the internal energy, since all types of energies are scalar quantities. Thus the energy of a polarized body in an electric field becomes:

$$u = -\int \mathbf{p} \cdot d\mathbf{E} = -\int p_\alpha \, dE_\alpha$$
$$= p_\alpha^{(0)} E_\alpha + \frac{1}{2} \chi_{\alpha\beta} E_\beta E_\alpha$$
$$+ \frac{1}{3} \chi_{\alpha\beta\gamma} E_\beta E_\gamma E_\alpha$$
$$+ \frac{1}{4} \chi_{\alpha\beta\gamma\delta} E_\beta E_\gamma E_\delta E_\alpha \quad (10)$$

This manipulation is a simple application of tensor calculus, and illustrates the simplification of the summation convention (summation signs have been omitted). The intrinsic symmetry of a property can reduce the number of independent components;

thus neglecting complications that may arise if electric fields of more than one frequency are present, the order of the field components in Eq. (10) is immaterial. Hence the properties $\chi_{\alpha\beta}$, $\chi_{\alpha\beta\gamma}$ and $\chi_{\alpha\beta\gamma\delta}$ must be symmetric with respect to interchange of suffixes, and this immediately reduces the number of independent components of the tensor properties to six for $\chi_{\alpha\beta}$, ten for $\chi_{\alpha\beta\gamma}$ and 15 for $\chi_{\alpha\beta\gamma\delta}$.

Another example of a second rank tensor property is electrical conductivity $\sigma_{\alpha\beta}$ which relates the current flow j_α in a particular direction to the electric field:

$$j_\alpha = \sigma_{\alpha\beta} E_\beta \quad (11)$$

This may also be written in terms of resistivity $\rho_{\alpha\beta}$ as:

$$E_\alpha = \rho_{\alpha\beta} j_\beta \quad (12)$$

and the Joule heating which a current generates in a sample is a scalar given by:

$$u = \mathbf{j} \cdot \mathbf{E} = j_\alpha E_\alpha = \rho_{\alpha\beta} j_\alpha j_\beta \quad (13)$$

Each tensor property has an intrinsic symmetry, which relates to the interchangeability of suffixes. However, the number of independent tensor components for a property also depends on the symmetry of the system it is describing. Thus the properties of an isotropic liquid, which has full rotational symmetry, can be defined in terms of a single independent coefficient. The number of independent components for a particular tensor property depends on the point group symmetry of the phase to which it refers. This is expressed by Neumann's principle which states that the symmetry elements of any physical property of a crystal must include the symmetry elements of the point group of the crystal.

Many properties of interest for liquid crystals are second rank tensors, and these have some special properties, since they can be represented as 3×3 matrices. Those second rank tensor properties which have an intrinsic symmetry with respect to interchange of suffixes are called symmetric and only have six independent components. A symmetric 3×3 matrix can always be diagonalized to give three principal components, which is equivalent to finding an axis system for which the off-diagonal components are zero. The principal axis system requires three angles to define it with respect to an arbitrary axis frame, so there is no loss of variables: three principal components and three angles being equivalent to the six independent components. However, if the material being described (i.e. the liquid crystal) has some symmetry, then the principal axes will be defined, and so it is possible to reduce the number of independent components of a second rank tensor to three (the principal values). For uniaxial liquid crystal phases of symmetry $D_{\infty h}$ such as N and SmA, as well as SmB and Col_{ho} and Col_{hd} phases, a unique symmetry axis can be defined, parallel to the director, and there are just two independent components of any second rank tensor property $\chi_{\alpha\beta}$: $\chi_{\parallel}$ (parallel) and $\chi_{\perp}$ (perpendicular) to the symmetry axis:

$$\chi_{\alpha\beta} = \begin{matrix} x \\ y \\ z \end{matrix}\begin{bmatrix} \chi_{\perp} & 0 & 0 \\ 0 & \chi_{\perp} & 0 \\ 0 & 0 & \chi_{\parallel} \end{bmatrix} \quad (14)$$

where the z-axis is defined as the symmetry axis. The values of the components in any other axis system can be obtained from the transformation law Eqs. (6) or (7). Two useful quantities defined in terms of these independent components are the anisotropy $\Delta\chi$ and the mean $\bar{\chi}$:

$$\Delta\chi = \chi_{\parallel} - \chi_{\perp}$$

$$\bar{\chi} = \frac{1}{3}(\chi_{\parallel} + 2\chi_{\perp}) \quad (15)$$

Another special feature of second rank tensor properties in three dimensions is that they can be represented by a property ellipsoid, such that the value of the property in a particular direction is represented by the length of the corresponding radius vector of the property ellipsoid. A three dimensional surface representing an ellipsoid can be defined by

$$C_{xx} x^2 + C_{yy} y^2 + C_{zz} z^2 \\ + 2 C_{xy} xy + 2 C_{xz} xz + 2 C_{yz} yz = 1 \quad (16)$$

where the coefficients $C_{\alpha\beta}$ behave as the components of a second rank tensor. If the ellipsoid is expressed in terms of principal axes, then the off diagonal terms in Eq. (16) are zero, and the equation of the ellipsoid becomes:

$$C_{xx} x^2 + C_{yy} y^2 + C_{zz} z^2 = 1 \quad (17)$$

where the lengths of the semi-axes are given by $[C_{ii}]^{-1/2}$. The value of this property in any direction (l) defined by direction cosines a_{lx}, a_{ly}, a_{lz} will be:

$$C = C_{xx} a_{lx}^2 + C_{yy} a_{ly}^2 + C_{zz} a_{lz}^2 \\ = (x^2 + y^2 + z^2)^{-1} \quad (18)$$

so the length of any radius of the property ellipsoid (Fig. 4) is equal to the reciprocal of the square root of the magnitude of the property in that direction.

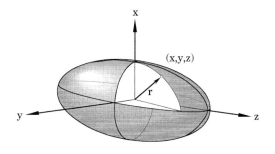

Figure 4. Property ellipsoid.

1.1   Macroscopic and Microscopic Properties

Liquid crystals as anisotropic fluids exhibit a wide range of complex physical phenomena that can only be understood if the appropriate macroscopic tensor properties are fully characterized. This involves a determination of the number of independent components of the property tensor, and their measurement. Thus a knowledge of refractive indices, electric permittivity, electrical conductivity, magnetic susceptibilities, elastic and viscosity tensors are necessary to describe the switching of liquid crystal films by electric and magnetic fields. Development of new and improved materials relies on the design of liquid crystals having particular macroscopic tensor properties, and the optimum performance of liquid crystal devices is often only possible for materials with carefully specified optical and electrical properties.

The anisotropy of liquid crystals stems from the orientational order of the constituent molecules, but the macroscopic anisotropy can only be determined through measurement of tensor properties, and macroscopic tensor order parameters can be defined in terms of various physical properties. The anisotropic part of a second rank tensor property can be obtained by subtracting the mean value of its principal components:

$$\chi_{\alpha\beta}^{(a)} = \chi_{\alpha\beta} - \overline{\chi}\,\delta_{\alpha\beta} \qquad (19)$$

where the Kronecker delta is equivalent to a unit second rank tensor or unit matrix, such that for $\alpha = \beta$, $\delta_{xx} = \delta_{yy} = \delta_{zz} = 1$, otherwise all other components are zero. Note that $\delta_{\alpha\alpha} = 3$, because of the repeated suffix convention. The advantage of using $\chi_{\alpha\beta}^{(a)}$ is that it is traceless, and the sum of the diagonal elements is zero. For an isotropic material $\chi_{\alpha\beta} = \chi\,\delta_{\alpha\beta}$, and so $\chi_{\alpha\beta}^{(a)} = 0$; thus the quan-

tity $\chi_{\alpha\beta}^{(a)}$ can be used as an order parameter. Alternatively a dimensionless order parameter tensor can be defined as

$$Q_{\alpha\beta} = \text{constant}\left(\chi_{\alpha\beta} - \overline{\chi}\,\delta_{\alpha\beta}\right) \qquad (20)$$

and without loss of generality the maximum value of a principal component (say Q_{zz}) of $Q_{\alpha\beta}$ can be set equal to one, so that:

$$Q_{zz} = 1$$
$$= \text{constant} \times \frac{2}{3}\left[\chi_{zz} - \frac{1}{2}\left(\chi_{xx} + \chi_{yy}\right)\right]$$
$$= \text{constant} \times \frac{2}{3}\left[\Delta\chi_{\max}\right] \qquad (21)$$

This provides the definition of the constant, and the macroscopic second rank tensor parameter becomes:

$$Q_{\alpha\beta} = \frac{3}{2}\left[\Delta\chi_{\max}\right]^{-1}\left(\chi_{\alpha\beta} - \frac{1}{3}\chi_{\gamma\gamma}\,\delta_{\alpha\beta}\right) \qquad (22)$$

The quantity $[\Delta\chi_{\max}]$ refers to the anisotropy of the tensor for a fully aligned state for which the order parameter is one. For a biaxial phase (i.e. a phase which has different properties along each of the three principal axes), the macroscopic order parameter in principal axes can be written as:

$$Q_{\alpha\beta} = \begin{bmatrix} -\frac{1}{2}(Q-P) & 0 & 0 \\ 0 & -\frac{1}{2}(Q+P) & 0 \\ 0 & 0 & Q \end{bmatrix} \qquad (23)$$

defining $Q = \left[\Delta\chi_{\max}\right]^{-1}\left(\chi_{zz} - \frac{1}{2}\left(\chi_{xx} + \chi_{yy}\right)\right)$ and $P = \frac{3}{2}\left[\Delta\chi_{\max}\right]^{-1}\left(\chi_{xx} - \chi_{yy}\right)$. Other definitions of $Q_{\alpha\beta}$ are possible, but the chosen one maintains the correspondence between the macroscopic order parameter and the microscopic one to be introduced next. One problem with the macroscopic order parameter as defined through Eq. (23) is that it has been assumed that the principal axes of the property $\chi_{\alpha\beta}$ and the order parameter tensor coincide. While this is necessarily true for uniaxial materials, it is not true for biaxial materials, the principal axes for which will be different for different properties.

The relationship between macroscopic properties and molecular properties is a major area of interest, since it is through manipulation of the molecular structure of mesogens, that the macroscopic liquid crystal properties can be adjusted towards particular values which optimize performance in applications. The theoretical connection between the tensor properties of molecules and the macroscopic tensor properties of liquid crystal phases provides a considerable challenge to statistical mechanics. A key factor is of course the molecular orientational order, but interactions between molecules are also important especially for elastic and viscoelastic properties. It is possible to divide properties into two categories, those for which molecular contributions are approximately additive (i.e. they are proportional to the number density), and those properties such as elasticity, viscosity, thermal conductivity etc. for which intermolecular forces are responsible, and so have a much more complex dependence on number density. For the former it is possible to develop a fairly simple theory using single particle orientational order parameters.

In the context of liquid crystals, single particle angular distribution functions are of major interest. They give the probability of a single molecule having a particular orientation with respect to some defined axis system, but they contain no information on pair correlations between molecules. A familiar example of single particle angular distribution functions is the hydrogenic s, p, d-orbitals, which are used via the square of the wave function to determine the probability distribution for a single electronic charge in

an atom. For a many-electron atom the wave function for the electrons is often written as the sum of component orbital contributions. In an analogous fashion, the distribution of molecular orientations in a liquid crystal can be represented as a sum of contributions of particular symmetries. Restricting attention to axially symmetric molecules having an axially symmetric distribution of orientations with respect to the director (z-axis), it is convenient to use Legendre functions $P_{\mathrm{L}}(\cos\theta)$ to describe the angular distribution function $f(\theta)$:

$$f(\theta) = \frac{1}{2}\left[\begin{array}{l}1 + a_1\,P_1(\cos\theta) + a_2\,P_2(\cos\theta) \\ + a_3\,P_3(\cos\theta) + a_4\,P_4(\cos\theta) + \dots\end{array}\right]$$
(24)

The expansion in Legendre polynomials or more generally spherical harmonics is chosen because they are orthogonal functions. The coefficients a_{L} in the expansion can be obtained by multiplying both sides of Eq. (24) by $P_{\mathrm{L}}(\cos\theta)$ and integrating over θ, with the result:

$$a_{\mathrm{L}} = (2\mathrm{L}+1)\frac{\int\limits_0^\pi P_{\mathrm{L}}(\cos\theta)\,f(\theta)\sin\theta\,d\theta}{\int\limits_0^\pi f(\theta)\sin\theta\,d\theta}$$
(25)

The ratio of integrals in Eq. (25) is the definition of the average value of $P_{\mathrm{L}}(\cos\theta)$ over the single particle angular distribution function, so the single particle distribution function can now be written as:

$$f(\theta) = \frac{1}{2}\left[\begin{array}{l}1 + 3\langle P_1(\cos\theta)\rangle\,P_1(\cos\theta) \\ + 5\langle P_2(\cos\theta)\rangle\,P_2(\cos\theta) \\ + 7\langle P_3(\cos\theta)\rangle\,P_3(\cos\theta) \\ + 9\langle P_4(\cos\theta)\rangle\,P_4(\cos\theta) \\ + \dots\end{array}\right]$$
(26)

The amplitudes of the coefficients in this expansion now have a special significance:

they are the order parameters for the distribution function. A knowledge of all order parameters will provide a complete description of the single particle angular distribution function, and the magnitude of each order parameter gives the contribution of a particular symmetry to the distribution function for the disordered structure. If we wish to consider the distribution function for a molecule requiring three angles to specify its orientation, $f(\Omega = \theta, \phi, \psi)$ must be expanded in terms of a set of orthogonal functions which span the orientation space of the Euler angles. A convenient set of such functions are the Wigner rotation matrices, denoted $\mathbf{D}_{m,n}^{L}(\theta, \phi, \psi)$, so the single particle angular distribution function becomes (see [3] for further details):

$$f(\Omega) = \frac{1}{8\pi^2} \sum_{L} \sum_{m,n=-L}^{L} \\ \cdot (2L+1) a_{L,m,n} D_{m,n}^{L}(\theta,\phi,\psi) \quad (27)$$

where as before the coefficients $a_{L,m,n}$ can be obtained from the orthogonality condition to give:

$$a_{L,m,n} = \left\langle D_{m,n}^{L}(\theta,\phi,\psi) \right\rangle \quad (28)$$

and the quantities $\langle D_{m,n}^{L}(\theta, \phi, \psi) \rangle$ are generalized orientational order parameters, the indices L, m, n relating to the angular variables θ, ϕ, ψ.

The symmetry of the constituent molecules and the symmetry of the liquid crystal phase provide some constraints on the terms which contribute to the distribution function. For example if a molecule has inversion symmetry, then only terms even in L will contribute; similarly for molecules with a C_2 rotation axis along the z-direction, only terms even in n will survive. A full set of symmetry operations and non-vanishing order parameters is given by Zannoni [4]. The number of independent order parameters

ters necessary to specify the angular distribution function will also be reduced by the symmetry of the liquid crystal phase, see for example [5]. Returning to the simple case of uniaxial molecules having inversion symmetry in a uniaxial phase, the distribution function can be written as:

$$f(\theta) = \frac{1}{2} \begin{bmatrix} 1 + 5\langle P_2(\cos\theta) \rangle P_2(\cos\theta) \\ + 9\langle P_4(\cos\theta) \rangle P_4(\cos\theta) + \ldots \end{bmatrix} \quad (29)$$

and the two leading order parameters are $\langle P_2(\cos\theta) \rangle$ and $\langle P_4(\cos\theta) \rangle$: the former is often referred to as S. One point to notice is that order parameters are multivalued in the sense that different distributions may give the same value for an order parameter, hence the value of information on more than one order parameter. For example, a distribution in which the molecular axes were at an average angle of about 55° to the z-axis would give $\langle P_2(\cos\theta) \rangle$ close to zero with a negative $\langle P_4(\cos\theta) \rangle$. An isotropic distribution of molecules is indicated by all order parameters being zero. Alternative but equivalent definitions of order parameters as tensor quantities are sometimes more convenient, particularly in relation to physical properties. These definitions are only usefully compact for systems in which either the molecules or the phase are uniaxial. For a uniaxial liquid crystal phase of biaxial molecules, the single particle angular distribution function can be written as [3]:

$$f(\theta,\phi) = \frac{1}{4\pi} \begin{bmatrix} 1 + 3S_\alpha l_\alpha + 5 S_{\alpha\beta} l_\alpha l_\beta \\ + 7 S_{\alpha\beta\gamma} l_\alpha l_\beta l_\gamma \\ + 9 S_{\alpha\beta\gamma\delta} l_\alpha l_\beta l_\gamma l_\delta + \ldots \end{bmatrix} \quad (30)$$

where $l_\alpha = l_x$, l_y, l_z are direction cosines of the director with respect to the molecular axes, and the quantities S_α, $S_{\alpha\beta}$, $S_{\alpha\beta\gamma}$ and $S_{\alpha\beta\gamma\delta}$ are ordering tensors of ranks 1, 2, 3, and 4, respectively, and the summation convention has been adopted. For phases with inversion

Table 1. Relationship between different definitions of order parameters.

	Uniaxial phase		Biaxial phase	
	Uniaxial molecule	Biaxial molecule	Uniaxial molecule	Biaxial molecule
Saupe ordering matrix $S_{\alpha\beta} = \left\langle \frac{1}{2}\left(3 l_\alpha l_\beta - \delta_{\alpha\beta}\right)\right\rangle$	$S = \left\langle \frac{1}{2}\left(3 l_z^2 - 1\right)\right\rangle$	$S_{zz} = \left\langle \frac{1}{2}\left(3 l_z^2 - 1\right)\right\rangle$ $D = S_{xx} - S_{yy} = \left\langle \frac{3}{2}\left(l_x^2 - l_y^2\right)\right\rangle$	$S_{zz}^Z = \left\langle \frac{1}{2}\left(3 l_{Z,z}^2 - 1\right)\right\rangle$ $P = S_{zz}^X - S_{zz}^Y = \left\langle \frac{3}{2}\left(l_{X,z}^2 - l_{Y,z}^2\right)\right\rangle$	$S_{zz}^Z = \left\langle \frac{1}{2}\left(3 l_{Z,z}^2 - 1\right)\right\rangle$ $D = S_{xx}^Z - S_{yy}^Z = \left\langle \frac{3}{2}\left(l_{Z,x}^2 - l_{Z,y}^2\right)\right\rangle$ $P = S_{zz}^X - S_{zz}^Y = \left\langle \frac{3}{2}\left(l_{X,z}^2 - l_{Y,z}^2\right)\right\rangle$ $C = \left(S_{xx}^X - S_{yy}^X\right) - \left(S_{xx}^Y - S_{yy}^Y\right)$ $= \left\langle \frac{3}{2}\left[\left(l_{X,x}^2 - l_{X,y}^2\right) - \left(l_{Y,x}^2 - l_{Y,y}^2\right)\right]\right\rangle$
Legendre functions $\langle P_L(\cos\theta)\rangle$ and spherical harmonics $\langle Y_{Lm}(\theta,\psi)\rangle$	$S = \langle P_2(\cos\theta)\rangle$ $= \left\langle \frac{1}{2}\left(3\cos^2\theta - 1\right)\right\rangle$	$S = \left(\frac{5}{4\pi}\right)^{1/2}\langle Y_{20}(\theta,\psi)\rangle$ $D = \sqrt{6}\left(\frac{5}{4\pi}\right)^{1/2}\langle Y_{22}(\theta,\psi)\rangle$ $= \frac{3}{2}\langle \sin^2\theta \cos 2\psi\rangle$		
Wigner rotation matrices $\langle D_{mn}^L(\theta,\phi,\psi)\rangle$	$S = \langle D_{0,0}^2\rangle$	$S = \langle D_{0,0}^2\rangle$ $D = \sqrt{\frac{3}{2}}\langle D_{0,2}^2 + D_{0,-2}^2\rangle$	$S = \langle D_{0,0}^2\rangle$ $P = \sqrt{\frac{3}{2}}\langle D_{2,0}^2 + D_{-2,0}^2\rangle$ $= \frac{3}{2}\langle \sin^2\theta \cos 2\phi\rangle$	$S = \langle D_{0,0}^2\rangle$ $D = \sqrt{\frac{3}{2}}\langle D_{0,2}^2 + D_{0,-2}^2\rangle$ $P = \sqrt{\frac{3}{2}}\langle D_{2,0}^2 + D_{-2,0}^2\rangle$ $C = \frac{3}{2}\langle D_{2,2}^2 + D_{-2,-2}^2 + D_{2,-2}^2 + D_{-2,2}^2\rangle$ $= \frac{3}{2}\left\langle \begin{array}{l}\left(1+\cos^2\theta\right)\cos 2\phi \cos 2\psi \\ -2\cos\theta \sin 2\phi \sin 2\psi\end{array}\right\rangle$

symmetry all ordering tensors of odd rank are zero, and the first nonvanishing tensor order parameter $S_{\alpha\beta}$ is sometimes known as the Saupe ordering matrix. Definitions of the ordering tensors as averages over direction cosines are:

$$S_\alpha = \langle l_\alpha \rangle \tag{31}$$

$$S_{\alpha\beta} = \left\langle \frac{1}{2}\left(3 l_\alpha l_\beta - \delta_{\alpha\beta}\right)\right\rangle$$

$$S_{\alpha\beta\gamma} = \left\langle \frac{1}{2}\left(\begin{array}{c} 5 l_\alpha l_\beta l_\gamma \\ -\left(l_\alpha \delta_{\beta\gamma} + l_\beta \delta_{\gamma\alpha} + l_\gamma \delta_{\alpha\beta}\right)\end{array}\right)\right\rangle$$

$$S_{\alpha\beta\gamma\delta} = \left\langle \begin{array}{c} \frac{1}{8}\left|\begin{array}{c} 35 l_\alpha l_\beta l_\gamma l_\delta \\ -5\left(\begin{array}{c} l_\alpha l_\beta \delta_{\gamma\delta} + l_\alpha l_\gamma \delta_{\beta\delta} \\ + l_\alpha l_\delta \delta_{\beta\gamma} + l_\beta l_\gamma \delta_{\alpha\delta} \\ + l_\beta l_\delta \delta_{\alpha\gamma} + l_\gamma l_\delta \delta_{\alpha\beta}\end{array}\right)\end{array}\right| \\ + \frac{1}{8}\left(\begin{array}{c}\delta_{\alpha\beta}\delta_{\gamma\delta} + \delta_{\alpha\gamma}\delta_{\beta\delta} \\ + \delta_{\beta\gamma}\delta_{\alpha\delta}\end{array}\right)\end{array}\right\rangle$$

These tensors are defined to be zero for an isotropic phase. For uniaxial molecules there is only one independent component for each of the tensors:

$$S_z = \langle \cos\theta \rangle = \langle P_1(\cos\theta)\rangle$$

$$S_{zz} = \left\langle \frac{1}{2}\left(3\cos^2\theta\right) - 1\right\rangle = \langle P_2(\cos\theta)\rangle$$

$$S_{zzz} = \left\langle \frac{1}{2}\left(5\cos^3\theta - 3\cos\theta\right)\right\rangle = \langle P_3(\cos\theta)\rangle$$

$$S_{zzzz} = \left\langle \frac{1}{8}\left(35\cos^4\theta - 30\cos^2\theta + 3\right)\right\rangle$$

$$= \langle P_4(\cos\theta)\rangle$$

One advantage of this representation of order parameters is that it readily describes nonuniaxial molecular order in a macroscopically uniaxial liquid crystal phase. If the reference axis frame has been chosen to diagonalize the ordering tensor, then its principal components are the order parameters of the different molecular axes with respect to a uniaxial director. If the distributions of the two shorter molecular axes (say x and y) are not identical, then $S_{xx} \neq S_{yy}$, which represents a local biaxial ordering of molecular axes. This may be expressed in terms of a new order parameter $D = S_{xx} - S_{yy}$, and the Saupe ordering matrix can be written as:

$$S_{\alpha\beta} = \begin{bmatrix} -\frac{1}{2}(S-D) & 0 & 0 \\ 0 & -\frac{1}{2}(S+D) & 0 \\ 0 & 0 & S \end{bmatrix} \tag{33}$$

If the Saupe ordering matrix is written in terms of the laboratory axis frame, but assuming now that the molecules are uniaxial, then phase biaxiality can be described in terms of the order parameter P, which is nonzero for tilted smectic phases and other intrinsically biaxial phases. For example the diagonal ordering matrix for the molecular long axis z can be written as:

$$S_{ij}^{(z)} = \left\langle \frac{1}{2}\left(3 l_{i,z} l_{j,z} - \delta_{ij}\right)\right\rangle$$

$$= \begin{bmatrix} -\frac{1}{2}(S-P) & 0 & 0 \\ 0 & -\frac{1}{2}(S+P) & 0 \\ 0 & 0 & S \end{bmatrix} \tag{34}$$

and the magnitude of P is a measure of the different probabilities of finding the z-molecular axis along the X and Y directions of the laboratory or phase reference frame.

The Cartesian tensor representation can be extended to describe the orientational ordering of biaxial molecules in biaxial phases by introducing [6] a fourth rank ordering tensor:

$$S_{\alpha\beta,ij} = \left\langle \frac{1}{2}\left(3 l_{i,\alpha} l_{j,\beta} - \delta_{ij}\delta_{\alpha\beta}\right)\right\rangle \tag{35}$$

where $l_{i,\alpha}$ is the cosine of the angle between the molecular axis α and the laboratory or phase axis i. The tensor $S_{\alpha\beta,ij}$ still describes second rank orientational ordering, and should not be confused with the ordering tensor $S_{\alpha\beta\gamma\delta}$ which refers to fourth rank orientational order. For a suitable choice of both sets of axes, the 81 components of $S_{\alpha\beta,ij}$ can be reduced to nine such that $i=j$ and $\alpha=\beta$. This is equivalent to defining three diagonal Saupe ordering matrices, one for each of the three axes, $i=X, Y, Z$:

$$S_{\alpha\beta}^{(i)} = \left\langle \frac{1}{2}\left(3 l_{i,\alpha} l_{i,\beta} - \delta_{\alpha\beta}\right)\right\rangle \tag{36}$$

Taking the diagonal components of these three matrices allows the construction of a 3×3 matrix:

$$S_{\alpha\beta}^{(i)} = \begin{bmatrix} S_{xx}^X & S_{yy}^X & S_{zz}^X \\ S_{xx}^Y & S_{yy}^Y & S_{zz}^Y \\ S_{xx}^Z & S_{yy}^Z & S_{zz}^Z \end{bmatrix} \tag{37}$$

and the generalized biaxial order parameters can be defined as follows. The long axis ordering is described by $S=S_{zz}^Z$, while the phase biaxiality for a uniaxial molecule is given by $P=S_{zz}^X-S_{zz}^Y$. For biaxial molecules in a uniaxial phase the biaxial order parameter is $D=S_{xx}^Z-S_{yy}^Z$, but it would be equally possible to define a biaxial order parameter with respect to the X-axis, such that $D'=S_{xx}^X-S_{yy}^X$, and this would equal $D''=S_{xx}^Y-S_{yy}^Y$ for uniaxial phases. However, if the phase is biaxial, then the biaxiality defined with respect to phase axes X and Y will be different, and this new form of biaxiality is described in terms of a new biaxial order parameter:

$$C = D' - D''$$
$$= \left(S_{xx}^X - S_{yy}^X\right) - \left(S_{xx}^Y - S_{yy}^Y\right) \tag{37a}$$

In terms of averages over Euler angles, these may be defined as:

$$S = \left\langle \frac{1}{2}\left(3\cos^2\theta - 1\right)\right\rangle$$

$$D = \left\langle \frac{3}{2}\left(\sin^2\theta \cos 2\psi\right)\right\rangle$$

$$P = \left\langle \frac{3}{2}\left(\sin^2\theta \cos 2\phi\right)\right\rangle$$

$$C = \left\langle \begin{array}{c} \frac{3}{2}\left[\left(1+\cos^2\theta\right)\cos 2\phi \cos 2\psi \right. \\ \left. - 2\cos\theta \sin 2\phi \sin 2\psi\right] \end{array}\right\rangle \tag{38}$$

Slightly different definitions have been adopted by some other authors with different numerical factors. The advantage of the definitions in Eq. (38) is that these order parameters are simply related to the components of the Saupe ordering matrices Eq. (34), as indicated in Table 1.

The order parameters introduced in the preceding paragraphs are sufficient to describe the orientational order/disorder of rigid molecules in liquid crystal phases. They will be used to relate molecular properties to macroscopic physical properties, but there are additional sources of order/disorder which may affect physical properties. For flexible molecules certain physical properties or responses may be sensitive to a particular group or bond within the molecule, and under these circumstances it is the order parameter of that moiety which determines the measured anisotropy. As an example, the degree of order of flexible alkyl chains attached to a rigid molecular core is reduced by internal rotation of the chain segments. Using selectively deuterated mesogens, deuterium magnetic resonance is able to measure the order parameters of different segments of a flexible chain, and it has been shown that as expected the orientational order decreases along the chain away from the rigid core of the molecule [7].

Local biaxial ordering of molecules in uniaxial liquid crystal phases can be detected by spectroscopic techniques such as linear dichroism and NMR. It is often more convenient to probe the biaxial order of solutes in liquid crystal hosts, for example a detailed analysis of the ^{2}D NMR of fully deuteriated anthracene d$_{10}$ in various liquid crystal solvents yields both S and D order parameters [8]. These are illustrated in Fig. 5, where they are compared to mean field calculations with the ratio (λ) of uniaxial and biaxial energies as an adjustable parameter. It is reasonable to assume that the orientational order parameters introduced will also be appropriate for smectic and columnar phases. However, there are additional contributions to the order/disorder, which can contribute to the measured anisotropy in physical properties. The characteristic structural feature of smectic and columnar phases is the presence of some translational order, and so the radial distribution function will have long range periodicity in certain directions, the amplitude of which will be determined by a suitable order parameter. Furthermore, there is the likelihood of coupled orientational and translational order: for example in a smectic A phase molecules will be more likely to be aligned parallel to the layer normal (the director) when their centres of mass coincide with the average layer position. For disordered smectics there is one dimensional positional order, and assuming uniaxial molecules in a uniaxial smectic phase, the corresponding single particle distribution function can be written as [9]:

$$f(z,\theta) = \frac{1}{2d} \begin{bmatrix} 1+2\left\langle\cos\left(\frac{2\pi z}{d}\right)\right\rangle\cos\left(\frac{2\pi z}{d}\right) \\ +5\langle P_2(\cos\theta)\rangle P_2(\cos\theta) \\ +10\left\langle P_2(\cos\theta)\cos\left(\frac{2\pi z}{d}\right)\right\rangle \\ \cdot P_2(\cos\theta)\cos\left(\frac{2\pi z}{d}\right)+... \end{bmatrix}$$

$$(39)$$

where translational (τ) and translational–rotational (σ) order parameters can be defined as:

$$\tau = \left\langle\cos\left(\frac{2\pi z}{d}\right)\right\rangle$$

$$\sigma = \left\langle P_2(\cos\theta)\cos\left(\frac{2\pi z}{d}\right)\right\rangle \qquad (40)$$

Columnar phases have two degrees of translational order, and the corresponding order parameters now require averages over periodic functions in two spatial dimensions; for completeness we include the distribution function for a uniaxial columnar phase consisting of uniaxial disc-like molecules:

$$f(x,y,\theta) = \frac{1}{2d_x d_y} \qquad (41)$$

$$\begin{bmatrix} 1+2\left\langle\cos\left(\frac{2\pi x}{d_x}\right)\cos\left(\frac{2\pi y}{d_y}\right)\right\rangle \\ \cdot\cos\left(\frac{2\pi x}{d_x}\right)\cos\left(\frac{2\pi y}{d_y}\right) \\ +5\langle P_2(\cos\theta)\rangle P_2(\cos\theta) \\ +20\left\langle P_2(\cos\theta)\cos\left(\frac{2\pi x}{d_x}\right)\cos\left(\frac{2\pi y}{d_y}\right)\right\rangle \\ \cdot P_2(\cos\theta)\cos\left(\frac{2\pi x}{d_x}\right)\cos\left(\frac{2\pi y}{d_y}\right)+... \end{bmatrix}$$

The contribution of translational order parameters to the anisotropy of physical properties of liquid crystals has not been studied in detail. Evidence suggests that there is a very small influence of translational ordering on the optical properties, but effects of translational order can be detected in the measurement of dielectric properties. There are strong effects in both elastic properties and viscosity, but the statistical theories of these properties have not been extended to include explicitly the effects of translational order.

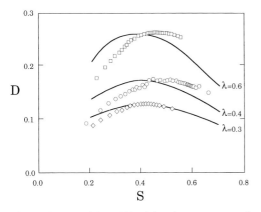

Figure 5. Uniaxial and biaxial order parameters for anthracene d_{10} in different liquid crystal solvents: ($\square$) – ZLI1167 (best fit $\lambda = 0.6$); ($\circ$) – E9 ($\lambda = 0.4$); ($\diamond$) – Phase 5 ($\lambda = 0.3$), reproduced from [8].

Having defined orientational order parameters, it is now possible to develop a general statistical theory which relates the physical properties of molecules to the macroscopic properties of liquid crystal phases. There are however many simplifying approximations which have to be introduced to give usable results. An important factor is that the nature of the property will determine the order parameters that will be included: in particular a property of tensor rank n will in general require order parameters up to tensor rank n to specify it.

If $\kappa_{\alpha\beta}$ is a molecular second rank tensor property, the principal components of which are $\kappa_{xx}^{(m)}$, $\kappa_{yy}^{(m)}$ and $\kappa_{zz}^{(m)}$ defined in a molecular axis system, then using the transformation rule for second rank tensors, the property in a laboratory frame is:

$$\kappa_{\alpha\beta} = a_{\alpha\gamma}\, a_{\beta\delta}\, \kappa_{\gamma\delta}^{(m)} \tag{42}$$

Ignoring the effect of molecular interactions, which is a gross assumption, the macroscopic response $\chi_{\alpha\beta}$ measured in a laboratory axis frame will be the molecular property multiplied by the number density, averaged over all possible orientations of the molecules:

$$\chi_{\alpha\beta} = N\langle\kappa_{\alpha\beta}\rangle = N\langle a_{\alpha\gamma}\, a_{\beta\delta}\rangle \kappa_{\gamma\delta}^{(m)} \tag{43}$$

The average over the products of direction cosine matrices contains the orientational order parameters, and in terms of the principal components of $\kappa_{\alpha\beta}^{(m)}$ the anisotropic part of the macroscopic tensor property becomes:

$$\chi_{\alpha\beta}^{(a)} = \frac{N}{3}\left\{\begin{array}{l}\langle 3a_{\alpha z}\, a_{\beta z} - \delta_{\alpha\beta}\rangle\\[4pt]\cdot\left[\kappa_{zz}^{(m)} - \frac{1}{2}\left(\kappa_{xx}^{(m)} + \kappa_{yy}^{(m)}\right)\right]\\[6pt]+\frac{1}{2}\langle 3a_{\alpha x}\, a_{\beta x} - 3a_{\alpha y}\, a_{\beta y}\rangle\\[4pt]\cdot\left(\kappa_{xx}^{(m)} - \kappa_{yy}^{(m)}\right)\end{array}\right\} \tag{44}$$

Those terms in Eq. (44) which have nonzero averages depend on the symmetry of $\chi_{\alpha\beta}^{(a)}$, which by Neumann's principle must contain the symmetry of the phase to which it relates. If the phase is uniaxial, the principal components of $\chi_{\alpha\beta}^{(m)}$ become:

$$\chi_{\parallel}^{(a)} = \frac{2N}{3}\left\{\begin{array}{l}S\left[\kappa_{zz}^{(m)} - \frac{1}{2}\left(\kappa_{xx}^{(m)} + \kappa_{yy}^{(m)}\right)\right]\\[6pt]+\frac{D}{2}\left(\kappa_{xx}^{(m)} - \kappa_{yy}^{(m)}\right)\end{array}\right\}$$

$$\chi_{\perp}^{(a)} = -\frac{N}{3}\left\{\begin{array}{l}S\left[\kappa_{zz}^{(m)} - \frac{1}{2}\left(\kappa_{xx}^{(m)} + \kappa_{yy}^{(m)}\right)\right]\\[6pt]+\frac{D}{2}\left(\kappa_{xx}^{(m)} - \kappa_{yy}^{(m)}\right)\end{array}\right\} \tag{45}$$

where the order parameters S and D are defined by Eq. (38). Comparison of Eq. (45) with the definition of the macroscopic order parameter Eq. (22) shows that:

$$Q_{zz} = \frac{N}{[\Delta\chi]_{max}}$$

$$\cdot\left\{\begin{array}{l}S\left[\kappa_{zz}^{(m)} - \frac{1}{2}\left(\kappa_{xx}^{(m)} + \kappa_{yy}^{(m)}\right)\right]\\[6pt]+\frac{D}{2}\left(\kappa_{xx}^{(m)} - \kappa_{yy}^{(m)}\right)\end{array}\right\} \tag{46}$$

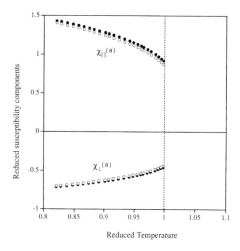

Figure 6. Reduced values $\chi_\parallel^{(a)}$ and $\chi_\perp^{(a)}$ as a function of reduced temperature plotted according to Eq. (45) using order parameters calculated using mean field theory. Open squares (□) assume no molecular biaxiality, so $D=0$; full circles (●) are for an assumed molecular biaxiality of 0.3 and a λ value of 0.3.

For perfect alignment $Q_{zz}=S=1$, and the macroscopic anisotropy $(\Delta\chi)_{max}=N[k_{zz}^{(m)}- 1/2\,k_{xx}^{(m)}+k_{yy}^{(m)}]$, which is simply N times the molecular anisotropy. Both order parameters S and D contribute to the anisotropy of second rank tensor properties, but they cannot be separated from a single measurement of the macroscopic anisotropy. The behavior of the macroscopic property components as a function of temperature is illustrated in Fig. 6 for a material of positive molecular anisotropy, for various values of the order parameters S and D calculated from mean field theory; for a negative anisotropy material the signs of the parallel and antiparallel components are interchanged.

If the liquid crystal phase is biaxial, then any second rank tensor property has three independent principal components. These are the diagonal elements of the anisotropic tensor $\chi_{\alpha\beta}^{(a)}$, and can be expressed in terms of the order parameters introduced for bi-

axial phases:

$$\chi_\parallel^{(a)} = \frac{2N}{3}\left\{ \begin{array}{l} S\left[\kappa_{zz}^{(m)} - \frac{1}{2}\left(\kappa_{xx}^{(m)} + \kappa_{yy}^{(m)}\right)\right] \\ + \frac{D}{2}\left(\kappa_{xx}^{(m)} - \kappa_{yy}^{(m)}\right) \end{array} \right\}$$

$$\chi_{\perp'}^{(a)} = -\frac{N}{3}\left\{ \begin{array}{l} (S+P)\left[\kappa_{zz}^{(m)} - \frac{1}{2}\left(\kappa_{xx}^{(m)} + \kappa_{yy}^{(m)}\right)\right] \\ + \frac{D+C}{2}\left(\kappa_{xx}^{(m)} - \kappa_{yy}^{(m)}\right) \end{array} \right\}$$

$$\chi_\perp^{(a)} = -\frac{N}{3}\left\{ \begin{array}{l} (S-P)\left[\kappa_{zz}^{(m)} - \frac{1}{2}\left(\kappa_{xx}^{(m)} + \kappa_{yy}^{(m)}\right)\right] \\ + \frac{D-C}{2}\left(\kappa_{xx}^{(m)} - \kappa_{yy}^{(m)}\right) \end{array} \right\}$$

$$(47)$$

It has been assumed that molecular properties contribute additively to the macroscopic tensor components, which are consequently proportional to the number density. If intermolecular interactions contribute to the physical property, then deviations from a linear dependence of the property on density are expected. Also the contribution of orientational order will be more complex, since the properties will depend on the degree of order of interacting molecules. Effects of molecular interactions contribute to the dielectric properties of polar mesogens, and are particularly important for elastic and viscoelastic properties. Molecular mean field theories of elastic properties predict that elastic constants should be proportional to the square of the order parameter; this result highlights the significance of pairwise interactions.

1.2 References

[1] J. F. Nye, *Physical Properties of Crystals*, Oxford University Press, Oxford, **1985**.
[2] S. Altmann, *Rotations, Quaternions and Double Groups*, Oxford University Press, Oxford, **1986**, p. 65.

[3] C. Zannoni in *Nuclear Magnetic Resonance of Liquid Crystals* (Ed.: J. W. Emsley), D. Reidel, Dordrecht, Netherlands, **1983**, p. 1.

[4] C. Zannoni in *Molecular Physics of Liquid Crystals* (Eds.: G. W. Gray, G. R. Luckhurst), Academic Press, London, **1979**, p. 51.

[5] Z. Luz, D. Goldfarb, H. Zimmermann in *Nuclear Magnetic Resonance of Liquid Crystals* (Ed.: J. W. Emsley), D. Reidel, Dordrecht, Netherlands, **1983**, p. 351.

[6] D. W. Allender, M. A. Lee, N. Hafiz, *Mol. Cryst. Liq. Cryst.* **1985**, *124*, 45–52.

[7] J. W. Emsley, G. R. Luckhurst, C. P. Stockley, *Proc. R. Soc. Lond.* **1982**, *A381*, 117.

[8] J. W. Emsley in *Nuclear Magnetic Resonance of Liquid Crystals* (Ed.: J. W. Emsley), D. Reidel, Dordrecht, Netherlands, **1983**, p. 379.

[9] C. Zannoni in *The Molecular Dynamics of Liquid Crystals* (Eds.: G. R. Luckhurst, C. A. Veracini), Kluwer, Dordrecht, Netherlands, **1994**, p. 34.

2 Magnetic Properties of Liquid Crystals

David Dunmur and Kazuhisa Toriyama

In Sec. 1 of this chapter it is shown that macroscopic anisotropy in liquid crystals can be related to molecular properties through appropriate microscopic orientational order parameters, as in Eq. (47) of that section. This relationship assumes that the macroscopic response of a liquid crystal is simply the sum of the individual molecular responses averaged over an orientational distribution function (i.e. interactions between molecules are ignored, except to the extent that they determine the orientational order). For most physical properties such an approximation is very crude; however, magnetic properties are only very weakly influenced by intermolecular interactions, and so it can be assumed that the magnetic response of liquid crystals is simply the aggregated molecular response. The weak interaction between molecules and magnetic fields is shown by the magnetic permeability relative to that of free space for nonferromagnetic materials, which is close to unity. The magnetic response of materials depends on their electronic structure, and the susceptibility may be negative, characteristic of diamagnetic compounds, positive denoting a paramagnetic response, or ferromagnetic which indicates a permanent magnetization resulting from coupling between electrons

on constituent atoms, ions or molecules; ferromagnetism is largely restricted to the solid state. Both diamagnetic and paramagnetic liquid crystals are known, and ferromagnetic liquid crystals have been prepared from colloidal suspensions of ferromagnetic materials in a liquid crystal host.

2.1 Magnetic Anisotropy

Like other tensor properties of liquid crystals, the magnetic susceptibility is anisotropic, and so magnetic fields can be used to control the alignment of liquid crystal samples. This is perhaps the single most useful application of the magnetic properties of liquid crystals, and the combination of magnetic field alignment with some other measurement forms the basis of many experimental investigations.

Macroscopically the magnetic susceptibility relates the induced magnetization M to the strength of the magnetic field, but assuming that local field effects are ignored, the susceptibility is usually defined in terms of magnetic induction B. For this chapter the magnetic induction will be referred to as the magnatic field, and the magnetization is given by:

$$M_\alpha = \mu_0^{-1} \chi_{\alpha\beta}^{\text{mag}} B_\beta \qquad (1)$$

where μ_0 is the permeability of free space. The magnetic contribution to the free energy density becomes:

$$
\begin{aligned}
g_{\text{mag}} &= -\int B_\alpha \, d M_\alpha \\
&= -\mu_0^{-1} \int B_\alpha \chi_{\alpha\beta} \, d B_\beta \\
&= g_0 - \frac{1}{2} \mu_0^{-1} \chi_{\alpha\beta} B_\alpha B_\beta \qquad (2)
\end{aligned}
$$

$\chi_{\alpha\beta}$ is a volume susceptibility, but a molar susceptibility $\chi_{\alpha\beta}^{\text{mol}}$ may be defined as

$$\chi_{\alpha\beta}^{\text{mol}} = \chi_{\alpha\beta} V^{\text{mol}} \qquad (3)$$

where V^{mol} is the molar volume. The susceptibility has the symmetry of the material, so expressing this in terms of the principal axes of χ gives for the free energy density:

$$
\begin{aligned}
g_{\text{mag}} &= g_0 - \frac{1}{2} \mu_0^{-1} \\
&\quad \cdot \left(\chi_\parallel B_\parallel^2 + \chi_{\perp'} B_{\perp'}^2 + \chi_\perp B_\perp^2 \right) \\
&= g_0 - \frac{1}{2} \mu_0^{-1} B^2 \\
&\quad \cdot \big(\chi_\parallel \cos^2 \theta + \chi_{\perp'} \sin^2 \theta \sin^2 \phi \\
&\quad + \chi_\perp \sin^2 \theta \cos^2 \phi \big) \qquad (4)
\end{aligned}
$$

where θ and ϕ are polar angles defining the orientation of B with respect to the principal axes of the susceptibility. For a uniaxial material $\chi_{\perp'} = \chi_\perp$ and:

$$
\begin{aligned}
g_{\text{mag}} &= g_0 - \frac{1}{2} \mu_0^{-1} B^2 \left(\chi_\perp + \Delta\chi \cos^2 \theta \right) \\
&= g_0 - \frac{1}{2} \mu_0^{-1} B^2 \chi_\perp \\
&\quad - \frac{1}{2} \mu_0^{-1} \Delta\chi (B \cdot n)^2 \qquad (5)
\end{aligned}
$$

where $\Delta\chi = \chi_\parallel - \chi_\perp$, and n is the director, which also defines the principal axis of the

susceptibility for uniaxial materials. From Eq. (5) it is clear that the sign of the anisotropy $\Delta\chi$ will determine the orientation of the director with respect to a magnetic field. In order to minimize the free energy, the director will align parallel to the magnetic field for a material having positive $\Delta\chi$, while for negative $\Delta\chi$ the director will be perpendicular to B; both situations can occur in practice.

As explained earlier, any anisotropic property can be used to define a macroscopic order parameter, and because it is largely unaffected by molecular interactions, the magnetic susceptibility is a particularly useful measure: definitions are given as Eqs. (21) and (22) of Sec. 1 of this chapter. The value of $\Delta\chi_{\text{max}}$ corresponding to perfect alignment can in principal be obtained from measurements on single crystals or from molecular susceptibilities. Eqs. (45) and (47) of Sec. 1 of this chapter relate a macroscopic susceptibility to a microscopic molecular property $\kappa_{\alpha\beta}$, and introduce appropriate order parameters. The molecular susceptibilities κ_{xx}, κ_{yy}, κ_{zz} are defined for the principal axes of the molecular susceptibility, which may not coincide with the axes that define the local orientational order; however, it is usually assumed that any differences can be neglected. Measurements of the magnetic susceptibility can provide a useful route to the order parameters of liquid crystals [1, 2], but require a knowledge of the molecular susceptibilities. These are not usually available for mesogens, but they can be obtained from single crystal measurements, provided full details are available for the crystal structure; the method for deriving molecular susceptibilities from crystal susceptibilities is explained in detail in [3].

2.2 Types of Magnetic Polarization

2.2.1 Diamagnetism

A diamagnetic response is the induction of a magnetic moment in opposition to an applied magnetic field, which thereby raises the free energy. Thus a diamagnetic material will be expelled from a magnetic field, or will adjust itself to minimize the diamagnetic interaction. Most liquid crystals are diamagnetic and this diamagnetism originates from the dispersed electron distribution associated with the molecular electronic structure. The diamagnetic susceptibility is a second rank tensor, and its principal components can be expressed as:

$$\kappa_{ll}^{dia} = -\frac{e^2 \mu_0}{4m_e}\left\langle m^2 + n^2 \right\rangle$$

$$\kappa_{mm}^{dia} = -\frac{e^2 \mu_0}{4m_e}\left\langle l^2 + n^2 \right\rangle$$

$$\kappa_{nn}^{dia} = -\frac{e^2 \mu_0}{4m_e}\left\langle m^2 + l^2 \right\rangle \quad (6)$$

where e is the electronic charge, m_e is the mass of an electron, l, m, n are the molecular axes, and the quantities $\langle m^2 + n^2 \rangle$ are averages over the electron distribution for a plane perpendicular to the component axis (l in this case see Fig. 1).

The induced diamagnetic moment depends on the extent of the electron distribution in a plane perpendicular to an applied magnetic field. In a molecule, delocalized charge makes a major contribution to κ^{dia}, and in particular the ring currents associated with aromatic units give a large negative component of diamagnetic susceptibility for directions perpendicular to the plane of the aromatic unit. It is for this reason that the diamagnetic anisotropy of most calamitic

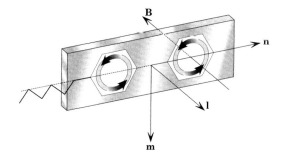

Figure 1. Diagram of molecular axes with representation of the perpendicular plane.

mesogens is positive since both components are negative, but $|\chi_\perp| > |\chi_\parallel|$.

2.2.2 Paramagnetism

Molecular paramagnetism is mostly connected with unpaired electron spins, which have associated magnetic moments. For paramagnetic mesogens the electron spin is introduced by metal centres, and one of the motivations for research into metal-containing mesogens is the desire to prepare paramagnetic liquid crystals. Orientation by an external magnetic field of the magnetic moment derived from an electron spin will induce a magnetization along the field direction, and so provides a positive contribution to the magnetic susceptibility:

$$\overline{\kappa}^{para} = \frac{\mu_0 g_e^2 \mu_B s(s+1)}{3k_B T} \quad (7)$$

where g_e is the electronic g-value, s is the total electron spin quantum number and μ_B is the Bohr magneton. κ^{para} is the isotropic molecular paramagnetic susceptibility, and the coefficient of 1/3 arises from the isotropic average of $\langle \cos^2 \theta \rangle$ over the spin orientation in a magnetic field. The temperature dependence of κ^{para} follows a simple Curie Law, and as written there is no anisotropy in κ^{para}, since in most molecules the unpaired electron spin is decoupled from the molecular structure and will align with an external magnetic field independently of the orientation of the molecule. Anisotropy in κ^{para} can be introduced if the g-value becomes anisotropic. The g-value is in fact a tensor quantity which describes the modification to the magnetic field experienced by the electron spin arising from the electron distribution in the molecule. It is analogous to the nuclear shielding in nuclear magnetic resonance, and it contributes to the magnetic internal energy as:

$$u_{\text{mag}} = -g^e_{\alpha\beta}\, \gamma_e\, s_\alpha\, B_\alpha \qquad (8)$$

where γ_e is the electronic magnetogyric ratio and s_α is the vector component of the electronic spin. Interaction between an electron spin and its local electronic environment changes g_e from its free-electron value of 2.0023. These interactions are termed spin–orbit interactions, since they arise from a coupling between the electron spin and the angular momentum of the molecular orbitals. The principal components of the molecular paramagnetic susceptibility can be written as:

$$\kappa^{\text{para}}_{ii} = \frac{\mu_0 \left(g^e_{ii}\right)^2 \mu_B^2 s(s+1)}{3 k_B T} \qquad (9)$$

Hence using Eq. (45) of Sec. 1 of this chapter for a uniaxial liquid crystal, and neglecting any biaxial local order gives for the anisotropic part of the paramagnetic susceptibility:

$$\chi^{(a)\text{para}}_{\parallel} = \frac{2N\,\mu_0\,\mu_B^2\,s(s+1)\,S}{3 k_B T}$$
$$\cdot\left[\left(g^e_{nn}\right)^2 - \frac{1}{2}\left\{\left(g^e_{ll}\right)^2 + \left(g^e_{mm}\right)^2\right\}\right] \qquad (10)$$

$$\chi^{(a)\text{para}}_{\perp} = -\frac{N\,\mu_0\,\mu_B^2\,s(s+1)\,S}{3 k_B T}$$
$$\cdot\left[\left(g^e_{nn}\right)^2 - \frac{1}{2}\left\{\left(g^e_{ll}\right)^2 + \left(g^e_{mm}\right)^2\right\}\right]$$

Thus for a uniaxial $\mathbf{g}_e$ tensor, the paramagnetic susceptibility anisotropy and mean susceptibility can be written as:

$$\Delta\chi^{(a)\text{para}} = \frac{N \mu_0 \mu_B^2 s(s+1) S}{k_B T}$$
$$\cdot\left[\left(g^e_{nn}\right)^2 - \left(g^e_{ll}\right)^2\right] \qquad (11)$$

$$\bar{\chi}^{\text{para}} = \frac{N \mu_0 \mu_B^2 s(s+1)}{9 k_B T}$$
$$\cdot\left[\left(g^e_{nn}\right)^2 + 2\left(g^e_{ll}\right)^2\right] \qquad (12)$$

Care is necessary in defining the principal axes of the $\mathbf{g}_e$ tensor, since they are determined by the local symmetry of the free-electron spin, and therefore usually differ from the molecular axes that define the orientational order of the mesogen. A consequence of Eq. (11) is that there can be competition between paramagnetic and diamagnetic contributions to the macroscopic anisotropy, so the alignment of a paramagnetic mesogen in a magnetic field will be determined by the larger of $\Delta\chi^{\text{dia}}$ and $\Delta\chi^{\text{para}}$.

There is a further contribution to the molecular paramagnetic susceptibility from magnetic field induced distortion of the orbital angular momentum: this is known as temperature independent paramagnetism (TIP), and is the precise magnetic analogue of the electronic polarizability that determines the response of a molecule to a high frequency electric field. The importance of $\kappa^{\text{TIP}}_{\alpha\beta}$ for mesogens is yet to be established: it is only likely to be significant for molecules with low-lying excited electronic states that are connected to the ground state by magnetically allowed dipole transitions; such states are also important in the circular dichroism spectra of molecules. Taking account of all contributions to the molecular susceptibility, the components of the macroscopic susceptibility for a uniaxial liquid crystal composed of uniaxial molecules can be written as:

$$\chi_{\parallel} = \bar{\chi} + \frac{2N\Delta\kappa^{\text{dia}} S}{3} + \frac{2N\Delta\kappa^{\text{TIP}} S}{3}$$
$$+ \frac{2N\mu_0 \mu_B^2 s(s+1) S}{9k_B T} \left[\left(g_{nn}^e\right)^2 - \left(g_{ll}^e\right)^2 \right]$$

$$\chi_{\perp} = \bar{\chi} - \frac{N\Delta\kappa^{\text{dia}} S}{3} - \frac{N\Delta\kappa^{\text{TIP}} S}{3}$$
$$- \frac{N\mu_0 \mu_B^2 s(s+1) S}{9k_B T} \left[\left(g_{nn}^e\right)^2 - \left(g_{ll}^e\right)^2 \right]$$

$$\tag{13}$$

2.2.3   Ferromagnetism

At sufficiently low reduced temperatures, and/or strong spin–spin interactions, spin magnetic moments can become ordered in a parallel array to give ferromagnetic materials, or ordered in an antiparallel fashion to give an antiferromagnetic structure. This behavior is rare in organic materials, for which ferromagnetic organization only occurs at low temperatures. There is a requirement for unpaired electron spins, and so in the context of liquid crystals, possible ferromagnetic materials will almost certainly require metal-containing mesogens. The contribution of ferromagnetic or antiferromagnetic coupling to the susceptibility can be described in terms of a modified Curie Law, known as the Curie–Weiss Law, and the isotropic susceptibility can be written as:

$$\bar{\chi}^{\text{para}} = \frac{N \mu_0 \, g_e^2 \, \mu_B^2 \, s(s-1)}{3k_B (T - \Theta)} \tag{14}$$

where the characteristic temperature Θ, the Curie temperature, is a measure of the ferromagnetic coupling, and marks the onset of permanent magnetization. If Θ is negative, then the local magnetic interactions are antiferromagnetic and Θ is called the Néel

temperature. Although no ferromagnetic liquid crystals have been discovered so far, ferromagnetic liquid crystals can be prepared by the dispersion of ferromagnetic particles in a liquid crystal host. In preparing such systems it is desirable to make the ferromagnetic particles very small, so that each particle has a permanent magnetic moment. Normally in the absence of an external magnetic field or special conditioning, the permanent magnetization characteristic of ferromagnetic materials forms in domains of opposing magnetic moments, so that the total magnetisation is cancelled. However for sufficiently small particle sizes, the domain wall energies become relatively too high to sustain, and single domain particles are preferred. This effect is known as super-paramagnetism or collective paramagnetism, and dispersed particles satisfying the requirements for single magnetic domains act in fluids as micromagnets: such systems are known as ferrofluids. The dispersion of single domain ferromagnetic particles in a liquid crystal host can form anisotropic ferrofluids or ferromagnetic liquid crystals.

2.3   Diamagnetic Liquid Crystals

Most liquid crystals are diamagnetic and their magnetic anisotropy arises from the electronic structure of the mesogens. Delocalisation of electronic charge will enhance the diamagnetic susceptibility and aromatic groups in particular make a large contribution to the diamagnetic susceptibility. In Table 1 are listed molecular susceptibility components for a number of molecules to indicate the likely contributions of various groups to mesogenic structures. These values have been obtained from susceptibility measurements on single crystals.

Table 1. Molecular susceptibility components from single crystal measurements.

Compound and molecular axes	Molecular susceptibility components			Ref.
	$N\kappa_l/10^{-9}$ m^3 mol^{-1}	$N\kappa_m/10^{-9}$ m^3 mol^{-1}	$N\kappa_n/10^{-9}$ m^3 mol^{-1}	
(benzene, axes 1, n, m)	-1.19	-0.44	-0.44	[3]
H_3C–O– ... –O–CH_3	-1.46	-0.99	-0.82	[3]
(benzoic acid dimer, O···HO / OH··O)	-1.38	-0.59	-0.69	[3]
(biphenyl)	-2.31	-0.78	-0.85	[3]
(anthracene)	-3.13	-0.96	-0.97	[3]
H_3CO– ... –N=N– ... –OCH_3 (O)	-3.08	-1.13	-1.33	[23]

Table 2. Diamagnetic susceptibilities for liquid crystals.

Compound and acronym	Diamagnetic susceptibilities		Ref.
	$\Delta\chi/10^{-9}$ m^3 kg^{-1}	$\chi/10^{-9}$ m^3 kg^{-1}	
C_5H_{11}– ... –CN	1.51	8.43	[2]
C_7H_{15}– ... –CN	1.37	8.66	[2]
C_5H_{11}– ... –CN	0.46		[4]
C_7H_{15}– ... –CN	0.42	9.32	[4]
C_7H_{15}– ... –CN	-0.38	8.87	[2]

Most thermotropic mesogens contain aromatic groups, and since the component of the diamagnetic susceptibility perpendicular to a benzene ring is greater than the in-plane component, liquid crystals composed of calamitic mesogens will have a positive diamagnetic anisotropy, while liquid crystals of disc-like molecules will have a negative diamagnetic anisotropy. Thus calamitic nematics and smectics will tend to align with their directors along the direction of an external magnetic field, while discotic liquid crystals will align with the director perpendicular to the field. Replacement of aromatic rings by saturated groups such as cyclohexyl, bicyclo-octyl or alkyl chains will reduce the anisotropy of the molecular core, so that liquid crystals based on the *trans-trans*-cyclohexylcyclohexyl core have a negative anisotropy due to the attached terminal groups. Some results for the magnetic susceptibilities of liquid crystals are given in Table 2.

2.4 Paramagnetic Liquid Crystals

Known paramagnetic liquid crystals are based on metal-containing mesogens, which have a variety of metal centres and co-ordination geometries [5, 6]. A requirement for paramagnetism is an unpaired spin, but to have an influence on the magnetic anisotropy, there must also be a significant **g**-tensor anisotropy. The effect of competi-

tion between diamagnetic and paramagnetic contributions to the susceptibility is illustrated by the behavior of salicyl-aldimine complexes of copper [7, 8]. These are formed from copper(II) having a d^9 electron configuration, which results in a square planar geometry around the metal centre. The **g**-tensor anisotropy is such that $\left(g_{nn}^e\right)^2 < \frac{1}{2}\left(\left(g_{ll}^e\right)^2 + \left(g_{mm}^e\right)^2\right)$ so that the paramagnetic contribution to the anisotropy is negative. For complexes with four benzene rings in the structure (Table 3, compounds 1 and 2), the paramagnetic term is larger than the diamagnetic term in the anisotropy, and so the complexes align with the major axis (*n*) perpendicular to the field direction: free rotation about the molecular long axis is assumed. Increasing the number of benzene rings to six (Table 3, compounds 3 and 4) causes the diamagnetic anisotropy to dominate, and the director aligns parallel to a magnetic field (see Fig. 2).

Electron paramagnetic resonance measurements on these liquid crystals give **g**-values of $g_{nn}^e = 2.053$ and $1/2\,(g_{ll}^e + g_{mm}^e) = 2.082$. By contrast the corresponding

Table 3. Structures, susceptibility anisotropies and alignment of salicylaldimine complexes of copper [7, 8].

X	Y	$\Delta\chi^{para}/10^{-9}$ m^3 mol^{-1}	$\Delta\chi^{dia}/10^{-9}$ m^3 mol^{-1}	Orientation to magnetic field
$C_7H_{15}O-$	⬡-OC$_{12}$H$_{25}$	−84.4	72.4	⊥
$C_7H_{15}O$-⬡-COO-	−C$_{12}$H$_{25}$	−91.5	79.5	⊥
$C_7H_{15}O-$	⬡-COO⬡-OC$_{12}$H$_{25}$	−74.7	117	∥
$C_7H_{15}O-$	⬡-⬡-OC$_{12}$H$_{25}$	−66.8	120	∥

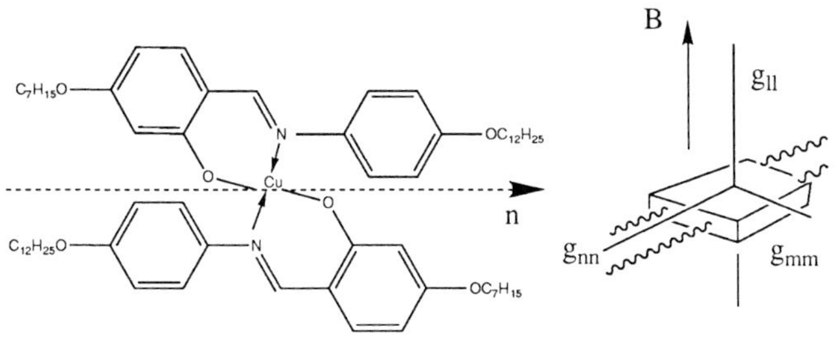

$\Delta\chi$ -ve, **n** perpendicular to field

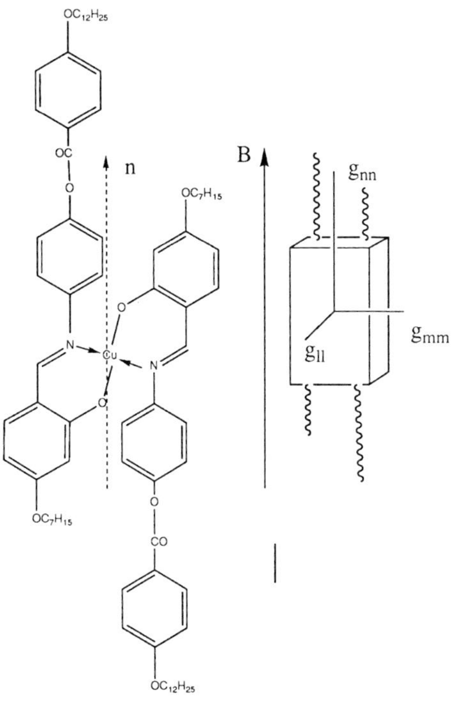

$\Delta\chi$ +ve **n** parallel to B

Figure 2. Alignment of salicyldimine complexes in a magnetic field.

vanadyl (VO) d^1 complexes have a reversed **g**-tensor anisotropy $g_{nn}^e = 1.987$ and $1/2(g_{ll}^e + g_{mm}^e) = 1.966$, and so these complexes always align with their molecular long axes along the magnetic field direction. Mesogenic paramagnetic salicylaldimine complexes of a number of rare earths have been reported showing SmA phases [9], and found to have large magnetic anisotropies. A similar result has been obtained [10] for a β-enaminoketone complex of dysprosium, but the corresponding gadolinium complex had a very small paramagnetic anisotropy.

2.5   Ferromagnetic Liquid Crystals

The possibility for a mesogenic material exhibiting ferromagnetism is at the present time remote. Organic ferromagnets have been prepared [11, 12], but they mostly have very low Curie temperatures, well below the melting points of the compounds. Since the origin of ferromagnetism is long range spin–spin interactions, it is unlikely that these will persist in a fluid liquid crystalline state, although there may be more chance of preparing metal-containing liquid crystal polymers having a potential for magnetic ordering [13]. A different approach to the preparation of ferromagnetic liquid crystals was proposed by Brochard and de Gennes [14] based on the dispersion of ferromagnetic particles in a liquid crystal host. As explained above, ferrofluids can be formed from dispersions of ferromagnetic materials that will form essentially single domain particles. Examples are colloidal suspensions of ferrite γ-Fe_2O_3, magnetite Fe_3O_4 or cobalt metal in either hydrocarbon or water-based solvents. Ferromagnetic particles may be imagined to couple with a liquid crystal host through interaction between the magnetic moment of the particle and the magnetic anisotropy of the surrounding liquid crystal [14]. Depending on the anisotropy of the liquid crystal the magnetisation of the particles should align parallel (positive $\Delta\chi$) or perpendicular (negative $\Delta\chi$) to the director.

Another mechanism for coupling the orientation of a magnetic particle to a liquid crystal is through elastic interactions. If the magnetic particles are anisotropic, then defining the director orientation at the surface of the particle with a suitable surfactant will cause a preferred alignment of the particle in a liquid crystal host. Chen and Amer [15]

succeeded in stabilizing a suspension of particles of length 0.35 µm and 0.04 µm diameter in MBBA. The particles were coated with a surfactant which defined the director orientation at the particle surface as perpendicular to the particle axis, and changes in the observed optical anisotropy in the presence of a magnetic field were consistent with the reorientation of the liquid crystal director perpendicular to the field.

Lyotropic liquid crystals doped with ferromagnetic particles have also been studied [16, 17], and the magnetic particles can be stabilised in either hydrophobic or hydrophilic regions. Changes in birefringence with magnetic fields have been observed, suggesting that the optical anisotropy of the liquid crystal has been coupled to the magnetic anisotropy of the dispersed particles. It is possible that such magnetic field effects in anisotropic ferrofluids may find application in the future.

2.6   Applications of Magnetic Properties

Since intermolecular forces scarcely affect the magnetic susceptibility, measurements of the magnetic anisotropy can provide a direct measure of the orientational order. Using Eq. (13), the anisotropy of susceptibility for a uniaxial liquid crystal phase formed from uniaxial mesogens can be written as:

$$\Delta\chi = N\Delta\kappa S \qquad (15)$$

where $\Delta\kappa$ contains contributions from diamagnetic, paramagnetic and temperature independent paramagnetic terms:

$$\Delta\kappa = \Delta\kappa^{dia} + \Delta\kappa^{TIP} \qquad (16)$$
$$+ \frac{N\mu_0\mu_B^2\, s(s+1)}{3k_B T}\left[\left(g_{nn}^e\right)^2 - \left(g_{ll}^e\right)^2\right]$$

$\Delta\kappa$ can be obtained from measurements on single crystals, and provided that accurate values are available for the density, the order parameter S can be obtained directly. An alternative way to determine S from measurements of the temperature dependence of the susceptibility is to fit values to a functional form for the variation of S with temperature. The simplest procedure known as the Haller extrapolation is described in the context of birefringence measurements in Sec. 3.2 of this chapter. The effects of local biaxial ordering on the measured susceptibility for cyanobiphenyls has been considered by Bunning, Crellin and Faber [2] using crystal data for biphenyl.

Magnetic properties have an importance in the NMR of liquid crystals [18,19], but the moments of the nuclear spins responsible for the NMR signal are far too small to make any contribution to magnetic susceptibilities. However, bulk susceptibility corrections to the NMR chemical shift of a standard immersed in the sample can be used to determine diamagnetic susceptibilities. The chemical shift of the standard is shifted to lower fields in a cylindrical sample due to the bulk magnetization, according to:

$$\sigma_{observed} = \sigma_{standard} - \frac{2\pi}{3}\chi_{ii} \qquad (17)$$

where χ_{ii} is the susceptibility component in the direction of the external magnetic field. Diamagnetic liquid crystals will align such that the smallest component of χ is along the magnetic field direction, and this causes a splitting in the NMR lines, which can be related to the order parameter. This technique is extremely useful for obtaining detailed information on the ordering of different segments of flexible molecules [18,19] and can also yield values for the local biaxial order parameters of molecules [18]. The method has been successfully applied to

both pure liquid crystals and to dopant molecules dissolved in liquid crystal hosts, which serve to orient the solute molecules. For these experiments the direction of alignment of the director with respect to the magnetic field is important, and since most liquid crystals have a positive susceptibility anisotropy, the director will align parallel to the magnetic field.

The standard method for measuring magnetic susceptibilities is to use a Faraday balance, which involves the measurement of the force on a sample in an inhomogeneous magnetic field [20]; other methods use a Gouy balance [3], or a SQUID magnetometer [21, 22]. All methods measure a single component of the susceptibility – the largest for paramagnetic samples, and the smallest for diamagnetic samples, assuming that the alignment of the sample liquid crystal is not constrained by other forces. In order to obtain the anisotropy, a second measurement is required, and this is usually taken as the mean susceptibility measured in the isotropic phase. Diamagnetic susceptibilities are independent of temperature, and it is reasonable to assume that the mean susceptibility in the liquid crystal phase is the same as in the isotropic phase: hence for positive materials, the susceptibility anisotropy is given by:

$$\Delta\chi = \frac{3}{2}(\chi_{\parallel} - \overline{\chi}) \qquad (18)$$

For liquid crystals having a negative susceptibility, the anisotropy is:

$$\Delta\chi = 3(\overline{\chi} - \chi_{\perp}) \qquad (19)$$

It is not possible to determine the sign of $\Delta\chi$ from magnetic measurements alone.

The ability of magnetic fields to control the alignment of liquid crystals is widely used, for example in X-ray structural studies of liquid crystals, or for optical measurements on aligned liquid crystal films.

An advantage of magnetic fields over electric fields for controlling alignment is that complications due to electrical conduction or electrohydrodynamic effects are not present. Competition between aligning fields has been used to obtain direct measurements of susceptibility anisotropies. The basis of the method can be understood from Eq. (5). A similar equation can be written for the free energy density of a liquid crystal in an electric field, such that:

$$\mathbf{g}^{e} = g_0 - \frac{1}{2}\varepsilon_0 E^2 \left(\varepsilon_\perp + \Delta\varepsilon \cos^2\theta\right) \qquad (20)$$

For balancing torques of the electric and magnetic fields on a liquid crystal:

$$\mu_0^{-1}\Delta\chi B^2 = \varepsilon_0 \Delta\varepsilon E^2 \qquad (21)$$

Thus by measuring the corresponding fields, and knowing the permittivity anisotropy, it is possible to determine $\Delta\chi$ directly [4].

2.7   References

[1] P. L. Sherrell, D. A. Crellin, *J. de Phys.* **1979**, *40*, C3-211.
[2] J. D. Bunning, D. A. Crellin, T. E. Faber, *Liq. Cryst.* **1986**, *1*, 37.
[3] J. W. Rohleder, R. W. Munn, *Magnetism and Optics of Molecular Crystals*, John Wiley, Chichester, **1992**, p. 27.
[4] H. Schad, G. Baur, G. Meier, *J. Chem. Phys.* **1979**, *70*, 2770.

[5] D. W. Bruce, *J. Chem. Soc. Dalton Trans.* **1993**, 2983.
[6] S. A. Hudson, P. M. Maitlis, *Chem. Rev.* **1993**, *93*, 861.
[7] J.-L. Serrano, P. Romero, M. Marcos, P. J. Alonso, *J. Chem. Soc., Chem. Commun.* **1990**, 859.
[8] I. Bikchantaev, Yu. Galyametdinov, A. Prosvirin, K. Griesar, E. A. Soto-Bustamente, W. Haase, *Liq. Cryst.* **1995**, *18*, 231.
[9] Yu. Galyametdinov, M. A. Athanassopoulou, K. Griesar, O. Kharitonova, E. A. Soto-Bustamente, L. Tinchurina, I. Ovchinnikov, W. Haase, *Chem. Mat.* **1996**, *8*, 922.
[10] I. Bikchantaev, Yu. Galyametdinov, O. Kharitonova, I. Ovchinnikov, D. W. Bruce, D. A. Dunmur, D. Guillon, B. Heinrich, *Liq. Cryst.* **1996**, *20*, 831.
[11] O. Kahn, Yu. Pie, Y. Journaux in *Inorganic Materials* (Ed.: D. W. Bruce, D. O'Hare), John Wiley, Chichester, **1992**, p. 61.
[12] D. Gatteschi, *Europhys. News* **1994**, *25*, 50.
[13] S. Takahashi, Y. Takai, H. Morimoto, K. Sonogashira, *J. Chem. Soc., Chem. Commun.* **1984**, 3.
[14] F. Brochard, P.-G. de Gennes, *J. de Phys.* **1970**, *31*, 691.
[15] S. H. Chen, N. M. Amer, *Phys. Rev. Lett.* **1983**, *51*, 2298.
[16] P. Fabre, C. Casagrande, M. Veyssie, V. Cabuil, R. Massart, *Phys. Rev. Lett.* **1990**, *64*, 539.
[17] C. Y. Matuo, F. A. Tourinho, A. M. Figueiredo-Neto, *J. Mag. Mag. Mat.* **1993**, *122*, 53.
[18] J. W. Emsley (Ed.), *Nuclear Magnetic Resonance of Liquid Crystals*, D. Reidel, Dordrecht, Netherlands, **1983**.
[19] R. Dong, *Nuclear Magnetic Resonance of Liquid Crystals*, Springer-Verlag, Berlin, **1994**.
[20] D. Jiles, *Introduction to Magnetism and Magnetic Materials*, Chapman & Hall, **1991**.
[21] J. S. Philo, W. M. Fairbank, *Rev. Sci. Instrum.* **1977**, *48*, 1529.
[22] C. Butzlaff, A. X. Trantwein, H. Winkler, *Meth. Enzym.* **1993**, *227*, 412.
[23] W. H. de Jeu, *Physical Properties of Liquid Crystalline Materials,* Gordon and Breach, New York, **1980**, p 31.

3   Optical Properties

David Dunmur and Kazuhisa Toriyama

The optical properties of liquid crystals determine their response to high frequency electromagnetic radiation, and encompass the properties of reflection, refraction, optical absorption, optical activity, nonlinear response (harmonic generation), optical waveguiding, and light scattering [1]. Most applications of thermotropic liquid crystals rely on their optical properties and how they respond to changes of the electric field, temperature or pressure. The optical properties can be described in terms of refractive indices, and anisotropic materials have up to three independent principal refractive indices defined by a refractive index ellipsoid.

Solution of Maxwell's equations for the propagation of a wave through an anisotropic medium gives three principal wave velocities for directions $i=1, 2, 3$ as:

$$v_i^2 = \left(\mu_{ii}^m\, \varepsilon_{ii}\right)^{-1} \tag{1}$$

where ε_{ii} and μ_{ii}^m are the principal components of the electric permittivity and magnetic permeability tensors, which are assumed to be diagonal in the same frame of axes. For other than ferromagnetic materials μ_{ii}^m is very close to unity, and comparing the velocities of the wave in a vacuum with the velocities of the wave in an anisotropic medium gives the principal refractive indices as:

$$n_i = (\varepsilon_{ii}/\varepsilon_0)^{1/2} \tag{2}$$

In fact, two waves of different velocity (having the same wave normal but orthogonal polarizations) can propagate through an optically anisotropic medium along two different directions. This results in the appearance of a double image of an object viewed through anisotropic crystals, and is termed double refraction. These two rays have different refractive indices: the *ordinary* ray propagates along the wave normal, and its direction obeys the normal Snell's law of refraction so that $n_o = \sin i/\sin r$, while for the other *extraordinary* ray, the ray direction and wave normal are not parallel. The two refractive indices for a particular wave normal can be obtained from the refractive index indicatrix [2]. This ellipsoid is defined by the equation:

$$\frac{x^2}{n_1^2}+\frac{y^2}{n_2^2}+\frac{z^2}{n_3^2}=1 \tag{3}$$

where n_1, n_2, and n_3 are termed the three principal refractive indices, and the directions x, y, z are the principal axes of the electric permittivity tensor (Fig. 1).

For any direction in a crystal (OP) in Fig. 1, the refractive indices of the two wave

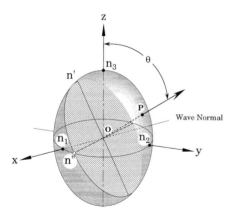

Figure 1. The optical indicatrix, where the principal refractive indices are labelled as n_1, n_2 and n_3. Refractive indices corresponding to the wave front normal OP are shown as n' and n''.

fronts normal to OP that can propagate are given by the semi-major and semi-minor axes of the ellipse perpendicular to OP. In the case of an indicatrix of revolution, which would represent a uniaxial liquid crystal, $n_1=n_2$ which is the ordinary refractive index n_o, and for light propagating along the z-direction (i.e. $\theta=0$), which is the symmetry axis of the indicatrix, the ordinary and extraordinary rays are coincident; the largest refractive index (n_3 in the figure) is called n_e. Materials having two equal principal refractive indices are referred to as uniaxial, and the unique direction z is the optic axis. The special feature of the optic axis is that light of any polarization travels along this axis without any change in its polarization (i.e. the material responds as an optically isotropic medium. The difference between the two independent principal refractive indices $\Delta n=n_3-n_1$ is termed the birefringence. For light propagating along a direction which is not the optic axis, the ordinary and extraordinary rays are not coincident and they travel with different velocities corresponding to different refractive

indices. For any direction, one ray has a refractive index of n_o, but the refractive index of the extraordinary ray depends on direction, such that:

$$n_e(\theta)^2 = \left(\frac{\cos^2\theta}{n_1^2} + \frac{\sin^2\theta}{n_3^2} \right)^{-1} \tag{4}$$

where the direction makes an angle θ with the z-axis. If $n_1 \neq n_2 \neq n_3$, there are two optic axes for which the perpendicular cross-section of the indicatrix is circular, such materials are biaxial, and there are two directions along which the material appears to be optically isotropic.

Many of the interesting properties of liquid crystals are a result of chirality or handedness, which is manifest in optical properties by optical activity. For isotropic materials or anisotropic materials viewed along their optic axes, optical activity causes the plane of polarization of propagating light to be rotated by an angle ϕ. This can be expressed in terms of a difference between refractive indices for left (n_l) and right (n_r) circularly polarized light:

$$\phi = \frac{\pi d}{\lambda} (n_l - n_r) \tag{5}$$

where d is the path length and λ is the vacuum wavelength of the light. For anisotropic materials including liquid crystals, the optical activity interacts with the linear birefringence, and the two propagating waves which correspond to a particular wave normal are elliptically polarized, the axes of the ellipses being perpendicular.

For an absorbing medium the refractive index can be represented as a complex quantity consisting of a real part (n_i') and an imaginary part (k_i):

$$n_i = n_i' - i k_i \tag{6}$$

and using the relation with the permittivity, we have:

$$n_i^2 = \varepsilon_o^{-1} \left[\varepsilon_{ii}^{(\text{real})} - i\,\varepsilon_{ii}^{(\text{imaginary})} \right]$$

$$n_i'^2 - k_i^2 = \varepsilon_o^{-1}\,\varepsilon_{ii}^{(\text{real})}$$

$$2n_i'k_i = \varepsilon_o^{-1}\,\varepsilon_{ii}^{(\text{imaginary})} \qquad (7)$$

The real part of the refractive index determines the speed of light through the medium, while the imaginary part measures the attenuation of its intensity, so k_{ii} is defined as the absorption coefficient. If the absorption coefficients for light plane-polarized along different directions are different, then the material is said to exhibit linear dichroism. Nonchiral materials can only be linearly dichroic, but chiral materials can also show circular dichroism, which arises from a difference between the imaginary parts (absorption coefficients) of the refractive indices for left and right circularly polarized light.

3.1 Symmetry of Liquid Crystal Phases and the Optical Indicatrix

The symmetry of liquid crystalline phases can be categorized in terms of their orientational and translational degrees of freedom. Thus nematic, smectic and columnar phase types have respectively three, two, and one degrees of translational freedom, and within each type there can be different phases depending on the orientational or point group symmetry. Their optics (uniaxial, biaxial, optically active) are determined by the point group symmetries, which are listed in Table 1 for common liquid crystal phases; the optical symmetries of variants of these phases can usually be established directly from their structures.

It will be seen that most nematics and orthogonal smectic and columnar phases are uniaxial, with two equal principal refractive indices, while the tilted smectic and columnar phases are biaxial. The optical symmetries of liquid crystal phases can be determined by conoscopic observation of aligned thin films [3] but the technique is difficult, and made more so by the small biaxiality of tilted liquid crystal phases. Principal refractive indices of liquid crystals range from 1.4 to 1.9, and uniaxial birefringences $\Delta n = n_e - n_o$, can be between 0.02 and 0.4; negative birefringences are associated with discotic versions of liquid crystal phases (e.g. discotic nematic or columnar phases). For biaxial liquid crystals, all three principal refractive indices are different, but usually one (n_3) is significantly greater (or less) than the other two, in which case the uniaxial birefringence can be defined as $\Delta n = n_3 - \frac{1}{2}(n_2 + n_1)$ and the biaxiality is $\partial n = n_2 - n_1$. The biaxiality of liquid crystals is small (≈ 0.01) [4, 5], which is a consequence of the small degree of structural biaxiality of these phases.

3.2 Molecular Theory of Refractive Indices

The characteristics of optical and electro-optical liquid crystal devices are determined by the refractive indices of the materials, thus an understanding of the relationship between refractive indices and molecular properties is necessary for the design of improved liquid crystal materials. In developing a molecular theory for any electrical or optical property, the problem of the internal or local electric field has to be addressed. This arises because the field experienced by a molecule in a condensed phase differs from that applied across the macroscopic sample. The internal field has

Table 1. Symmetries of common liquid crystal phase types: only those phases with well-established phase structures are included. Crystal smectic phases, including the cubic D phase have been omitted, as have recently discovered twist grain boundary phases, and structurally modulated variants of smectic phases.

Liquid crystal phase	Point group and translational degrees of freedom – $T(n)$	Optical symmetry; uniaxial – u(+) $n_3 > n_2 = n_1$ u(−) $n_3 < n_2 = n_1$; biaxial (b); helicoidal (h)
Achiral calamitic, micellar nematic N or N_u	$D_{\infty h} \times T(3)$	u(+)
Achiral nematic discotic (N_D), columnar nematic (N_{Col})	$D_{\infty h} \times T(3)$	u(−)
Chiral nematic (cholesteric) N^*	$D_{\infty} \times T(3)$	b, h, locally biaxial but globally u(−)
Biaxial nematic (N_b) (only micellar confirmed)	$D_{2h} \times T(3)$	b
Achiral calamitic orthogonal smectic or lamellar phases (SmA)	$D_{\infty h} \times T(2)$	u(+)
Achiral tilted smectic phase (SmC)	$C_{2h} \times T(2)$	b
Chiral tilted smectic phase (SmC*)	$C_2 \times T(2)$	b, h
Orthogonal and lamellar hexatic phase (SmB)	$D_{6h} \times T(1)$ locally, $D_{6h} \times T(2)$ globally	u(+)
Tilted and lamellar hexatic phases (SmF and SmI)	$C_{2h} \times T$ (1 or 2 see above)	b
Chiral tilted and lamellar hexatic phases (SmF* and SmI*)	$C_2 \times T$ (1 or 2 see above)	b, h
Discotic columnar: hexagonal order of columns, ordered or disordered within columns (Col_{ho} or $_{hd}$)	$D_{6h} \times T(1)$	u(−)
Rectangular array of columns (Col_{ro} or $_{rd}$)	$D_{2h} \times T(1)$	b
Molecules tilted within columns (Col_{to} or $_{td}$)	$C_{2h} \times T(1)$	b

a special significance for anisotropic materials such as liquid crystals. For isotropic media the Lorentz local field is used

$$\left(E_{loc} = \frac{(\varepsilon + 2)}{3} E \right)$$ where ε is the mean permittivity, and this results in the Lorenz–Lorentz expression relating the refractive index to the mean molecular polarizability:

$$\frac{n^2 - 1}{n^2 + 2} = \frac{N\alpha}{3\varepsilon_o} \qquad (8)$$

where N is the number density and α is the mean polarizability. For low density gases, this will be the molecular polarizability, but in condensed fluids it is an effective or

dressed property, which takes account of short range intermolecular interactions. In anisotropic liquid crystals it is reasonable to adopt the isotropic model for the internal field, and their principal refractive indices can be written as:

$$\frac{n_i^2 - 1}{n^2 + 2} = \frac{N\langle \alpha_{ii} \rangle}{3\varepsilon_o} \qquad (9)$$

The polarizability component $\langle \alpha_{ii} \rangle$ is the average value along the direction of the principal refractive index n_i, and $n^2 = \frac{1}{3}\left(n_1^2 + n_2^2 + n_3^2\right)$ is a mean refractive index.

Using the general results for the transformation of second rank tensor properties, these polarizabilities can be expressed in terms of molecular components and orientational order parameters. For liquid crystal phases of uniaxial symmetry the optic axis coincides with the average alignment direction of the molecules, termed the director, and from Eq. (45) of Sec. 1 of this chapter we obtain for components parallel ($\langle \alpha_\parallel \rangle$) and perpendicular ($\langle \alpha_\perp \rangle$) to the director:

$$\langle \alpha_\parallel \rangle = \alpha + \frac{2}{3} \left\{ \begin{matrix} S\left[\alpha_{nn} - \frac{1}{2}(\alpha_{ll} + \alpha_{mm}) \right] + \\ \frac{1}{2} D(\alpha_{ll} - \alpha_{mm}) \end{matrix} \right\}$$

$$\langle \alpha_\perp \rangle = \alpha + \frac{1}{3} \left\{ \begin{matrix} S\left[\alpha_{nn} - \frac{1}{2}(\alpha_{ll} + \alpha_{mm}) \right] + \\ \frac{1}{2} D(\alpha_{ll} - \alpha_{mm}) \end{matrix} \right\} \quad (10)$$

The principal axes of the molecular polarizability tensor are labelled l, m, n, as shown in Fig. 2. Thus the importance of order parameters in determining the anisotropy of optical properties is clearly demonstrated. Both order parameters S and D contribute to the anisotropy of second rank tensor properties even in uniaxial liquid crystals, but they cannot be separated from a single measurement of the birefringence.

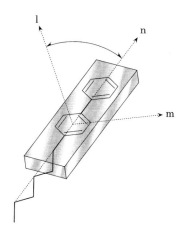

Figure 2. The principal axes of a molecular polarizability tensor.

If the liquid crystal phase is biaxial, as with SmC phases, then any second rank tensor property has three independent principal components and the average polarizabilities corresponding to the three refractive indices can be expressed in terms of the orientational order parameters introduced for biaxial phases:

$$\langle \alpha_{33} \rangle = \\ \alpha + \frac{2}{3} \left\{ \begin{matrix} S\left[\alpha_{nn} - \frac{1}{2}(\alpha_{ll} + \alpha_{mm}) \right] + \\ \frac{1}{2} D(\alpha_{ll} - \alpha_{mm}) \end{matrix} \right\}$$

$$\langle \alpha_{22} \rangle = \\ \alpha - \frac{1}{3} \left\{ \begin{matrix} (S+P)\left[\alpha_{nn} - \frac{1}{2}(\alpha_{ll} + \alpha_{mm}) \right] + \\ \frac{1}{2}(D+C)(\alpha_{ll} - \alpha_{mm}) \end{matrix} \right\}$$

$$\langle \alpha_{11} \rangle = \\ \alpha - \frac{1}{3} \left\{ \begin{matrix} (S-P)\left[\alpha_{nn} - \frac{1}{2}(\alpha_{ll} + \alpha_{mm}) \right] + \\ \frac{1}{2}(D-C)(\alpha_{ll} - \alpha_{mm}) \end{matrix} \right\} \quad (11)$$

Refractive indices of liquid crystals may be measured by a variety of methods, but all require a well-aligned thin film. The simplest method is to use a refractometer with a suitably coated prism surface to give an aligned sample, and use of a polarizer permits the separation of the two refracted rays. This works well for N and SmA phases, but it is not usually possible to align both directors for SmC phases; there are also limitations of temperature. Wedge cells have also been used to obtain refractive indices of liquid crystals, relying on external magnetic fields for alignment [6, 7]. Guided-mode methods can be used to obtain refractive indices [8, 9], while the Z-scan method [10, 11] can be used to obtain information on the field dependence of refractives indices. It is easier to measure birefringence changes of liquid crystals directly using interferometric methods, and high precision can be achieved [12–14]. The wavelength dependence of refractive indices can be obtained from so-called channelled spectra, which are interference bands observed from thin films in spectrophotometers [15–17].

Not only are refractive indices needed to optimise device and materials design, but also they provide a simple route to order parameters. Extrapolation methods can be used to determine the uniaxial order parameter S, if contributions from molecular biaxiality (D) are ignored [18–20]. For molecules having axial symmetry ($\alpha_{ll} = \alpha_{mm}$) Eq. (9) can be written as:

$$\frac{n_{\parallel}^2 - n_{\perp}^2}{n^2 - 1} = \frac{S\Lambda\alpha}{\alpha} \qquad (12)$$

where $\Delta\alpha = (\alpha_{nn} - \alpha_{ll})$. If it is assumed that the order parameter can be written in a simple form as:

$$S = \left(1 - \frac{T}{T_{\text{N-I}}}\right)^b \qquad (13)$$

where $T/T_{\text{N-I}}$ is a reduced temperature then a plot of $\log\left(\frac{n_{\parallel}^2 - n_{\perp}^2}{n^2 - 1}\right)$ against $\log\left(1 - \frac{T}{T_{\text{N-I}}}\right)$ will give a straight line of intercept $\frac{\Delta\alpha}{\alpha}$, using which the order parameters S can be calculated from Eq. (12). It is also possible to

Table 2. Molecular polarizabilities and refractive indices for a selection of mesogens.

Mesogen	Polarizability		Refractive index; reduced $T_R = 0.95$		λ (nm)	Ref.
	$\Delta\alpha/10^{-40}$ $C^2 J^{-1} m^2$	$\alpha/10^{-40}$ $C^2 J^{-1} m^2$	Δn	$\bar{n}$		
C₅H₁₁⟨⟩⟨⟩-CN	19.4	37.5	0.194 / 0.194	1.589 / 1.595	633 / 589	(a, f)
C₅H₁₁⟨⟩⟨⟩-CN	16.0	36.2	0.125	1.533	589	(b, f)
C₇H₁₅⟨⟩⟨⟩-CN	11.1	40.4	0.045 [a]	1.471 [a]	589	(b, c)
H₃CO⟨⟩-C=N-⟨⟩-C₄H₉	31.2	41.1	0.21 / 0.22	1.61 / 1.62	633 / 589	(g, h)
C₈H₁₇⟨⟩-C=C-⟨⟩-OC₂H₅	33.4	52.8	0.172 / 0.179	1.523 / 1.528	633 / 589	(d)
C₅H₁₁⟨⟩-C-O-⟨⟩-OC₆H₁₃	26.4	43.9	0.121	1.530	589	(e)
C₅H₁₁⟨⟩-C-S-⟨⟩-CN	25.3	37.0	0.197	1.592	589	(e)

[a] $T_R = 0.965$.
(a) R. G. Horn, *J. de Phys.* **1978**, *39*, 105; or Ref. [19].
(b) M. M. M. Abdoh, S. N. C. Shivaprakash, J. S. Prasad, *J. Chem. Phys.* **1982**, *77*, 2570.
(c) I. H. Ibrahim, W. Haase, *Mol. Cryst. Liq. Cryst.* **1981**, *66*, 189.
(d) Reference [18].
(e) I. H. Ibrahim, W. Haase, *J. de Phys.* **1979**, *40*, 191.
(f) D. A. Dunmur, A. E. Tomes, *Mol. Cryst. Liq. Cryst.* **1983**, *97*, 241.
(g) M. Mitra, B. Majumdar, R. Paul, S. Paul, *Mol. Cryst. Liq. Cryst.* **1990**, *180B*, 187.
(h) I. Haller, H. A. Huggins, M. J. Freiser, *Mol. Cryst. Liq. Cryst.* **1972**, *16*, 53.

fit refractive index data to a mean field expression for the order parameter using a numerical procedure [6]. The refractive indices also provide a possible route to determining molecular polarizabilities, and design of molecules with specifically high or low polarizability is important for particular applications.

3.3 Optical Absorption and Linear Dichroism

The attenuation of the intensity of a beam of light on passing through an absorbing medium can be measured by the absorption coefficient, which is the imaginary part of the refractive index. A more usual measure is the optical absorbance (A) or molar extinction coefficient ε, which is defined in terms of the Beer–Lambert law as:

$$\log_{10}(I_0/I) = \varepsilon C l \tag{14}$$

where C is the concentration in moles per m^3, l is the optical path length in the sample, and I_0/I is the ratio of the incident intensity I_0 to the transmitted intensity I. The extinction coefficient is a function of the frequency of the light, and integrating over the absorption band $i(\omega_i \pm \Gamma_{on}/2)$, centered on the frequency of maximum absorption ω_i gives the optical absorbance as:

$$A = \frac{\log_e 10}{2\pi} \int_{-\Gamma_{on}/2}^{\Gamma_{on}/2} \varepsilon(\omega) \, d\omega \tag{15}$$

Finally for a narrow absorption, the absorption coefficient, k, can be related to the molar extinction coefficient by:

$$k(\omega) = \frac{6c\varepsilon(\omega) C \log_e 10}{2\pi n \omega_i} \tag{16}$$

where c is the velocity of light and n is the real part of the refractive index. The real and imaginary parts of the refractive index are related through a Kramers–Kronig relation, such that:

$$k(\omega) = \frac{2}{\pi} \int_0^\infty \frac{n(\omega) - 1}{\omega - \omega_i} \, d\omega \tag{17}$$

Equations (14)–(17) apply to isotropic media. In an orientationally ordered material the extinction coefficient becomes dependent on the angle between the alignment axis and the polarization direction of the incident light, and has the characteristics of a second rank tensor. At a microscopic level, the optical absorption depends on the angle between the molecular transition dipole moment μ_i for the particular absorption band, and the electric field of the light wave. Restricting attention to uniaxial systems, an effective order parameter (S_{op}) for optical absorption can be defined as:

$$S = \frac{A_{\parallel} - A_{\perp}}{A_{\parallel} + 2A_{\perp}} \tag{18}$$

in which $A_{\parallel}$ and $A_{\perp}$ are the extinction coefficients for light polarized parallel or perpendicular to the director; the difference $A_{\parallel} - A_{\perp}$ is known as the linear dichroism. These extinction coefficients can be related to the transition dipole moments using the general result Eq. (45) of Sec. 1 of this chapter, such that:

$$A_{\parallel} = A_0 + B\left\{ \frac{2}{3} S\left[(\mu_i)_n^2 - \frac{1}{2}\left((\mu_i)_l^2 + (\mu_i)_m^2\right)\right] \right.$$
$$\left. + \frac{1}{3} D\left((\mu_i)_l^2 - (\mu_i)_m^2\right)\right\} \tag{19}$$

$$A_{\perp} = A_0 - B\left\{ \frac{1}{3} S\left[(\mu_i)_n^2 - \frac{1}{2}\left((\mu_i)_l^2 + (\mu_i)_m^2\right)\right] \right.$$
$$\left. + \frac{1}{6} D\left((\mu_i)_l^2 - (\mu_i)_m^2\right)\right\} \tag{20}$$

where $A_0 = \frac{1}{3}(A_{\parallel} + 2A_{\perp})$ is the mean optical absorbance or the optical absorbance of the isotropic fluid. The factor B contains various fundamental constants, and μ_{il}, μ_{im} and μ_{in} are components of the transition moment along the principal axes of the molecule. Using the angles defined in Fig. 3 a simple manipulation of Eqs. (18)–(20) leads to the result [21]:

$$S_{op} = \frac{1}{2}\left(3\cos^2\beta - 1\right)S_{\mu}$$
$$- \frac{1}{2}(\sin^2\beta \cos 2\alpha) D_{\mu} \qquad (21)$$

where S_{μ} and D_{μ} are order parameters for the transition moment. A more complex result can be derived for biaxial samples using Eq. (47) of Sec. 1 of this chapter. S_{op} defined above is the order parameter for an optically absorbing mesogen; often the chromophore in a liquid crystal is not the mesogen but a solute dye molecule, which may or may not be mesogenic itself. There can still be dichroism because the liquid crystal host orders the dye molecule, but the order parameter of the dye may be substantially different from that of the host. In a mixture, the order parameters for the chromophore can in principle be related to the order parameters of the host material using the mean field theory of mixtures [23, 24].

It is sometimes useful to express the dichroism in terms of the dichroic ratio

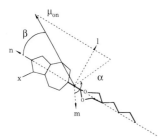

Figure 3. Orientation of the transition moment μ_{on} with respect to molecular axes for a dichroic mesogen [22].

$R = A_{\parallel}/A_{\perp}$. Assuming that there is no local biaxial order, the order parameter can be written in terms of R and the angle between the transition dipole and the ordered axis of the absorbing molecule:

$$R = \frac{1 + S\left(3\cos^2\beta - 1\right)}{1 - \frac{1}{2}S\left(3\cos^2\beta - 1\right)} \qquad (22)$$

For the special cases where the transition dipole is parallel or perpendicular to the molecular axis, the relationships between the dichroic ratio and the order parameter reduce to simple forms:

$$R(\beta = 0°) = \frac{1 + 2S}{1 - S}$$

$$R(\beta = 90°) = \frac{1 - S}{1 + \frac{1}{2}S} \qquad (23)$$

A particularly useful aspect of dichroic measurements is the chance to probe orientational order using more than one electronic transition in a molecule. Thus optical order parameters can be determined for different absorption bands, and if the transition moment directions are known, it is possible to determine both order parameters S and D. If it is assumed that there is a relationship between the uniaxial and biaxial order parameters, as given in Fig. 5 of Sec. 1 in this chapter, then it is possible to obtain both order parameters from the polarized spectra from a single absorption band [25]. This method has been applied [26] to the determination of order parameters of rigid aromatic probes, such as azulene, phenanthrene, and anthracene and related compounds. Dichroism measurements on impurity molecules in liquid crystal solvents have also been used [27, 28] to study intermolecular interactions, and their influence on electronic absorption bands. Polarization effects of the type described above for simple optical absorption spectroscopy can be observed and interpreted in a similar manner for many other types of spectroscopy such as Raman scattering [29] and resonance Raman scattering [30].

The Kramers–Kronig relationship between the real and imaginary parts of the refractive index shows that materials having a strong electronic absorption will tend to have a high refractive index. Conjugated mesogens or polarizable mesogens will therefore have relatively large refractive indices, and the birefringence will be determined by the polarization of the electronic absorptions. Some typical values for refractive indices of a range of different liquid crystals are given in Table 2. Materials with electronic absorptions in the UV at wavelengths less than 200 nm such as substituted bicyclohexanes will have small refractive indices and usually small birefringences. Changes in refractive indices with wavelength are also determined by the electronic absorptions for particular mesogens, and a polarized UV/visible spectrum for a standard liquid crystal is illustrated in Fig. 4. The dispersion in the corresponding refractive indices can be readily obtained from the following equations based on Drude's theory of optical dispersion:

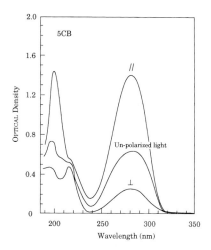

Figure 4. UV/visible spectrum of a mesogen (5CB).

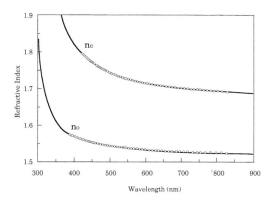

Figure 5. Fitted dispersion of refractive indices for 5CB.

$$n_e = 1 + n_{0e} + b_{1e}\left(\frac{1}{\lambda_1^2} - \frac{1}{\lambda^2}\right)^{-1} +$$

$$b_{2e}\left(\frac{1}{\lambda_2^2} - \frac{1}{\lambda^2}\right)^{-1} + \ldots$$

$$n_o = 1 + n_{0o} + b_{1o}\left(\frac{1}{\lambda_1^2} - \frac{1}{\lambda^2}\right)^{-1} +$$

$$b_{2o}\left(\frac{1}{\lambda_2^2} - \frac{1}{\lambda^2}\right)^{-1} + \ldots \tag{24}$$

Measured refractive indices for 5CB fitted to these equations are given in Fig. 5 [31].

3.4   Refractive Indices and Liquid Crystal Phase Structure

In Eqs. (9)–(11) the electronic polarizability is independent of temperature, so the temperature dependence of refractive indices is determined primarily by the order parameter, and to a lesser extent by changes in the density: the latter may be important at phase transitions. The variation of refractive indices of *n*-pentyloxyphenyl *trans*-4-noctylcyclohexanoate with temperature is given in Fig. 6, where the effect of phase changes is clearly seen. The changes mostly reflect changes in the order parameter at the transitions.

Thin films of oriented liquid crystals can act as optical wave guides, and examination of the eigenmodes of thin liquid crystal films can be used to obtain values for the real and imaginary parts of the refractive indices, as well as giving information on the director configuration in thin films [32]. The liquid crystal film is contained between two metallized (silver) reflecting surfaces, and the reflectivity is measured as a function of the angle and polarization of incident monochromatic light. The intensities of the reflected beam for light polarized in the plane of incidence (p) and perpendicular to the plane of incidence (s) are measured, and can be fitted to a model for the refractive indices, film thickness and director configuration. An example of experimental and fitted results is shown in Fig. 7 for the SmA phase of a commercial liquid crystal mixture Merck (UK) SCE3 [33].

A significant advance provided by guided mode experiments is the ability to meas-

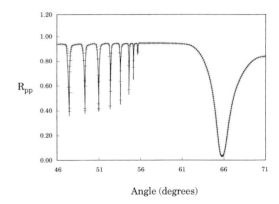

Figure 7. Reflectivity as a function of angle with fitted function for SCE3 at 72.8 °C (SmA); fitting parameters $\varepsilon_\parallel = 2.760$ and $\varepsilon_\perp = 2.208$.

ure not only the optical parameters of a liquid crystal film, but also the director configuration in complex geometries, such as the chevron structure in a SmC phase (see Fig. 8). The analysis of the optical response of complex liquid crystal structures is most conveniently achieved using the methods of matrix optics. There are a number of variants which can be used, depending on the particular problem, but the basic method is to represent an optical element as a matrix, which acts on the incident light, represented as a column vector, to give a resultant vector characteristic of the transmitted light. The approach is particularly suited to the geometries encountered in liquid crystal systems, since a complex optical structure can be split up into a series of elements each having its own characteristic matrix. Within each element it is assumed that the director is uniform, so the optical properties can be simply described in terms of principal refractive indices, the absorption coefficients for absorbing materials and the orientation of the optical indicatrix. The resultant response of the complex structure is then given by successive multiplication of the matrices for each element. The simplest method is that due to Jones, where the light wave

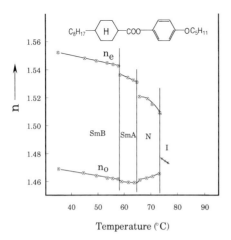

Figure 6. Refractive indices against temperature, including phase transitions [31]

is represented as a 2×1 column vector consisting of the electric field components E_x and E_y in a plane perpendicular to the propagation direction z. Thus the transmitted wave (E_{tx}, E_{ty}) through a birefringent element of thickness d and principal refractive indices n_e and n_o making an angle of α with the x axis can be written as:

$$\begin{bmatrix} E_{tx} \\ E_{ty} \end{bmatrix} = \begin{bmatrix} \cos\alpha & -\sin\alpha \\ \sin\alpha & \cos\alpha \end{bmatrix}$$
$$\begin{bmatrix} e^{-2\pi i n_e d/\lambda} & 0 \\ 0 & e^{-2\pi i n_o d/\lambda} \end{bmatrix}$$
$$\begin{bmatrix} \cos\alpha & \sin\alpha \\ -\sin\alpha & \cos\alpha \end{bmatrix} \begin{bmatrix} E_{ix} \\ E_{iy} \end{bmatrix}$$

or

$$\mathbf{E}_t = \mathbf{R}^{\text{trans}} \, \mathbf{BRE}_i \tag{25}$$

where the incident wave is represented by E_{ix}, E_{iy}, $\mathbf{R}$ is the rotation matrix and the matrix $\mathbf{B}$ represents the birefringent element. Extra elements such as polarizers and other birefringent elements can be included as appropriate matrix multipliers, and the transmitted light intensity is given by $I = E_{tx}^2 + E_{ty}^2$.

As an example of the application of the Jones matrices, the director configuration in a thin liquid crystal film of a SmC material has been investigated [34] by measuring the transmission of polarized light as a function of wavelength. Knowing the refractive indices of the material, the experimental results can be fitted to a model for the director configuration, as illustrated in the Fig. 8.

Another analytical method which had been extensively applied to liquid crystal structures was developed by Berreman [35]. It is based on a 4×4 matrix representation of optical elements [36], where the incident and transmitted light are described by a 4×1 column vector consisting of components of both the electric and magnetic field associated with the electromagnetic wave.

A widely used technique to study the properties of thin films is ellipsometry, and it has been used to investigate the structure of free-standing films of smectic liquid crystals consisting of only a few layers [37]. The method involves measuring the polarization characteristics of a transmitted or reflected beam of monochromatic light for different angles of incidence. Writing the phase difference between the s- and p-polarizations of the transmitted beam as $\Delta = \phi_p - \phi_s$, the value of Δ will depend on the integrated optical path difference across the smectic layers for light of the two polarizations. The technique has been used to probe the structure of ferroelectric and antiferro-

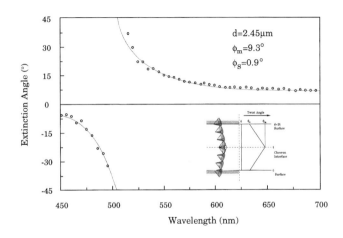

Figure 8. Fitted transmission against wavelength for a chevron structure [34].

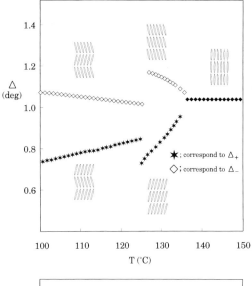

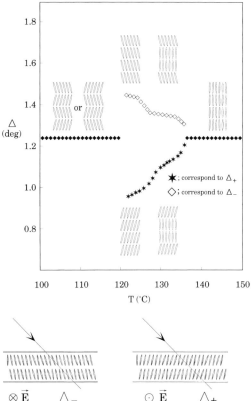

Figure 9. Measured optical path difference for three and four layer smectic structures of antiferroelectric, ferrielectric and ferroelectric SmC phases [37]. Bottom figure indicates angle of incidence for Δ_+ and Δ_- measurements.

electric smectic liquid crystals. For smectic films with alternating tilt directions, as in antiferroelectric or ferrielectric chiral SmC phases, the optical path difference will differ for odd or even numbers of layers. This is illustrated in Fig. 9 for a 3 and 4 layer film of MHPOBC, and the observed values for Δ measured in the antiferroelectric phase for 2 angles of incidence are consistent with the proposed structure for antiferroelectric phases.

3.5   Optics of Helicoidal Liquid Crystal Structures

In simple fluids of chiral molecules (molecules having structures such that mirror images are not superposable) the molecular chirality can be identified through the associated optical activity (i.e. the rotation of the plane of incident plane polarized light). If the light is of a frequency that corresponds with an electronic or vibrational absorption, then the fluid can exhibit differential absorption for left and right circularly polarized light, and the transmitted light is elliptically polarized. Optical activity can also be observed in differential scattering of left or right circularly polarized light. Chiral liquid crystals as well as exhibiting optical activity have a number of characteristic chiral properties, the most important of which is a tendency for the phases to develop helicoidal structures. Thus optical effects due to chirality in liquid crystals can be due to molecular chirality or a consequence of the helicoidal structure. This may seem to be an unnecessary distinction, since the helicoidal structures of chiral liquid crystal phases are a result of chiral interactions between molecules. However there is a difference when it comes to calculating the optical response of chiral liquid crystals, which can usually be modelled by twisted layers of linearly birefringent material; the intrinsic molecular optical activity of a liquid crystal is normally neglected in modelling the optical properties of chiral liquid crystals. For example the specific rotation measured in isotropic solution of a typical liquid crystal material such as CE6 $(C_{10}H_{21}OC_6H_4COOC_6H_4CH_2CH(CH_3)C_2H_5)$ is $3\,°cm^{-1}$, while the optical rotation of an aligned film of the chiral nematic phase of such a material, which arises primarily from the helicoidal structure, is around $10^4\,°cm^{-1}$.

One optical feature of helicoidal structures is the ability to rotate the plane of incident polarized light. Since most of the characteristic optical properties of chiral liquid crystals result from the helicoidal structure, it is necessary to understand the origin of the chiral interactions responsible for the twisted structures. The continuum theory of liquid crystals is based on the Frank–Oseen approach to curvature elasticity in anisotropic fluids. It is assumed that the free energy is a quadratic function of curvature elastic strain, and for positive elastic constants the equilibrium state in the absence of surface or external forces is one of zero deformation with a uniform, parallel director. If a term linear in the twist strain is permitted, then spontaneously twisted structures can result, characterized by a pitch p, or wave-vector $\mathbf{q}=2\pi p^{-1}\mathbf{i}$, where $\mathbf{i}$ is the axis of the helicoidal structure. For the simplest case of a nematic, the twist elastic free energy density can be written as:

$$g = -k_2(\mathbf{n}\cdot\nabla\times\mathbf{n}) + k_{22}(\mathbf{n}\cdot\nabla\times\mathbf{n})^2 \quad (26)$$

and the pitch of the corresponding chiral nematic is given by $p^{-1}=k_2/2\pi k_{22}$. The optical properties of such twisted structures have been determined using a model of twisted layers of linearly birefringent material, which predicts a variety of optical response depending on the value of $p\,\Delta n$ in

comparison with the wavelength λ. For long pitch materials such that $p\,\Delta n \gg 1$, the structure behaves as a rotator, and linear polarized modes rotate with the helix: this is the mode utilized in twisted nematic displays. For shorter pitches the optical response is more complicated, but analytical results can be obtained for normal incidence, leading to the de Vries equation [38] for the optical rotation per unit length $\rho = \phi/d$:

$$\rho = -\frac{\pi}{4p}\left(\frac{n_e^2 - n_o^2}{n_e^2 + n_o^2}\right)^2\left(\frac{\lambda_o}{\lambda}\right)^2\left(1-\left(\frac{\lambda}{\lambda_o}\right)^2\right)^{-1} \quad (27)$$

This equation predicts that the optical rotation diverges at a critical wavelength $\lambda_o=np$. The variation of ρ with wavelength is illustrated schematically in Fig. 10.

However the de Vries equation is not valid in the region of λ_o, which corresponds to a total reflection of circularly polarized light having the same sense as the helical pitch. This is often referred to as Bragg reflection, by analogy with X-ray diffraction, but only first order reflections are allowed for nor-

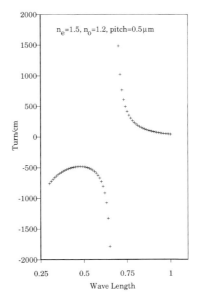

Figure 10. Variation of ρ with $\lambda/\mu m$ according to Eqn. 27.

mal incidence. The band width for total reflection centered on a wavelength of λ_o is $\Delta\lambda = p\,\Delta n$, and the sign of the optical rotation reverses on crossing the band at λ_o. It is observed that the wavelength of reflected light varies as a function of the angle of incidence, but an analytical solution of the optic response for this situation is not possible. Away from normal incidence, all orders of reflection are permissible [39], but an approximate result for the angular dependence of the wavelength of reflected light is

$$\lambda(\theta) = \lambda(0)\cos\left(\sin^{-1}\frac{\theta}{n}\right) \qquad (28)$$

From the above it is clear that the optical response of helicoidal structures of liquid crystal molecules depends on the pitch, and in order to relate these optical properties to molecular structures, the dependence of pitch on molecular structure must be considered.

Chiral liquid crystal phases readily form in mixtures of chiral and nonchiral materials, and for mixtures of a chiral dopant and a nonchiral host it is convenient to define a twisting power b, which is a measure of the pitch induced per unit concentration of chiral dopant:

$$b = \frac{dp^{-1}}{dc} \qquad (29)$$

The twist induced by different molecular species can be qualitatively related to the molecular structure of the chiral species [40]. The twist induced in a nonchiral liquid crystal solvent by a chiral dopant also depends on the nature of the solvent, and it has been proposed that chiral dopants can preferentially promote chiral conformations of the solvent molecules [41]. This effect has also been observed in isotropic solutions, where an enhanced optical rotation in solutions of a chiral biaryl in a cyanobiphenyl solvent was attributed [42] to an induction of chirality via preferential interactions

between solute and solvent conformations of the same chirality. Attempts have been made to relate the twisting power to some molecular measure of chirality based on a variety of geometric indices [43], but with only limited success.

Chiral interactions in the isotropic phase of liquid crystals are clearly seen in the pretransitional increase in optical activity observed at the isotropic to chiral nematic or chiral smectic phase transition. This was first observed by Cheng and Meyer [44] and explained by them in terms of fluctuations in the off-diagonal elements of the correlation function for fluctuations in the ordering matrix [45]. There are five fluctuation modes which can contribute to the pretransitional optical activity, and the experimentally observed behaviour depends on the relative amplitudes of these modes and coupling between them [46]. For some systems of high chirality there is a reversal in the sign of the pretransitional optical activity which has been attributed to mode-coupling; similar results have been obtained in the pretransitional region of smectic phases [47]. Scattering of circularly polarized light has also been used [46] to probe different chiral fluctuation modes, and by selection of combinations of incident right or left and scattered left or right circularly polarized light it is possible to identify scattering from three of the five individual modes.

The imaginary part of the complex refractive indices for left and right circularly polarized light relates to circular dichroism, that is differential absorption for light of different circular polarizations. It is treated in a similar manner to linear dichroism, except that the definition of principal components follows a different convention. For linear birefringence and dichroism the principal values of the complex refractive index relate to the electric field polarization direction which is transverse to the propagation direction (TE); principal components of the circular dichroism tensor are defined for the propagation direction. Thus components of the transition moments will contribute differently to corresponding principal values of the linear dichroism and circular dichroism tensors [48].

The circular dichroism (CD) ΔA_{CD} is defined as the difference in the optical absorption for left and right circularly polarized light, and for small values of the dichroism this gives rise to a corresponding ellipticity ψ for transmitted plane polarized light:

$$\Delta A_{CD} = k_l - k_r \qquad (30)$$

$$\psi = \frac{(k_l - k_r)\omega d}{2c} \qquad (31)$$

where ω is the frequency of the light in radians/sec, d is the pathlength and c the velocity of light.

The use of CD as a probe of liquid crystalline properties has been rather limited. A helicoidal structure will induce circular dichroism at an absorption band of a nonchiral chromophore, and the magnitude of the induced CD absorption depends on the pitch of the helix, the sign of the CD changing if pitch inversion occurs. This technique has been used to investigate phase transitions between ferrielectric, ferroelectric and antiferroelectric smectic C* states of MHPOBC ($C_8H_{17}OC_6H_4COOC_6H_4COOCH(CH_3)C_6H_{13}$) [49]. Spectra were recorded in the visible for a dissolved nonchiral dye molecule, and a small induced CD was detected in the UV originating in the intrinsic absorption of MHPOBC. A change in sign of the CD of the dye molecule was observed at the phase transition between ferroelectric and ferrielectric states. Measurements were made on 100 μm cells homeotropically aligned along the helix axis, to avoid complications from linear birefringence or dichroism effects.

3.6 References

[1] I.-C. Khoo, S.-T. Wu, *Optics and Nonlinear Optics of Liquid Crystals*, World Scientific, Singapore, 1993.
[2] J. F. Nye, *Physical Properties of Crystals*, Oxford University Press, Oxford, **1985**.
[3] N. H. Hartshorne, *Microscopy of Liquid Crystals*, Microscope Publications, London, **1974**.
[4] T. E. Lockhart, D. W. Allender, E. Gelerinter, D. C. Johnson, *Phys. Rev.* **1979**, *A20*, 1655.
[5] F. Yang, J. R. Sambles, *Liq. Cryst.* **1993**, *13*, 1.
[6] D. A. Dunmur, D. A. Hitchen, Xi-Yun Hong, *Mol. Cryst. Liq. Cryst.* **1986**, *140*, 303.
[7] V. P. Arora, S. A. Prakash, V. K. Garwal, B. Bahadur, *Ind. J. Appl. Phys.* **1992**, *30*, 406.
[8] K. R. Welford, J. R. Sambles, *Appl. Phys. Lett.* **1987**, *50*, 871.
[9] C. R. Lavers, *Liq. Cryst.* **1992**, *11*, 819.
[10] M. Sheik-Bahae, A. A. Said, T.-H. Wei, D. J. Hagan, E. W. Van Stryland, *IEEE J. Quant. Electron.* **1990**, *26*, 760.
[11] P. Palffy-Muhoray, H. J. Yuan, L. Li, *Mol. Cryst. Liq. Cryst.* **1991**, *199*, 223.
[12] P. Palffy-Muhoray, D. A. Balzarini, *Can. J. Phys.* **1981**, *59*, 515.
[13] K. C. Lim, J. T. Ho, *Mol. Cryst. Liq. Cryst.* **1978**, *47*, 173.
[14] D. W. Bruce, D. A. Dunmur, P. M. Maitlis, M. R. Manterfield, R. Orr, *J. Mater. Chem.* **1991**, *1*, 288.
[15] M. Warenghem, C. P. Grover, *Rev. Phys. Appl.* **1988**, *23*, 1169.
[16] M. Warenghem, G. Joly, *Mol. Cryst. Liq. Cryst.* **1991**, *207*, 205.
[17] M. Boschmans, *Mol. Cryst. Liq. Cryst.* **1991**, *199*, 267.
[18] I. Haller, H. A. Huggins, H. R. Lillenthal, T. R. McGuire, *J. Phys. Chem.* **1973**, *77*, 950.
[19] R. G. Horn, *J. Phys.* **1978**, *39*, 167.
[20] R. G. Horn, T. E. Faber, *Proc. Roy. Soc. Lond.* **1979**, *A368*, 199.
[21] E. H. Korte, *Mol. Cryst. Liq. Cryst.* **1983**, *100*, 127.
[22] R. Brettle, D. A. Dunmur, S. Estdale, C. M. Mason, *J. Mater. Chem.* **1993**, *3*, 327.
[23] P. Palffy-Muhoray, D. A. Dunmur, D. A. Balzarini, *Ordered Fluids and Liquid Crystals* (Ed.: A. C. Griffin), Plenum Press, New York, **1984**, *4*, p. 615.
[24] E. M. Averyanov, *Nuovo Cim* **1990**, *12*, 1281.
[25] H.-G. Kuball, R. Memmer, A. Straus, M. Junge, G. Scherowsky, A. Schonhofer, *Liq. Cryst.* **1989**, *5*, 969.
[26] H.-G. Kuball, M. Junge, B. Schulteis, A. Scholhofer, *Ber. Bunsenges. Phys. Chem.* **1991**, *95*, 1219.
[27] E. M. Averyanov, V. M. Muratov, V. G. Rumyantsev, *Sov. Phys. JETP* **1985**, *61*, 476.

[28] E. M. Averyanov, V. M. Muratov, V. G. Rum-
     yantsev, V. A. Churkina, *Sov. Phys. JETP* **1986**,
     63, 57.

[29] A. A. Minko, V. S. Rachevich, S. Ye. Yakoven-
     ko, *Liq. Cryst.* **1989**, *4*, 1.

[30] S. Yakovenko, R. Ignatovich, *Mol. Cryst. Liq.
     Cryst.* **1990**, *179*, 93.

[31] G. Pelzl in *Liquid Crystals* (Ed.: H. Stege-
     meyer), Sternkopff, Darmstadt, Germany, **1994**,
     p. 51.

[32] E. L. Wood, J. R. Sambles, S. J. Elston, *J. Mod.
     Opt.* **1991**, *38*, 1385.

[33] S. J. Elston, J. R. Sambles, M. G. Clark, *J. Mod.
     Opt.* **1989**, *36*, 1019.

[34] M. H. Anderson, J. C. Jones, E. P. Raynes, M. J.
     Tower, *J. Phys. D.: Appl. Phys.* **1991**, *24*,
     338.

[35] D. W. Berremen, *J. Opt. Soc. Am.* **1973**, *63*,
     1374.

[36] R. M. A. Azzam, N. M. Bashara, *J. Opt. Soc.
     Am.* **1972**, *62*, 1252.

[37] Ch. Bahr, D. Flieger, *Phys. Rev. E* **1993**, *70*,
     1842.

[38] G. Vertogen, W. H. de Jeu, *Thermotropic Liquid
     Crystals, Fundamentals 1988*, Springer-Verlag,
     Berlin, Heidelberg, p. 133.

[39] R. Dreher, G. Meier, *Phys. Rev.* **1973**, *A8*, 1616.

[40] G. Solladie, R. G. Zimmerman, *Angew. Chem.,
     Int. Ed. Engl.* **1984**, *23*, 348.

[41] G. Gottarelli, *Liq. Cryst.* **1986**, *1*, 29.

[42] G. Gottarelli, M. A. Osipov, G. P. Spada,
     J. Phys. Chem. **1991**, *95*, 3879.

[43] L. A. Kutulya, V. E. Kuzmin, I. B. Stelmakh,
     T. V. Handrimailova, P. P. Shtifanyuk, *J. Phys.
     Org. Chem.* **1992**, *5*, 308.

[44] J. Cheng, R. B. Meyer, *Phys. Rev. Lett.* **1972**, *29*,
     1240.

[45] J. Cheng, R. B. Meyer, *Phys. Rev. A* **1974**, 2744.

[46] P. R. Battle, J. D. Miller, P. J. Collings, *Phys.
     Rev.* **1987**, *A36*, 369.

[47] K. C. Frame, J. L. Walker, P. J. Collings, *Mol.
     Cryst. Liq. Cryst.* **1991**, *198*, 91.

[48] H.-G. Kuball, S. Neubrech, A. Schonhofer,
     J. Spectroscopy **1989**, *10*, 91.

[49] J. Watanabe, *J. Phys. Condens. Matter* **1990**, *2*,
     SA271–SA274.

4 Dielectric Properties

David Dunmur and Kazuhisa Toriyama

The purpose of this Section is to describe the dielectric properties of liquid crystals, and relate them to the relevant molecular properties. In order to do this, account must be taken of the orientational order of liquid crystal molecules, their number density and any interactions between molecules which influence molecular properties. Dielectric properties measure the response of a charge-free system to an applied electric field, and are a probe of molecular polarizability and dipole moment. Interactions between dipoles are of long range, and cannot be discounted in the molecular interpretation of the dielectric properties of condensed fluids, and so the theories for these properties are more complicated than for magnetic or optical properties. The dielectric behavior of liquid crystals reflects the collective response of mesogens as well as their molecular properties, and there is a coupling between the macroscopic polarization and the molecular response through the internal electric field. Consequently, the molecular description of the dielectric properties of liquid crystals phases requires the specification of the internal electric field in anisotropic media which is difficult.

4.1 Dielectric Response of Isotropic Fluids

The various factors which influence the dielectric properties of a liquid crystal can be identified by recalling the results for isotropic fluids. The Onsager equation (1) for an isotropic fluid relates the permittivity to the mean polarizability ($\bar{\alpha}$) and molecular dipole moment (μ):

$$(\varepsilon - 1) = \frac{NFh}{\varepsilon_0}\left[\bar{\alpha} + \frac{\mu^2 F}{3k_B T}\right] \qquad (1)$$

where F and h are reaction field and cavity field factors which account for the field dependent interaction of a molecule with its environment, and N is the number density. Molecular contributions to the permittivity are approximately additive, since $(\varepsilon - 1)$ is proportional to N, but the internal field factors for the reaction field (F) and cavity field (h) are also density dependent. For isotropic fluids, the internal field factors are given by:

$$F = (1 - \alpha f)^{-1} \text{ and } f = \frac{2(\varepsilon - 1)}{4\pi\varepsilon_0 a^3(2\varepsilon + 1)}$$

$$h = \frac{3\varepsilon}{(2\varepsilon + 1)} \qquad (2)$$

where a is the radius of the spherical cavity which accommodates the molecule. Using the Lorenz–Lorentz equation for the isotropic polarizability, Eq. (8) of Sec. 3 of this chapter gives for the reaction field factor:

$$F = \frac{(2\varepsilon + 1)(n^2 + 2)}{3(2\varepsilon + n^2)} \tag{3}$$

Specific pairwise dipole–dipole interactions can be accounted for by introducing the Kirkwood correlation factor g_1, such that the mean square dipole moment is replaced by an effective mean square moment defined by:

$$\mu^2_{\text{effective}} = g_1 \mu^2 \tag{4}$$

and this correlation factor g_1 can be related to the spatial dipole correlation function $G(\mathbf{r})$ in the fluid:

$$g_1 = 1 + V^{-1} \int G_1(\mathbf{r}) d\mathbf{r} \text{ and}$$

$$G_1(\mathbf{r}) = \frac{\langle \mu(0)\mu(\mathbf{r}) \rangle}{\langle \mu(0)\mu(0) \rangle} \tag{5}$$

The Kirkwood–Frohlich equation incorporates this factor, and enables the mean square effective dipole moment to be deduced from measurements of the electric permittivity, refractive index and number density of a fluid:

$$\frac{(\varepsilon - n^2)(2\varepsilon + n^2)}{\varepsilon(n^2 + 2)^2} = \frac{N g_1 \mu^2}{9\varepsilon_0 k_B T} \tag{6}$$

The electric permittivity determines the polarization (dipole moment per unit volume) induced in a material by an electric field. If the applied field varies with time, then the frequency dependence of the permittivity is an additional property of the material. A complication with any time-dependent response is that it may not be in-phase with the applied field. Thus to describe the frequency-dependent dielectric response of a

material, the amplitude and phase of the induced polarization must be measured. A convenient way of representing phase and amplitude is through complex notation, so that ε' (real) measures the in-phase response, and ε'' (imaginary) measures the 90° out-of-phase response:

$$\varepsilon^*(\omega) = \varepsilon'(\omega) - i\varepsilon''(\omega) \tag{7}$$

and the phase angle is $\tan^{-1}(\varepsilon''/\varepsilon')$. The effective response of a molecule to an alternating field of frequency ω, assuming a single molecular dipole relaxation and neglecting the contribution from the polarizability, can be described through the complex permittivity as:

$$\varepsilon^*(\omega) - 1 = (1 + i\omega\tau)^{-1} \frac{N\mu^2}{3\varepsilon_0 k_B T} \tag{8}$$

which gives (assuming $g_1 = 1$):

$$\varepsilon'(\omega) - 1 = \left[1 + \omega^2\tau^2\right]^{-1} \frac{N\mu^2}{3\varepsilon_0 k_B T}; \tag{9}$$

$$\varepsilon''(\omega) = \omega\tau\left[1 + \omega^2\tau^2\right]^{-1} \frac{N\mu^2}{3\varepsilon_0 k_B T} \tag{9}$$

The time τ is the relaxation time for dipole reorientation in an electric field of frequency ω (radians s^{-1}). For real systems there may be a number of contributions to the electric permittivity, each relaxing at a different frequency, for example due to internal dipole motion in flexible molecules or collective dipole motion. If these contributions to the electric permittivity are at sufficiently different frequencies, they can be separated in the dielectric spectrum, and it is possible to apply Eq. (9) to each relaxation process. At low frequencies ($\omega \to 0$), the orientation polarization contribution to the permittivity is $\dfrac{N\mu^2}{3\varepsilon_0 k_B T}$, neglecting any internal field effects, while at high frequencies ($\omega \to \infty$), the molecular dipoles do not rotate fast enough to contribute to the dielectric response. More generally the real and imaginary parts of the permittivity can be expressed as:

$$\varepsilon^*(\omega) - \varepsilon'(\infty) = \frac{[\varepsilon'(0) - \varepsilon'(\infty)]}{[1 + i\omega\tau]}$$

$$\varepsilon'(\omega) - \varepsilon'(\infty) = \frac{[\varepsilon'(0) - \varepsilon'(\infty)]}{[1 + \omega^2\tau^2]};$$

$$\varepsilon''(\omega) = \frac{\omega\tau[\varepsilon'(0) - \varepsilon'(\infty)]}{[1 + \omega^2\tau^2]} \qquad (10)$$

where $\dfrac{N\mu^2}{3\varepsilon_0 k_B T} = [\varepsilon'(0) - \varepsilon'(\infty)]$.

These equations, due to Debye, can be used to describe any relaxation process in a material, but in such cases the frequencies $\omega = 0$ and $\omega = \infty$ refer to frequencies below and above the relaxation frequency $\omega_0 = \tau^{-1}$, as illustrated in Fig. 1.

If the variable $\omega\tau$ is eliminated from Eq. (10), we obtain:

$$\varepsilon''(\omega)^2 + \left\{\varepsilon'(\omega) - \frac{1}{2}[\varepsilon'(0) + \varepsilon'(\infty)]\right\}^2$$

$$= \frac{1}{4}[\varepsilon'(0) - \varepsilon'(\infty)]^2 \qquad (11)$$

which is the equation of a circle of radius $\frac{1}{2}[\varepsilon(0) + \varepsilon(\infty)]$ centered on the point $\varepsilon''(\omega)$

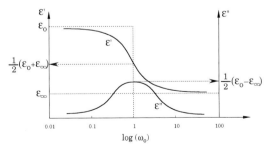

Figure 1. Schematic plot of the real and imaginary parts of the complex permittivity as a function of relative frequency $\omega_0 = \omega\tau$.

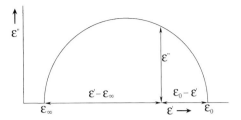

Figure 2. Cole–Cole plot of the real and imaginary parts of the permittivity.

$= 0$, $\varepsilon'(\omega) = \frac{1}{2}[\varepsilon(0) + \varepsilon(\infty)]$. This representation of the real and imaginary parts of the electric permittivity is known as a Cole–Cole plot, and is illustrated in Fig. 2.

In plotting experimental data it is sometimes more convenient to use the Debye–Pellat equations which are obtained by rearranging Eq. (10) thus:

$$\varepsilon'(\omega) = \varepsilon'(\infty) + \frac{\varepsilon''(\omega)}{\omega\tau}$$

$$\varepsilon'(\omega) = \varepsilon'(0) - \omega\tau\varepsilon''(\omega) \qquad (12)$$

since these provide a simple way to determine the low $\varepsilon'(0)$ and high $\varepsilon'(\infty)$ frequency contributions to the real part of the permittivity.

The measured frequency dependences of $\varepsilon''(\omega)$ and $\varepsilon'(\omega)$ in real fluids do not always fit the Debye–Pellat equations, and many methods [1] have been proposed to analyse skewed or displaced Cole–Cole plots. Debye's theory of dipole relaxation assumes that rotational motion can be described in terms of a single relaxation time. In a real system, fluctuations in the local structure of a molecule or its environment may result in a distribution of relaxation times about the Debye value, and such a situation can be described by a modification to Eq. (10)

$$\varepsilon^*(\omega) - \varepsilon'(\infty) = \frac{\varepsilon'(0) - \varepsilon'(\infty)}{1 + (i\omega\tau)^{1-\alpha}} \qquad (13)$$

where α is a parameter introduced by Cole and Cole [2]. The effect of α is to produce a semicircular Cole–Cole plot, the center of which is depressed below the abscissa. Asymmetric or skewed plots of $\varepsilon''(\omega)$ against $\varepsilon'(\omega)$ can sometimes be fitted by the Cole–Davidson equation:

$$\varepsilon^*(\omega) - \varepsilon'(\infty) = \frac{\left[\varepsilon'(0) - \varepsilon'(\infty)\right]}{\left[1 + (i\omega\tau)\right]^{\beta}} \quad (14)$$

where β is another parameter which is a measure of an asymmetric distribution of relaxation times. Another empirical functional form which is sometimes used to describe a non-Debye like relaxation is due to Fuoss and Kirkwood. This function, Eq. (15), gives a symmetric Cole–Cole plot, but depressed with respect to the result for a single relaxation: (α here is not the same as the α parameter in Eq. (13))

$$\varepsilon''(\omega) = \frac{1}{2}\alpha\big(\varepsilon(0) - \varepsilon(\infty)\big)\,\mathrm{sech}\,\alpha(\ln\omega\tau) \quad (15)$$

In some isotropic liquids, and quite generally in polar liquid crystals, the plot of $\varepsilon''(\omega)$ or $\varepsilon'(\omega)$ against frequency shows evidence of two or more separate relaxation times. In some circumstances these may be well separated in frequency giving distinct Cole–Cole arcs, but more usually they overlap to give a composite Cole–Cole arc. The simplest analysis of such measurements is to assume that the relaxation processes contribute additively to the permittivity so that the complex permittivity can be written:

$$\varepsilon^*(\omega) - \varepsilon'(\infty) = \sum_{j} \frac{x_j}{\left[1 + i\omega\tau_j\right]} \quad (16)$$

where x_j is a weighting factor for each relaxation centred at frequency $\omega_j = \tau_j^{-1}$. Using this, nonsemi-circular Cole–Cole plots can be analysed in terms of a sum of contributions (see Fig. 3).

In deriving the macroscopic equations from the microscopic result, Eq. (8) the effect of the environment on a rotating molec-

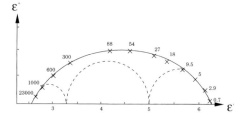

Figure 3. Cole–Cole plot of the perpendicular component of the permittivity for the nematic phase of 4-heptyl-4-cyanobiphenyl [3].

ular dipole has been neglected. This does not invalidate the Debye–Pellat equations or Cole–Cole plot, but requires a different interpretation of the relaxation time. Thus $\omega_j = \tau_j^{-1}$ is no longer the angular velocity of a rotating dipole, but is a macroscopic relaxation frequency. For an isotropic fluid an approximate relationship between the macroscopic and microscopic relaxation times is [4]:

$$\tau_{\mathrm{macroscopic}} = \frac{3\varepsilon(0)}{2\varepsilon(0) + \varepsilon(\infty)}\tau_{\mathrm{microscopic}} \quad (17)$$

but this result is not applicable to liquid crystals.

If the rotating molecule is assumed to be rigid and axially symmetric, then the dipole will lie along the symmetry axis, and the reorientation of the dipole in an electric field will be governed by the rotational motion of the molecule about an axis perpendicular to the symmetry axis. This in turn can be related to a rotational diffusion constant $D_\perp$:

$$\tau_{\mathrm{molecular}} = (2D_\perp)^{-1} = I(2k_{\mathrm{B}} T \tau_J)^{-1} \quad (18)$$

where τ_J is the relaxation time for the angular momentum about the short axis, and I is the moment of inertia. Assuming an isotropic distribution of angular momenta, Debye showed that τ_J could be simply related to the moment of inertia, an effective molecular radius (a) and a microscopic viscosity (η), giving:

$$\tau_{\text{molecular}} = \frac{4\pi \eta a^3}{k_B T} \tag{19}$$

Various extensions of this to other molecular shapes have been reported [5].

In the above discussion of the frequency dependent permittivity, the analysis has been based on either the single particle rotational diffusion model of Debye, or empirical extensions of this model. A more general approach can be developed in terms of time correlation functions [6], which in turn have to be interpreted in terms of a suitable molecular model. While using the correlation function approach does not simplify the analysis, it is useful, since experimental correlation functions can be compared with those deduced from approximate theories, and perhaps more usefully with the results of molecular dynamics simulations. Since the use of correlation functions will be mentioned in the context of liquid crystals, they will be briefly introduced here. The dipole–dipole time correlation function $C(t)$ is related to the frequency dependent permittivity through a Laplace transform such that:

$$\frac{\varepsilon^*(\omega) - n^2}{\varepsilon(0) - n^2} \left[\frac{\left(2\varepsilon'(\omega) + n^2\right)\varepsilon'(0)}{\left(2\varepsilon'(0) + n^2\right)\varepsilon'(\omega)} \right]$$

$$= 1 - i\omega \int_0^\infty C(t) e^{i\omega t} \, dt \tag{20}$$

where $C(t)$ is the dipole time correlation function defined by:

$$C(t) = \frac{\langle \boldsymbol{\mu}(0) \cdot \boldsymbol{\mu}(t) \rangle}{\langle \boldsymbol{\mu}(0) \cdot \boldsymbol{\mu}(0) \rangle} \tag{21}$$

4.2 Dielectric Properties of Anisotropic Fluids

In an orientationally ordered fluid the electric permittivity becomes a second rank tensor $\varepsilon_{\alpha\beta}$, and an electric susceptibility can be defined which relates the induced polarization $\boldsymbol{P}$ to the applied electric field $\boldsymbol{E}$:

$$P_\alpha = \varepsilon_0 (\varepsilon_{\alpha\beta} - \delta_{\alpha\beta}) E_\beta$$

$$= \varepsilon_0 \, \chi_{\alpha\beta}^{\text{electric}} E_\beta \tag{22}$$

Thus the number of independent components of the permittivity tensor will depend on the symmetry of the liquid crystal phase. The frequency dependence of the permittivity is described in terms of real and imaginary parts, and these also will be tensor quantities. Apart from complications of anisotropic internal fields, the static or low frequency part of the permittivity tensor can be related to the molecular polarizability and dipole moment averaged over the appropriate orientational distribution functions.

4.2.1 The Electric Permittivity at Low Frequencies: The Static Case

4.2.1.1 Nematic Phase

In this section we wish to consider all the possible contributions to the electric permittivity of liquid crystals, regardless of the time-scale of the observation. Conventionally this permittivity is the static dielectric constant (i.e. it measures the response of a system to a d.c. electric field) in practice experiments are usually conducted with low frequency a.c. fields to avoid conduction and space charge effects. For isotropic dipolar fluids of small molecules, the permittivity is effectively independent of frequency below 100 MHz, but for liquid crystals it may be necessary to go below 1 kHz to measure the static permittivity; polymer liquid crystals can have relaxation processes at very low frequencies.

The electric susceptibility for anisotropic materials is a second rank tensor, and we can use the general results of Sec. 1 of this chapter. Firstly we will only consider the anisotropic part of the permittivity:

$$\chi^{(a)}_{\alpha\beta} = \chi_{\alpha\beta} - \frac{1}{3}\chi_{\gamma\gamma}\,\delta_{\alpha\beta} \tag{23}$$

The principal components of this can be evaluated in terms of order parameters and molecular properties as given in Eq. (47) of Sec. 1 of this chapter. All that remains is to evaluate the components of the molecular susceptibility tensor $\kappa^{(\text{electric})}_{\alpha\beta}$. This contains contributions from (1) the dipole induced by

the internal electric field $E^{(\text{int})}$, and (2) the orientation polarization arising from the partial alignment of dipoles by the directing field $E^{(\text{dir})}$. Thus in molecular axes the principal components of the microscopic electric susceptibility tensor will be obtained from:

$$\kappa^{(\text{electric})}_{ii}E_i = \alpha_{ii}\,E_i^{(\text{int})} + \mu_i \tag{24}$$

If a molecule is freely rotating then the average dipole moment along any axis will be zero. The effect of an electric field is to break the $\pm$ symmetry of the axis, and the value of μ_i is the average over the $+i$ and $-i$ directions weighted by the Boltzmann energy associated with the directing electric field, so that:

$$\bar{\mu}_i = \frac{\mu_{+i}\exp-\left(\dfrac{u_0 + \mu_i\,E_i^{(\text{dir})}}{k_B\,T}\right) + \mu_{-i}\exp-\left(\dfrac{u_0 - \mu_i\,E_i^{(\text{dir})}}{k_B\,T}\right)}{\exp-\left(\dfrac{u_0 + \mu_i\,E_i^{(\text{dir})}}{k_B\,T}\right) + \exp-\left(\dfrac{u_0 + \mu_i\,E_i^{(\text{dir})}}{k_B\,T}\right)} \tag{25}$$

where $\mu_{-i} = -\mu_{+i}$. Assuming that $\mu_i\,E_i^{(\text{dir})} \ll k_B\,T$, expanding the exponentials gives:

$$\bar{\mu}_i = \frac{\mu_i^2\,E_i^{(\text{dir})}}{k_B\,T} \tag{26}$$

and using this in Eq. (24) along with the results for $E^{(\text{int})}$ and $E^{(\text{dir})}$ obtained for isotropic liquids gives:

$$\kappa^{(\text{electric})}_{ii} = F h\left[\alpha_{ii} + \frac{\mu_i^2\,F}{k_B\,T}\right] \tag{27}$$

This can be substituted into Eq. (47) of Sec. 1 to obtain the anisotropic part of the electric susceptibility tensor. The isotropic part of $\chi^{(\text{electric})}_{\alpha\beta}$ is the result obtained for isotropic fluids:

$$\frac{1}{3}\chi^{(\text{electric})}_{\gamma\gamma} = \frac{NFh}{\varepsilon_0}\left[\bar{\alpha} + \frac{\mu^2\,F}{3k_B\,T}\right] \tag{28}$$

so the final result for the principal components of the permittivity tensor can be obtained from Eqs. (27) of this section and (47) of Sec. 1 to give:

$$\chi^{(\text{electric})}_{\parallel} = \varepsilon_\parallel - 1 = \varepsilon_0^{-1}\,NFh\left\{\bar{\alpha} + \frac{2}{3}\alpha_1\,S + \frac{1}{3}\alpha_2\,D\right.$$
$$\left. + \frac{Fg_1^{(\parallel)}}{3k_B\,T}[\mu_z^2(1+2S) + \mu_y^2(1-S-D) + \mu_x^2(1-S+D)]\right\} \tag{29a}$$

$$\chi_{\perp'}^{(electric)} = \varepsilon_{\perp'} - 1 = \varepsilon_0^{-1} NFh \left\{ \bar{\alpha} - \frac{1}{3}\alpha_1 (S+P) - \frac{1}{6}\alpha_2 (D+C) \right. \tag{29b}$$

$$\left. + \frac{Fg_1^{(\perp')}}{6k_B T} [2\mu_z^2 (1-S-P) + \mu_y^2 (2+S+P+D+C) + \mu_x^2 (2+S+P-D-C)] \right\}$$

$$\chi_{\perp}^{(electric)} = \varepsilon_{\perp} - 1 = \varepsilon_0^{-1} NFh \left\{ \bar{\alpha} - \frac{1}{3}\alpha_1 (S-P) - \frac{1}{6}\alpha_2 (D-C) \right. \tag{29c}$$

$$\left. + \frac{Fg_1^{(\perp)}}{6k_B T} [2\mu_z^2 (1-S+P) + \mu_y^2 (2+S-P+D-C) + \mu_x^2 (2+S-P-D+C)] \right\}$$

These are the Maier and Meier equations [7] for the low frequency components of the permittivity extended [8] to include all orientational order parameters. They predict that the mean value of the permittivity

$$\bar{\varepsilon} = \frac{1}{3}(\varepsilon_{\parallel} + \varepsilon_{\perp'} + \varepsilon_{\perp})$$ should be independent

of the orientational order, and apart from changes in density it is expected to be continuous through all liquid crystal phase changes. This prediction is not always confirmed by experiment, and for polar mesogens there are often detectable changes in $\bar{\varepsilon}$ at phase transitions, including those to the isotropic liquid.

A simplified form of Eq. (29) may be used to write down the permittivity anisotropy of a uniaxial liquid crystal consisting of uniaxial molecules having an off-axis dipole which makes an angle of β with the principal molecular axis (see Fig. 4):

$$\Delta\varepsilon = \frac{NhFS}{\varepsilon_0} \left[\Delta\alpha + \frac{\mu^2 F}{2k_B T} (3\cos^2 \beta - 1) \right] \tag{30}$$

Figure 4. Schematic of a mesogen with an off-axis dipole moment making an angle of β with the molecular long axis.

For values of β less than $54.7°$, the dipolar term is positive, while for angles greater than this it becomes negative, and may result in an overall negative dielectric anisotropy. For a particular combination of molecular properties, the polarisability anisotropy and the dipolar terms in Eq. (30) may cancel at a particular temperature. This has been observed in certain fluorinated cyclohexyl ethanyl biphenyls [9], which change the sign of their anisotropy from negative at low temperatures to positive at high temperatures. There is, however, a basic inconsistency with the model described by Eq. (30), since any molecule with an off-axis dipole is necessarily biaxial; rotation about the molecular long axis may result in the biaxial order parameter D averaging to zero.

Dielectric measurements on liquid crystal phases probe the dipole organization of molecules, and changes in the permittivity components as a liquid crystal undergoes transitions from nematic to various smectic phases will primarily reflect changes in orientational order and symmetry changes. Different degrees of translational order will only influence the permittivity components indirectly through macroscopic internal field corrections and through short range dipole–dipole interactions. The effect of macroscopic anisotropy on the dielectric properties of a material has been calculated for the model of a polarized sphere immersed in a dielectric continuum, with the result: [10]

$$\frac{\left(\varepsilon_i - n_i^2\right)\left[\varepsilon_i + \Omega_i\left(1 - \varepsilon_i\right)\right]^2}{\varepsilon_i\left[\varepsilon_i + \left(n_i^2 - \varepsilon_i\right)\Omega_i\right]} = \frac{\langle M_i^2\rangle}{\varepsilon_0 k_B T} \tag{31}$$

where $\langle M_i^2\rangle$ is the mean square dipole moment of the sphere in the i-direction, and the factor Ω_i accounts for the depolarization field associated with a sphere in an anisotropic medium. Ω_i is defined in terms of the components of the permittivity tensor by:

$$\Omega_i = \frac{1}{2}\varepsilon_i \int_0^\infty \frac{\left(s\varepsilon_{ii} + 1\right)^{-1} ds}{\sqrt{\left(s\varepsilon_{xx} + 1\right)\left(s\varepsilon_{yy} + 1\right)\left(s\varepsilon_{zz} + 1\right)}} \tag{32}$$

and equals 1/3 for an isotropic dielectric. To proceed beyond Eq. (32) it is necessary to model both the anisotropic local field acting on molecules and the short range interactions between molecular dipoles in the sphere. The former depends on the long range anisotropy in the radial distribution function, while dipole–dipole correlations can be described in terms of anisotropic Kirkwood g-factors defined for different directions in the sample. These are most usefully defined in terms of the appropriate dipole correlation functions as:

$$g_1^i = 1 + V^{-1}\int G_1^i(\mathbf{r})\,d\mathbf{r}$$

$$G_1^i(\mathbf{r}) = \frac{\langle \mu_i(0)\mu_i(\mathbf{r})\rangle}{\langle \mu_i(0)\mu_i(0)\rangle} \tag{33}$$

where (i) refers to the parallel and perpendicular directions. Thus $g_1^\parallel$ and $g_1^\perp$ measure the extent to which the projections of molecular dipole components are correlated along the principal axes of an anisotropic fluid, assumed to be uniaxial in this case. If the rotational motion of molecules was isotropic, then clearly the correlation factors along the axes would be the same. This anisotropic dipole correlation is illustrated in

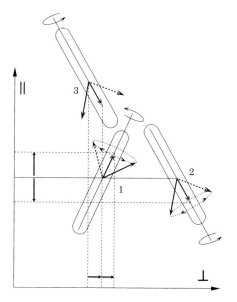

Figure 5. Anisotropic correlation of molecular dipole components.

Fig. 5, where the projected parallel components of the dipoles on molecules (1) and (2) are opposed ($g_1^\parallel < 1$), while the corresponding dipoles projected on to the perpendicular axis, molecules (1) and (3) are reinforcing ($g_1^\perp$).

Evaluating the mean square moment of a sphere (Eq. (31)), and using the definition of the correlation factor above gives the anisotropic version of the Kirkwood–Frohlich equation:

$$\frac{\left(\varepsilon_i - n_i^2\right)\left[\varepsilon_i + \Omega_i\left(n_i^2 - \varepsilon_i\right)\right]}{\varepsilon_i\left[n_i^2 + 2\right]^2} = \frac{N g_1^i\langle \mu_i^2\rangle}{9\varepsilon_0 k_B T} \tag{34}$$

This equation can be written as:

$$\left(\varepsilon_i - n_i^2\right) = \frac{N h_i F_i^2 g_1^i\langle \mu_i^2\rangle}{\varepsilon_0 k_B T} \tag{35}$$

where $h_i = \varepsilon_i\left[\varepsilon_i + \Omega_i\left(n_i^2 - \varepsilon_i\right)\right]^{-1}$ is the anisotropic cavity field factor, and $F_i = \frac{1}{3}\left(n_i^2 + 2\right)$

is an approximation to the anisotropic reaction field factor. While some progress has been made in describing local dipole correlations in partially ordered phases, the formulation of the internal electric field for a macroscopically anisotropic dipolar fluid remains a formidable problem. Various approximate models have been described [11–13], but it is difficult to assess the relative merits of such approaches. In liquid crystals, it has been proposed [14] that the long range anisotropy in the radial distribution function can be modelled by assuming an ellipsoidal cavity of dimensions a, b and c which depend on molecular shape. However er comparison of local field tensors calculated for a continuum model with the results of dipole–dipole lattice sums have led to the conclusion [15] the contribution of shape anisotropy to the local field in anisotropic fluids can be neglected, and the assumption of isotropic internal field factors is justified. Under these circumstances, Eq. (35) can be written as more familiar Maier and Meier equations in terms of these effective mean square dipole moments:

$$\varepsilon_\parallel - 1 = \varepsilon_0^{-1} NLF$$
$$\cdot \left\{ \bar{\alpha} + \frac{2}{3} \Delta\alpha S + \frac{F}{k_B T} \left[\mu_{\text{eff}}^\parallel \right]^2 \right\} \quad (36)$$

$$\varepsilon_\perp - 1 = \varepsilon_0^{-1} NLF$$
$$\cdot \left\{ \bar{\alpha} - \frac{1}{3} \Delta\alpha S + \frac{F}{k_B T} \left[\mu_{\text{eff}}^\perp \right]^2 \right\} \quad (37)$$

where the effective mean square dipole moments can be written in terms of the order parameter and the longitudinal (μ_l) and transverse (μ_t) components of the molecular dipole as:

$$\left[\mu_{\text{eff}}^\parallel \right]^2 = \frac{g_1^\parallel}{3} \left[\mu_l^2 (1 + 2S) + \mu_t^2 (1 - S) \right] \quad (38)$$

$$\left[\mu_{\text{eff}}^\perp \right]^2 = \frac{g_1^\perp}{3} \left[\mu_l^2 (1 - S) + \mu_t^2 \left(1 + \frac{1}{2} S \right) \right] \quad (39)$$

A number of studies of dipole association in liquid crystalline systems have been reported [16–18], and it is clear that the orientation of the molecular dipole with respect to the molecular axis has a large influence on the local dipole organisation. A mean field theory of short range dipole–dipole correlation between interacting hard ellipsoids with embedded dipoles has been developed [19], and this predicts that prolate ellipsoidal molecules (rod-like) with longitudinal dipoles will exhibit local antiferroelectric order in ordered fluids, while oblate ellipsoids with dipoles along the shortest axis will order ferroelectrically. These studies can aid the development of new materials, for which carefully tailored dielectric properties are required [20], but are also of relevance in the research on anisotropic fluids having long range dipole organization.

As was pointed out in Sec. 1 of this chapter the symmetry of the phases will determine the number of independent components of the second rank electric permittivity; furthermore the point group symmetry of the phase and the constituent molecules will fix the orientational order parameters that contribute to a microscopic expression for the permittivity. In order to complete the description of the low frequency or static electric permittivity of liquid crystals, it is necessary to consider the additional effects of chirality, and the translational order associated with smectic phases.

The chiral nematic phase is characterized by a helical structure, and so the electric permittivity is biaxial, with three independent components along the principal axes, which are the local director axis, the helix and a third orthogonal axis. Since the pitches of chiral nematics are usually many molecular diameters, chiral nematics are locally uniaxial, and the pitch does not affect the symmetry or the magnitude of the permittivity.

4.2.1.2   The Smectic Phases

Experimentally there are changes in the components of the permittivity at nematic/smectic and smectic/smectic phase transitions, as illustrated in Figs. 6 and 7.

These changes reflect the molecular reorganization that takes place at transitions between different liquid crystal phases. The interpretation of the dielectric properties of smectic phases can be carried out using Eq. (34). Differences in orientational order in smectic phases are accounted for through the appropriate orientational order parameters given in Eq. (29), while other influences of the translational smectic order will affect the internal field factors and short range dipole–dipole interactions. For strongly

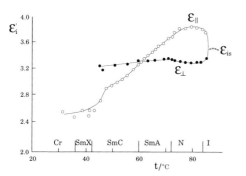

Figure 6. Dielectric permittivities for 95S showing effect of smectic phase transitions [21].

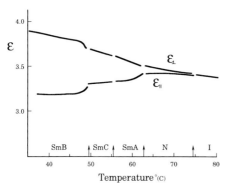

Figure 7. Dielectric permittivities of 4-pentyloxyl-benzylidene-4′-heptylaniline [22].

polar mesogens dipole–dipole association is an important contributor to the physical properties of liquid crystal phases, and this is particularly important in smectic phases, where translational order can affect the dipole–dipole correlation factors. Dielectric studies have been made on materials exhibiting a number of different SmA phases (SmA$_1$, SmA$_2$, SmÃ), which are characterized by different degrees of dipole–dipole organization [23, 24]. The effect of different local interactions can be measured through the dipole correlation factors $g_1^{(i)}$, but evaluation of the anisotropic Kirkwood correlation factors either requires a detailed microscopic model for the liquid crystal, or it can be calculated from computer simulation. One model approach [25, 26] is to assume perfect orientational order (i.e. $S=1$) so that the molecules are constrained to be parallel or antiparallel to the director axis. The problem is then reduced to a two-state model, and if the dipole moment is assumed to be along the molecular axis, (i.e. parallel or antiparallel to the layer normal (z-axis), the net correlation is given by the relative probabilities for parallel or antiparallel dipole organization. Thus the dipole–dipole correlation factor can be written as:

$$g_1^{(\parallel)} = 1 + \frac{n\langle\mu_{1z}\,\mu_{2z}\rangle}{\langle\mu_{1z}^2\rangle} = 1 + n\langle p_+ - p_-\rangle \quad (40)$$

where n is the number of neighbors. The probabilities for parallel (+) or antiparallel (−) dipole orientation are determined by the dipole–dipole energy $u(\mu_1, \mu_2)$ (Eq. 41), and the average in Eq. (40) must be evaluated over the microscopic structure of the liquid crystal:

$$u(\mu_1,\mu_2) = -\frac{\mu^2}{4\pi r^3 \varepsilon_0}$$
$$\begin{pmatrix}2\cos\theta_1\cos\theta_2 - \\ \sin\theta_1\sin\theta_2\cos\phi\end{pmatrix} \quad (41)$$

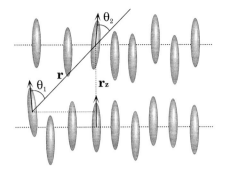

Figure 8. Dipole correlation in smectic phases.

The angles θ_1 and θ_2 are defined in Fig. 8 where the dihedral angle ϕ has been set equal to 0. Writing the probabilities in terms of the dipole–dipole energy:

$$g_1^{(\parallel)} = 1 + \tag{42}$$

$$n \frac{\left\langle \exp-\left(\dfrac{u_{parallel}}{k_B T}\right) - \exp-\left(\dfrac{u_{antiparallel}}{k_B T}\right) \right\rangle}{\left\langle \exp-\left(\dfrac{u_{parallel}}{k_B T}\right) + \exp-\left(\dfrac{u_{antiparallel}}{k_B T}\right) \right\rangle}$$

expanding the exponentials, and assuming that $u(\mu_1, \mu_2) \ll k_B T$, gives:

$$g_1^{(\parallel)} = 1 + n \frac{\langle u_{antiparallel} - u_{parallel} \rangle}{k_B T} \tag{43}$$

and for the model depicted in Fig. 8, this can be evaluated to give:

$$g_1^{(\parallel)} = 1 - \frac{n\mu^2 \left\langle 3\left(\dfrac{r_z}{r}\right)^2 - 1 \right\rangle}{4\pi \varepsilon_0 r^3 k_B T} \tag{44}$$

In a smectic phase, the average separation perpendicular to the layers (r_z) is likely to be greater than the in-plane separation, and this results in a $g_1^{(\parallel)} < 1$, while $g_1^{(\perp)}$ would be unity because the perpendicular component of the dipole is zero. This simple model can be extended to molecules with a molecular dipole inclined at an angle β to the molecu-

lar alignment axis, in which case:

$$g_1^{(\parallel)} = 1 - \frac{n\mu^2 \cos^2 \beta \left\langle 3\left(\dfrac{r_z}{r}\right)^2 - 1 \right\rangle}{4\pi \varepsilon_0 r^3 k_B T} \tag{45}$$

$$g_1^{(\perp)} = 1 - \frac{n\mu^2 \sin^2 \beta \left\langle 3\left(\dfrac{r_x}{r}\right)^2 - 1 \right\rangle}{8\pi \varepsilon_0 r^3 k_B T} \tag{46}$$

If $\langle r_x^2 \rangle$ is less than $\langle r^2/3 \rangle$, the perpendicular dipole correlation factor will be greater than one, indicating a preferred parallel alignment of dipoles in the smectic layer. This model has been used [26] to explain the change in sign of the dielectric anisotropy from positive to negative on passing from the nematic to the smectic A phase of p-heptylphenylazoxy-p'-heptylbenzene.

Low frequency dielectric studies on smectic C, F and I phases are complicated by the intrinsic biaxiality of these phases. It is possible to use dielectric measurements to determine the tilt angle in SmC materials [27], but of more interest is the direct determination of the three principal components of the dielectric tensor, since such measurements can give additional information on the local molecular organization from Eq. (29). The orientation of the principal axes for tilted smectic phases is not determined by symmetry, except that one principal axis coincides with the C_2 rotation axis perpendicular to the tilt-plane. It is assumed that a further principal axis lies along the tilt direction, and this appears to be justified by experiment; the orientation of the axes are indicated in Fig. 9.

Most recent studies [28, 29] of the dielectric properties of the SmC phase have focussed on the ferroelectric chiral smectic C phase, because of its importance in applications. The molecular interpretation of the principal permittivities is contained in Eqs. (27), with appropriate correlation fac-

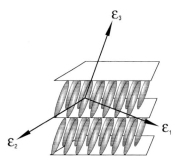

Figure 9. Definition of principal axes for SmC phases.

Specimen results for SCE13(R), a commercial racemic host, are given in Fig. 11 [28].

The reduced symmetry of chiral phases results in additional contributions to the low frequency permittivity. Tilted chiral phases such as smectic C*, F* and I* lack a centre of symmetry, and it is possible for these materials to be ferroelectric. The resulting spontaneous polarization P_S is directed along the C_2 symmetry axis, and is perpendicular to the tilt plane; it also depends di-

tors to account for the tilted layered structure [25], but the experimental problem is to find suitable alignment geometries that allow the independent measurement of the principal components of the permittivity, denoted as ε_1, ε_2, and ε_3 in Fig. 9. One approach [28], which is equally applicable to nonchiral and chiral tilted phases is to measure the permittivity of a homeotropically aligned sample in which the measurement direction is along the smectic layer normal. This gives a result for ε_{homo}, defined by:

$$\varepsilon_{homo} = \varepsilon_1 + (\varepsilon_3 - \varepsilon_1)\cos^2\theta \qquad (47)$$

where $(\varepsilon_3 - \varepsilon_1)$ is defined as the dielectric anisotropy (ε_3 is equivalent to $\varepsilon_\parallel$ and ε_1 is equivalent to $\varepsilon_\perp$). A second permittivity can be measured for the so-called planar state, for which a chevron structure is assumed (Fig. 10) with a layer tilt angle of δ: the corresponding permittivity is:

$$\varepsilon_{planar} = \varepsilon_2 - \partial\varepsilon\frac{\sin^2\delta}{\sin^2\theta} \qquad (48)$$

the quantity $\partial\varepsilon = (\varepsilon_2 - \varepsilon_1)$ is defined as the dielectric biaxiality.

In order to obtain the three components, it is assumed that the mean permittivity extrapolated from higher smectic and nematic phases may be used, such that:

$$\bar{\varepsilon} = \frac{1}{3}(\varepsilon_1 + \varepsilon_2 + \varepsilon_3) \qquad (49)$$

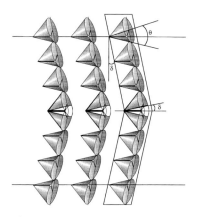

Figure 10. Chevron structure on a SmC phase.

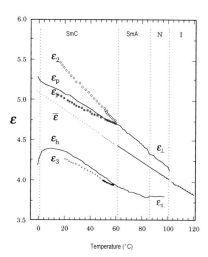

Figure 11. Dielectric results for SCE13(R) taken from [28].

rectly on the tilt angle θ:

$$P_S = P_0 \sin \theta \qquad (50)$$

In the absence of any constraints, the direction of P_S rotates from one smectic layer to the next, with a period equal to the smectic C* pitch, and so the average polarization for a sample would be zero. However, surface treatment or application of a field can cause the helix to untwist, resulting in a permanently polarized sample. The spontaneous polarization arises from a preferred alignment of molecular dipole components which are perpendicular to the molecular long axis, but it behaves differently from the ferroelectric and ferromagnetic polarization characterised for crystals. The liquid crystalline ferroelectric phases identified so far are *improper* ferroelectrics, since the spontaneous polarization results from a symmetry constraint, whereas in *proper* ferroelectrics the polarization results from dipole – dipole interactions. The Curie – Weiss law for proper ferroelectrics predicts a second order phase transition at the Curie temperature from the high temperature paraelectric state to a permanently polarized ferroelectric state:

$$P_S = \frac{N\mu^2 E}{3k_B (T - T_c)} \qquad (51)$$

However in chiral tilted liquid crystal smectic phases, the polarization is driven by the tilt angle, and the phase transition will not necessarily be of second order.

The helical structure which can develop in thick cells of chiral smectic C phases having planar surface alignment conditions can be used to obtain measurements of the components of the dielectric permittivity tensor [29], but the technique is restricted to chiral smectic phases. Measurements are made (see Fig. 9) of the homeotropic state, as above, and additionally the helical state (Fig. 12), and the uniformly-tilted state ob-

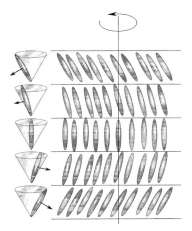

Figure 12. Alignment states for the helical state of chiral SmC phase.

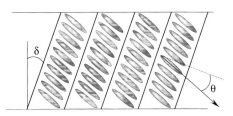

Figure 13. Alignment of the unwound chiral SmC phase.

tained by applying a field to unwind the helix (Fig. 13):

$$\varepsilon_{\text{helix}} = \frac{1}{2} \cos^2 \delta \left(\varepsilon_1 \cos^2 \theta + \varepsilon_3 \sin^2 \theta + \varepsilon_2 \right) + \sin^2 \delta \left(\varepsilon_1 \sin^2 \theta + \varepsilon_3 \cos^2 \theta \right)$$

$$\varepsilon_{\text{unwound}} = \varepsilon_2 \cos^2 \delta + \sin^2 \delta \left(\varepsilon_1 \sin^2 \theta + \varepsilon_3 \cos^2 \theta \right) \qquad (52)$$

The use of Eqs. (47) and (52) then allows the evaluation of three principal permittivity components.

The electric susceptibilities of chiral smectic A and chiral smectic C, I and F contain terms related to the permanent dipole polarization which can develop in tilted smectic phases (ferroelectricity), or which can be induced in orthogonal smectic phases through an induced tilt (electroclinism) by electric fields in the plane of the layers. The broken symmetry associated with these phases is along an axis perpendicular to the director, and the layer normal for smectics, and so the extra dielectric contributions are to a single perpendicular component of the permittivity. The origin of the polarization contributions to the electric permittivity of chiral smectic A and smectic C phases is illustrated in Fig. 14 [30].

Orientation of the transverse molecular dipole moments become biased along the y-axis (corresponding to ε_2 above) when a tilt develops either induced by an electric field along the y-axis (orthogonal smectics) or spontaneously in tilted smectics. This is known as the soft mode contribution to the electric susceptibility, and as well as contributing to all smectic phases there is a detectable effect in chiral nematic phases, close to phase transitions. Tilted chiral smectic phases can develop helicoidal structures, the helix axis being perpendicular to the layers. In the unperturbed state the polarization associated with the spontaneously aligned transverse dipole components rotates with the helix, however application of an electric field perpendicular to the helix axis gives a contribution to the electric polarization, and hence to the electric permittivity component: this is known as the Goldstone mode. In terms of Fig. 14 the Goldstone mode describes polarization resulting from changes in the azimuthal angle (ϕ) of the director, while the soft mode is polarization from changes in the tilt angle θ. A Goldstone mode contribution will only be measured if there is a helicoidal structure, and so will be absent in surface-stabilized chiral smectic C structures, such as the planar and unwound states described above.

The theory of the dielectric properties of chiral smectic liquid crystals is far from complete, particularly with respect to a molecular statistical approach. Simple Landau theory [31] gives expressions for the contributions of soft modes (χ_S) and Goldstone modes (χ_G) to the low frequency permittivity as:

$$\chi_S = \frac{\varepsilon_{\infty y}^2 \mu_p^2}{kq_0^2 + 2a(T_c - T)}$$

and

$$\chi_G = \frac{\varepsilon_{\infty y}^2 \mu_p^2}{2kq_0^2} \tag{53}$$

where μ_p is the piezoelectric coefficient in the Landau free energy, which measures the coupling between the director and the polarization, and $\varepsilon_{\infty y}$ is the high frequency part

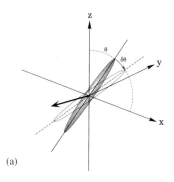

(a)

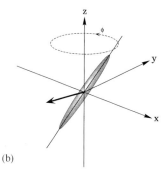

(b)

Figure 14. Contributions of soft mode (a) and Goldstone mode (b) deformations to the electric susceptibility.

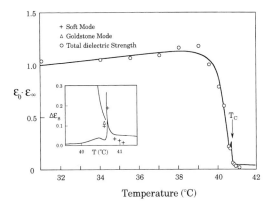

Figure 15. Soft mode and Goldstone mode contributions to the electric permittivity for the mixture BAHABAC [32]; the full lines are theoretical results from a generalized Landau theory.

of the permittivity component along the y-axis, k is an effective elastic constant for the helicoidal distortion, $q_0 = 2\pi/p$ is the wave vector of the helical pitch p and a is the coefficient in the Landau free energy of the term quadratic in the primary order parameter. The temperature dependence of the soft mode and Goldstone mode contributions to the electric permittivity is illustrated schematically in Fig. 15.

4.2.2   Frequency Dependence of the Electric Permittivity: Dielectric Relaxation

A full theory of the frequency dependence of the electric permittivity of liquid crystals cannot yet be given. It is a formidable problem since it requires proper account to be taken of the influence of orientational order on molecular motion, as well as the effects of macroscopic dielectric anisotropy. The dielectric response of a rigid dipolar molecule rotating in an isotropic fluid can be described in terms of a single relaxation time or rotational diffusion constant. In a liquid crystal phase, the rotational motion is no longer isotropic, and for a rigid rod-like or disc-like mesogen in a uniaxial phase two rotational diffusion constants can be defined, parallel to or perpendicular to the unique inertial axis. In general a dipole moment will not be along the inertial axis of a molecule, and there is likely to be a complicated relationship between the motion of the dipole, which determines the frequency dependence of the permittivity, and the rotational motion of the molecule. For rigid molecules, an anisotropic rotational diffusion model in which the isotropic motion of the molecule is modified by the orienting potential of the phase is sufficient to describe the dielectric response of nematic. For smectic phases translational order can affect the reorientation of molecules, and hence the dielectric relaxation, although this seems to have a minor influence for SmA and SmC phases. In tilted smectic phases the molecular tilt causes biaxiality, and provides an additional environmental constraint on molecular rotation; such effects are sometimes detectable in the dielectric properties. The increased in-plane order associated with hexatic liquid crystals and more ordered crystal smectic phases can be detected through measurements of the permittivity components. For chiral liquid crystal phases such as chiral SmA* SmC*, SmF* and SmI*, there are new contributions to the permittivity which arise from the alignment and collective motion of molecular dipoles, and these collective relaxation modes contribute to the dielectric behaviour of these phases.

The origin of a frequency dependent permittivity is molecular motion associated with a dipole moment. In an oriented fluid, induced or permanent dipole moments contribute differently to the components of the permittivity tensor; similarly the effects of molecular motion as reflected by the frequency dependence of the permittivity will also be different for different components.

Both real and imaginary parts of the permittivity tensor are frequency dependent, and there is a relationship between them known as a Kramers–Kronig relation:

$$\varepsilon_i(\omega)' = \varepsilon_i(\infty)' + \frac{2}{\pi} \int_0^\infty \frac{\omega \varepsilon_i(\omega)''}{\left(\omega^2 - \omega_0^2\right)} \, d\omega$$

$$\varepsilon_i(\omega)'' = \frac{2\omega_0}{\pi} \int_0^\infty \frac{\varepsilon_i(\omega)'}{\left(\omega^2 - \omega_0^2\right)} \, d\omega \qquad (54)$$

where the integral excludes the singular point at $\omega = \omega_0$. It is normal to measure and analyse both the real and imaginary parts of $\varepsilon(\omega)^*$, but any model which violates Eq. (54) must be open to doubt.

An idealized picture of the frequency dependence of $\varepsilon_\parallel$ and $\varepsilon_\perp$ may be constructed by examining Eqs. (36) and (37) for a uniaxial liquid crystal having no local biaxial order. The effects of induced moments can be removed by subtracting the high frequency part of the permittivity or square of the refractive index, giving for the real parts of the permittivity:

$$\varepsilon_\parallel(\omega) - n_\parallel^2 = \frac{NLF^2 g_\parallel^\parallel}{3\varepsilon_0 k_B T}$$
$$\left[\mu_l^2(1+2S) + \mu_t^2(1-S) \right]$$

$$\varepsilon_\perp(\omega) - n_\perp^2 = \frac{NLF^2 g_\perp^\perp}{3\varepsilon_0 k_B T}$$
$$\left[\mu_l^2(1-S) + \mu_t^2\left(1+\frac{1}{2}S\right) \right] \quad (55)$$

Each component of $\varepsilon(\omega)'$ contains two contributions from the molecular dipole moment, and each can have different relaxation times or frequencies. Thus frequency scans of $\varepsilon_\parallel(\omega)^*$ and $\varepsilon_\perp(\omega)^*$ are each expected to show two relaxation regions. The characteristic frequencies or relaxation times will be related to the rotational motion of a molecule in an anisotropic environment, and for a uniaxial molecule with two inde-

pendent moments of inertia, the dynamics can be approximately described in terms of three rotational modes. There is some arbitrariness in choosing these modes, but for illustration we assert that the rotation of a molecule can be broken down into contributions from end-over-end rotation (ω_1), precessional motion (ω_2) about the director and rotation (ω_3) about its own long molecular axis. These motions are illustrated in Fig. 16.

A dipole component will contribute to a principal permittivity if there is a mechanism for that component to follow an electric field applied along the particular principal direction of the permittivity. The manner in which the rotational modes allow different dipole components to reverse in particular directions is seen in Fig. 16. In a nematic environment, it is expected that the magnitudes of the characteristic frequencies for the rotational modes will be in the order $\omega_1 \ll \omega_2 < \omega_3$. Contributions to the permittivity from different dipole components will be lost above frequencies corresponding to ω_1, ω_2 and ω_3, and the variation of $\varepsilon_\parallel(\omega)^*$ and $\varepsilon_\perp(\omega)^*$ with frequency is shown in Fig. 17.

This over simplified model matches the experimental measurements obtained for a

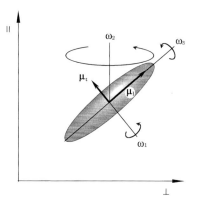

Figure 16. Molecular rotational modes that contribute to the dielectric relaxation.

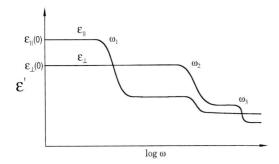

Figure 17. Schematic plot of $\varepsilon'_{\parallel}$ and $\varepsilon'_{\perp}$ against frequency.

number of nematic materials [33], except that the relaxations associated with ω_2 and ω_3 are not separated in the frequency spectrum. The low frequency absorption measured in most liquid crystals can usually be fitted very accurately by a semicircular Cole–Cole plot, but higher frequency relaxations tend to be broader, indicating a range of relaxation times. If other dipolar contributions are included in the permittivity expression, for example those arising from local biaxial order, then these would be expected to relax at different frequencies, and the corresponding dielectric spectrum would be more complicated.

For rigid molecules the frequency dependence of the orientational polarization in isotropic liquids can be calculated using Debye's model for rotational diffusion. This may be modified to describe rotational diffusion in a liquid crystal potential of appropriate symmetry, but the resulting equation is no longer soluble in closed form. Martin, Meier and Saupe [34] obtained numerical solutions for a nematic pseudopotential of the form:

$$u = -bS \cos^2 \theta \qquad (56)$$

where θ is the angle between the long axis of a uniaxial molecule and the director. They assumed that the nematic potential had no

effect on molecular rotation about the long axis of the molecule, so that relaxation processes involving μ_t are not influenced by orientational order.

The solutions to the anisotropic diffusion equation can be written as a series expansion, each term of which can be associated with a particular relaxation time. For a harmonic perturbation of the rotational distribution function, as occurs in a dielectric relaxation experiment with an ac electric field, it was found that a single relaxation time was sufficient to describe the relaxation of μ_l, and this could be expressed in terms of the relaxation time (τ_0) for μ_l in the absence of a nematic potential by:

$$\tau_i = j_i \tau_0 \qquad (57)$$

The subscript $i = \parallel$ or $\perp$ identified the component of the permittivity, and the quantity j_i is a retardation factor calculated numerically from the model, which depends on the coefficient b of the pseudopotential, (j_i is often written as g_i, which we have avoided because of confusion with the Kirkwood dipole correlation factor $g_1^{(i)}$). An approximate result for the retardation factor was calculated [35] such that:

$$j_{\parallel} = \frac{k_B T}{bS} \left[\exp\left(\frac{bS}{k_B T} \right) - 1 \right] \qquad (58)$$

where bS is identified as the height of the potential barrier to end-over-end rotation of a molecule.

The full solution of the rotational diffusion equation including a general single particle potential of $D_{\infty h}$ symmetry has been investigated [36], and it is found that the dipole correlation function, which can be related to the permittivity as a function of frequency, is a sum over many exponential terms each characterised by a different relaxation time. Extending Eq. (20) for an anisotropic fluid gives:

$$\frac{\varepsilon_i(\omega)^* - n_i^2}{\varepsilon_i(0) - n_i^2} \left[\frac{\left(2\varepsilon_i(\omega) + n_i^2\right)\varepsilon_i(0)}{\left(2\varepsilon_i(0) + n_i^2\right)\varepsilon_i(\omega)}\right]$$

$$= 1 - i\omega \int_0^{\infty} C_i(t) e^{i\omega t}\, dt \qquad (59)$$

where $C_i(t)$ is the dipole time correlation function defined by:

$$C_i(t) = \frac{\langle \mu_i(0)\,\mu_i(t)\rangle}{\langle \mu_i(0)\,\mu_i(0)\rangle} \qquad (60)$$

It was observed that in many cases $C_i(t)$ could be approximated by the first term, and this gives rise to four relaxation times corresponding to the four dipole contributions appearing in Eq. (55). Resolving the dipole into its longitudinal and transverse components, as above, enables the correlation functions for directions parallel and perpendicular to the director to be written as [36, 37]:

$$C_{\parallel}(t) = \frac{1}{3\mu_0^2}\left[\begin{array}{l}\mu_l^2(1+2S)\Phi_{00}(t) +\\ \mu_t^2(1-S)\Phi_{01}(t)\end{array}\right] \qquad (61)$$

$$C_{\perp}(t) = \frac{1}{3\mu_0^2}\left[\begin{array}{l}\mu_l^2(1-S)\Phi_{10}(t) +\\ \mu_t^2\left(1+\frac{1}{2}S\right)\Phi_{11}(t)\end{array}\right] \qquad (62)$$

where $\Phi_{kl}(t)$ describe the time dependence of different angular functions representing different relaxation modes for the molecular dipole in an anisotropic environment [38]. In the rotational diffusion model each $\Phi_{kl}(t)$ can be written in terms of a single relaxation time related to a particular rotational mode. For example the low frequency end-over-end rotational is given by Φ_{00}, and can be accurately represented by a single exponential, so that:

$$\Phi_{00}(t) = e^{-\frac{t}{\tau_{00}}} = \langle\cos\theta(0)\cos\theta(t)\rangle \qquad (63)$$

The relaxation times τ_{ij} depend on the parameters of the assumed nematic pseudopo-

tential and the anisotropy in the rotational diffusion constants $D_{\parallel}$ and $D_{\perp}$, which are related to the molecular shape and the local viscosity. If $D_{\parallel}$ and $D_{\perp}$ are assumed to be equal, then $\Phi_{01}(t) = \Phi_{10}(t)$, and the rotational modes can be represented as shown in Fig. 16 and the effect of an ordering potential on the relaxation times is shown schematically in Fig. 18.

These results indicate that the effect of the liquid crystal ordering potential is to decrease the relaxation frequency for end-over-end rotation (τ_{00}^{-1}) and increase it for rotation about the molecular long axis. If anisotropy in the rotational diffusion constants is included, then the relaxation time τ_{00} is further retarded.

Experimental measurements of dielectric relaxation confirm qualitatively the predictions of the rotational diffusion model, in that $\varepsilon_{\parallel}$ has a low and high frequency relaxation, while $\varepsilon_{\perp}$ only shows relaxations at higher frequencies. Unfortunately there are few liquid crystal systems that have been studied over wide frequency ranges, and measurements at high frequencies >50 MHz on aligned samples are difficult. Some typical results are shown in Fig. 19.

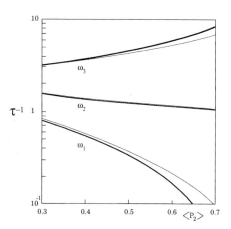

Figure 18. Calculated relaxation frequencies in reduced units plotted as a function of order parameter for para-azoxyanisole [36].

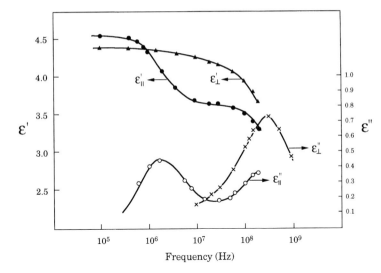

Figure 19. Complex permittivity components for 4-pentylphenyl 4-propylbenzoate [39].

In the study of dielectric relaxation, temperature is an important variable, and it is observed that relaxation times decrease as the temperature increases. In Debye's model for the rotational diffusion of dipoles, the temperature dependence of the relaxation is determined by the diffusion constant or microscopic viscosity. For liquid crystals the nematic ordering potential contributes to rotational relaxation, and the temperature dependence of the order parameter influences the retardation factors. If rotational diffusion is an activated process, then it is appropriate to use an Arrhenius equation for the relaxation times:

$$\tau = A \exp \frac{E_a}{k_B T} \qquad (64)$$

where E_a is the activation energy or barrier to dipole reorientation, and A is another parameter of the model. This would seem to be a useful way to describe end-over-end rotation in liquid crystals [40].

It is observed experimentally that the activation energies for dipole reorientation often change at liquid crystal phase transitions. Changes occur due to differences in the degree of order and the local viscosity,

and it is expected that the activation energies will be higher in smectic phases. In practice is often observed that the end-over-end rotation has a lower activation energy in the smectic A phase than in the higher temperature nematic phase, and this casts some doubt on the role of the order parameter and viscosity, which are both higher in smectic phases than the nematic phase. Benguigui [41] has satisfactorily explained the results for τ_{00} measured for nematic and SmA phases in terms of a free-volume model due to Vogel and Fulcher. Their result for for the relaxation time is:

$$\tau = B \exp \frac{E_a}{k_B (T - T_0)} \qquad (65)$$

where T_0 is a hypothetical glass transition temperature. At phase transitions from disordered smectics (A and C) to those with hexatic order (B, F, and I) there is an order of magnitude increase in relaxation times, but activation energies for end-over-end dipole reorientation are similar in all smectic phases. This is illustrated in Fig. 20 using results for 4-pentylphenyl 4'-heptylbiphenyl-4-carboxylate [42]. The relaxation frequencies decrease by nearly a decade on go-

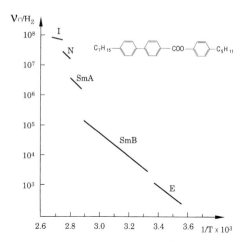

Figure 20. Relaxation frequencies plotted on a logarithmic scale as a function of inverse temperature (K) for 4-pentylphenyl 4′-heptylbiphenyl-4′-carboxylate.

ing from the nematic phase or the smectic A phase, and from the smectic A phase to the smectic B phase, but the activation energies, given by the slope of the lines in Fig. 20 are approximately the same.

Models for the interpretation of the low frequency relaxation in liquid crystals are often based on a single particle relaxation process, but spectroscopic probes of molecular motion such as magnetic resonance, neutron scattering and time-resolved fluorescence depolariastion, suggest that reorientation times for mesogens are of the order of 10^{-9} s to 10^{-10} s in isotropic, nematic and disordered smectic phases. Thus interpretation of dielectric relaxation processes at MHz or even kHz frequencies in terms of single molecule rotation is not likely to be correct. The low frequency relaxations observed in liquid crystals are the result of collective molecular motion, although the models outlined above are useful in analysing results and comparing materials.

It has been demonstrated that the different dipolar contributions to the permittivity components in Eq. (55) cease to contribute at different frequencies, and if these frequencies are well-separated the Debye equations can be applied separately to each term. The dielectric anisotropy changes as a function of frequency, and for particular relative values of longitudinal and transverse dipole moments, it is possible for $\Delta\varepsilon$ to change sign from positive at low frequencies to negative at high frequencies [39]. Liquid crystals exhibiting this behaviour are known as dual-frequency or two-frequency materials, because of potential applications in fast-switched liquid crystal devices, which are addressed by both low frequency ($\Delta\varepsilon_{lf} + ve$) and high frequency ($\Delta\varepsilon_{hf} - ve$) signals. Application of the Debye Equation (10) for $\varepsilon_{\parallel}$ gives the frequency at which $\Delta\varepsilon(\omega)$ changes sign:

$$\omega(\Delta\varepsilon = 0) = \tau_{\parallel}^{-1} \sqrt{\frac{\Delta\varepsilon_{lf}}{-\Delta\varepsilon_{hf}}} \qquad (66)$$

The frequency dependence of the permittivity component of chiral smectic phases along the director is similar to that observed for nonchiral materials, but for dielectric measurements perpendicular to the layer normal, ferroelectric polarization results in additional contributions to the electric susceptibility. Both the soft mode and Goldstone mode contributions are frequency dependent, and the latter gives rise to a low frequency relaxation for helicoidal smectic structures corresponding to the rotation of the polarization about the helix axis. This relaxation frequency is approximately independent of temperature, indicating that rotation of the polarization about the helical axis has zero activation energy, and so is identified as a Goldstone process. There are in principle four relaxation processes that can contribute to $\varepsilon_{\perp}$ in chiral smectic phases: two at low frequency are associated with the motion of the director, but there are also two high frequency modes which relate to changes in the polarization for a fixed director orientation. The former low frequency process exists for the orthogonal smectic phases. Typical relaxation frequencies for these modes are $\omega_S \approx 1 \to 10^4$ kHz and $\omega_G \approx 10 \to 200$ Hz. The high frequency polarization modes (ω_{PS}, ω_{PG}) are essentially the same as the dipole relaxation modes for $\varepsilon_\perp$ already discussed in the context of non-chiral phases, except they relate to a linear transverse dipolar contribution to the permittivity; they are degenerate in orthogonal smectic phases, and have relaxation frequencies in the region of 500 MHz. The temperature dependence of these relaxation processes is illustrated schematically in Fig. 21.

Contributions to the permittivity from fluctuations in the amplitude of the tilt angle are expected to be small away from phase transitions, but this process is strongly temperature dependent, and the relaxation frequency tends to zero at a phase transition: this relaxation is known as the soft mode in common with similar behaviour in crystals. Both the Goldstone mode and the soft mode relaxation processes are a result of the cooperative motion of molecular dipoles, and there is no proper molecular theory for them. The real and imaginary parts of the permittivity fit semicircular Cole–Cole plots, and so each mode can be characterized by a single relaxation frequency, although the dielectric absorptions for the two processes may overlap in the frequency spectrum.

The dielectric response of smectic C* liquid crystals can be derived from a Landau model [43] with the result:

$$\varepsilon_\perp(\omega)^* - \varepsilon_\perp(\infty)'$$

$$= \frac{\chi_G}{1 + i\omega\tau_G} + \frac{\chi_S}{1 + i\omega\tau_S}$$

$$+ \frac{\chi_{PG}}{1 + i\omega\tau_{PG}} + \frac{\chi_{PS}}{1 + i\omega\tau_{PS}} \tag{67}$$

where χ_i are the increments in the electric susceptibility associated with the Goldstone and soft modes for the director and the polarization having relaxation frequencies (τ_i). The dielectric increment for the Goldstone and soft modes is given in Eq. (53). χ_G can also be expressed in terms of the spontaneous polarization as:

$$\varepsilon_0 \chi_G = \frac{1}{2k}\left(\frac{P_S}{q\sin\theta}\right)^2 \tag{68}$$

As explained earlier, both the Goldstone and soft modes contribute to the perpendicular permittivity component in the smectic C* phase, although away from T_c the Goldstone mode dominates in twisted structures.

4.3 References

[1] C. J. F. Bottcher, P. Bordewijk, *Theory of Electric Polarization*, **1978**, Vol. 2, Elsevier, Amsterdam, Netherlands, p. 61.
[2] K. S. Cole, R. H. Cole, *J. Chem. Phys.* **1941**, 9, 341.
[3] D. Lippens, J. P. Parneix, A. Chapoton, *J. de Phys.* **1977**, 38, 1465.
[4] J. M. Deutch, *Faraday Symposium of the Chemical Society* **1977**, p. 26.
[5] See Ref. [1], p. 206.
[6] D. D. Klug, D. E. Kranbuehl, W. E. Vaughan, *J. Chem. Phys.* **1969**, 50, 3904.
[7] W. Maier, G. Meier, *Z Naturforsch.* **1961**, 16 a, 262.
[8] K. Toriyama, D. A. Dunmur, S. E. Hunt, *Liquid Crystals* **1988**, 5, 1001.

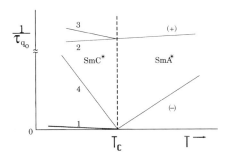

Figure 21. Schematic variation of the frequency of relaxation modes for chiral SmA and SmC phases [30].

[9] D. A. Dunmur, D. A. Hitchen, Xi-Jun Hong, *Mol. Cryst. Liq. Cryst.* **1986**, *140*, 303.

[10] See Ref. [1], p. 444.

[11] A. A. Minko, V. S. Rachevich, S. Ye Yakovenko, *Liq. Cryst.* **1989**, *4*, 1.

[12] E. M. Averyanov, M. A. Osipov, *Sov. Phys. Usp.* **1990**, *33*, 365.

[13] P. Palffy-Muhoray, D. A. Balzarini, D. A. Dunmur, *Mol. Cryst. Liq. Cryst.* **1984**, *110*, 315.

[14] P. Bordewijk, W. H. de Jeu, *J. Chem. Phys.* **1978**, *68*, 116.

[15] D. A. Dunmur, R. W. Munn, *Chem. Phys.* **1983**, *76*, 249.

[16] K. Toriyama, D. A. Dunmur, *Mol. Phys.* **1985**, *56*, 478.

[17] D. A. Dunmur, K. Toriyama, *Mol. Cryst. Liq. Cryst.* **1995**, *264*, 131.

[18] K. Toriyama, S. Sugimari, K. Moriya, D. A. Dunmur, R. Hanson, *J. Phys. Chem.* **1996**, *100*, 307.

[19] D. A. Dunmur, P. Palffy-Muhoray, *Mol. Phys.* **1992**, *76*, 1015.

[20] D. A. Dunmur, R. H. H. Okada, H. Onnagawa, S. Sugimori, K. Toriyama, *Proceedings of 15th Int. Display Res. Conference*, Hamamatsu, Japan, **1995**, p. 563.

[21] J. Chrusciel, B. Gestblom, M. Makrenek, W. Haase, M. Pfeiffer, S. Wrobel, *Liq. Cryst.* **1993**, *14*, 565.

[22] L. Benguigui, *J. de Phys.* **1980**, *41*, 341.

[23] C. Druon, J. M. Wacrenier, F. Hardouin, N. H. Tinh, H. Gasparoux, *J. de Phys.* **1983**, *44*, 1195.

[24] B. R. Ratna, C. Nagabhushana, V. N. Raja, R. Shashidhar, S. Chandrasekhar, G. Heppke, *Mol. Cryst. Liq. Cryst.* **1986**, *138*, 245.

[25] L. Benguigui, *J. de Phys.* **1979**, *40*, 705.

[26] W. H. de Jeu, W. J. A. Goosens, P. Bordewijk, *J. Chem. Phys.* **1974**, *61*, 1985.

[27] L. Benguigui, D. Cahib, *Phys. Stat. Sol.* **1978**, *a47*, 71.

[28] C. Jones, E. P. Raynes, *Liq. Cryst.* **1992**, *11*, 199.

[29] F. Gouda, W. Kuczynski, S. T. Lagerwall, M. Matuszczyk, K. Sharp, *Phys. Rev. A* **1992**, *46*, 951.

[30] B. Zeks, R. Blinc, *Ferroelectric Liquid Crystals* (Ed.: G. W. Taylor), Gordon and Breach, Philadelphia, USA **1991**, p. 395.

[31] See Ref. [30], p. 372.

[32] T. Carlsson, B. Zeks, C. Filipic, A. Levstik, *Ferroelectrics* **1988**, *84*, 223.

[33] C. Druon, J. M. Wacrenier, *Mol. Cryst. Liq. Cryst.* **1982**, *88*, 99.

[34] A. J. Martin, G. Meier, A. Saupe, *Faraday Society Symposium No 5*, **1971**, 119.

[35] G. Meier, A. Saupe, *Mol. Cryst. Liq. Cryst.* **1996**, *1*, 515.

[36] P. L. Nordio, G. Rigatti, U. Segre, *Mol. Phys.* **1973**, *25*, 129.

[37] W. Otowski, W. Demol, W. van Dael, *Mol. Cryst. Liq. Cryst.* **1993**, *226*, 103.

[38] G. S. Attard, *Mol. Phys.* **1986**, *58*, 1087.

[39] M. F. Bone, A. H. Price, M. G. Clark, D. G. McDonnell, *Liquid Crystals and Ordered Fluids*, Vol. 4 (Eds.: A. C. Griffin, J. P. Johnson), Plenum, New York **1984**, p. 799.

[40] W. H. de Jeu, Th. W. Latouwers, *Mol. Cryst. Liq. Cryst.* **1974**, *26*, 225.

[41] L. Benguigui, *Mol. Cryst. Liq. Cryst.* **1984**, *114*, 51.

[42] A. Buka, L. Bata, *Cryst. Res. Tech.* **1981**, *16*, 1439.

[43] T. Carlsson, A. Levstik, B. Zeks, R. Blinc, F. Gouda, S. T. Lagerwall, K. Skarp, C. Filipic, *Phys. Rev.* **1988**, *A38*, 5833.

5 Elastic Properties

David Dunmur and Kazuhisa Toriyama

Elasticity is a macroscopic property of matter defined as the ratio of an applied static stress (force per unit area) to the strain or deformation produced in the material; the dynamic response of a material to stress is determined by its viscosity. In this section we give a simplified formulation of the theory of torsional elasticity and how it applies to liquid crystals. The elastic properties of liquid crystals are perhaps their most characteristic feature, since the response to torsional stress is directly related to the orientational anisotropy of the material. An important aspect of elastic properties is that they depend on intermolecular interactions, and for liquid crystals the elastic constants depend on the two fundamental structural features of these mesophases: anisotropy and orientational order. The dependence of torsional elastic constants on intermolecular interactions is explained, and some models which enable elastic constants to be related to molecular properties are described. The important area of field-induced elastic deformations is introduced, since these are the basis for most electro-optic liquid crystal display devices.

5.1 Introduction to Torsional Elasticity

An important aspect of the macroscopic structure of liquid crystals is their mechanical stability, which is described in terms of elastic properties. In the absence of flow, ordinary liquids cannot support a shear stress, while solids will support compressional, shear and torsional stresses. As might be expected the elastic properties of liquid crystals are intermediate between those of liquids and solids, and depend on the symmetry and phase type. Thus smectic phases with translational order in one direction will have elastic properties similar to those of a solid along that direction, and as the translational order of mesophases increases, so their mechanical properties become more solid-like. The development of the so-called continuum theory for nematic liquid crystals is recorded in a number of publications by Oseen [1], Frank [2], de Gennes and Prost [3] and Vertogen and de Jeu [4]; extensions of the theory to smectic [5] and columnar phases [6] have also been developed. In this section it is intended to give an introduction to elasticity that we hope will make more detailed accounts accessible: the importance of elastic properties in determining the behavior of the mesophases will also be described.

The starting point for a discussion of elastic properties is Hooke's Law, which states that the relative extension of a wire (strain) is proportional to the force per unit area (stress) applied to the wire, and the constant of proportionality is the elastic constant:

$$\frac{F}{A} = k\frac{x}{l} \tag{1}$$

The corresponding elastic energy can be obtained by integrating with respect to strain to give:

$$U = \int F dx \quad \text{or} \quad \frac{U}{V} = \frac{1}{2}ke^2 \tag{2}$$

where $e = x/l$ is the strain, and U/V is the energy density. To describe the relationship between stress and strain in three dimensions requires a tensorial representation of the elements of stress and strain. If a body is subjected to a general stress, then the relative positions of points within the body will change.

The change in positions of particles at points r_1 and r_2 (see Fig. 1) due to some applied stress can be written as:

$$r_1' = r_1 + u$$
$$r_2' = r_2 + v \tag{3}$$

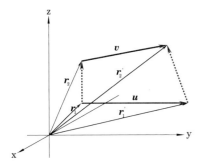

Figure 1. Strain in a body.

for small strains we can write using the tensor suffix notation:

$$v_\alpha = u_\alpha + \left(\frac{du_\beta}{dr_\alpha}\right)r_\beta \tag{4}$$

where $r_\alpha = r_{2\alpha} - r_{1\alpha}$, so that the change in relative positions of r_1 and r_2 becomes:

$$(r_{2\alpha}' - r_{1\alpha}') - (r_{2\alpha} - r_{1\alpha}) = \left(\frac{du_\beta}{dr_\alpha}\right)r_\beta \tag{5}$$

The tensor $e_{\alpha\beta} = \left(\dfrac{du_\beta}{dr_\alpha}\right)$ is known as the strain tensor, the diagonal elements of which measure extensional strain along the x, y, z axes, while off-diagonal elements are a measure of the shear strains (i.e. the change in the angle between the position vectors r_1 and r_2 in the strained state). For small strains, the change in this angle (θ_{12} defined in radians) projected onto the xy, yz and zx planes is the sum of appropriate off-diagonal elements:

$$\gamma_{xy} = e_{xy} + e_{yx}$$
$$\gamma_{yz} = e_{yz} + e_{zy} \tag{6}$$
$$\gamma_{zx} = e_{zx} + e_{xz}$$

Clearly if the strain tensor is antisymmetric such that $e_{\alpha\beta} = -e_{\beta\alpha}$, then the shear angles are zero, which corresponds to a whole body rotation without distortion. It is usual, therefore, to redefine the strain tensor in terms of its symmetric and antisymmetric parts, so that:

$$e_{\alpha\beta} = \frac{1}{2}\left(e_{\alpha\beta} + e_{\beta\alpha}\right)_{sym} + \frac{1}{2}\left(e_{\alpha\beta} - e_{\beta\alpha}\right)_{antisym} \tag{7}$$

The symmetric part of the strain tensor can be associated with changes in the relative positions of particles within a strained sample. For incompressible materials this is zero, and such an assumption is normally applied to nematic liquid crystals. However, in smectic and columnar phases the translational order results in some nonzero components of the symmetric part of the strain tensor.

We have defined the stress applied to a body as the force per unit area: the force may be perpendicular to the unit area, as with normal stress or pressure, or it can be in the plane of the unit area when it is known as shear stress. For any particular direction defining the normal to an element of area A, there will be a single component of normal stress and two shear stress components. Thus a system of forces acting on a body can be described in terms of the nine components of the stress tensor $\sigma_{\alpha\beta}$, defined as:

$$\sigma_{\alpha\beta} = \frac{n_\alpha F_\beta}{A} \tag{8}$$

where F_α is a force acting on an element of area A, the direction of which is defined by its normal n_α. Diagonal components of the stress are pressure, while off-diagonal elements refer to shear stress; for a stressed body to be in mechanical equilibrium, the normal forces on opposite faces must be equal, and the turning moment represented by off-diagonal elements must be zero (see Fig. 2).

Clearly the strain is a consequence of stress, and for small strains there is a linear relationship between them. Since both stress and strain are second rank tensors, the material property that links them must in general be a fourth rank tensor, so that:

$$\sigma_{\alpha\beta} = \sum_{\gamma,\delta=x,y,z} c_{\alpha\beta,\gamma\delta}\, e_{\gamma\delta} \tag{9}$$

The intrinsic symmetry of $\sigma_{\alpha\beta}$ and $e_{\gamma\delta}$ reduces the number of components of $c_{\alpha\beta,\gamma\delta}$ to 36: the elasticity tensor $c_{\alpha\beta,\gamma\delta}$ is also symmetric with respect to interchange of pairs of suffixes $\alpha\beta$ and $\gamma\delta$ which further limits the number of independent components to 21. Phase symmetry also lowers the numbers of elasticity components that have to be independently measured.

For small strains, the elastic energy density is second order in the strain so that:

$$u = \frac{1}{2} c_{\alpha\beta,\gamma\delta}\, e_{\alpha\beta}\, e_{\gamma\delta} \tag{10}$$

and for isothermal strains this is a direct contribution to the free energy density of the system; in Eq. (10) summation over all suffixes is assumed, representing 81 terms. There is an alternative notation widely used in elasticity theory, which enables the elasticity to be expressed in a more compact way. For homogeneous strain, and in the absence of a turning moment, both $\sigma_{\alpha\beta}$ and $e_{\alpha\beta}$ are symmetric tensors having six independent components. New elastic constants can be defined by:

$$\sigma_i = \sum_{j=1,6} c_{ij}\, e_j \tag{11}$$

but care must be exercised in manipulating the components with reduced indices, since they no longer transform as tensors.

Nematic liquid crystals having no translational order will not support extensional or shear strains, but they will support torsional strain which results from application of a torque. The torsional strain is conveniently represented in terms of the angular

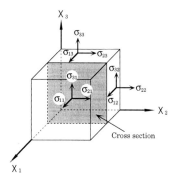

Figure 2. Elements of stress tensor.

displacement of a unit vector (the director) at position r from its equilibrium orientation at the origin, and by analogy with Eq. (4) the torsional strain tensor $e_{\alpha\beta}$ can be defined:

$$n'_\alpha = n_\alpha + \left(\frac{dn_\beta}{dr_\alpha}\right) r_\beta$$

or

$$n'_\alpha = n_\alpha + e_{\alpha\beta}\, r_\beta \tag{12}$$

As before the energy density can be written as:

$$g = \frac{1}{2} k_{\alpha\beta,\gamma\delta}\, e_{\alpha\beta}\, e_{\gamma\delta} \tag{13}$$

where we have now used free energy for an isothermal strain, and the tensor $k_{\alpha\beta,\gamma\delta}$ is the torsional elasticity. If n is chosen to be along the z-axis at the origin, for small strains, the components $e_{\beta z}$ can be neglected, and the nonvanishing elements of the strain tensor can be identified as follows with splay, twist and bend deformations of the director:

$$\text{splay}: \quad e_{xx} = \frac{dn_x}{dx};\ e_{yy} = \frac{dn_y}{dy}$$

$$\text{twist}: \quad e_{xy} = \frac{dn_y}{dx};\ e_{yx} = \frac{dn_x}{dy} \tag{14}$$

$$\text{bend}: \quad e_{zx} = \frac{dn_x}{dz};\ e_{zy} = \frac{dn_y}{dz}$$

these are illustrated in Fig. 3.

The condensed notation for the elements of the torsional elasticity tensor is normally used, and the torsional strain elements are written as a column vector with the compo-

nents:

$$e_i(i = 1 - 6) = e_{xx}, e_{yx}, e_{zx}, e_{xy}, e_{yy}, e_{zy} \tag{15}$$

so the free energy density can now be written as:

$$g = \frac{1}{2} \sum_{i,j=1,6} k_{ij}\, e_i\, e_j \tag{16}$$

The strain tensor must conform to the symmetry of the liquid crystal phase, and as a result, for nonpolar, nonchiral uniaxial phases there are ten nonzero components of k_{ij}, of which four are independent (k_{11}, k_{22}, k_{33} and k_{24}). These material constants are known as torsional elastic constants for splay (k_{11}), twist (k_{22}), bend (k_{33}) and saddle-splay (k_{24}); terms in k_{24} do not contribute to the free energy for configurations in which the director is constant within a plane, or parallel to a plane. The simplest torsional strains considered for liquid crystals are one dimensional, and so neglect of k_{24} is reasonable, but for more complex director configurations and at surfaces, k_{24} can contribute to the free energy [7]. In particular k_{24} is important for curved interfaces of liquid crystals, and so must be included in the description of lyotropic and membrane liquid crystals [8]. Evaluation of Eq. (16) making the stated assumptions, leads to [9]:

$$g = \frac{1}{2} k_{11}\left(e_{xx} + e_{yy}\right)^2 + \frac{1}{2} k_{22}\left(e_{xy} - e_{yx}\right)^2$$
$$+ \frac{1}{2} k_{33}\left(e_{zx} + e_{zy}\right)^2$$
$$- \left(k_{22} - k_{24}\right)\left(e_{xx}\, e_{yy} - e_{xy}\, e_{yx}\right) \tag{17}$$

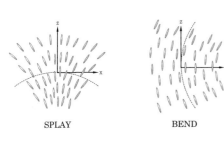

SPLAY   BEND

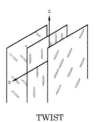

TWIST

Figure 3. Principal torsional elastic deformations.

This expression is often written in a more compact vector notation as:

$$g = \frac{1}{2}\left[k_{11}(\nabla \cdot n)^2 + k_{22}(n \cdot \nabla \times n)^2\right.$$
$$+ k_{33}(n \times \nabla \times n)^2 - (k_{22} - k_{24})$$
$$\left.\cdot(\nabla \cdot \{n\nabla \cdot n + n \times \nabla \times n\})\right] \quad (18)$$

In the above we have assumed that the lowest energy state is one of uniform parallel alignment of the director; however, it is possible to modify these expressions for situations where the lowest energy state might be one of uniform splay, bend or twist. Of these, the last is important because it describes the helical liquid crystal phases that result from chiral molecules. Such states arise if there are terms in the free energy that are linear in the strain, and for a chiral nematic the free energy density becomes:

$$g = -k_2(e_{xy} - e_{yx}) + \frac{1}{2}k_{11}(e_{xx} + e_{yy})^2 \quad (19)$$
$$+ \frac{1}{2}k_{22}(e_{xy} - e_{yx})^2 + \frac{1}{2}k_{33}(e_{zx} + e_{zy})^2$$

Defining the torsional twist strain as $t=(e_{xy}-e_{yx})$, and minimising Eq. (19) with respect to t results in a stabilized helical structure having a finite twist strain $t_0=k_2/k_{22}$, and the free energy density can be written as:

$$g = \frac{1}{2}k_{11}(e_{xx} + e_{yy})^2$$
$$+ \frac{1}{2}k_{22}(e_{xy} - e_{yx} - t_0)^2$$
$$+ \frac{1}{2}k_{33}(e_{zx} + e_{zy})^2 - \frac{1}{2}k_{22}t_0^2 \quad (20)$$

The lowest energy state for this structure is where the director describes a helix along one of the axes. Thus assuming that there is no bend or splay strain energy, a chiral nematic has a director configuration such that:

$$n_x = \cos(2\pi z/p); \quad n_y = \sin(2\pi z/p); \quad n_z = 0$$

where p is the pitch of the helix, as illustrated in Fig. 4.

Using this director distribution, the nonvanishing terms in the free energy expression Eq. (20) are:

$$g = \frac{1}{2}k_{22}\left[n_y\left(\frac{\partial n_x}{\partial n_z}\right) - n_x\left(\frac{\partial n_y}{\partial n_z}\right) - t_0\right]^2$$
$$- \frac{1}{2}k_{22}t_0^2 \quad (21)$$

which is a minimum when $t_0 = 2\pi/p$, i.e. the lowest energy state is one of uniform pitch.

It is also possible to envisage minimum energy structures with non-zero splay or bend strain, when other terms linear in strain contribute to the free energy. However, a uniformly splayed or bent structure will no longer have the reversal symmetry $+n=-n$ and so will be associated with permanently polarized structures; a uniformly bent structure would have to be biaxial with a polar structure perpendicular to the major axis. These structures, illustrated in Fig. 5, are associated with the phenomenon known as flexoelectricity, where polarization is coupled to elastic strain.

It is not essential that the molecular asymmetry is linked to an electric polarization, although symmetry suggests that the two are

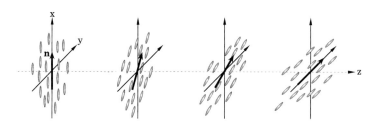

Figure 4. Director helix in a chiral nematic.

Figure 5. Flexo-electric structures.

coupled. Equivalent structures to those illustrated may be obtained on the basis of shape asymmetry, the so called steric dipole.

So far only torsional contributions to the elastic strain energy have been considered, but smectic and columnar liquid crystals having translational order may support extensional and shear strain like solids along certain directions. For a uniaxial SmA phase with one degree of translational freedom, the only homogeneous strain supported is an extension or compression along the axis of translational order (i. e. perpendicular to the layers). Bend and twist strains involve changes in layer spacing, and so are likely to be of very high energy; they will couple with layer compression but will be high order contributions to the free energy, and are neglected for small strains. Thus the elastic free energy of a SmA phase reduces to:

$$g = \frac{1}{2} B \left(\frac{\mathrm{d}u_z}{\mathrm{d}z} \right)^2 + \frac{1}{2} k_{11} (\boldsymbol{\nabla} \cdot \boldsymbol{n})^2 \qquad (22)$$

where B is a compression elastic constant for the smectic layers; $(\mathrm{d}u_z/\mathrm{d}z)$ is the layer strain along the layer normal, and k_{11} is the splay elastic constant.

The description of the elastic properties of columnar phases, biaxial smectics, chiral

smectics and more ordered liquid crystal phases is being developed [10]. Since columnar phases are two dimensional solids, it is expected that compression elasticity in the plane perpendicular to the columns will be important, and the high strain energy associated with in-plane deformations prevents splay or twist torsional distortion [11]. It is however possible to bend the columns, while maintaining a constant in-plane separation of the columns, so bend distortions can be expected in columnar liquid crystals: the corresponding free energy density for a uniaxial columnar phase becomes:

$$g = \frac{1}{2} B \left(\frac{\mathrm{d}u_x}{\mathrm{d}x} + \frac{\mathrm{d}u_y}{\mathrm{d}y} \right)^2$$

$$+ \frac{1}{2} C \left[\left(\frac{\mathrm{d}u_x}{\mathrm{d}x} - \frac{\mathrm{d}u_y}{\mathrm{d}y} \right)^2 + \left(\frac{\mathrm{d}u_x}{\mathrm{d}y} + \frac{\mathrm{d}u_y}{\mathrm{d}x} \right)^2 \right]$$

$$+ \frac{1}{2} k_{33} (\boldsymbol{n} \times \boldsymbol{\nabla} \times \boldsymbol{n})^2 \qquad (23)$$

Results for biaxial smectics and columnar phases have additional compressional terms, but tilted smectic phases can support additional torsional distortions. Such phases are conveniently described in terms of two directors, one along the tilt direction, corresponding to the nematic director $\boldsymbol{n}$, and the projection of $\boldsymbol{n}$ on the layer plane, known as the $\boldsymbol{c}$-director (see Fig. 6).

In practice it is mathematically more convenient to define a director (**a**) normal to the

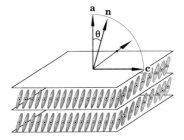

Figure 6. Definition of **a** and **c** directors.

layer such that:

$$n = a \cos\theta + c \sin\theta \qquad (24)$$

Torsional distortions can now be written in terms of derivatives of a and c, and it is found [10] that nine torsional elastic constants are required for the smectic C phase. Mention should be made of the biaxial smectic C* phase, which has a twist axis along the normal to the smectic layers. This helix is associated with a twist in the c-director, and so elastic strain energy associated with this can be described by terms similar to those evaluated for the chiral nematic phase.

5.2 Director Distribution, Defects and Fluctuations

The development of Eq. (18) was based on the idea that torsional strain resulted from a director that is a function of position. Provided that the first nonvanishing term in the free energy is quadratic in torsional strain, the minimum energy configuration of a liquid crystal will be that in which the torsional strain is zero and the director is everywhere parallel to a symmetry axis. Linear contributions to the strain energy result in equilibrium structures in which the director is not uniformly parallel, but has some particular spatial dependence, as with the chiral nematic phase. External influences will also affect the director distribution in space, and surface interaction, external fields and thermal fluctuations give rise to structures in which the director is a function of position. Neglecting fluctuations, the director distribution will be that which minimizes the free energy given by:

$$g = \frac{1}{2} \int \left[k_{11} (\nabla \cdot n)^2 + k_{22} (n \cdot \nabla \times n)^2 \right.$$
$$\left. + k_{33} (n \times \nabla \times n)^2 + g_{ext} \right] dr \qquad (25)$$

here g_{ext} contains contributions to the free energy from external forces.

Experimental observations indicate that it is possible to define the director orientation by surface treatment or external electric or magnetic fields, and a simple example of the effect of surfaces on the director distribution in a nematic is illustrated in Fig. 7.

Two rubbed plates are held at a distance l apart, such that the alignment directions include an angle θ. The director orientation of a nematic between these plates will be defined by the rubbing directions at the plate surfaces, but in the bulk the director distribution will be that which minimizes Eq. (25). De Gennes [12] has shown that for this simple twist deformation, the director orientation varies linearly with position (y) as indicated in the figure. Under these circumstances, the director can be written in terms of:

$$n_x = \sin\left(\frac{\theta y}{l}\right); \quad n_y = 0; \quad n_z = \cos\left(\frac{\theta y}{l}\right) \qquad (26)$$

and the corresponding free energy density per unit wall area becomes:

$$g = \frac{1}{2} k_{22} \frac{\theta^2}{l} \qquad (27)$$

The torque t is given by:

$$t = -\frac{dg}{d\theta} = -\frac{k_{22}\theta}{l} \qquad (28)$$

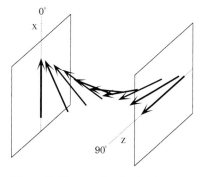

Figure 7. Director distribution between twisted plates for $\theta = 90°$.

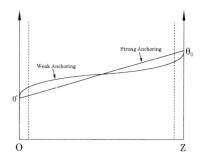

Figure 8. Plot of director orientation for strong and weak anchoring.

and is constant through the sample. Thus plate 1 exerts a torque of $k_{22}\theta/l$ on plate 2, while plate 2 exerts a torque of $-k_{22}\theta/l$ on plate 1.

This calculation assumes that the torques of opposing plates are insufficient to disturb the orientation of the director at the surface: such a condition is known as strong anchoring. In practice the strength of surface interactions may have to be considered, and in the example given, weak anchoring would cause the director orientation close to the boundary surfaces to depart from a linear dependence on position, as illustrated in Fig. 8.

For weak anchoring there is a competition between the torque in the bulk due to one plate, and the torque resulting from the other surface; this is usually confined to a boundary region as indicated in the figure. External electric and magnetic fields will also affect the equilibrium director distribution, and this is the basis of many liquid crystal applications; these effects will be considered later.

5.2.1　Defects in Liquid Crystals

It has been explained how nonuniform director distributions can arise owing to surface interactions or external fields, and these can be detected optically or by other physical methods. In many liquid crystal samples, large inhomogeneities of director orientation associated with structural defects can be observed by polarized light microscopy, and these clearly involve elastic strain energy. In this section we will outline the elastic theory of these defects: reviews of this topic have been given by Chandrasekhar [13] and Kleman [14]. As an illustration we will consider the simplest type of fluid defect observed in nematics named disclination lines by Frank.

In two dimensions and setting all the torsional elastic constants equal, the free energy density expression Eq. (17) can be written as:

$$g = \frac{1}{2}k\left[\left(e_{xx}+e_{yy}\right)^2 + \left(e_{xy}-e_{yx}\right)^2\right] \qquad (29)$$

If the director is confined to a plane, then its components may be represented by:

$$n_x = \cos\theta(r) \quad \text{and} \quad n_y = \sin\theta(r)$$

where $\theta(r)$ is a function of x and y in the plane. The free energy density then becomes:

$$g = \frac{1}{2}k\left[\left(\frac{d\theta}{dx}\right)^2 + \left(\frac{d\theta}{dy}\right)^2\right] \qquad (30)$$

This is minimized, corresponding to an equilibrium structure when:

$$\frac{d^2\theta}{dx^2} + \frac{d^2\theta}{dy^2} = 0 \qquad (31)$$

The defect-free solution to Eq. (31) is obviously when θ is independent of position, but other solutions corresponding to disclination lines are given by:

$$\theta = s\tan^{-1}\left(\frac{y}{x}\right) + \theta_0 \qquad (32)$$

where s is the strength of the disclination.

For a fixed distance from the origin, the director orientation as given by Eq. (32) changes by $s\pi$ as y/x varies from $+\infty$ to $-\infty$; contours of solutions for different s are given in Fig. 9. These director distributions give rise to characteristic images of disclination lines, which are readily observed for thin films in a polarizing microscope.

The elastic strain energy associated with a disclination line can now be calculated from Eq. (30) as:

$$g/\text{unit length} = s^2 \pi k \ln\left(\frac{r_{max}}{r_{min}}\right) \quad (33)$$

r_{max} is the distance from the centre of the disclination at which the strain energy becomes effectively zero, while r_{min} defines the size of the core of the disclination, the energy of which cannot be calculated from the above equations, which apply to small strains. The quadratic dependence of the strain energy on s explains why usually only disclinations of low s are observed in low molecular weight liquid crystals: higher strength disclinations have been observed in

polymer liquid crystals [14], since these can support higher strain energies.

In real samples there will be a number of disclinations corresponding to different magnitudes and signs of s. The resultant effect of a number of disclinations on the director orientation at some point in the fluid can be obtained by combining the corresponding director angles. The corresponding free energy shows that disclinations of similar sign will repel each other, while those of opposite sign will be mutually attracted.

This simple treatment of liquid crystalline defects is only applicable to nematics, and the detailed appearance of disclination lines will differ from the simple structures described above because of differences between the elastic constants for splay, twist and bend. In smectic phases, defects associated with positional disorder of layers will also be important, and some smectic phase defects such as edge dislocations have topologies similar to those described for crystals. The defect structures of liquid crystals contribute to the characteristic optical tex-

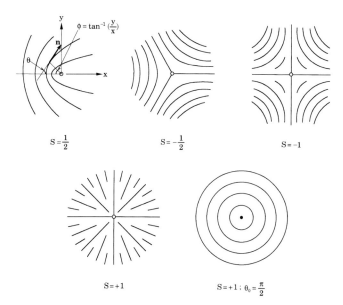

Figure 9. Disclinations obtained by solution of Eq. (32). The lines represent contours of the director.

tures, which are used to identify different liquid crystal phases, and it is apparent that there is a wealth of information on elastic properties contained in the observed optical textures.

5.2.2 Fluctuations

Neglecting static distortions which may result from interactions with surfaces or external fields, the lowest energy director configuration is predicted to be one of uniform director orientation. However, even in the absence of external influences, locally the director fluctuates about its lowest energy configuration, and this local director orientation disorder can contribute to liquid crystal properties. For example the characteristic turbidity of liquid crystalline phases is due to fluctuations in the refractive indices on the scale of the order of the wavelength of light. The origin of this nonuniform alignment is thermal excitation of librational modes associated with the director. The energy is assumed to be a quadratic function of the torsional strain; but since the torsional elastic constants are very small, little thermal energy is required to excite a torsional libration and so disturb the uniform director configuration. In reality there are many torsional modes that can be simultaneously excited, and the long-range orientational structure of the liquid crystal can be extensively disordered. This picture of director disorder assumes that the molecular organization is unaffected by the long-range macroscopic disorder, although it must be remembered that the director has no independent existence, and is itself defined by the molecular orientations averaged over an appropriate volume.

The director orientational disorder can be formally represented in exactly the same way as the molecular disorder by defining a director order parameter as:

$$Q_{\alpha\beta} = \frac{1}{2}\langle 3n_\alpha(r)n_\beta(r) - \delta_{\alpha\beta}\rangle \tag{34}$$

where $n_\alpha(r)$ are the components of the director as a function of position.

The spatial variation of the director can be expressed in terms of Fourier components of wave vector q, such that:

$$n_\alpha(q) = V^{-1}\int n_\alpha(r)\exp(i\,q\cdot r)\mathrm{d}^3r \tag{35}$$

$$n_\alpha(r) = V(2\pi)^{-3}\int n_\alpha(q)\exp(-i\,q\cdot r)\mathrm{d}^3q$$

Small distortions from a director axis z can be expressed in terms of $n_x(q)$ and $n_y(q)$; however these are not the normal coordinates for a mode of wavelength $\lambda = 2\pi/q$, and using the following transformation, the free energy can be written as the sum of quadratic contributions from the normal modes:

$$\begin{bmatrix} n_1(q) \\ n_2(q) \end{bmatrix} = \frac{1}{\sqrt{(q_x^2 + q_y^2)}}\begin{bmatrix} q_x & q_y \\ -q_y & q_x \end{bmatrix}\begin{bmatrix} n_x \\ n_y \end{bmatrix} \tag{36}$$

For each q in a uniaxial phase there are two normal modes corresponding to a splay–bend distortion $n_1(q)$ and a twist–bend distortion $n_2(q)$: biaxial liquid crystal phases have five normal modes for each value of q. The free energy density can be written in terms of the normal coordinates for torsional displacement in a uniaxial nematic as:

$$g = \frac{1}{2}\sum_q\left[\lambda_+(q)n_1(q)^2 + \lambda_-(q)n_2(q)^2\right] \tag{37}$$

The coefficients λ_+ and λ_- are given by:

$$\lambda_+ = k_{11}\left(q_x^2 + q_y^2\right) + k_{33}\,q_z^2$$

$$\lambda_- = k_{22}\left(q_x^2 + q_y^2\right) + k_{33}\,q_z^2 \tag{38}$$

Because the torsional elastic constants are small, the terms in the expression for the free energy density are treated classically so that according to the equipartition theorem the average contribution of each mode is supposed to be $k_B T/2$, giving:

$$\langle n_1(\boldsymbol{q})^2 \rangle = \frac{k_B T}{V\lambda_+(\boldsymbol{q})}$$

$$\langle n_2(\boldsymbol{q})^2 \rangle = \frac{k_B T}{V\lambda_-(\boldsymbol{q})} \qquad (39)$$

It is now possible to determine the mean square fluctuation of the director from its aligned state along the z-axis, such that:

$$\langle n_x(\boldsymbol{r})^2 + n_y(\boldsymbol{r})^2 \rangle$$

$$= \frac{V}{(2\pi)^3} \int_{q_{min}}^{q_{max}} \langle n_1(\boldsymbol{q})^2 + n_2(\boldsymbol{q})^2 \rangle \, d^3\boldsymbol{q} \qquad (40)$$

There is one problem remaining which is to fix the limits for the integration over q [15]. The lower limit for q is determined by the volume of the sample, since it is unrealistic to have distortion modes of wavelength longer than the maximum dimensions of the sample container: for an infinite sample $q_{min} = 0$. The upper limit for q corresponds to the fluctuation mode of shortest wavelength, which is likely to be of the order of molecular dimensions.

If the elastic constants for splay twist and bend are assumed to be independent of q, and for simplicity are set equal to k, then evaluation of Eq. (40) enables the director order parameter to be determined as follows:

$$\langle Q_{zz}(\boldsymbol{r}) \rangle = \frac{1}{2}\langle 3n_z(\boldsymbol{r})^2 - 1 \rangle$$

$$= 1 - \frac{3}{2}\langle n_x(\boldsymbol{r})^2 + n_y(\boldsymbol{r})^2 \rangle$$

$$= 1 - \frac{3k_B T q_{max}}{2\pi^2 k} \qquad (41)$$

Fluctuations in the director orientation will modulate the anisotropy of physical properties such as the electric permittivity tensor,

which can be written as:

$$\varepsilon_{\alpha\beta} = \bar{\varepsilon} + \frac{\Delta\varepsilon}{3}\left(3n_\alpha(\boldsymbol{r})\,n_\beta(\boldsymbol{r}) - \delta_{\alpha\beta}\right) \qquad (42)$$

where $\bar{\varepsilon}$ is the average permittivity and $\Delta\varepsilon = \varepsilon_{\parallel} - \varepsilon_{\perp}$ is the permittivity anisotropy. Light scattering arises because of spatial and time-dependent fluctuations of the local dielectric tensor:

$$\delta\varepsilon_{\alpha\beta}(\boldsymbol{r}, t) = \varepsilon_{\alpha\beta}(\boldsymbol{r}, t) - \varepsilon_{\alpha\beta}(0) \qquad (43)$$

and if these fluctuations are expressed in terms of Fourier components of wave-vector $\boldsymbol{q}$, the intensity of scattered light can be derived as [16]:

$$I_{i,f}(\boldsymbol{q},\omega) = \frac{const}{2\pi}$$

$$\cdot \int_{-\infty}^{+\infty} \langle \delta\varepsilon_{if}(\boldsymbol{q},0)\,\delta\varepsilon_{if}(\boldsymbol{q},t)\rangle \exp - i\omega t \, dt \quad (44)$$

The subscripts i, f refer to the polarization directions of the incident and scattered light, $\boldsymbol{q}$ is the wave-vector of the scattered light, and the constant is given by:

$$const = \frac{\pi^2 I_0}{\lambda^4 R^2 \varepsilon_0^2} \qquad (45)$$

where I_0 is the incident light intensity, wavelength λ, and R is the distance from the scattering volume to the detector. Choice of suitable scattering geometries [17] and measurement of the intensity as a function of scattering angle allow the determination of the elastic constant ratios k_{33}/k_{11} and k_{22}/k_{11} [18].

5.3   Curvature Elasticity of Liquid Crystals in Three Dimensions

As pointed out for Eq. (17) there is a contribution to the elastic energy from saddle-splay distortions, which can be neglected in some circumstances. However for distortions such that the director is not constant in one plane, the term in k_{24} must be included in the free energy. Liquid crystal structures in which there is curvature in two dimensions can support saddle-splay distortion, and such curvature is often associated with fluid interfaces in liquid crystals. The main area of importance for interfacial contributions to the elastic energy is lyotropic liquid crystals formed by molecular association and segregation in materials having a minimum of two chemical constituents. A typical lyotropic system consisting of water and an amphiphile exhibits lamellar (smectic), hexagonal (columnar) and occasionally nematic liquid crystal phases, and in the phase diagram these are often separated by or bounded by narrow regions of cubic phases. The molecular organization in lyotropics represents an extreme example of amphiphilic liquid crystal phases in which the spatial separation of molecular subunits is stabilized by a second component, usually water, to produce a partitioning of space into polar and nonpolar regions. Inverse lyotropic phase structures can also be prepared where the spatial separation of alkyl chains is stabilized by a nonpolar solvent such as hexane. Spatial organization of molecular subunits can also occur in nonlyotropic liquid crystals resulting in a variety of modulated phase structures, and the formation of curved interfaces is of especial importance for biological membranes.

The macroscopic topology of lyotropic or liquid crystal phases involving segregation is determined by the curvature of the interface; a lamellar structure has zero curvature, while micellar phases or hexagonal phases exhibit interfacial curvature. An interface is defined by the segregation of different molecules or molecular subunits. Deformation of this interface may occur in a variety

of ways: tangential stress will cause a planar stretch, while molecular tilting will generate a stress normal to the interface. Both have a high associated elastic energy, but deformation through interfacial curvature involves much lower elastic energies, and so is the preferred mechanism through which topological changes occur. The formation of spherical micelles requires curvature strain in three dimensions, and a director defined along the normal to the interface is no longer uniform within a plane or parallel to a plane. Bearing this in mind the curvature elastic energy per unit area (w) can be written as (c.f. Eq. (17)):

$$w = \frac{1}{2} k_{11}^{(a)} \left(e_{xx} + e_{yy} - c_0 \right)^2$$
$$+ k_{24}^{(a)} \left(e_{xx} e_{yy} - e_{xy} e_{yx} \right) \tag{46}$$

where c_0 allows for equilibrium structures having a non-zero splay. Since the deformation is confined to the interface and the director n remains everywhere normal to the interface, there are no bend or twist contributions to the elastic energy, and $e_{yx} = e_{xy} = 0$. The first term in Eq. (46) is clearly a splay elastic energy, while the second term was designated by Frank as a saddle-splay energy. From the definition of curvature strain elements $e_{\alpha\beta}$ (Eq. 14), it can be seen that for small strains:

$$e_{xx} = \frac{\mathrm{d}n_x}{\mathrm{d}x} = \frac{1}{R_1}$$

and

$$e_{yy} = \frac{\mathrm{d}n_y}{\mathrm{d}y} = \frac{1}{R_2} \tag{47}$$

where R_1 and R_2 are the radii of curvature of the surface in the zx and zy planes respectively (see Fig. 10), and c_0 is the spontaneous curvature of the surface in the absence of strain. Thus the elastic free energy per unit area can now be written in terms of the

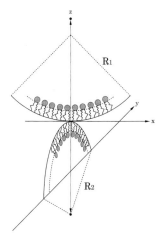

Figure 10. Principal radii of curvature in a saddle-splay distortion.

principal radii of curvature as:

$$w = \frac{1}{2} k_m^{(a)} \left(\frac{1}{R_1} + \frac{1}{R_2} - c_0 \right)^2 + k_g^{(a)} \frac{1}{R_1 R_2} \quad (48)$$

with $k_m^{(a)}$ and $k_g^{(a)}$ defined as elastic constants for the mean curvature $\frac{1}{2}\left(R_1^{-1} + R_2^{-1}\right)$ and gaussian curvature $(R_1 R_2)^{-1}$ respectively.

Different types of surface can be categorized in terms of their mean and gaussian curvatures. For example, a sphere of radius R has a mean curvature of R^{-1} and a gaussian curvature of R^{-2}, while a cylinder has one principal radius of curvature equal to infinity, so the gaussian curvature is zero. Surfaces with zero mean curvature such that $R_1^{-1} = -R_2^{-1}$ are known as minimal surfaces and have been proposed as structures for some cubic phases [19].

5.4 Electric and Magnetic Field-induced Deformations

Competition between two competing elastic torques results in a director distribution that minimises the free energy. The origin of the torque may be mechanical as repre-

sented in Fig. 7, or it may arise through a coupling between an external field and the corresponding susceptibility anisotropy. For a uniaxial material the internal energy density in the presence of a field (F) can be written as:

$$u = -\frac{F^2}{2}\left[\bar{\chi} + \frac{\Delta\chi}{3}\left(3\sin^2\theta - 1\right)\right] \quad (49)$$

where $(90 - \theta°)$ is the angle between the field and the director (see Fig. 11), and so the torque is given by:

$$t = -\frac{du}{d\theta} = -\Delta\chi F^2 \cos\theta \sin\theta$$
$$= -\Delta\chi(\boldsymbol{F}\cdot\boldsymbol{n})(\boldsymbol{F}\times\boldsymbol{n}) \quad (50)$$

The torque is zero when the field is parallel or perpendicular to the director, but depending on the sign of $\Delta\chi$, one of these states is in stable equilibrium, while the other is in unstable equilibrium. For positive $\Delta\chi$, increasing a field perpendicular to n will raise the energy and hence destabilize this state: eventually the increase in energy due to the field exceeds the elastic energy, and so the liquid crystal adopts a new director configuration. Reorientation of the liquid crystal does not necessarily occur as soon as the field is applied (because the torque is zero), and can be a threshold phenomenon occur-

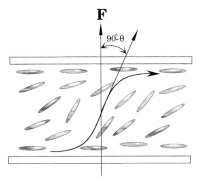

Figure 11. Director reorientation by a field.

ring above a critical field strength. The magnitude of the threshold field can be estimated as follows; if the director is deformed over a distance l, the strain is of order l^{-1}, and the elastic free energy density is k/l^2. The angle-dependent contribution to the free energy density from the field is $\approx \Delta \chi F^2$, and the threshold field for director reorientation is when these energies become equal:

$$\Delta \chi F^2 \approx \frac{k}{l^2}$$

i.e. $F_{\text{threshold}} \approx \frac{1}{l} \sqrt{\frac{k}{\Delta \chi}}$ (51)

the length l is typically of order of the sample thickness. This reorientation of the director by a field is known as a Fréedericksz transition [20], and forms the basis of most display applications of nematic liquid crystals. If the director is not perpendicular (or parallel) to the field because of misalignment or surface tilt of the director, then there is a torque for vanishingly small field strengths, and deformation occurs without a threshold.

5.4.1 Director Distribution in Magnetic Fields

Before outlining the effect of fields on the orientation of the director, it must be emphasized that surface interactions and boundary conditions are usually of importance. The models applied here assume that there is a well-defined director distribution in the absence of any external field, and in practice this can only be provided by suitable treatment of boundary surfaces. The strength of surface interactions must also be considered as this will influence the equilibrium director configuration in the presence of a field. For the simplest description of field effects in liquid crystals it is usual to assume an infinite anchoring energy for

the director at the surface: this is the strong anchoring limit, but it is possible to include finite surface coupling energies [21 – 24]. It is found that there can still be a threshold response to external fields, but the critical field strength is reduced in comparison with the strong anchoring limit.

It is easier to model the deformation induced by magnetic fields than electric fields, because for nonferromagnetic liquid crystals the relative volume magnetic susceptibility anisotropy is very small ($\approx 10^{-6}$) in comparison with the corresponding dielectric anisotropy (≈ 10). This means that the induced magnetization is always parallel to the field direction, in contrast with the electric polarization, which in general makes some angle with the electric field. The magnetic contribution to the free energy of a liquid crystal can be written as (see Eq. 49):

$$g_{\text{mag}} = -\frac{1}{2\mu_0} \int \left(\chi_\perp + \Delta \chi \sin^2 \theta \right) B^2 \, dV \quad (52)$$

where $90 - \theta$ is the angle between the director and the magnetic field. Thus the total free energy including any elastic deformation becomes:

$$g = \frac{1}{2} \int k_{11} (\nabla \cdot \boldsymbol{n})^2 + k_{22} (\boldsymbol{n} \cdot \nabla \times \boldsymbol{n})^2$$
$$+ k_{33} (\boldsymbol{n} \times \nabla \times \boldsymbol{n})^2$$
$$- \mu_0^{-1} \left(\chi_\perp B^2 + \Delta \chi (\boldsymbol{n} \cdot \boldsymbol{B})^2 \right) d\boldsymbol{r}^3 \quad (53)$$

and the equilibrium director distribution $\boldsymbol{n}(\boldsymbol{r})$ is that which minimizes g. To proceed further, it is necessary to specify the geometry of the system more closely, and it is usual to consider three standard configurations (Fig. 12) where the applied field is perpendicular to the director of a uniformly aligned sample of positive susceptibility anisotropy.

These configurations define Fréedericksz transitions from undeformed to deformed states for which the threshold fields are sep-

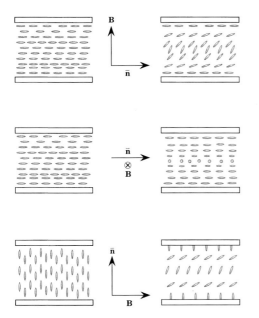

Figure 12. Field induced splay, twist and bend deformations.

arately related to the elastic constants for splay twist and bend. Only deformations for which the director is uniform in a plane are considered (so k_{24} is unimportant), and applying the Euler–Lagrange equations [25] to minimize g (Eq. (53) gives the following results for the three cases in Fig. 12.

(1) Splay:

$$\frac{d}{dz}\left[\left\{k_{11}\cos^2\theta + k_{33}\sin^2\theta\right\}\left(\frac{d\theta}{dz}\right)^2 + \mu_0^{-1}\Delta\chi B^2\sin^2\theta\right] = 0$$

(2) Twist:

$$\frac{d}{dz}\left[k_{22}\left(\frac{d\phi}{dz}\right)^2 + \mu_0^{-1}\Delta\chi B^2\sin^2\phi\right] = 0$$

(3) Bend:

$$\frac{d}{dz}\left[\left\{k_{33}\cos^2\theta + k_{11}\sin^2\theta\right\}\left(\frac{d\theta}{dz}\right)^2 + \mu_0^{-1}\Delta\chi B^2\sin^2\theta\right] = 0 \tag{54}$$

where the deformation plane is defined by the directions $z-n$. The case (1) will be considered in a little more detail [26], and results will be quoted for other boundary conditions.

Eq. (54) implies that:

$$\left[k_{11}\cos^2\theta + k_{33}\sin^2\theta\right]\left(\frac{d\theta}{dz}\right)^2$$
$$+ \mu_0^{-1}\Delta\chi B^2\sin^2\theta = \text{constant } C \tag{55}$$

and this constant may be evaluated by recognizing that for the selected geometry, θ will be a maximum at the mid-point of the cell such that $z=d/2$. Setting this condition in Eq. (55) gives that:

$$C = \mu_0^{-1}\Delta\chi B_2\sin^2\theta_m \tag{56}$$

Substitution in Eq. (55) and rearranging now gives a differential equation for the equilibrium director distribution in terms of the angle θ, the field strength and the position in the cell:

$$\frac{d\theta}{dz} = \left(\mu_0^{-1}\Delta\chi B^2\right)^{1/2}$$
$$\cdot\left[\frac{\sin^2\theta_m - \sin^2\theta}{k_{11}\cos^2\theta + k_{33}\sin^2\theta}\right]^{1/2} \tag{57}$$

or in integral form:

$$\int_0^{\theta_m}\left[\frac{1+k'\sin^2\theta}{\sin^2\theta_m - \sin^2\theta}\right]^{1/2}d\theta$$
$$= \int_0^{d/2}\left(\frac{\mu_0^{-1}\Delta\chi}{k_{11}}\right)^{1/2}B\,dz \tag{58}$$

where a reduced elastic constant $k' = (k_{33}-k_{11})/k_{11}$ has been introduced. It is now convenient to change the variable θ to ψ using:

$$\sin\theta = \sin\theta_m\sin\psi, \text{ and}$$

$$\frac{d\theta}{d\psi} = \left[\frac{\sin^2\theta_m - \sin^2\theta}{1-\sin^2\theta}\right]^{1/2}$$

so that when $\theta = \theta_m$, $\psi = \pi/2$ and $z = d/2$, hence Eq. (58) becomes:

$$\int_0^{\pi/2} \left[\frac{1 + k' \sin^2 \theta_m \sin^2 \psi}{1 - \sin^2 \theta_m \sin^2 \psi} \right]^{1/2} d\psi$$

$$= \int_0^{d/2} \left(\frac{\mu_0^{-1} \Delta \chi}{k_{11}} \right)^{1/2} B \, dz \qquad (59)$$

The right hand side of Eq. (59) is simply $\dfrac{Bd}{2} \left(\dfrac{\mu_0^{-1} \Delta \chi}{k_{11}} \right)^{1/2}$, but the left hand side is an elliptic integral of the third kind and must be evaluated numerically. At the threshold $\theta_m = 0$, the left hand side reduces to $\pi/2$ giving the result for the Fréedericksz threshold magnetic field (B_c) as:

$$B_c = \frac{\pi}{d} \left(\frac{k_{11}}{\mu_0^{-1} \Delta \chi} \right)^{1/2} \qquad (60)$$

for fields above threshold:

$$\frac{B}{B_c} = \frac{2}{\pi} \int_0^{\pi/2} \left[\frac{1 + k' \sin^2 \theta_m \sin^2 \psi}{1 - \sin^2 \theta_m \sin^2 \psi} \right]^{1/2} d\psi \quad (61)$$

The director distribution through the sample may be obtained for a particular value of B/B_c by finding θ_m from Eq. (61), and the value of $\theta(z)$ at position z within the sample is obtained (Eq. 58) from:

$$\int_0^{\theta(z)} \left[\frac{1 + k' \sin^2 \theta}{\sin^2 \theta_m - \sin^2 \theta} \right]^{1/2} d\theta$$

$$= z B \left(\frac{\mu_0^{-1} \Delta \chi}{k_{11}} \right)^{1/2} = \frac{z \pi B}{d B_c} \qquad (62)$$

Results for $\theta(z)$ at various values of $\dfrac{B}{B_c}$ are plotted in Fig. 13 for an assumed value of $k' = 1$, corresponding to $k_{33} = 2 k_{11}$.

Similar derivations can be applied to the other geometries illustrated in Fig. 12 with the results:

(2) Twist:

$$\frac{B}{B_c} = \frac{2}{\pi} \int_0^{\pi/2} \left[\frac{1}{1 - \sin^2 \phi_m \sin^2 \psi} \right]^{1/2} d\psi$$

$$\text{and} \quad B_c = \frac{\pi}{d} \left(\frac{k_{22}}{\mu_0^{-1} \Delta \chi} \right)^{1/2} \qquad (63)$$

where the right hand side is an elliptic integral of the first kind.

The result for a bend deformation is the same as that given for splay, except the elastic constants k_{11} and k_{33} are interchanged; thus the threshold field is given by:

$$B_c = \frac{\pi}{d} \left(\frac{k_{33}}{\mu_0^{-1} \Delta \chi} \right)^{1/2} \qquad (64)$$

In the above treatment, splay and bend deformations can be regarded as limiting cases of the effect of a field on a uniformly tilted structure with a zero-field tilt $\theta_0 = 0°$ (splay) and $\theta_0 = 90°$ (bend). For cell configurations with a uniform tilt between $0°$ and $90°$, there will still be a threshold response provided that the field is perpendicular or parallel (depending on the susceptibility an-

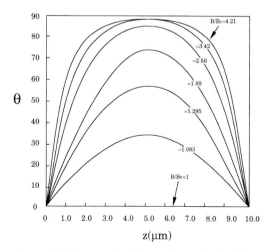

Figure 13. Director distribution above threshold.

isotropy) to the initial alignment direction; the corresponding threshold field is [24]:

$$B_c = \frac{\pi}{d} \left\{ \frac{k_{11}}{\mu_0^{-1} \Delta \chi} \left[1 + k' \sin^2 \theta_0 \right] \right\}^{1/2} \quad (65)$$

For the twisted (TN) geometry widely encountered in displays, the director distribution in three dimensions has to be evaluated in terms of both $\theta(z)$ and $\phi(z)$. This has been done [27] and the corresponding threshold field is:

$$B_c = \frac{\pi}{d} \quad (66)$$

$$\cdot \left\{ \frac{1}{\mu_0^{-1} \Delta \chi} \left[k_{11} + (k_{33} - 2k_{22}) \left(\frac{\phi_0}{\pi} \right)^2 \right] \right\}^{1/2}$$

where ϕ_0 is the zero-field twist angle. In these simple derivations the condition of strong anchoring has been assumed; however effects of weak anchoring where a surface interaction term is included have been examined [24].

5.4.2 Director Distribution in Electric Fields

The complication associated with electric fields is due to the large anisotropy of the electric permittivity, which means that above threshold the induced electric polarization is no longer parallel to the applied field. In a deformed sample the director orientation is inhomogeneous through the cell, and as a consequence the electric field is also nonuniform. An additional problem can arise with conducting samples, for which there is a contribution to the electric torque from the conductivity anisotropy. Neglecting this, the expressions for threshold electric fields are similar to those obtained for magnetic fields:

(1) Splay

$$E_c = \frac{\pi}{d} \left(\frac{k_{11}}{\varepsilon_0 \Delta \varepsilon} \right)^{1/2}$$

(2) Twist

$$E_c = \frac{\pi}{d} \left(\frac{k_{22}}{\varepsilon_0 \Delta \varepsilon} \right)^{1/2}$$

(3) Bend

$$E_c = \frac{\pi}{d} \left(\frac{k_{33}}{\varepsilon_0 \Delta \varepsilon} \right)^{1/2} \quad (67)$$

and the corresponding threshold voltages are independent of sample thickness.

In order to calculate the electric response above threshold, it is necessary to start from the free energy:

$$g = \frac{1}{2} \int k_{11} (\boldsymbol{\nabla} \cdot \boldsymbol{n})^2 + k_{22} (\boldsymbol{n} \cdot \boldsymbol{\nabla} \times \boldsymbol{n})^2 \quad (68)$$
$$+ \; k_{33} (\boldsymbol{n} \times \boldsymbol{\nabla} \times \boldsymbol{n})^2 - \varepsilon_0 \boldsymbol{D} \cdot \boldsymbol{E} \; dr^3$$

$$= \frac{1}{2} \int k_{11} (\boldsymbol{\nabla} \cdot \boldsymbol{n})^2 + k_{22} (\boldsymbol{n} \cdot \boldsymbol{\nabla} \times \boldsymbol{n})^2$$
$$+ \; k_{33} (\boldsymbol{n} \times \boldsymbol{\nabla} \times \boldsymbol{n})^2$$
$$- \; \varepsilon_0 \left(\varepsilon_\perp E^2 + \Delta \varepsilon (\boldsymbol{n} \cdot \boldsymbol{E})^2 \right) dr^3$$

In the absence of charge $\nabla \cdot \boldsymbol{D} = 0$, so that D_z is constant; by symmetry E will only have a component along the z-direction, but its magnitude will be a function of z. Considering for a moment just the electric field contribution to the energy per unit area, this can be written as:

$$g_{elec} / area = -\frac{1}{2} D_z \int E_z dz$$

$$= -\frac{1}{2} D_z V \quad (69)$$

where the voltage V is just the integral of the field. The electric displacement $D_z = \varepsilon_0 \varepsilon_{zz} E_z$, so from Eq. (69), D_z can be written as

$$D_z = V / \int (\varepsilon_0 \varepsilon_{zz})^{-1} dz$$

For an initially planar cell (case (1) above), $\varepsilon_{zz} = \varepsilon_{\parallel} \sin^2\theta + \varepsilon_{\perp} \cos^2\theta$, so the free energy per unit area becomes:

$$
\begin{aligned}
g_{\text{elec}}/\text{area} \qquad\qquad\qquad\qquad (70)\\
= \frac{1}{2} \int \left(k_{11}\cos^2\theta + k_{33}\sin^2\theta\right)\left(\frac{d\theta}{dz}\right)^2 dz \\
- \frac{1}{2}\varepsilon_0 V^2 \left[\int \left(\varepsilon_{\parallel}\sin^2\theta + \varepsilon_{\perp}\cos^2\theta\right)^{-1} dz\right]^{-1}
\end{aligned}
$$

Applying variational calculus to this expression for the free energy gives an integral equation (c.f. Eq. (58) for the director distribution:

$$
\frac{V}{V_c} = \frac{2}{\pi}(1 + \gamma \sin^2\theta_m)^{1/2} \qquad (71)
$$

$$
\cdot \int_0^{\theta_m} \left[\frac{1 + k'\sin^2\theta}{(1+\gamma\sin^2\theta)(\sin^2\theta_m - \sin^2\theta)}\right]^{1/2} d\theta
$$

where $\gamma = \Delta\varepsilon/\varepsilon_{\perp}$, with a similar expression for the bend deformation (case (3) above), except that $\varepsilon_{\parallel}$ and $\varepsilon_{\perp}$ and k_{11} and k_{33} are interchanged: the case (2) for a twisted deformation is more complicated [26]. This treatment has neglected any coupling between elastic deformation and electric polarization, but such flexoelectricity can contribute to Fréedericksz transitions [28].

5.4.3   Fréedericksz Transitions as a Method for Measuring Elastic Constants

As well as providing the basis for electro-optic displays, the Fréedericksz transitions can be used to determine the elastic constants of liquid crystals. Any physical technique that is sensitive to a change in the director distribution can be used to obtain elastic constants, but the most common methods rely on measurement of capacitance or birefringence changes during a Fréedericksz transition. As before the simplest configuration to consider is the planar to homeotropic transition observed in materials having a positive electric or magnetic susceptibility anisotropy.

5.4.3.1   Capacitance Method

For a particular applied field, the measured capacitance of a sample will be an integral over the permittivity component ε_{zz} across the cell. If the cell is imagined as a series of thin slices each of which acts as a parallel plate capacitor, the addition theorem for series capacitors gives the cell capacitance C as:

$$
C^{-1} = \int \frac{dz}{\varepsilon_0 A \varepsilon_{zz}(z)}
$$

$$
= \frac{2}{A\varepsilon_0\varepsilon_{\perp}} \int_0^{d/2} \left(1 + \gamma\sin^2\theta\right)^{-1} dz \qquad (72)
$$

It is convenient to change the variable z in Eq. (72) to one involving the director deformation θ, and for a magnetic field-induced Fréedericksz transition, Eq. (57) can be used for $\left(\frac{d\theta}{dz}\right)$ to give:

$$
\begin{aligned}
C^{-1} &= \frac{2\left(\mu_0^{-1}\Delta\chi B^2\right)^{-1/2}}{A\varepsilon_0\varepsilon_{\perp}} \int_0^{\theta_m} \left[\frac{k_{11}\cos^2\theta + k_{33}\sin^2\theta}{\sin^2\theta_m - \sin^2\theta}\right]^{1/2}\left(1+\gamma\sin^2\theta\right)^{-1} d\theta \\
&= \frac{2B_c}{\pi C_0 B}\int_0^{\theta_m}\left[\frac{1+k'\sin^2\theta}{\sin^2\theta_m - \sin^2\theta}\right]^{1/2}\left(1+\gamma\sin^2\theta\right)^{-1}d\theta \qquad (73)
\end{aligned}
$$

where $C_0 = \dfrac{A\varepsilon_0\varepsilon_\perp}{d}$ is the zero-field capacitance. For excitation by an electric field the result is:

$$\frac{C_0}{C} = \frac{\displaystyle\int_0^{\theta_m} \left[\frac{1+k'\sin^2\theta}{(1+\gamma\sin^2\theta)(\sin^2\theta_m - \sin^2\theta)}\right]^{1/2} d\theta}{\displaystyle\int_0^{\theta_m} \left[\frac{(1+\gamma\sin^2\theta)(1+k'\sin^2\theta)}{\sin^2\theta_m - \sin^2\theta}\right]^{1/2} d\theta} \tag{74}$$

5.4.3.2   Birefringence Method

This technique usually measures the change in birefringence as a cell is switched from planar to homeotropic with an electric or magnetic field. Initially the birefringence is $(n_e - n_0)$, which at infinite field becomes zero. The change (Δ) in the birefringence can be written as:

$$(n_e - n_0) - (n_{eff} - n_0) = (n_e - n_{eff}) = \Delta \tag{75}$$

where n_{eff} is the effective refractive index along the field direction, and is given by:

$$n_{eff} = \frac{1}{d}\int_0^d n(z)\,dz \tag{76}$$

$n(z)$ depends on the director orientation, and can be obtained from Eq. (4) of Sec. 3 of this chapter, but note that the angle θ used in the earlier equation is $(90° - \theta)$ used here. Thus $n(z) = n_e n_0 (n_e^2 \sin^2\theta + n_0^2 \cos^2\theta)^{-1/2}$ $= n_e(1 + v\sin^2\theta)^{-1/2}$, and $v = (n_e^2 - n_0^2)/n_0^2$, and the birefringence change Δ can be obtained by transforming Eq. (76) to an integral over θ using Eq. (57), so that:

$$\Delta = n_e - \frac{2n_e\left(\mu_0^{-1}\Delta\chi B^2\right)^{-1/2}}{d}$$
$$\cdot \int_0^{\theta_m} \left[\frac{k_{11}\cos^2\theta + k_{33}\sin^2\theta}{(1+v\sin^2\theta)(\sin^2\theta_m - \sin^2\theta)}\right]^{1/2} d\theta \tag{77}$$

and using the result for the threshold field, this becomes:

$$\frac{\Delta}{n_e} = 1 - \frac{2B_c}{\pi B} \tag{78}$$

$$\cdot \int_0^{\theta_m} \left[\frac{1+k'\sin^2\theta}{(1+v\sin^2\theta)(\sin^2\theta_m - \sin^2\theta)}\right]^{1/2} d\theta$$

It will be useful to express this in terms of the new angular variable ψ introduced earlier, to give:

$$\frac{\Delta}{n_e} = 1 - \frac{2B_c}{\pi B}$$
$$\cdot \int_0^{\pi/2} \left[\frac{1+k'\sin^2\theta_m\sin^2\psi}{1+v\sin^2\theta_m\sin^2\theta}\right]^{1/2}$$
$$\cdot \left(1 - \sin^2\theta_m\sin^2\psi\right)^{-1} d\theta \tag{79}$$

The usual way to observe birefringence changes in a nematic slab undergoing a Fréedericksz transition is to illuminate the sample with polarized light, the plane of which makes an angle of 45° with the director axis. Increasing the field above threshold and measuring the intensity transmitted through a crossed polariser gives rise to a series of maxima and minima (fringes). These can be related to the expression (79) by the result:

$$\frac{I}{I_0} = \frac{1}{4}\left(1 - \cos\left(\frac{2\pi\,d\Delta}{\lambda}\right)\right) \tag{80}$$

where the ratio I/I_0 refers to changes in the intensity of unpolarized incident light (I_0).

The result for electric field excitation is more complex for the reasons given earlier, but has a similar form to Eq. (79)

$$
\frac{\Delta}{n_e} = 1 - \frac{\int_0^{\pi/2} \left[\dfrac{\left(1 + k' \sin^2 \theta_m \sin^2 \psi\right)\left(1 + \gamma \sin^2 \theta_m \sin^2 \psi\right)}{\left(1 + v \sin^2 \theta_m \sin^2 \theta\right)\left(1 - \sin^2 \theta_m \sin^2 \theta\right)} \right]^{1/2} d\theta}{\int_0^{\pi/2} \left[\dfrac{\left(1 + k' \sin^2 \theta_m \sin^2 \psi\right)\left(1 + \gamma \sin^2 \theta_m \sin^2 \psi\right)}{1 - \sin^2 \theta_m \sin^2 \theta} \right]^{1/2} d\theta}
\tag{81}
$$

Numerical methods have been developed [29–31] to fit experimental values of $\frac{C}{C_0}$ or $\frac{\Delta}{n_e}$ to the theoretical expressions with B_c, k', v, V_c and γ as adjustable parameters appropriate to the particular experiment being considered. Another way of using these expressions is to develop low-field or high-field expansions, so that material parameters can be obtained directly from a linear or polynomial expression. For example the low and high field expansions for electric field excitation have been derived [32, 33] as:

$$
\left(\frac{C - C_0}{C_0}\right)_{E \to 0} = 2\gamma \left(1 + k' + \gamma\right)^{-1} \left(\frac{V - V_c}{V_c}\right)
$$

$$
\left(\frac{C - C_0}{C_0}\right)_{E \to \infty} = \gamma - 2\gamma \, \pi^{-1} \left(1 + \gamma\right)^{1/2} \frac{Vc}{V}
\tag{82}
$$

Care is required in using these expressions, since they are usually only valid close to the limiting field values where measurements are difficult and liable to error.

In this section the effects of magnetic and electric fields have been considered for a few standard geometries for materials having positive susceptibility anisotropies, but there are many possible variations with negative anisotropies or simultaneous excitation with both electric and magnetic fields of materials having both positive, both negative or different signs of electric and magnetic susceptibility anisotropies. Another

variation which may be introduced is to have different boundary conditions for the containing surfaces of the liquid crystal film [34], so-called hybrid-aligned cells. The theoretical treatment outlined here has excluded the possibility of defect formation, although this can in principle be described by the elasticity theory already developed. Defect structures can be formed as a result of deformation by electric or magnetic fields. In some situations they may arise from natural degeneracies in the sample: for example in a twist cell, states of opposite twist will be of equal energy, and so may form distinct regions separated by a disclination. In real device cells the structures are designed to avoid the formation of defects.

5.4.4   Fréedericksz Transition for Chiral Nematics

Field effects on chiral nematics can be interpreted by adding a pitch term to the free energy, so it might be expected that the Fréedericksz transitions observed for chiral nematics will be similar to those described above for achiral nematics. In reality this is not the case because the helical structure in chiral phases prevents the formation of uniformly aligned films, and so defects and defect-modulated structures are unavoidable in many field-induced orientational changes. The effects of external fields on chiral nematic films have been described [35, 36], and some of the associated phenomena form the basis for display devices. In most cases a theoretical description of the induced deformation is only possible by numerical solution of the Euler–Lagrange equations, but one simple effect that has an analytical solution is the so-called field-induced cholesteric to nematic phase transition. An external field applied perpendicular to the helix axis of a material having a positive susceptibility anisotropy will cause the helix to unwind and the pitch to increase. A treatment similar to that given for the twist Fréedericksz transition shows that the critical field for divergence of the pitch to infinity is:

$$F_c = \frac{\pi \, q_0}{2} \left(\frac{k_{22}}{\Delta \chi \, \mu_0^{-1}} \right)^{1/2} \tag{83}$$

This results contrasts with the threshold field for a nematic Fréedericksz transition, which is thickness dependent: also F_c marks the end of the deformation rather than the beginning which defines the normal threshold fields.

predicts a threshold field which is inversely proportional to the square-root of the sample thickness [37] e. g. for a magnetic field:

$$B_c = \left(\frac{2\pi \, k_{11}}{\Delta \chi \, \mu_0^{-1} \lambda d} \right)^{1/2} \tag{84}$$

where d is the cell thickness and $\lambda = \sqrt{k_{11}/B}$ (see Fig. 14).

External field distortions in SmC and chiral SmC phases have been investigated [38], but the large number of elastic terms in the free-energy, and the coupling between the permanent polarization and electric fields for chiral phases considerably complicates the description. In the chiral smectic C phase a simple helix unwinding Fréedericksz transition can be detected for the *c* director. This is similar to the chiral nematic–nematic transition described by Eq. (83), and the result is identical for the SmC* phase. Indeed it appears that at least in interactions with magnetic fields in the plane of the layers, SmC and SmC* phases behave as two dimensional nematics [39].

5.4.5   Fréedericksz Transitions for Smectic Phases

The simplest elasticity theory for SmA phases includes two elastic contants, one for splay and one for layer compressibility. It might therefore be expected that a Fréedericksz transition for splay deformation should be observed corresponding to an initial deformation of layers in a planar to homeotropic transition. This is not observed, and field induced deformations in smectic A phases are accompanied by defect formation. The Helfrich–Hurault mechanism for the homeotropic to planar transition via the formation of undulations

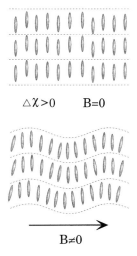

Figure 14. Helfrich–Hurault mechanism for field deformation of smectic layers.

5.5   Molecular Aspects of Torsional Elasticity

5.5.1   van der Waals Theory

Torsional elasticity is of special interest from a microscopic viewpoint since it is a property characteristic of liquid crystals, which distinguishes them from ordinary liquids. The elastic properties contribute to many physical phenomena observed for liquid crystals, and a molecular theory of torsional elasticity should enable the identification of particular molecular properties responsible for many aspects of liquid crystalline behavior.

The principal elastic constants for a nematic liquid crystal have already been defined in Sec. 5.1 as splay (k_{11}), twist (k_{22}) and bend (k_{33}). In this section we shall outline the statistical theory of elastic constants, and show how they depend on molecular properties. The approach follows that of the generalised van der Waals theory developed by Gelbart and Ben-Shaul [40], which itself embraces a number of earlier models for the elasticity of nematic liquid crystals. Corresponding theories for smectic, columnar and biaxial phases have yet to be developed.

Elastic constants are defined in terms of the deformation free energy of a liquid crystal subjected to torsional strain. Statistical models for liquid crystals result in equations for the free energy in an undistorted state: thus to calculate elastic constants it is necessary to obtain a statistical expression for the free energy of a strained liquid crystal. In developing a statistical theory it is easier to use the Helmholtz free energy to calculate elastic constants, although Frank originally defined them in terms of the Gibbs free energy, corresponding to strain at constant external pressure. We shall be considering torsional strain at constant volume, for which changes in both the internal energy and entropy of a liquid crystal will contribute to the elastic constants.

The most widely used statistical model for fluids is that due to van der Waals, which includes a mean attractive potential with a hard particle excluded volume. For such a model the Helmholtz free energy can be written as [41, 42]:

$$A\big[f(\boldsymbol{\Omega}_1, \boldsymbol{\Omega}_2, \boldsymbol{R}_1, \boldsymbol{R}_2)\big] \qquad (85)$$
$$= A\big[f(\boldsymbol{\Omega}_1, \boldsymbol{\Omega}_2, \boldsymbol{R}_1, \boldsymbol{R}_2)\big]_{\mathrm{hp}}$$
$$- \frac{1}{2}\,\rho^2 \int f(\boldsymbol{\Omega}_1, \boldsymbol{R}_1) f(\boldsymbol{\Omega}_2, \boldsymbol{R}_2)$$
$$\cdot u_{12}\big(\boldsymbol{R}_{12}, \boldsymbol{\Omega}_1, \boldsymbol{\Omega}_2\big)$$
$$\cdot g_{\mathrm{hp}}\big(\boldsymbol{R}_{12}, \boldsymbol{\Omega}_1, \boldsymbol{\Omega}_2\big)\mathrm{d}\boldsymbol{\Omega}_1, \mathrm{d}\boldsymbol{\Omega}_2, \mathrm{d}\boldsymbol{R}_1, \mathrm{d}\boldsymbol{R}_2$$

where $f(\boldsymbol{\Omega}_1, \boldsymbol{R}_1)$ is a single particle distribution function for molecule 1, $u_{12}(\boldsymbol{R}_{12}, \boldsymbol{\Omega}_1, \boldsymbol{\Omega}_2)$ is the attractive part of the pair potential and $g_{\mathrm{hp}}(\boldsymbol{R}_{12}, \boldsymbol{\Omega}_1, \boldsymbol{\Omega}_2)$ is the pair distribution function for the hard particle interactions in the isotropic state.

To proceed we need to know how the functional $A[f(\boldsymbol{\Omega}_1, \boldsymbol{\Omega}_2, \boldsymbol{R}_1, \boldsymbol{R}_2)]$ varies when the equilibrium state of the liquid crystal is elastically distorted. A macroscopic strain will not influence u_{12} or g_{hp}, since these are dependent only on molecular parameters of the model: the free energy changes because the single particle distribution functions change. We assume that for the small distortions described by the Frank elastic constants, the single particle orientational distribution function, defined with respect to a local director axis, is also independent of strain i.e. elastic torques do not change the molecular order parameters. The product of distribution functions $f(\boldsymbol{\Omega}_1, \boldsymbol{R}_1) f(\boldsymbol{\Omega}_2, \boldsymbol{R}_2)$ will change with strain because the director orientations at $\boldsymbol{R}_1$ and $\boldsymbol{R}_2$ will differ, and the evaluation of the strain dependence of the

Helmholtz free energy has reduced to determining the strain derivatives of the single particle distribution functions. Both the terms in Eq. (85) will contribute to the elastic free energy, but the first term, which describes hard particle contributions to the Helmholtz free energy will only appear in the entropy of the system, since the internal energy of a hard particle fluid is zero. Furthermore it can be shown [40] that the rotational hard particle entropy is independent of strain, provided that the single particle orientational distribution defined with respect to the local director does not depend on strain. Thus the hard particle repulsion contributes to the elastic strain energy in two ways: firstly through the orientation-dependent excluded volume, which affects the translational entropy, and secondly because the integration over the pair attractive potential energy is convoluted with the hard particle distribution function. A particularly convenient form for the hard particle translational entropy is provided by the 'y-expansion' [41], and using this the Helmholtz free energy becomes:

$$A\big[f(\boldsymbol{\Omega}_1, \boldsymbol{\Omega}_2, \boldsymbol{R}_1, \boldsymbol{R}_2)\big] \tag{86}$$

$$= T\mathbf{S}_{\mathrm{hp}}^{\mathrm{rot}} - \frac{kT\rho^2}{2(1 - v_0\rho)}$$

$$\cdot \int f(\boldsymbol{\Omega}_1, \boldsymbol{R}_1) f(\boldsymbol{\Omega}_2, \boldsymbol{R}_2)$$

$$\cdot \big\{g_{\mathrm{hp}}^{(0)} - 1\big\} \mathrm{d}\boldsymbol{\Omega}_1, \mathrm{d}\boldsymbol{\Omega}_2, \mathrm{d}\boldsymbol{R}_1, \mathrm{d}\boldsymbol{R}_2$$

$$- \frac{1}{2}\rho^2 \int f(\boldsymbol{\Omega}_1, \boldsymbol{R}_1) f(\boldsymbol{\Omega}_2, \boldsymbol{R}_2)$$

$$\cdot u_{12}(\boldsymbol{R}_{12}, \boldsymbol{\Omega}_1, \boldsymbol{\Omega}_2) g_{\mathrm{hp}}^{(0)} \mathrm{d}\boldsymbol{\Omega}_1, \mathrm{d}\boldsymbol{\Omega}_2, \mathrm{d}\boldsymbol{R}_1, \mathrm{d}\boldsymbol{R}_2$$

$$= -T\mathbf{S}_{\mathrm{hp}}^{\mathrm{rot}} - \frac{1}{2}\rho^2$$

$$\cdot \int f(\boldsymbol{\Omega}_1, \boldsymbol{R}_1) f(\boldsymbol{\Omega}_2, \boldsymbol{R}_1 + \boldsymbol{R}_{12})$$

$$\cdot \left\{u_{12}(\boldsymbol{R}_{12}, \boldsymbol{\Omega}_1, \boldsymbol{\Omega}_2) g_{\mathrm{hp}}^{(0)} - \frac{kT}{(1 - v_0\rho)}\big(g_{\mathrm{hp}}^{(0)} - 1\big)\right\}$$

$$\cdot \mathrm{d}\boldsymbol{\Omega}_1, \mathrm{d}\boldsymbol{\Omega}_2, \mathrm{d}\boldsymbol{R}_1, \mathrm{d}\boldsymbol{R}_{12}$$

where $g_{\mathrm{hp}}^{(0)} = \exp{-u_{\mathrm{hp}}/k_{\mathrm{B}}T}$ is the pair correlation function for the hard particles, v_0 is the particle volume, u_{hp} is the hard particle potential, and $\mathbf{S}_{\mathrm{hp}}^{\mathrm{rot}}$ is the hard particle rotational entropy. In order to obtain the elastic-distortion-free energy from Eq. (86), we assume that molecule 1 is located at some arbitary origin in the fluid. The orientational distribution function for molecule 2 at position $\boldsymbol{R}_{12}$ only depends on the orientation of the director at $\boldsymbol{R}_{12}$ with respect to the director at the origin. Thus in the undeformed state the director at $\boldsymbol{R}_{12}$ is parallel to that at the origin (at least for non-chiral liquid crystals), but in the deformed state the director at $\boldsymbol{R}_{12}$ makes an angle of $\theta(\boldsymbol{R}_{12})$ to that at the origin; see Fig. 15.

The distortion free energy density is then the difference in free energies given by Eq. (86) between the distorted and undistorted states, that is:

$$\frac{A_{\mathrm{distorted}}}{V} = \frac{1}{2}\rho^2 \int f(\boldsymbol{\Omega}_1, 0) \tag{87}$$

$$\cdot \Big\{f\big(\boldsymbol{\Omega}_2, \theta(\boldsymbol{R}_{12})\big) - f(\boldsymbol{\Omega}_2, 0)\Big\}$$

$$\cdot \left\{u_{12}(\boldsymbol{R}_{12}, \boldsymbol{\Omega}_1, \boldsymbol{\Omega}_2) g_{\mathrm{hp}}^{(0)} - \frac{kT}{(1 - v_0\rho)}\big(g_{\mathrm{hp}}^{(0)} - 1\big)\right\}$$

$$\cdot \mathrm{d}\boldsymbol{\Omega}_1, \mathrm{d}\boldsymbol{\Omega}_2, \mathrm{d}\boldsymbol{R}_{12}$$

The change in the single particle orientational distribution function for small defor-

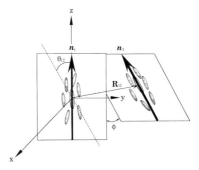

Figure 15. Molecular distributions in a deformed nematic liquid crystal.

mations can be obtained from a Taylor expansion, such that:

$$f(\Omega,\theta(\mathbf{R})) - f(\Omega,0)$$

$$= \left(\frac{\mathrm{d}f(\Omega,\theta(\mathbf{R}))}{\mathrm{d}\theta}\right)_0 \theta$$

$$+ \frac{1}{2}\left(\frac{\mathrm{d}^2 f(\Omega,\theta(\mathbf{R}))}{\mathrm{d}\theta^2}\right)_0 \theta^2 + \dots \qquad (88)$$

The way in which $\theta(\mathbf{R})$ varies with position depends on the form of the torsional deformation applied to the liquid crystal, and in order to calculate the principal elastic constants, it makes sense to calculate the free energy density for 'normal mode' deformations, i.e. those which correspond to splay, twist and bend. These can be achieved easily by confining the director to a plane, and assuming the undisturbed director at the origin to be along the z-axis. $\mathbf{q}$ is the wavevector of the deformation, and for $\mathbf{q}$ constrained to the x, z plane, the components of the director as a function of position become:

$$n_x = \sin\mathbf{q}\cdot\mathbf{R}; \quad n_y = 0; \quad n_z = \cos\mathbf{q}\cdot\mathbf{R}$$

Director configurations corresponding to pure splay, twist and bend are illustrated in Fig. 16 and to the lowest order of approximation can be described in terms of long

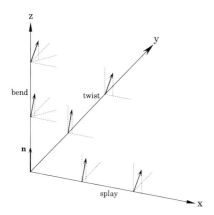

Figure 16. Deformations of director for splay, twist, bend.

wavelength deformations parallel to the x-axis for splay, parallel to the y-axis for twist and parallel to the z-axis for bend. Under these circumstances, the angle between the director at the origin and that at position $\mathbf{R}$ becomes:

$$\theta(\mathbf{R}) = \begin{cases} qx & \text{splay} \\ qy & \text{twist} \\ qz & \text{bend} \end{cases} \qquad (89)$$

Gelbart and Ben Shaul have shown that a more consistent director distribution giving rise to normal mode elastic deformations is:

$$\tan\theta(\mathbf{R}) = \begin{cases} qx(1+qz)^{-1} & \text{splay} \\ qy & \text{twist} \\ qz(1-qx)^{-1} & \text{bend} \end{cases}$$

$$\text{or } \theta(\mathbf{R}) = \begin{cases} qx(1-qz) & \text{splay} \\ qy & \text{twist} \\ qz(1+qx) & \text{bend} \end{cases} \qquad (90)$$

Substituting these results into Eq. (88) gives:

$$f(\Omega,\theta(\mathbf{R})) - f(\Omega,0)$$

$$= \left(\frac{\mathrm{d}f(\Omega,\theta(\mathbf{R}))}{\mathrm{d}\theta}\right)_0 \begin{cases} (qx - q^2\,xz) \\ qy \\ (qz + q^2\,zx) \end{cases}$$

$$+ \frac{1}{2}\left(\frac{\mathrm{d}^2 f(\Omega,\theta(\mathbf{R}))}{\mathrm{d}\theta^2}\right)_0 \begin{cases} q^2 x^2 \\ q^2 y^2 \\ q^2 z^2 \end{cases} \qquad (91)$$

The macroscopic expression for the elastic free energy density (Eq. 25) is:

$$g - g_0 = \frac{1}{2V}\int k_{11}(\nabla\cdot\mathbf{n})^2 + k_{22}(\mathbf{n}\cdot\nabla\times\mathbf{n})^2$$

$$+ k_{33}((\mathbf{n}\cdot\nabla)\mathbf{n})^2\,\mathrm{d}\mathbf{R} \qquad (92)$$

Using the expressions for $\theta(\mathbf{R})$ corresponding to pure splay, twist and bend deformations gives the result:

$$g - g_0 = \begin{cases} \dfrac{1}{2} k_{11} q^2 & \text{splay} \\ \dfrac{1}{2} k_{22} q^2 & \text{twist} \\ \dfrac{1}{2} k_{33} q^2 & \text{bend} \end{cases} \qquad (93)$$

Combining Eqs. (91) and (87) and equating the result for the free energy density to the macroscopic expression Eq. (93) gives:

$$\left.\begin{array}{c} k_{11} \\ k_{22} \\ k_{33} \end{array}\right\} = \rho^2 \int f(\Omega_1, 0)$$

$$\cdot \left\{ \left(\frac{df(\Omega_1, \theta(\mathbf{R}))}{d\theta} \right)_0 \begin{Bmatrix} -xz \\ 0 \\ +zx \end{Bmatrix} \right.$$

$$+ \frac{1}{2} \left(\frac{d^2 f(\Omega_1, \theta(\mathbf{R}))}{d\theta^2} \right)_0 \begin{Bmatrix} x^2 \\ y^2 \\ z^2 \end{Bmatrix} \Bigg\}$$

$$\cdot \left[u_{12}(\mathbf{R}_{12}, \Omega_1, \Omega_2) g_{hp}^0 - \frac{kT}{(1 - v_0 \rho)}(g_{hp}^0 - 1) \right]$$

$$\cdot d\Omega_1, d\Omega_2, d\mathbf{R}_{12} \qquad (94)$$

The two terms in the square brackets of Eq. (94) can be identified as a temperature independent internal energy term, and a temperature dependent entropy term resulting from the hard particle pair distribution function. From this equation it can be seen that the calculation of the principal elastic constants of a nematic liquid crystal depends on the first and second derivatives with respect to the angle θ of the single particle orientational distribution function. Any appropriate angular function may be used for $f(\Omega_1, \theta(\mathbf{R}))$, but the usual approach is to use an expansion in terms of spherical harmonics. The necessary mathematical manipulations are complicated, but give relatively compact results. Thus the ingredients of a molecular calculation of torsional elastic constants within the van der Waals

theory, are a single particle angular distribution function, an attractive intermolecular potential and a hard particle pair distribution function. An immediate result of the above theory is that since the elastic constants depend on the product of single particle orientational distribution functions, they will depend on the product of order parameters.

5.5.2 Results from Lattice Models

For the simplest distribution function, only the term involving the second derivative in Eq. (94) is nonzero, and the torsional elastic constants are given by an average over the square of the intermolecular distances x, y and z. Since macroscopic uniaxiality is assumed, the averages over x and y, perpendicular to the undisturbed director, will be equal, with the result [43]:

$$k_{11} = k_{22} = \text{constant } \rho^2 S^2 b \langle x^2 \rangle \quad \text{and}$$
$$k_{33} = \text{constant } \rho^2 S^2 b \langle z^2 \rangle \qquad (95)$$

where b is an energy parameter. If a lattice model is assumed, then the averages over x and z will relate to the unit cell dimensions, or the dimensions of an 'interaction volume'. The result, Eq. (95), fails to account for the observed difference between splay and twist elastic constants, and it fails to provide a useful basis for investigating the effects of molecular structure on elastic properties.

5.5.3 Mean Field and Hard Particle Theories

The first attempt at a molecular theory of elastic constants due to Saupe and Nehring [44] assumed an attractive pair potential of the form $u_{12} = -b r^{-6} P_2(\cos \theta_{12})$, and set

$g_{\text{(hp)}}^{(0)}$ equal to that for hard spheres of diameter σ, with the result that:

$$\begin{Bmatrix} k_{11} \\ k_{22} \\ k_{33} \end{Bmatrix} = \frac{12\pi\rho^2 S^2 b}{5\sigma} \begin{pmatrix} 5 \\ 11 \\ 5 \end{pmatrix} \qquad (96)$$

This suggests that the elastic constants for splay and bend should be equal. The difference between k_{11} and k_{33} has been identified as due to a term in $\langle P_2(\cos\theta)\rangle$ $\langle P_4(\cos\theta)\rangle$, and while this usually leads to $k_{33} > k_{11}$, particular choices of intermolecular potential can result in $k_{33} < k_{11}$ or indeed a change in sign of the quantity $\left(\dfrac{k_{33}}{k_{11}} - 1\right)$ as a function of temperature [45].

Various authors [46, 47] have reported calculations of torsional elastic constants for hard spherocylinders. These neglect any attractive interactions, and so only give entropic contributions to the elastic free energy; their results can be summarized as:

$$\begin{Bmatrix} \dfrac{k_{11}-k}{k} \\[2mm] \dfrac{k_{22}-k}{k} \\[2mm] \dfrac{k_{33}-k}{k} \end{Bmatrix} = \begin{Bmatrix} \Delta - 3\Delta'\dfrac{\langle P_2(\cos\theta)\rangle}{\langle P_4(\cos\theta)\rangle} \\[3mm] -2\Delta - \Delta'\dfrac{\langle P_2(\cos\theta)\rangle}{\langle P_4(\cos\theta)\rangle} \\[3mm] \Delta + 4\Delta'\dfrac{\langle P_2(\cos\theta)\rangle}{\langle P_4(\cos\theta)\rangle} \end{Bmatrix} \qquad (97)$$

where $k = \dfrac{1}{3}(k_{11} + k_{22} + k_{33})$, and Δ and Δ' depend on the details of the potential. For hard spherocylinders, the parameters Δ and Δ' are:

$$\Delta = \frac{2R^2 - 2}{7R^2 + 20} \quad \text{and} \quad \Delta' = \frac{9(3R^2 - 8)}{16(7R^2 + 20)} \qquad (98)$$

where $R + 1 = (l/w)$; l is the length and w is the width of the spherocylinder. Calculations of the principal elastic constants for various molecular models, with both attractive and repulsive interactions have been reported by various authors [48–52].

5.5.4   Computer Simulations

Some progress has been made in computer simulation of torsional elastic properties [53, 54]. A variety of methods is available for modelling elastic distortions, but the most used technique is based on the direct calculation of the amplitude of director fluctuations. In Sec. 5.2.2 of this Chapter the background to long range director fluctuations was outlined, and it was found that the mean square amplitude of the director component along z could be expressed in terms of the splay and bend elastic constants and the wave vector for a particular deformation mode (it is assumed that for the undistorted liquid crystal, the director lies along the z axis). In order to allow for the influence of local molecular disorder, the fluctuations in the director order parameter are calculated, assuming that for long wavelength fluctuations, the molecular order parameter simply scales the director order parameter. Thus the Fourier components of the order parameter are written as:

$$Q_{\alpha\beta}(\boldsymbol{q})$$
$$= \langle P_2(\cos\theta)\rangle \left(\frac{1}{2}(3n_\alpha(\boldsymbol{q})n_\beta(\boldsymbol{q}) - \delta_{\alpha\beta})\right) \qquad (99)$$

Using the results Eqs. (38) and (39) give expressions for the fluctuation in the order parameter as [53]:

$$\langle Q_{xz}(\boldsymbol{q})Q_{xz}(-\boldsymbol{q})\rangle = \frac{9}{4}\langle P_2(\cos\theta)\rangle^2 \frac{k_B T}{V(k_{11}q_\perp^2 + k_{33}q_\parallel^2)}$$

$$\langle Q_{yz}(\boldsymbol{q})Q_{yz}(-\boldsymbol{q})\rangle = \frac{9}{4}\langle P_2(\cos\theta)\rangle^2 \frac{k_B T}{V(k_{22}q_\perp^2 + k_{33}q_\parallel^2)} \qquad (100)$$

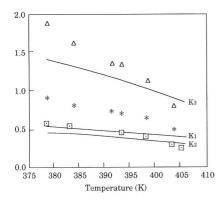

Figure 17. Comparison of calculated and theoretical elastic constants for *p*-azoxyanisole [53] (units are 10^{-11} N).

where $q_{\parallel}$ and $q_{\perp}$ are the components of the distortion mode wave vector parallel and perpendicular to the undisturbed director. Comparisons between theory and experiment are encouraging (see Fig. 17), although sometimes at the expense of assuming rather high values for the local molecular order parameters; this is attributed to a weakness in the theory which neglects long-range correlations [53].

5.6 Experimental Aspects of Elastic Constants and Comparison with Theoretical Predictions

The statistical theories of elasticity have shown that the principal elastic constants depend on the single particle distribution functions and the intermolecular forces. The former can be accounted for in terms of order parameters, but intermolecular parameters are more difficult to interpret in terms of molecular properties. Results for hard particle potentials relate the elastic constants to particle dimensions, but the dependence of elastic properties on details of molecular structure is more obscure. Measured values for the torsional elastic constants of nematics are of the order of 10^{-12} N, corresponding to $kT(4 \times 10^{-21}$ J)$/l(10^{-9}$ m), where $T \approx 300$ K and l is a molecular dimension. The effect of elastic distortions on molecular interactions is very small, so that the assumption of strain-independent distribution functions is justified. For example the elastic energy in a 1 μm liquid crystal film in a $\pi/2$ twisted nematic cell is only $\approx 10^{-27}$ J/molecule, which compares with an intermolecular energy of about 10^{-20} J/molecule. The angular displacement per molecule in such a cell is 10^{-3} rad, and the small energies associated with such torsional strain mean that direct mechanical measurement of elastic constants in liquid crystals is difficult.

Indirect experimental measurements of torsional elastic constants indicate that for many materials $k_{33} > k_{11} > k_{22}$. It is usual to consider ratios of elastic constants, and most measured values lie in the ranges $0.5 < k_{33}/k_{11} < 3.0$ and $0.5 < k_{22}/k_{11} < 0.8$. Simple mean field theory predicted that $k_{33} = k_{11} < k_{22}$, which is clearly in error. For rigid molecules increasing the molecular length increases k_{33}/k_{11}, while increasing the width decreases k_{33}/k_{11}, in qualitative agreement with the results of hard particle theories. However within a homologous series, increasing the length of an alkyl chain causes the ratio k_{33}/k_{11} to decrease, which is likely to be a result of increased flexibility rather than any change in effective shape. For SmA phases the torsional elastic constants for twist and bend are expected to be very large, and it is observed that k_{33} and sometimes k_{22} measured for a nematic phase increase dramatically with decreasing temperature as an underlying smectic phase is approached. Short-range smectic-like ordering will influence the measured elastic constants; long alkyl chains promote smectic phase formation, and as a result k_{33}/k_{11} may increase in a homologous series with increasing chain length.

The main temperature dependence of elastic constants is due to the order parameter. Following the predictions of mean field theory Eq. (96), reduced elastic constants c_i have been introduced, which should be independent of temperature, defined by:

$$c_i = \frac{k_{ii}}{S^2 \rho^{7/3}} \qquad (101)$$

Experimentally the c_is do change with temperature, and this can be attributed to the influence of repulsive interactions on the elastic properties.

5.7   References

[1]   C. W. Oseen, *Trans. Faraday Soc.* **1933**, *29*, 883.
[2]   F. C. Frank, *Discuss. Faraday Soc.* **1958**, *25*, 19.
[3]   P. G. de Gennes, J. Prost, *The Physics of Liquid Crystals,* Oxford University Press, Oxford, **1993**, p. 98.
[4]   G. Vertogen, W. H. de Jeu, *Thermotropic Liquid Crystals, Fundamentals* **1988**, p. 74.
[5]   Ref. [3], p. 341.
[6]   Ref. [3], p. 349.
[7]   R. D. Polak, G. P. Crawford, B. C. Kostival, J. W. Doane, S. Zumer, *Phys. Rev. E* **1994**, *49*, R978.
[8]   J. Charvolin, J. F. Sadoc, *J. Phys. Chem.* **1988**, *92*, 5787.
[9]   J. Nehring, A. Saupe, *J. Chem. Phys.* **1971**, *54*, 337.
[10]   F. M. Leslie, I. W. Stewart, M. Nakagawa, *Mol. Cryst. Liq. Cryst.* **1991**, *198*, 443.
[11]   Ref. [3], p. 351.
[12]   Ref. [3], p. 74.
[13]   S. Chandrasekhar, G. S. Ranganath, *Adv. Phys.* **1986**, *35*, 507.
[14]   M. Kleman, *Liq. Cryst.* **1989**, *5*, 399.
[15]   T. E. Faber, *Liq. Cryst.* **1991**, *9*, 95.
[16]   B. J. Berne, R. Pecora, *Dynamic Light Scattering,* John Wiley, New York, **1976**, p. 27.
[17]   S. Chandrasekhar, *Liquid Crystals,* Cambridge University Press, Cambridge, **1992**, p. 167.
[18]   J. P. van der Meulen, R. J. J. Zijlstra, *J. de Phys.* **1984**, *45*, 1627.
[19]   J. H. Seddon, R. H. Templer, *Phil. Trans. R. Soc. Lond. A* **1993**, *334*, 377.
[20]   V. Fréedericksz, V. Zolina, *Trans. Faraday Soc.* **1933**, *29*, 919.
[21]   J. Nehring, A. Saupe, *J. Chem. Phys.* **1971**, *54*, 337.
[22]   A. J. Derzhanski, A. G. Petrov, M. D. Mitov, *J. de Phys.* **1978**, *39*, 272.
[23]   T. Motooka, A. Fukuhara, K. Suzuki, *Appl. Phys. Lett.* **1979**, *34*, 305.
[24]   T. Motooka, A. Fukuhara, *J. Appl. Phys.* **1979**, *50*, 6907.
[25]   Ref. [4], p. 94.
[26]   H. J. Deuling, *Sol. State Phys. Suppl.* 14 (Liquid Crystals), Ed.: L. Liebert, Academic Press, London 1978, p. 77.
[27]   F. M. Leslie, *Mol. Cryst. Liq. Cryst.* **1970**, *12*, 57.
[28]   H. J. Deuling, *Mol. Cryst. Liq. Cryst.* **1974**, *26*, 281.
[29]   C. Maze, *Mol. Cryst. Liq. Cryst.* **1978**, *48*, 273.
[30]   M. J. Bradshaw, E. P. Raynes, *Mol. Cryst. Liq. Cryst.* **1983**, *91*, 145.
[31]   H. J. Deuling, *Sol. Stat. Commun.* **1974**, *14*, 1073.
[32]   M. G. Clark, E. P. Raynes, R. A. Smith, R. J. A. Tough, *J. Phys. D* **1980**, *13*, 2151.
[33]   T. Uchida, Y. Takahashi, *Mol. Cryst. Liq. Cryst.* **1981**, *72*, 133.
[34]   G. Barbero, *Z. Naturforsch.* **1984**, *39a*, 575.
[35]   L. M. Blinov, *Electro-optical and Magneto-optical Properties of Liquid Crystals,* J. Wiley, Chichester, **1983**, p. 240.
[36]   V. G. Chigrinov, *Kristallografiya* **1988**, *33*, 260.
[37]   Ref. [3], p. 364.
[38]   Ref. [3], pp. 347, 376.
[39]   S. P. A. Gill, F. M. Leslie, *Liq. Cryst.* **1993**, *14*, 1901.
[40]   W. M. Gelbart, A. Ben-Shaul, *J. Chem. Phys.* **1982**, *77*, 916.
[41]   W. M. Gelbart, *J. Chem. Phys.* **1979**, *71*, 3053.
[42]   M. A. Cotter, *J. Chem. Phys.* **1977**, *66*, 4710.
[43]   H. Gruler, *Z. Naturforsch.* **1975**, *30a*, 230.
[44]   J. Nehring, A. Saupe, *J. Chem. Phys.* **1972**, *56*, 5527.
[45]   W. M. Gelbart, J. Stecki, *Mol. Phys.* **1981**, *41*, 1451.
[46]   R. Priest, *Phys. Rev.* **1973**, *A7*, 720.
[47]   J. P. Straley, *Phys. Rev.* **1973**, *A8*, 2181.
[48]   A. Ponerwierski, J. Stecki, *Mol. Phys.* **1979**, *38*, 1931.
[49]   H. Kimura, M. Hoshino, H. Nakano, *Mol. Cryst. Liq. Cryst.* **1981**, *74*, 55.
[50]   B. W. van der Meer, F. Postma, A. J. Dekker, W. H. de Jeu, *Mol. Phys.* **1982**, *45*, 1227.
[51]   Y. Singh, S. Singh, K. Rajesh, *Phys. Rev.* **1992**, *A44*, 974.
[52]   M. A. Osipov, S. Hess, *Mol. Phys.* **1993**, *78*, 1191.
[53]   B. Tjipto-Margo, G. T. Evans, M. P. Allen, D. Frenkel, *J. Phys. Chem.* **1992**, *96*, 3942.
[54]   M. A. Osipov, S. Hess, *Liq. Cryst.* **1994**, *16*, 845.

6 Phase Transitions

6.1 Phase Transition Theories

Philippe Barois

6.1.1 Introduction

The understanding of continuous phase transitions and critical phenomena has been one of the important breakthrough in condensed matter physics in the early seventies. The concepts of scaling behavior and universality introduced by Kadanoff and Widom and the calculation of non-gaussian exponents by Wilson and Fisher are undeniably brilliant successes of statistical physics in the study of low temperature phase transitions (normal to superconductor, normal to superfluid helium) and liquid–gas critical points.

But no other field in condensed matter physics has shown such a rich variety of continuous or weakly first-order phase transitions than liquid crystals: order parameters of various symmetrics, anisotropic scaling behaviors, coupled order parameters, multicritical points, wide critical domains, defect mediated transitions, spaces of low dimensionality, multiply reentrant topologies are currently found in liquid crystals, at easily accessible temperatures. Beside their famous technical applications in optics

and electronic displays, liquid crystals can certainly be regarded as a paradise of the physics of phase transitions.

The main features of the most general phase transitions encountered in liquid crystals are presented in this section.

6.1.2 The Isotropic–Nematic Transition

6.1.2.1 Mean Field Approach (Landau–de Gennes)

The nematic phase being the liquid crystal of highest symmetry, its condensation from the isotropic liquid should be the simplest to describe. Indeed, molecular theories convincingly explain the natural onset of nematic ordering in a population of anisotropic molecules with excluded volume interaction (Onsager) or in mean field theory (Maier–Saupe). Regarding the effect of symmetry on the isotropic to nematic (I–N) phase transition, the phenomenological approach is useful too.

It was explained in Chapter III of this volume that the order parameter of the nematic phase is a symmetric traceless tensor Q_{ij} that can be constructed from any macroscopic tensor property such as the magnetic susceptibility χ_{ij}. For a uniaxial nematic for instance, the order parameter can be defined as [1]:

$$Q_{ij} = G\left(\chi_{ij} - \frac{1}{3}(\chi_{\parallel} + 2\chi_{\perp})\delta_{ij}\right) \quad (1)$$

in which $\chi_{\parallel}$ and $\chi_{\perp}$ are susceptibilities parallel and perpendicular to the director respectively and δ_{ij} is the Kronecker unity tensor. The normalization constant G is generally chosen to ensure $Q_{zz} = 1$ for a perfectly ordered nematic. Such normalization suggests that Q_{zz} is similar to the scalar order parameter S defined in microscopic theories from the statistical distribution of molecular axes (see Chap. III, Sec. 1 of this volume). This is not exactly true, however, since macroscopic quantities χ_{ij} may not be a simple sum of uncorrelated microscopic susceptibilities (think of dielectric polarizability). Consequently, macroscopic nematic order parameters Q_{zz} defined from different macroscopic tensor properties (such as magnetic susceptibilities and refractive indices) may not show the same temperature behaviour. Experimental differences are, however, very weak [2].

In the eigenframe, the Q_{ij} matrix reads [1]:

$$Q = \begin{pmatrix} -\dfrac{q+\eta}{2} & 0 & 0 \\ 0 & -\dfrac{q-\eta}{2} & 0 \\ 0 & 0 & q \end{pmatrix} \quad (2)$$

The phenomenological scalar order parameter $q = Q_{zz}$ is non-zero in nematic phases and vanishes in the isotropic phase. η is non-zero in biaxial nematics only and referred to as the biaxial order parameter. The symmetric invariants of Q_{ij} may be written as [1, 3]:

$$\sigma_1 - Q_{ij} = 0$$

$$\sigma_2 = \frac{2}{3}Q_{ij}Q_{ij} = q^2 + \frac{\eta^2}{3}$$

$$\sigma_3 = 4Q_{ij}Q_{jk}Q_{ki} = q(q^2 - \eta^2) \quad (3)$$

The second and third order invariants can be regarded as independent in the biaxial phase with the constraint $\sigma_3^2 < \sigma_2^3$. Equality $\sigma_3^2 = \sigma_2^3$ is reached in the uniaxial nematic.

A Landau–de Gennes free energy may now be expanded in powers of the two invariants σ_2 and σ_3 [1, 3–5]:

$$\Delta F = F_N - F_{Iso} \quad (4)$$

$$= a\sigma_2 + b\sigma_3 + \frac{1}{2}c\sigma_2^2 + d\sigma_2\sigma_3 + \frac{1}{2}e\sigma_3^2$$

Uniaxial Nematic

The most common case of uniaxial nematics can be treated first: the free energy expansion reduces then to [1]:

$$\Delta F_{Uniaxial} = \frac{1}{2}A(T)q^2 + \frac{1}{3}Bq^3$$
$$+ \frac{1}{4}Cq^4 + O(q^5) \quad (5)$$

As usual with Landau theories, the phase transition is governed by the coefficient of the quadratic term $A(T) = a(T - T^*)$. Other coefficients B and C are supposed to have weak temperature dependance. The presence of a third order term, imposed by the symmetry, drives the transition first order [1, 6] at a Landau temperature T_{N-I} determined by $2B^2 = 9aC(T_{N-I} - T^*)$. It must be emphasized that such third order term reflects the physical relevance of the sign of q: think about a nematic arrangement of long rods with director **n** parallel to reference axis $\hat{z}$, $q > 0$ corresponds to long axes lining up along $\hat{z}$ whereas $q < 0$ corresponds to rods parallel to the (x, y) plane with no angular order in this plane. If the first case is, of course, physical for rods, the second is more appropriate for discs. Corresponding uniaxial nematic phases are often called cylindric and discotic respectively [4, 5].

Below T_{N-I}, the order parameter varies as:

$$q = -\frac{B}{2c}\left(1 + \frac{4aC}{B^2}(T^{**} - T)^{1/2}\right) \quad (6)$$

in which T^{**} is the absolute limit of superheating of the nematic phase. The order parameter at the transition is:

$$q_c = -\frac{3a}{B}(T_{N-I} - T^*) \quad (7)$$

Experiments do confirm the first order character of the I–N transition: discontinuities are observed but are weak enough to justify the relevance of the Landau expansion.

Biaxial Nematic

The stability of biaxial nematics is calculated by minimizing the free energy (Eq. 4). The stability of the model requires $c > 0$, $e > 0$ and $ce > d^2$ (higher order terms have to be considered in the expansion otherwise). The absolute minimum of ΔF is straightforwardly obtained at [1, 3, 5]:

$$\tilde{\sigma}_2 = \frac{bd - ae}{ce - d^2} \quad \text{and} \quad \tilde{\sigma}_3 = \frac{ad - bc}{ce - d^2} \quad (8)$$

The phase diagram is calculated in the (a, b) plane, as usual with Landau theories. The various possible transition lines are obtained as follows:

– The first order isotropic–uniaxial nematic corresponds to the set of constraints $\Delta F = 0$ and $\sigma_2^{1/2} = \sigma_3^{1/3} = q$. At first order in b, its equation is:

$$a_{I-N} = \frac{b^2}{2c}\left(1 + \frac{8bd}{c^2} + O(b^2)\right) \quad (9)$$

– The isotropic–biaxial nematic line is determined by $\Delta F = 0$ and independent σ_2 and σ_3 which leads to the trivial equation:

$$a_{I-N_b} = \pm b\sqrt{\frac{c}{e}} \quad (10)$$

– Finally, the transition lines from uniaxial nematic to biaxial are determined Eq. (8) with the constraint $\tilde{\sigma}_2^{1/2} = \tilde{\sigma}_3^{1/3}$. At lowest order in b, the slope of the two lines at their meeting point $a = b = 0$ is given by:

$$a_{N_\pm - N_b} = b\frac{cd}{2ce - d^2} + O_\pm(b^2) \quad (11)$$

The complete phase diagram is obtained with the selection rule $\tilde{\sigma}_2^3 \geq \tilde{\sigma}_3^2$ and reproduced in Fig. 1 [1]. Note that a direct transition I–N_b occurs at a single point $a = b = 0$ on the line (Eq. 10) [7]. The biaxial nematic region separates two uniaxial nematics N_+ and N_- of opposite sign. $N_\pm - N_b$ transitions are second order since the condition $\tilde{\sigma}_2^{1/2} = \tilde{\sigma}_3^{1/3}$ can be approached continuously from the biaxial phase.

Experiments on lyotropic systems [4, 5] do confirm the existence of a biaxial nematic phase separating two uniaxial nematics of opposite birefringence. Experimental features are in good agreement with the Landau–de Gennes mean field approach.

If the condition $ce > d^2$ is not satisfied, the minimum energy is clearly obtained with σ_2 as large as possible within the limit

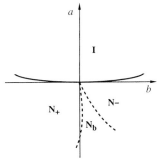

Figure 1. Phase diagram in the a, b plane resulting from minimization of Eq. (4) with biaxiality in the case $ce < d^2$. N_+ and N_- are uniaxial nematics with positive and negative order parameter respectively. N_b is a biaxial nematic. Solid lines are first-order transformations, dashed lines are second-order transformations.

$\tilde{\sigma}_2^3 \geq \tilde{\sigma}_3^2 \cdot \tilde{\sigma}_2 = \tilde{\sigma}_3$ is then the physical limit in this case and the biaxial domain is replaced by a direct first order transition from a positive to negative nematic [1].

6.1.2.2 Fluctuations

Like for any weakly first order phase transition, the mean field approach is expected to fail close to the temperature of transition $T_{N–I}$ when fluctuations of the order parameter reach correlation lengths much larger than molecular sizes. Spatial variations of the order parameter are classically accounted for by adding a gradient term $\frac{1}{2} L (\nabla q)^2$ to the expansion (Eq. 5) and taking the new expression as a local Hamiltonian (L is a positive Landau coefficient). Up to second order in q, the model is exactly soluble (Gaussian model) but has no condensed phase (it becomes unstable for $A(T) < 0$). In the isotropic phase, the Gaussian correlation length is then $\xi_G = (a/L(T-T^*))^{-1/2}$ for $T > T_{N–I} > T^*$.

The first step beyond mean field and Gaussian approximations is to expand the partition function in powers of the interaction (i.e. terms of order higher than 2 in q) and integrate over Gaussian fluctuations. It is a standard result that such perturbation expansion presents infrared divergences (i.e. for wavenumber $k \to 0$) in three dimensions [8]. It can be used, however, to estimate the critical domain $T_{Ginz} - T^*$, T_{Ginz} being the Ginzburg temperature at which the lowest order contribution of the interaction is of order the mean field value [1, 9, 10]:

$$T_{Ginz} - T^* \approx \frac{L}{a}\left(\frac{7B^2 k_B T}{4\pi^2 L^3}\right)^{2/3} \tag{12}$$

Experimental estimates of B lead to a noticeable critical domain ($T_{Ginz} - T^*$ of order 1 to 10 K [1]). Significant deviations from mean field predictions are indeed commonly observed [11]. For instance, the differ-

ence $(T^{**} - T_{N–I})$ is smaller than its mean field value $B^2/36\,aC$ which shows that fluctuations decrease the effective value of B. Note that this leads to underestimate the critical domain by setting the experimental value of B in Eq. (12).

The divergences of the perturbation expansion can be overcome by length scale transformations of the renormalization group method [12]. Each step of the process consists in integrating out fluctuations of large wavenumber k close to the upper cutoff Λ, say $\Lambda/b < k < \Lambda$ with $0 < b < 1$ and rescaling all lengths ($k \to bk$) to restore original momentum space. High k integrations are safe with respect to infrared divergences whereas length rescaling decreases the correlation length and thus brings the system away from the critical point towards 'safer' (i.e. less diverging) regions. Iterations of this process lead to recursion relations which define trajectories of the Landau–Ginzburg coefficients $B(b)$ and $C(b)$. Stable fixed points of these trajectories are linked to second order phase transitions whereas unstable flows denote first order transitions. Given that the Landau rule predicts a first order N–I transition, the renormalization flow is expected to hit such unstable domain for large enough correlation length. It is interesting to notice that besides this first order behavior, the existence of an unexpected stable fixed point has been reported [13]. This fixed point should be physically accessible for small values of the ratio B^2/C, i.e. close to the biaxial nematic. Although conjectural though, this result suggests the possibility of a second order isotropic to nematic transition, despite the third order invariant!

6.1.2.3 Isotropic–Nematic Transition in Restricted Geometries

The size or shape of the liquid crystal sample has not been specified so far: phase transition theories implicitely deal with infinite samples, that is samples of macroscopic size D much larger than any relevant physical length such as the correlation length ξ of the order parameter. Liquid crystals, however, are commonly used in restricted geometries such as thin films or dispersions in a polymer matrix (PDLC) or a solid labyrinth (aerogels). There are cases where the restriction of the available space plays a role on the phase transition. We will discuss, as an example, the influence of a solid interface that develops different microscopic interactions with molecules in the isotropic and the nematic phase. A macroscopic way of considering this difference is to assume different wetting of the solid by the two phases. This problem was addressed by Poniewiersky and Sluckin [14] who showed that the isotropic to nematic transition is shifted towards lower temperatures if the isotropic phase wets the solid surface more than the nematic phase. The slope of the coexistence curve is given by a generalized Clausius-Clapeyron equation:

$$\frac{dT_{N-I}}{dD} = \frac{2}{D^2}\frac{(\gamma_N - \gamma_I)}{L}T_{N-I}(D) \qquad (13)$$

in which D is the thickness of the liquid crystal film, γ_N and γ_I the surface tension of the nematic and isotropic phase relative to the substrate and L the latent heat of the N–I transition.

It could happen, on the contrary, that a solid substrate promotes local nematic ordering in the isotropic phase which amounts to setting $(\gamma_N - \gamma_I) < 0$ in Eq. (13) (this was experimentally observed for mica surfaces in lyotropic nematics [15]). The nematic phase may then form above T_{N-I} for thin enough liquid crystal films. This is the so-called capillary condensation.

Geometrical constraints may affect the N–I transition in different ways: strong anchoring conditions at the boundaries of lim-

ited volumes (PDLC or aerogels) produce topological defects in the nematic phase. The N–I transition may then be decreased to compensate for the extra energetical cost of the defects.

6.1.3   The Nematic–Smectic A Transition

6.1.3.1   Definition of an Order Parameter

After the isotropic to nematic transition, the next step towards more ordered mesophases is the condensation of SmA order when the continuous translational symmetry is broken along the director. The theoretical description of the N–SmA transition begins with the identification of an order parameter. Following de Gennes and McMillan [1, 16], we notice that the layered structures of a SmA phase is characterized by a periodic modulation of all the microscopic properties along the direction $\hat{z}$ perpendicular to the layers. The electron density for instance, commonly detected by X-ray scattering can be expanded in Fourier series:

$$\rho(\mathbf{r}) = \rho_0 + \sum_{n=1}^{\infty} \Psi_n\, e^{iq_{Sm}\cdot r} + \Psi_n^*\, e^{-iq_{Sm}\cdot r} \qquad (14)$$

in which q_{Sm} is the smectic wavenumber of modules $2\pi/d$. The smectic period d (commonly referred to as layer thickness) is of the order of a molecular length in most thermotropic smectics and varies essentially as the inverse fraction Φ^{-1} of the material constituting the lamellae in lyotropic smectics. The average electron density ρ_0 is about the same in the nematic and smectic A phases. The fundamental term Ψ_1 of the Fourier expansion is obviously zero in the nematic phase. It is a natural choice for the N–SmA order parameter. As a complex, it has two independent components:

$$\Psi_1(\mathbf{r}) = \rho_1(\mathbf{r})e^{i\Phi_1(\mathbf{r})} \qquad (15)$$

The modulus $\rho_1(\mathbf{r})$ is a measure of the strength of the local SmA ordering: the more the molecules are segregated into well defined layers, the higher ρ_1. Note that unlike the nematic scalar order parameter S, ρ_1 cannot be defined on an absolute scale that would assign 1 to perfect SmA ordering for instance. The amplitude of the electron density wave depends on each particular system. Variations of the phase $\Phi_1(\mathbf{r})$ are related to the local displacements of the layers commonly denoted as $u(\mathbf{r})$ with:

$$\Phi_1(\mathbf{r}) = -i\,q_{sm}\,u(\mathbf{r}) \qquad (16)$$

The basic elastic distortions of the layers, namely compression and curvature correspond to $\nabla_\parallel u$ and $\Delta_\perp u$ respectively (subscripts $\parallel$ and $\perp$ denote directions parallel and perpendicular to the director respectively).

6.1.3.2 Mean Field Description (McMillan, de Gennes)

Mean field approximation neglects all spatial inhomogeneities. The order parameter is thus considered as constant all over the sample. In the vicinity of the N–SmA phase transition, a Landau free energy density may be expanded in invariant combinations of the order parameter Ψ (the subscript 1 referring to the first term of the Fourier expansion will be omitted in the following). Translational invariance requires that the phase Φ does not enter the free energy. The Landau expansion thus reduces to:

$$\Delta F_1 = \frac{r}{2}|\Psi|^2 + \frac{u_0}{4}|\Psi|^4 + \dots \qquad (17)$$

With $r = a\,(T - T_{N-A})$ and $u_0 > 0$, the N–SmA transition is second order at the mean field temperature T_{N-A}. The modulus $|\Psi|$ grows as $(r/u_0)^{1/2}$ below T_{N-A}. The critical exponent β of the order parameter is thus 1/2 as usual with mean field theories.

This picture is slightly more complicated if the coupling with the nematic order parameter $S_0(T) = 1/2\langle 3\cos^2\theta_i - 1\rangle$ is included (θ_i denotes the angle between the long axis of the ith molecule and the director $\mathbf{n} = \hat{z}$ and the brackets $\langle\dots\rangle$ represent an average over the volume of the sample).

If δS measures the deviation of the microscopic alignment from its equilibrium value S_0 in the nematic phase (calculated from the Maier–Saupe theory [17] for instance), the free energy density reads:

$$\qquad\qquad\qquad\qquad\qquad\qquad (18)$$
$$\Delta F_2 = \frac{r}{2}|\Psi|^2 + \frac{u_0}{4}|\Psi|^4 + \frac{1}{2\chi}\delta S^2 - C|\Psi|^2\delta S$$

$\chi(T)$ is a response function (susceptibility) and C is a generally positive constant (the onset of the smectic order usually increases the average attraction between the molecules and hence reinforces the alignment).

Minimization with respect to δS yields $\delta S = \chi C|\Psi|^2$ (>0) and the new coefficient of the fourth order term is now:

$$u = u_0 - 2C^2\chi \qquad (19)$$

If the nematic susceptibility χ is low (i.e. far enough from T_{N-I}, temperature of the isotropic–nematic transition), u is positive and the N–SmA transition is second order at T_{N-A} again.

If χ is larger (i.e. T_{N-A} close to T_{N-I}, u is negative. A sixth order term $v/6|\Psi|^6$ must be added to Eq. (18) and the N–SmA transition is first order at a temperature $T_{N-A} + 3\,u^2/16\,a\,v > T_{N-A}$.

$\chi = u_0/2\,C^2$ (i.e. $u = 0$) defines a tricritical point on the N–SmA line.

McMillan estimated that the tricritical point corresponds to a value 0.87 of the ratio T_{N-I}/T_{N-A} [18] which agrees reasonably well with many experimental data.

6.1.3.3 Analogy with Superconductors

Because the smectic order is one dimensional, the fluctuations of the layers described by the local displacement field $u(\mathbf{r})$ are known to play an important role, even far away from any phase transition: the square amplitude $\langle u^2(\mathbf{r}) \rangle$ diverges like the logarithm of the size of the sample (this is the well known Landau–Peierls instability [19, 20]) hence killing a true long range order.

Close to the N–SmA transition, the vanishing of the elastic constant of compression of the layers amplifies the fluctuations of the phase whereas critical fluctuations of the amplitude $|\Psi(\mathbf{r})|$ are expected to be important too.

Fluctuations are accounted for in a Landau–Ginzburg expansion of a local Hamiltonian. Once again even powers of $|\psi|$ only are permitted. Including gradient terms and fluctuations of the nematic director $\delta\mathbf{n}_\perp = \mathbf{n}(\mathbf{r}) - \mathbf{n}_0 \, (\mathbf{n}_0 = \hat{z})$ yields the following Landau–Ginzburg functional:

$$F_S = \frac{1}{2}\int d^3 r \left\{ r|\psi|^2 + \frac{u}{2}|\psi|^4 \right.$$
$$+ C_\parallel |\nabla_z \psi|^2 + C_\perp |(\nabla_\perp - i q_{sm}\,\delta\mathbf{n}_\perp)\psi|^2$$
$$+ K_1 (\mathrm{div}\,\delta\mathbf{n}_\perp)^2 + K_2(\hat{z}\cdot\mathbf{curl}\,\delta\mathbf{n}_\perp)^2$$
$$\left. + K_3(\nabla_z\,\delta\mathbf{n}_\perp)^2 \right\} \qquad (20)$$

The first two terms are the Landau part, Eq. (17). Because of the nematic anisotropy, the gradient terms exhibit anisotropic coefficients ($C_\parallel \neq C_\perp$) along directions parallel and perpendicular to the director $\mathbf{n}$. With the notation $\nabla_z = \partial/\partial z$ and $\nabla_\perp = (\partial/\partial x, \partial/\partial y)$ and at lowest relevant order in ($\delta\mathbf{n}_\perp$), these gradients have the form Eq. (20). The last three terms are the usual Frank–Oseen elastic energy of the nematic [21].

If one forgets about the fluctuations of the director (i.e. set $\delta\mathbf{n}_\perp = 0$ in (20)) the N–SmA problem becomes equivalent to the condensation of superfluid helium (XY model, $d = 3$) since the smectic order parameter has two independent components (the anisotro-

py of the elastic coefficients $C_\parallel$ and $C_\perp$ can be removed by a simple anisotropic rescaling).

With non-zero $\delta\mathbf{n}_\perp$, expression (Eq. (20)) is very similar to the Landau–Ginzburg functional describing the normal–superconductor transition [1, 22, 23]:

$$F_{SC} = \frac{1}{2}\int d^3 r \left\{ a|\psi|^2 + \frac{u}{2}|\psi|^4 \right.$$
$$+ \frac{1}{4m}\left|\left(\hbar\nabla - i\frac{2e}{c}\mathbf{A}\right)\psi\right|^2$$
$$\left. + \frac{1}{8\pi\mu}(\mathbf{curl}\,\mathbf{A})^2 \right\} \qquad (21)$$

Here, ψ is the superconductor gap order parameter. It corresponds to the wave function of the superconducting pair in BCS theory and has the XY symmetry of the smectic order parameter. The magnetic vector potential $\mathbf{A}$ comes analogous to the director $\mathbf{n}$ (m and e are the mass and charge of a single electron, $\hbar$ Planck's constant, c the velocity of light and μ the magnetic permittivity).

In liquid crystals, $\mathbf{curl}\,\mathbf{n}$ is then analogous to the magnetic field $\mathbf{B} = \mathbf{curl}\,\mathbf{A}$. Twist ($\mathbf{n}\cdot\mathbf{curl}\,\mathbf{n}$) and bend ($\mathbf{n}\times\mathbf{curl}\,\mathbf{n}$) are components parallel and perpendicular to the director respectively (they correspond to components of $\mathbf{B}$ parallel and perpendicular to $\mathbf{A}$ in superconductors). The anisotropy of the field $\mathbf{curl}\,\mathbf{n}$ follows from $K_2 \neq K_3$.

Interesting behaviors of the smectic state can be deduced from this analogy.

Just as superconductors expel magnetic fields $\mathbf{curl}\,\mathbf{A}$ (Meisner effect [24]) smectics expel bend and twist. The bend and twist moduli K_2 and K_3 should therefore diverge upon approaching the smectic state from the nematic.

Two important lengths characterize the superconductors: the order parameter coherence length $\xi = (m/|r|)^{1/2}$ over which ψ can vary and the London penetration depth of a magnetic field $\lambda = (m\,c^2/2\,\mu\,e^2|\psi|^2)^{1/2}$.

For smectics, two order parameter coherence lengths

$$\xi_{\parallel,\perp} = \left(\frac{C_{\parallel,\perp}}{|r|} \right)^{\frac{1}{2}} \tag{22}$$

and four penetration lengths associated with twist and bend:

$$\lambda_{\parallel,\perp}^{2,3} = \left(\frac{K_{2,3}\,u}{2C_{\parallel,\perp}\,q_{Sm}^2\,|r|} \right)^{\frac{1}{2}} \tag{23}$$

can be identified, which precludes a simple classification.

The superconductor analogy suggests, however, two distinct behaviors:

- Type I ($\lambda\sqrt{2} > \xi$). The superconducting state is observed with perfect Meisner effect below a critical field H_c.
- Type II ($\lambda\sqrt{2} < \xi$). Two critical values of the field are found. Vortices bearing a quantum flux $\phi_0 = h/q$ penetrate the system for $H < H_{c1}$ whereas the normal–type I superconductor transition occurs at $H_{c2} > H_{c1}$.

The analog of the magnetic intensity **H** (produced by external currents) would be an external field coupled to the two components of **curl n**. A local microscopic source of bend [23] is not easy to imagine. On the other hand, chirality is naturally coupled to twist: chiral mesogens develop a spontaneous twist in the nematic phase. The cholesteric is thus analogous to a normal metal in a magnetic field.

The smectic analog of a vortex is a dislocation. Type II smectics would therefore develop dislocations when submitted to a bending or twisting stress whereas type I would resist until a critical stress eventually induces the nematic state.

The observation of edge dislocation arrays in a wedge (i.e. bending stress) [25] and of screw dislocation arrays in chiral compounds [26] (the so-called twist grain

boundary (TGB) phase anologous to the Abrikosov flux phase in superconductors) suggests that type II smectics do exist, but type I behaviors have also been reported [27]. The N–SmA transition in chiral smectics and the ability of TGB phases will be discussed in Sec. 6.1.5 of this chapter.

Finally, the bare smectic coherence length at 0 temperature $\xi_0 = (C_\perp / a\,T_{N-A})^{1/2}$ with $r = a\,(T - T_{N-A})$ is significantly shorter than its superconductor equivalent (10–20 Å instead of 5000 Å) because of the higher value of T_{N-A}. An interesting consequence is that the critical domain is expected to be much larger (i.e. more easily accessible) in the smectic case.

Important differences exist, however, about gauge invariance and the absence of true smectic long range order. How severe these differences are is not fully understood yet.

The superconductor Hamiltonian (Eq. 21) is invariant under the following gauge transformation:

$$\begin{cases} \mathbf{A}' = \mathbf{A} - \nabla L \\ \psi' = \psi' \exp\left(-\frac{q}{c}\,L \right) \end{cases} \tag{24}$$

so that any gauge choice is physically acceptable. The Coulomb gauge div $\mathbf{A} = 0$ is generally used.

In the smectic case, a gauge transformation reads:

$$\begin{cases} \delta\mathbf{n}' = \delta\mathbf{n} - \nabla L \\ \psi' = \psi\,\exp(-i q_{Sm}\,L) \end{cases} \tag{25}$$

but the Frank–Oseen energy is not gauge invariant because of the splay term K_1. The lack of gauge invariance reflects the fact that the director field **n** is a physical observable in liquid crystals (unlike its magnetic analog **A**). There is in fact only one physical gauge, namely $\delta n_z = 0$ (remember that the condition $n^2 = 1$ implies that $\delta\mathbf{n}$ lies in the

x, y plane). In the unphysical Coulomb gauge (div $\delta\mathbf{n}=0$), the splay term disappears and it can be shown that true long range smectic order is not killed by the layer fluctuations [28]. If the physical $\delta\mathbf{n}$ is expressed in the Coulomb gauge via Eq. (25), the splay term reads:

$$\frac{K_1}{2}\int(\Delta L)^2 d^3x \tag{26}$$

and with the condition $\delta n_z=0$ (i.e. $A_z(q)-iq_zL(q)=0$):

$$\frac{K_1}{2}\int\frac{q^4}{q_z^2}\left|A_z(q)\right|^2\frac{d^3q}{(2\pi)^3} \tag{27}$$

The splay term is non-analytic, which implies that K_1 cannot be renormalized by the fluctuations of ψ at any order in perturbation theory. In a renormalization flow, K_1 will simply evolve according to simple power counting.

6.1.3.4   Critical Exponents and Scaling

Experiments suggest the existence of anisotropic scaling laws: $\xi_\parallel\propto t^{-\nu_\parallel}$ and $\xi_\perp\propto t^{-\nu_\perp}$ with $\nu_\parallel\neq\nu_\perp$ for instance for the correlation lengths parallel and perpendicular to the director $\mathbf{n}$ ($t=(T-T_{N-A})/T_{N-A}$ is the reduced temperature).

Anisotropic homogeneous functions can indeed be defined to describe the scaling behavior of the correlation functions of the order parameters:

$$\langle\psi(q)\,\psi^*(q)\rangle = G(q_\parallel, q_\perp, t) \tag{28}$$
$$= t^{-(2-\eta_\perp)\nu_\perp}G(t^{-\nu_\parallel}q_\parallel, t^{-\nu_\perp}q_\perp)$$

$$\langle\delta n_\alpha(q)\delta n_\beta^*(q)\rangle = G_{\alpha\beta}(q_\parallel, q_\perp, t) \tag{29}$$
$$= t^{-(2-\eta_n)\nu_\perp}G_{\alpha\beta}(t^{-\nu_\parallel}q_\parallel, t^{-\nu_\perp}q_\perp)$$

G and $G_{\alpha\beta}$ are the two point correlation functions of the smectic order parameter ψ and of the director $\delta n_\perp$ respectively (indices α, β can be x or y).

The exponent γ of the susceptibility ($G(q=0)\approx t^{-\gamma}$) comes immediately from

Eq. (28):

$$\gamma = (2-\eta_\perp)\nu_\perp = (2-\eta_\parallel)\nu_\parallel \tag{30}$$

The free energy density $f(t)$ classically scales like k_BT over the volume of a 'block' of condensed phase ξ^3. In the smectic case, this yields:

$$f(t) \sim \frac{k_BT}{\xi_\parallel\xi_\perp^2} \sim t^{2\nu_\perp+\nu_\parallel} \sim t^{2-\alpha} \tag{31}$$

in which α is the critical exponent of the singular part of the specific heat δC_P.

The critical behavior of the twist and bend elastic constants on the nematic phase follows from the equipartition theorem

$$\begin{cases}\langle\delta n_1(q)\delta n_1(-q)\rangle = \dfrac{k_BT}{K_1 q_\perp^2 + K_3 q_\parallel^2} \\[2ex] \langle\delta n_2(q)\delta n_2(-q)\rangle = \dfrac{k_BT}{K_2 q_\perp^2 + K_3 q_\parallel^2}\end{cases} \tag{32}$$

Axes 1 and 2 in Eq. (32) are the eigendirections of the Frank–Oseen elastic energy in the Fourier plane (q_x, q_y). The corresponding unit vectors $\mathbf{e}_1$ and $\mathbf{e}_2$ are such that $\mathbf{e}_2$ is normal to $\mathbf{q}$ and $\mathbf{e}_1$ is normal to $\mathbf{e}_2$ [1]. With the structure (Eq. 29), one gets:

$$K_2 \sim t^{-\eta_n\nu_\perp}, \quad K_3 \sim t^{2(\nu_\perp-\nu_\parallel)-\eta_n\nu_\perp} \tag{33}$$

On the smectic side:

$$\langle\delta n_\perp(q)\delta n_\perp(-q)\rangle = \frac{k_BT q_\perp^2}{Bq_z^2 + K_2q_\perp^4} \tag{34}$$

Equations (29) and (34) yield:

$$B \approx t^{(4-\eta_n)\nu_\perp-2\nu_\parallel} \tag{35}$$

The final scaling laws are then:

$$\begin{cases}\text{(a)}\ \ \xi_\parallel = \xi_0\,t^{-\nu_\parallel} \qquad \xi_\perp = \xi_0\,t^{-\nu_\perp} \\ \text{(b)}\ \ \delta K_2 \propto \xi_\perp^2/\xi_\parallel \qquad \delta K_3 \propto \xi_\parallel \\ \text{(c)}\ \ B \propto \xi_\parallel/\xi_\perp^2 \\ \text{(d)}\ \ 2-\alpha = \nu_\parallel+2\nu_\perp \quad \gamma = (2-\eta_\perp)\nu_\perp\end{cases} \tag{36}$$

Equations (36 b) and (36 d) are valid above and below T_{N-A}.

6.1.3.5   Renormalization Group Procedures

According to the classical Wilson–Fisher renormalization process [12], the parameters of the Hamiltonian (Eq. 20) will evolve under an anisotropic rescaling of the wavevectors of the form ($q_\perp \to e^l q_\perp$ and $q_z \to e^{v_\parallel/v_\perp l} q_z$). The recursion relations defining the renormalization flow of the physical parameters are controlled by simple dimensional analysis (power counting) and Gaussian integration of the fluctuations of large wavevectors (perturbation theory).

The splay constant K_1 plays a particular role since its flow depends on power counting only (see Sec. 6.1.3.3 of this volume) which yields:

$$\frac{dK_1}{dl} = -\eta_n K_1(l) = -\frac{(2v_\perp - v_\parallel)}{v_\perp} K_1(l) \quad (37)$$

There are therefore only three possibilities for the fixed point:

(1) $K_1^* = 0$: for $2v_\perp > v_\parallel$, the K_1 eigendirection is stable. The problem reduces exactly to the normal–superconductor transition [29] which implies $v_\perp = v_\parallel$. There is no such stable fixed point, however, in the smectic case ($n = 2$, $d = 3$) [30].
(2) $K_1^* = \infty$: the fixed point can be stable if $2v_\perp < v_\parallel$ only, but none is found [29].
(3) K_1^* finite and $2v_\perp - v_\parallel = 0$: this anisotropy is too large to fit any experimental behavior.

Classical theories fail to find a stable fixed point (not even isotropic) and consequently cannot corroborate the experimental evidence that the transition can be second order. Two possible situations can be imagined upon moving away from the mean field tricritical point towards negative values of the Landau coefficient, u: either the transition is still first order but the discontinuities are too small to be detected (this would agree with the lack of stable fixed point of the renormalization flow) or beyond some original multicritical point the transition becomes truly second order. The theory of a transition induced by dislocation loops provides an example that fits in this last situation.

6.1.3.6   Dislocation Loops Theory

Dislocations are the most elementar defects of a SmA phase. Since they cannot end in the smectic material, they must form closed loops (at least in the thermodynamic limit of an infinite sample). The elastic energy of a dislocation line is proportional to its length [31]. It is then characterized by a line tension γ_0. The total free energy per unit length is:

$$\gamma = \gamma_0 - \frac{k_B T}{x} \quad (38)$$

k_B/x measures the entropy contribution that favors spontaneous nucleation of dislocation loops (x has the dimension of a length). γ becomes negative above some temperature so that the density of dislocations becomes finite. Helfrich [32] and Nelson and Toner [33] have shown that such proliferation of dislocations destroys smectic order: the elastic response of a SmA with a finite density of unbound dislocations is equivalent to that of a nematic [34] (unbound dislocation loops are those thermally excited defects that can move around freely under the effect of thermal fluctuations). Nelson and Toner [33], Dasgupta and Halperin [35] and Toner [34] have formalized this problem: the main result is the existence of a stable fixed point which for the first time gave the theoretical possibility of a true second order N–SmA transition.

This fixed point is isotropic and belongs to the so-called inverted XY universality class. Inverted refers to the inversion of the high and low temperature sides of the transition since the melting is described by a 'disorder' parameter ψ (density of dislocations). $|\psi| \neq 0$ corresponds to the nematic phase. The universal ratio of the amplitudes A^+ and A^- of the heat capacity singularity for instance ($\delta C_p = A^\pm t^{-\alpha}$) is the inverse of the superfluid helium value.

6.1.4 The Smectic A–Smectic C Transition

6.1.4.1 The Superfluid Helium Analogy

The SmC state differs from the SmA by a tilt θ of the director **n** with respect to the direction $\hat{z}$ normal to the layers. The director **n** is totally specified by θ and the azimuthal angle φ. The SmC order can thus be described by two real angles θ and φ or equivalently by the complex order parameter:

$$\psi(\mathbf{r}) = \theta(\mathbf{r}) e^{i\varphi(\mathbf{r})} \qquad (39)$$

A modulus $\theta = 0$ corresponds to the SmA state. A Landau–Ginzburg functional similar to Eq. (20) with $\delta\mathbf{n} = 0$, and therefore to the superfluid–normal helium problem, can be constructed to describe the SmA–SmC transition.

The straightforward consequence of this analogy is that the SmA–SmC transition may be continuous at a temperature $T_{SmC-SmA}$ with XY critical exponents. Below $T_{SmC-SmA}$ the tilt angle θ for instance should vary as $\theta = \theta_0 |t|^\beta$ with $\beta = 0.35$. Above $T_{SmC-SmA}$ an external magnetic field can induce a tilt θ proportional to the susceptibility $\chi \approx t^{-\gamma}$ with $\gamma = 1.33$.

Experiments (heat capacity measurements in particular [36]) rather show a mean field behavior which may be due to the narrowness of the critical domain or to the influence of a close by tricritical point.

The width of the XY critical domain can be estimated from the usual Ginzburg cri-

terion: equating the mean field heat capacity discontinuity and the contribution of the fluctuations in the Gaussian approximation leads to a Ginzburg temperature T_{Ginz} [1, 37]:

$$T_{Ginz} - T_{SmC-SmA} \approx T_{SmC-SmA}/64\pi^2 \qquad (40)$$

so that the critical regime extends over a fraction of a degree. It may easily switch from observable to non-observable with a slight change of the roughly estimated numerical constant that lead to Eq. (40). Furthermore, the crossover regime may be different from one observable to another. Recent estimates of the Ginzburg criterion, for instance, indicate that elastic constants measurements are one hundred times more sensitive to fluctuations than heat capacity ones [37]. Experiments tend to confirm this point: a mean field behavior of the heat capacity has been reported [36] whereas fluctuations seem to be important in some tilt, susceptibility or bulk modulus measurements [38, 39].

Finally, a first order SmA–SmC transition is always possible.

6.1.4.2 The N–SmA–SmC Point

The existence of a N–SmA–SmC multicritical point (i.e. a point where the N–SmA, SmA–SmC and N–SmC lines meet) was demonstrated in the late seventies [40, 41]. Various theories have been proposed to describe the N–SmA–SmC diagram [42–45]. The phenomenological model of Chen and Lubensky [46] (referred to as the N–SmA–SmC model) captures most of the experimental features. The starting point is the observation that the X-ray scattering in the nematic phase in the vicinity of the N–SmA transition shows strong peaks at wavenumber $\mathbf{q}_A = \pm q_0 \mathbf{n}$. Near the N–SmC transition, these two peaks spread out into two rings at $\mathbf{q}_C = (\pm g_\parallel, \, q_\perp \cos\varphi, \, q_\perp \sin\varphi)$.

Once again, the order parameter is defined from the mass density wave $\rho(\mathbf{r})$. In the smectic phases, $\rho(\mathbf{r})$ becomes periodic with fundamental wave number $\mathbf{q}_0 = (\pm q_\parallel, q_\perp)$ ($q_\perp = 0$ in the SmA phase). Chen and Lubensky defined a covariant Landau–Ginzburg Hamiltonian including fluctuations of the director through the Frank free energy. A mean field description of the N–SmA–SmC model is given here. The director is chosen along z: $\mathbf{n} = \mathbf{n}_0$ $(0,0,1)$. Like in the SmA phase, the order parameter Ψ is chosen as the part of ρ with wave numbers in the vicinity of $\mathbf{q}_0$:   (41)

$$\Psi(\mathbf{r}) = \psi \exp i(\mathbf{q}_0 \cdot \mathbf{r}) = \psi \exp i(q_\parallel z + \mathbf{q}_\perp \cdot \mathbf{x}_\perp)$$

$\mathbf{x}_\perp$ and $z_\perp$ are two dimensional vectors in the plane (x, y) and (q_x, q_y) respectively. The mean field free energy density reads:

$$\Delta F_{\text{N–SmA–SmC}}$$
$$= \frac{1}{2}\Big\{ a|\psi|^2 + D_\parallel |\nabla_z^2 \psi|^2$$
$$- C_\parallel |\nabla_z \psi|^2 + \frac{C_\parallel}{4D_\parallel}|\psi|^2$$
$$- C_\perp |\nabla_\perp \psi|^2 + D_\parallel |\nabla_\perp^2 \psi|^2 \Big\} + u|\psi|^2 \quad (42)$$

with $C_\parallel$, $D_\parallel$, $D_\perp$, and $u > 0$ and $a =$

$$a'\left(\frac{T - T_{\text{N–SmA}}}{T_{\text{N–SmA}}} \right).$$

In Fourier space, $\Delta F_{\text{N–SmA–SmC}}$ reads:

$$\Delta F_{\text{N–SmA–SmC}}$$
$$= \frac{1}{2}\Big\{ a + D_\parallel\big(q_z^2 - Q_\parallel^2\big)^2 + C_\perp q_\perp^2 + D_\perp q_\perp^4 \Big\}$$
$$\times \psi^2 + u\psi^4 \quad (43)$$

where $Q_\parallel^2 = C_\parallel / 2D_\parallel$.

The free energy (Eq. 43) is minimized the usual way:

– For $C_\perp > 0$, the minimum is reached for $q_\perp = 0$ and $q_z 0 \pm q_\parallel$. A second order N–SmA transition occurs at the mean field temperature $T = T_{\text{N–SmA}}$.

– For $C_\perp < 0$, $\Delta F_{\text{N–SmA–SmC}}$ can be rewritten as:

$$\Delta F_{\text{N–SmA–SmC}}$$
$$= \frac{1}{2}\Big\{ \tilde{a} + D_\parallel\big(q_z^2 - Q_\parallel^2\big)^2 + D_\perp\big(q_\perp^2 - Q_\perp^2\big)^2 \Big\}$$
$$\times \psi^2 + u\psi^4 \quad (44)$$

with $Q_\perp^2 = |C_\perp|/2D_\perp$ and
$\tilde{a} = a'(T - T_{\text{N–SmC}})/T_{\text{N–SmA}}$,
$T_{\text{N–SmC}} = T_{\text{N–SmA}} + C_\perp^2 T_{\text{N–SmA}}/4D_\perp a'$.

The minimum corresponds to $q_\perp = \pm Q_\perp$, $q_z = \pm Q_\parallel$ and a N–SmC transition occurs at the mean field temperature $T_{\text{N–SmC}}$. The mean field N–SmA–SmC diagram is shown in Fig. 2.

The scattered X-ray intensity $I(q)$ can be estimated in the nematic phase from Gaussian fluctuations about the mean field solution $\psi = 0$:

$$I(q) \approx \langle \rho(q)\rho(-q) \rangle \quad (45)$$
$$\approx \frac{1}{a + D_\parallel\big(q_\parallel^2 - Q_\parallel^2\big) + C_\perp q_\perp^2 + D_\perp q_\perp^4}$$

For $C_\perp > 0$, $I(q)$ has peaks at $q_\parallel = \pm Q_\parallel$ corresponding to fluctuations into the SmA

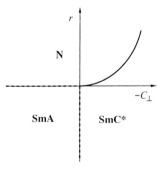

Figure 2. N–SmA–SmC phase diagram from Chen and Lubensky [46]. All transitions are second order in mean field but fluctuations lead to a first order N–SmC transition [47]. The line $C_\perp = 0$ separates two regions in the nematic phase with diffuse X-ray scattering centered about $q_\perp = 0$ ($C_\perp > 0$) and $q_\perp \neq 0$ ($C_\perp < 0$). The point $C_\perp = 0$, $r = 0$ is a Lifschitz point.

phase. When $C_\perp$ is negative (see Eq. 44) $I(q)$ is maximum on two rings ($\pm q_\parallel$, $q_\perp \cos \varphi$, $q_\perp \sin \varphi$) as observed.

The N–SmA–SmC problem bears strong similarities with the transition from paramagnetic to helimagnetic states [47]. Fluctuations are maximum at a finite m-dimensional vector $q_\perp$ in a d-dimensional space. Mean field N–SmA–SmC theory predicts a second oder N–SmC transition as shown. Fluctuations, however, are believed to lead to a first-order transition when $m = d$ or $d - 1$. The N–SmC transition with $d = 3$ and $m = 2$ is therefore expected to be first order.

Finally, we notice that the N–SmA–SmC point is an example of Lifschitz point defined as the place where some of the coefficients of the gradient terms ($C_\perp$ here) vanish [48, 49].

Complications arise from the vanishing of the N–SmC latent heat at the N–SmA–SmC point and from the difficulties connected to the smectic state (Landau–Peierls instability) the N–SmA transition (lack of gauge invariance) and the SmA–SmC transition (proximity of a tricritical point).

The description of the Chen–Lubensky model is reasonably well borne out by experiment [50, 51]: the universal topology of the phase diagram and the existence of a $C_\perp = 0$ line in particular are well established. An interesting possibility pointed out by Grinstein and Toner [52] with a model of dislocation unbinding is the existence of a biaxial nematic phase: if the N–SmA and SmA–SmC lines are second order with XY critical exponents, the N–SmA–SmC point should be tetracritical [53] and a mixed phase (i.e. a biaxial nematic) should show up. It has however not been observed so far.

6.1.5   The Smectic A–Hexatic Smectic B Transition

The *crystalline* smectic B differs from the SmA phase by the existence of a regular hexagonal packing of the molecules in the plane of the layers. The B–SmA transition corresponds to the melting of a two-dimensional crystalline order in a three-dimensional material whereas the layered structure along the third direction switches from true to quasi-long range order. In two dimensions (i.e. in a hypothetic isolated single layer) a Kosterlitz–Thouless [54] mechanism of dislocations unbinding may happen to make the transition continuous with no singularities. In three dimensions, a direct transition B–SmA is expected to be first order.

An intermediate layered structure SmB, however, exists in which the two-dimensional ordering of the molecules within a layer is organized as follows: the correlations of the position of the center of mass of the molecules decay exponentially (like in a liquid) but the direction of the bonds linking two adjacent molecules exhibit a long range order with six-fold symmetry. Halperin, Nelson and Young [55, 56] have pointed out that dislocations can create such a two dimensional hexatic order. The existence of this hexatic order in smectics is a beautiful success of the Halperin–Nelson–Young concept of bond ordering.

Accounting for the six-fold symmetry of the B and SmB phases, the modulation of the mass density within the (x, y) plane of the layers can be expanded in a Fourier series:

$$\rho(\mathbf{r}) - \bar{\rho} = \rho_0(r) + \rho_6(r) \cos 6(\theta - \phi)$$
$$+ \text{higher harmonics} \tag{46}$$

$\mathbf{r}$ is a 2-d vector of cartesian coordinates $x = r \cos \theta$ and $y = r \sin \theta$. The correlation function of the radial distribution $\rho_0(r)$ falls off exponentially in the SmB and SmA phases. The amplitude $\rho_6(r)$ describing the six-fold order vanishes in the SmA phase only. A possible order parameter of the SmA–SmB transition is therefore:

$$\psi_6 = \rho_6 \, e^{6i\phi} \tag{47}$$

The phase ϕ accounts for spatial fluctuations of the orientation of the local six-fold axes.

The order parameter ψ_6 has two components and the SmA–SmB transition is expected to belong to the XY universality class.

Experiments do confirm the second order nature of the transition but high resolution a.c. calorimetry gives values of the specific heat exponent α about 0.6 [57] inconsistent wih the XY class ($\alpha = -0.06$). High sensitivity heat capacity measurements on freely suspended thin films (down to four layer) show a crossover from bulk ($\alpha = 0.59$) to a two dimensional cusp-like behavior with $\alpha = -0.26$ [58].

6.1.6   Phase Transitions in Chiral Liquid Crystals

6.1.6.1   Chirality in Nematic and Smectic Liquid Crystals

Experiments show that the introduction of chirality (chiral mesogen or chiral dopant) in a nematic phase generates a spontaneous twist of the director, $\mathbf{n} = (\cos 2\pi z/P, \sin 2\pi z/P, 0)$ for instance for a helical twist of pitch P along the direction z. The higher the chirality (i.e. fraction of chiral dopant for instance), the higher the twist (i.e. the shorter the pitch). It is thus clear that twist is a structural response coupled to the microscopic constraint chirality. How can these two physical quantities be linked in a quantitative way? The covariant expression of the local twist, commonly used in the Frank energy, is $\mathbf{n} \cdot \mathbf{curl}\, \mathbf{n}$ ($=-2\pi/P$ in the example of a simple cholesteric helix). A natural extension of the Frank energy density to chiral nematics is [1, 59]:

$$\Delta F_{\text{Frank}}^{\text{Chiral}} = \frac{K_1}{2}(\nabla \cdot \mathbf{n})^2 + \frac{K_2}{2}(\mathbf{n} \cdot \mathbf{curl}\, \mathbf{n})^2$$
$$+ \frac{K_3}{2}(\mathbf{n} \times \mathbf{curl}\, \mathbf{n})^2 - h\,\mathbf{n} \cdot \mathbf{curl}\, \mathbf{n} \tag{48}$$

in which the field h coupled with the twist is a measure of the chirality. In the superconductor analogy, h is analogous to the magnetic field $\mathbf{H}$ imposed by external currents (or more exactly, to the component of $\mathbf{H}$ along $\mathbf{A}$; other components are coupled to bend).

In cholesterics, no other energy depends on twist and Eq. (48) is minimum ($=-h^2/2K_2$) for the helical solution of pitch $P = 2\pi K_2/h = 2\pi/k_0$: $\mathbf{n} = (\sin 2\pi z/P, \cos 2\pi z/P, 0)$ which satisfies $\nabla \cdot \mathbf{n} = 0$, $\mathbf{n} \times \mathbf{curl}\, \mathbf{n} = 0$ and $\mathbf{n} \cdot \mathbf{curl}\, \mathbf{n} = h/K_2$. Note that the pitch then gives a measure of the chirality. The sign of h reflects the handedness of the helix.

In defect-free smectic phases of constant layer thickness d, twisted arrangement of the layers are forbidden as can be shown easily: let A and B be two points in a smectic phase and $\mathbf{N}(r)$ the (oriented) layer normal at position $\mathbf{r}$. The (algebraic) number of layers crossed along a path going from A to B is $\int_A^B \mathbf{dl} \cdot \left(\dfrac{\mathbf{N}}{d}\right)$ independent of the path AB. This property implies that there exists a potential ϕ such as $\mathbf{N}/d = \nabla\phi$ which in turn leads to $\mathbf{curl}\,(\mathbf{N}/d) = 0$.

In a SmA phase, the layer normal $\mathbf{N}$ identifies with the director field $\mathbf{n}$ so that $\mathbf{curl}\,\mathbf{n} = 0$: twist and bend cannot develop on a macroscopic scale. The penetration of twist requires either non-uniform layer thickness $d(\mathbf{r})$ (which can be achieved on a limited scale: the twist penetration depth λ_2, Eq. (23)) or defects such as dislocations which make path-dependent the integral $\int_A^B \mathbf{dl} \cdot \left(\dfrac{\mathbf{N}}{d}\right)$. In mean field, $\mathbf{n} \cdot \mathbf{curl}\ \mathbf{n} = 0$ everywhere in the SmA phase with no defect and Eq. (48) implies that the free energy density of the cholesteric phase is decreased by the twist term $-h^2/2\,K_2$ with respect to the untwisted SmA. The cholesteric to SmA transition becomes first order at a temperature $T_{N^*-A} = T_{N-A} - \sqrt{2\,u_0/K_2}\,|h|/a$ lower than T_{N-A} (u_0 and a are defined from Eq. (17)).

In a SmC phase, the layer normal $\mathbf{N}$ no longer coincides with the director field $\mathbf{n}$. Non-zero twist and bend (i.e. $\mathbf{curl}\ \mathbf{n} \neq 0$) are permitted. The well known structure of the director helix is $\mathbf{n}(\mathbf{r}) = (\sin\theta \cos 2\pi z/P,\ \sin\theta \sin 2\pi z/P,\ \cos\theta)$ for a layer normal along z and a SmC tilt θ ($\mathbf{N}\cdot\mathbf{n} = \cos\theta$). The twist and bend terms of the Frank energy are:

$$|\mathbf{n}\cdot\mathbf{curl}\ \mathbf{n}| = \frac{2\pi}{P}\sin^2\theta$$

$$|\mathbf{n}\times\mathbf{curl}\ \mathbf{n}| = \frac{2\pi}{P}\sin\theta\cos\theta \qquad (49)$$

Note that the gain in twist energy associated with the SmC* helix is coupled to a loss in bend energy. Phase transitions between chiral phases will be described below in the frame of the mean field approximation.

6.1.6.2 Mean Field Chiral N–SmA–SmC model

The chiral version of the N–SmA–SmC model of Chen and Lubensky was investigated by Lubensky and Renn [59]. The mean field free energy density is:

$$(50)$$

$$\Delta F_{\text{N–SmA–SmC*}} = \Delta F_{\text{N–SmA–SmC}} + \Delta F_{\text{Frank}}^{\text{Chiral}}$$

in which $\Delta F_{\text{N–SmA–SmC}}$ is the smectic part of the N–SmA–SmC model and $\Delta F_{\text{Frank}}^{\text{Chiral}}$ is given by Eq. (48). In the simplified version of the mean field model (Eq. 41) of the smectic order parameter, $\Delta F_{\text{N–SmA–SmC}}$ reduces to Eq. (42). It is more convenient to define the z axis along the layer normal in the smectic phases:

$$(51)$$

$$\Psi(\mathbf{r}) = \psi_0\, e^{i\mathbf{k}\cdot\mathbf{r}} = \psi_0\, e^{ikz} \quad \text{with} \quad \mathbf{k} = (0, 0, k)$$

and the director $\mathbf{n}$ as:

$$\mathbf{n} = (\sin\theta \cos\phi,\ \sin\theta \sin\phi,\ \cos\theta) \qquad (52)$$

ψ_0, k, θ, and ϕ are then treated as variational parameters. The important parameter is $\cos\theta = \mathbf{n}\cdot\mathbf{k}/k$ independent of the choice of the coordinate system. The three phases of interest are thus characterized by:

$$\psi_0 = 0, \quad \theta = \frac{\pi}{2}, \quad \nabla_z\phi = k_0$$
    for the N* phase

$$\psi_0 \neq 0, \quad \theta = 0, \quad k = q_0$$
    for the SmA phase

$$\psi_0 \neq 0, \quad \theta \neq 0, \frac{\pi}{2}, \quad k \neq q_0$$
    for the SmC* phase $\qquad (53)$

where $k_0 = h/K_2 = 2\pi/P$ and q_0 is the wavenumber of the SmA phase. Using Eqs. (51) and (52) in the free energy density, Renn and Lubensky obtain:

$$\Delta F_{\text{N*–SmA–SmC*}} \qquad\qquad (54)$$

$$= a|\psi_0|^2 + \frac{1}{2}u|\psi_0|^4$$

$$+ C_\parallel (k\cos\theta - q_0)^2 |\psi_0|^2$$

$$+ (C_\perp k^2 \sin^2\theta + D_\perp k^4 \sin^4\theta)|\psi_0|^2$$

$$+ \frac{1}{2}K(\theta)\sin^2\theta(\nabla_z\phi)^2 + h\sin^2\theta(\nabla_z\phi)$$

where $a = a'(T - T_{\text{N–SmA}})$ and

$$K(\theta) = K_3 \cos^2\theta + K_2 \sin^2\theta \qquad (55)$$

Equation (54) forms the basis for the derivation of the N*–SmA–SmC* phase diagram. Rather than using k and θ as independent parameters, it is convenient to use $k_\parallel = k\cos\theta$ and $k_\perp = k\sin\theta$. Straightforward minimization with respect to k and $\nabla_z\phi$ yields:

$$k_\parallel = q_0 \quad \text{and} \quad \nabla_z \phi = -\frac{h}{K(\theta)} \qquad (56)$$

and the free energy density is finally a function of only two variational parameters $\tan\theta$ and ψ_0:

$$\Delta F_{\text{N*–SmA–SmC*}}$$

$$= a|\psi_0|^2 + \frac{1}{2}u|\psi_0|^4$$

$$+ (C_\perp q_0^2 \tan^2\theta + D_\perp q_0^4 \tan^4\theta)|\psi_0|^2$$

$$- \frac{1}{2}\frac{h^2}{K_3}\frac{\tan^2\theta}{1+(K_2/K_3)\tan^2\theta} \qquad (57)$$

The final minimization is somewhat tedious but presents no difficulties [59]. The equations for the phase boundaries are

$$a_{\text{N*–SmA}} = -\left(\frac{u}{K_2}\right)^{1/2} h \qquad (58\,\text{a})$$

$$a_{\text{N*–SmC*}} = -\left(\frac{u}{K_2}\right)^{1/2} h$$

$$- \frac{1}{4D_\perp q_0^4}\left[-C_\perp q_0^2 + \frac{1}{2}\left(\frac{K_2 u}{K_3^2}\right)^2\right]^2 \qquad (58\,\text{b})$$

$$a_{\text{SmA–SmC*}} = -\frac{u}{2K_3 C_\perp q_0^2}h^2 \qquad (58\,\text{c})$$

The N*–SmA–SmC* phase diagram is shown in Fig. 3. The N*–SmA and N*–SmC* lines are first order whereas the

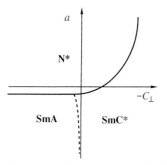

Figure 3. Chiral N*–SmA–SmC* phase diagram from Lubensky and Renn [59]. Solid lines are first order and dashed lines second order.

SmA–CmC* line is second order: the director tilt θ grows continuously at finite twist $\nabla_z \phi = 2\pi/P = -h/K_3$ across the SmA–SmC* transition (Eq. 56 with $\theta=0$).

It is interesting to calculate the value of the Ginzburg parameter $\kappa = \lambda_2/\xi$ that controls the transition from type I to type II behavior. With the coherence length ξ and the twist penetration depth λ_2 given by Eqs. (22) and (23), one gets:

$$\kappa = \frac{\lambda_2}{\xi} = \frac{1}{C_\perp q_0}\left(\frac{u K_2}{2}\right)^2 \qquad (59)$$

κ diverges in the vicinity of the N–SmA–SmC point where $C_\perp$ vanishes. The type II condition is thus expected to be fulfilled close to the N–SmA–SmC point. In presence of chirality, this is precisely the place where a liquid crystal analog of the Abrikosov flux phase should show up. The structure of this new liquid crystalline state will be described in subsequent section.

6.1.6.3   Twist Grain Boundary Phases

Structural Properties

The most spectacular outcome of the analogy with superconductors is undoubtedly the identification of twist grain boundary smectic phases (TGB for short) as liquid crystal analogs of the Abrikosov flux phase in type II superconductors. The existence and structural properties of the TGB phases were first predicted theoretically by Renn and Lubensky [60] (RL) in 1988 and discovered and characterized experimentally by Goodby and coworkers shortly after [26, 61].

The highly dislocated structure of a TGB phase is shown in Fig. 4. Slabs of SmA materials of thickness l_b are regularly stacked in a helical fashion along an axis $\hat{\mathbf{x}}$ parallel to the smectic layers. Adjacent slabs are continuously connected via a grain boundary constituted of a grid of parallel equi-

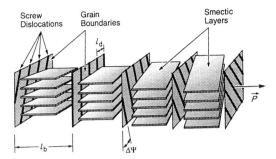

Figure 4. Structure of the TGB phase (from [60, 62]).

spaced screw dislocation lines analogous to magnetic vortices. The finite twist angle of each grain boundary is $\Delta\theta = 2 \sin^{-1}(d/2 l_d)$ where d is the smectic period and l_d the distance between parallel screw dislocations within a grain boundary. Twist penetrates the smectic structure just as magnetic induction penetrates the type II superconducting phase via the Abrikosov lattice of magnetic vortices. The difference lies in the lattice structure: two dimensional hexagonal packing of parallel lines for superconductors, twisted array of rows in TGBs.

RL pointed out that the crystallographic nature of a TGB stack depends on the ratio of the slab thickness to the helical pitch l_b/P (or equivalently $\Delta\theta/2\pi$). If the ratio l_b/P is rational ($=p/q$ with p, q mutually prime integers) the structure is periodic of period $pP = q l_b$, with q-fold symmetry about the $\hat{x}$ axis. Such TGB stack is referred to as commensurate. Reciprocal space is formed of q equispaced Bragg spots distributed on rings of radius $2\pi/d$ in planes q_y, q_z. If q differs from crystallographic values 2, 3, 4 or 6, the commensurate TGB structure is quasi-crystalline rather than crystalline. If on the other hand l_b/P is irrational, the TGB stack is non-periodic (or incommensurate). The scattering is dense on a ring of radius $2\pi/d$ and axis q_x. Reciprocal spaces of the commensurate and incommensurate TGB phases are consequently qualitatively different.

On the experimental point of view, a TGB structure is characterized by five parameters (d, P, l_b, l_d and $\Delta\theta$) linked by two structural relations:

$$\Delta\theta = 2\sin^{-1}(d/2l_d) \quad \text{and} \quad \frac{l_p}{P} = \Delta\theta/2\pi \tag{60}$$

Full experimental characterization requires measurement of three independent parameters. The layer spacing d (of the order of 3 to 4 nm) and the helical pitch P (a few µm) are easily deduced from X-ray and optical measurements. The third parameter can be the twist boundary angle $\Delta\theta$ which can be measured in the case of commensurate TGB only. For incommensurate TGB [61], other parameters are inferred from the RL theoretical estimate $l_b \approx l_d$ which yields l_b and l_d of the order of a few tens of nm and $\Delta\theta$ about $15°$.

Critical Fields h_{c1} and h_{c2}

The thermodynamic critical field h_c above with the SmA phase becomes globally unstable to the N* phase at some temperature T_{N*-SmA} lower than T_0 follows from Eq. (58a):

$$h_c = -a_{N*-SmA}\left(\frac{K_2}{u}\right)^{1/2}$$

$$= a'(T_0 - T_{N*-SmA})\left(\frac{K_2}{u}\right)^{1/2} \tag{61}$$

The critical fields h_{c1} and h_{c2} of the SmA–TGB and TGB–N* transitions were calculated by RL from the superconductor model.

The lower critical field h_{c1} is reached when the SmA phase becomes unstable to the formation of TGB, that is when the gain in twist energy compensates the cost in formation of grain boundaries. On the other hand, the upper critical field h_{c2} corresponds to the stable growth of a non-zero smectic order parameter in the N* phase, that is the inverse susceptibility of the smectic order parameter changes sign.

The result is [60]:

$$h_{c1} = \frac{h_c}{\kappa\sqrt{2}}\ln\kappa$$

$$h_{c2} = \sqrt{2}\,\kappa\,h_c\left(1 + O(k_0/q_0)\right) \qquad (62)$$

For large enough κ (i.e. type II condition $\kappa > 1/\sqrt{2}$), $h_{c1} < h_c < h_{c2}$ and a stable TGB domain opens up in between the SmA and N* phases.

Experimental Situation

Most features of the RL model were confirmed by experiment. The expected Gaussian shape of the scattering along the pitch direction was confirmed [61]. Networks of screw dislocations were observed by freeze fracture experiments [62]. The link between the existence of TGB phases and the proximity of the N*–SmA–SmC* region was demonstrated by Nguyen and coworkers [63] with the discovery of a TGB phase with local SmC order (so called TGBc also predicted by RL). Quasi-crystalline (i.e. commensurate) TGB structures are commonly observed in TGBc systems [64]. Furthermore, a supercritical transformation within the cholesteric domain above the TGB phase was detected by high resolution calorimetry [65]. The low temperature cholesteric may be identified with the vortex liquid phase appearing in high T_c superconductors [66].

Although a number of experimental question remain, like the origin of commensurability of TGBc, the analogy of smectic liquid crystals with type II superconductors is now clearly established.

6.1.7 Frustrated Smectics

6.1.7.1 Polar Smectics

Liquid crystal molecules with long aromatic cores and strongly polar head groups like –CN or NO_2 exhibit a rich SmA (and SmC) polymorphism. Since the discovery of the first SmA–SmA transition by Sigaud, Hardouin and Achard in 1979 [67], seven different smectic phases have been identified in pure compounds or in binary mixtures of polar molecule.

Extensive experimental studies have been carried out about the structures, phase diagrams and physical properties of these thermotropic liquid crystals [68, 69].

The variety of structures arises from the asymmetry of the molecules: in addition to the classical N–SmA–SmC polymorphism, the long range organization of the position of the polar heads generates new phases. An antiferroelectric stacking of polarized layers for instance generates the bilayer SmA called SmA_2. If the dipoles are randomly oriented, the asymmetry can be forgotten and a monolayer SmA_1 phase is obtained. X-ray diffraction patterns clearly show the doubling of the lattice spacing at the SmA_1–SmA_2 transition.

Another experimental characteristic of polar mesogens is the intrinsic incommensurability of their structures. Nematic phases of polar compounds often exhibit diffuse X-ray scattering corresponding to a short range smectic order. Two sets of diffuse spots centered around incommensurate wavevectors $\pm q_1$ and $\pm q_2$ with $q_1 < q_2 < 2q_1$ are usually found. The wavevector q_2 associated with the classical monolayer order is clearly of order $2\pi/l$ where l is the length of a molecule in its most extended configuration. The wavevector q_1 associated with the head to tail association of the polar molecules reveals the existence of another natural length l' such as $l<l'<2l$. l' is commonly identified with the length of a pair of antiparallel partially overlapping molecules, although microscopic approaches suggest that the emergence of l' involves more than just two molecules [70, 71]. Condensed smectic phases also exhibit incommensurate behaviours: the so-called partially bilayer smectic phase SmA_d with a lattice period $d \cdot l$ ($l<d<2$) is commonly observed.

More rarely encountered, but definitely revealing of the incommensurate nature of polar smectics, are the incommensurate SmA phases SmA_{inc} in which the phase of the bilayer modulation shifts periodically (with a period Z) with respect to the monolayer order [72, 73]. The structure is truly incommensurate if the ratio Z over the period of the underlying smectic lattice (Z/l) is irrational. If it is rational ($Z/l=m/n$), the structure is rather modulated and has a period nZ.

The four structures described above are uniaxial since all their modulations have collinear wavevectors but polar mesogens can form biaxial structures too.

The smectic antiphase $Sm\tilde{A}$, first discovered by Sigaud et al. [74] exhibits a periodic modulation of the antiferroelectric order along a direction x parallel to the plane of the layers. The incommensurate wavevectors tilt over to lock in in two dimensions (Fig. 5).

The tilted antiphase (or ribbon phase) $Sm\tilde{C}$ [75] arises from an asymmetric 2-d lock in of the wavevectors. The denomination C emphasizes the fact that both the layers and the antiferroelectric modulations are tilted with respect to the director **n**.

At last, careful studies of the SmA_2–$Sm\tilde{A}$ and SmA_2–$Sm\tilde{C}$ transitions have revealed the existence of the so-called crenelated SmA_{cre} phase over a very narrow range of temperature [76, 77]. SmA_{cre} exhibits the basic transverse modulation of $Sm\tilde{A}$ but with non-equal up and down domains in the plane of the layers.

The experimental phase diagrams of polar compounds are usually represented in axes temperature–pressure or temperature–concentration in binary mixtures. Although the whole set of structures described above is not found in one single system, most phase diagrams fit in a common topology: N, SmA_1, SmA_2 and SmA_d form the generic phase diagram of polar systems [67, 78] (Fig. 6). If biaxial phases ($Sm\tilde{A}$ and $Sm\tilde{C}$) are present, their domain opens up between SmA_1 and SmA_d [77, 79]. Tricritical points are observed on the N–SmA_1 and N–SmA_2 lines [78] whereas re-entrant behavior is often associated with the triple (or multicritical) point N–SmA_1–SmA_2 [80]. The SmA_2–SmA_d line may end up on a critical point beyond which no transition is detected [81].

6.1.7.2 The Model of Frustrated Smectics (Prost)

Prost showed that the properties and structures of frustrated smectics can be described by two order parameters [72, 82]. The first $\rho(\mathbf{r})$ measures mass density modulation familiar in SmA phases [1]. The second $P_z(\mathbf{r})$, often referred to as a polarization wave, describes long range head-to-tail correlations of asymmetric molecules along the z axis

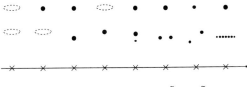

| N | SmA₁ | SmA₂ | SmA_d | SmA_inc | SmÃ | SmC̃ | SmA_cre |

Figure 5. Sketches of the X-ray scattering patterns for various phases appearing in polar smectics. Axes q_x and q_z are horizontal and vertical respectively. Note incommensurate diffuse spots in the nematic phase at natural wavenumbers q_1 and q_2.

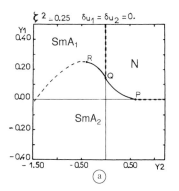

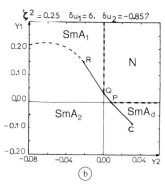

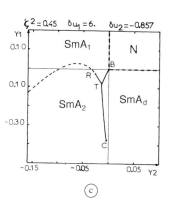

Figure 6. Mean field phase diagram obtained from the model of frustrated smectics for different values of the incommensurability parameter ζ. (a) Very weak incommensurability: N, SmA$_1$ and SmA$_2$ form the generic phase diagram of frustrated smectics. First calculated by Prost [82] this diagram is similar to the experimental one observed on the mixture DB5-TBBA [67]. (b) For slightly larger incomensurability, a SmA$_d$ domain appears in between the N and SmA$_2$ phases. The SmA$_2$–SmA$_d$ line terminates at an isolated critical point C. This diagram reproduces well the behavior of DB6–TBBA mixtures [78] including the order of the transition and the tricritical point R, but not C. (c) For large incommensurability, a new SmA$_1$–SmA$_d$ line and a mean field bicritical point B appear. The critical point C was discovered in this topology by Hardouin et al. [93].

(normal to the smectic layers). Although emphasized by the notation P_z, the antiferroelectric nature of this modulation is not a necessary input of the model.

In the absence of coupling between ρ and P_z, ρ would develop spatial modulation along the z axis at wavevector $q_2 = 2\pi/l$ where l is of the order of a molecular length whereas P_z would develop modulations at wavevector $q_1 = 2\pi/l'$ where l' is identified with the length of a pair of antiparallel partially overlapping molecules as discussed in the previous section.

To describe the appearance of modulated order, two complex fields ψ_1 and ψ_2 are introduced:

$$P_z(\mathbf{r}) = \mathrm{Re}(\psi_1(\mathbf{r}))$$
and $(\psi_1(\mathbf{r}) = |\psi_1|\, e^{i(\mathbf{Q}_1 \cdot \mathbf{r} + \varphi_1)}$

$$\rho(\mathbf{r}) = \mathrm{Re}(\psi_2(\mathbf{r}))$$
and $(\psi_2(\mathbf{r}) = |\psi_2|\, e^{i(\mathbf{Q}_2 \cdot \mathbf{r} + \varphi_2)}$ (63)

In terms of these fields, the Landau free energy of the model reads:

$$\Delta F = \frac{1}{V} \int_V dr \left\{ \frac{A_1}{2} |\psi_1|^2 + \frac{D_1}{2} |(\Delta + q_1^2)\psi_1|^2 \right.$$

$$+ \frac{C_1}{2} |\nabla_\perp \psi_1|^2 + \frac{U_1}{2} |\psi_1|^4$$

$$+ \frac{A_2}{2} |\psi_2|^2 + \frac{D_2}{2} |(\Delta + q_2^2)\psi 2|^2$$

$$+ \frac{C_2}{2} |\nabla_\perp \psi_2|^2 + \frac{U_2}{2} |\psi_2|^4$$

$$\left. + \frac{U_{12}}{2} |\psi_1|^2 |\psi_2|^2 - w\,\mathrm{Re}\left(\psi_1^2 \psi_2^*\right) \right\}$$
                                                      (64)

where V is the volume, $A_1 = a_1(T - T_{c1})$ and $A_2 = a_2(T - T_{c2})$ measure the temperatures from the non-interacting mean field transition temperatures T_{c1} and T_{c2} of the fields ψ_1 and ψ_2. $\nabla_\perp$ is a derivative in the plane perpendicular to the director $\mathbf{n}$. The terms in D_1 and D_2 favor $Q_1^2 = q_1^2$ and $Q_2^2 = q_2^2$

respectively whereas the coupling term w $\mathrm{Re}(\psi_1^2 \psi_2^*)$ favors lock-in conditions $2\mathbf{Q}_1 = \mathbf{Q}_2$ in the case of a weak overlap (l' close to $2l$). A linear coupling term $w'\mathrm{Re}(\psi_1 \psi_2^*)$ would be more appropriate in the strong overlapping limit $l' \approx l$ [72].

Frustration arises from the impossibility to satisfy simultaneously all these tendencies.

Note that the fluctuations of the director and the coupling with the nematic order are not included.

6.1.7.3 The Mean Field Model

To study the different structures and the phase diagrams in mean field, the free energy (Eq. 64) has to be minimized with respect to the smectic amplitudes $|\psi_1|$ and $|\psi_2|$ and the wavevectors $\mathbf{Q}_1$ and $\mathbf{Q}_2$. The following phases are expected:

(1) The nematic phase (N) with $|\psi_1| = |\psi_2| = 0$. The director $\mathbf{n}$ defines the z axis.
(2) Uniaxial layered structures with $\mathbf{Q}_1$ and $\mathbf{Q}_2$ along z. $|\psi_1| = 0$, $|\psi_2| \neq 0$ and $Q_2 = q_2$ defines the monolayer smectic phase SmA$_1$. $|\psi_1| \neq 0$, $|\psi_2| \neq 0$ and $Q_2 = 2\,Q_1$ defines the bilayer antiferroelectric smectic phase SmA$_2$.
(3) Biaxial layered structures: $|\psi_1| \neq 0$, $|\psi_2| \neq 0$ and at least $\mathbf{Q}_1$ non collinear with $\mathbf{Q}_2$.
(4) Uniaxial modulated structures: $|\psi_1| \neq 0$, $|\psi_2| \neq 0$ and modulated phases $\varphi_1(z)$ and $\varphi_2(z)$.

The number of Landau parameters in Eq. (64) can be reduced by an apropriate straightforward rescaling of variables:

with complex order parameters:

$$\begin{cases} \theta_1 = x_1 \exp[i(k_0 z + k_x x + \alpha_1(z))] \\ \theta_2 = x_2 \exp[i(2k_0 z + k_x' x + \alpha_2(z))] \end{cases} \tag{66}$$

The phase diagrams are classically calculated in the plane y_1, y_2.

The difference $\zeta = k_1^2 - k_2^2/4$ of the reduced wavevectors k_1 and k_2 turns out to be an important parameter. The physical significance of ζ is clear with the original parameters $\zeta \propto (q_1^2 - q_2^2/4)w$ measures incommensurability over coupling strength. The frustration is thus essentially controlled by ζ, referred to as the incommensurability parameter. Incommensurate structures for instance are expected at high values of ζ.

The elastic coefficients of tilt γ_1 and γ_2 control the appearance of the biaxial phases SmÃ and SmC̃.

The last two coefficients δu_1 and δu_2 account for anisotropic fourth order terms. Although they can have a significant effect on the shape of the phases diagrams in the plane y_1, y_2 they do not change the qualitative features of their topology.

The most significant phase diagrams calculated from this model are given in Figs. 6 and 7.

Uniaxial Structures

For a small incommensurability parameter ζ and symmetric fourth order coefficients ($\delta u_1 = \delta u_2$) the phase diagram shown in Fig. 6a is similar to the very first diagram calculated by Prost in which incommensurability was not considered [82]. A second order N–SmA$_1$ line terminates at a mean

$$\left[\frac{2w^2}{u_{12}D_1}\right]^{3/4} \frac{u_{12}^3}{64w^4} \frac{D_1}{D_2} \frac{\Delta F}{V} = \Delta F[\theta_1, \theta_2]$$

$$= \frac{1}{V} \int_V d^3x \{y_1 \,|\theta_1|^2 + |(\Delta + k_1^2)\theta_1|^2 + \gamma_1 \,|\nabla_\perp \theta_1|^2 + \delta u_1 \,|\theta_1|^4 + y_2 \,|\theta_2|^2 + |(\Delta + k_2^2)\theta_2|^2$$
$$+ \gamma_2 \,|\nabla_\perp \theta_2|^2 + \delta u_2 |\theta_2|^4 + (|\theta_1|^2 + |\theta_2|^2)^2 - (\theta_1^2 \theta_2^* + \theta_1^{2*}\theta_2)\} \tag{65}$$

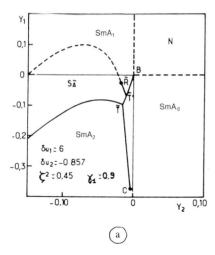

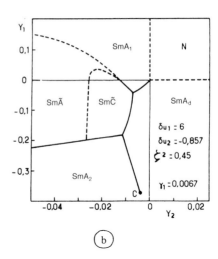

(a)

(b)

Figure 7. Mean field phase diagrams with biaxial phases. (a) For weak incommensurability, the antiphase SmÃ shows up in between the SmA₁ and SmA₂ phases. The SmA₁–SmÃ transition is second order in mean field but first order with fluctuations [47].

(b) For larger incommensurability, the tilted ribbon phase SmC̃ is stable instead of SmÃ close to SmA$_d$. These topologies again compare well with experimental situations [74, 75, 77, 93].

field critical end point Q where the N, SmA₁ and SmA₂ phases meet. A second order N–SmA₂ line terminates at a tricritical point P. The N–SmA₂ line QP is first order and continues into the smectic region as a first order SmA₁–SmA₂ line. Beyond a tricritical point R, the SmA₁–SmA₂ line is second order. Because of the coupling term w, the phase of the bilayer order parameter ψ_1 is locked in the monolayer smectic SmA₁ so that the amplitude x_1 only is critical at the SmA₁–SmA₂ transition. It is therefore expected to be in the Ising universality class [82, 83].

For higher incommensurability parameter ζ (and/or asymmetric fourth order terms $\delta u_1 > \delta u_2$) [84] a new phase boundary separating two SmA₂ phases appears (Fig. 6b). The two SmA₂ phases are distinguished by different values of the amplitudes of the order parameters x_1 and x_2 and therefore of the wavevector k_0. For $x_1 > x_2$, k_0 is of order k_1 (i.e. the smectic period is close to $l' = d \times l < 2\,l$) and the modulus x_1 is much

larger than x_2. The SmA₂ phase is identified with the partially bilayer SmA$_d$ phase.

The new SmA₂–SmA$_d$ phase boundary terminates at a critical point C where the jump in wavevector goes to zero. Although the fluctuating parameter is a scalar (layer thickness) it will be shown in this chapter, Sec. 6.1.7.4 that the new critical point C is not expected to belong to the Ising universality class [85].

When the incommensurability parameter ζ is further increased, a new SmA₁–SmA$_d$ line appears (Fig. 6c) terminating at a mean field bicritical point B where the N, SmA₁ and SmA$_d$ phases meet, and a triple point T where the SmA₁, SmA₂, and SmA$_d$ phases coexist.

Biaxial Structures

Two different structures will be investigated:

$$\begin{cases} \theta_1 = x_1 \exp[i(k_0 z + k_x x)] \\ \qquad + x_1 \exp[i(k_0 z - k_x x)] \\ \theta_2 = x_2 \exp[2 i k_0 z] \end{cases} \quad (67)$$

describes the smectic antiphase SmÃ [86] and

$$\begin{cases} \theta_1 = x_1 \exp[i((k_0 + \delta k_0)z + k_x x)] \\ \quad + x_1' \exp[i((k_0 - \delta k_0)z - (k_x' - k_x)x)] \\ \theta_2 = x_2 \exp[i(2k_0 z - k_x' x)] \end{cases} \quad (68)$$

the tilted antiphase SmC̃ [87].

Note that the name smectic is improperly but commonly used to denote these two dimensionally ordered phases.

Figure 7a shows the phase diagram in the case of weak/medium incommensurability parameter. A biaxial domains opens up for $\gamma_1 < 2\zeta$. The SmA_2–SmÃ transition is always first order since the free energies cannot link up continuously when k_x goes to zero. The SmA_1–SmÃ line is found second or first order in mean field with a tricritical point R̃. It meets the first order SmA_1–SmA_2 line at a triple point T̃. If variations of the elastic coefficient γ_1 with temperature are allowed for, the second order SmA_1–SmÃ line may meet the second order SmA_1–SmA_2 line at a SmA_1–SmA_2–SmÃ Lifshitz point [88]. Because of the continuous degeneracy of the fluctuations of the SmÃ order parameter in reciprocal space, the SmA_1–SmÃ transition (like N–SmC) is, however, expected to be always first order [47].

At higher incommensurability parameter, the biaxial domain reaches the SmA_d region. The tilted antiphase SmC̃ is stable in between SmA_d and SmÃ so that two new lines are found: a first order SmC̃–SmA_d and a second order SmÃ–SmC̃ (Fig. 7b).

Uniaxial Modulated Structures

Incommensurate smectics with modulated phases $\alpha_1(z)$ and $\alpha_2(z)$ were among the first to be predicted by Prost [72]. An exact functional minimization of Eq. (65) with respect to the phases $\alpha_1(z)$ and $\alpha_2(z)$ is possible in terms of elliptic integrals [72, 89, 90]. The problem is in fact isomorphous to the cholesteric–nematic transition induced by a magnetic field [91]. In the modulated smectic, the phase difference $\alpha_1(z) - \alpha_2(z)$ undergoes π jumps (discommensurations or solitons) with a period Z. Such modulated structures (or SmA_{inc}) are found to be stable in the vicinity of the SmA_1–SmA_2–SmA_d point [90, 92]. The SmA_2–SmA_{inc} and SmA_d–SmA_{inc} lines are first order, whereas the SmA_1–SmA_{inc} line is second order. The period Z of the discommensurations does not diverge except at the SmA_1–SmA_2–SmA_{inc} mean field bicritical point. Further away from this point, Z is not much larger than the layer thickness so that the phase difference $\alpha_1(z) - \alpha_2(z)$ is close to a linear function of z (weak coupling limit) [90]. An unfortunate consequence of this remark is that X-ray diffraction experiments may not easily distinguish SmA_{inc} from a simple coexistence of SmA_2 and SmA_d.

Conclusion

Experimental phase diagrams turn out to compare remarkably well to the calculated ones. Actually most of the experimental topologies exhibiting uniaxial phases only fall in one of the theoretical schemes of Figs. 6 and 7.

Binary mixtures of TBBA (terephthal bis butyl aniline) and homologous compounds of the DBn (alkylphenyl cyanobenzoyloxy benzoate) series for instance illustrate well cases in Fig. 6a and b. The DB6–TBBA system in particular [78] exhibits the SmA_d phase, the tricritical SmA_1–SmA_2 point and the order of all the phase transitions agrees well with Fig. 6b.

The DBn–C_5 stilbene binary mixtures [77] and the nOPCBOB (alkoxyphenyl cyanobenzyloxy benzoate) series [93] reproduce well the theoretical diagram of Fig. 7a and b respectively.

The SmA_2–SmA_d critical point was first unambiguously identified by Shashidar et al. [81].

The existence of coexisting incommensurate smectic modulations was reported by Brownsey and Leadbetter [94] and demonstrated by Mang et al. [73].

The critical properties of the various transitions were also investigated by high resolution X-ray diffraction or a.c. calorimetry. 3-d XY critical exponents with non-inverted temperatures are measured for the N–SmA_1 transition [95].

The Ising nature of the SmA_1–SmA_2 transition is confirmed [78, 96, 97] with Fisher-renormalized exponents [98].

The SmA_1–$Sm\tilde{A}$ transition is found to be weakly first order with unusually large pre-transitional fluctuations [99].

The stability of the crenelated phase in the general theoretical diagram has not been discussed yet. SmA_{cre} can be roughly described by the following order parameter:

$$\begin{cases} \theta_1 = \left(x_0 + x_1 \exp[ik_x\,x] + x_1 \exp[-ik_x\,x]\right) \\ \qquad \times \exp[ik_0\,z] \\ \theta_2 = x_2 \exp[2ik_0\,z] \end{cases} \quad (69)$$

If the structure of Eq. (69) is inserted in the free energy density Eq. (65), SmA_{cre} is found to be always unstable with respect to both SmA_2 and $Sm\tilde{A}$. If a more realistic profile of the polarization wave along the layers is considered (i.e. include higher harmonics or compute the profile of lowest energy) SmA_{cre} is found to be only marginably stable at the SmA_2–$Sm\tilde{A}$ transition: it has exactly the same free energy as SmA_2 and $Sm\tilde{A}$ right on the SmA_2–$Sm\tilde{A}$ line [100]. Whether the experimental stability is due to higher order terms in the free energy density [101] or to fluctuations is still an open question. The mean field result suggests anyway that the SmA_{cre} domain should be narrow, as indeed observed [76, 77].

The mean field analysis of the Prost's model of frustrated smectics describes most of the reported experimental observations on polar smectics. Situations where fluctuations are important are however not correctly described: the understanding of multiply re-entrant behavior, the appearance of nematic bubbles and the scaling properties of the SmA_2–SmA_d critical point for instance require more elaborate analysis. Some of these points will be discussed in the following sections.

6.1.7.4　Critical Properties of the Isolated SmA–SmA Critical Point

The behavior of the smectic wavevector q_0 in the vicinity of the SmA_2–SmA_d critical point C [61, 81] clearly suggests a close similarity with a liquid-gas critical point.

In order to describe the fluctuations around C, it is convenient to express the smectic order parameter denoted $m(\mathbf{r})$ as follows:

$$m(\mathbf{r}) = \left[S_{01} + a_1\,S(\mathbf{r})\right] e^{i\theta} e^{iq_0[z+u(\mathbf{r})]}$$
$$+ \left[S_{02} - a_2\,S(\mathbf{r})\right] e^{2i\theta} e^{2iq_0[z+u(\mathbf{r})]} \quad (70)$$

where S_{01} and S_{02} are the amplitudes of ψ_1 and ψ_2 at the critical point. The equilibrium wavevector is $q_0 = q_{0c}(1 + \langle \nabla_z u \rangle)$. $S(\mathbf{r})$ measures the deviation of the amplitudes from the critical values.

The part $H_{Am}[S(\mathbf{r}')]$ of the Landau–Ginzburg Hamiltonian depending on the amplitude of order parameter $S(\mathbf{r})$ has the liquid–gas form:

$$H_{Am}[S] = \int d^d r \left\{ h_s S + \frac{1}{2}\left[rS^2 + c(\nabla S)^2\right] \right.$$
$$\left. + \frac{w_s}{3!}S^3 + \frac{u_s}{4!}S^4 \right\} \quad (71)$$

where $r = a_s\,(T - T_s)$. In absence of coupling, H_{Am} is critical when r vanishes.

The energy of elastic deformations in SmA phases $H_{Sm}[u(\mathbf{r})]$ depends on the phase $u(\mathbf{r})$ only. It can be developed likewise with the constraint that it must be invariant with respect to both rigid translations and rigid rotations. The former invariance is assured if H_{Sm} is a function only of the gradients of $u(\mathbf{r})$, the latter by requiring that the compressional term appears only in the rotationally invariant combination $E[u(\mathbf{r})] = \nabla_z u + 1/2 (\nabla u)^2$ [102]. The elastic Hamiltonian is thus:

$$H_{Sm}[u] = \int d^d r \{ h_0 E(u) + B_0 E^2(u) $$

$$+ \frac{w_0}{3!} E^3(u) + \frac{v_0}{4!} E^4(u)$$

$$+ \frac{1}{2} [K_1^0 (\nabla_\perp^2 u)^2 + K_2^0 (\nabla_z^2 u)^2 $$

$$+ K_3^0 (\nabla_z \nabla_\perp u)^2] \} \qquad (72)$$

The classical physics of SmA phases is recovered by expanding Eq. (72) in powers of the gradients of $u(r)$. The quadratic terms in $(\nabla_z u)^2$ and $(\nabla_\perp^2 u)^2$ are responsible for the Landau–Peierls instability [1, 6]. The non-harmonic terms in $(\nabla_z u)(\nabla_\perp u)^2$ and $(\nabla_\perp u)^4$ lead to a break-down of conventional elasticity: $B(q)$ and $K_1(q)$ respectively vanishes and diverges as powers of $\ln(q)$ at small wavevectors q [102]. Other anharmonic terms are irrelevant in the renormalization group sense in the smectic phase.

The Hamiltonian coupling S to u is:

$$H_{compl.} = \int d^d r \{ \lambda_{11} SE + \lambda_{12} S^2 E $$

$$+ \lambda_{21} SE^2 + \lambda_{22} S^2 E^2 \} \qquad (73)$$

The final Hamiltonian is $H[S, u] = H_{Am} + H_{Sm} + H_{compl.}$. The physics of the critical point can be understood from the Gaussian truncation of H in reciprocal space:

$$H_G[S, u] = \frac{1}{2} \int \frac{d^d q}{(2\pi)^d} \{ (B_1 q_z^2 + K_1 q_\perp^4 $$

$$+ K_2 q_z^4 + K_{12} q_z^2 q_\perp^2) |u(q)|^2$$

$$+ (r + Cq^2) |S(q)|^2$$

$$- \lambda_{11} (i q_z u(q) S(-q)$$

$$+ \text{conj. complex}) \} \qquad (74)$$

with $B_1 = B_0 + h_0$.

The linear coupling of S and u in Eq. (74) can be removed via the transformation:

$$S(q) = \sigma(q) + \frac{i \lambda_{11} q_z u(q)}{r + Cq^2} \qquad (75)$$

The Gaussian correlation functions of the new variables are:

$$\begin{cases} G_{uu}^{(2)}(q, r) \qquad\qquad (76) \\[2mm] = \dfrac{1}{B_1 \dfrac{(r - r_c)}{r} q_z^2 + K_1 q_\perp^2 + K_2' q_z^4 + 2K_{12}' q_\perp^4} \\[4mm] G_{\sigma\sigma}^{(2)}(q, r) = \dfrac{1}{r + Cq^2} \end{cases}$$

with $r_c = \lambda_{11}^2 > 0$. The layer compression is controlled by an effective elastic constant $B_{eff}(q)$

$$B_{eff}(q = 0) = B_1 \frac{(r - r_c)}{r} \qquad (77)$$

which goes to zero for a positive value of r. $G_{\sigma\sigma}^{(2)}$ is therefore not critical at C and a Gaussian integration of the fluctuations of σ is acceptable. $u(r)$ is the critical variable and the critical point is reached when B_{eff} vanishes. This behavior is to be compared to the divergence of the compressibility in the liquid–vapor problem: $1/B$ is analogous to a smectic compressibility along the direction z.

The critical Landau–Ginzburg Hamiltonian is then:

$$H = \int d^d r \{ h \nabla_z u + \frac{B_1}{2} (\nabla_z u)^2 + \frac{B_2}{2} (\nabla_\perp u)^2 $$

$$+ \frac{1}{2} [K_1 (\nabla_z u)^2 + K_2 (\nabla_\perp u)^2 $$

$$+ 2 K_{12} (\nabla_z \nabla_\perp u)^2]$$

$$+ \frac{w_1}{3!} (\nabla_z u)^3 + \frac{w_2}{3!} (\nabla_z u)(\nabla_\perp u)^2$$

$$+ \frac{v_1}{4!} (\nabla_z u)^4 + \frac{v_1}{4!} (\nabla_\perp u)^4$$

$$+ \frac{v_{12}}{12} (\nabla_z u)^2 (\nabla_\perp u)^2 \} \qquad (78)$$

In the absence of external field, $h=B_2=0$. In mean field theory, $\langle\nabla_\perp u\rangle$ is zero, the order parameter is $\langle\nabla_z u\rangle$ and the critical point is found for $B_1=w_1=0$.

When fluctuations are included, $\langle\nabla_\perp u\rangle$ cannot be ignored. In the liquid–gas Hamiltonian, there is only one third-order invariant ϕ^3 which can be removed by shifting ϕ. This implies that the liquid–gas transition belongs to the same universality as the Ising model with additional higher order irrelevant potentials. In the SmA–SmA Hamiltonian (Eq. 78) there are two distinct third order potentials $(\nabla_z u)^2$ and $\nabla_z u(\nabla_\perp u)^2$. They cannot be removed by shifting the order parameter $\langle\nabla_z u\rangle$. A renormalized theory should therefore take them into account explicitly. The general ϕ^3-field theory suggests that the critical dimension is $d_c=6$.

The complete perturbation theory of the Hamiltonian (Eq. 78) has been developed by Park et al. [85]. The inverse susceptibility $\tilde{B}$ is found to be related to the coefficients of Eq. (78) as follows:

$$B_1 - B_{1c} = \tilde{B}\left(1 - w_{2c}^2\,\tilde{B}^{d/2-3} + v\,\tilde{B}^{d/2-2} + \ldots\right)$$
(79)

since $w_{2c}\neq0$, the expansion (Eq. 79) breaks down for spatial dimensions d lower than 6.

For $4<d<6$, fourth order terms are irrelevant in Eq. (78). An anisotropic ϕ^3-renormalized theory was constructed [85]. The Gaussian and Ising fixed point are unstable, as expected. A stable non-trivial fixed point exists and anisotropic critical exponents were calculated in $6-\varepsilon$ dimensions.

Although a number of theoretical questions remain (unknown equation of state, non analytic corrections to the third order vertex for $6<d<8$) the main result is that the SmA–SmA critical point is expected to belong to a new universality class. Experiments confirm this point: a high resolution calorimetric study on a binary mixture gives

a critical exponent γ/Δ of 0.6 ± 0.2 [103] whereas elastic constant measurements show a vanishing of B with a value 0.4 ± 0.2 of the same exponent γ/Δ [104]. These numbers are only marginally consistent with one another but definitely different from the Ising value 0.79. On the other hand, a mean-field behavior ($\gamma/\Delta=0.67$) would not explain the asymmetry of the calorimetric and X-ray data.

The mean field theory of frustrated smectics in the limit of a strong overlap ($l'\approx l$) also predicts several critical points [105]. All of them (including the present SmA$_2$–SmA$_d$ point) are in fact of SmA$_d$–SmA$_d$ type (with $1\leq d\leq2$) and belong to the same universality class. It is worth emphasizing the unusual possibility of going from SmA$_1$ to SmA$_2$ via a second-order transition, a first-order jump or without any phase transition at all by a continuous growth of the smectic wavevector since all SmA phases have the same symmetry. Second order transitions may only occur with exact doubling (or tripling) of the period by continuous condensation of a subharmonic modulation.

6.1.7.5 The Re-entrant Phenomenon

Although not specific of liquid crystals, the re-entrant phenomenon (i.e. re-appearance of the phase of higher symmetry upon cooling) is often observed in frustrated smectics [80, 106]. Theoretical analyses suggest that there is no universal explanation for re-entrance in the N–SmA problem.

Single re-entrance may occur in mean field: the coupling of two order parameters may generate a curved N–SmA phase boundary and thus produce a re-entrant phenomenon if the physical temperature axis is a suitable combination of the two Landau control parameters.

Experiments suggest that double re-entrance is associated with the bi- or tetracritical N–SmA$_1$–SmA$_d$ point. Prost and Lubensky [107, 108] have argued that the correlation function of the smectic order parameter in both SmA$_1$ and SmA$_d$ obeys a scaling relation such as:

$$G(t,p) = \tilde{t}^{-\gamma} f\left(\frac{\tilde{p}}{\tilde{t}^{x}}\right) \qquad (80)$$

where t, p are the reduced temperature and pressure in the vicinity of the critical point (defined by $t=p=0$) $\tilde{t}$ and $\tilde{p}$ are linear combinations of t and p. Equation (80) implies that the phase boundaries obey a relation

$$\tilde{p} = w \pm \tilde{t}^{x} \qquad (81)$$

which gives a doubly re-entrant behavior with suitable rotation of the physical axes p, t with respect to $\tilde{p}$ and $\tilde{t}$.

Beside these phenomenological approaches that say nothing about the microscopic origin of re-entrance, molecular theories have been proposed. The frustrated spin glass model of Indekeu and Berker [70, 71] has been particularly successful.

Re-entrance may also occur as closed nematic domains (or nematic 'bubbles') deep in the smectic region. Such bubbles seem to be closely related to the existence of a SmA$_2$–SmA$_d$ or SmA$_1$–SmA$_d$ critical point [105, 109].

As explained in Sec. 6.1.7.4 of this chapter, the compressional elastic constant B vanishes at the SmA–SmA critical point so that the system is very close to a nematic. An interesting consequence is that the energy of dislocations becomes very weak and their proliferation may lead to a destruction of smectic order. Prost and Toner [109] have shown that depending on bare parameters of a particular system, either the nematic bubble or the critical point could be observed.

Beside this fluctuation corrected mean field theory, an exact model has been developed by Pommier and Prost [105] in the limit of an infinite number of components n

of the order parameter (recall $n=2$ for the smectic case). With the same physics of frustration, exact phase diagrams have been calculated that reproduce fairly well the continuous inclusion of the re-entrant nematic phase that leads to a closed island.

6.1.8   Conclusions

The study of phase transitions in thermotropic liquid crystals has made considerable progress since 1980. Strong theoretical efforts coupled with high resolution and high sensitivity experimental techniques have provided a good understanding of the physics of the phase transformations that occur in mesophases. Clear experimental universal behaviors have been identified and most exceptions have received a reasonable theoretical explanation (Ginzburg criterion for the SmA–SmC transition, crossover towards tricritical behavior, analogy with superconductors).

A number of open questions remain: the N–SmA transition for instance is almost understood but not quite which suggests more efforts have to be done.

Transitions involving disk-like molecules, mesogenic polymers or lyotropic micelles have certainly not received such a great deal of attention and still constitute an active field of research.

Acknowledgments

I wish to thank the editors and particularly Professor Demus for infinite patience and constant encouragement for writing this chapter. I am also indebted to C. Coulon, C. W. Garland, F. Hardouin, A.-M. Levelut, T. C. Lubensky, J. P. Marcerou, F. Nallet, L. Navailles, H. T. Nguyen, Y. Park, R. Pindak, J. Pommier, J. Prost, R. Shashidar and G. Sigaud for helpful discussions about the work presented in this chapter.

6.1.9   References

[1] P. G. de Gennes, J. Prost, *The Physics of Liquid Crystals*, 2nd ed., Oxford Science Publications Clarendon Press, Oxford, **1993**.

[2] B. Deloche, B. Cabane, D. Jérôme, *Mol. Cryst. Liq. Cryst.* **1971**, *15*, 197.

[3] M. J. Freiser, *Phys. Rev. Lett.* **1970**, *24*, 1041; C. S. Shih, R. Alben, *J. Chem. Phys.* **1972**, *57*, 3055; Y. Rabin, W. E. Mullen, W. M. Gelbart, *Mol. Cryst. Liq. Cryst.* **1982**, *89*, 67.

[4] A. Saupe, P. Boonbrahm, L. J. Yu, *J. Chim. Phys.* **1983**, *80*, 7.

[5] Y. Galerne, J. P. Marcerou, *Phys. Rev. Lett.* **1983**, *51*, 2109.

[6] L. D. Landau in *Collected Papers* (Ed.: D. Ter Haar), Gordon and Breach, New York, **1965**.

[7] R. Alben, *J. Chem. Phys.* **1973**, *59*, 4299.

[8] D. J. Wallace in *Phase Transitions and Critical Phenomena*, Vol. 6 (Eds.: C. Domb, M. S. Green), Academic Press, London, **1976**.

[9] E. F. Gramsbergen, L. Longa, W. H. De Jeu, *Phys. Rep.* **1986**, *135*, 195.

[10] C. P. Fan, M. J. Stephen, *Phys. Rev. Lett.* **1970**, *25*, 500.

[11] J. Thoen, H. Marynissen, W. Van Dael, *Phys. Rev. A* **1982**, *A26*, 2886.

[12] K. G. Wilson, J. Kogut, *Phys. Rep.* **1974**, *12*, 77.

[13] A. L. Korzhenevskii, B. N. Shalaev, *Sov. Phys. JETP* **1979**, *49*, 1094.

[14] A. Poniewiersky, T. J. Sluckin, *Liq. Cryst.* **1987**, *2*, 281.

[15] P. Richetti, L. Moreau, P. Barois, P. Kékicheff, *Phys. Rev. E* **1996**, *54*, 1749.

[16] W. L. McMillan, *Phys. Rev.* **1971**, *A4*, 1238.

[17] W. Maier, A. Saupe, *Z. Naturforsch.* **1959**, *14A*, 882; **1960**, *15A*, 287.

[18] W. L. McMillan, *Phys. Rev. A* **1972**, *A6*, 936; *Phys. Rev.* **1973**, *A7*, 1673.

[19] L. D. Landau, *Phys. Z. Sowj. Un.* **1937**, *2*, 26.

[20] R. E. Peierls, *Annls Inst. H. Poincaré* **1935**, *5*, 177.

[21] C. W. Oseen, *Trans. Faraday Soc.* **1933**, *29*, 883; F. C. Frank, *Disc. Faraday Soc.* **1958**, *25*, 19.

[22] V. L. Ginzburg, L. D. Landau, *J. Exptl. Theoret. Phys. (USSR)*, **1950**, *20*, 1064.

[23] P. G. de Gennes, *Solid State Commun.* **1972**, *10*, 753.

[24] W. Meisner, R. Ochsenfeld, *Naturwiss.* **1933**, *21*, 787.

[25] S. T. Lagerwall, R. B. Meyer, B. Stebler, *Ann. Phys. (France)* **1978**, *3*, 249.

[26] J. Goodby, M. Waugh, S. Stein, R. Pindak, J. Patel, *Nature* **1988**, *337*, 449.

[27] P. E. Cladis, S. Torza, *J. Applied Phys. (USA)* **1975**, *46*, 584.

[28] F. Jähnig, F. Brochard, *J. Phys. (France)* **1974**, *35*, 301.

[29] T. C. Lubensky, Jing-Huei Chen, *Phys. Rev. A* **1978**, *17*, 366.

[30] B. Halperin, T. C. Lubensky, Shang-Keng Ma, *Phys. Rev. Lett.* **1974**, *32*, 292.

[31] C. E. Williams, M. Kléman, *J. Phys. Colloq.* **1975**, *36*, C1–315; M. Kléman, *Points, Lignes, Parois,* Les Editions de Physique, Orsay, France, **1977**.

[32] W. Helfrich, *J. Phys. (Paris)* **1978**, *39*, 1199.

[33] D. R. Nelson, J. Toner, *Phys. Rev. B* **1981**, *B24*, 363.

[34] J. Toner, *Phys. Rev. B* **1982**, *B26*, 462.

[35] C. Dasgupta, B. I. Halperin, *Phys. Rev. Lett.* **1981**, *47*, 1556.

[36] M. A. Anisimov, V. P. Voronov, A. O. Kulkov, F. Kholmurodov, *J. Phys. (France)* **1985**, *46*, 2157.

[37] L. Benguigui, P. Martinoty, *Phys. Rev. Lett.* **1989**, *63*, 774.

[38] Y. Galerne, *J. Phys. (France)* **1985**, *46*, 733.

[39] D. Collin, J. L. Gallani, P. Martinoty, *Phys. Rev. Lett.* **1988**, *61*, 102.

[40] G. Sigaud, F. Hardouin, M. F. Achard, *Solid State Commun.* **1977**, *23*, 35.

[41] D. Johnson, D. Allender, D. Dehoff, C. Maze, E. Oppenheim, R. Reynolds, *Phys. Rev. B* **1977**, *B16*, 470.

[42] P. G. de Gennes, *Mol. Cryst. Liq. Cryst.* **1973**, *21*, 49.

[43] W. L. McMillan, *Phys. Rev. A* **1973**, *8*, 1921; R. B. Meyer, W. McMillan, *Phys. Rev. A* **1974**, *9*, 899.

[44] A. Wulf, *Phys. Rev. A* **1975**, *A11*, 365.

[45] R. J. Priest, *J. Phys. (Paris)* **1975**, *36*, 437.

[46] Jing-Huei Chen, T. C. Lubensky, *Phys. Rev. A* **1976**, *A14*, 1202.

[47] S. A. Brazowskii, *Soviet Phys. JETP* **1975**, *41*, 85.

[48] L. D. Landau, E. M. Lifshitz, *Statistical Physics*, Pergamon, New York, **1968**.

[49] R. M. Hornreich, M. Luban, S. Shtrikman, *Phys. Rev. Lett.* **1975**, *35*, 1678.

[50] C. R. Safinya, L. J. Martinez-Miranda, M. Kaplan, J. D. Litster, R. J. Birgeneau, *Phys. Rev. Lett.* **1983**, *50*, 56.

[51] L. J. Martinez-Miranda, A. R. Kortan, R. J. Birgeneau, *Phys. Rev. Lett.* **1986**, *56*, 2264; *Phys. Rev. A* **1987**, *A36*, 2372.

[52] G. Grinstein, J. Toner, *Phys. Rev. Lett.* **1983**, *51*, 2386.

[53] M. E. Fisher, D. R. Nelson, *Phys. Rev. Lett.* **1974**, *32*, 1350.

[54] J. M. Kosterlitz, D. J. Thouless, *J. Phys. (France)* **1973**, *C6*, 1181.

[55] B. I. Halperin, D. R. Nelson, *Phys. Rev. Lett.* **1978**, *41*, 121; D. R. Nelson, B. I. Halperin, *Phys. Rev. B* **1979**, *B19*, 2457.

[56] A. P. Young, *Phys. Rev. B* **1979**, *B19*, 1855.

[57] C. C. Huang, J. M. Viner, R. Pindak, J. W. Goodby, *Phys. Rev. Lett.* **1981**, *46*, 1289.

[58] R. Geer, C. C. Huang, R. Pindak, J. W. Goodby, *Phys. Rev. Lett.* **1989**, *63*, 540.

[59] T. C. Lubensky, S. R. Renn, *Phys. Rev. A* **1990**, *41*, 4392.

[60] S. R. Renn, T. C. Lubensky, *Phys. Rev. A* **1988**, *A38*, 2132.

[61] G. Srajer, R. Pindak, M. A. Waugh, J. W. Goodby, J. S. Patel, *Phys. Rev. Lett.* **1990**, *64*, 1545.

[62] K. J. Ihn, J. A. N. Zasadzinski, R. Pindak, A. J. Slaney, J. W. Goodby, *Science* **1992**, *258*, 275.

[63] H. T. Nguyen, A. Bouchta, L. Navailles, P. Barois, N. Isaert, R. J. Twieg, A. Maaroufi, C. Destrade, *J. Phys. (France)* **1992**, *2*, 1889.

[64] L. Navailles, P. Barois, H. T. Nguyen, *Phys. Rev. Lett.* **1993**, *71*, 545.

[65] T. Chan, C. W. Garland, H. T. Nguyen, *Phys. Rev. E* **1995**, *52*, 5000.

[66] R. D. Kamien, T. C. Lubensky, *J. Phys. I (France)* **1994**, *3*, 2123.

[67] G. Sigaud, F. Hardouin, M. F. Achard, *Phys. Lett.* **1979**, *72A*, 24.

[68] F. Hardouin, A. M. Levelut, M. F. Achard, G. Sigaud, *J. Chim. Phys.* **1983**, *80*, 53.

[69] R. Shashidar, B. R. Ratna, *Liq. Cryst.* **1989**, *5*, 421.

[70] J. O. Indekeu, A. N. Berker, *J. Phys. (France)* **1988**, *49*, 353.

[71] A. N. Berker, J. S. Walker, *Phys. Rev. Lett.* **1981**, *47*, 1469.

[72] J. Prost, *Proceedings of the Conf. on Liq. Cryst. of One and Two Dimensional Order, Garmisch Partenkirchen,* Springer Verlag, Berlin, **1980**.

[73] J. T. Mang, B. Cull, Y. Shi, P. Patel, S. Kumar, *Phys. Rev. Lett.* **1995**, *74*, 21.

[74] G. Sigaud, F. Hardouin, M. F. Achard, A. M. Levelut, *J. Phys. (France)* **1981**, *42*, 107.

[75] F. Hardouin, H. T. Nguyen, M. F. Achard, A. M. Levelut, *J. Phys. Lett. (France)* **1982**, *43*, L-327.

[76] A. M. Leelut, *J. Phys. Lett. (France)* **1984**, *45*, L-603.

[77] G. Sigaud, M. F. Achard, F. Hardouin, *J. Phys. Lett. (France)* **1985**, *46*, L-825.

[78] K. K. Chan, P. S. Pershan, L. B. Sorensen, F. Hardouin, *Phys. Rev. Lett.* **1985**, *54*, 1694; *Phys. Rev. A* **1986**, *A34*, 1420.

[79] H. T. Nguyen, C. Destrade, *Mol. Cryst. Liq. Cryst. Lett.* **1984**, *92*, 257.

[80] P. E. Cladis, *Phys. Rev. Lett.* **1975**, *35*, 48.

[81] R. Shashidar, B. R. Ratna, S. Krishna, S. Somasekhar, G. Heppke, *Phys. Rev. Lett.* **1987**, *59*, 1209.

[82] J. Prost, *J. Phys. (France)* **1979**, *40*, 581.

[83] Wang Jiang, T. C. Lubensky, *Phys. Rev. A* **1984**, *A29*, 2210.

[84] P. Barois, J. Prost, T. C. Lubensky, *J. Phys. (France)* **1985**, *46*, 391.

[85] Y. Park, T. C. Lubensky, P. Barois, J. Prost, *Phys. Rev. A* **1988**, *A37*, 2197.

[86] P. Barois, C. Coulon, J. Prost, *J. Phys. Lett. (France)* **1981**, *42*, L-107.

[87] J. Prost, P. Barois, *J. Chim. Phys. (France)* **1983**, *80*, 66.

[88] L. G. Benguigui, *J. Phys. (France)* **1983**, *44*, 273.

[89] P. Barois, J. Prost, *Ferroelectrics* **1984**, *58*, 193.

[90] P. Barois, J. Pommier, J. Prost, *Solitons in Liquid Crystals* (Ed.: Lui Lam, J. Prost), Springer Verlag, Berlin **1991**, Chapter 6.

[91] P. G. de Gennes, *Solid State Commun.* **1968**, *6*, 163.

[92] P. Barois, *Phys. Rev. A* **1986**, *A33*, 3632.

[93] F. Hardouin, M. F. Achard, C. Destrade, H. T. Nguyen, *J. Phys. (France)* **1984**, *45*, 765.

[94] G. J. Brownsey, A. J. Leadbetter, *Phys. Rev. Lett.* **1980**, *44*, 1608.

[95] C. W. Garland, G. Nounesis, K. J. Stine, G. Heppke, *J. Phys. (France)* **1989**, *50*, 2291; C. W. Garland, G. Nounesis, K. J. Stine, *Phys. Rev. A* **1989**, *A39*, 4919.

[96] C. Chiang, C. W. Garland, *Mol. Cryst. Liq. Cryst.* **1985**, *122*, 25.

[97] C. W. Garland, C. Chiang, F. Hardouin, *Liq. Cryst.* **1986**, *1*, 81.

[98] D. A. Huse, *Phys. Rev. Lett.* **1985**, *55*, 2228; M. A. Anisimov, A. V. Voronel, Gorodetskii, *Sov. Phys. JETP* **1971**, *33*, 605.

[99] K. Ema, C. W. Garland, G. Sigaud, H. T. Nguyen, *Phys. Rev. A* **1989**, *A39*, 1369.

[100] J. Pommier, P. Barois, unpublished.

[101] L. G. Benguigui, *Phys. Rev. A* **1986**, *A33*, 1429.

[102] G. Grinstein, R. A. Pelcovits, *Phys. Rev. A* **1982**, *A26*, 915.

[103] Y. H. Jeong, G. Nounesis, C. W. Garland, R. Shashidar, *Phys. Rev. A* **1989**, *A40*, 4022.

[104] J. Prost, J. Pommier, J. C. Rouillon, J. P. Marcerou, P. Barois, M. Benzekri, A. Babeau, H. T. Nguyen, *Phys. Rev. B* **1990**, *B42*, 2521.

[105] J. Pommier, Thèse de l'Université de Bordeaux 1, n° 274, **1989**.

[106] H. T. Nguyen, F. Hardouin, C. Destrade, *J. Phys. (France)* **1982**, *43*, 1127.

[107] J. Prost, *Adv. Phys.* **1984**, *33*, 1.

[108] T. C. Lubensky, unpublished.

[109] J. Prost, J. Toner, *Phys. Rev. A* **1987**, *A36*, 5008.

6.2 Experimental Methods and Typical Results

6.2.1 Thermal Methods

Jan Thoen

6.2.1.1 Introduction

Liquid crystals exhibit a rich variety of phases and phase transitions. The phase transitions can either be first-order or second-order and critical fluctuations often play an important role. In order to elucidate the nature of these phase transitions the application of high-resolution measuring techniques are essential.

Calorimetric studies have played a significant role in providing information on energy effects near many liquid crystal phase transitions and complimented structural information from X-ray investigations. The vast majority of thermal information concerns calorimetric data on the static thermal quantities enthalpy, H, or specific heat capacity, C_p. However, recently high-resolution results for thermal transport properties have also been obtained.

The rest of this section is divided in subsections: Section 6.2.1.2 gives some general thermodynamic aspects of phase transitions as well as on the crucial role played by critical fluctuations. Section 6.2.1.3 describes several high-resolution techniques and the way in which they allow to extract information on static and/or dynamic ther-

mal quantities. Section 6.2.1.4 deals with calorimetric results obtained for a variety of phase transitions with the purpose of illustrating the possibilities of several of the measuring techniques in arriving at information on static thermal quantities. In Section 6.2.1.5 the capability of photoacoustic and photopyroelectric methods to obtain static and dynamic thermal quantities is illustrated.

6.2.1.2 Theoretical Background

Thermal Characteristics of Phase Transition

Phase transitions can be first-order or second-order (or continuous), and critical energy fluctuations quite often have a significant impact on thermal parameters. First-order transitions are characterized by discontinuous jumps in the first derivatives of the free energy, resulting in finite density ρ and enthalpy H differences between two distinct coexisting phases at the transition temperature T_{tr}. For a second-order transition there are no discontinuities in the density or the enthalpy but the specific heat capacity C_p will exhibit either a discontinuous-jump (for mean-field regime) or a critical anomaly (critical fluctuation regime). In the left hand part of Fig. 1 characteristic behavior for H as a function of temperature near T_{tr} is shown schematically for first-order as well as for second-order transitions. The right hand part of Fig. 1 gives the corresponding temperature behavior of the specific heat capacity $C_p = (\partial H/\partial T)_p$.

Figure 1(a) gives the temperature dependence of H near a strongly first-order transition with a large latent ΔH_L at T_{tr}. The variation of H with T is nearly linear above and below T_{tr} resulting in almost temperature independent C_p values in the low and high temperature phases. Figure 1(b) represents the case of a weakly first-order transition with only a small latent heat $\Delta H_L \neq 0$, but H shows substantial pretransitional temperature variation, which show up as anomalous pretransitional increases in the corresponding C_p behavior. The total enthalpy change associated with the phase transition can be written as the sum of two terms:

$$\Delta H_L + \delta H = \Delta H_L + \int \Delta C_p dT \qquad (1)$$

with $\Delta C_p = C_p - C_p^b$ the excess specific heat capacity (above the background C_p^b) associated with changes in ordering.

Figure 1(c–e) schematically represent three commonly encountered cases for second-order phase transitions. At a second-order phase transition the latent heat $\Delta H_L = 0$ and the specific heat capacity C_p (the temperature derivative of H) shows singular behavior at the critical point (CP). Figure 1(c) is the case of a mean-field second-order transition with a normal linear behavior above the critical temperature $T_c = T_{tr}$ and a rapid variation of H below T_c due to changes in long-range order with temperature. This is reflected in a rapid change of C_p below T_c on approaching T_c and a discontinuous jump at T_c. Both cases given in Fig. 1(d and e) are critical fluctuations dominated second-order phase transitions with pretran-

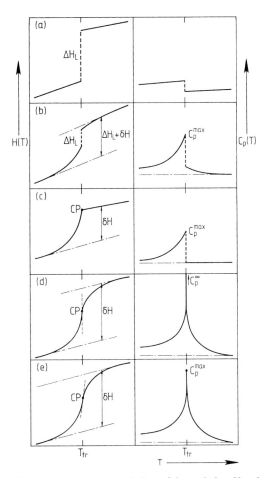

Figure 1. Schematic variation of the enthalpy H and specific heat capacity C_p with temperature T for transitions (at T_{tr}) that are: (a) strongly first-order, (b) weakly first-order with pretransitional fluctuation behavior, (c) mean-field second-order (CP indicates the critical point on the enthalpy curve for the Landau second-order transition temperature $T_c = T_{tr}$), (d) and (e) are critical fluctuation dominated second-order transitions with a diverging (d) or large but finite (e) specific heat capacity at the critical temperature $T_c = T_{tr}$. For the first-order transitions the latent heats ΔH_L correspond with the steps in $H(T)$ at $T = T_{tr}$. δH represent the fluctuation induced enthalpy change associated with the phase transition.

sitional enthalpy variations and anomalies in the specific heat capacity above and below T_c. The main difference between the two cases in the difference is the slope $(\partial H/\partial T)_p$

at the critical point, resulting in a divergence of the specific heat capacity C_p to infinity in Fig. 1(d), or in a large finite value at T_c (Fig. 1e). In all three of these second-order cases the total enthalpy change associated with the transition contains the δH contribution, which is indicated in the figures.

All these types of phase transitions have been encountered in the large amount of high-quality calorimetric studies during the last two decades. In Section 6.2.1.4 a series of typical cases will be considered in detail. More extensive overviews can be found in reviews that have appeared recently [1–6].

The calorimetric measurements provide information only on the static thermal quantities H and C_p. The dynamic thermal quantities such as the thermal diffusivity D and thermal conductivity κ, which are linked to the dynamics of the fluctuations near phase transitions, are substantially more difficult to measure. Only recently have high-resolution techniques with sufficiently small thermal gradients been used for the study of liquid crystal phase transitions.

Fluctuation Effects

In order to be able to make, even qualitatively, a determination that a transition is first-order or second-order one has to apply a measuring technique probing the temperature dependence of the enthalpy H. Very important is also the magnitude of the latent heat ΔH_L relative to the pretransitional enthalpy change δH (see Fig. 1). Quite often it is also important to know how this ratio changes with some physical parameter like pressure or concentration (for mixtures). As a function of these parameters the order of the transition may, indeed, change at a tricritical point or reach an isolated critical point (liquid–gas type transition in simple fluids).

Accurate determination of the temperature dependence of C_p above and below the transition temperature is also very important because the direct link with critical energy fluctuations. This allows one to find out whether mean-field or critical fluctuation theories are relevant for the transition under investigation [7–9].

If the intermolecular interactions responsible for ordering in the system are sufficiently long-range, the system can be described by a Landau mean-field model. Mean-field models are approximations in which only the long-range aspects of enthalpy fluctuations are taken into account and the short-range fluctuations in the order are ignored. The Landau model assumes the Gibbs free energy, G, to be analytic and a function of a long-range order parameter θ, given by

$$G = G_0 + a\varepsilon\theta^2 + b\theta^4 + c\theta^6 \tag{2}$$

where G_0 is the free energy of the disordered phase and $\varepsilon = (T_c - T)/T_c$ the reduced temperature with T_c the critical temperature. The coefficients a and c are positive, but b can be positive (second-order), zero (tricritical), or negative (first-order). In Eq. (2) odd powers of θ are omitted under the symmetry assumption $G(\theta) = G(-\theta)$. This assumption is, however, not always valid and a third-order term has to be included in some cases (see further the nematic-isotropic transition). The presence of a θ^3 term will always cause the transition to be (weakly) first-order [7, 9].

On the basis of Eq. (2) one arrives at

$$C_p = C_p^0, \quad \text{for} \quad T > T_c \tag{3}$$

$$C_p = C_p^0 + A\frac{T}{T_c}\left(\frac{T_m - T_c}{T_m - T}\right)^{1/2}$$
$$\text{for} \quad T < T_c \quad \text{and} \quad b > 0 \tag{4}$$

$$C_p = C_p^0 + 2A\frac{T}{T_c}\left(\frac{T_m - T_1}{T_m - T}\right)^{1/2}$$
$$\text{for} \quad T < T_c \quad \text{and} \quad b < 0 \tag{5}$$

where $A = |a^2/2 b T_c|$, $T_m = T_c + (b^2 T_c/3 ac)$ and C_p^0 the background heat capacity arising from the regular part G_0 of the free energy. When $b < 0$, one has a first-order transition at T_1 with a specific heat capacity jump given by $2A$. For a continuous transition, one must have $b \geq 0$ and a, $c > 0$. One has an ordinary second-order transition for $b > 0$ with a specific heat capacity jump (see Fig. 1c) given by A. If $b = 0$, a tricritical point is observed at $T_m = T_c$ and below T_m a divergent C_p is predicted with a critical exponent of 1/2.

In liquid crystals intermolecular interactions are short-range and C_p is dominantly related to short-range order fluctuations. Since fluctuations in local order occur above and below the transition, one expects, in contrast to the mean-field situation, anomalous C_p behavior above as well as below T_c. Unfortunately models taking short-range order parameter fluctuations into account are much more difficult to handle and analytic solutions are only available for some special cases. However, sophisticated renormalization group (RG) analyses have been carried out for several types of n vector order parameter models of which the three-dimensional cases are of particular relevance for liquid crystals [7, 10, 11].

In the renormalization group theory one obtains for the anomaly in the specific heat capacity an expression of the following form:

$$C_p^\pm = A^\pm |\varepsilon|^{-\alpha} \left(1 + D_1^\pm |\varepsilon|^{\Delta_1} + D_2^\pm |\varepsilon|^{\Delta_2} \right) + B \quad (6)$$

The superscripts $\pm$ indicate the quantities above and below the transition temperature. A^+ and A^- are the critical amplitudes. The constant term B contains a regular as well as a critical background contribution. $D_1^\pm$ and $D_2^\pm$ are the coefficients of the first- and second-order correction-to-scaling terms, while Δ_1 and Δ_2 are the correction exponents (typically $\Delta_1 \approx 0.5$ and $\Delta_2 \approx 2 \Delta_1$). The critical exponent α has a unique theoretical value for a given universality class (e.g. $\alpha = 0.11$ for the three-dimensional $n = 1$ Ising model). Although the amplitudes in Eq. (6) are not universal, the ratios A^-/A^+ and D_1^-/D_1^+ have fixed values for each universality class.

The fact that many liquid crystalline compounds exhibit several phase transitions of different types in a limited temperature range makes them very suitable for testing phase transition theories. However, in several cases the closeness of two phase transitions may cause coupling between different order parameters and result in complex crossover between different kinds of critical behavior. An important and complicated example of crossover is the N–SmA transition, where order parameter coupling results in an evolution from second-order to first-order via a tricritical point by varying pressure or the composition of a binary mixture. In many experimental systems an intermediate crossover behavior with effective critical exponents which differ from the ones theoretically expected for the uncoupled cases, is often encountered. Crossover from mean-field behavior at large reduced temperatures to critical fluctuation behavior close to T_c is another important case. There are also situations where critical fluctuation results are expected but mean-field behavior is observed in the experimentally accessible reduced temperature range. This can be understood on the basis of the Ginzburg criterion [12] relating the size of the critical region to the range of correlated fluctuations.

In many cases one studies binary mixtures of liquid crystals in order to change or avoid crossover effects or follow a given phase transition as it evolves from one type to another. In principle, however, one should then apply Eq. (6) to $C_{p\phi}$ as a function of $\varepsilon_\phi = |(T - T_c)/T_c|$, where ϕ designates a path of constant chemical potential difference, and not to C_{px} as measured along a line of constant concentration. The distinction between measurements along paths of constant x or ϕ is not very important as long as dT_c/dx is not too large. Fisher [13] found that approaching a critical point via a constant-composition path results in exponent renormalization. For sufficiently large dT_c/dx values one may experimentally observe this Fisher renormalization with, for example, a renormalized critical exponent $\alpha_R = -\alpha/(1-\alpha)$, which results in a non-divergent specific heat capacity for $\alpha > 0$.

6.2.1.3   Experimental Methods

Differential Scanning Calorimetry

Differential scanning calorimetry (DSC) is by far the most commonly used thermal technique for studying liquid crystals. It is a very useful and sensitive survey technique for discovering new phase transitions and for determining the qualitative magnitude of thermal features. However, DSC is not well suited for making detailed quantitative measurements near liquid crystal phase transitions.

The reason for this can be understood by considering Fig. 2. In Fig. 2(a) representative enthalpy curves for a first-order and second-order transition are given. In part (b) the corresponding DSC responses (for heating runs) are compared. In a DSC measurement, a constant heating (or cooling) rate is imposed on a reference sample, which results in a constant and rapid temperature ramp (see Fig. 2). A servosystem forces the sample temperature to follow that of the reference by varying the power input to the sample. What is recorded then is the differential power dH/dt (between reference and sample), and the integral $\int (dH/dt)\,dt$ for a DSC peak approximates the enthalpy change associated with the corresponding

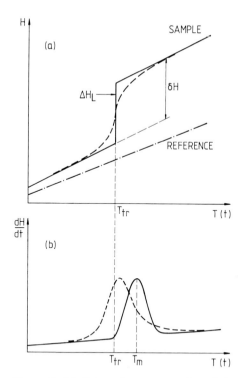

Figure 2. (a) Schematic representation of the temperature dependence of the enthalpy H of a liquid crystal sample near a first-order (solid curve) or second-order (dashed curve) phase transition at T_{tr}. The dashed-dotted line gives the enthalpy of a (DSC) reference material with a nearly constant heat capacity and without a phase transition. (b) Corresponding DSC responses (from heating runs) for the first-order (solid curve) and second-order (dashed curve) cases of part (a).

transition. However, as can be seen in Fig. 2(b) the DSC response is very similar for a first-order transition with latent heat ΔH_L and a second-order transition with a comparable integrated enthalpy change δH. The broadening of the DSC signal for the first-order case is due to the fact that the latent heat cannot be delivered to the sample instantaneously. The DSC response will be at a maximum for a temperature T_m well above T_{tr} for a heating run and well below T_{tr} for a cooling run. The difference between T_m and T_{tr} depends on the scan rate that is chosen. The DSC response for a second-order transition, given by the dashed curve in Fig. 2(b) is qualitatively the same. It is thus very difficult to distinguish between first- and second-order transitions and almost impossible to detect the difference between the latter ones and a weak first-order transition with significant pretransitional specific heat capacity variation. Furthermore, there is additionally a problem of thermodynamic equilibrium with DSC instruments. Although slow rates of the order of 0.1 K min^{-1} are possible, the best operational conditions of DSC machines are realized for fairly rapid scan rate (above 1 K min^{-1}). However, many liquid crystal phase transitions require scan rates which are typically two to three orders of magnitude slower than slow DSC scans.

Traditional Adiabatic Calorimetry

The classical method for measuring the specific heat capacity of a sample as a function of temperature is called adiabatic calorimetry and carried out by the stepwise addition of accurately measured small increments of electrically supplied heat [14, 15]. The heat capacity (at constant pressure p), C_p, of a sample can be obtained from the measured total heat capacity

$$C_1 = C_p + C_h = \frac{\Delta Q}{\Delta T} = \frac{\Delta H}{\Delta T} \qquad (7)$$

provided the heat capacity of the sample holder, C_h, has been independently determined in a calibration experiment. In the experiment a known amount of heat ΔQ is thus applied and the corresponding temperature rise ΔT measured. This ΔT step has of course to be sufficiently small compared to the curvature of the $H(T)$ curve at a given (average) temperature T, especially near a transition, which makes the method somewhat tedious and slow. Anisimov and coworkers [2,16] have mainly used this method for the investigations of liquid crystals.

Adiabatic Scanning Calorimetry

Much of the intrinsic difficulties with DSC measurements and the tedious data collecting process of traditional adiabatic calorimetry can be avoided by adiabatic scanning calorimetry. In this technique a measured heating power is continuously applied to (or extracted from) the sample and sample holder. It was used in the 1970s for the study of liquid–gas [17] and liquid–liquid critical points [18] and first applied to liquid crystals by us [19] and later also by Anisimov et al. [20]. In the dynamic modes the total heat capacity C_1 is now given by:

$$C_1 = C_p + C_h = \frac{P_1}{\dot{T_1}} = \frac{P_1^e + P_1^l}{\dot{T_1}} \qquad (8)$$

In Eq. (8) the total heating power P_1 has been divided in two parts: P_1^e the power applied electrically to a heater, heating the sample and sample holder, and P_1^l representing leaks with an (adiabatic) shield surrounding the sample holder. For cooling runs P_1^l has to be given a controlled negative value. By keeping P_1 or $\dot{T_1}$ constant, combined with increasing or decreasing the sample temperature, four practical modes of operation are obtained [3, 21]. These modes require different settings for the servo-systems controlling the temperature and maintaining adiabatic conditions, or a controlled heat transfer between sample holder and shielding. The most interesting operating conditions are the ones with constant heating or cooling power (P_1 constant). As will be pointed out, this has distinct advantages at first-order transitions. On the other hand, in modes in which $\dot{T_1}$ is kept constant, one runs basically into the same kind of problems as in case of the DSC technique, even if one uses substantially lower scanning rates. It is sufficient to consider here only the constant heating mode (the constant cooling mode is analogous). In this case $P_1 =$

P_1^e and P_1^l is kept negligibly small. In order to obtain the heat capacity of the sample one has to measure P_1^e, $\dot{T}_1$ and C_h. P_1^e is easily obtainable from a measurement of the d.c. current through and the voltage drop across the heating resistor on the sample holder. The rate $\dot{T}_1$ has to be obtained by numerical differentiation of the carefully measured time dependence of T with the sensor in close thermal contact with the sample. Values of $C_h(T)$ can be derived from calibration runs without the sample or with calibration fluids in the sample holder. In fact, the numerical differentiation is not necessary because the time dependence of temperature can be easily converted into a more basic result, namely, the enthalpy versus temperature. Indeed, after subtracting the contribution due to the sample holder, one gets $H(T)-H(T_s)=P(t-t_s)$, the index s refers to the starting conditions of the run. $P=P_1-P_h$, with P_h the power needed to heat the sample holder. Here, T is the sample temperature at time t. In this manner, the latent heat can also be obtained. If a first-order transition occurs at a certain temperature T_{tr}, a step change will occur in $H(T)$ in the interval $\Delta T = t_f - t_i$ until the necessary (latent) heat ΔH_L has been supplied to cross the transition. One then can write:

$$H(T) - H(T_s) = P(t_i - t_s) + P(t_f - t_i)$$

$$+P(t - t_f) = \int_{T_s}^{T_{tr}} C_p dT + \Delta H_L(T_{tr}) + \int_{T_{tr}}^{T} C_p dt \quad (9)$$

From the above considerations, it should be clear that running an adiabatic scanning calorimeter in the constant heating (or cooling) modes makes it possible to determine latent heats when present and distinguish between first-order and second-order phase transitions. On the basis of $C_p = P/\dot{T}$, it is also possible to obtain information on the pretransitional heat capacity behavior, provided one is able to collect sufficiently detailed and ac-

curate information on the temperature evolution $T(t)$ during a scan [19, 21]. Details of the mechanical construction and alternative operating modes of previously used adiabatic scanning calorimeters, can be found elsewhere [3, 21]. Figure 3 gives a schematic representation of a recently constructed wide temperature range version.

A. C. Calorimetry

The a.c. calorimetric technique is very well suited to measure pretransitional specific heat capacities. Indeed it measures C_p directly and not the enthalpy H. This method has been extensively used for the investigation of liquid crystals by Johnson [22], Huang [23–25], Garland [4, 26, 27] and their coworkers.

For a.c. calorimetry an oscillating heating power $P_{ac} = P_0(1 + \cos \omega t)$ is supplied to a sample (in a sample holder), usually loosely coupled to a heat bath at a given temperature T_0, and from the amplitude ΔT_{ac} of the resulting temperature oscillation (see Fig. 4) the specific heat capacity of the sample can be derived. From Fig. 4, it should also be clear that for a first-order transition one can not determine the latent heat ΔH_L in this way.

In a basic one-dimensional heat flow model [28, 25] for a planar sample with thermal conductivity κ_s, the sample is assumed to be thermally coupled to the bath at T_0 with a finite thermal conductance K_b, but the coupling medium (gas) is assumed to have zero heat capacity. The following results are obtained:

$$\Delta T_{ac} \quad (10)$$

$$= \frac{P_0}{\omega C_p} \left[1 + \frac{1}{(\omega \tau_1)^2} + (\omega \tau_2)^2 + \frac{2K_b}{3K_s} \right]^{-1/2}$$

$$T_a = T_0 + \Delta T_{dc} = T_0 + \frac{P_0}{K_b} \quad (11)$$

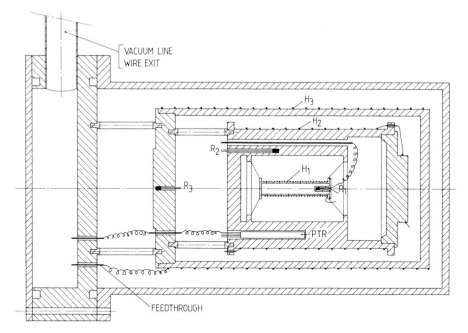

Figure 3. Schematic diagram of an adiabatic scanning calorimeter. Electric heaters and thermistors are denoted by H and R. PTR is a platinum resistance thermometer. The whole calorimeter is placed in a hot air temperature controlled oven. Details on a sample holder with stirring capabilities have been given elsewhere [21].

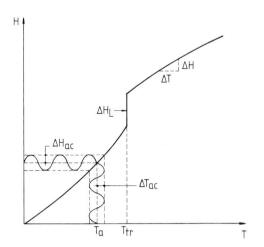

Figure 4. Enthalpy H curve for a first-order transition at T_{tr} and schematic representation of the a.c. calorimetric and standard adiabatic methods for determining the heat capacity C_p. ΔH_{ac} and ΔT_{ac} are, respectively the amplitude of the enthalpy variation and the corresponding temperature change. T_a is given by Eq. (11). In the standard adiabatic method, $C_p = (\Delta H / \Delta T)_p$ is obtained from the temperature increase ΔT resulting from the heat input ΔH.

In Eq. (11) T_a is the average sample temperature (see Fig. 4) and ΔT_{dc} is the offset between bath and sample temperature. In Eq. (10) K_s is the thermal conductance for heat flow perpendicular to the surface of the planar sample. τ_1 is the sample bath relaxation time and τ_2 is a combined (internal) relaxation time for the sample and addenda (heater, temperature sensor, and sample holder when present). τ_1 and τ_2 result, respectively, in a low and high frequency cut-off. In the normal a.c. calorimetric mode one chooses a frequency ω so that $\omega \tau_2 \ll 1$ in order to avoid temperature gradients in the sample. In Eq. (10) the third term can then be neglected. If one further assume the thermal conductance of the sample K_s to be much larger than K_b also the last term can be omitted. In addition to ΔT_{ac} one may also observe a phase shift $(\varphi - \pi/2)$ between $T(t)$ and $P(t)$, with φ given by [6]:

$$\tan \varphi = (\omega \tau_1)^{-1} = K_b (\omega C_p)^{-1} \qquad (12)$$

Eliminating τ_1 between Eqs. (12) and (10) (without the last two terms) results in:

$$C_p = \frac{P_0}{\omega \Delta T_{ac}} \cos\varphi \qquad (13)$$

When ω is well above the low frequency cutoff ($\omega\tau_1 \gg 1$), which corresponds to the normal operating conditions, one can set $\cos\varphi = 1$ as a good approximation in many cases. If necessary, one can correct for small non-zero φ values [6].

Satisfying the above normal operating conditions ($\omega\tau_1 \gg 1 \gg \omega\tau_2$) can be achieved by limiting the thermal coupling between the sample and bath and by using flat thin sample holders. Several designs for a.c. calorimeters have been described in the literature [22–27]. Figure 5 shows the mechanical and thermal design of a calorimeter capable of both a.c.-mode and relaxation-mode (see further) operation [6].

In Huang's group a special purpose a.c. calorimeter was built with the aim of studying free-standing liquid crystal films [25]. This calorimeter contains a special constant-temperature oven that allows the manipulations required for spreading smectic films, and a thermocouple detector placed very near ($\approx$10 μm) but not touching the film. Periodic heating of the sample is achieved by utilizing a chopped laser beam. Thick smectic films ($\approx$100 layers) as well as films as thin as two smectic layers have been studied [25, 29, 30].

Relaxation Calorimetry

In conventional relaxation calorimetry the bath temperature is held constant at a value T_0 and a step function d.c. power is supplied to the sample cell [31–33]. For heating runs, P is switched from 0 at time $t = 0$ to a con-

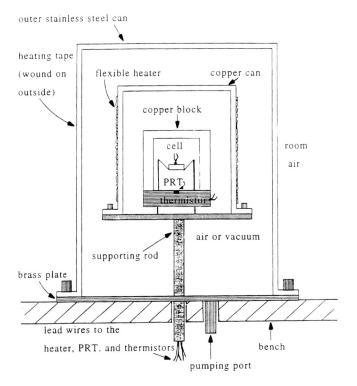

outer stainless steel can

heating tape (wound on outside)

flexible heater   copper can

copper block

cell

PRT

thermistor

room air

air or vacuum

supporting rod

brass plate

lead wires to the heater, PRT. and thermistors

bench

pumping port

Figure 5. Mechanical/thermal design of a computer-controlled calorimeter capable of operating in the a.c. mode or in relaxation modes including non-adiabatic scanning [34, 35].

stant value P_0, while for cooling runs the power is switched from P_0 to 0. In the absence of latent heat effects the cell temperature $T(t)$ relaxes exponentially from $T(0)$ to $T(\infty)$. For the heating regime one has:

$$T(t) = T_0 + \Delta T_\infty [1 - \exp(-t/\tau_1)] \tag{14}$$

For the cooling case one has:

$$T(t) = T_0 + \Delta T_\infty \exp(-t/\tau_1) \tag{15}$$

In both equations $\tau_1 = C/K_b$ and $\Delta T_\infty = P_0/K_b$, with C the heat capacity of the cell and K_b the thermal conductance between cell and bath. In deriving Eqs. (14) and (15) it is assumed that K_b and C are constant over the (narrow) range from $T(0)$ to $T(\infty)$. From a fit of the $T(t)$ data with Eq. (14) or Eq. (15) both τ_1 and ΔT_∞ are obtained. Provided P_0 is known, this results in $C = \tau_1 K_b$.

If a first-order phase transition occurs between $T(0)$ and $T(\infty)$, there will be a non-exponential $T(t)$ variation due to latent heat effects. This situation can be handled [32, 33] by defining a time-dependent heat capacity $C(t) = \dot{P}/\dot{T}$, where $\dot{P} = P_0 - (T - T_0) K_b$. In this procedure it is possible to obtain the latent heat using:

$$\Delta H_L = \int_0^\infty [C(t) - C(0)] \dot{T}(t) \, dt \tag{16}$$

Nonadiabatic Scanning Calorimetry

Recently a new type of relaxation calorimetry has been developed [34] in which the heater power is linearly ramped. This new method has been called nonadiabatic scanning calorimetry [6, 34]. For a heating run, $P = 0$ for $t < 0$, $P = \dot{P}t$ for $0 \leq t \leq t_1$ where $\dot{P} = dP/dt$ is a constant, and $P = P_0 = \dot{P}t_1$ for $t > t_1$. The initial ($t \leq 0$) sample temperature is equal to the bath temperature T_0 and the final ($t \gg t_1$) sample temperature is $T(\infty) = T_0 + P_0 K_b^{-1}$. For a cooling run a reversed power profile is used: $P = P_0$ for $t < 0$,

$P = P_0 + \dot{P}t$ with $\dot{P}$ negative and constant for $0 \leq t \leq t_1$, and $P = 0$ for $t > t_1$. Now the initial temperature is $T(\infty)$ and the final one T_0. For the heating run the time dependence of the temperature of the cell over the time regime $0 \leq t \leq t_1$ is given by [6, 34]:

$$T(t) = T_0 + K_b^{-1} \dot{P}(t - \tau_1) + \tau_1 K_b^{-1} \dot{P} \exp(-t/\tau_1) \tag{17}$$

For a cooling run ($\dot{P}$ negative) a similar expression is obtained:

$$T(t) = T(\infty) + K_b^{-1} \dot{P}(t - \tau_1) + \tau_1 K_b^{-1} \dot{P} \exp(-t/\tau_1) \tag{18}$$

The thermal conductance can be derived from:

$$K_b = P_0 (T_\infty - T_0)^{-1} \tag{19}$$

and the heat capacity in both cases is given by:

$$C(T) = \frac{dH}{dT} = \frac{P - (T - T_0) K_b}{dT/dt} \tag{20}$$

Equation (20) is identical to Eq. (8) if one identifies P_1^1 with $K_b(T_0 - T)$. In normal adiabatic scanning calorimetry one imposes $P_1^1 = 0$ (or $T = T_0$) for heating runs or P_1^1 constant for a cooling run [3, 21]. Here, however, $(T - T_0)$ and P are time dependent, because $P(t')$ is the power at time t' corresponding to the cell temperature $T(t')$ lying in the interval T_0 to $T(\infty)$. dT/dt is obtained by fitting $T(t)$ data over a short time interval centered at t'. In comparison with the conventional transient method, this method has the advantage of avoiding large transient disturbances (by step increases or decreases of P) and also an optimal nearly linear behavior of dT/dt, except for regions with very rapid H variation with T.

This new method also allows a determination of the latent heat of a first-order transition with a two-phase region [35]. In the two-phase region Eq. (20) represents an effective heat capacity C_e. Outside the two-phase region C_e is identical to $C(T)$. One identifies a first-order transition by observ-

ing anomalous behavior of C_e and the occurrence of hysteresis. The latent heat can
be derived from the following expression
[35]:

$$\Delta H_L = \int_{t_1}^{t_2}[P - K_b(T - T_0)]dt - \int_{T_1}^{T_2}[C_h + C_c)dt \qquad (21)$$

where two-phase coexistence exists between $T_1(t_1)$ and $T_2(t_2)$, C_h is the heat
capacity of the sample holder, and $C_c = X_a C_a$
$+ X_b C_b$ (with X the mole fraction) the heat
capacity of the two coexisting phases that
would be observed in the absence of phase
conversion.

By integrating Eq. (20) between $T_1(t_1)$
and $T(t)$, and assuming $\dot{P}$, K_b and T_0 constant one also arrives at the following expression for the temperature dependence of
the enthalpy:

$$H(t) - H(t_1) = \frac{\dot{P}}{2}(t^2 - t_1^2) \qquad (22)$$

$$- K_b\int_{t_1}^{t}T(t)dt + K_b T_0(t - t_1)$$

The calorimeter developed at MIT [6, 35]
and shown in Fig. 5, can also be operated in
the new relaxation mode.

The Photoacoustic Method

Photoacoustics is well established [36] for
the investigation of optical and thermal
properties of condensed matter, but it has
only rather recently been applied to liquid
crystals [37–41]. The photoacoustic technique is based on the periodic heating of a
sample, induced by the absorption of modulated or chopped (electromagnetic) radiation. In the gas microphone detection configuration the sample is contained in a gas-
tight cell (Fig. 6). The thermal wave produced in the sample by the absorbed radiation couples back to the gas above the sample and periodically changes the tempera-

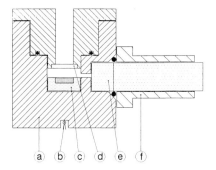

Figure 6. Schematic representation of a photoacoustic measuring cell with microphone detection:
(a) copper cell body, (b) thermistor, (c) gold plated
sample holder with sample, (d) quartz window,
(e) condenser microphone, (f) Teflon microphone
holder, solid dots are O-rings; heating elements for
temperature control are not shown.

ture of a thin gas layer (with a thickness determined by the thermal diffusion length μ_g)
above the sample surface. This will result in
a periodic pressure change in the gas cell
which can be detected by a microphone. A
general one-dimensional theoretical model
has been developed by Rosencwaig and
Gersho [42].

Assuming a sinusoidal variation, part
$I = I_0/2 \exp(i \omega t)$ of the radiation intensity
will give rise to a time-dependent har-
monic surface temperature variation
$\theta = \theta_0 \exp(i \omega t)$. Starting with this temperature variation one arrives at the following
result for the corresponding pressure variation in the gas [42]:

$$\delta p = \left[\gamma_g P_0 \mu_g / \left(\sqrt{2} T_0 l_g\right)\right]\theta e^{-i\pi/4}$$

$$\equiv Q e^{i(\omega t - \pi/4)} \qquad (23)$$

where γ_g is the ratio of the specific heat capacities at constant pressure and constant
volume of the gas, P_0 and T_0 the static pressure and temperature, μ_g the thermal diffusion length of the gas and l_g the gas length
between the sample and the light window in
the planar geometric configuration. Rosencwaig and Gersho [42] derived the follow-

ing explicit relation for $\theta_0(\omega)$ for a homogeneous sample on top of a backing material:

$$\theta_0 = \frac{I_0\beta}{2\kappa(\beta^2 - \sigma^2)} \quad (24)$$

$$\cdot \left[\frac{(r-1)(b+1)e^{\sigma l} - (r+1)(b-1)e^{-\sigma l} + 2(b-r)e^{\beta l}}{(g+1)(b+1)e^{\sigma l} - (g-1)(b-1)e^{-\sigma l}} \right]$$

In this equation β is the optical absorption coefficient of the sample, l is the sample thickness and $\sigma = (1+i)a$ with $a = 1/\mu$ the thermal diffusion coefficient and μ the thermal diffusion length of the sample. One further has: $b = \kappa_b a_b \mu \kappa^{-1}$, $g = \kappa_g a_g \mu \kappa^{-1}$, $r = (1-i)\beta\mu/2$. Here, and also in Eq. (23), quantities without subscript refer to the sample and the subscripts g and b refer to the cell gas and the sample backing material.

For measurements in liquid crystals, one usually works in the optically and thermally thick regime ($e^{\sigma l} \gg e^{-\sigma l}$, $e^{-\beta l}$), resulting in substantially simplified expressions. This means that β^{-1} and μ of the sample have to be much smaller than the sample thickness, l. The contribution of the backing material then disappears from the expressions. One then obtains for the photoacoustic microphone signal $Q = q e^{-i\psi}$, with amplitude q and phase ψ (with respect to the radiation modulation) [38]:

$$q = \frac{\gamma_g P_0 I_0 t (2t^2 + 2t + 1)^{-1/2}}{2\sqrt{2T_0 l_g}(1+s)\kappa_g a_g^2} \quad (25)$$

$$\tan\psi = 1 + 1/t \quad (26)$$

with $t = \mu\beta/2$ and $s = a\kappa(\kappa_g a_g)^{-1}$.

The Eqs. (25) and (26) allow the simultaneous determination of the heat capacity per unit volume (ρC_p) and the thermal conductivity κ of the sample by solving for t and s. This gives

$$\rho C_p = \beta \kappa_g a_g st^{-1}\omega^{-1} \quad (27)$$

$$\kappa = 2\kappa_g a_g st\beta^{-1} \quad (28)$$

However, one should arrange the measuring conditions in such a way that $1/t = 2/(\mu\beta)$ is not too small compared to l in Eq. (26). This can to some extent be done by choosing the proper modulation frequency (changing μ) and/or changing β by choosing an appropriate wavelength λ for the modulated light source. In any case one also has to measure (in a separate experiment) $\beta(\lambda)$. Moreover, Eqs. (25) and (26) for q and ψ contain several other quantities related to the cell characteristics, the cell gas and the light source, which have to be known in order to arrive at quantitative results for ρC_p and κ. This problem is usually solved by simultaneous or separate calibration runs with a sample of known optical and thermal parameters [39, 40].

In photoacoustics, as in a.c. calorimetry, relatively small periodic temperature variations (in the mK range) are used, and only small amounts of samples are needed. The small temperature variations allow measurements very close to phase transitions but latent heats of first-order phase transitions can also not be obtained in this way.

One of the important advantages of photoacoustical techniques is the possibility of simultaneously measuring the heat capacity and the thermal conductivity in small samples. These methods also offer the possibility of investigating anisotropic thermal transport properties and to carry out depth profiling in thermally inhomogeneous samples [40]. Studying the thermal conductivity anisotropy requires homogeneous samples with known director orientation. This can be achieved by placing the photoacoustic cell and/or sample in an external magnetic field.

It should also be pointed out that with the extension of the standard a.c. calorimetric technique by Huang et al. [24], one can also measure simultaneously the heat capacity and the thermal conductivity. In this case a.c. calorimetric measurements have to be extended to high frequencies into the regime where the thermal diffusion length $\mu = [2\,\kappa/(\rho\,C_p\,\omega)]^{1/2}$ of the sample becomes comparable to the sample thickness in which case the second term in Eq. (10) can no longer be neglected.

The Photopyroelectric Method

If we consider a pyroelectric transducer, with thickness l_p and surface area A, in a one-dimensional configuration, a change of the temperature distribution $\theta(x, t)$ relative to an initial reference situation $\theta(x, t_0)$ will cause a change of polarization. This in turn induces an electric charge given by:

$$q(t) = \frac{pA}{l_p} \int_0^{l_p} [\theta(x,t) - \theta(x,t_0)]dx = \frac{pA}{l_p}\theta(t) \quad (29)$$

with p the pyroelectric coefficient of the transducer. Consequently also a current is produced:

$$i_p = \frac{pA}{l_p}\frac{d\theta(t)}{dt} \quad (30)$$

Usually the electric signal from the pyroelectric transducer is detected by a lock-in amplifier. Thus only the a.c. component of the temperature variation gives rise to the detected signal. The signal depends, of course, on the impedance of the transducer and of the detecting electronics. The pyroelectric element can be represented by an ideal current source with a parallel leakage resistance and capacitance, while the detection electronics can be described by an input capacitance and a parallel load resistance [43]. Circuit analysis results then in a general expression for the signal $V(\omega)$ [44, 45].

Under properly chosen experimental conditions [44], however, it is possible to arrive at a simultaneous determination of the specific heat capacity and the thermal conductivity of a sample in thermal contact with the

pyroelectric detector. A suitable setup for liquid crystal samples is given in Fig. 7. The wavelength of the modulated light, the modulation frequency, sample and detector thicknesses l_s and l_p are chosen in such a way that the sample and transducer are optically opaque, the detector thermally very thick ($\mu_p \ll l_p$) and the sample quasithermally thick ($\mu_s \leq l_s$). One then obtains for the signal amplitude and phase [46]:

$$|V(\omega)| = \frac{I_0\,\eta_s\,A\,R\,p\,e_p}{l_p[1+(\omega\tau)^2]^{1/2}\,\rho_p\,C_p}$$
$$\cdot\frac{\exp[-(\omega/2\alpha_s)^{1/2}l_s]}{e_s(e_m/e_s+1)(e_p/e_s+1)} \quad (31)$$

$$\phi(\omega) = \tan^{-1}(\omega\tau) - (\omega/2\alpha_s)^{-1/2}l_s \quad (32)$$

where subscripts s, p and m, respectively, refer to the sample, the pyroelectric transducer and to the medium in contact with the sample front surface. I_0 is the nonreflected light source intensity, η_s is the nonradiative conversion efficiency, ρ is the density and $e = (\rho\,C\,\kappa)^{1/2}$ is the thermal effusivity. R and τ are a circuit equivalent resistance and time constant. From Eq. (32) it is possible to determine the sample thermal diffusivity α_s. Inserting α_s in Eq. (31) yields the sample effusivity. The thermal conductivity and specific heat capacity are then given by:

$$C_s = e_s\,\rho_s^{-1}\,\alpha_s^{-1/2} \quad \text{and} \quad \kappa_s = e_s\,\alpha_s^{-1/2} \quad (33)$$

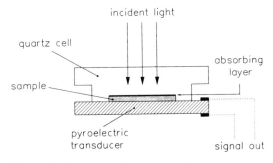

Figure 7. Schematic diagram of a photopyroelectric cell for liquid crystal samples [46].

6.2.1.4 Calorimetric Results

High resolution calorimetry has been extensively used to study the many important and often puzzling phase transitions in thermotropic liquid crystals of rodlike molecules. Discotic liquid crystals seem to be, from the calorimetric point of view, less exciting because most phase transitions are first order with little pretransitional fluctuation effects [47]. Although lyotropic liquid crystal systems show complex phase behavior, they have not been studied calorimetrically in great detail, because calorimetric signatures of phase transitions tend to turn out quite small [48]. No attempt will be made to give an exhaustive treatment of any of the transitions discussed further. Thermal results will be presented to show what can be accomplished for typical cases with high-resolution calorimetric experiments, careful analysis and theoretical interpretation. More extensive overviews can be found elsewhere [1–6]. A detailed account of calorimetric studies of free standing liquid crystal films can be found in a review article by Huang [49].

The I–N Transition

The only difference between the nematic phase and the isotropic phase is the orientational order. A proper description of this orientational order requires the introduction of a tensor of the second rank [7, 8]. This tensor can be diagonalized and for anisotropic liquids with uniaxial symmetry, the nematic phase can be described by only one scalar order parameter. The thermodynamic behavior in the vicinity of the N–I transition is usually described in terms of the mean-field Landau–de Gennes theory [7]. For the uniaxial nematic phase one can obtain the expansion of the free energy G in terms of the modulus of an order parameter Q.

$$G - G_0 \tag{34}$$

$$= \frac{1}{2}AQ^2 - \frac{1}{3}BQ^3 + \frac{1}{4}CQ^4 + \frac{1}{6}DQ^6 + \ldots$$

In the isotropic phase $Q=0$ and in the nematic phase $Q\neq0$. In Eq. (34) one has $A = a(T-T^*)/T_{N-I}$ and $B>0$. The presence of the cubic term, which does not disappear at T_{N-I}, leads to a first-order transition with a finite discontinuity in the order parameter ($Q_{N-I} = 2B/3C$). T^* is the stability limit of the isotropic phase. For $B=0$, a normal second-order transition at T_{N-I} is expected. The excess heat capacity in the nematic phase is given by [16]:

$$C_p = -aQ\left(\frac{\partial Q}{\partial T}\right)_p \tag{35}$$

$$= \frac{a^2}{2CT_{N-I}}\left[1 + \frac{B}{2(aC)^{1/2}}\left(\frac{T^{**}-T}{T_{N-I}}\right)^{-1/2}\right]$$

with T^{**} the stability limit of the nematic phase. One, thus, has an anomalous contribution with an exponent $\alpha = 1/2$ in the nematic phase, resulting in a jump in C_p at T_{N-I} equal to $\Delta C_p = 2a^2/CT_{N-I}$. In the case of a mean field second-order ($B=0$ and $C>0$) transition the singular contribution (see Fig. 1c) follows from the mean-field behavior (with the critical exponent $\beta = 1/2$) of the order parameter [16].

For the enthalpy discontinuity at T_{N-I} one obtains $H_I - H_N = 2aB^2/9C^2$. When $B=0$, there is no enthalpy jump or latent heat and one has a critical point (Landau point) on an otherwise first-order line. Because of the presence of this (small) cubic term in Eq. (34), the N–I transition should (normally) be weakly first order. This is in agreement with experimental observations. In Fig. 8, part of the enthalpy curve near T_{N-I} is shown for hexylcyanobiphenyl (6CB), a compound of the alkylcyanobiphenyl (nCB) homologous series [5]. The nearly vertical part of the enthalpy curve corresponds to the

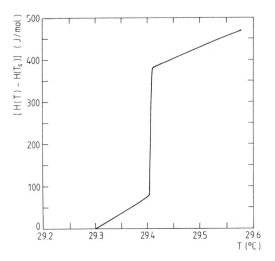

Figure 8. Temperature dependence of the enthalpy near the nematic to isotropic transition for hexylcyanobiphenyl (6CB) [5].

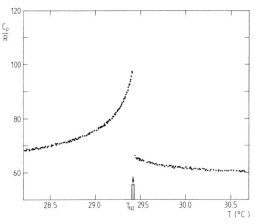

Figure 9. Temperature dependence of the reduced heat capacity per mole (R is the gas constant) near the nematic to isotropic (N–I) transition for hexylcyanobiphenyl (6CB). The width of the arrow on the temperature axis represents the two-phase region [5].

latent heat of the N–I transition. It is not perfectly vertical because of the small (≈ 20 mK wide) impurity-induced two-phase region. ΔH_L for the transitions in 6CB is (293 ± 3) J/mol. This latent heat is indeed quite small, about two orders of magnitude smaller than the latent heat at the melting transitions in these cyanobiphenyl compounds [19, 50].

Because the latent heat at the N–I transition is very small and since the orientational interactions between liquid crystal molecules are short range, one might expect many physical properties to display critical-like behavior, described by power laws with the appropriate critical exponents. These fluctuation effects are, indeed, observed in many properties including the specific heat capacity. Figure 9 gives the temperature dependence of C_p near the N–I transition for the same substance as in Fig. 8. No data points are displayed in the coexistence region, whose width is indicated by the narrow box around the arrow marking T_{N-I}.

Since the Landau–de Gennes theory is a mean-field theory, one can only expect a qualitative description of the specific heat capacity anomaly [2, 16]. Attempts to determine a critical exponent α have been largely unsuccessful due to the first-order character of the N–I transition. Fits with Eq. (6) must be made separately for data above and below T_{N-I} with different effective critical temperatures (T^* and T^{**}), which both are different from the first-order transition temperature. For $T < T_{N-I}$, typical α_{eff} values lie in the range 0.3–0.4 but depend strongly on the fitting range. For $T > T_{N-I}$, an even wider range of α_{eff} values between 0.1 and 0.5 is observed. Furthermore $(T^* - T_{N-I})$ and $(T^{**} - T_{N-I})$ values obtained from C_p fits are consistently about one-tenth the magnitude of those from several other properties [2, 16]. The experimental behavior of the N–I transition is quite well characterized, but theoretical understanding is still rather poor.

Blue Phase Transitions

In the case of optically active molecules one arrives at a special situation for the nematic phase. In addition to the long range orientational order there is a spatial variation of the director leading to a helical structure. The local preferred direction of alignment of the molecules is slightly rotated in adjacent planes perpendicular to the pitch axis. This phase is usually called chiral nematic (N*) or cholesteric phase. The helical pitch in the N* phase is substantially larger than molecular dimensions and varies with the type of molecules. For a long pitch one has a direct first-order transition between the isotropic phase and the chiral nematic phase with characteristics very similar to the normal N–I transition in nonchiral compounds. For chiral nematics with short pitch (typically less than 0.5 μm), a set of intermediate blue phases (BP) are observed between the isotropic and the N* phase [51–53]. In order of increasing temperature these phases are denoted BP_I, BP_{II} and BP_{III}. The first two blue phases have three-dimensional cubic defect structures, whereas BP_{III}, which is also called the fog phase appears to be amorphous [51–53].

Different phase transitions involving blue phases are expected theoretically to be first-order [54] except the BP_{III} to isotropic transition which could become second order [54] in an isolated critical point at the termination of a first-order line [6, 55]. Figure 10 shows C_p results of cholesteryl non-nanoate (CN) with adiabatic scanning calorimetry [56]. From Fig. 10 it is clear that there are substantial pretransitional heat capacity effects associated with the BP_{III}–I transition, which means that a large amount of energy is going into changing the local nematic order. The other transitions appear as small, narrow features on the BP_{III}–I transition peak. From the inspection of the en-

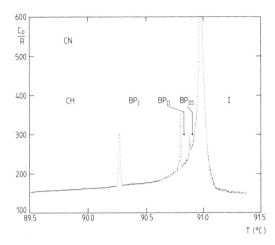

Figure 10. General overview for the reduced heat capacity per mole C_p/R for a temperature range covering all phase transitions involving the blue phases in cholesteryl nonanoate (CN) [56].

thalpy behavior, it was found that these small features correspond to first-order transitions with very small latent heats [56]. In CN the BP_{III}–I transition was also observed to be first order. This can be seen in Fig. 11 where the enthalpy versus temperature is given for a T-range of 0.2 K below and 0.2 K above the transition.

Further calorimetric studies by Voets and Van Dael [57, 58] have shown that the latent heat for BP_{III}–I decreases as the chirality of the molecule increases and that its magnitude varies considerably as a function of composition in binary mixtures of R- and S-enantiomers. Recently an a.c. and non-adiabatic scanning calorimetry investigation of the chiral compound S,S-(+)-4″-(2-methylbutylphenyl)-4′-(2-methylbutyl)-4-biphenyl-4-carboxylate (S,S-MBBPC), which has very short pitch, yielded the results given in Fig. 12 [55]. By combining these two calorimetric techniques one can show that there are sharp first-order transitions with small latent heats at the N*–BP_I and the BP_I–BP_{III} transitions (the BP_{II} phase is absent in this compound). However, no thermodynamic

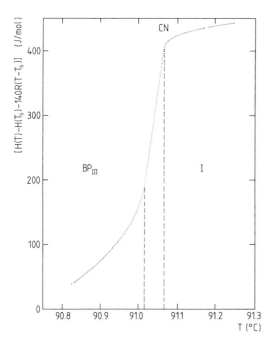

Figure 11. Detailed plot of the enthalpy near the phase transition from the blue phase III to the isotropic phase in cholesteryl nonanoate (CN). Note that for clarity a large linear background 140 R $(T-T_s)$ with $T_s = 90.73°C$, has been subtracted from the direct data. The two vertical dashed lines indicate the width of the two-phase region [56].

BP$_{III}$–I transition occurs in this S,S-enantiomer of MBBPC. One only observes a supercritical evolution. As pointed out by Kutnyak et al. [55] an investigation of S,S- and R,R-enantiomers of MBBPC should allow one to find the BP$_{III}$–I critical point and determine its critical exponent. The observed continuous supercritical evolution implies that the BP$_{III}$ and I phases must have the same macroscopic symmetry allowing, as in the case of the liquid–gas transition of simple fluids, the first-order line to terminate in an isolated critical point.

The N–SmA Transition

This transition has been the most extensively studied of all liquid crystal phase transi-

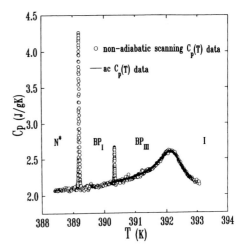

Figure 12. The specific heat capacity variation in the N*–BP$_I$–BP$_{III}$–I transition region for S,S-MBBPC. The sharp peaks from non-adiabatic scanning through the N*–BP$_I$ and BP$_I$–BP$_{III}$ regions are due to latent heat effects at the first-order transitions. No such effects are seen for the BP$_{III}$–I transition, because this is only a continuous supercritical evolution not a true phase transition in this material [55].

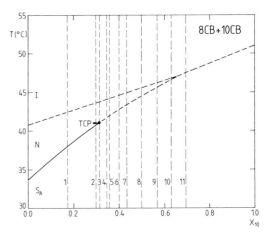

Figure 13. Phase diagram of 8CB + 10CB mixtures. Heavy solid and dashed lines are, respectively, second-order and first-order transitions. TCP is the N–A tricritical point. Vertical dashed lines indicate the measured mixtures [64].

tions. In spite of vigorous experimental and theoretical efforts, many aspects of the critical behavior of this simple kind of one-dimensional freezing, are not yet understood, making this transition an intriging and challenging problem in the statistical mechanics of condensed matter.

The SmA liquid crystalline phase results from the development of a one-dimensional density wave in the orientationally ordered nematic phase. The smectic wave vector q is parallel to the nematic director (along the z-axis) and the SmA order parameter $\psi = |\psi| e^{i\phi}$ is introduced by $\rho(r) = \rho_0 [1 + \text{Re } \psi e^{iqz}]$. Thus the order parameter has a magnitude and a phase. This led de Gennes to point out the analogy with superfluid helium and the normal-superconductor transition in metals [7, 59]. This would than place the N–SmA transition in the three-dimensional XY universality class. However, there are two important sources of deviations from isotropic 3D-XY behavior. The first one is crossover from second-order to first-order behavior via a tricritical point due to coupling between the smectic order parameter ψ and the nematic order parameter Q. The second source of deviation from isotropic 3D-XY behavior arises from the coupling between director fluctuations and the smectic order parameter, which is intrinsically anisotropic [60–62].

The first-order to second-order crossover can be qualitatively understood on the basis of a mean-field approximation for the excess smectic free energy (above the nematic one) as formulated by de Gennes [7]:

$$\Delta G = a|\psi|^2 + b_0|\psi|^4 + (\delta Q)^2/2\chi - C|\psi|^2 \delta Q \tag{36}$$

where $a = a_0(T - T_{N-A})/T_{N-A} = a_0 \varepsilon$; b_0, $C > 0$; and δQ the change in the nematic order induced by the formation of the smectic layers, and χ the temperature dependent nematic susceptibility [7] whose value at T_{N-A} depends on the width of the nematic range. Minimization of ΔG with respect to δQ yields:

$$\Delta G = a|\psi|^2 + b|\psi|^4 + c|\psi|^6 \tag{37}$$

with $b = b_0 - C^2 \chi/2$. For narrow nematic ranges $\chi(T_{N-A})$ is large and $b < 0$, resulting in a first-order transition ($c > 0$ for stability reasons). For wide nematic ranges $\chi(T_{N-A})$ is small and $b > 0$, the transition is then second-order. In this mean-field approach the C_p behavior would then be given in Fig. 1(c). A tricritical point occurs for $b = 0$. The specific heat capacity in the low temperature side then diverges with a critical exponent $\alpha = 1/2$. In Fig. 1(c) one then has $C_p^{\text{max}} = \infty$.

The crossover from first- to second-order behavior and the disappearance of measurable latent heats with increased widths of the nematic ranges has been shown on the basis of adiabatic scanning calorimetry of the 4′-n-alkyl-4-cyano-biphenyl (nCB) binary mixtures 9B + 10CB [63] and 8CB + 10CB [64]. The investigation of 8CB + 10CB mixtures with nematic ranges from 7 K to zero (see the phase diagram in Fig. 13) resulted in a partly first- and partly second-order SmA line with a tricritical point at $X_{10CB} = 0.314$. In Fig. 14 the enthalpy curves near T_{N-A} are given for several mixtures of 8CB + 10CB with first-order transitions.

Along the second-order part of the transition line one would expect 3D-XY critical behavior at least in the limit of the critical point. In real experiments, one observes crossover between XY critical and tricritical behavior, resulting in α_{eff} values where $\alpha_{XY} < \alpha_{\text{eff}} < \alpha_{TC}$ because the temperature range for most experimental data is limited to $10^{-5} < \varepsilon < 10^{-2}$. In addition to that the coupling between director fluctuations and the smectic order parameter intervenes as a second source of deviation from isotropic behavior and influences the behavior of the smectic susceptibility and the correlation lengths (parallel and perpendicular to the director) much more than the specific heat. Here also a broad crossover should be ob-

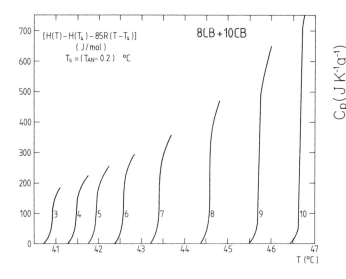

Figure 14. Enthalpy versus temperature near T_{N-A} for the tricritical mixture (no. 3) and several other 8CB + 10CB mixtures with a first-order N–A-transition. A regular part 85 R $(T-T_s)$ has been subtracted for display reasons [64].

served from isotropic XY to a weakly anisotropic regime to the strong coupling limit with highly anisotropic correlation behavior. Narrow nematic ranges would result in highly anisotropic behavior and (very) wide nematic ranges should show isotropic or weakly anisotropic behavior. Figure 15

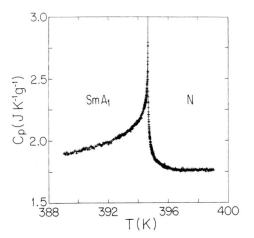

Figure 15. Heat capacity of 8OPCBOB near the nematic to smectic A_1 phase transition. The smooth curve represents a fit to the data with Eq. (6) based on critical parameters in agreement with the three-dimensional XY model. The index 1 in SmA refers to the monolayer structure of this frustrated smectic compound [65].

shows a typical example of 3D-XY specific heat capacity behavior for the compound 4-n-octyloxyphenyl-4-(4-cyanobenzyloxy)-benzoate (8OPCBOB) with a large nematic range [65]. The evolution from second-order to tricritical to first-order is illustrated in Fig. 16 by the behavior of mixtures of 4O.8 and 6O.8 (butyl- and hexyloxybenzylidene octylaniline) [66]. Pure 4O.8 has a nematic range of 14.72 K and an $\alpha_{eff} = 0.13$ is obtained. Pure 6O.8 has a nematic range of only 0.87 K and the transition is strongly first-order with a latent heat $\Delta H_L = 3700$ J mol^{-1}. Fits to the data in Fig. 16 yield the following α_{eff} values: 0.13, 0.22, 0.30, 0.45, 0.50 for X_1 to X_5 respectively. The mixtures with X_6 and X_7 are weakly first-order.

More extensive general discussions of the N–SmA problem can be found elsewhere [3, 5, 7, 9, 67] and detailed theoretical treatments are also available [60–62, 68].

The SmA–SmC Transition

The SmC phase differs from the SmA phase by a tilt angle θ of the director with respect to the layer normal. However, in order to fully specify the director one also needs the az-

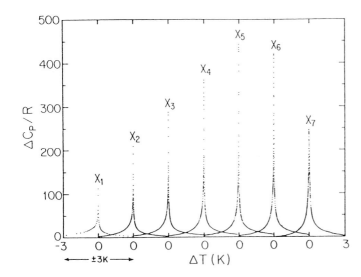

Figure 16. Reduced excess molar heat capacity $\Delta C_p/R$ for the N–A transition in mixtures of 4O.8 +6O.8. In each case data are shown over the range $-3\,\mathrm{K} < T - T_c < 3\,\mathrm{K}$. The values of the mole fraction X of 6O.8 are $X_1 = 0$, $X_2 = 0.10$, $X_3 = 0.20$, $X_4 = 0.35$, $X_5 = 0.35$, $X_6 = 0.40$, and $X_7 = 0.50$ [66].

imuthal angle φ. This results in an order parameter $\varphi = \theta\,e^{i\varphi}$ with two components, and one would expect that the second-order AC transition should belong to the XY universality class. However, AC transitions in nonpolar nonchiral materials are second-order transitions that are very well described by the Landau theory with a large sixth-order coefficient c in Eq. (2). An example of this kind of behavior in N-(4-n-heptyloxybenzylidene)-4′-n-heptylaniline (7O.7) is shown in Fig. 17 [69]. The explanation for this mean-field behavior is the fact that the Ginsburg [12] criterion indicates that the critical region is extremely small ($\varepsilon_{\mathrm{crit}} < 10^{-5}$). It should, however, also be noted that the magnitude and sharpness of the Landau heat capacity peak for the SmA–SmC transition varies greatly from one material to another. In contrast to the monolayer SmA–SmC transitions (as e.g. in 7O.7) bilayer SmA_2–SmC_2 transitions show a steplike C_p Landau behavior with $c \approx 0$ [70].

If the constituent molecules are optically active, the chiral SmC* phase will be observed instead of the normal SmC phase. There will be a helical precession of the di-

rection of the director tilt with respect to the layer normal. Chiral compounds can exhibit strongly first-order SmA–SmC* transitions. By varying the composition of a bi-

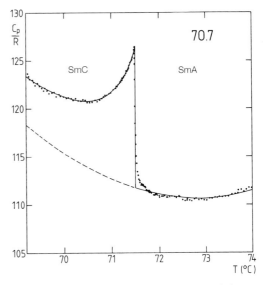

Figure 17. The temperature dependence of the reduced heat capacity per mole for 7O.7 near the SmA–SmC phase transition. The solid line represents a fit with Eq. (4) combined with the background behavior given by the dashed line. The curvature of the background is caused by the nearby SmC–B transition [69].

nary mixture the first-order can crossover via a tricritical point to a second-order transition [71, 72].

The N–SmA–SmC Multicritical Point

In some mixtures of liquid crystals exhibiting N, SmA and SmC phases, the N–SmA, SmC–SmC, and N–SmC transition lines meet at the N–SmA–SmC multicritical point. Fig. 18 gives the phase diagram with an N–SmA–SmC point for mixtures of $\overline{5}O\overline{8}+\overline{6}O\overline{8}$ (4-n-alkyloxyphenyl-4′-n-alkyloxybenzoate) compounds [73]. The N–SmA–SmC point has been the subject of extensive theoretical and experimental studies during the past decade. The nature of the point is, however, still not clearly established.

Since the first discovery [74, 75] of multicritical points in 1977, several systems have been investigated experimentally, also

calorimetrically [5]. The calorimetric data show essentially a similar behavior near the multicritical point for all materials investigated in detail. The N–SmC transitions are first-order with a latent heat becoming zero at or very near the multicritical point (see e.g. Fig. 19). SmA–SmC/N–SmA transitions are found to be second-order. The heat capacity anomalies along the SmA–SmC line as well as along the N–SmC line become larger and sharper on approaching the multicritical point, thus suggesting a mean-field tricritical character of the point. On the other hand, high resolution X-ray scattering results [76] are in general agreement with the point being a Lifshitz point [77]. The multicritical point seems to simultaneously exhibit the characteristics of a Lifshitz point and of a tricritical point.

When chiral compounds are involved the phase diagram near the intersection of the N*–SmA, SmA–SmC and N*–(SmC*) transition lines is significantly more complicated. Several kinds of TGB phases have

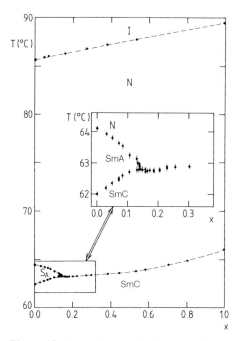

Figure 18. Phase diagram for the binary liquid crystal system $\overline{5}O.\overline{8}$ (left) – $\overline{6}O.\overline{8}$ (right), with X the weight fraction of $\overline{6}O.\overline{8}$ in the mixture [73].

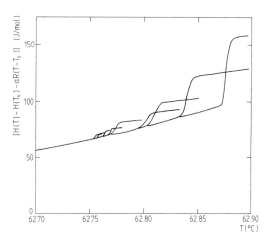

Figure 19. Temperature dependence of the enthalpy for several mixtures of $\overline{5}O.\overline{8}+\overline{6}O.\overline{8}$ with direct N–SmC transitions. For clarity reasons a large linear background $C_p^b = a R$ has been subtracted from the direct experimental data. The curve with the smallest steplike increase is closest to the N–SmA–SmC point, and has a latent heat of (1.7 ± 0.3) J/mol [73].

been theoretically predicted [78, 79] and experimentally observed [35, 80].

6.2.1.5   Photoacoustic and Photopyroelectric Results

As pointed out above the photoacoustic and photopyroelectric techniques permit the simultaneous measurement of static as well as transport thermal properties. In comparison with the vast amount of high-quality calorimetric information on the static thermal quantities specific heat capacity and enthalpy, there is rather limited information on thermal transport properties of liquid crystals in general and near phase transitions in particular. In early studies of the 1970s, and later on conventional steady-state gradient and transient techniques have been used to detect the expected anisotropy (in oriented samples) of the thermal diffusivity or the thermal conductivity [81, 82]. However, the techniques usually applied were not very well suited for measurements very close to phase transitions, because large samples and sizable temperature gradients are usually needed. Although in the photoacoustic and photothermal method, one works with very small samples and small temperature gradients, only rather recently these methods have been applied for liquid crystal studies. In fact, high-resolution investigations of thermal transport near liquid crystal phase transitions were pioneered about ten years ago by Huang and coworkers [24] by a high-frequency extension of the standard a.c. calorimetric technique (see Section for a.c. calorimetry). Instead of using Eq. (10) for ΔT_{ac} in a ω regime such that $\omega \tau_1 \gg 1$ and $\omega \tau_2 \ll 1$, one measures ΔT_{ac} as a function of ω in the region $\omega \tau_1 \gg 1$ including the high frequency cutoff $\omega_H \tau_2 = 1$. From this quantity the sample thermal diffusivity can be deduced. By also simultaneously measuring the heat capacity in the normal low ω, a.c.

calorimetric regime, one can derive also the thermal conductivity κ. A number of phase transitions have been studied in this way by Huang and coworkers. An overview and discussion of the major results can be found elsewhere [5].

The first high-resolution photoacoustic investigations of phase transitions have been carried out by Zammit et al. [41] for a series of samples with N–SmA transitions. These investigations were carried out on samples of 9CB, 8CB and a mixture of 8CB + 7CB with a mole fraction $X = 0.76$ of 8CB. No effort was made to prepare homogeneously aligned samples. It is, however, likely that at least a substantial part of the sample near the free surface is homeotropically aligned, because these cyanobiphenyls strongly prefer homeotropic alignment at free surfaces [40]. The results showed a critical anomaly both in the specific heat capacity C_p and the thermal conductivity κ. The results for 9CB were subsequently also confirmed by measurements (also for non-aligned samples) in a photopyroelectric set-up [46]. The values of the heat capacity critical exponents from fits to power laws of the type of Eq. (6), are fully consistent with previous results obtained by Thoen et al. [19, 63] by means of adiabatic scanning calorimetry (see above). For the critical exponent values which have been obtained from power law fits to the thermal conductivity, no agreement with theory or a systematic trend could be obtained [41]. However, in our photoacoustic measurements on homeotropically aligned 8CB samples [83] we did not see any evidence for a critical increase of the thermal conductivity at the N–SmA transition.

Very recently M. Marinelli et al. [84] carefully remeasured photopyroelectrically properly homeotropically as well as planar aligned 8CB samples. In Fig. 20 the specific heat capacity for both configurations is

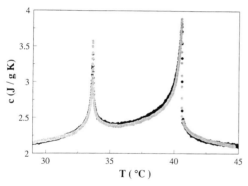

Figure 20. The specific heat capacity of 8CB as a function of temperature for planar (gray dots) and homeotropic (black dots) aligned samples [84].

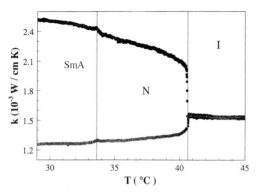

Figure 21. The thermal conductivity of 8CB as a function of temperature for planar (gray dots, lower curve) and homeotropic (black dots, upper curve) aligned samples [84].

given. Corresponding data for the thermal conductivity are given in Fig. 21. Contrary to their previous results [41, 46] a critical anomaly in κ is virtually absent. Whether this difference has to do with the fact that in the new results aligned samples have been used is not entirely clear, because also recently no evidence for a thermal conductivity anomaly at the N–SmA transition was found for nonaligned samples of octylcyanothiolbenzoate by the same group [85]. A possible explanation of the unusual κ behavior of the older results could be the presence of thermal gradients [84, 85].

6.2.1.6   Acknowledgements

The author wishes to thank C. W. Garland for many helpful discussions. He also thanks him, M. Marinelli and U. Zammit for the permission to reproduce figures from their publications.

6.2.1.7   References

[1] C. W. Garland in *Geometry and Thermodynamics* (Ed.: J.-C. Tolédano), NATO ASI Ser. B 229, Plenum, New York **1990**, pp. 221–254.
[2] M. A. Anisimov, *Critical Phenomena in Liquids and Liquid Crystals*, Gordon and Breach, Philadelphia **1990**, Chap. 10.
[3] J. Thoen in *Phase Transitions in Liquid Crystals* (Eds.: S. Martellucci, A. N. Chester), NATO ASI Ser. B 290, Plenum, New York **1992**, Chap. 10.
[4] C. W. Garland in *Phase Transitions in Liquid Crystals* (Eds.: S. Martellucci, A. N. Chester), NATO ASI Ser. B 290, Plenum, New York **1992**, Chap. 11.
[5] J. Thoen, *Int. J. Mod. Phys. B* **1995**, 9, 2157–2218.
[6] C. W. Garland in *Liquid Crystals: Physical Properties and Phase Transitions* (Ed.: S. Kumar), Oxford University Press **1997**, Chap. 10 (in preparation).
[7] P. G. de Gennes and J. Prost, *The Physics of Liquid Crystals*, 2nd edn, Clarendon Press, Oxford **1993**.
[8] G. Vertogen, W. de Jeu, *Thermotropic Liquid Crystals, Fundamentals*, Springer-Verlag, Berlin **1988**.
[9] P. M. Charkin, T. C. Lubensky, *Principles of Condensed Matter Physics*, Cambridge University Press, **1995**.
[10] C. Bagnuls, C. Bervillier, *Phys. Rev. B* **1985**, 32, 7209–7231.
[11] C. Bagnuls, C. Bervillier, D. I. Meiron, B. G. Nickel, *Phys. Rev. B* **1987**, 35, 3585–3607.
[12] V. I. Ginzburg, *Sov. Phys. Solid State* **1961**, 2, 1824–1829.
[13] M. E. Fisher, *Phys. Rev.* **1968**, 176, 257–272.
[14] M. A. Anisimov, A. V. Voronel, T. M. Ovodova, *Sov. Phys. JETP* **1972**, 35, 536–539.
[15] M. Sorai, S. Asahina, C. Destrade, N. H. Tinh, *Liq. Cryst.* **1990**, 7, 163–180.
[16] M. A. Anisimov, *Mol. Cryst. Liq. Cryst.* **1988**, 162A, 1–96.
[17] J. A. Lipa, C. Edwards, M. J. Buckingham, *Phys. Rev. Lett.* **1970**, 25, 1086–1090; *Phys. Rev. A* **1977**, 15, 778–789.
[18] E. Bloemen, J. Thoen, W. van Dael, *J. Chem. Phys.* **1980**, 73, 4628–4635.
[19] J. Thoen, H. Marijnissen, W. Van Dael, *Phys. Rev. A* **1982**, 26, 2886–2905.
[20] M. A. Anisimov, V. P. Voronov, A. O. Kulkov, F. Kholmurodov, *J. Phys. (Paris)* **1985**, 46, 2137–2143.

[21] J. Thoen, E. Bloemen, H. Marijnissen, W. Van Dael, *Proceedings of the 8th Symposium on Thermophysical Properties*, Nat. Bur. Stand. Maryland, 1981, Am. Soc. Mech. Eng., New York **1982**, 422–428.

[22] C. A. Schantz, D. L. Johnson, *Phys. Rev. A* **1978**, *17*, 1504–1512.

[23] J. M. Viner, D. Lamey, C. C. Huang, R. Pindak, J. W. Goodby, *Phys. Rev. A* **1983**, *28*, 2433–2441.

[24] C. C. Huang, J. M. Viner, J. C. Novak, *Rev. Sci. Instrum.* **1985**, *56*, 1390–1393.

[25] R. Geer, T. Stoebe, T. Pitchford, C. C. Huang, *Rev. Sci. Instrum.* **1991**, *62*, 415–421.

[26] G. B. Kasting, K. J. Lushington, C. W. Garland, *Phys. Rev. B* **1980**, *22*, 321–331.

[27] C. W. Garland, *Thermochim. Acta* **1985**, *88*, 127–142.

[28] P. Sullivan, G. Seidel, *Phys. Rev.* **1968**, *173*, 679–685.

[29] A. J. Jin, T. Stoebe, C. C. Huang, *Phys. Rev. E* **1994**, *49*, R4791–R4791.

[30] T. Stoebe, C. C. Huang, *Phys. Rev. E* **1994**, *50*, R32–R35.

[31] D. Djurek, J. Baturic-Rubcic, K. Franulovic, *Phys. Rev. Lett.* **1974**, *33*, 1126–1129.

[32] K. Ema, T. Uematsu, A. Sugata, H. Yao, *Jpn. J. Appl. Phys.* **1993**, *32*, 1846–1850.

[33] K. Ema, H. Yao, I. Kawamura, T. Chan, C. W. Garland, *Phys. Rev. E* **1993**, *47*, 1203–1211.

[34] H. Yao, T. Chan, C. W. Garland, *Phys. Rev. E* **1995**, *51*, 4585–4597.

[35] T. Chan, PhD Thesis in Physics, Massachusetts Institute of Technology **1995**.

[36] A. Rosencwaig, *Photoacoustics and Photothermal Spectroscopy*, Wiley & Sons, New York **1980**.

[37] G. Louis, P. Peretti, G. Billard, *Mol. Cryst. Liq. Cryst.* **1985**, *122*, 261–267.

[38] M. Marinelli, U. Zammit, F. Scudieri, S. Martellucci, J. Quartieri, F. Bloisi, L. Vicari, *Nuovo Cimento D* **1987**, *9*, 557–563.

[39] C. Glorieux, E. Schoubs, J. Thoen, *Mat. Sci. and Eng. A* **1989**, *122*, 87–91.

[40] J. Thoen, C. Glorieux, E. Schoubs, W. Lauriks, *Mol. Cryst. Liq. Cryst.* **1990**, *191*, 29–36.

[41] U. Zammit, M. Marinelli, R. Pizzoferrato, F. Scudieri, S. Martellucci, *Phys. Rev. A* **1990**, *41*, 1153–1155.

[42] A. Rosencwaig, A. Gersho, *J. Appl. Phys.* **1976**, *47*, 64–69.

[43] H. J. Coufal, R. K. Grygier, D. E. Horne, J. E. Fromm, *J. Vac. Technol. A* **1987**, *5*, 2875–2889.

[44] A. Mandelis, M. M. Zver, *J. Appl. Phys.* **1985**, *57*, 4421–4429.

[45] C. Christofides, *Crit. Rev. Sol. State Mater. Sci.* **1993**, *18*, 113–174.

[46] M. Marinelli, U. Zammit, F. Mercuri, R. Pizzoferrato, *J. Appl. Phys.* **1992**, *72*, 1096–1100.

[47] S. Shandrasekhar, *Liquid Crystals*, 2nd edn, Cambridge University Press **1992**.

[48] S. T. Shin, S. Kumar, D. Finotello, S. S. Keast, M. E. Neubert, *Phys. Rev. A* **1992**, *45*, 8683–8692.

[49] T. Stoebe, C. C. Huang, *Int. J. Mod. B* **1995**, *9*, 147–181.

[50] H. Marijnissen, J. Thoen, W. Van Dael, *Mol. Cryst. Liq. Cryst.* **1983**, *97*, 149–161.

[51] P. P. Crooker, *Liq. Cryst.* **1989**, *5*, 751–775.

[52] D. J. Wright, N. D. Mermin, *Rev. Mod. Phys.* **1989**, *61*, 385–432.

[53] T. Seideman, *Rep. Prog. Phys.* **1990**, *53*, 659–705.

[54] H. Grebel, R. M. Hornreich, S. Shtrikman, *Phys. Rev. A* **1984**, *30*, 3264–3278.

[55] Z. Kutnyak, C. W. Garland, J. L. Passmore, P. J. Collings, *Phys. Rev. Lett.* **1995**, *74*, 4859–4862.

[56] J. Thoen, *Phys. Rev. A* **1988**, *37*, 1754–1759.

[57] G. Voets, W. Van Dael, *Liq. Cryst.* **1993**, *14*, 617–627.

[58] G. Voets, PhD Thesis, Katholieke Univ. Leuven, Belgium **1993**.

[59] P. G. de Gennes, *Sol. St. Commun.* **1972**, *10*, 753–756.

[60] T. C. Lubensky, *J. Chim. Phys.* **1983**, *80*, 31–43.

[61] B. R. Patton, B. S. Andereck, *Phys. Rev. Lett.* **1992**, *69*, 1556–1559.

[62] B. S. Andereck, B. R. Patton, *Phys. Rev. E* **1994**, *49*, 1393–1403.

[63] J. Thoen, H. Marijnissen, W. Van Dael, *Phys. Rev. Lett.* **1984**, *52*, 204–207.

[64] H. Marijnissen, J. Thoen, W. Van Dael, *Mol. Cryst. Liq. Cryst.* **1985**, *124*, 195–203.

[65] C. W. Garland, G. Nounesis, K. J. Stine, G. Heppke, *J. Phys. (Paris)* **1989**, *50*, 2291–2301.

[66] K. J. Stine, C. W. Garland, *Phys. Rev. A* **1989**, *39*, 3148–3156.

[67] C. W. Garland, G. Nounesis, *Phys. Rev. E* **1994**, *49*, 2964–2971.

[68] J. Prost, *Adv. Phys.* **1984**, *33*, 1–46.

[69] J. Thoen, G. Seynaeve, *Mol. Cryst. Liq. Cryst.* **1985**, *127*, 229–256.

[70] Y. H. Jeong, K. J. Stine, C. W. Garland, N. H. Tinh, *Phys. Rev. A* **1988**, *37*, 3465–3468.

[71] T. Chan, Ch. Bahr, G. Heppke, C. W. Garland, *Liq. Cryst.* **1993**, *13*, 667–675.

[72] H. Y. Liu, C. C. Huang, Ch. Bahr, G. Heppke, *Phys. Rev. Lett.* **1988**, *61*, 345–348.

[73] J. Thoen, R. Parret, *Liq. Cryst.* **1989**, *5*, 479–488.

[74] G. Sigaud, F. Hardouin, M. F. Achard, *Sol. St. Commun.* **1977**, *23*, 35–36.

[75] D. L. Johnson, D. Allender, R. DeHoff, C. Maze, E. Oppenheim, R. Reynolds, *Phys. Rev. B* **1977**, *16*, 470–475.

[76] L. J. Martinez-Miranda, A. R. Kortan, R. J. Birgeneau, *Phys. Rev. Lett.* **1986**, *56*, 2264–2267; *Phys. Rev. A* **1987**, *36*, 2372–2383.

[77] J. Chen, T. C. Lubensky, *Phys. Rev. A* **1976**, *14*, 1202–1207.

[78] S. R. Renn, T. C. Lubensky, *Phys. Rev. A* **1988**, *38*, 2132–2147; *Phys. Rev. A* **1990**, *41*, 4392–4401.
[79] S. R. Renn, *Phys. Rev. A* **1992**, *45*, 953–976.
[80] T. Chan, C. W. Garland, H. T. Nguyen, *Phys. Rev. E* **1995**, *52*, 5000–5003.
[81] R. Vilanove, E. Guyon, G. Mitescu, P. Pieranski, *J. Phys. (Paris)* **1974**, *35*, 153–162.
[82] T. Akahane, M. Kondoh, K. Hashimoto, M. Nagakawo, *Jpn. J. Appl. Phys.* **1987**, *26*, L1000–1005.

[83] J. Thoen, E. Schoubs, V. Fagard in *Physical Acoustics: Fundamentals and Applications* (Eds. O. Leroy, M. Breazeale), Plenum Press, New York **1992**, p. 179–187.
[84] M. Marinelli, F. Mercuri, S. Foglietta, U. Zammit, F. Scudieri, *Phys. Rev. E* **1996**, *54*, 1604–1609.
[85] M. Marinelli, F. Mercuri, U. Zammit, F. Scudieri, *Phys. Rev. E* **1996**, *53*, 701–705.

6.2.2   Density

Wolfgang Wedler

The publication of density and specific volume studies on mesogens started soon after the discovery of liquid crystalline phases. In 1898/99 and 1905 Schenck [1, 2] reported specific volume data for PAA and *p*-methoxycinnamic acid. Until 1958/1960, with the publication of the Maier–Saupe theory [3], specific volume and density data on mesophases were the topics of only a few publications, and were always combined with results of other measurements [4–15]. A brief historical review of this aspect of liquid crystal research has been given by Bahadur [16].

6.2.2.1   Instrumentation

Density and specific volume of mesophases are measured under atmospheric pressure with pycnometers [17–22], dilatometers [16, 23–25] (e.g., differential scanning dilatometers, DSD [26–31]), with a buoyancy method [32], and with a vibrating device, known as a digital precision density meter system, or densitometer (Anton Paar K.G.) [33–38]. Devices for high-pressure measurements of specific volumes have been described by Kuss [39, 40] and Dörrer et al. [41], but were also mentioned earlier [42]. More recently, a device for simultaneous

scans of heat capacities and in-plane densities of smectic free-standing films, using reflectivity measurements, was reported by Stoebe et al. [43].

6.2.2.2   General Conclusions from Density Studies on Liquid Crystals

Specific volume and density are, as enthalpy changes, among the important parameters that indicate the order of a phase transition [44–47]. In fact, parallel development of enthalpy and relative volume changes, although not exactly proportional, is frequently observed [48–51]. The volume effects at mesophase–mesophase or mesophase–isotropic transitions are considerably smaller than those at melting into the mesophases, which have rarely been reported. For example, for the transition Cr/N in PAA, Bahadur [45] found fractional volume changes of 7.74% and a smaller thermal expansion coefficient in the crystalline phase $(4.11 \times 10^{-4} \text{ K}^{-1})$ than in the nematic phase $(9.26 \times 10^{-4} \text{ K}^{-1})$. Similar results have been obtained for chiral substances [52–56]. Despite the small magnitude of the observed discontinuities, first-order (e.g., Is/A [57], N/A [46, 47, 58, 301], A/B [57], and B/G [47]) weakly first-order (e.g., N/A [57, 59],

A/B [60], and B/F [61]), and second-order (e.g., N/A [62–65], A/C [47, 66–68], I/F [69, 70]) phase transitions, as well as singularities in the critical behavior [71], have been inferred. Lists of selected values of density and specific volume changes at different phase transitions are given in Tables 1 and 2. Collections of data and reviews have been given by Bahadur [16], Beguin et al. [72], Pisipati et al. [73], and Tsykalo [74].

From the temperature dependence of volume and density, the thermal expansion coefficient α and the thermal expansivity β [34] can easily be obtained. These quantities make phase transitions more apparent than does the density [33, 48]. Figures 1 and 2 demonstrate this with two examples. Densities, molar volumes, and thermal expansion coefficients show systematic behavior in homologous series. It has been shown that the magnitude of density changes at N/Is transitions [51, 57, 75–80, 301] and N/A transitions [75, 77–79, 301], and the absolute values of densities [79–81], are subject to even–odd effects with respect to the length of longitudinally attached alkyl

Table 1. Fractional volume changes (in %, upper right from diagonal, maximum and minimum reported values), and related information (lower left from diagonal) for phase transitions in mesophases.

Phase I → / Phase II ↓	Is	N	A	C
Is	–	0.51 [301] (7O.7) 0.0711 [112] (NPOOB)	2.00 [117] Diethyl (4,4′-azoxybenzoate) 0.22 [73] (4O.10)	1.94 [123] (OOAB) 0.55 [69] (TBDA)
N	First-order, ΔV= 0.95 ± 0.15 cm^3 mol^{-1} (8CB) [146]	–	0.65 [301] (7O.2) 0–0.08 [37]	0.15 [299] (HOAB) 0.01–0.023 [37]
A	First-order (Two-phase coexistence [117])	First-order [138] (7O.4) Weakly first-order (CBOOA) [44] Second-order (CBOOA) [31] ΔV=0.14 $\pm$0.04 cm^3 mol^{-1} (8CB) [146]	–	0.146 [48] (DOBACA) 0–0.005 [37]
C	First-order	First-order [299]	Second-order	–
B	No data available	First-order	First-order Weakly first-order [60]	First-order (Two phase coexistence [116])
I	No data available	No data available	No data available	First-order [69, 70]
F	First-order	No data available	First-order	First-order
G	No data available	First-order	No data available	First-order
H	No data available	No data available	No data available	First-order $\Delta V/V$=5.37% (DOBAMBC) [162]

Table 1. (continued)

Phase I / Phase II	B	I	F	G
Is	No data available	No data available	2.14 [302] (100.14)	No data available
N	1.50 [223] (HBT) 1.36 [285] (PMMA)	No data available	No data available	1.16 [139] (60.2) 0.38 [37] (50.2)
A	1.59 [73] (70.2) 0.05 [60] (70.1)	No data available	1.44 [308] (90.4) 1.42 [283] (90.4)	No data available
C	1.23 [123] (OOAB) 0.41 [79] (40.7)	1.00 [309] (TBAA9) 0.29 [70] (TBAA12)	1.07 [73] (70.6) 0.39 [65] (50.5)	>0.7 [73] 0.14 [101] (70.4)
B	–	No data available	0.03 [65] (50.6) 0.02 [61] (50.6)	0.06 [115] (80.4)
I	No data available	–	0.00 [69,70]	No data available
F	Weakly first-order Second-order	Second-order [69, 70]	–	0.04 [302] (100.14) 0.03 [61] (50.6)
G	First-order	No data available	First-order (50.6) [120] Weakly first-order (100.14) [302] Second-order (TBOA) [118]	
H	First-order, measured, no data reported [46]	No data available	No data available	No data available

chains in homologous series. The discontinuities in the absolute volume, like those in the enthalpy, increase with the length of the alkyl chains [50, 51, 82, 301], as can be seen for the two homologous series of 4O.m and 5O.m compounds in Table 3. This also was found for the Cr/N* transition [83]. The discontinuities are directly proportional to the molar mass, as can be seen when different substances are compared [48]. Absolute densities tend to decrease, and specific and molar volumes to increase, with chain length in a series of crystalline, smectic, and nematic phases [51, 84–88]; see the example in Fig. 3. A predicted inverse relation between molecular flexibility (due mainly to aliphatic chains) and the magnitude of the fractional volume change at the N/I transition [89] was verified by experiments in homologous series [23, 90–93] for longitudinal and lateral chain attachment. Furthermore, an increase in the thermal expansion coefficient with alkyl chain length has been found for N, A, and B phases [50, 94].

Table 2. Fractional volume changes (in %, upper right) and related information (lower left) for phase transitions in mesophases of chiral compounds.

Phase II \ Phase I	Is	Blue phase	N*	A*
Is	–	No data reported	0.17 [56] (cholesteryl stearate) 0.10 [27] (cholesteryl nonanoate)	No data reported
Blue phase	$\Delta V = 7.5 \times 10^{-4}$ cm^3 g^{-1} (cholesteryl oleate) [28]	–	0.004 [27] (cholesteryl nonanoate)	No data reported
N*	First-order [28] $\Delta V = 4 \times 10^{-5}$ cm^3 g^{-1} (cholesteryl nonanoate) [28]		–	0.14 [53] (cholesteryl myristate) 0.00 (cholesteryl [107] linolenate)
A*			$\Delta V = 154 \times 10^{-5}$ cm^3 g^{-1} (cholesteryl myristate) [28]	–

Based on density measurements, molar volume increments for methylene chain units, separated from the contribution of the aromatic part [95, 96], have been determined. The literature mentions mostly increments for longitudinal chain position [50, 51, 57, 60, 75, 79, 85, 86, 90, 94–97]. They agree well with those found for lateral chain position [91] and columnar mesophases [87]; see Table 4. On the basis of

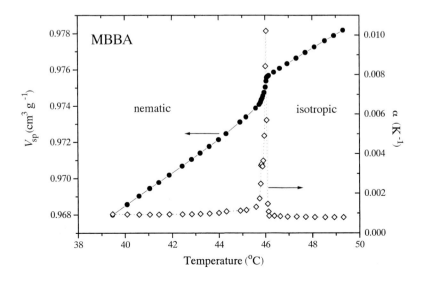

Figure 1. Temperature dependence of specific volume V_{sp}, and thermal expansion coefficient α for MBBA. Adapted from data set 2 of Gulari and Chu [33].

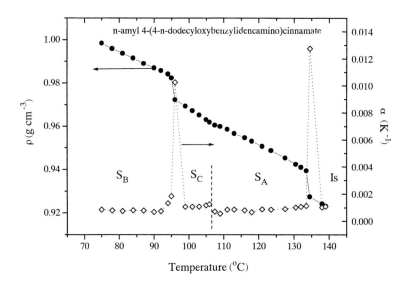

Figure 2. Temperature dependence of density, and thermal expansion coefficient for *n*-amyl 4-(4-*n*-dodecyloxybenzylidene-amino)cinnamate. Adapted from Demus and Rurainski [48].

these increments, conclusions have been drawn about the packing of the alkyl groups, locating the packing density in the nematic phase closely to that in the isotropic phase, with slightly higher values in the nematic phase [60]. An investigation of the influence of aromatic ring number on the packing density in homologous series can be found elsewhere [51, 91].

Table 3. Fractional volume and density changes (in %) in the homologous series *n*O.*m*: Is/N and N/A phase transitions.

$$C_n H_{2n+1} O - \bigcirc - CH=N - \bigcirc - C_m H_{2m+1}$$

m	*n* = 4		*n* = 5	
	$\frac{\Delta V}{V}$	$\frac{\Delta \rho}{\rho}$	$\frac{\Delta \rho}{\rho}$	$\frac{\Delta \rho}{\rho}$
	at Is/N	at N/A	at Is/N	at N/A
4	0.28 [79]		0.22 [88]	
5	0.298 [79]		0.34 [88]	
6	0.22 [79]		0.30 [88]	0.10 [88]
7	0.35 [79]		0.33 [88]	
8	0.31 [79]		0.25 [88]	0.08 [88]
9	0.39 [79]			
10	0.21 [144]	0.02 [88]	0.46 [88]	
12	0.39 [88]	0.02 [88]	0.32 [88]	0.06 [88]

6.2.2.3 Studies of Calamitic Compounds

Density and Order of Phase Transitions

A systematic development of density and thermal expansion coefficients in calamitic compounds is frequently observed when neighboring mesophases are compared. At the N/Is transition, for example, the thermal expansion coefficient was found on average to be larger in the nematic than in the isotropic phase [45–47, 55, 57, 98–104]. This was also confirmed for the N*/Is transition [53, 105] (with a few exceptions [106]). For example in PAA, Price and Wendorff [55] reported 9.4×10^{-4} K^{-1} in the nematic and 8.4×10^{-4} K^{-1} in the isotropic phase. Similar data were given by Bahadur [45] and by Rao and co-workers. These observations were attributed to the rapid development of high order in the nematic phase. It was further suggested that the discontinuities at the N*/Is transition are slightly lower than those at the N/Is transition [107, 108] (compare Tables 1 and 2), and that the thermal expansion coefficient in N* phases is smaller than in N phases [109]. The temperature dependence of the thermal expansion coefficient

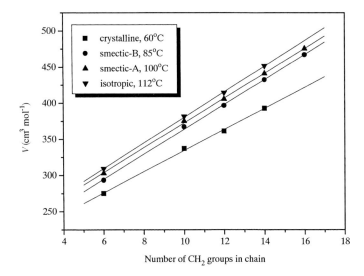

Figure 3. Molar volume development in the crystalline, smectic B, smectic A, and isotropic phases within the homologous series of 4-bromo-*N*-(4-*n*-alkyloxybenzylidene)anilines. Adapted from Seurin et al. [94].

Table 4. Selected molar volume increments for one methylene unit at different molecular positions in the isotropic and nematic phases.

ΔV_{CH_2} (cm^3 mol^{-1})	Substance(s)	Reference
(a) Rod-shaped molecules, longitudinal position		
14.6–18.66 (isotropic phase)	4O.*m*	[79]
17.14±0.44 (isotropic phase, $T_{NI}+3$ K)	*n*O.1	[60]
16.75±0.35 (nematic phase, $T_{NI}-3$ K)	*n*O.1	[60]
(b) Rod-shaped molecules, lateral position		
18.7–22.3	1.4-bis[4-*n*-hexylbenzyloxy]-2-substituted benzenes	[91]
(c) Columnar phases		
16.9±0.3	Copper(II) alkanoates	[87]

in the isotropic state was found to be linear [84, 100, 110], and has been treated by linear regression [19, 92]. In mesophases like the nematic phase, mostly nonlinear behavior is observed [111]. Consequently, some authors [19, 92, 112] have used polynomials to describe the temperature dependence in the nematic and isotropic state outside the pretransitional zone. Regarding other phase transitions, N phases [101, 113] and one blue phase [114] were shown to have higher thermal expansion coefficients than the A phases, or the respective N* phase, into which they transform at lower temperatures.

However, contrary reports are also found [57, 66]. In a similar way, different results have been found for the A/Is transition. For example, OBBA [115] or 7O.8 and 8O.8 [116] were shown to have higher thermal expansion coefficients in the A phase, whereas the opposite was shown for diethyl 4,4′-azoxybenzoate [117]. Further, it should be mentioned that the C phase reportedly has higher thermal expansion coefficients than the adjacent A [66, 68, 79] and B phases, which was attributed to higher flexibility and rotational freedom [101]. However, there are also reports of the opposite case

[118]. Furthermore, in TBDA and TBAA12, it was found that the C phase has higher values than the adjacent I phase, whereas the I phase in turn has higher values than the low-temperature F phase [69, 70] with respect to the thermal expansion coefficient. Also, the F phase was found to have higher values than an adjacent A phase [308]. This was explained by chain stiffening and increasingly closer packing in the smectic layers during cooling. It is noteworthy here that the difficulty of density data acquisition by dilatometric/pycnometric methods due to high viscosity in the smectic phases with higher order in the layer structure is frequently mentioned [79].

Dilatometric investigations of materials with more complex polymorphisms revealed a second-order character for the F/I [69], C/A [68], C/N [37], and N*/blue phase [114] transitions; the last result was confirmed by dilatometric and ultrasonic experiments [119]. Uncertainties about the G/F transition [61, 73, 118, 120, 129] are noted in Table 1. Further results on polymorphous systems have been reported [48, 121–124]. A measurable density change at the cybotactic nematic/ordinary nematic transition [125, 126], occurring at 59.6°C in BBBA is doubtful [46]. Also, densitometric scanning in two mixtures showed a continuous change at the A_1/A_2 transition, and a slight increase in density (less than 0.02%) at the A_d/A_1 transition [127]. Only a slight change in slope of specific volume versus temperature was observed in an experiment scanning the hexatic B/crystal B transition [128].

The order of the A/N transition was the subject of an extended theoretical discussion during the early 1970s [130–135], which has been surveyed with respect to density measurements by Rao et al. [79, 103] and Potukuchi et al. [136]. Extensive density measurements on $nO.m$ compounds [57, 58, 60, 61, 65, 73, 79, 88, 101–104, 113, 115, 136–145, 301] were employed, along with other methods, to test these theories, and to find a predicted A/N transition order change within homologous series. The A/N transition occurs over a remarkable temperature interval, and the possible existence of a tricritical point was the subject of density studies, also on other materials, for example 8CB [146], CBOOA [31, 44], NPOOB [112], nCBs [34], HOPDOP [124], TBBA [66], side-group polysiloxanes [147], H_xBPAA [148], 8OCB, 8S5 [64], and 9CB/10CB blends ([149], successfully using a method described previously [150–153]). Based on the evidence of density measurements and other data, influences of molecular architecture (e.g., the dominant influence of the alkyloxy chains in $nO.m$ compounds [88]) and polarity, formation of cybotactic groups, and mono/bilayer organization in the smectic phase on the A/N transition order were found [60, 65, 66, 73, 79, 146, 154]. Several dilatometric studies have indicated a change from first to second order of the N*/A transition in a mixture of 62 mol% cholesteryl oleyl carbonate and 38 mol% cholesteryl chloride [155, 156], and blends of cholesteryl myristate with cholesteryl propionate [119]. The opposite influence of cholesterol was also studied [157]. Dilatometric testing in cholesteryl myristate revealed an untwisting of the cholesteric helix when the volume discontinuity at the N*/A transition vanishes [36], after high-pressure studies [158].

Further specific volume measurements with chiral mesogens, cholesteric, and blue phases have been made [26–28, 159–161]. The influence of chiral dopants on C/A and C/N phase transitions has been studied [37, 162]. Earlier studies, particularly on cholesteric mesophases and their mixtures, have been published, for example, for cholesteryl acetate [163], cholesteryl myristate [53], cholesteryl nonanoate [54], cholesteryl stearate [56], and a variety of other compounds [27, 106, 164–166]. In particular, Price et al. have published Avrami exponents from kinetic studies of density changes on two compounds, yielding values of 2 for the cholesteric/smectic and isotropic/cholesteric transitions [53, 56].

Pretransition Behavior

Precise measurements have shown that the density is subject to pronounced pretransition behavior for a variety of phase transitions. This may depend upon purity, and also upon the cooling/heating rate [60]. The N/Is and N*/Is transitions exhibit such effects only on the low-temperature side [53–55], which was theoretically explained and compared with enthalpy changes by Chandrasekhar and Sashidar [167, 168]. In contrast to mesophase/mesophase transitions, on the high-temperature side post-transitional behavior was detectable only with high-sensitivity testing [28, 67]. Several workers pointed out that post-transitional effects vanish in accordance with the predictions of the Maier–Saupe theory [3] when the samples are carefully purified [51, 53, 112]. The density jump at this specific phase transition was shown to occur in a temperature interval of 0.1–0.5 K, depending upon the purity of the sample [19, 92]. In the special case of ultra-high purity materials, Dunmur and Miller [34] and Demus et al. [112] found interval widths of less than 0.1 and 0.02 K, respectively. Evidence of the influence of domain size on the discontinuity at the transition was found [44]. Pretransitional effects pose a main problem for the exact evaluation of changes in molar volume, specific volume, and density at phase transitions. A precise method was devised for dilatometric and calorimetric studies on PAA by Klement and Cohen [169]. It was further developed by Dunmur and Miller [34] for *n*CBs, and used in other work

[57, 170]. The method uses plots of $1/\rho$ versus $|T - T_t|^{0.5}$, and linearly extrapolates to $T = T_t$ (where T_t is the phase transition temperature) for the low- and high-temperature sides. Other functional forms for the modelling of density in the nematic/isotropic pretransitional region have been given [31, 79, 171–173]. Based on the Landau–de Gennes theory [174], the expansion $\Delta V = B\,\Delta T + E\,[(\Delta T + D)^{1-\alpha} - D^{1-\alpha}]$, in terms of the critical exponent α, was applied for CBOOA [31], MBBA and PAA [29], and several cholesterol derivatives [28], and used to describe pretransition behavior for other transitions (e.g., N/C in HOAOB [30]). The influence of the order parameter was taken into account by an expansion given by Chang [172]. Bendler [175, 176], using high-precision data on MBBA [33], attempted to model expansion coefficients in isotropic phases very close to the N/Is transition in terms of order parameter fluctuations within the framework of the de Gennes theory. This was extended to the pretransitional behavior on the nematic side, and applied to nematic mixtures [177] with good coincidence between theory and experiment. Note finally that the equation given by Chang and Gyspers [178], $\alpha_N = [e + f(T_{NI} - T)^h] \times 10^{-3}$, enjoys widespread popularity for data evaluation in the nematic state.

Molecular Interaction, Order Parameter, and Density

The influence of attractive and repulsive interactions between molecules on volumetric behavior, packing density, and the resulting phase stability have been studied by several authors [19, 20, 35, 51, 91–93, 111]. In mixtures, strong attractive forces due to dipole–dipole interactions or EDA complex formation [179–183] result in the formation of induced phases [184–186]. Sadowska et al. [35] reported results of density measurements for two blends, indicating a measurable shrinking effect. EDA complex formation was investigated by Demus et al. [20], who obtained complex stability constants, complex formation enthalpies, and complex concentrations as temperature functions by dilatometry and calorimetry. An interpretation was given using the theories of Cotter and Gelbart [187, 188]. Further density investigations on blends with nematic reentrant phases [189, 190] and induced smectic phases [191], were reported, and related to anomalies in pressure–temperature diagrams [192–194]. Aleksandriiskii and coworkers [111] assessed the influence of hydrogen bonds on phase transition characteristics. Enhanced values of fractional volume changes at the N/Is transition, compared with those for other polar compounds, were found, which is consistent with observations by Bahadur [16]. Obviously, hydrogen bond formation and also polar interactions [195] favor the formation of cybotactic groups, and lead to the nonlinear behavior that is also observed. However, differing results were found in a systematic study of longitudinally polar and nonpolar substituted liquid crystal blend systems [196], using a method described by Press and Arrott [32]. The dominant role of steric repulsion in the stabilization of the nematic phase was suggested on the basis of high-pressure experiments on PAA, BMAB, and EBBA by Stishov et al. [197] and Kachinskii et al. [198], and demonstrated by density studies [19, 91–93] under atmospheric pressure on laterally branched compounds [199]. The results are in qualitative agreement with molecular statistical theories based on hard rod models [187, 200, 201] and with van der Waals theories [202]. There has been discussion of repulsive forces [203–205].

Density studies help to assess the influence of molecular interaction, cybotactic groups, and the order parameter in nematic phases, in combination with refractive index measurements and other methods for pure substances [51, 76, 206–212] and binary blends [213–216]. This has been demonstrated, using different theoretical concepts for the inner field [217–222], also for smectic phases [59, 223, 224]. There have been additional density studies of binary systems [81, 170, 225–228].

Relation between Density and
Other Properties in Calamitic Compounds

Combination with calorimetric data, and use of the Clausius–Clapeyron equation or the Ehrenfest equations (e.g., for the A/C transition [67]), yields slopes in the p–T phase diagram. These can be verified by high-pressure measurements of transition temperatures (e.g. [229], compared with density data [70]) or metabolemetric scans [230] (see Sec. 6.2.3 of this Chapter). On the other hand, direct measurements of densities or volumes at high pressures were also obtained [39–42, 158, 198, 231–240], yielding isothermal compressibilities and permitting conclusions about phase transition mechanisms. For example, using temperature as a parameter, Dörrer et al. [41] combined data gained from two blends with high-pressure data of the rotational viscosity γ_1, and constructed master curves, introducing reduced density coordinates.

After initial tests with N-(4-n-nonyloxy-benzylidene)-4-toluidine [241], it was demonstrated in a variety of work on azomethine compounds of the nO.m type [58, 98, 100, 103, 104, 116, 137, 140–142, 148, 242–244] and other compounds [119, 148, 160, 177] that, with the ultrasonic velocity v, temperature T, and molar weight M, parameters like adiabatic compressibility $\beta_{ad} = \dfrac{V_{sp}}{v^2}$,

Rao number (molar sound velocity) [245]

$R_n = M V_{sp} v^{1/3}$, and Wada constant (molar compressibility) [246] $B = \dfrac{M V_{sp}}{\beta_{ad}^{1/7}}$ can be calculated from the specific volume V_{sp} and compared with theoretical concepts [3, 132, 247, 248].

6.2.2.4 Discotics

Few density measurements on discotic phases have been reported. Data have been given for the Cr/D_{ho} and D_{ho}/Is transitions [249] and the D_{hd}/Is [18], D_{rd}/Is [21], and Cr/columnar [87] transitions. The last reference mentions data in older publications [250, 251]. Similar fractional volume changes and pretransitional effects, but higher transition enthalpies and entropies, compared with the nematic/isotropic transition, were found [18] – in some disagreement with theory [252, 253]. The transition from the high-viscosity D_{rd} phase into the isotropic phase was reported to show an increase in density [21]. The density within homologous series decreases with increasing alkyl chain length [87]. Further data are available [254].

6.2.2.5 Lyotropics and Anhydrous Soaps

Density behavior at phase transitions in lyotropic or micellar mesophases [110, 255–257] are found to be less pronounced than in thermotropic mesophases (for example, there is no discontinuity at the N/Is transition in DACl/NH$_4$Cl/H$_2$O [257], and there is a 0.012% volume change at the N/Is transition in CsPFO/H$_2$O [110]). Several publications [38, 110, 258–260] have discussed the system cesium pentadecafluorooctanoate (CsPFO)/water, where decreases in volume were found at the lamellar/nematic and nematic/isotropic transitions [258, 259]. This was interpreted in terms of significant changes in the micellar structure during the transitions. Several explanations

for this observation were attempted [38, 260]. Effects of additional electrolytes on the density in these systems have been discussed [110]. The older literature also refers to density studies in anhydrous surfactants [261–270], for example, in alkali metal stearates, laurates, and palmitates. Finally, we mention reports of density measurements on mesophasic systems involving liposomes and membranes [271, 272].

6.2.2.6 Polymeric Liquid Crystals

Density measurements in liquid crystalline polymeric systems have rarely been reported. Density scans of A/N transitions in liquid crystalline side-chain polysiloxanes [147] and of N/Is transitions in side-chain polysiloxanes and polymethacrylates [273] have been published. Increased densities and lower expansion coefficients compared with corresponding low molecular weight model systems were found, and a van der Waals analysis has been given [273]. One main-chain polybenzoate was tested in solution [274].

6.2.2.7 Further Studies

Density and its changes at phase transitions have been the subject of molecular theories [275–280]. It is also an important parameter in molecular dynamics (MD) studies. Two examples are an MD study of the density dependence of diffusion coefficients in nematic phases [281] and simulations of the density dependence of orientational order in nematics using a Gay–Berne potential [282]. Further explanations and references are given in Chap. III.

Investigations of the density and expansion coefficients in mesogens, in connection with data from other experiments, and especially in the transition regions have been described [61, 82, 108, 129, 234, 235, 283–309].

6.2.2.8   References

[1] F. R. Schenck, (a) Z. Phys. Chem. **1898**, *25*, 337–352; (b) Z. Phys. Chem. **1898**, *27*, 167–171; (c) Z. Phys. Chem. **1899**, *28*, 280–288; *Royal Society of London, Catalogue of Scientific Papers* **1923**, *18*, 503.

[2] F. R. Schenck, *Kristallinische Flüssigkeiten und Flüssige Kristalle*, Engelmann, Leipzig **1905**.

[3] W. Maier, A. Saupe, (a) Z. Naturforsch. **1958**, *13a*, 564–566; (b) Z. Naturforsch. **1959**, *14a*, 882–889; (c) Z. Naturforsch. **1960**, *15a*, 287–292.

[4] E. Eichwald, Untersuchungen über flüssige Kristalle, Dissertation, Marburg **1905**.

[5] F. Dickenschied, Untersuchungen über Dichte, Reibung, Kapillarität kristalliner Flüssigkeiten, Dissertation, Halle (Saale) **1908**.

[6] F. Conrat, *Phys. Z.* **1909**, *10*, 202 [*Chem. Abstr.* **1909**, *3*, 1609].

[7] F. Schäfer, Dichte, Reibung und Kapillarität kristallinischer Flüssigkeiten, Dissertation, Halle (Saale) **1911**.

[8] F. M. Jaeger, Z. Anorg. Chem. **1917**, *101*, 152 [*Chem. Abstr.* **1918**, *12*, 875].

[9] D. Vorländer, *Phys. Z.* **1930**, *31*, 428–435 [*Chem. Abstr.* **1930**, *24*, 4433].

[10] E. Bauer, J. Bernamont, *J. Phys. Radium* **1936**, *7*, 19–22 [*Chem. Abstr.* **1936**, *30*, 5088^1].

[11] G. Becherer, W. Kast, *Ann. Physik (5. Folge)* **1942**, *41*, 355–374 [*Chem. Abstr.* **1942**, *36*, 6390^5].

[12] F. Linsert, Thesis, Halle (Saale) **1945**.

[13] I. Gabrielli, L. Verdini, *Nuovo Cim. (Sr. 10)* **1955**, *2*, 526–541 [*Chem. Abstr.* **1956**, *50*, 6122b].

[14] W. A. Hoyer, A. W. Nolle, *J. Chem. Phys.* **1956**, *24*, 803–811.

[15] W. R. Runyan, A. W. Nolle, *J. Chem. Phys.* **1957**, *27*, 1081–1087.

[16] B. Bahadur, *J. Chim. Phys., Phys.-Chim. Biol.* **1976**, *73*, 255–267 [*Chem. Abstr.* **1976**, *85*, 54670s].

[17] H. Sackmann, F. Sauerwald, Z. Phys. Chem. **1950**, *195*, 295–312 [*Chem. Abstr.* **1951**, *45*, 8314d].

[18] T. H. Smith, G. R. van Hecke, Proceedings of the 8th International Liquid Crystal Conference, Kyoto, Part D, *Mol. Cryst. Liq. Cryst.* **1981**, *68*, 23–28.

[19] A. Hauser, R. Rettig, C. Selbmann, W. Weissflog, J. Wulf, D. Demus, *Cryst. Res. Technol.* **1984**, *19*, 261–270.

[20] D. Demus, A. Hauser, G. Pelzl, U. Böttger, S. Schönburg, *Cryst. Res. Technol.* **1985**, *20*, 381–390.

[21] K. A. Lawler, G. R. Van Hecke, *Liq. Cryst.* **1991**, *10*, 341–346.

[22] ASTM Standard D 1481-93, Standard Test Method for Density and Relative Density (Specific Gravity) of Viscous Materials by Lipkin Bicapillary Pycnometer, *Annual Book of ASTM Standards* **1993**, Vol. 05.01, pp. 521–525.

[23] D. Guillon, A. Skoulios, *Mol. Cryst. Liq. Cryst.* **1977**, *39*, 139–157.

[24] D. Grasso, *J. Calorim., Anal. Therm. Thermodyn. Chim.* **1986**, *17*, 318–320 [*Chem. Abstr.* **1987**, *107*, 49926y].

[25] D. Grasso, *Liq. Cryst.* **1987**, *2*, 557–560.

[26] D. Armitage, F. P. Price, *Bull. Am. Phys. Soc.* **1975**, *20*, 886.

[27] D. Armitage, F. P. Price, *J. Appl. Phys.* **1976**, *47*, 2735–2739.

[28] D. Armitage, F. P. Price, *J. Chem. Phys.* **1977**, *66*, 3414–3417.

[29] D. Armitage, F. P. Price, *Phys. Rev. A* **1977**, *15*, 2496–2500.

[30] D. Armitage, F. P. Price, *Phys. Rev. A* **1977**, *15*, 2069–2071.

[31] D. Armitage, F. P. Price, *Mol. Cryst. Liq. Cryst.* **1977**, *38*, 229–237.

[32] M. J. Press, A. S. Arrott, *Phys. Rev. A* **1973**, *8*, 1459–1465.

[33] E. Gulari, B. Chu, *J. Chem. Phys.* **1975**, *62*, 795–797.

[34] D. A. Dunmur, W. H. Miller, *J. Phys. (Paris) Colloque C3* **1979**, *40*, 141–146.

[35] K. W. Sadowska, A. Zywocinski, J. Stecki, R. Dabrowski, *J. Phys.* **1982**, *43*, 1673–1678.

[36] P. Pollmann, K. Schulte, *Ber. Bunsenges. Phys. Chem.* **1985**, *89*, 780–786.

[37] R. Kiefer, G. Baur, *Liq. Cryst.* **1990**, *7*, 815–837.

[38] N. Boden, K. W. Jolley, *Phys. Rev. A* **1992**, *45*, 8751–8758.

[39] E. Kuss, *Mol. Cryst. Liq. Cryst.* **1978**, *47*, 71–83.

[40] E. Kuss, *Mol. Cryst. Liq. Cryst.* **1981**, *76*, 199–210.

[41] H. L. Dörrer, H. Kneppe, F. Schneider, *Liq. Cryst.* **1992**, *11*, 905–915.

[42] Y. B. Kim, K. Ogino, *Phys. Lett. A* **1977**, *61*, 40–42.

[43] T. Stoebe, C. C. Huang, J. W. Goodby, *Phys. Rev. Lett.* **1992**, *68*, 2944–2947.

[44] S. Torza, P. E. Cladis, *Phys. Rev. Lett.* **1974**, *32*, 1406–1409.

[45] B. Bahadur, *Mol. Cryst. Liq. Cryst.* **1976**, *35*, 83–89.

[46] J. V. Rao, N. V. S. Rao, V. G. K. M. Pisipati, C. R. K. Murty, *Ber. Bunsenges. Phys. Chem.* **1980**, *84*, 1157–1160.

[47] J. V. Rao, K. R. K. Rao, L. V. Choudary, P. Venkatacharyulu, *Cryst. Res. Technol.* **1986**, *21*, 1245–1249.

[48] D. Demus, R. Rurainski, Z. Phys. Chem. (Leipzig) **1973**, *253*, 53–67.

[49] D. Demus, S. Diele, S. Grande, H. Sackmann in *Advances in Liquid Crystals* Vol. 6 (Ed.: G. H. Brown), Academic Press, New York **1983**, pp. 1–107.

[50] G. Poeti, E. Fanelli, G. Torquati, D. Guillon, *Nuovo Cim. D* **1983**, *2*, 1335–1346.

[51] A. Hauser, C. Selbmann, R. Rettig, D. Demus, *Cryst. Res. Technol.* **1986**, *21*, 685–695.

[52] E. McLaughlin, M. A. Shakespeare, A. R. Ubbelohde, *Trans. Faraday Soc.* **1964**, *60*, 25–32.

[53] F. P. Price, J. H. Wendorff, *J. Phys. Chem.* **1971**, *75*, 2839–2849.

[54] F. P. Price, J. H. Wendorff, *J. Phys. Chem.* **1972**, *76*, 276–280.

[55] F. P. Price, J. H. Wendorff, *J. Phys. Chem.* **1972**, *76*, 2605–2608.

[56] F. P. Price, J. H. Wendorff, *J. Phys. Chem.* **1973**, *77*, 2342–2346.

[57] V. G. K. M. Pisipati, N. V. S. Rao, M. V. V. N. Reddy, C. G. Rama Rao, G. Padmavathi, *Cryst. Res. Technol.* **1991**, *26*, 709–716.

[58] V. G. K. M. Pisipati, N. V. S. Rao, M. K. Rao, D. M. Potukuchi, P. R. Alapati, Proceedings of the 11th International Liquid Crystal Conference, Berkeley, CA 1986, Part C, *Mol. Cryst. Liq. Cryst.* **1987**, *146*, 89–96.

[59] R. K. Sarna, V. G. Bhide, B. Bahadur, *Mol. Cryst. Liq. Cryst.* **1982**, *88*, 65–79.

[60] N. V. S. Rao, P. V. Datta Prasad, V. G. K. M. Pisipati, *Mol. Cryst. Liq. Cryst.* **1985**, *126*, 175–186.

[61] M. Takahashi, S. Kondo, *Tokyo Kogyo Koto Senmon Gakko Kenkyu Hokokusho (Research Report of Tokyo Technical College)* **1987**, *19*, 19–23 [*Chem. Abstr.* **1988**, *109*, 220081m].

[62] L. Longa, *J. Chem. Phys.* **1986**, *85*, 2974–2985.

[63] L. Longa, *Z. Phys. B* **1986**, *64*, 357–361.

[64] A. Zywocinski, S. A. Wieczorek, J. Stecki, *Phys. Rev. A* **1987**, *36*, 1901–1907.

[65] P. R. Alapati, D. M. Potukuchi, N. V. S. Rao, V. G. K. M. Pisipati, A. S. Paranjpe, U. R. K. Rao, *Liq. Cryst.* **1988**, *3*, 1461–1479.

[66] N. V. S. Rao, V. G. K. M. Pisipati, *Mol. Cryst. Liq. Cryst.* **1984**, *104*, 301–306.

[67] A. Zywocinski, S. A. Wieczorek, *Phys. Rev. A* **1985**, *31*, 479–482.

[68] A. Zywocinski, S. A. Wieczorek, *Mol. Cryst. Liq. Cryst.* **1987**, *151*, 399–410.

[69] P. R. Alapati, D. M. Potukuchi, N. V. S. Rao, V. G. K. M. Pisipati, D. Saran, Proceedings of the 11th International Liquid Crystal Conference, Berkeley, CA, 1986, Part C, *Mol. Cryst. Liq. Cryst.* **1987**, *146*, 111–119.

[70] N. V. S. Rao, V. G. K. M. Pisipati, P. R. Alapati, D. M. Potukuchi, *Mol. Cryst. Liq. Cryst.* **1988**, *162B*, 119–125.

[71] J. Stecki, A. Zywocinski, S. A. Wieczorek, *Phys. Rev. A* **1983**, *28*, 434–439.

[72] A. Beguin, J. Billard, F. Bonamy, J. M. Buisine, P. Cuvelier, J. C. Dubois, P. Le Barny, Sources of Thermodynamic Data on Mesogens, *Mol. Cryst. Liq. Cryst.* **1984**, *115*, 1–326.

[73] V. G. K. M. Pisipati, N. V. S. Rao, A. Alapati, *Cryst. Res. Technol.* **1989**, *24*, 1285–1290.

[74] A. L. Tsykalo, *Thermophysical Properties of Liquid Crystals*, Gordon and Breach, New York **1991**.

[75] P. Adomenas, V. A. Grozhik, *Vestsi Akad. Navuk BSSR, Ser. Khim. Navuk* **1977**, *2*, 39–42 [*Chem. Abstr.* **1977**, *87*, 29178w].

[76] W. H. de Jeu, W. A. P. Claasen, *J. Chem. Phys.* **1978**, *68*, 102–108.

[77] M. Takahashi, S. Mita, S. Kondo, *Mol. Cryst. Liq. Cryst.* **1987**, *147*, 99–105.

[78] M. Takahashi, S. Mita, S. Kondo, *Phase Transitions* **1987**, *9*, 1–9.

[79] N. V. S. Rao, D. M. Potukuchi, V. G. K. M. Pisipati, *Mol. Cryst. Liq. Cryst.* **1991**, *196*, 71–87.

[80] O. N. Puchkov, V. A. Molochko, *Zh. Prikl. Khim. (S.-Peterburg)* **1992**, *65*, 825–829 [*Chem. Abstr.* **1993**, *118*, 212851h].

[81] W. Labno, J. Jadzyn, *Pr. Kom. Mat.-Przyr., Poznan. Tow. Przyj. Nauk, Fiz. Dielektr. Radiospektrosk.* **1981**, *12*, 75–84 [*Chem. Abstr.* **1983**, *98*, 99290c].

[82] R. A. Orwoll, V. J. Sullivan, G. C. Campbell, *Mol. Cryst. Liq. Cryst.* **1987**, *149*, 121–140.

[83] N. A. Nedostup, V. V. Gal'tsev, (a) *Russ. J. Phys. Chem.* **1977**, *51*, 121–122; (b) *Zh. Fiz. Khim.* **1977**, *51*, 214–216.

[84] R. Dabrowski, K. Kenig, Z. Raszewski, J. Kedzierski, K. Sadowska, *Mol. Cryst. Liq. Cryst.* **1980**, *61*, 61–78.

[85] G. Albertini, E. Fanelli, D. Guillon, S. Melone, G. Poeti, F. Rustichelli, G. Torquati, *J. Chem. Phys.* **1983**, *78*, 2013–2016.

[86] G. Albertini, E. Fanelli, D. Guillon, S. Melone, G. Poeti, F. Rustichelli, G. Torquati, *J. Phys.* **1984**, *45*, 341–346.

[87] H. Abied, D. Guillon, A. Skoulios, A. M. Giroud-Godquin, P. Maldivi, J. C. Marchon, *Colloid Polym. Sci.* **1988**, *266*, 579–582.

[88] V. G. K. M. Pisipati, N. V. S. Rao, D. M. Potukuchi, P. R. Alapati, P. B. Rao, *Mol. Cryst. Liq. Cryst.* **1989**, *167*, 167–171.

[89] A. Wulf, A. G. De Rocco, *J. Chem. Phys.* **1971**, *55*, 12–27.

[90] I. Haller, H. A. Huggins, H. R. Lilienthal, T. R. McGuire, *J. Phys. Chem.* **1973**, *77*, 950–954.

[91] D. Demus, A. Hauser, C. Selbmann, W. Weissflog, *Cryst. Res. Technol.* **1984**, *19*, 271–283.

[92] D. Demus, A. Hauser, A. Isenberg, M. Pohl, C. Selbmann, W. Weissflog, S. Wieczorek, *Cryst. Res. Technol.* **1985**, *20*, 1413–1421.

[93] D. Demus, S. Diele, A. Hauser, I. Latif, C. Selbmann, W. Weissflog, *Cryst. Res. Technol.* **1985**, *20*, 1547–1558.

[94] P. Seurin, D. Guillon, A. Skoulios, *Mol. Cryst. Liq. Cryst.* **1981**, *65*, 85–110.

[95] D. Guillon, A. Skoulios, *J. Phys. (Paris)* **1976**, *37*, 797–800.

[96] D. Guillon, A. Skoulios, *J. Phys. (Paris) C3* **1976**, *37*, 83–84.

[97] J. H. Ibrahim, W. Haase, *Z. Naturforsch. A* **1976**, *31*, 1644–1650.

[98] B. Bahadur, S. Chandra, *J. Phys. C* **1976**, *9*, 5–9.

[99] M. N. Rao, V. G. K. M. Pisipati, N. V. S. Rao, J. V. Rao, C. R. K. Murty, *Phase Transitions* **1981**, *2*, 231–238.

[100] K. R. K. Rao, J. V. Rao, L. V. Choudary, C. R. K. Murty, *Acta Phys. Pol. A* **1983**, *63*, 419–424 [*Chem. Abstr.* **1983**, *98*, 189509k].

[101] N. V. S. Rao, V. G. K. M. Pisipati, *Phase Transitions* **1983**, *3*, 317–327.

[102] N. V. S. Rao, V. G. K. M. Pisipati, P. V. Datta Prasad, P. R. Alapati, *Phase Transitions* **1985**, *5*, 187–195.

[103] K. R. K. Rao, J. V. Rao, P. Venkatacharyulu, *Acta Phys. Pol. A* **1986**, *69*, 261–265 [*Chem. Abstr.* **1984**, *104*, 159999v].

[104] J. V. Rao, L. V. Choudary, K. R. K. Rao, P. Venkatacharyulu, *Acta Phys. Pol. A* **1987**, *72*, 517–522 [*Chem. Abstr.* **1988**, *108*, 66511x].

[105] N. M. Sakevich, (a) *Zh. Fiz. Khim.* **1968**, *42*, 2930; (b) *Russ. J. Phys. Chem.* **1968**, *42*, 1555–1557.

[106] L. E. Hajdo, A. C. Eringen, J. Giancola, A. E. Lord Jr., *Lett. Appl. Eng. Sci.* **1975**, *3*, 61 [*Chem. Abstr.* **1975**, *83*, 69402p].

[107] A. E. Lord Jr., F. E. Wargocki, L. E. Hajdo, A. C. Eringen, *Lett. Appl. Eng. Sci.* **1975**, *3*, 125–132 [*Chem. Abstr.* **1975**, *83*, 88976v].

[108] R. S. Porter, J. F. Johnson, *J. Appl. Phys.* **1963**, *34*, 55–59.

[109] T. Matsumoto, *Kogakuin Daigaku Kenkyu Hokoku (Research Reports of the Kogakuin University)* **1974**, *36*, 1–8 [*Chem. Abstr.* **1975**, *83*, 171116k].

[110] S. Plumley, M. R. Kuzma, *Mol. Cryst. Liq. Cryst.* **1991**, *200*, 33–41.

[111] V. V. Aleksandriiski, V. A. Burmistrov, O. I. Koifman, (a) *Zh. Fiz. Khim.* **1993**, *67*, 1623–1625; (b) *Russ. J. Phys. Chem.* **1993**, *67*, 1456–1458.

[112] D. Demus, H.-J. Deutscher, S. König, H. Kresse, F. Kuschel, G. Pelzl, H. Schubert, C. Selbmann, W. Weissflog, A. Wiegeleben, J. Wulf, *Wiss. Beitr. Martin-Luther-Univ. Halle-Wittenberg* **1978**, *21*, 9–20 [*Chem. Abstr.* **1978**, *89*, 189209b].

[113] N. V. S. Rao, V. G. K. M. Pisipati, *J. Phys. Chem.* **1983**, *87*, 899–902.

[114] D. Demus, H. G. Hahn, F. Kuschel, *Mol. Cryst. Liq. Cryst.* **1978**, *44*, 61–70.

[115] V. G. K. M. Pisipati, N. V. S. Rao, *Phase Transitions* **1983**, *3*, 169–175.

[116] V. G. K. M. Pisipati, N. V. S. Rao, Y. Gouri Sankar, J. S. R. Murty, *Acustica* **1986**, *60*, 163–168.

[117] L. E. Hajdo, A. C. Eringen, A. E. Lord Jr., *Lett. Appl. Eng. Sci.* **1974**, *2*, 367–371 [*Chem. Abstr.* **1975**, *83*, 19359j].

[118] N. V. S. Rao, V. G. K. M. Pisipati, Y. Gouri Sankar, *Mol. Cryst. Liq. Cryst.* **1985**, *131*, 237–243.

[119] A. K. George, A. R. K. L. Padmini, *Mol. Cryst. Liq. Cryst.* **1981**, *65*, 217–226.

[120] Y. Thiriet, J. A. Schulz, P. Martinoty, D. Guillon, *J. Phys.* **1984**, *45*, 323–329.

[121] D. Demus, S. Diele, M. Klapperstück, V. Link, H. Zaschke, *Mol. Cryst. Liq. Cryst.* **1971**, *15*, 161–174.

[122] D. Demus, R. Rurainski, *Mol. Cryst. Liq. Cryst.* **1972**, *16*, 171–174.

[123] D. Demus, H. König, D. Marzotko, R. Rurainski, *Mol. Cryst. Liq. Cryst.* **1973**, *23*, 207–214.

[124] D. Demus, M. Pohl, S. Schönberg, L. Weber, A. Wiegeleben, W. Weissflog, *Wiss. Beitr. Martin-Luther-Univ. Halle-Wittenberg* **1983**, *41* (Forsch. Flüss. Krist.), 18–28 [*Chem. Abstr.* **1984**, *101*, 178680p].

[125] A. De Vries, *Mol. Cryst. Liq. Cryst.* **1970**, *10*, 31–37.

[126] A. De Vries, *Liquid Crystals: Proceedings of International Conference 1973, Pramana Suppl.* **1975**, *1*, 93–113 [*Chem. Abstr.* **1976**, *84*, 114666m].

[127] B. R. Ratna, C. Nagabhushana, V. N. Raja, R. Shashidar, G. Heppke, *Mol. Cryst. Liq. Cryst.* **1986**, *138*, 245–257.

[128] G. Poeti, E. Fanelli, D. Guillon, *Mol. Cryst. Liq. Cryst. Lett.* **1982**, *82*, 107–114.

[129] P. Bhaskara Rao, N. V. S. Rao, V. G. K. M. Pisipati, D. Saran, *Cryst. Res. Technol.* **1989**, *24*, 723–731.

[130] K. Kobayashi, *Phys. Lett. A* **1970**, *31*, 125–126.

[131] K. K. Kobayashi, *J. Phys. Soc. Jpn.* **1970**, *29*, 101–105 [*Phys. Abstr.* **1970**, *73*, 55591].

[132] W. L. McMillan, (a) *Phys. Rev. A* **1971**, *4*, 1238–1246; (b) *Phys. Rev. A* **1972**, *6*, 936–947; (c) *Phys. Rev. A* **1973**, *7*, 1419–1422.

[133] P. G. de Gennes, *Solid State Commun.* **1972**, *10*, 753–756.

[134] R. Alben, *Solid State Commun.* **1972**, *13*, 1783–1785.

[135] B. I. Halperin, T. C. Lubensky, S. K. Ma, *Phys. Rev. Lett.* **1974**, *32*, 292–295.

[136] D. M. Potukuchi, K. Prabhakar, N. V. S. Rao, V. G. K. M. Pisipati, D. Saran, *Mol. Cryst. Liq. Cryst.* **1989**, *167*, 181–189.

[137] L. V. Choudary, J. V. Rao, P. N. Murty, C. R. K. Murty, *Z. Naturforsch. A* **1983**, *38*, 762–764.

[138] N. V. S. Rao, S. M. Rao, V. G. K. M. Pisipati, *Phase Transitions* **1983**, *3*, 159–167.

[139] V. G. K. M. Pisipati, N. V. S. Rao, *Phase Transitions* **1984**, *4*, 91–96.

[140] V. G. K. M. Pisipati, N. V. S. Rao, P. V. Datta Prasad, P. R. Alapati, *Z. Naturforsch. A* **1985**, *40*, 472–475.

[141] A. K. Jaiswal, G. L. Patel, *Acta Phys. Pol. A* **1986**, *69*, 723–726 [*Chem. Abstr.* **1986**, *104*, 234839y].

[142] K. R. K. Rao, J. V. Rao, P. Venkatacharyulu, V. Baliah, *Acta Phys. Pol. A* **1986**, *70*, 541–547 [*Chem. Abstr.* **1987**, *106*, 59321r].

[143] N. V. S. Rao, V. G. K. M. Pisipati, Y. Gouri Sankar, D. M. Potukuchi, *Phase Transitions* **1986**, *7*, 49–57.

[144] N. V. S. Rao, D. M. Potukuchi, P. B. Rao, V. G. K. M. Pisipati, *Cryst. Res. Technol.* **1989**, *24*, 219–225.

[145] S. Lakshminarayana, C. R. Prabhu, D. M. Potukuchi, N. V. S. Rao, V. G. K. M. Pisipati, *Liq. Cryst.* **1993**, *15*, 909–914.

[146] A. J. Leadbetter, J. L. A. Durrant, M. Ruyman, *Mol. Cryst. Liq. Cryst. Lett.* **1977**, *34*, 231–235.

[147] M. F. Achard, F. Hardouin, G. Sigaud, M. Mauzac, *Liq. Cryst.* **1986**, *1*, 203–207.

[148] K. R. K. Rao, J. V. Rao, P. Venkatacharyulu, V. Baliah, *Mol. Cryst. Liq. Cryst.* **1986**, *136*, 307–316.

[149] V. N. Raja, S. Krishna Prasad, D. S. Shankar Rao, S. Chandrasekhar, *Liq. Cryst.* **1992**, *12*, 239–243.

[150] C. W. Garland, G. B. Kasting, K. J. Lushington, *Phys. Rev. Lett.* **1979**, *43*, 1420–1423.

[151] C. Rosenblatt, J. T. Ho, *Phys. Rev. A* **1982**, *26*, 2293–2296.

[152] H. Marynissen, J. Thoen, W. Van Dael, *Mol. Cryst. Liq. Cryst.* **1983**, *97*, 149–161.

[153] T. Pitchford, G. Nounesis, S. Dumrongrattana, J. M. Viner, C. C. Huang, J. W. Goodby, *Phys. Rev. A* **1985**, *32*, 1938–1940.

[154] V. G. K. M. Pisipati, S. B. Rananavare, *Liq. Cryst.* **1993**, *13*, 757–764.

[155] W. U. Müller, H. Stegemeyer, *Chem. Phys. Lett.* **1974**, *27*, 130–132.

[156] W. U. Müller, H. Stegemeyer, *Ber. Bunsenges. Phys. Chem.* **1974**, *78*, 880–883.

[157] H. Stegemeyer, W. U. Müller, *Naturwissenschaften* **1976**, *63*, 388.

[158] V. K. Semenchenko, V. M. Byankin, V. Yu. Baskakov, (a) *Sov. Phys. Crystallogr.* **1975**, *20*, 111–113; (b) *Kristallografija* **1975**, *20*, 187–191.

[159] A. W. Neumann, L. J. Klementowski, R. W. Springer, *J. Colloid Interface Sci.* **1972**, *41*, 538–541.

[160] J. R. Otia, A. R. K. L. Padmini, *Mol. Cryst. Liq. Cryst.* **1976**, *36*, 25–39.

[161] M. Nakahara, Y. Yoshimura, J. Osugi, *Nippon Kogakukai (Bull. Chem. Soc. Jpn.)* **1981**, *54*, 99–102 [*Chem. Abstr.* **1981**, *94*, 74987q].

[162] D. M. Potukuchi, N. V. S. Rao, V. G. K. M. Pisipati, *Ferroelectrics* **1993**, *141*, 287–296.

[163] F. P. Price, J. H. Wendorff, *J. Phys. Chem.* **1971**, *75*, 2849–2853.

[164] S. N. Mochalin, P. P. Pugachevich, *Uch. Zap., Ivanov. Gos. Pedagog. Inst.* **1972**, *99*, 200–207. From: Ref. Zh., Fiz., E. **1972**, Abstr. No. 10E138 [*Chem. Abstr.* **1973**, *78*, 152444w].

[165] S. N. Mochalin, P. P. Pugachevich in *Sb. Dokladov II. nauchnoi konferentsii po zhidkim kristallam (Collection of Contributions to the 2nd Conference on Liquid Crystal Science)*, Ivanovo **1972**, p. 200.

[166] S. N. Mochalin, *Uch. zap. Ivanov., un-t* **1974**, *128*, 86–89. *Ref. Zh., Khim.* **1975**, Abstr. No. 413795, only title translated [*Chem. Abstr.* **1975**, *83*, 152689h].

[167] S. Chandrasekhar, R. Shashidar, N. Tara, *Mol. Cryst. Liq. Cryst.* **1971**, *12*, 245–250.

[168] S. Chandrasekhar, R. Sashidar, *Mol. Cryst. Liq. Cryst.* **1972**, *16*, 21–32.

[169] W. Klement, L. H. Cohen, *Mol. Cryst. Liq. Cryst.* **1974**, *27*, 359–373.

[170] G. A. Oweimreen, A. K. Shihab, K. Halhouli, S. F. Sikander, *Mol. Cryst. Liq. Cryst.* **1986**, *138*, 327–338.

[171] H. Imura, K. Okano, *Chem. Phys. Lett.* **1972**, *17*, 111–113.

[172] R. Chang, *Solid State Commun.* **1974**, *14*, 403–406.

[173] G. R. Van Hecke, J. Stecki, *Phys. Rev. A* **1982**, *25*, 1123–1126.

[174] P. G. de Gennes, *The Physics of Liquid Crystals*, Clarendon, Oxford **1974**.

[175] J. T. Bendler, Theory of pretransitional effects in the i phase of liquid crystals, Dissertation Yale University **1974**.

[176] J. Bendler, *Mol. Cryst. Liq. Cryst.* **1977**, *38*, 19–30.

[177] A. K. George, R. A. Vora, A. R. K. L. Padmini, *Mol. Cryst. Liq. Cryst.* **1980**, *60*, 297–310.

[178] R. Chang, J. C. Gyspers, *J. Phys. (Paris) C1* **1975**, *36*, 147–149.

[179] G. Pelzl, D. Demus, H. Sackmann, *Z. Phys. Chem.* **1968**, *238*, 22–32.

[180] J. W. Park, C. S. Bak, M. M. Labes, *J. Am. Chem. Soc.* **1975**, *97*, 4398–4400.

[181] J. W. Park, M. M. Labes, (a) *J. Appl. Phys.* **1977**, *48*, 22–24; (b) *Mol. Cryst. Liq. Cryst. Lett.* **1977**, *34*, 147–152.

[182] A. C. Griffin, T. R. Britt, N. W. Buckley, R. F. Fisher, S. J. Havens, D. W. Goodman in *Liquid Crystals and Ordered Fluids*, Vol. 3 (Eds.: J. F. Johnson, R. S. Porter), Plenum Press, New York **1978**, pp. 61–73.

[183] N. K. Sharma, G. Pelzl, D. Demus, W. Weissflog, *Z. Phys. Chem.* **1980**, *261*, 579–584.

[184] L. Longa, W. H. De Jeu, *Phys. Rev. A* **1982**, *26*, 1632–1647.

[185] M. Domon, J. Billard, *J. Phys. Colloq.* **1979**, *40*, 413–418.

[186] F. Schneider, N. K. Sharma, *Z. Naturforsch.* **1981**, *36a*, 62–67.

[187] M. Cotter in *The Molecular Physics of Liquid Crystals* (Eds.: G. R. Luckhurst, G. W. Gray), Academic Press, London **1979**, p. 181.

[188] W. M. Gelbart, *J. Phys. Chem.* **1982**, *86*, 4298–4307.

[189] Y. Guichard, G. Sigaud, F. Hardouin, *Mol. Cryst. Liq. Cryst. Lett.* **1984**, *102*, 325–330.

[190] F. R. Bouchet, P. E. Cladis, *Mol. Cryst. Liq. Cryst.* **1980**, *64*, 81–87.

[191] V. V. Belyaev, T. P. Antonyan, L. N. Lisetski, M. F. Grebenkin, G. G. Salshchova, V. F. Petrov, *Mol. Cryst. Liq. Cryst.* **1985**, *129*, 221–233.

[192] R. Shashidar, H. D. Kleinhans, G. M. Schneider, *Mol. Cryst. Liq. Cryst. Lett.* **1981**, *72*, 119–126.

[193] H. D. Kleinhans, G. M. Schneider, R. Shashidar, *Mol. Cryst. Liq. Cryst.* **1982**, *82*, 19–24.

[194] H. D. Kleinhans, G. M. Schneider, R. Shashidar, *Mol. Cryst. Liq. Cryst.* **1983**, *103*, 255–259.

[195] A. E. White, P. E. Cladis, S. Torza, *Mol. Cryst. Liq. Cryst.* **1977**, *43*, 13–31.

[196] R. Kiefer, G. Baur, *Mol. Cryst. Liq. Cryst.* **1990**, *188*, 13–24.

[197] S. M. Stishov, V. A. Ivanov, V. N. Kachinskii, (a) *Pis'ma Zh. Eksp. Teor. Fiz.* **1976**, *24 (6)*, 329–332; (b) *JETP Lett.* **1976**, *24 (6)*, 297–300.

[198] V. N. Kachinski, V. A. Ivanov, A. N. Zisman, S. M. Stishov, (a) *Sov. Phys. JETP* **1978**, *48*, 273–277; (b) *Zh. Eksp. Teor. Fiz.* **1978**, *75*, 545–553.

[199] W. Weissflog, D. Demus, Proceedings of the 10th International Liquid Crystal Conference, York, U.K. 1984, Part E, *Mol. Cryst. Liq. Cryst.* **1985**, *129*, 235–243.

[200] M. Cotter, (a) *Phys. Rev. A* **1974**, *10*, 625–636; (b) *Mol. Cryst. Liq. Cryst.* **1976**, *35*, 33–70.

[201] R. Pynn, *J. Chem. Phys.* **1974**, *60*, 4579–4581.

[202] B. A. Baron, W. M. Gelbart, *J. Chem. Phys.* **1978**, *67*, 5795–5801.

[203] D. Demus, *Mol. Cryst. Liq. Cryst.* **1988**, *165*, 45–84.

[204] D. Demus, A. Hauser in *Selected topics in Liquid Crystal Research* (Ed.: H.-D. Koswig), Akademie-Verlag, Berlin **1990**, Chap. 2 [*Chem. Abstr.* **1991**, *115*, 61027d].

[205] D. Demus, A. Hauser, M. Keil, W. Wedler, *Mol. Cryst. Liq. Cryst.* **1990**, *191*, 153–161.

[206] A. P. Kovshik, Yu. I. Denite, E. I. Ryumtsev, V. N. Tsvetkov, (a) *Kristallografiya* **1975**, *20*, 861–864; (b) *Sov. Phys. Crystallogr.* **1975**, *20*, 532–534.

[207] H. S. Subramhanyan, J. Shashidara Prasad, *Mol. Cryst. Liq. Cryst.* **1976**, *37*, 23–27.

[208] J. Shashidara Prasad, H. S. Subramhanyam, *Mol. Cryst. Liq. Cryst.* **1976**, *33*, 77–82.

[209] W. H. de Jeu, P. Bordewijk, *J. Chem. Phys.* **1978**, *68*, 109–115.

[210] F. Leenhouts, W. H. De Jeu, A. J. Dekker, *J. Phys. (Paris)* **1979**, *40*, 989–995.

[211] N. C. Shivaprakash, M. M. M. Abdoh, S. and J. Shashidara Prasad, *Mol. Cryst. Liq. Cryst.* **1982**, *80*, 179–193.

[212] A. K. Garg, G. K. Gupta, V. P. Arora, V. K. Agarwal, B. Bahadur, Proceedings of Nuclear Physics and Solid State Physics Symposium 1981, **1982**, *24C*, 343–344 [*Chem. Abstr.* **1982**, *97*, 102189p].

[213] C. Cabos, J. Sicard, *C. R. Acad. Sci. Paris, Sér. B* **1975**, *281*, 109–111.

[214] S. Denprayoonwong, P. Limcharoen, O. Phaovibul, I. M. Tang, *Mol. Cryst. Liq. Cryst.* **1981**, *69*, 313–326.

[215] O. Phaovibul, K. Chantanasmit, I. M. Tang, *Mol. Cryst. Liq. Cryst.* **1981**, *71*, 233–247.

[216] O. Phaovibul, S. Denprayoonwong, I. M. Tang, *Mol. Cryst. Liq. Cryst.* **1981**, *73*, 71–79.

[217] M. F. Vuks, *Opt. Spectrosc.* **1966**, *20*, 361–364 [*Chem. Abstr.* **1966**, *65*, 3174e].

[218] H. E. J. Neugebauer, (a) *Phys. Rev.* **1952**, *88*, 1210 [*Chem. Abstr.* **1953**, *47*, 3076b]; (b) *Can. J. Phys.* **1954**, *32*, 1–8 [*Phys. Abstr.* **1954**, *57*, 3442].

[219] E. M. Aver'yanov, V. F. Shabanov, (a) *Kristallografiya* **1979**, *24*, 184–186; (b) *Sov. Phys. Cryst.* **1979**, *24*, 107–109.

[220] E. M. Aver'yanov, V. F. Shabanov, (a) *Kristallografiya* **1979**, *24*, 992–997; (b) *Sov. Phys. Cryst.* **1979**, *24*, 567–570.

[221] P. Palffy-Muhoray, D. A. Balzarini, *Can. J. Phys.* **1981**, *59*, 375–377.

[222] P. Palffy-Muhoray, D. A. Balzarini, D. A. Dunmur, *Mol. Cryst. Liq. Cryst.* **1984**, *110*, 315–330.

[223] R. K. Sarna, B. Bahadur, V. G. Bhide, *Mol. Cryst. Liq. Cryst.* **1979**, *51*, 117–136.

[224] B. Bahadur, R. K. Sarna, V. G. Bhide, *Mol. Cryst. Liq. Cryst. Lett.* **1982**, *72*, 139–145.

[225] A. I. Pirogov, I. V. Novikov, *Zh. Prikl. Khim. (Leningrad)* **1988**, *61*, 6, 1382–1384 [*Chem. Abstr.* **1988**, *109*, 139708z].

[226] O. Phaovibul, K. Pongthana-Ananta, I. Ming Tang, *Mol. Cryst. Liq. Cryst.* **1980**, *62*, 25–32.

[227] G. A. Oweinreen, M. Hasan, *Mol. Cryst. Liq. Cryst.* **1983**, *100*, 357–371.

[228] J. Jadzyn, W. Labno, *Chem. Phys. Lett.* **1980**, *73*, 307–310.

[229] A. Bartelt, H. Reisig, J. Herrmann, G. M. Schneider, *Mol. Cryst. Liq. Cryst. Lett.* **1984**, *102*, 133–138.

[230] J. M. Buisine, R. Cayuela, C. Destrade, N. H. Tinh, Proceedings of the 11th International Liquid Crystal Conference, Berkeley, CA, 1986, Part B, *Mol. Cryst. Liq. Cryst.* **1987**, *144*, 137–160.

[231] V. K. Semenchenko, N. A. Nedostup, V. Yu. Baskakov, 1(a) *Russ. J. Phys. Chem.* **1975**, *49*, 909–912; 1(b) *Zh. Fiz. Khim.* **1975**, *49*, 1543–1547; 2(a) *Russ. J. Phys. Chem.* **1975**, *49*, 912–914; 2(b) *Zh. Fiz. Khim.* **1975**, *49*, 1547–1550.

[232] A. C. Zawisza, J. Stecki, *Solid State Commun.* **1976**, *19*, 1173–1175.

[233] N. A. Nedostup, V. K. Semenchenko, (a) *Russ. J. Phys. Chem.* **1977**, *51*, 958–959; (b) *Zh. Fiz. Khim.* **1977**, *51*, 1628–1631.

[234] A. P. Kapustin, *Eksperimental'nye issledovaniya zhidkikh kristallov (Experimental Study of Liquid Crystals)*, Nauka, Moscow **1978**, p. 368 [*Chem. Abstr.* **1979**, *90*, B 95971n].

[235] E. A. S. Lewis, H. M. Strong, G. H. Brown, *Mol. Cryst. Liq. Cryst.* **1979**, *53*, 89–99.

[236] T. Shirakawa, T. Inoue, T. Tokuda, *J. Phys. Chem.* **1982**, *86*, 1700–1702.

[237] T. Shirakawa, M. Arai, T. Tokuda, *Mol. Cryst. Liq. Cryst.* **1984**, *104*, 131–139.

[238] C. S. Johnson, P. J. Collings, *J. Chem. Phys.* **1983**, *79*, 4056–4061.

[239] S. N. Nefedov, A. N. Zisman, S. M. Stishov, *Zh. Eksp. Teor. Fiz.* **1984**, *86*, 125–132 [*Chem. Abstr.* **1984**, *100*, 94933q].

[240] T. Shirakawa, Y. Kikuchi, T. Seimiya, *Thermochim. Acta* **1992**, *197*, 399–405.

[241] A. P. Kapustin, G. E. Zvereva, (a) *Kristallografiya* **1965**, *10*, 723–726; (b) *Sov. Phys. Cryst.* **1966**, *10*, 603–606.

[242] B. Bahadur, J. Prakash, K. Tripathi, S. Chandra, *Acustica* **1975**, *33*, 217–219.

[243] B. Bahadur, *Z. Naturforsch. A* **1975**, *30*, 1093–1096.

[244] V. G. K. M. Pisipati, N. V. S. Rao, *Z. Naturforsch. A* **1984**, *39*, 696–699.

[245] M. R. Rao, *J. Chem. Phys.* **1941**, *9*, 682–685.

[246] Y. Wada, *J. Phys. Soc. Jpn.* **1949**, *4*, 280–283 [*Chem. Abstr.* **1950**, *44*, 6703c].

[247] J. Frenkel, *Kinetic Theory of Liquids*, Dover, New York **1955**.

[248] F. T. Lee, H. T. Tan, Y. M. Shin, C. N. Woo, *Phys. Rev. Lett.* **1973**, *31*, 1117–1120.

[249] H. Gasparoux, M. F. Achard, F. Hardouin, G. Sigaud, *C. R. Acad Sci. Paris II* **1981**, *293*, 1029–1032 [*Chem. Abstr.* **1982**, *97*, 47743n].

[250] P. A. Spegt, A. E. Skoulios, *Acta Crystallogr.* **1963**, *16*, 301–306.

[251] P. A. Spegt, A. E. Skoulios, *Acta Crystallogr.* **1964**, *17*, 198–207.

[252] W. M. Gelbart, B. Barboy, *Mol. Cryst. Liq. Cryst.* **1979**, *55*, 209–226.

[253] W. M. Gelbart, B. Barboy, *Acc. Chem. Res.* **1980**, *13*, 290–296.

[254] V. N. Raja, R. Shashidar, S. Chandrasekhar, R. E. Boehm, D. E. Martire, *Pramana* **1985**, *25*, L119–L122 [*Chem. Abstr.* **1985**, *103*, 113798e].

[255] J. S. Clunie, J. M. Corkill, J. F. Goodman, *Proc. R. Soc. Lond. Ser. A* **1965**, *285*, 520–533.

[256] S. Yano, K. Tadano, K. Aoki, *Mol. Cryst. Liq. Cryst. Lett.* **1983**, *92*, 99–104.

[257] M. Stefanov, A. Saupe, *Mol. Cryst. Liq. Cryst.* **1984**, *108*, 309–316.

[258] P. Photinos, A. Saupe, *J. Chem. Phys.* **1989**, *90*, 5011–5015.

[259] P. Photinos, A. Saupe, *Phys. Rev. A* **1990**, *41*, 954–959.

[260] A. A. Barbosa, A. V. A. Pinto, *J. Chem. Phys.* **1993**, *98*, 8345–8346.

[261] A. S. C. Lawrence, *Trans. Faraday Soc.* **1938**, *34*, 660–677 [*Chem. Abstr.* **1938**, *32*, 4033^1].

[262] R. D. Vold, M. J. Vold, *J. Am. Chem. Soc.* **1939**, *61*, 808–816 [*Chem. Abstr.* **1939**, *33*, 9108^7].

[263] R. D. Vold, F. B. Rosevear, R. H. Ferguson, *Oil Soap* **1939**, *16*, 48–51.

[264] M. J. Vold, M. Macomber, R. D. Vold, *J. Am. Chem. Soc.* **1941**, *63*, 168–175 [*Chem. Abstr.* **1941**, *35*, 1658^2].

[265] W. Gallay, I. E. Puddington, *Can. J. Res. B* **1943**, *21*, 202–210.

[266] R. D. Vold, M. J. Vold, *J. Phys. Chem.* **1945**, *49*, 32–42 [*Chem. Abstr.* **1945**, *39*, 1588^6].

[267] F. W. Southam, I. E. Puddington, *Can. J. Res. B* **1947**, *25*, 121–124.

[268] G. Stainsby, R. Farnand, I. E. Puddington, *Can. J. Chem.* **1951**, *29*, 838–842 [*Chem. Abstr.* **1952**, *46*, 3300i].

[269] D. P. Benton, P. G. Howe, J. R. Farnand, I. E. Puddington, *Can. J. Chem.* **1955**, *33*, 1798–1805 [*Chem. Abstr.* **1956**, *50*, 11688i].

[270] K. U. Ingold, I. E. Puddington, *J. Inst. Petrol.* **1958**, *44*, 41–44 [*Chem. Abstr.* **1958**, *52*, 6774i].

[271] A. G. MacDonald, *Biochim. Biophys. Acta* **1978**, *507*, 26–37.

[272] K. Ohki, K. Tamura, I. Hatta, *Biochim. Biophys. Acta* **1990**, *1028*, 215–222.

[273] M. Wolf, J. H. Wendorff, Proceedings of the 5th European Winter Liquid Crystal Conference, Borovets, Bulgaria, 1987, *Mol. Cryst. Liq. Cryst.* **1987**, *149*, 141–162.

[274] M.-J. Gonzalez-Tejera, J. M. Perena, A. Bello, I. Hernandez-Fuentes, *Polym. Bull. (Berlin)* **1993**, *31*, 111–115.

[275] S. K. Ghosh, S. Amadesi, *Phys. Lett. A* **1976**, *59*, 282–284.

[276] J. G. Ypma, G. Vertogen, *Phys. Lett. A* **1977**, *61*, 45–47.

[277] L. Feijoo, V. J. Rajan, Chia-Wei Woo, *Phys. Rev. A* **1979**, *19*, 1263–1271.

[278] K. Singh, S. Singh, *Mol. Cryst. Liq. Cryst.* **1984**, *108*, 133–148.

[279] A. V. Belik, V. A. Potemkin, Yu. N. Grevtseva, *Dokl. Akad. Nauk* **1994**, *336*, 361–364 [*Chem. Abstr.* **1994**, *121*, 242457p].

[280] A. V. Belik, V. A. Potemkin, Yu. N. Grevtseva, (a) *Zh. Fiz. Khim.* **1995**, *69*, 101–105; (b) *Russ. J. Phys. Chem.* **1995**, *69*, 91–94.

[281] M. P. Allen, *Phys. Rev. Lett.* **1990**, *65*, 2881–2884.

[282] J. W. Emsley, G. R. Luckhurst, W. E. Palke, D. J. Tildesley, *Liq. Cryst.* **1992**, *11*, 519–530.

[283] P. R. Alapati, D. M. Potukuchi, P. Bhaskara Rao, N. V. S. Rao, V. G. K. M. Pisipati, A. S. Paranjpe, *Liq. Cryst.* **1989**, *5*, 545–551.

[284] V. V. Aleksandriiski, V. A. Burmistrov, O. I. Koifman, *Izv. Vyssh. Uchebn. Zaved., Khim.*

Khim. Tekhnol. **1988**, *31*, 111–114 [*Chem. Abstr.* **1988**, *109*, 30485y].

[285] D. Demus, M. Klapperstück, R. Rurainski, D. Marzotko, *Z. Phys. Chem.* **1971**, *246*, 385–395.

[286] D. Guillon, A. Skoulios, *C. R. Acad. Sci., Sér. C* **1974**, *278*, 389–391.

[287] D. Guillon, A. Skoulios, *J. Phys. (Paris)* **1977**, *38*, 79–83.

[288] D. Guillon, A. Skoulios, *Mol. Cryst. Liq. Cryst.* **1979**, *51*, 149–160.

[289] S. D. Lotke, S. B. Desai, R. N. Patil, Proceedings of the 9th International Liquid Crystal Conference, Bangalore, 1982, Part C, *Mol. Cryst. Liq. Cryst.* **1983**, *99*, 267–277.

[290] A. P. Kapustin, *Elektroopticheskije i akusticheskie svoistva zhidkikh kristallov (Electro-Optic and Acoustic Properties of Liquid Crystals)*, Nauka, Moscow **1973**, p. 232.

[291] R. Kiefer, G. Baur, Proceedings of the 17th Freiburger Arbeitstagung Flüssigkristalle, Freiburg, 1987.

[292] T. Matsumoto, *Kogakuin Daigaku Kenkyu Hokoku (Research Reports of the Kogakuin University)* **1973**, *34*, 1–6 [*Chem. Abstr.* **1974**, *80*, 149676j].

[293] R. Paul, B. Jha, D. A. Dunmur, *Liq. Cryst.* **1993**, *13*, 629–636.

[294] A. I. Pirogov, I. V. Novikov, *Izv. Vyssh. Uchebn. Zaved. Khim. Khim. Tekhnol.* **1987**, *30*, 3, 63–68 [*Chem. Abstr.* **1987**, *107*, 145296p].

[295] A. I. Pirogov, I. V. Novikov, *Izv. Vyssh. Uchebn. Zaved. Khim. Khim. Tekhnol.* **1987**, *30*, 10, 63–67 [*Chem. Abstr.* **1988**, *108*, 66525e].

[296] A. I. Pirogov, N. Kodabakas, *Zh. Fiz. Khim.* **1987**, *61*, 1754–1760 [*Chem. Abstr.* **1987**, *107*, 209494m].

[297] A. I. Pirogov, N. Kodabakas, *Zh. Fiz. Khim.* **1989**, *63*, 368–372 [*Chem. Abstr.* **1989**, *110*, 203502c].

[298] R. S. Porter, J. F. Johnson, *J. Appl. Phys.* **1963**, *34*, 51–54.

[299] N. V. S. Rao, V. G. K. M. Pisipati, D. Saran, *Phase Transitions* **1984**, *4*, 275–279.

[300] N. V. S. Rao, D. M. Potukuchi, P. V. Sankar Rao, V. G. K. M. Pisipati, *Liq. Cryst.* **1992**, *12*, 127–135.

[301] P. Bhaskara Rao, D. M. Potukuchi, J. S. R. Murty, N. V. S. Rao, V. G. K. M. Pisipati, *Cryst. Res. Technol.* **1992**, *27*, 839–849.

[302] N. V. S. Rao, G. Padmaja Rani, D. M. Potukuchi, V. G. K. M. Pisipati, *Z. Naturforsch.* **1994**, *49a*, 559–562.

[303] P. I. Rose, *Mol. Cryst. Liq. Cryst.* **1974**, *26*, 75–85.

[304] P. Seurin, D. Guillon, A. Skoulios, *Mol. Cryst. Liq. Cryst.* **1980**, *61*, 185–190.

[305] A. G. Shashkov, I. P. Zhuk, V. A. Karolik, *High Temp.–High Press.* **1979**, *11*, 485–490 [*Chem. Abstr.* **1980**, *93*, 58566a].

[306] R. Somashekar, M. S. Madhava, *Mol. Cryst. Liq. Cryst.* **1987**, *147*, 79–84.

[307] C. I. Venkatamana Shastry, J. Shashidara Prasad, Proceedings of the 11th International Liquid Crystal Conference, Berkeley, CA 1986, Part A, *Mol. Cryst. Liq. Cryst.* **1986**, *141*, 191–200.

[308] D. M. Potukuchi, P. B. Rao, N. V. S. Rao, V. G. K. M. Pisipati, *Z. Naturforsch. A* **1989**, *44*, 23–25.

[309] N. V. S. Rao, V. G. K. M. Pisipati, J. S. R. Murthy, P. Bhaskara Rao, P. R. Alapati, *Liq. Cryst.* **1989**, *5*, 539–544.

6.2.3 Metabolemeter

Wolfgang Wedler

In 1983, Buisine and coworkers [1] described a device which, in a small cell with constant volume, permits the detection of pressure differences in condensed, especially liquid-crystalline, phases as a function of the temperature. This device was designed to gain quickly comprehensive information about the phase behaviour with very small quantities of material. It produces plots of pressure difference–temperature which were named thermobarograms. Clearly, it relates two intensive properties. The authors were able to show that the only occurrence of intensive properties conveniently permits a significant miniaturization of the apparatus. Furthermore, a theoretical estimate of the pressure changes in the chamber when heated or cooled beyond first-order phase transitions was carried out. This was combined with first experimental data from scans of MBBA, EBBA and Octylcyanobiphenyl, which indicated that the sensitivity of the method was sufficiently high to detect phase transitions which proceeded with only 0.06% fractional volume change. The authors suggested to name this apparatus metabolemeter, from μεταβολη (transformation) and μετρον (to measure).

Several publications showed its usefulness by describing phase transitions in mesogenic materials with a rich polymorphism [2–4]. Liquid-crystalline polymorphism of chiral compounds, having multicritical points [5, 6], as well as mesogens with discotic or pyramidic molecular geometry, and re-entrant behaviour [7–11] were subject to thermobarometric analysis (TBA). Furthermore, the metabolemeter proved its usefulness in studies of binary mixtures of mesogens [12], and in polymer research [13, 14].

An overview of the construction is given in Fig. 1. Additionally, Fig. 2 shows a schematic representation of the sample cell. The

following description bases on the informations given by the authors [1].

To verify simultaneous recording of sample pressure and temperature, a rigid but dilatable cell is used. This cell can be heated and contains a pressure transducer and a thermometer. The pressure transducer used is a HEM 375-20000-Kulite International. It has a flushing sensible metallic membrane, a working temperature range between −55°C and 260°C, and a maximum sustainable pressure of 1700 bar. The cell is composed of a crucible (3) in which is machined a cavity (4) giving a sample volume of 5.97 mm^3. The pressure transducer (1) covers the upper part of the measurement cell. For successful measurements, sensor and crucible have to be aligned parallel to each other in order to tightly close the chamber for measurement. The hemispheric shape of the crucible together with its placement on

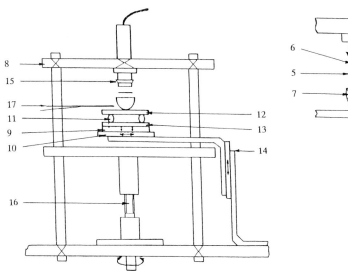
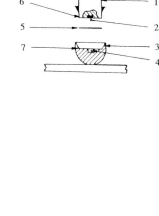

Figure 1. Schematic representation of the metabolemeter (initial design). The following parts are indicated by numbers: (1) pressure transducer (Kulite International, HEM 375 20000), (2) flushing sensible membrane, (3) crucible (17–4–PH stainless steel), (4) chamber (6 mm^3), (5) tin joint, (6) crowned insensible surface of the head transducer, (7) plane surface of the crucible, (8) pressure sensor support, (9, 10) horizontal translation movements, (11) steel balls, (12, 13) aluminium plates, (14) vertical translation movement, (15) centring cone, (16) screw, (17) temperature sensor. (Reproduced from [1] by permission of Gordon and Breach Science Publishers, Inc).

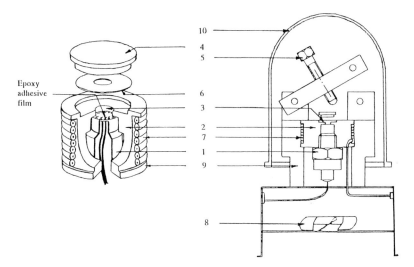

Epoxy
adhesive
film

Figure 2. Schematic representation of the scanning numerical metabolemeter metallic sample cell. (1) pressure transducer, (2) steel crucible, (3) cavity, (4) steel cover, (5) set screw, (6) annular joint, (7) heater (8) fan, (9) steel stand, (10) glass housing. (Reproduced from [10] by permission of Société Française Physique).

ball bearings (11) and aluminium plates (12) and (13), is intended to solve this problem. When the lower part is moved towards the stationary upper part, the centring cone (15) moves the crucible into the right position, without sliding. Another essential point is to guarantee a homogenous deformation of the chamber during measurements. Therefore, the cell is constructed of the same material as the pressure sensor (17–4–PH stainless steel). A tin joint (5) insures tightness when exposed to pressures. This material also introduces an upper limit to the available temperature, being at 230°C, beyond which the tin melts. Later studies extended the temperature range to 270°C, marking the maximum working temperature of the pressure transducer, by using a gold joint [3]. Also, attempts were made to obtain tightness of the cell up to 2500 bar [13]. The apparatus is hermetically closed by vertical motion (14) of the lower part by means of a screw (16), located at the base of the apparatus. The cell then can be heated with an oven, and the temperature is measured with a platinum resistance thermometer (17). Configured in the described way, heating rates of 5 K min^{-1}, and cooling rates of

1 K min^{-1} were achieved [3]. The signals which come from the thermometer and the pressure transducer, are respectively transmitted to the X- and Y-inputs of a recorder, or can be fed into a computer, giving the thermobarograms.

The apparatus has been further developed into a scanning numerical metabolemeter [11, 15, 16].

6.2.3.1   Thermobarograms

Plots which are produced by TBA, so-called thermobarograms, are similar and related to the pressure–temperature phase diagrams which are commonly used to express equilibrium between phases. However, the first and main difference between such a phase diagram and a thermobarogram is that the latter also shows out-of-equilibrium states. An estimate [1] from literature data leads to the result, that in between phase transitions, the slope is often less than 13 bar K^{-1}, whereas it is higher than 26 bar K^{-1} at first-order phase transitions for most cases. Hence, these phase transitions express themselves in a thermobarogram by noticable slope changes. The second difference between a phase diagram and a thermobarogram is the choice of the pressure variable. Phase diagrams depict the absolute pressure at equilibrium as a function of phase transition temperature. In contrast, thermobarograms show pressure differences evolving with the temperature, which is not necessarily the phase transition temperature at equilibrium. Therefore, experiments can be started at different temperatures: close to, or far from phase transitions. The zero pressure difference then always starts with ambient pressure, but the trajectory in the phase diagram is different for every case. When crossing phase transitions in this way, one finally is able to piece together the phase diagram from several different thermobarograms. Figure 3 illustrates this procedure for the example of HBPD [3].

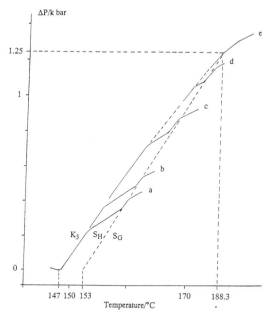

Figure 3. Generation of a pressure–temperature phase diagram from thermobarograms of HBPD. Reproduced from [3] by permission of Gordon and Breach Science Publishers, Inc. Sequence of four thermobarograms (a–d) showing the K_3–S_H, and S_H–S_G phase transitions. The fifth curve indicates the triple point K_2–K_3–S_G (HBPD was shown to have three different solid phases)

It was shown [1] that an assessment of the isothermal pressure increment $(\Delta P)_T$, occuring at phase transitions in the thermobarograms, is possible. Assuming fractional volume changes $\Delta V/V_m$ of 6%, 0.6%, and 0.06%, which are realistic (see Sec. 6.2.2 of this chapter), together with reasonable assumptions about the other parameters give pressure increments of 1600 bar maximum, and 12 bar minimum, which are well in the range of the sensitivity of the pressure transducer. Furthermore, the influence of respiration (incurvation variation of the membrane) as a source of error has been discussed [1]. Respiration increases the cavity volume v_c, and decreases the observed pressure increment, setting a lower limit to sample cell minimization. It was found that for 6% relative volume change, and a cavity volume of 6 mm³, independence of $(\Delta P)_T$ from v_c occured. A drop in cavity size to 0.4 mm³ leaves the pressure increment still higher than 90% of the maximum value for large cavity volumes. During data evaluation, the expression for the pressure increment provides the possibility to calculate fractional volume changes and phase transition enthalpies [2].

In blends, incompatibility between two mesogenic compounds A and B in the crystalline state, decrease of the melting temperature and eutectic point formation cause more complexity in the thermobarograms. For the case of the perfect solution Buisine [17], and Buisine and Billard [12] derived mathematical relations which allow calculation and understanding of thermobarograms of blends and estimation of the detectability of phase transition effects. Using these relations, pressure increments of 238 bar and 768 bar were estimated, which are well in the range of detectability of the metabolemeter, for the eutectic melting and the dissolution of the majority component, respectively. According to [12], the eutectic melting is reflected in the thermobarogram by a slope of approximately 40 bar K^{-1} which is similar to the melting of the pure components. The subsequent dissolution starts with a drop of slope to 15 bar K^{-1}, and proceeds by increasing up to around 33 bar K^{-1}. Further equations were given [12] which model pressure increments and slopes for fluid – fluid phase transitions in blends. It could be shown that, for example, the slope always has to be positive. For the simple case of horizontal curves of coexistence in the isobaric temperature–concentration phase diagram, the slope becomes definitely bigger than the out-of-transition slope. Also, its value stays always smaller than that of the corresponding transitions in the pure compounds. Deviations from the horizontal shape of the coexistence curve in the phase diagram, and a consequent widening of the spindle, decrease the slope value. Estimates for the pressure increments at several different fluid–fluid transitions are compiled from [12], and summarized in Table 1. In some cases, new phases were discovered and identified. Also, phase transitions of weakly first order [3] or second order [6], and glass transitions [13] were detected. A general introduction into theory and practice of TBA has been given by Buisine [17, 18].

Table 1. Estimated typical values for the thermobarometric pressure increments at different fluid–fluid phase transitions (compiled from [12]).

Transition	ΔP_I (bar K^{-1})
Smectic-A/nematic	20
Nematic/isotropic, or smectic-C/nematic	30
Highly organized mesophase/ smectic-A (C), or highly organized mesophase/nematic	40
Highly organized mesophase/ highly organized mesophase	150

As an example we mention the TBA results for the binary system 4-methoxy-4′-nonyltolane (MNT)/4-methoxy-4′-ethyltolane (MET). This system has only one nematic phase, and has been described previously [19]. Figure 4 shows the thermobarogram, as obtained on heating from a blend, containing 11.9 wt% MET. The melting at 32°C under atmospheric pressure, and T_{N-I} at 63.5°C under 500 bar, are visible. The dissolution process of excess MNT crystals starts with a slope of 6.2 bar K^{-1}, and continously develops to a slope of 13.6 bar K^{-1}. The nematic dilatation begins with a faint change of slope at 60°C, under 450 bar.

Finally, we note that the further development of TBA was not restricted exclusively to the phase characterization of liquid crystals. Since 1988, a more exended use in

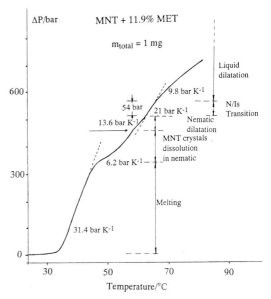

Figure 4. Thermobarogram on heating for the mixture MET/MNT with 11.9 wt% MET. Sample weight was 1 mg. Sequence of phase transitions: eutectic melting, crystal dissolution of excess MNT, nematic/isotropic phase transition. Reproduced from [12] by permission of Gordon and Breach Science Publishers, Inc. and OPA Ltd

polymer science, paraffin and bitumen research, and in the characterization of reacting systems has been reported [20], underlining the versatility of the apparatus.

6.2.3.2 References

[1] J. M. Buisine, B. Soulestin, J. Billard, *Mol. Cryst. Liq. Cryst.* **1983**, *91*, 115–127.

[2] J. M. Buisine, B. Soulestin, J. Billard, *Proceedings of the 9th International Liquid Crystal Conference*, Bangalore, India, Part D, **1982**, *Mol. Cryst. Liq. Cryst.* **1983**, *97*, 397–406.

[3] J. M. Buisine, *Mol. Cryst. Liq. Cryst.* **1984**, *109*, 143–157.

[4] J. M. Buisine, *C. R. Acad. Sci. Paris, Sér. II*, **1983**, *297*, 323–326.

[5] C. Legrand, N. Isaert, J. Hmine, J. M. Buisine, J. P. Parneix, H. T. Nguyen, C. Destrade, (a) *Ferroelectrics* **1991**, *121*, 21–31. (b) *J. Phys. II* **1992**, *2*, 1545–1562.

[6] A. Anakkar, A. Daoudi, J. M. Buisine, N. Isaert, T. Delattre, H. T. Nguyen, C. Destrade, *J. Therm. Anal.* **1994**, *41*, 1501–1513.

[7] J. M. Buisine, R. Cayuela, C. Destrade, N. H. Tinh, *Proceedings of the 11th International Liquid Crystal Conference*, Berkeley, Part B, 1986. *Mol. Cryst. Liq. Cryst.* **1987**, *144*, 137–160.

[8] J. M. Buisine, J. Malthête, C. Destrade, N. H. Tinh, *Physica B* **1986**, *139/140*, 631–635.

[9] J. M. Buisine, M. Domon, *C. R. Acad. Sc. Paris, Sér. II*, **1986**, *303*, 1769–1772.

[10] J. M. Buisine, B. Soulestin, *Rev. Phys. Appl.* **1987**, *22*, 1211–1214.

[11] J. M. Buisine, H. Zimmermann, P. Poupko, Z. Luz, J. Billard, *Mol. Cryst. Liq. Cyst.* **1987**, *151*, 391–398.

[12] J. M. Buisine, J. Billard, *Proceedings of the 10th International Liquid Crystal Conference*, York, United Kingdom, Part D, 1984. *Mol. Cryst. Liq. Cryst.* **1985**, *127*, 353–379.

[13] J. M. Buisine, P. Le Barny, J. C. Dubois, *J. Polym. Sci.: Polym. Lett.* **1984**, *22*, 149–152.

[14] C. Lahmamssi, X. Coqueret, J. M. Buisine, C. Gors, *Calorim. Anal. Therm.* **1992**, *23*, 359–366.

[15] The scanning numerical metabolemeter was manufactured and distributed by Micro Technique Métropole LEADER, Moulin 1, 2 Rue de la Créativité 59650 Villeneuve d'Ascq, France, under the reference MAB 02 (Information from Ref. [11]).

[16] J. M. Buisine, J. L. Bigotte, M. T. M. Leader, *Calorim. Anal. Therm.* **1987**, *18*, 387–391.

[17] J. M. Buisine, *Thèse*, Lille, France, **1984**.

[18] J. M. Buisine, *Calorim. Anal. Therm.* **1988**, *19*, C24.1-C24.8.

[19] J. Malthête, M. Leclercq, M. Dvolaitzky, J. Gabard, J. Billard, V. Pontikis, J. Jacques, *Mol. Cryst. Liq. Cryst.* **1973**, *23*, 233–260.

[20] The metabolemeter was also used in experiments with different, non-liquid-crystalline systems. Publications regarding these studies are listed below. (a) Bitumen, Paraffins: J. M. Buisine, C. Such, A. Eiadlani, *Calorim. Anal. Therm.* **1988**, *19*, P 21.1-P 21.8. J. M. Buisine, C. Such, A. Eiadlani, *Prepr.-Am. Chem Soc., Div. Pet. Chem.* **1990**, *35*, 320–329. J. M. Buisine, C. Such, A. Eiadlani, *Fuel Sci. Technol. Int.* **1992**, *10*, 835–853. G. Joly, F. Farcas, A. Eiadlani, C. Such, J. M. Buisine, *Calorim. Anal. Therm.* **1992**, *23*, 343–349. D. Lourdin, A. H. Roux, J. P. E. Grolier, J. M. Buisine, *Thermochim. Acta* **1992**, *204*, 99–110. J. Li. Tamarit, B. Legendre, J. M. Buisine, *Mol. Cryst. Liq. Cryst. Sci. Technol., Sect. A* **1994**, *250*, 347–358.

(b) Polymers, Polymeric Blends, Filled Systems: J. M. Buisine, P. Cuvelier, B. Addadi, N. Elbounia, *Calorim. Anal. Therm.* **1991**, *22*, 185–192. D. Lourdin, J. R. Quint, A. H. Roux, J. P. E. Grolier, *Calorim. Anal. Therm.* **1992**, *23*, 225–232. P. Cuvelier, B. Haddadi, J. M. Buisine, N. Elbounia, *Thermochim. Acta* **1992**, *204*, 123–135.

(c) Reacting Systems: A. Squalli, L. Montagne, P. Vast, G. Palavit, J. M. Buisine, *J. Therm. Anal.* **1991**, *37*, 1673–1678. A. Squalli, J. M. Buisine, *Calorim. Anal. Therm.* **1992**, *23*, 401–408.

6.2.4 High Pressure Investigations

P. Pollmann

6.2.4.1 Introduction

By application of hydrostatic pressure the liquid crystalline range of existence can be varied: it can be increased, decreased (in the extreme case suppressed) or induced at all

(Fig. 1). Liquid crystalline phases can disappear with increasing pressure and appear again at still higher pressures. Thus there exists a valuable tool for influencing the phase behavior of liquid crystals.

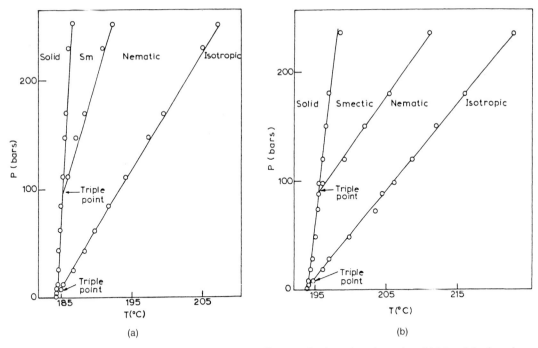

Figure 1. Experimental pressure–temperature phase diagrams for 4-methoxybenzoic acid (a) and 4-ethoxyben-zoic acid (b) showing the solid–nematic–isotropic and solid–smectic–nematic triple points. (From [19], repro-duced by permission of Indian Academy of Sciences.)

Pressure plays an important role in the field of critical phenomena in liquid crystal systems. Contrary to investigations at atmospheric pressure where, for instance, the composition of a mixture must be varied, single compounds can be observed only by increasing the pressure.

The measurement of the pressure dependence of a physical property of a liquid crystalline phase at constant temperatures offers the advantage of studying only the influence of density without temperature effects.

6.2.4.2 Phase Transitions at High Pressures

Phase transitions involving liquid crystalline phases have been the subject of experimental work for many years. By 1899 Hulett [1] had already investigated 4,4'-bis-methoxy-azoxybenzene, 4,4'-bis-ethoxy-azoxybenzene and cholesteryl benzoate up to 30 MPa. After a long period of inactivity

in this field, in 1926 Puschin and Greben-schtschikow [2] studied 4,4'-bis-methoxy-azoxybenzene up to 250 MPa, and in 1938 Robberecht [3] studied cholesteryl pentano-ate and cholesteryl hexanoate up to 76 MPa.

For an indication of a discontinuous phase transition each physical quantity can be used in principle which is changed sensitively enough by the phase transition also at higher pressures. The detection methods for a continuous phase transition are naturally restricted.

Experimental Techniques

Differential Thermal Analysis (DTA)/
Differential Scanning Calorimetry (DSC)

DTA and DSC are often utilized for detecting liquid crystalline phase transitions, which are associated with low transition enthalpies.

After the first measurements of Garn [4] up to 1 MPa, Reshamwala and Shashidhar [5] designed a DTA device with a coaxial cell working up to 750 MPa. The sample was enclosed in Teflon and thus totally separated from the pressure-transmitting medium.

Herrmann and coworkers [6] developed a DTA apparatus with a diamond anvil cell for measurements up to 300 MPa. For measuring the temperature difference between sample and reference cell the thermocouples are soldered into two blind plugs. The samples are encapsulated in lead, indium or nickel cells.

Sandrock, Bartelt and Schneider [7, 8] later on used a microcomputer-assisted apparatus with a pressure range up to 1000 MPa. It consisted of a twin autoclave. The sample is encapsulated in a lead or indium cup.

Investigations of phase transitions with DSC [9] are advantageous when, aside from the phase boundary lines, enthalpy changes of the transitions shall be determined. Using the transition enthalpies and the slopes of the transition curves the volume changes accompanying the transitions can additionally be calculated by means of the Clausius–Clapeyron equation.

Garland [10] developed an a.c. calorimeter technique for the study of second-order phase transitions. The device works up to 300 MPa and requires only a small sample (50–100 mg). In [10] two versions of this calorimeter are described: a manually operated calorimeter with a computerized data-acquisition system, and a fully computerized calorimeter which can work in a scanning mode.

Optical Methods

A simple method for detecting phase transitions is the optical transmission technique [11]. The transition is indicated by a discontinuity in the intensity of transmitted light or a change in its pressure dependence. The measuring substance is contained in an optical high pressure cell which is usually sealed by two sapphire windows. In many cases the sample is filled into a special cuvette which in the pressure chamber is surrounded by the pressure transmitting not light absorbing fluid [12, 13]. The pressure then is transmitted to the contents of the cuvette by a variety of devices (e.g. bellows, shrinking hose, thin-walled tube).

A diamond anvil cell can also serve as an optical high pressure cell [14]. Transitions between liquid crystalline phases which exhibit optical activity can be indicated by jumps in the angle of optical rotation [15].

The characteristic behavior of the selective light reflection (e.g. of a cholesteric phase near a smectic A phase) can be used to observe the transition of phases. In some cases a distinction between a discontinuous and a continuous smectic A–cholesteric phase transition is possible [16].

Often a phase transition between two liquid crystalline phases is associated with a change in texture, which is characteristic for a special phase. The observation of the sample in the high pressure optical cell by a polarizing microscope then offers the advantage not only of detecting the phase transition but also of identifying the type of the involved liquid crystalline phases [13, 17].

The apparatus described in [18] allows not only the observation of the sample with a polarizing microscope but also includes the application of optical spectroscopy electrooptic and electrical measurements.

A multipurpose and a constant sample thickness cell can be used alternatively as specimen cells. In the first case the pressure transmitting medium is argon. Samples which are affected by argon then are contained between two sheets of glass sealed with PTFE. The latter arrangement, however, has the disadvantage of a layer thickness which varies with pressure. If the sample is not affected by argon, it can be placed in an open cell which consists of two panes of glass with 10 μm thick stainless steel spacers or even it can be observed as a free-standing film.

The pressure transmitting medium in the case of the constant sample thickness cell is oil. The sample is contained in a small reservoir, the deformable bottom (copper membrane) of which transmits the pressure to the specimen. The observation is performed in such a way that the layer thickness is not influenced by the deformation of the copper membrane.

Very special methods of detecting phase transitions include the following.

Normal Behavior (including pressure-induced and suppressed behavior)

For facilitating the survey of the numerous publications concerning phase transitions the latter are classified by the method of examination. The letter after the reference gives the types of the liquid crystalline phases which are involved in the transitions investigated. The different smectic phases are all listed under the symbol S; if only transitions between smectic phases were observed see letter **k**. In the cases of twisted-grain-boundary phases (TGB), cholesteric phases (N*) and blue phases (BP) their symbols are positioned beside the reference number of the literature and the referring letter. The letters mean the following combinations of phase transitions: (a) N–I; (b) S–I; (c) S–N–I; (d) Cr–N–I; (e) Cr–S–I; (f) Cr–S–N–I; (g) S–N; (h) Cr–N; (i) Cr–S–N; (j) Cr–S; (k) S–S.

Differential Thermal Analysis (DTA)

[19] – c, f (N*): Whereas the nematic range of 4,4'-bis-n-(heptyloxy)-azoxybenzene is extended by increasing pressure, that of the SmC one disappears at a pressure of about 668 MPa.

The cholesteric range of cholesteryl oleyl carbonate increases with pressure. There is evidence of a tricritical point. In the case of methoxy- and ethoxybenzoic acid a pressure-induced nematic and a smectic phase are observed.

[20] – f: The DTA measurements of 4-*n*-hexyloxyphenyl-4-*n*-decyloxybenzoate are combined with microscopic observations of the sample in a diamond anvil cell. A SmA–SmA phase transition is supposed to occur. Monotropic SmB and SmE phases are observed.

[21] – e, f: The SmC phases of the three compounds di-(4'-n-octyloxyphenyl)-trans-cyclohexane-1,4'-dicarboxylate), 4-*n*-nonyloxybenzoic acid and *n*-pentyl-4(4'-n-decyloxybenzylideneamino) cinnamate exhibit a different temperature dependence of the tilt angle in all three cases but the same pressure behavior: the SmC phases are suppressed at higher pressures.

[22] – d, f: The experimental values for the slopes of the nematic–isotropic phase boundary lines dT/dp of the first six homologs of the 4,4'-di-*n*-alkoxyazoxybenzene are in good agreement with those calculated from the Clausius–Clapeyron equation. The fifth and sixth homologs show in contrast to the other ones additionally a smectic phase with an unusual pressure behavior.

[23] – c (N*), d, f (N*): The monotropic nematic phase of 4-cyanophenyl-4-n-propyloxy-α-methyl cinnamate becomes enantiotropic at 58 MPa and 86 °C. The cholester-

ic phase of cholesteryl nonanoate is suppressed at 285 MPa and 196 °C. All transitions of cholesteryl tetradecanoate remain enantiotropic up to 160 MPa. 4,4'-bis-n-butyl-azobenzene and 4-biphenyl-4-ethylbenzoate exhibit a pressure-induced mesophase.

[24] – d, f: The pressure dependence of the phase transition temperatures of the pentyl to octyl homologs of the 4'-n-alkyl-4-cyanobiphenyls has been determined. The slopes of the nematic-isotropic phase boundary lines dT/dp exhibit an alternation with the number of carbon atoms in the alkyl chain. This is not the case for the Cr–N (or Cr–S)transition. The SmA–N transition of 4'-n-octyl-4-cyanobiphenyl becomes tricritical at 268 MPa and 92.5 °C.

[25] – d, e, f: 4-cyanophenyl-trans-4-n-butylcyclohexane-1-carboxylate, 4-n-pentylphenyl-trans-4-n-pentylcyclohexane-1-carboxylate, trans-4,4'-di-n-propyl-1,1-bicyclohexyl-cis-4-carbonitrile, trans-4-methoxy-4'-propyl-1,1'-bicyclohexane, and trans-4-methoxy-4'-n-butyl-1,1'-bicyclohexane were investigated in the temperature range 300–500 K up to 800 MPa. Some transitions of these compounds undergo a change from monotropic to enantiotropic behavior at higher pressures. An extended Simon equation reflects the influence of the chain length on the phase behavior of the homologous series.

[26] – c, d: The phase transitions of N-(4-methoxybenzylidene)-4-n-butylaniline, N-(4-ethoxybenzylidene)-4-n-butylaniline, N-(4-pentoxybenzylidene)-4-n-butylaniline, 4-ethoxybenzylidene-4-aminobenzonitrile, 4,4'-dimethoxyazoxybenzene, and 4,4'-diethoxyazoxybenzene were followed to 300 MPa in the temperature range 273–550 K. From the experimentally determined transition entropies at atmospheric pressure and the slopes of the obtained transition curves the transition volumes were evaluated by means of the Clausius–Clapeyron equation.

[27] – d, f: The influence of saturation with helium, nitrogen, argon or argon/carbondioxid mixtures on the phase behavior of 4-n-hexyloxy- and 4-n-decyloxy-4'-cyanobiphenyl was studied up to 270 MPa. The latter compound exhibits a pressure-induced nematic phase above 35 MPa.

The nematic and the smectic phases are destabilized by the saturation with the gases mentioned above. The pressure-induced nematic phase does not appear in the presence of the gases with the exception of helium.

[28] – e, f: The phase transitions of terephthaldiylidene-bis-(4-n-octyl)aniline and terephthaldiylidene-bis-(4-n-dodecyl)aniline were observed in a diamond anvil cell up to 300 MPa in the temperature range 300–600 K. For identifying the different smectic phases (A, C, F, G, I) and detecting phase transitions of higher order a polarizing microscope was used. For the SmA–SmC transition line of the octyl member there seems to be a tricritical point at 110 MPa and 512 K. At 23 MPa and 488 K a pressure-induced nematic phase appears. The dodecyl member shows a pressure-induced phase at 80.5 MPa and 492 K, which is probably a SmA phase. For the butyl and pentyl member of this homologous series see [29] – **f**.

[6] – d, f: The phase behavior of eight members of the homologous series of the 4,4'-bis-(n-alkoxybenzylidene)-1,4-phenylenediamine (butoxy-, pentyloxy-, hexyloxy, heptyloxy, octyloxy, dodecyloxy-, tridecyloxy-, tetradecyloxy-) are observed up to 300 MPa. The nematic phase of all members is stabilized by increasing pressure.

This is confirmed of a pressure-induced nematic phase in the case of the tri- and tetra-decyloxy members. The pressure behavior of the particular smectic ranges (C, G, H, I) is very complex. Some smectic phases vanish at higher pressures. Two compounds investigated additionally: 4-*n*-octyloxy-4'-cyanobiphenyl and 4-cyanobenzylidene-4'-*n*-octyloxy aniline and their mixtures show pressure-induced nematic re-entrant behavior.

[30] – d: 4,4'-bis-methoxy-azoxybenzene was studied up to 200 MPa. The slopes of the phase boundaries obtained are compared with those calculated from the Clausius–Clapeyron equation.

[31] – d: The clearing temperatures of numerous nematic 2-substituted hydroquinone-bis-(4-substituted benzoates) could be measured up to 500 MPa. The N–I phase transition curves are strongly nonlinear and can be described by a modified Simon–Glatzel equation. A preferred orientation of the lateral substituents parallel to the molecular long axis was expected to be strengthened by increasing pressure. The latter effect should manifest itself in enhanced values of the slopes of the transition lines. However, despite the strong deviation of the compounds from the ideal rod-like shape, the data are similar to those of classical nematics.

[32] – d: The DTA measurements were combined with a method which allows recording of dielectric losses. The pressure dependence of the clearing and melting temperatures of 4,4'-bis-methoxy-azoxybenzene and *N*-(4-methoxybenzylidene)-4-*n*-butylaniline could be determined up to 700 MPa.

[33]: Pressure-temperature phase diagrams of the disc-like mesogens benzene-hexa-*n*-

hexanoate and benzene-hexa-*n*-octanoate are presented up to 200 MPa and 110 °C. Both compounds exhibit one mesophase. The mesophase of the latter compound is bounded about 140 MPa.

[34] – d: The phase diagram of 4-(trans-4-pentyl-cyclohexyl)-benzonitrile is presented up to 260 MPa and 97 °C. Dielectric studies were performed in the nematic phase in the pressure range 0.1–140 MPa, the frequency range 1 kHz–13 MHz and the temperature range 38–77 °C (in the isotropic phase up to 45 MPa).

Differential Scanning Calorimetry (DSC)

[35] – f (N*): In order to examine if there is a tricritical point on the SmA–N* transition line of cholesteryl tetradecanoate the corresponding phase transition enthalpy was measured up to 300 MPa in the temperature range 300–600 K. The pressure–temperature phase diagram was obtained by DTA.

[36] – d: Numerical data for the transition temperatures, enthalpies, entropies and volumes are presented for the N–I and two Cr–N phase transitions of N-(4'-ethoxy-benzylidene)-4-*n*-butylaniline. The measurements were performed up to 250 MPa.

[37]: Simultaneous measurements of the rate of heat evolution and volume changes by phase transitions were carried out with a pressure–volume–temperature controlled scanning calorimeter up to 175 MPa. No pressure–temperature phase diagram for the compound under test 4-*n*-pentylphenyl-4-*n*-decyloxythiobenzoate, which has a nematic and three smectic phases, is given.

Optical Transmission Technique

[38] – i: The effect of pressure on the phase transitions in the seventh to tenth homologs

of 4-alkoxy-4-(4'-nitrobenzoyloxy)-benzo-ates (DB.*k*O.NO$_2$) was investigated up to 260 MPa and 200°C. The pressure does not cause any peculiarities for the $k = 7$ and 8 homologs, which have a SmA and SmC phase. Only both SmC phases are suppressed at higher pressures. The $k = 9$ and 10 homologs, however, exhibit a very complex re-entrant phase behavior.

[39] – i, f: The phase transitions of four pure compounds with strongly polar terminal groups (CN or NO$_2$) and two binary mixtures of components with these groups are observed up to 400 MPa and 190 °C. In all cases the partially bilayer SmA$_d$ is bounded, while the nematic phase is stabilized with higher pressures. Furthermore some compounds show re-entrant nematic behavior.

[40] – f, g, i: The pressure–temperature phase diagrams of 4-*n*-pentylphenyl-4-*n*-heptyloxy-thiobenzoate and 4-*n*-pentylphe-nyl-4-*n*-octyloxy-thiobenzoate and their mixtures are given up to 350 MPa and 130 °C. While in all cases the stability of the SmA phase is increased by increasing pressure, the SmC phase of the pure compounds gets bounded at higher pressures.

[41] – b, e, k: The phase behavior of three ferroelectric compounds, the eighth and tenth homologs of 2-methylbutyl-N-(4-*n*-alkoxybenzylidene)-4-amino-cinnamate and 2-chloro-propyl-N-(4-*n*-hexyloxyben-zylidene)-4-amino-cinnamate was studied up to 400 MPa and 175 °C. All the three compounds possess three smectic phases: SmA, SmC* and SmI*. The chiral C phases of both homologs are suppressed at higher pressures. The pressures and temperatures where these phases disappear are 170 MPa, 115 °C and 380 MPa, 162 °C, respectively. They are quite different, although their transition temperatures at at-mospheric pressure are nearly the same. A suppression of the SmC* phase of the third compound is supposed only about 800 MPa.

[11] – a (N*), c, d: The pressure dependence of the melting and clearing temperatures of numerous 4.4′ disubstituted azobenzenes, azoxybenzenes and benzylideneanilines are determined up to 250 MPa and 150 °C. Only the clearing temperatures are given for cholesteryl oleyl carbonate and cholesteryl geranyl carbonate up to 300 MPa and 150 °C. For the mathematical description of the pressure dependence of the phase transition temperatures the melting point equation of Simon–Glatzel in the modified form of Kraut and Kennedy is successfully applied.

[42] – c (N*): The effect of pressure on the phase transition temperatures of cholesteryl oleate, cholesteryl linoleate and cholesteryl linolenate was studied up to 100 MPa and 80 °C. The transition volume of the Sm–N* transition increases with the number of the double bonds in the molecule, while that of the N*–I transition shows little effect.

[20] – f (see DTA)

[43] – a: Turbidity measurements in the isotropic phases of N-(4-methoxybenzyli-dene)-4-*n*-butylaniline, N-(4-ethoxybenzy-lidene)-4-*n*-butylaniline, and N-(4-cyano-benzylidene)-4-*n*-nonylaniline were per-formed up to 120 MPa and 110 °C. N–I tran-sition parameters for the three compounds as clearing pressure, temperature, and order parameter are presented in tabular form.

[44] – d (N*): The N*–I and Cr–I tran-sitions of cholesteryl stearate are observed up to 80 MPa and 103 °C. The transitions are monotropic with respect to the solid phase.

[45] – a (N*, BP): The BP_{III}–I, BP_I–BP_{III} and N*–BP_I transitions of 4-(2-methylbutylphenyl)-4′-(2-methylbutyl)-4-biphenylcarboxylate were recorded up to 120 MPa and 170 °C. The phase boundaries in the pressure–temperature diagram are linear.

Optical Microscopy

[20] – f: The optical microscopy studies were combined with DTA measurements. Compounds and pressure–temperature range are therefore described in the DTA subsection.

An opposed diamond anvil cell was used as a high pressure optical cell. For getting good textures a 0.1 mm thick gasket made of hardened steel was used. A Sm–N transition is identified by a change from the Schlieren to the focal conic texture with ellipses. The monotropic SmB phase shows a mosaic texture.

[14] – f: A diamond anvil cell was used as a high pressure optical cell. The pressure–temperature phase diagram of trans-4-n-propyl-1-(4-n-butoxyphenyl)cyclohexane is given up to 1600 MPa and 277 °C. The compound exhibits a pressure-induced N and SmB phase. The phases were identified by their textures.

[46] – e, f: Pressure–temperature phase diagrams of N-(4-n-butyloxybenzylidene)-4-n-octylaniline and n-hexyl-4-n-pentyloxy-biphenyl-4′-carboxylate are presented up to 300 MPa and 150 °C. Both compounds possess in addition to a SmA phase a SmB phase. One B phase is liquid (hexatic B) the other one crystalline in nature. The neighboring SmA phase of the latter one is bounded above 194 MPa, while the other A phase remains stable with increasing pressure. To explain this differing behavior a model is suggested.

[47] – c: The phase transitions of a racemic composition of 4-(2′-methylbutyl)phenyl-4′-n-nonyloxybiphenyl-4-carboxylate were observed up to 220 MPa and 200 °C. The phase boundaries are extremely linear. The compound under test has five smectic phases: A, C, I, G and H.

[48] – f: The phase diagrams of 4,4′-bis-n-decyloxytolane and 3β-n-tetradecyl-(5α)-cholestane are presented up to 400 MPa and 160°C. While in the case of the former compound the SmA and nematic phase disappear at higher pressures, these phases are stabilized by increasing pressure in the case of the latter compound. This contrary effect of pressure is related to the different dipole moments of both compounds.

[49] – f: The phase transitions of 4-n-octyl-4-cyanobiphenyl and 4-n-octyloxy-4-cyanobiphenyl are observed up to 500 MPa and 140 °C. The SmA phase of the latter compound is a bilayer SmA and is destabilized by increasing pressure.

[50] – d: The pressure–temperature phase diagrams of the 4-n-butyl-, 4-n-heptyl- and 4-n-nonyl-4-methoxy-tolanes are presented up to 400 MPa and 140 °C. The compounds exhibit no peculiarities.

[18] – a, b (TGB): For testing their apparatus the pressure–temperature phase diagrams of 4′-n-pentyl-4-cyanobiphenyl and 1-methylheptyl-4′-(4-n-tetradecyloxyphenyl-propioloyloxy)biphenyl-4-carboxylate are determined up to 60 MPa and 370 K. The latter compound exhibits a TGB_A phase which is suppressed at 25 MPa. The TGB_A–I phase boundary line shows a negative gradient. The measurements could be confirmed on the free-standing film.

Barometric Method

This method allows study of first- and second-order phase transitions of pure compounds. The mesogenic sample is enclosed in a rigid but dilatable metallic cell. The sample pressure and temperature are simultaneously recorded. The thermobarograms obtained exhibit a clear change of slope at the phase transition. The measuring apparatus is called a metabolmeter and requires only a very small amount of mesogen.

[51] – d, e: For testing the barometric method the pressure dependence of the phase transition temperatures of 4-n-octyl-4′-cyanobiphenyl and methoxy- and N-(4-ethoxybenzylidene)-4-n-butylaniline are determined.

[52] – f: For further testing the barometric method the SmA–N transition of 4-n-octyl-4′-cyanobiphenyl is additionally observed. This transition, associated with a very small transition volume, is detected as well.

[53] – e, j: The pressure–temperature phase diagrams of bis-(4-n-heptyloxybenzylidene)-1,4-phenylenediamine and terephthalylidene-bis-(4-n-decylaniline) are presented up to 120 MPa and 200 °C. In the case of the former compound the SmI–SmG and SmI–SmC transitions can be individually observed. The SmG–SmF, SmI–SmF, SmC–SmA and SmA–isotropic transitions of the latter compound are all detectable.

[54] – d, f: The method is extended to phase transitions of binary mixtures of 4-methoxy-4′-ethyltolane and 4-methoxy-4′-n-nonyltolane as well as of terephthaldiylidene-N,N′-bis-(1-methyl-heptyl-4-aminocinnamate) and N,N′-bis-(4-n-heptyloxybenzylidene)-phenylene-1,4-diamine.

[55] – f (TGB, N*): The phase transitions of two homologs ($n = 16$ and $n = 18$) of the

4-(3-fluoro-4(R)- or (S)-methylheptyloxy)-4-(4-fluorobenzoyloxy)-tolanes were studied up to 140 MPa and 125 °C. Both compounds have a TGB$_A$ phase which is stabilized by increasing pressure. The $n = 18$ homolog exhibits a pressure-induced cholesteric phase resulting in a TGB$_A$–N*–I triple point.

[56] – discotic: The pressure–temperature phase diagrams of two disc-like mesogens are presented up to 90 MPa and 170 °C. The compounds investigated are (–)-2.3.6.7.10.11-hexa-[S-(3-methyl)-n-nonanoyloxy]-triphenylene and 2.3.6.7.10.11-hexa-(n-dodecanoyloxy) triphenylene. The transitions between both columnar mesophases of the first compound and between the D_0, hexagonal and rectangular columnar mesophases of the second one can be detected.

Selective Light Reflection

Chiral mesogens can reflect light within a narrow region of wavelength. The reflection is represented by the wavelength of maximum light reflection, which shows a characteristic pressure and temperature behavior near and at a phase transition point. The light reflection causes a quasiabsorption and in the most cases therefore is measured by an absorption spectrophotometer.

[57] – a (N*, BP): The measurements were performed with cholesteryl nonanoate up to 115 MPa and 137 °C. The BP$_{II}$–BP$_{III}$ transition is detected by a disappearance of the reflection band, the BP$_I$–BP$_{II}$ and N*–BP$_I$ transitions by jumps in the wavelength of maximum reflection.

[58] – g (N*): The SmA–N* phase transitions of four cholesteryl n-alkanoates, the cholesteryl decanoate, tridecanoate, pentadecanoate and heptadecanoate are observed

up to 260 MPa and 130 °C. Since the wavelength of maximum light reflection is directly proportional to the pitch of the cholesteric phase, the pressure behavior of the reflection wavelength corresponds to that of the pitch. If the n-alkyl chain is not too short, the cholesteric phase transforms into the SmA phase with a finite pitch. With increasing transition pressure this transition pitch is shifted to higher values till at a definite transition pressure the pitch is infinite. For the tri-, penta- and heptadecanoate this pressure is about 100 MPa.

[59] – d (N*): The pressure–temperature phase diagrams of cholesteryl n-pentanoate and n-hexanoate are presented up to 190 MPa and 150 °C. The pressure–temperature behavior of the wavelength of maximum light reflection of both compounds reveals no indication of a pressure-induced SmA phase, which was supposed for the pentanoate by other authors. For both compounds separate crystallization and melting curves were found.

[60] – g (N*): The SmC*–N* phase transition of 4-n-hexyloxyphenyl-4'-(2-methylbutyl)biphenyl-4-carboxylate was studied up to 200 MPa and 115 °C. For the SmC* phase also light reflection measurements were performed. Probably because of the high viscosity of this phase the reproducibility was low. No crossing over of the phase transition from first- to second-order was observed at higher pressures.

X-ray

[61] – j: The high pressure X-ray investigations were performed with a high pressure X-ray camera equipped with two cone shaped X-ray windows made of boron single crystals. The apparatus works up to 500 MPa and 120 °C. The pressure dependence of the crystalline-SmG* melting temperature of 4-(2'-methylbutyl)phenyl-4'-n-octylbiphenyl-4-carboxylate is given (see also [62]).

[63] – k: The high pressure X-ray measurements were carried out with an X-ray diffraction system equipped with a rotating anode and a bent quartz crystal. The patterns are registered with a stable position-sensitive X-ray detector. The sample is contained in a quartz capillary, which is located in a beryllium cylinder. The pressure is transmitted to the sample by a flexible Teflon bellows, which is connected to the upper part of the cylinder (for details see [64]). The thickness of the smectic layers (SmA, SmC*, SmI*) is studied as a function of pressure at different temperatures. While the layer spacing of a SmC phase with a nematic phase as neighbor is independent of pressure, that of a chiral C phase with a SmA phase as neighbor changes strongly as the pressure increases from the SmC*–SmA transition pressure. The compounds investigated are 4,4'-bis(heptyloxy)azoxybenzene (no SmA phase) and 2-methylbutyl-N-(4-n-decyloxybenzylidene)-4-aminocinnamate (with SmA phase).

[65] – f (TGB, N*): The phase behavior of N-[4-((6-cholesteryloxycarbonyl)pentyloxy)benzylidene]-4-n-butylaniline was investigated by a wide-angle X-ray scattering apparatus up to 100 MPa and 200 °C. The compound exhibits a complex phase sequence with a TGB phase at atmospheric pressure. Only a few phase transitions (without TGB phase) can be observed at higher pressures. For details of the apparatus see [66].

Nuclear Magnetic Resonance (NMR)

[67] – d: The high pressure vessel was machined of nonmagnetic Cu–Be and

Cu–Ni alloys. The equipment works up to 700 MPa. The pressure transmitting medium was purified helium gas. The NMR coil and the sample are contained inside a heating tube. The N–I and the Cr–N transitions of 4,4'-bis-methoxy-azoxybenzene were observed up to 400 MPa and 210 °C. The order parameter of the nematic phase could be determined from a characteristic doublet of the NMR CW spectrum. The order parameter at the N–I transition is found to be constant up to 300 MPa.

Optical Activity

[15] – a (N*, BP): The optical rotation was determined by a half-shade polarimeter. The high pressure optical cell is similar to that of the optical transmission technique.

All phase transitions involving blue phases (BP_I, BP_{II}, BP_{III}) of S-(+)-4'-(2-methylbutyl)phenyl-4-n-decyloxy and -dodecyloxy benzoate could be observed up to 280 MPa and 103 °C. The decyloxy homolog exhibiting all three blue phases at atmospheric pressure looses BP_{II} at 120 MPa. The dodecyloxy homolog only with BP_I shows a pressure-induced BP_{II} already at lower pressures. A correlation between the pretransitional behavior of the optical activity in the isotropic liquid phase and the phase behavior of the blue phases at high pressures is found.

Re-entrant Behavior

In 1975 Cladis discovered the sequence of phases nematic, smectic, and again nematic at atmospheric pressure. The lower-temperature nematic phase was designated as the 're-entrant nematic' phase (N_{re}). By 1977 Cladis [68] was successful in giving evidence of a pressure-induced 're-entrant nematic' phase (Fig. 2). The investigated compounds were cyano Schiff bases and cyanobiphenyls with terminal n-alkyl or

n-alkoxy chains, which all were known (or suspected) to exhibit a bilayer SmA phase.

The transitions were observed with a high pressure optical microscope stage working up to 1000 MPa.

The most important results are the following (Fig. 2):
1. The re-entrant phase exists only in the supercooled region of the liquid.
2. There is a maximum pressure p_m above which the bilayers are destabilized.

It turned out that p_m increases with increasing transition enthalpy of the SmA–N transition determined at atmospheric pressure. p_m decreases with decreasing number of methylene groups interacting within a layer.

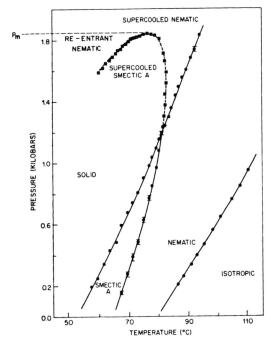

Figure 2. Pressure–temperature phase diagram for 4-cyano-4'-octyloxybiphenyl. Data taken on the re-entrant nematic–smectic A transition in the supercooled liquid are shown as squares. P_m is the maximum pressure at which the smectic exists for this compound. (From [68], reproduced by permission of American Physical Society.)

The studied compounds all have a terminal cyano group which means a very strong dipole. Cladis et al. [69] proposed a structural model of the bilayer smectic A phase for this kind of molecules (Fig. 3). The molecules are assumed to be associated in antiparallel pairs, which results in a weak interacting between the different polar parts of the pairs and a less dense packing of the molecules. Thus a transition of such bilayer SmA phase to a re-entrant nematic phase seems to be evident, because in this phase empty spaces of the structure are filled up more efficiently. Probably for similar reasons Pollmann et al. [70–72] found a pressure-induced re-entrant cholesteric phase behavior for ternary mixtures of cholesteryl n-alkanoates which, however, are terminally nonpolar (see Fig. 4). The mixtures of

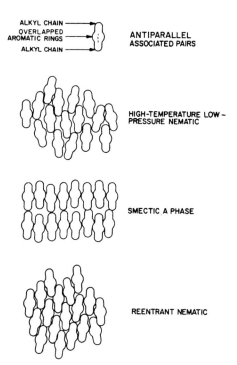

Figure 3. Schematic arrangement of antiparallel associated pairs in the nematic, smectic A, and re-entrant nematic phase. (From [64], reproduced by permission of American Physical Society.)

cholesteryl tetradecanoate (Ch-14), nonanoate (Ch-9) and propionate (Ch-3) possess one component which has a considerably shorter n-alkylcarboxy chain than the other components. This probably leads to 'holes' in the smectic structure. With increasing pressure the molecules are increasingly forced into these 'holes' until the layer structure of the SmA phase finally is destroyed and the re-entrant cholesteric phase appears.

Guillon et al. [73] investigated the pressure dependence of the SmA phase layer spacing of N-(4-cyanobenzylidene)-4-n-octyloxyaniline by X-ray-diffraction measurements. While the layer spacing decreases, when the SmA phase is pressurized towards the solid phase, it remains constant when this smectic phase is pressurized towards the re-entrant nematic phase.

Shashidhar and Rao [74] performed high pressure X-ray studies on liquid crystals with re-entrant behavior with an opposed diamond anvil cell. They found that the layer spacing of the SmA phase of 4-n-octyloxy-4'-cyanobiphenyl first decreases more or less linearly with increasing pressure up to 140 MPa, then increases at still higher pressures. Since this compound shows re-entrant nematic behavior at high pressures, this result confirms the prediction of Cladis et al. that the occurrence of a re-entrant nematic phase is associated with an expansion of the SmA phase layer spacing.

Shashidhar et al. [75] studied the influence of pressure on the SmA- (re-entrant) nematic and N–I phase boundaries of mixtures of 4-n-hexyloxy- and 4-n-octyloxy-4'-cyanobiphenyl. The maximum pressure where the SmA and re-entrant nematic phase, respectively, still exist, decreases with increasing mole fraction, x, of the hexyloxy homolog till at $x \approx 0.30$ the SmA phase disappears. Just in this mole fraction region the slope of the N–I transition

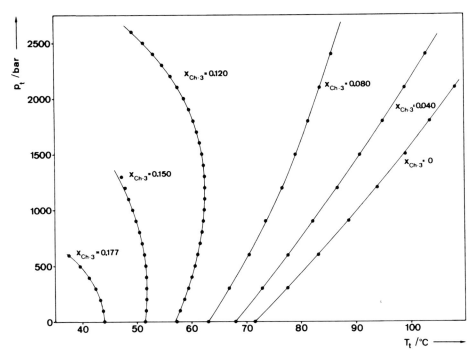

Figure 4. Pressure–temperature smectic A–cholesteric phase boundary lines of the ternary system Ch-9/Ch-14/Ch-3; $x_{Ch-9}/x_{Ch-14} = 2.11$ (x = mole fraction). (From [71], reproduced by permission of Taylor & Francis Ltd.)

(at 0.1 MPa) exhibits a marked anomaly. Corresponding investigations of mixtures of 4-n-octyloxy-4′-cyanobiphenyl and N-(4-n-cyanobenzylidene)-4′-octyloxyaniline were carried out by Herrmann [76].

Prasad et al. [77] studied the phase transition behavior of 4-n-octyloxy- and 4-n-nonyloxybenzoyloxy-4′-cyanoazobenzene. While the pressure–temperature diagram of the eighth homolog shows no peculiarities, the ninth one shows already at atmospheric pressure a SmA (partially bilayer)–N, N_{re}–SmA (partially bilayer) and a re-entrant SmA (monolayer)–N_{re} phase transition. The partially bilayer SmA–(re-entrant) N phase boundary shows an elliptic shape and the partially bilayer SmA is suppressed at higher pressures. The re-entrant monolayer SmA–re-entrant N phase boundary is a straight line at all pressures.

Similar investigations of eight pure compounds were performed by Kalkura et al. [78] and of further five pure compounds and some mixtures by Prasad et al. [79]. While pressure-induced re-entrant cholesteric phase behavior of mixtures of terminally nonpolar components is already known [70], Ratna et al. [80] found, that a mixture of the terminally nonpolar components n-dodecyl-N-4-(4-ethoxybenzylideneamino)-α-methylcinnamate and 4-n-heptyloxyphenyl-4-(4-ethylcyclohexanoyloxybenzoate) not only shows a re-entrant N phase behavior at atmospheric pressure but also at higher pressures.

Illian et al. [81] observed a closed loop re-entrant N–SmA phase boundary in a temperature–mole fraction diagram at atmospheric pressure. At higher pressures the SmA region of existence decreases until at a critical pressure of 145 MPa this region is reduced to a point. The measurements were carried out with mixtures of 4-*n*-heptyl-oxy-4′-cyanobiphenyl and a mixture of *N*-4-cyanobenzylidene-4′-*n*-octyloxyaniline and 4-*n*-pentylphenyl-4-*n*-octyloxybenzoate.

The pressure-induced re-entrant phase behavior is not restricted to one re-entrant N phase and not only to N phases. Raja et al. [38] studied the effect of high pressure on the phase transitions in the seventh to tenth homologs of 4-alkoxyphenyl-4′-nitrobenzoyloxybenzoates. The decyloxy member, and only this one, exhibits a conspicuous phase diagram: besides two pressure-induced SmA phases and a bounded SmC phase a 'quadruple re-entrance' is observed: SmA–N–SmA$_d$–N–SmA$_d$–N.

Illian et al. [82] studied the pressure effect on the phase transition behavior of binary mixtures of terminally polar and nonpolar components which exhibit induced SmA phases. The re-entrant N phase is stabilized by increasing pressure and at about 101 MPa the SmA–re-entrant N phase boundary meets the N–SmA one. At higher pressures a nematic gap appears and finally the SmA phase on the polar side of the temperature–mole fraction diagram (175 MPa) disappears.

Daoudi et al. [83] applied the thermodynamic approach 'equal Gibbs energy analysis' to the pressure–composition phase diagrams of binary mixtures exhibiting re-entrant phase behavior. Three different solution models are tested. The experimental data for the 4-*n*-hexyloxy- and 4-*n*-octyloxy-4′-cyanobiphenyl system are successfully described by the regular solution hypothesis.

6.2.4.3   Critical Phase Transitions Under Pressure

The different types of liquid crystals are closely related to one another in a thermodynamic sense so that a great potential for multicritical phenomena can be supposed.

Tricritical Behavior

Following the terminology of Landau and Lifshitz [84] order parameters and quantities of state, respectively, for instance the enthalpy, entropy and volume, change discontinuously at a phase transition of the first kind, and continuously at a phase transition of the second kind. More commonly the terms 'first order' and 'second order' are used.

When a line of first-order phase transitions changes to a line of second-order phase transitions the corresponding point in the phase diagram is called a critical point of the second kind (Tisza [85]) or a tricritical point (Griffiths [86]). In the following the latter expression will be taken. In 1973 Keyes et al. [87] could find such a tricritical point (TCP) for the SmA–N* phase transition of cholesteryl oleylcarbonate at 266 MPa and 60.3 °C. The TCP was indicated by a sharp discontinuity in the transmitted light. Shashidhar et al. [19] could confirm this discovery by DTA measurements (p (TCP) = 267 MPa) and later in Shichijyo et al. [88] by volume and DTA investigation (p (TCP) = 300 MPa) (see also [89]). The TCP p–T coordinates obtained in this way are probably essentially too high since the entire disappearance of the DTA peak is taken as indication of the TCP. In the case of cholesterogens following de Gennes [90, 91] Pollmann et al. [92] determined by selective light reflection measurements that transition pressure as the tricritical pressure, where the pitch of the cholesteric helix turns to infinity according to the divergence of the characteristic coherence length. At this tri-critical pressure clear transition volume [92] and enthalpy effects [35] remain, which can be understood as pretransitional effects. The latter disappear only at still higher pressures. The role of the pretransitional enthalpy of the SmA–N* transition can be seen in [16], where the tricritical pressures of nine cholesteryl *n*-alkanoates (nonanoate until heptadecanoate) were determined. For 4'-*n*-octyl-4-cyanobiphenyl, Shashidhar et al. [24] found a TCP of this transition at 268 MPa and 92.5 °C.

Garland [10] studied second-order SmA–N and SmC–SmA phase transitions by very precise heat capacity measurements up to 300 MPa. Similar measurements of the critical heat capacity near the SmA–N transition were performed by Kasting et al. [93, 94]. McKee et al. [95] carried out orientational order determinations near a possible SmA–N TCP ($p = 289$ MPa, 140 °C) by NMR. From McMillan's theory [96] in the case of a TCP of the SmA–N (N*) transition at atmospheric pressure a value of 0.866 (model parameter $\delta = 0$) for T(SmA–N)/T(N–I) follows. Thus a rough test by the corresponding transition temperatures at higher tricritical pressures is possible.

Bartelt et al. [28] suppose a TCP also for the SmC–SmA transition. The first-order SmC–SmA transition of terephthaldiylidene-bis-(4-*n*-octylaniline) probably changes to second-order at 110 MPa and 239 °C.

Multicritical Behavior

Re-entrant N–SmC–SmA and
N–SmA–SmC Multicritical Behavior

The re-entrant N–SmC–SmA multicritical point and the N–SmA–SmC multicritical point are, by definition, points in the temperature–concentration or pressure–temperature plane at which three second order phase boundaries meet. At this point all in-volved three phases become indistinguishable. After no N–SmA–SmC point for a single component liquid crystal system could be detected by studying the pressure–temperature phase behavior of N-(4-*n*-pentyloxybenzylidene)-4'-*n*-hexylaniline up to 800 MPa [97], Shashidhar et al. [98, 99] were successful in finding another kind of multicritical point at 52 MPa and 86.2 °C: a N_{re}–SmC–SmA point for N-4-(4-*n*-decyloxybenzoyloxy)-benzylidene-4'-cyano-aniline.

Shortly after finding this point Shashidhar et al. also observed a N–SmA–SmC point under high pressures [100]. The N–SmA–SmC point of 4-*n*-heptacylphenyl-4'-(4''-cyanobenzoyloxybenzoate) occurs at 30.4 MPa and 149.9 °C. The most remarkable feature of the obtained p–T phase diagram is its topology near the N–SmA–SmC point which is nearly the same as obtained for binary mixtures in a temperature-concentration diagram at atmospheric pressure (Fig. 5a and b). A quantitative description of the three phase boundaries of the pure compound near the N–SmA–SmC point agrees so closely with that of the binary mixtures that the universal behavior of the N–SmA–SmC point of the latter at atmospheric pressure can be extended to such a point at higher pressures for single component systems. This is not valid for the behavior of the N_{re}–SmC–SmA point under high pressure of the above mentioned compound which is different from that at atmospheric pressure in the case of binary mixtures. The N_{re}–SmC–SmA point at atmospheric pressure, however, can on certain conditions exhibit the same universal behavior as the N–SmA–SmC point [101]. There is no satisfying theoretical explanation for the uniqueness in the topology of the phase transition lines near the N–SmA–SmC and N_{re}–SmC–SmA multicritical points.

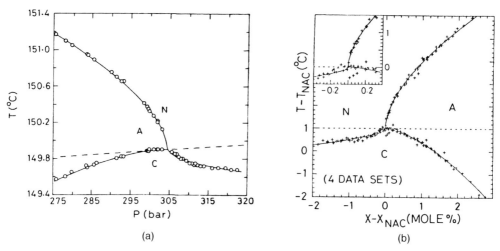

Figure 5. (a) High resolution temperature–pressure diagram in the vicinity of the N–A–C multicritical point in 4-*n*-heptacylphenyl-4′-(4″-cyanobenzoyloxybenzoate). The solid lines are computer fits of data evaluated with equations representing the NA, NC, and AC phase boundaries, respectively. (b) The universal temperature–concentration plot showing the data for four binary liquid crystal systems. (From [100], reproduced by permission of American Physical Society.)

Legrand et al. [102] report an electroclinic effect in the cholesteric phase near a N*–SmA–SmC* multicritical point of the $n = 8$ homolog of a biphenyl benzoate series. This 'chiral' analogon to the N–SmA–SmC point is found at 15 MPa and 144 °C.

Smectic C–Smectic A -TGB$_A$*
and Smectic A-TGB$_A$–Cholesteric
Multicritical Behavior

Experimental evidence under high pressures is given by Anakkar et al. [103] for two new kinds of multicritical points involving Twist Grain Boundary (TGB) phases: SmC*–SmA-TGB$_A$ and SmA-TGB$_A$-N* points for single component systems. The investigated compound was the $n = 12$ homolog of the series 3-fluoro-4((*R*) or (*S*)-methylheptyloxy)-4′-(4-alkyloxy-3-fluorobenzoyloxy)tolanes. The *p–T* coordinates of the multicritical points are: SmC*–SmA-TGB$_A$: 67.5 MPa, 91.5 °C; SmA-TGB$_A$-N*: 106 MPa, 113 °C. The result is in good agreement with the Renn–Lubensky theory.

Anakkar et al. [104] also investigated the $n = 10$ and $n = 11$ members of the above tolane series which both exhibit the TGB$_A$ phase. For $n = 11$ also experimental evidence is given for a SmC*–SmA-TGB$_A$ multicritical point ($p = 13$ MPa, $T = 367$ K).

Bicritical Behavior

A bicritical point is a point in the temperature–concentration or pressure–temperature plane at which two second-order phase boundaries and one first-order phase boundary meet. In order to find such a bicritical point in the pressure–temperature plane Bahr et al. [105] studied the pressure–temperature phase behavior of binary mixtures of 4-(4-*n*-butyloxybenzoyloxy)-4′-nitroazo-benzene and 4-(4-*n*-nonyloxybenzoyloxy)-4′-cyanoazobenzene. From the *p–T* phase diagrams of the pure components which are also presented a bicritical behavior of their mixtures can be expected. Two mixtures with molar fractions of about 0.80 for the cyano component show an abrupt change of the slope of the SmA–N phase boundary at 30 MPa. Here the transition of the partially bilayer SmA (A_d) phase to the nematic phase changes to a transition of the monolayer SmA phase (A_1) to the nematic phase. Just at this crossover point the SmA_1–SmA_d phase boundary, which, however, could not be detected by the used optical transmission technique, should meet the SmA–N phase boundary.

Since the SmA_1–N and SmA_d–N transitions are expected to be second order (low T_{A-N}/T_{N-I}) values [97]) and the SmA_1–SmA_d transition first-order (symmetry arguments), the meeting point should be a bicritical point. Moreover the topology of the p–T phase diagram obtained resembles that of a diagram exhibiting known bicritical points [106].

Prasad et al. [107] studied the critical character of the N*–TGB–SmA and TGB–SmA–SmC* meeting points in the pressure–temperature phase diagram of mixtures of 4-(2-methylbutyl)phenyl-4′-*n*-octyl-biphenyl-4-carboxylate and 4-*n*-dodecyloxy-biphenyl-4′-yl-4-(2′-methylbutyl)-benzoate. For a weight fraction, $x = 0.25$, of the benzoate there is a pressure-induced TGB phase between the N* and the SmA phase. Analysis of the topology of the pressure–temperature diagram in the vicinity of the resulting N*–TGB–SmA meeting point more points to a critical than to a bicritical point. At $x = 0.64$ there appears a pressure-induced SmA phase between the TGB and the SmC* phase so that a TGB–SmA–SmC* meeting point is observed. The topology of this phase diagram suggests that the latter point is a bicritical point.

6.2.4.4 Pressure-Volume-Temperature Behavior

The physical properties of a substance depend on the intermolecular distance and

since the change of the specific volume or density with pressure and temperature means a change of the mean distance between the molecules, p–V–T data are required to understand the pressure–temperature behavior of physical properties.

Experimental Techniques

(a) Tungsten Tip Surface Scanning by Screw Spindle [108]

The apparatus allows the measurement of p–V–T data of liquid crystal compounds up to 400 MPa.

The sample is enclosed in a glass piezometer and separated from the pressure transmitting fluid by mercury. As the sample volume decreases with increasing pressure the level of the mercury, which is contained in a calibrated glass tube, sinks and thus opens the electric contact to a tungsten tip. By means of a screw spindle the contact of the tungsten tip with the mercury level is made again. The number of revolutions needed for it is a measure for the decrease in volume of the sample and can be measured by an electronic altimeter outside the autoclave. The reproducibility of the density is ± 0.0003 g cm^{-3}.

(b) The Bridgman Flexible Bellows with Slide Wire [109]

In this experimental set up the liquid crystal is contained in a stainless-steel bellows which is placed inside a pressure-vessel-double-oven arrangement. The oven temperature is regulated by a proportional servomechanism using a sensor thermocouple. The Bridgman bellows with slide-wire technique [110] is used for the measurement of the sample volume. As the sample is compressed the flexible bellows contracts and moves a slide-wire under a contact. The fixed end does not move with respect to the contact.

The position of the slide-wire from which the change in volume of the liquid crystal sample can be determined is measured by means of an a.c. bridge arrangement. The accuracy in determination of the specific volume is ± 0.00015 cm^3 g^{-1}.

(c) Piston Displacement [111]

A steel cylinder is used to contain the sample under investigation. The sample is compressed by a steel piston, the displacement of which is measured by means of a differential transformer. From the displacement of the piston the volume decrease of the liquid crystal sample can be evaluated.

The pressure is determined from the pressure dependence of the electrical resistance of a manganin wire which is placed inside the pressure chamber with the sample. The accuracy of the measurements of the volume changes is within $\pm 0.3\%$.

Experimental Results and Theoretical Description

In the following only those publications which contain important theoretical descriptions are specified. The p–V–T investigated compounds of the publications not mentioned here, are however, in 'the list of compounds' with the referring citation.

The letter (a–c) after the following citations means the experimental method

which has been used (see 'experimental techniques').

[112] (a): p–V–T and viscosity data of the methoxy (MBBA) and ethoxy (EBBA) homologs of the N-(4-alkyloxybenzylidene)-4'-n-butylaniline series are presented up to 200 MPa.

The p–V–T behavior of both compounds can be well described by the semiempirical Tait equation:

$$\kappa = -\frac{1}{V_0}\left(\frac{\partial V}{\partial p}\right)_T = \frac{C}{B+p} \tag{1}$$

here used in the form

$$\frac{V_0 - V}{V_0} = C \cdot \log\frac{B+p}{(B+p_0)} \tag{2}$$

The constants B and C were evaluated from the experimental data and are given together with the compressibility coefficient κ at 1 bar for MBBA in Table 1.

C can be assumed to be independent of temperature. B is correlated with the molecular structure of the substance.

[108] (a): The p–V–T data for three trans-4-n-alkyl-1-(4-cyanophenyl)-cyclohexanes (propyl, pentyl, heptyl) and one eutectic mixture were measured up to 200 MPa.

For isotherms in the nematic phase beginning at high N–I transition pressures and temperatures (p_1, V_1), the V_0 of the Tait equation (see [112]) is not known with suf-

Table 1. The compressibility coefficient and the coefficients of the Tait equation in both phases of MBBA.

| | $\kappa \cdot 10^6$ [bar^{-1}] | | | | B [bar] | |
| | 1 bar | Transition point | | C | | |
°C	isotropic	isotropic	nematic	for all temperatures	isotropic	nematic
60	54.55	45.88	43.96	isotropic	2116.2	2087.4
70	55.94	41.77	40.13	0.2658	2063.7	2034.5
80	57.35	38.92	36.65	nematic	2012.5	1977.9
90	59.10	36.60	34.38	0.2619	1947.0	1908.7

ficient accuracy. The equation is therefore modified to:

$$\frac{V_1 - V}{V_1} = \frac{C \cdot V_0}{V_1} \cdot \log\left(\frac{B + p}{B + p_1}\right)$$

$$= C^* \cdot \log\left(\frac{B + p}{B + p_1}\right) \quad (3)$$

The constants B, C, and C^* (not independent of temperature) are listed in Tables 2 and 3 for the isotropic and nematic phase, respectively. An influence of the length of the alkyl chain is obvious.

[109] (b): p–V–T data were obtained for the nematic range of the first six members of the homologous series of 4,4′-bis-alkoxyazoxybenzenes up to 123 MPa.

Measurements performed at constant volume showed that the temperature range of the nematic phase is considerably larger at constant volume than at constant pressure.

Table 2. Tait parameter for the isotropic range, B [bar].

| | Temperature (°C) | | | |
	60	70	80	90
Propyl (30%)	1867	1812	1756	1701
Pentyl (40%)	1847	1793	1738	1683
Heptyl (30%)	1822	1767	1713	1657
Eutectic mixture	1834	1779	1724	1670

$C = 0.2850$ for all compounds and temperatures.

Measurements at constant temperature established a decrease in N–I transition volume and fractional volume of a substance as one moves up to higher transition pressures and temperatures. The volume at the N–I transition point also decreases. The results in this work, especially the influence of the length of the flexible tail of the molecules on the data, are compared with the predictions of numerous theories.

[113] (b): The p–V–T behavior of the non-re-entrant nematic (N) 4,4′-bis-n-heptyloxy-azoxybenzene (7OAB) and the re-entrant nematic (N$_{\text{re}}$) 4-n-octyloxy-4′-cyanobiphenyl (8OCB) were compared with one another. The N–I transition volumes of these two compounds are similar at low transition temperatures and pressures but exhibit quite a different transition temperature dependence.

The SmC–N transition volumes are an order of magnitude larger than those of the N$_{\text{re}}$–SmA transition. For theoretical considerations the isothermal compressibility coefficient and the parameter

$$\gamma = -\frac{\partial (\ln T/T_0)}{\partial (\ln V/V_0)} \quad (4)$$

are used, where T_0 and V_0 represent a point on the phase equilibrium curve. The latter parameter is sensitive to the relative

Table 3. Tait parameter for the nematic range, B [bar].

| | $C^* \times 10^4$ Temperature (°C) | | | | B [bar] Temperature (°C) | | | |
	60	70	80	90	60	70	80	90
Propyl	1834	1862	1889	1917	890	814	715	622
Pentyl	1818	1845	1872	1899	969	865	761	657
Heptyl	1807	1834	1861	1889	984	879	774	669
Eutectic mixture	1827	1854	1882	1907	959	855	751	647

$C = 0.1800$ for all compounds and temperatures.

strength of steric repulsions as compared to attractive potentials.

[114] (b): $p–V–T$ measurements were used to investigate short-range smectic fluctuations in the N–I transitions of the seventh to ninth members of the homologous series of 4,4'-bis-n-alkylazoxybenzenes.

The experimental results confirm the theoretical prediction that the N–I transition volume increases when the N–I transition gets closer to a SmA–N transition. This is due to short-range smectic order. This influence of the nematic range on the N–I transition volume can be very obviously seen for the ninth member of the series which exhibits a SmA–N–I triple point at 92 MPa and 103 °C.

[111] (c): Pressure–volume isotherms of the isotropic and nematic phase are presented in the temperature range 57–83 °C up to 90 MPa. Whereas the N–I transition volume remains constant with increasing transition pressure that of the Cr–N transition clearly decreases.

According to an equation of Hiwatari and Matsuda [115] the $p–V–T$ data for the nematic state yields an $1/r^{10.8}$ dependence of the repulsive potential energy on the intermolecular distance r or $\rho^{3.6}$, when the density ρ is used. McColl [116] had found $\rho^{3.7}$ for 4,4'-bis-methoxy-azoxybenzene.

[117] (c): The pressure–volume isotherms of 4-(trans-4-pentylcyclohexyl)benzonitrile were measured near its clearing transition. For the parameter $\gamma = -\partial \ln T_C / \partial \ln V_C$ a value of +5.24 is determined and related to the intermolecular potential energy. The value of γ' depends on the number of carbons in the alkyl chain.

[118] (b): $p–V–T$ measurements were carried out on N-(4-methoxybenzylidene)-4-n-butylaniline (MBBA), N-(4-cyanobenzyli-

dene)-4-n-octyloxyaniline (CBOOA), 4,4'-bis-n-heptyloxy-azoxybenzene (HAB) and cholesteryl nonanoate (CN).

The variations in the N–I transition volumes and the isothermal compressibility coefficients of the different compounds can be attributed to a large extent to the effects of end chain flexibility. The N(N*)–I transition volumes of CBOOA and CN decrease more rapidly with increasing transition temperature than that of the relatively inflexible MBBA.

The transition densities of all four compounds increase with increasing transition temperatures, for MBBA the slope is the smallest, for HAB the largest. The experimental result for MBBA is compared with the predictions of three theoretical models. The nematic ordering seems to be almost entireley a result of the repulsive forces.

[119] (b): The $p–V–T$ equation of state of 4-n-octyloxy-4'-cyanobiphenyl was measured to test the hypothesis of Pershan and Prost [120] that the re-entrant nematic phase occurs because the SmA phase can exist only near an optimum density. The experimental results are in substantial disagreement with the hypothesis.

[121–124] (c): The $p–V–T$ behavior of 4-n-octyl-4'-cyanobiphenyl was observed near the N–I transition. While the N–I transition volume was almost constant with increasing transition temperature (T_{tr}) the volume of the nematic phase at the transition temperature V_{tr} decreases.

The plot of $\ln T_{tr}$ vs. $\ln V_{tr}$ yields a straight line with a slope of -4.3. This value agrees with the prediction of the Pople–Karasz theory not with that of Maier–Saupe. An important role of repulsive forces in the N–I transition is suggested.

Analogous measurements with the pentyl, hexyl, and heptyl members of the series result -7.62, -6.10, and -5.15 for the above plot, hence a decrease with the length of the alkyl chain [122]. The larger values of these compounds compared to that of the octyl member make it obvious that in this case the volume-dependent part of the intermolecular potential for the nematic state is harder than the repulsive part of the Lennard–Jones potential but softer than the hard-rod potential.

In a p–V–T study on 4-fluorophenyl-trans-4-n-heptylcyclohexanecarboxylate $d\ln T_{tr}/d\ln V_{tr} = -2.98$ was obtained. This value is noticeably lower than those of the former compounds and attributed to a larger intramolecular potential at the phase transition point (see also 4′-mono-4-propyl-cyanophenylcyclohexane with a value of -8.29 [124]).

The results are discussed on the basis of the molecular structure, particularly of the molecular rigidity.

6.2.4.5 Appendix

List of Compounds

4-n-alkyl-4′-cyanobiphenyl (pentyl to heptyl) [122]
trans-4-alkyl-(4-cyanophenyl)-cyclohexane (propyl, pentyl, heptyl) [108]
4-(trans-4-n-butylcyclohexyl)benzonitrile [127]
cholesteryl butyrate [133]
cholesteryl hexadecanoate (palmitate) [130]
cholesteryl hexanoate (capronate) [129]
cholesteryl nonanoate (pelargonate) [118, 131]
cholesteryl octadecanoate (stearate) [137]
cholesteryl oleyl carbonate [88]
cholesteryl pentanoate (valerate) [128]
cholesteryl propionate [134]
cholesteryl tetradecanoate (myristate) [131]
α,ω-bis[4-cyanobiphenyl-4′-yl]alkane (nonane, decane) [126]

4,4′-bis-alkyloxy-azoxybenzene (methyl to hexyl) [109]
4,4′-bis-alkyl-azoxybenzene (heptyl to nonyl) [114]
4,4′-bis-methoxy-azoxybenzene [109, 135, 136]
4,4′-bis-n-octyloxy-azoxybenzene [132]
N-(4-ethoxybenzylidene)-4-n-butylaniline [112]
4-fluorophenyl-trans-4-heptylcyclohexanecarboxylate [123]
4,4′-bis-n-heptyloxy-azoxybenzene [113, 118, 131]
N-(4-methoxybenzylidene)-4-n-butylaniline [112, 118, 125]
4-n-octyl-4′-cyanobiphenyl [121]
4-n-octyloxy-4′-cyanobiphenyl [113, 119]
4-(trans-4-n-pentylcyclohexyl)benzonitrile [108, 117]
N-(4-n-pentyloxybenzylidene)-4-n-butylaniline [111]
4′-mono-4-propyl-cyanophenyl-cyclohexane [108, 124]

Physical Properties under Pressure

Property	Refs.
Viscosity	[112, 138, 139]
Dielectricity	[140–145, 146–148, 152, 153]
Ferroelectricity	[150]
Refraction	[149, 151]
Elasticity (elastic constants)	[151]
Ultrasound	[88, 154]
Electrooptic (elastic constant/ dielectric anisotropy ratio)	[155]
Acoustic	[156, 157]
Acoustooptic	[158]

6.2.4.6   References

[1] G. A. Hulett, *Z. Phys. Chem.* **1899**, *28*, 629–672.

[2] N. A. Puschin, I. W. Grebenschtschikow, *Z. Phys. Chem.* Stöchiometrie und Verwandtschaftslehre (Leipzig) **1926**, *124*, 270–276.

[3] J. Robberecht, *Bull. Soc. Chim. Belg.* **1938**, *47*, 597–639.

[4] P. D. Garn, R. J. Richardson, *Thermal Analysis* **1971**, *3*, 123–130.

[5] A. S. Reshamwala, R. Shashidhar, *J. Phys. E* **1977**, *10*, 180–183.

[6] J. Herrmann, H. D. Kleinhans, G. M. Schneider, *J. de Chimie Physique* **1983**, *80*, 111–117.

[7] R. Sandrock, Doctoral Thesis, **1982**, Ruhr-Universität Bochum, Germany.

[8] A. Bartelt, G. M. Schneider, *Rev. Sci. Instrum.* **1989**, *60*, 926–929.

[9] M. Kamphausen, *Rev. Sci. Instrum.* **1975**, *46*, 668–669.

[10] C. W. Garland, *Thermochimica Acta* **1985**, *88*, 127–142.

[11] M. Feyz, E. Kuss, *Ber. Bunsenges. Phys. Chem.* **1974**, *78*, 834–842.

[12] P. H. Keyes, H. T. Weston, W. B. Daniels, *Phys. Rev. (Lett.)* **1973**, *31*, 628–630.

[13] P. Pollmann, *J. Phys. E* **1974**, *7*, 490–492.

[14] A. Rothert, G. M. Schneider, *High Press. Res.* **1990**, *3*, 285–287.

[15] P. Pollmann, E. Voß, *Liq. Cryst.* **1997**, *23*, 299–307.

[16] P. Pollmann, B. Wiege, *Liq. Cryst.* **1989**, *6*, 657–666.

[17] G. Illian, H. Kneppe, F. Schneider, *Liq. Cryst.* **1989**, *4*, 643–652.

[18] C. Carboni, H. F. Gleeson, J. W. Goodby, A. J. Slaney, *Liq. Cryst.* **1993**, *14*, 1991–2000.

[19] R. Shashidhar, S. Chandrasekhar, *J. de Physique Col. C 1* **1975**, *36*, 49–51.

[20] R. Shashidhar, H. Herrmann, H. D. Kleinhans, *Mol. Cryst. Liq. Cryst.* **1982**, *72*, 177–182.

[21] A. N. Kalkura, R. Shashidhar, G. Venkatesh, M. E. Neubert, J. P. Ferrato, *Mol. Cryst. Liq. Cryst.* **1983**, *99*, 177–183.

[22] G. Venkatesh, R. Shashidhar, D. S. Parmar, *Proceedings of Int. Liq. Crystals Conference* **1979**, Bangalore.

[23] R. Shashidhar, *Mol. Cryst. Liq. Cryst.* **1977**, *43*, 71–81.

[24] R. Shashidhar, G. Venkatesh, *J. de Physique Col. C 3* **1979**, *40*, 396–400.

[25] J. Rübesamen, G. M. Schneider, *Liq. Cryst.* **1993**, *13*, 711–719.

[26] W. Spratte, G. M. Schneider, *Ber. Bunsenges. Phys. Chem.* **1976**, *80*, 886–891.

[27] R. Krombach, G. M. Schneider, *Thermochimica Acta* **1994**, *231*, 169–175.

[28] A. Bartelt, H. Reisig, J. Herrmann, G. M. Schneider, *Mol. Cryst. Liq. Cryst. (Lett.)* **1984**, *102*, 133–138.

[29] J. Herrmann, A. Bartelt, H. D. Kleinhans, H. Reisig, G. M. Schneider, *Mol. Cryst. Liq. Cryst. (Lett.)* **1983**, *92*, 225–230.

[30] W. Klement, L. H. Cohen, *Mol. Cryst. Liq. Cryst.* **1974**, *27*, 359–373.

[31] C. Rein, D. Demus, *Liq. Cryst.* **1993**, *15*, 193–202.

[32] N. A. Tikhomirova, L. K. Vishin, V. N. Nosow, *Sov. Physics – Crystallography* **1973**, *17*, 878–880.

[33] S. Chandrasekhar, B. K. Sadashiva, K. A. Suresh, N. V. Madhusudana, S. Kumar, R. Shashidhar, G. Venkatesh, *J. de Physique C 3* **1979**, *40*, 120–124.

[34] T. Brückert, T. Büsing, A. Würflinger, S. Urban, *Mol. Cryst. Liq. Cryst. Sci. Technol. Sect. A* **1995**, *262*, 1497–1508.

[35] J. Herrmann, R. Sandrock, W. Spratte, G. M. Schneider, *Mol. Cryst. Liq. Cryst. (Lett.)* **1980**, *56*, 183–188.

[36] R. Sandrock, M. Kamphausen, G. M. Schneider, *Mol. Cryst. Liq. Cryst.* **1978**, *45*, 257–265.

[37] S. L. Randzio, *J. Thermal. Anal.* **1992**, *38*, 1989–1993.

[38] V. N. Raja, B. R. Ratna, R. Shashidhar, G. Heppke, C. Bahr, J. F. Marko, J. O. Indeken, A. N. Berker, *Phys. Rev. A* **1989**, *39*, 4341–4344.

[39] S. K. Prasad, S. Pfeiffer, G. Heppke, R. Shashidhar, *Z. Naturforsch.* **1985**, *40a*, 632–635.

[40] A. N. Kalkura, S. K. Prasad, R. Shashidhar, *Mol. Cryst. Liq. Cryst.* **1983**, *99*, 193–202.

[41] S. K. Prasad, B. R. Ratna, R. Shashidhar, V. Surendranath, *Ferroelectrics* **1984**, *58*, 101–105.

[42] M. Nakahara, K. Maeda, J. Osugi, *Bull. Chem. Soc. Jpn.* **1980**, *53*, 2499–2501.

[43] W. J. Lin, P. H. Keyes, *J. Physique* **1980**, *41*, 633–638.

[44] K. Maeda, M. Nakahara, K. Hara, J. Osugi, *Rev. Phys. Chem. Jpn.* **1979**, *49*, 85–90.

[45] J. Hollmann, P. Pollmann, P. J. Collings, *Liq. Cryst.* **1993**, *15*, 651–658.

[46] P. E. Cladis, J. W. Goodby, *Mol. Cryst. Liq. Cryst. (Lett.)* **1982**, *72*, 307–312.

[47] P. E. Cladis, J. W. Goodby, *Mol. Cryst. Liq. Cryst. (Lett.)* **1982**, *72*, 313–317.

[48] L. Liebert, W. B. Daniels, J. Billard, J. Malthete, *C. R. Acad. Sci. Paris* **1977**, *285*, Série C 451–453.

[49] L. Liebert, W. B. Daniels, *J. de Physique (Lett.)* **1977**, *38*, L 333–335.

[50] L. Liebert, W. B. Daniels, J. Billard, *Mol. Cryst. Liq. Cryst.* **1977**, *41*, 57–62.

[51] J. M. Buisine, B. Soulestin, J. Billard, *Mol. Cryst. Liq. Cryst.* **1983**, *91*, 115–127.

[52] J. M. Buisine, B. Soulestin, J. Billard, *Mol. Cryst. Liq. Cryst.* **1983**, *97*, 397–406.

[53] J. M. Buisine, *Mol. Cryst. Liq. Cryst.* **1984**, *109*, 143–157.

[54] J. M. Buisine, J. Billard, *Mol. Cryst. Liq. Cryst.* **1985**, *127*, 353–379.

[55] A. Daoudi, A. Anakkar, J. M. Buisine, F. Bougrioua, N. Isaert, H. T. Nguyen, *J. Thermal. Anal.* **1996**, *46*, 337–345.

[56] J. M. Buisine, J. Malthete, C. Destrade, N. H. Tinh, *Physica (Amsterdam)* **1986**, *139/140 B*, 631–635.

[57] P. Pollmann, G. Scherer, *Z. Naturforsch.* **1979**, *34a*, 255–256.

[58] F. Pakusch, P. Pollmann, *Mol. Cryst. Liq. Cryst.* **1982**, *88*, 255–271.

[59] P. Pollmann, B. Wiege, *Mol. Cryst. Liq. Cryst. (Lett.)* **1982**, *72*, 271–276.

[60] P. Pollmann, K. Schulte, *Liq. Cryst.* **1987**, *2*, 701–706.

[61] J. Przedmojski, S. Gierlotka, R. Wisniewski, B. Pura, W. Zajac, *Ferroelectrics* **1989**, *92*, 345–348.

[62] J. Przedmojski, S. Gierlotka, B. Pura, R. Wisniewski, *Cryst. Res. Technol.* **1988**, *23*, K72–K73.

[63] D. Guillon, J. Stamatoff, P. E. Cladis, *J. Chem. Phys.* **1982**, *76*, 2056–2063.

[64] D. Guillon, P. E. Cladis, D. Aadsen, W. B. Daniels, *Phys. Rev. A* **1980**, *21*, 658–665.

[65] Y. Maeda, Y. Yun, J. Jin, *Mol. Cryst. Liq. Cryst.* **1996**, *280*, 85–90.

[66] Y. Maeda, N. Tanigaki, A. Blumstein, *Mol. Cryst. Liq. Cryst.* **1993**, *237*, 407–418.

[67] B. Deloche, B. Cabane, D. Jerome, *Mol. Cryst. Liq. Cryst.* **1971**, *15*, 197–209.

[68] P. E. Cladis, R. K. Bogardus, W. B. Daniels, G. N. Taylor, *Phys. Rev. (Lett.)* **1977**, *39*, 720–723.

[69] P. E. Cladis, R. K. Bogardus, D. Aadsen, *Phys. Rev. A* **1978**, *18*, 2292–2306.

[70] P. Pollmann, B. Wiege, *Mol. Cryst. Liq. Cryst. (Lett.)* **1984**, *102*, 119–124.

[71] P. Pollmann, B. Wiege, A. Rothert, *Liq. Cryst.* **1988**, *3*, 225–233.

[72] P. Pollmann, B. Wiege, *Liq. Cryst.* **1988**, *3*, 1203–1213.

[73] D. Guillon, P. E. Cladis, D. Aadsen, W. B. Daniels, *Phys. Rev. A* **1980**, *21*, 658–665.

[74] R. Shashidhar, K. V. Rao, *Proceedings of Int. Liq. Crystals Conference* **1979**, Bangalore.

[75] R. Shashidhar, H. D. Kleinhans, G. M. Schneider, *Mol. Cryst. Liq. Cryst. (Lett.)* **1981**, *72*, 119–126.

[76] J. Herrmann, *Mol. Cryst. Liq. Cryst.* **1982**, *72*, 219–224.

[77] S. K. Prasad, R. Shashidhar, K. A. Suresh, A. N. Kalkura, G. Heppke, R. Hopf, *Mol. Cryst. Liq. Cryst.* **1983**, *99*, 185–191.

[78] A. N. Kalkura, R. Shashidhar, N. Subramanya Raj Urs, *J. Physique* **1983**, *44*, 51–55.

[79] S. K. Prasad, S. Pfeiffer, G. Heppke, R. Shashidhar, *Z. Naturforsch.* **1985**, *40a*, 632–635.

[80] B. R. Ratna, R. Shashidhar, V. N. Raja, C. Nagabhushan, *Mol. Cryst. Liq. Cryst.* **1989**, *167*, 233–237.

[81] G. Illian, H. Kneppe, F. Schneider, *Ber. Bunsenges. Phys. Chem.* **1988**, *92*, 776–780.

[82] G. Illian, H. Kneppe, F. Schneider, *Liq. Cryst.* **1989**, *4*, 643–652.

[83] A. Daoudi, A. Anakkar, J. Buisine, *Thermochimica Acta* **1994**, *245*, 219–229.

[84] L. D. Landau, E. M. Lifschitz in *Lehrbuch der theoretischen Physik V*, Akademie Verlag, Berlin **1971**, p. 642.

[85] L. Tisza, *Ann. Phys.* **1961**, *13*, 1–92.

[86] R. B. Griffiths, *Phys. Rev. (Lett.)* **1970**, *24*, 715–717.

[87] P. H. Keyes, H. T. Weston, W. B. Daniels, *Phys. Rev. (Lett.)* **1973**, *31*, 628–630.

[88] S. Shichijyo, T. Okamoto, T. Takemura, *Jpn. J. Appl. Phys.* **1982**, *21*, 1260–1267.

[89] K. J. Lushington, G. B. Kasting, C. W. Garland, *Phys. Lett.* **1979**, *74A*, 143–145.

[90] P. G. de Gennes, *Solid State Commun.* **1972**, *10*, 753–756.

[91] P. G. de Gennes in *The Physics of Liquid Crystals*, Clarendon Press, Oxford, **1975**, p. 331.

[92] P. Pollmann, K. Schulte, *Ber. Bunsenges. Phys. Chem.* **1985**, *89*, 780–786.

[93] G. B. Kasting, C. W. Garland, K. J. Lushington, *J. Physique* **1980**, *41*, 879–884.

[94] G. B. Kasting, K. J. Lushington, C. W. Garland, *Phys. Rev. B* **1980**, *22*, 321–331.

[95] T. J. McKee, J. R. McColl, *Phys. Rev. (Lett.)* **1975**, *34*, 1076–1080.

[96] W. L. McMillian, *Phys. Rev. A* **1971**, *4*, 1238–1246.

[97] R. Shashidhar, A. N. Kalkura, S. Chandrasekhar, *Mol. Cryst. Liq. Cryst. (Lett.)* **1980**, *64*, 101–107.

[98] R. Shashidhar, A. N. Kalkura, S. Chandrasekhar, *Mol. Cryst. Liq. Cryst. (Lett.)* **1982**, *82*, 311–316.

[99] R. Shashidhar, S. K. Prasad, S. Chandrasekhar, *Mol. Cryst. Liq. Cryst.* **1983**, *103*, 137–142.

[100] R. Shashidhar, B. R. Ratna, S. K. Prasad, *Phys. Rev. (Lett.)* **1984**, *53*, 2141–2144.

[101] S. Somasekhara, R. Shashidhar, B. R. Ratna, *Phys. Rev. A* **1986**, *34*, 2561–2563.

[102] C. Legrand, N. Isaert, J. Hmine, J. M. Buisine, J. P. Parneix, H. T. Nguyen, C. Destrade, *Ferroelectrics* **1991**, *121*, 21–31.

[103] A. Anakkar, A. Daoudi, J. M. Buisine, N. Isaert, F. Bougrioua, H. T. Nguyen, *Liq. Cryst.* **1996**, *20*, 411–415.

[104] A. Anakkar, A. Daoudi, J. M. Buisine, N. Isaert, T. Delattre, H. T. Nguyen, C. Destrade, *J. Thermal Anal.* **1994**, *41*, 1501–1513.

[105] Ch. Bahr, G. Heppke, R. Shashidhar, *Z. Naturforsch.* **1985**, *40a*, 1311–1315.

[106] J. Prost, *Adv. Physics* **1984**, *33*, 1–46.

[107] S. K. Prasad, G. G. Nair, S. Chandrasekhar, J. W. Goodby, *Mol. Cryst. Liq. Cryst.* **1995**, *260*, 387–394.

[108] E. Kuss, *Mol. Cryst. Liq. Cryst.* **1981**, *76*, 199–210.

[109] R. V. Transfield, P. J. Collings, *Phys. Rev. A* **1982**, *25*, 2744–2749.

[110] W. A. Steele, W. Webb in *High Pressure Physics and Chemistry* (Ed. R. S. Bradley), Academic Press, New York, **1963**, Vol. 1, p. 145.

[111] C. Hanawa, T. Shirakawa, T. Tokuda, *Chem. Lett. Chem. Soc. Jpn.* **1977**, 1223–1226.

[112] E. Kuss, *Mol. Cryst. Liq. Cryst.* **1978**, *47*, 71–83.

[113] C. S. Johnson, P. J. Collings, *J. Chem. Phys.* **1983**, *79*, 4056–4061.

[114] M. W. Lampe, P. J. Collings, *Phys. Rev. A* **1986**, *34*, 524–528.

[115] Y. Hiwatari, H. Matsuda, *Progr. Theor. Phys.* **1972**, *47*, 741–764.

[116] J. R. McColl, *Physics Lett.* **1972**, *38A*, 55–57.

[117] H. Ichimura, T. Shirakawa, T. Tokuda, T. Seimiya, *Bull. Chem. Soc. Jpn.* **1983**, *56*, 2238–2240.

[118] P. H. Keyes, W. B. Daniels, *J. Physique C 3* **1979**, *40*, 380–383.

[119] R. Shashidhar, P. H. Keyes, W. B. Daniels, *Mol. Cryst. Liq. Cryst.* **1986**, *3*, 169–175.

[120] P. S. Pershan, J. Prost, *J. Physique (Lett.)* **1979**, *40*, L-27–30.

[121] T. Shirakawa, T. Inoue, T. Tokuda, *J. Phys. Chem.* **1982**, 1700–1702.

[122] T. Shirakawa, T. Hayakawa, T. Tokuda, *J. Phys. Chem.* **1983**, *87*, 1406–1408.

[123] T. Shirakawa, H. Eura, H. Ichimura, T. Ito, K. Toi, T. Seimija, *Thermochim. Acta* **1986**, *105*, 251–256.

[124] T. Shirakawa, M. Arai, T. Tokuda, *Mol. Cryst. Liq. Cryst.* **1984**, *104*, 131–139.

[125] E. A. S. Lewis, H. M. Strong, G. H. Brown, *Mol. Cryst. Liq. Cryst.* **1979**, *53*, 89–99.

[126] A. Abe, S. Y. Nam, *Macromol.* **1995**, *28*, 90–95.

[127] T. Shirakawa, H. Ichimura, I. Ikemoto, *Mol. Cryst. Liq. Cryst.* **1987**, *142*, 101–106.

[128] V. Ya. Baskakov, V. K. Semenchenko, *ZhETF Pis. Red.* **1973**, *17*, 580–583; *JETP Lett.* **1973**, *17*, 414.

[129] V. M. Byankin, V. K. Semenchenko, V. Ya. Baskakov, *Zh. Fiz. Khim.* **1974**, *48*, 1250–1253.

[130] N. A. Nedostup, V. K. Semenchenko, *Zh. Fiz. Khim.* **1977**, *51*, 1628–1631.

[131] V. K. Semenchenko, V. M. Byankin, V. Ya. Baskakov, *Sov. Phys. Crystallogr.* **1975**, *20*, 187–191.

[132] V. Ya. Baskakov, V. K. Semenchenko, V. M. Byankin, *Zh. Fiz. Khim.* **1976**, *50*, 200–202.

[133] V. K. Semenchenko, N. A. Nedostup, V. Ya. Baskakov, *Zh. Fiz. Khim.* **1975**, *49*, 1547–1550.

[134] V. K. Semenchenko, V. M. Byankin, V. Ya. Baskakov, *Zh. Fiz. Khim.* **1974**, *48*, 2353–2355.

[135] V. Ya. Baskakov, V. K. Semenchenko, N. A. Nedostup, *Kristallogr.* **1974**, *19*, 185–187.

[136] V. Ya. Baskakov, V. K. Semenchenko, V. M. Byankin, *Zh. Eksp. Teor. Fiz.* **1974**, *66*, 792–797; *Sov. Phys.-JETP* **1974**, *39*, 383–385.

[137] V. K. Semenchenko, N. A. Nedostup, Y. Ya. Baskakov, *Zh. Fiz. Khim.* **1975**, *49*, 1543.

[138] A. C. Diogo, *Solid State Commun.* **1984**, *50*, 895–897.

[139] H. Doerrer, H. Kneppe, F. Schneider, *Liq. Cryst.* **1992**, *11*, 905–915.

[140] M. Ozaki, N. Yasuda, K. Yoshino, *Jpn. J. Appl. Phys.* **1987**, *26*, L 1927–L 1929.

[141] T. Brückert, A. Würflinger, *Ber. Bunsenges. Phys. Chem.* **1993**, *97*, 1209–1213.

[142] H.-G. Kreul, S. Urban, A. Würflinger, *Phys. Rev. A* **1992**, *45*, 8624–8631.

[143] A. Würflinger, *Internat. Rev. Phys. Chem.* **1993**, *12*, 89–121.

[144] T. Brückert, T. Büsing, A. Würflinger, S. Urban, *Mol. Cryst. Liq. Cryst. Sci. Technol. Sect. A* **1995**, *262*, 1497–1508.

[145] T. Brückert, S. Urban, A. Würflinger, *Ber. Bunsenges. Phys. Chem.* **1996**, *100*, 1133–1137.

[146] S. Urban, B. Gestblom, T. Brückert, A. Würflinger, *Z. Naturforsch. A: Phys.-Sci.* **1995**, *50*, 984–990.

[147] S. Urban, *Z. Naturforsch. A: Phys.-Sci.* **1995**, *50*, 826–830.

[148] N. Yasuda, S. Fujimoto, S. Funado, *J. Phys. D: Appl. Phys.* **1985**, *18*, 521–530.

[149] R. G. Horn, *J. Physique* **1978**, *39*, 167–172.

[150] S. M. Khened, S. K. Prasad, V. N. Raja, S. Chandrasekhar, B. Shivkumar, *Ferroelectrics* **1991**, *121*, 307–318.

[151] P. L. Sherrell, J. D. Bünning, T. E. Faber, *Liq. Cryst.* **1987**, *2*, 3–19.

[152] T. Uemoto, K. Yoshino, Y. Inuishi, *Mol. Cryst. Liq. Cryst.* **1981**, *67*, 137–152.

[153] T. Uemoto, K. Yoshino, Y. Inuishi, *Jpn. J. Appl. Phys.* **1980**, *19*, 1467–1472.

[154] D. Ergashev, *Mezhvuz. Sb. Nauch. Tr. Vses. Zaoch. Mashinostr. Int.* **1982**, *34*, 129–133.

[155] V. I. Kireev, S. V. Pasechnik, V. A. Balandin, *Pis'ma Zh. Tekh. Fiz.* **1989**, *15*, 88–90.

[156] A. S. Lagunov, A. N. Larionov, *Akust. Zh.* **1984**, *30*, 344–351.

[157] S. V. Pasechnik et al., *J. Phys. (Les Ulis. Fr.)* **1984**, *45*, 441–449.

[158] S. V. Pasechnik, V. A. Balandin, V. I. Kireev, *Pis'ma Zh. Tekh. Fiz.* **1988**, *14*, 1756–1760.

6.3 Fluctuations and Liquid Crystal Phase Transitions

P. E. Cladis

6.3.1 Introduction

Many liquid crystal phase transitions involve broken continuous symmetries in real space and their interactions on a molecular scale are short range [1]. As a result, fluctuations have long been known to be an important feature of liquid crystal phase transitions: even weakly first order (discontinuous) ones. Compared to major advances in our understanding of fluctuation controlled second-order (continuous) phase transitions, relatively little is known about fluctuation phenomena (critical phenomena) at first-order phase transitions such as the nematic–isotropic transition.

The central concept of critical phenomena is "universality". Simply put, this means that fluctuation dominated continuous phase transitions are controlled by a unique length, $\xi \equiv \xi_0/\varepsilon^\nu$, where ξ is a measure of the length over which order parameter fluctuations are correlated. $\varepsilon \equiv |T - T_c|/T_c$, with T_c the second-order phase transition temperature. ν is the correlation length exponent that is completely defined by the symmetry of the system (i.e. the number of components of its order parameter) and the dimensionality, d, of the space in which the material is embedded [2]. ξ_0, the 'bare' correlation length, is a measure of what fluctuations have to beat to become critical (i.e. to take control of the phase transition). In low-temperature superconductors, $\xi_0 \approx 200$ nm while in liquid crystals, $\xi_0 \approx 0.5$–1 nm. Because ξ_0 is so large, low temperature superconductors have only been studied in a 'mean field' limit. In liquid crystals, a mean field limit with a cross-over to a critical regime as $T \rightarrow T_c$ can also be observed.

A surprising result is that the mean field limit is exact for spatial dimension, d, greater than or equal to a critical dimension, d_c. For phase transitions far from a tricritical point, $d_c = 4$. For phase transitions in the vicinity of a tricritical point, $d_c = 3$: exponents in the vicinity of a tricritical point are mean field [2].

In the vicinity of fluctuation dominated phase transitions, the temperature dependence of thermodynamic parameters such as the specific heat at constant pressure, $C_p \approx \varepsilon^\alpha$, and the order parameter, $\psi \approx \varepsilon^\beta$, are all related to ξ through a free energy density giving rise to scaling relations. For example: $\alpha = 2 - \nu d$ and $\beta = (d - 2)\,\nu/2$ [2]. Despite the variety of their continuous broken symmetries, most liquid crystal phase transitions are expected to fall in the 3D-XY (helium) universality class with, $\alpha \approx -0.01$, $\nu \approx 0.67$ and $\beta \approx 0.33$.

Here we give an overview of fluctuation dominated thermotropic liquid crystal phase transitions with a few hints of emerging aspects. From this perspective, the situation may be fairly summarized by noting that while analogies to phase transition models in spin-space (e.g. XY model with two components for the order parameter or Ising model with one) or momentum space (superconductivity) are a powerful tool to predict qualitative behavior for fluctuation dominated, real space, high temperature liquid crystal phase transitions, there is a significant gap between several quantitative (and qualitative) expectations and experimental measurements.

One reason may be that liquid crystals have first-order phase transitions that are so weakly discontinuous, their first-order nature escapes detection by traditional methods such as adiabatic calorimetry [3]. Recently, a macroscopic qualitative test of phase transition order [4] revealed that even immeasurably small discontinuities at first-order phase transitions [5] using static tests, have a distinct dynamic signature in interface (front) propagation compared to second-order or continuous phase transitions [4]. It is important to know the order of a phase transition because for universality to apply at all levels of its hierarchy [2], ξ must approach infinity continuously: there can be no discontinuities at T_c. If there are, 'all bets are off' [6].

A salient feature of molecular materials, including liquid crystals, is that they are liquid well above $T(K) \equiv 0$. As a result they have generic long range correlations even far from critical points or hydrodynamic instabilities [7] that could make it difficult to access critical regimes before being finessed by a first-order phase transition. At these high temperatures, externally supplied noise [8] (e.g. expressed by random thermal fluctuations) can *suppress* the onset of mac-

roscopic instabilities such as spatial turbulence far from any phase transition [9]. Dynamic correlation functions decay with long time tails supporting nonequilibrium steady states and contribute to divergences in transport coefficients that cannot be accounted for by a 'static' theory of phase transitions [7]. Liquid crystal contributions to dynamic critical phenomena [10] are beginning to emerge [11].

The richness implicit in what we have learned and can learn from fluctuations in liquid crystals seems endless. Here we give only the merest hint of the enormous volume of information contained in a vast, and still growing, body of research.

6.3.2   The Nematic–Isotropic Transition

Liquid crystal materials in the isotropic liquid state are transparent. As temperature decreases towards the nematic liquid state, the material becomes more and more turbid (i.e. scatters more and more light). The picture is that nematic droplets are forming in the isotropic liquid to scatter the light. As the temperature is lowered, more and more droplets of the ordered state appear until the whole system is nematic. In liquid crystals, such droplets, called 'cybotatic groups', were an early precursor of the notion of 'fluctuations'. The idea of droplets of a disordered state appearing in an ordered state, and vice versa, as phase transitions are approached is still a useful picture for explaining pretransitional changes in light scattering data at first-order phase transitions, for example. While similar to classical explanations of critical opalescence as the liquid–gas critical point is approached, the difference is that in liquid crystals, droplets in the nematic state have a different symmetry from the disordered state, for example isotropic liquid [12]. Because there is no surface energy between the ordered and disordered states at continuous or second-order phase transitions, the picture of compact structures driven by surface tension is unsatisfactory. At second-order phase transitions, it has turned out that a more useful concept is to think in terms of statistical objects termed fluctuations with a characteristic length over which ordering is correlated [1].

When there are no external magnetic or electric fields, the nematic and isotropic liquids do not have the same symmetry. The effect of an applied field is to induce orientational order in the isotropic phase that grows with increasing field intensity. Eventually, with increasing external fields, the jump at the transition vanishes at a field induced critical point. The increase in orientational order in the isotropic liquid results in an enhancement of the nematic–isotropic transition temperature, δT_{N-I}. In analogy to paramagnetism, the isotropic liquid with field induced orientational order is called paranematic. In low molecular weight liquid crystals, $\delta T_{N-I} \approx 1$ mK even in the most intense fields currently available. However, as can be seen in another chapter of this volume [13], in liquid crystal elastomers, $\delta T_{N-I} \approx 10$ K (i.e. is large for even quite modest mechanical fields). As a result, it is now possible to pass through the field induced critical point and observe the state beyond where nematic and paranematic states are indistinguishable.

Fluctuation effects are large in polymeric liquid crystals even far from phase transition temperatures. For example, in a novel liquid crystalline elastomer with a SmA–I transition, under an external mechanical stress, it was found in a mean field limit that, well in the isotropic phase, nematic fluctuations dominate with a cross-over temperature closer to the transition where SmA fluctuations become more important [14].

6.3.3 The Uniaxial–Biaxial Nematic Transition

Both uniaxial positive (rod-like) and uniaxial negative (disc-like) nematics exist. The nematic order parameter, $Q_{\alpha\beta}$, distinguishes between the two possibilities [15]. If $Q_{\alpha\beta}$ refers to uniaxial positive nematics (rod-like), then, $-Q_{\alpha\beta}$ describes uniaxial negative nematics (disc-like). As a result, the N–I transition is necessarily first order for geometric reasons [15].

While both side-on side-chain liquid crystal polymers [16] and low molecular weight liquid crystals [17] have been reported to have biaxial nematic phases, only in a lyotropic liquid crystal system [18], has the uniaxial–biaxial nematic transition been studied in detail from the point of view of critical phenomena. This transition is found to be second order. So far, it is the one liquid crystal system where earlier theoretical expectations [19] of fluctuation dominated phase transitions and later experimental results are most fully in agreement with respect to both static [20] and dynamic [21] aspects of critical phenomena. In particular, static critical phenomena predicts 3D-XY exponents which have been observed (with irrelevant corrections to scaling) by Saupe et al. [19]. Transport parameters were not expected [19] to show any singularities at the transition [19] as later verified by Roy et al. [21] because the dynamics of the biaxial order parameter is nonconserved (Model A) [19].

6.3.4 Type I Smectic A's: Nematic–Smectic A Transition

Following McMillan's prediction of the conditions under which the N–SmA transition could be second order [22] and de Gennes' analogy between the N–SmA transition and the normal-metal–superconducting transition [23], many experiments followed to test new scaling ideas at phase transitions. This was because de Gennes also pointed out that ξ_o was small in liquid crystals, so fluctuations did not have to work so hard to take over the transition (i.e. liquid crystal phase transitions could have observable critical regimes). At the normal-metal–superconducting transition in Type I superconductors, magnetic field lines ($\mathbf{H} = \text{curl}\mathbf{A}$, where $\mathbf{A}$ is the magnetic vector potential) are expelled (Meissner effect). Type I superconductors are perfect diamagnets. In a similar way, at the transition to an ordered Sm A phase from the nematic phase, twist and bend deformations contained in curl$\mathbf{n}$, with $\mathbf{n}$ the liquid crystal director, are expelled [23].

The most recent extensive overview of the experimental situation for the N–SmA transition has been given by Garland and Nounesis [24]. The overall picture is that exponents for the correlation length parallel ($v_{\parallel}$) and perpendicular ($v_{\perp}$) to the director are non-universal and different. Hyperscaling (substituting $v = (v_{\parallel} + 2 v_{\perp})/3$ in scaling relations e.g. $\alpha = 2 - vd$) approximately works. McMillan's number, $M = T_{\text{N-A}}/T_{\text{N-I}}$, emerges as a robust measure of a relatively sharp cross-over in the specific heat exponent form the 3D-XY value when $M < 0.93$ with a systematic transition to the tricritical value ($\alpha = 0.5$) as $M \to 1$. As the authors emphasize, the 3D-XY model cannot account for the observation $v_{\parallel} \neq v_{\perp}$.

Many of the compounds in the large list of Garland and Nounesis have N–SmA phase transitions determined to be continuous by calorimetry and X-ray diffraction and discontinuous using the more powerful dynamic test of phase transition order [4]. In particular, the compounds known as 8CB and 9CB, with second order N–SmA phase transitions by the standard tests [25], were found to exhibit dynamic behavior consistent with a first-order phase transition. This test further enabled a determination of the small cubic term (the HLM effect) [26] predicted to be a feature of the N–SmA transition in the vicinity of a tricritical point, an appropriate limit for this result [26] to apply [5]. The conclusion is that while the N–SmA transition is intrinsically first order, the discontinuity is 'small'. Indeed, precise measurements of changes in sound speed and the elastic constants deduced from these measurements find evidence for 3D-XY fluctuations even far from the transition temperature [27].

The theoretical picture for the N–SmA transition has been succinctly summarized by Lubensky. [28] He explains that there are several important differences between the normal-metal–superconducting and the N–SmA transitions where K_1, the splay elastic constant, emerges as a 'dangerous irrelevant parameter' [28].

First, unlike superconductors, SmA order is not long range (Peierls' argument). This introduces an additional length, l. As a result, correlations in director fluctuations and SmA order parameter fluctuations have *different* lengths. Where director orientational fluctuations are correlated on a length, ξ_{thermo}, SmA order parameter fluctuations are correlated on a length ξ_{eff} where $1/\xi_{\text{eff}} = 1/\xi_{\text{thermo}} + 1/l$. The temperature dependence (i.e. critical exponents) of l depends on splay fluctuations which in turn depend on K_1.

Second, in the N–SmA case, 'nature has so arranged it that measurable quantities are in a gauge where fluctuations are most violent' [28]. Measurable quantities in superconductors are gauge invariant, (i.e. independent of div$\mathbf{A}$). In the N–SmA case, this corresponds to $K_1 \equiv 0$. The Coulomb gauge, div$\mathbf{A} \equiv 0$ in superconductors, corresponds to div$\mathbf{n} \equiv 0$ or $K_1 \to \infty$ [29]. In both these cases $v_\parallel \equiv v_\perp$. However, in the $K_1 \to \infty$ limit, gauge transformation theory predicts that the 3D-XY fixed point is unstable leaving only the stable fixed point at $K_1 \equiv 0$.

Dislocation-loop melting theory [30], taking into account entropic effects, introduced two new fixed points on the K_1 axis, one stable and the other unstable, where $v_\parallel = 2v_\perp$ is 'built-in' [28]. As the value of the unstable fixed point is at smaller K_1 than the stable one, K_1 has to be larger than its unstable value for fluctuations to converge to dislocation-loop theory's fixed point [31]. Recent self-consistent one loop calculations by Andereck and Patton [32] and the persistent anisotropy in the critical behavior of ξ have renewed interest in this still mysterious aspect of the N–SmA transition [24]. However, Lubensky emphasizes that the only way theoretically to obtain the larger $K_1 \neq 0$ stable fixed point is by ignoring l [28]. If one includes l, the only stable fixed point is at $K_1 = 0$ where $v_\parallel = v_\perp$.

It may be that mixtures can be found [33] to tune K_1 sufficiently well to locate the stable fixed point where $K_1 \neq 0$. For example, some early experiments [34] measuring the elastic constants of K_2 and K_3 in mixtures of CBOOA (cyanooctyloxybiphenyl) and ortho-MBBA (o-methoxybenzilidene butylaniline), found at sufficiently large concentration of o-MBBA that $K_3 \approx \xi_\parallel$ diverged with an exponent $v_\parallel \approx 1$. In these same mixtures, $K_2 \approx \xi_\perp^2/\xi_\parallel$ showed no divergence [34] consistent with $v_\parallel = 2v_\perp$. In view of significant advances in materials and measuring techniques since these early days, redoing these experiments, including precise measurements of K_1 [35], in a system where this 'dangerous irrelevant parameter' can be varied through a large range of values, may be a way to find the $K_1 \neq 0$ stable fixed point.

A first indication that long range dynamic correlations [7] existed in liquid crystals appeared with the discovery of divergences in the Ericksen–Leslie viscous coefficients, α_1, α_3 and α_6 [36]. Brochard's result was found using Kawasaki's mode coupling theory [37]. One consequence of divergences in these viscous coefficients is that 'flow alignment' breaks down in the nematic phase on approaching the N–SmA transition [38]. While this result has inspired, and is continuing to inspire, the study of macroscopic instabilities in liquid crystals under well-controlled and well-defined conditions [39], more traditional features of dynamic critical phenomena are beginning to emerge, especially at other liquid crystal phase transitions [40]. In particular, it has been shown that dynamic behavior in the vicinity of the N–SmA transition depends on material parameters specific to a compound. The implication is that the universal behavior expected for dynamic critical phenomena will only be observed in certain compounds (e.g. those with relatively stiff and incompressible layers in the case of SmA) and not others [11].

6.3.5 Type II Smectic A's: Cholesteric–Smectic A Transition

The parameter distinguishing between Type I and Type II superconducting behavior in a magnetic field [41] is the Ginzburg–Landau parameter, κ. When $\kappa < 1/\sqrt{2}$, the superconductor is Type I (Meissner effect) and when $\kappa > 1/\sqrt{2}$, it is Type II (flux lines penetrate but the penetration is not complete). The analogue of studying superconductors in a magnetic field ($\mathrm{curl}A \neq 0$) is to turn on spontaneous twist deformations ($\mathbf{n}\mathrm{curl}\mathbf{n}$, chirality, a property of cholesteric liquid crystals) in the liquid crystal case [23]. The transition is analogous to a Type I superconductor and the N*–SmA transition, in a high enough chirality limit, is analogous to Type II superconductors exhibiting the analogue of a vortex lattice in what are now known as twist grain boundary phases (TGB) [42].

Renn and Lubensky proposed a model for the analogue of the vortex lattice [43] for the N*–SmA transition [42]. Simultaneously and independently such a phase was discovered by direct observation, supported by X-ray analysis as well as freeze-fracture, by Goodby et al. [44] between the isotropic liquid and a SmC* phase. The first TGBA phase found to exist between N* and SmA was studied in a dynamic experiment [45]. In the Renn-Lubensky model, uniform sheets of SmA of extent $\xi_\perp$, separated by parallel planes of screw dislocations, twist relative to each other [46].

When Dasgupta and Halperin included fluctuations in the HLM effect [47], they found that the second order nature was restored in the high κ limit of superconductors. Recently in a dynamic experiment [45], there was no interface between cholesteric and TGBA in a material for which κ was estimated to be ≈3 times larger than $1/\sqrt{2}$. In addition, the N phase region did not propagate (i.e. either advance or retreat): TGBA grew as SmA melted and was squeezed out by an advancing SmA phase. These observations [45] support a second order TGBA–N* phase transition [4].

High resolution specific heat measurements raise the possibility of another interesting scenario [48]. These measurements show a large broad feature well above the TGB–N* transition and only a tiny (if any) latent heat at the N*–TGB transition. A similar behavior was observed at the first TGBA–N* transition [45] in a different compound making it a possible generic feature of TGB–N* transitions. In this case [45], the size of the heat signature at the TGBA–N* transition was time dependent. Chan et al. [48] suggest that the broad pretransitional heat feature was consistent with

the formation of a liquid of screw dislocations condensing into either a glassy or an ordered TGB phase [49]. They point out [48] the resemblance between their phase diagram and a theoretical one for a Type II superconductor with strong thermal fluctuations where, with decreasing temperature, a vortex liquid state condenses into either a glassy or an ordered state [50]. Whether the second order nature of the N*-TGBA transition [45] is a result of fluctuations obliterating the small HLM singularity or a proposed [50] (but not yet experimentally verified) [51] liquid–glass transition calls for more work. Materials with TGB phases with a large temperature range [52] may help clarify this question relevant not only to liquid crystal and high T_c superconducting materials, but also other complex materials such as polymers, electrorheological fluids and ferrofluids [42].

6.3.6   Transitions between Tilted and Orthogonal Smectic Phases

In the SmC phase, the director tilts relative to the layer normal at an angle, $\theta_T \neq 0$, breaking the continuous rotational symmetry in the plane of SmA layers. The order parameter for this transition is θ_T, the angle between the layer normal and the director. Guillon and Skoulios [53] suggested that a direct measure of θ_T is given by cos $\theta_T = d_C/d_A$ where d_C is the layer spacing measured in the C phase and d_A is its (maximum) value at $T_{SmA-SmC}$. According to theory [54], this transition is in the 3D-XY universality class. Although measurements by the MIT group found mean field exponents and unobservably small critical regimes ($\xi_0 \approx 700$ nm) [55], a compound studied by Delaye [56] in light scattering, as well as optical measurements in other compounds by Ostrovskii et al. [57] and Galerne [58], and

X-ray layer spacing measurements of Keller et al. [59], observable critical regimes were found with 3D-XY exponents as expected by theory [54].

Recently Ema et al. [60] found that, with 'significant corrections-to-scaling', their high resolution specific heat measurements agreed with theoretical expectations [54] at the SmC_α^*–SmA transition. The SmC_α^* phase is composed of alternating layers of equal and opposite tilt [61]. Ema et al. [60] give a nice summary of these transitions in the context of critical phenomena as well as a sense of the detailed scrutiny high quality data are subjected to in such studies.

The controversy over the nature of the SmA–SmC transition was essentially resolved by Benguigui and Martinoty [62]. These authors addressed the question of why fluctuation effects can be observed on long lengths scales with ultra-sound propagation but not by specific heat or X-ray measurements. In particular in $\bar{8}S5$, where specific heat, X-ray and dilatometric measurements found mean-field behavior, ultrasound damping measurements showed significant fluctuation effects [63]. Most recently [64], they further elucidated the question of how certain compounds do show mean-field behavior while others show critical behavior at long wavelengths. Their explanation rests on the Andereck–Swift theory of the SmA–SmC transition [65] that couples the SmC order parameter to gradients in the density and gradients in the layer spacing.

6.3.7   B–SmB–SmA Transitions

The SmB phase is another example of a phase where entropic effects are important. SmA is a two dimensional isotropic liquid in the plane of its layers. B is a three dimensional hexagonal crystal. SmB in-plane ordering is between these two with long range hexagonal bond orientational order and short range translational order that is not correlated from layer to layer [66]. A simple picture is that in the SmB phase, molecules are delocalized on a hexagonal grid [59]. X-ray diffraction patterns of B are resolution limited while those of SmB are not [67]. Although a first order SmB–SmC* transition has been observed [59], most of the work has concentrated on the expected fluctuation dominated SmB–SmA transition [68].

So far, none of the exponents observed at this transition fit a recognizable universality class. Nounesis et al. measured the divergence in the thermal conductivity and found dynamic critical exponents consistent with tricritical values [69]. As they had already established that $\alpha \approx 0.6$, very different from expectations of static critical phenomena (i.e. 3D-XY) they suggested that the SmB–SmA transition, at least in the compound they were studying (65OBC) [70], was driven towards a tricritical point because of a coupling between hexatic order and a short range ordering field such as a 'herringbone' structure from a nearby, lower temperature crystal E phase [1]. However, later extensive work on different compounds along a SmB–SmA transition line, with SmB temperature ranges between 0.8 K and 22 K, found $\alpha \approx 0.6$ along the entire line [71]. These authors point out that only a tricritical point (*not* a line) is compatible with a coupling between bond-orientational order and short range positional order.

Gorecka et al. [72] studied a compound in thick freely suspended films exhibiting a B–SmB–SmA phase sequence. They found no significant difference in hexatic properties in this system and those with SmB–E transitions [71]. The exponent for the hexatic order parameter was found to be $\beta = 0.15$

±0.03 which the authors point out is consistent with the three state Potts model in three dimensions [68]. They could not exclude the possibility that the SmB–A transition in their material was first order. The suggestion was that the hexatic order parameter did not fully cross-over from mean field to 3D-XY. It was also found that while there is a sharp discontinuity in the in-plane positional correlation length at the SmB–B transition, suggesting a strongly first-order transition, there was no measurable enthalpy change.

Fluctuation effects at the more fluid of liquid crystal phase transitions have been sensitively probed using techniques exploiting long length scales. Indeed, ultrasonic wavelengths are highly sensitive to hexatic fluctuations in SmA. Gallani et al. [73] found evidence of hexatic fluctuations at the SmA–SmB transition resulting in a strong damping and velocity anomaly in the A phase of an ultrasonic wave. Their data suggests that the in-plane hexatic ordering couples to SmA layer undulations which in turn have a nonlinear coupling to the sound velocity field.

6.3.8 Fluctuations at Critical Points

A line of first order phase transitions between two phases with the same symmetry may end at a critical point where fluctuations are expected to dominate. The classic example of this is the liquid–gas critical point.

6.3.8.1 BP_III–Isotropic

The analogue of the pressure–temperature plane in a liquid–gas system is the temperature–chirality plane for blue phase transitions. Along the lines of the classic liq-

uid–gas example, Keyes [74] suggested that as the BP_{III} and isotropic phases have the same symmetry, the line of first order phase transitions could end at a critical point. Voigts and van Dael found, [75] later verified by others, that, with increasing fraction of a chiral component, the latent heat at the BP_{III}–I transition decreased. More recently, Kutjnak et al. [76] located a BP_{III}-I critical point and characterized its behavior with high resolution calorimetry and measurements of optical activity. They report that both measurements are consistent with mean field behavior instead of the theoretically expected [77] Ising fluctuation behavior. They suggest that mean field behavior is observed because ξ_o is large. However, measured values [74] of $\xi_o \approx 1.5$ nm (i.e. a molecular length scale).

6.3.8.2 SmA_d and SmA_I

These two smectic phases have the same symmetry but their layer spacings are different and incommensurate. In SmA_I, the layer spacing is about the molecular length while in SmA_d it is 1.2–1.3 times the molecular length. As the critical point is approached, fluctuations become so large the critical point explodes into a bubble of a disordered state–cholesteric if the compounds are chiral and nematic if they are not–embedded in SmA. [78] On the other hand, a line of first order phase transitions between SmA_d and SmA_2, for which the layer spacing is about twice the molecular length, ends at a critical point [79]. Prost and Toner [80] developed a dislocation melting theory for a line of first-order SmA–SmA phase transitions ending at a N–SmA_d–SmA, triple point. The first-order nature of this triple point for a nematic bubble has been verified by Wu et al. [81]. Furthermore, it was predicted that this line could continue linearly into the nematic region to end at a N–N critical point [80]. At the triple point on a nematic bubble [78], such a line would have to bend nearly 90°. But, at a nematic 'estuary' connected to a nematic 'ocean' (i.e. not surrounded by SmA), evidence has been found for a critical point terminating a first-order line of N–N transitions. [82]

6.3.8.3 NAC Multicritical Point

A point where three fluctuation dominated phase transition lines meet in a 2-dimensional parameter space is also expected to exhibit universal features. An extensively studied liquid crystal candidate was the N–SmA–SmC point in mixtures [83], in a pure compound under pressure [84] and at the re-entrant N–SmA–SmC multicritical point [85]. The situation may be summarized as follows. The systems studied showed qualitative and quantitative similarities. However, the exponents exhibited were not in the expected universality class for three second order phase transition lines meeting at a point. This is likely because, in the N–SmA–SmC case, the N–SmC transition line is first order [86] as is the N–SmA transition line, [26] leaving only the SmA–SmC second-order phase transition line.

6.3.9 Conclusion

Given the rich variety to their broken symmetries, enormous strides in perfecting low molecular weight organic liquid crystal materials for applications in industry, as well as important advances in measurement technologies of material properties under well-controlled conditions, liquid crystals emerge as useful materials to explore the intricate and beautiful interplay in nature between symmetry and spatial dimensionality at phase transitions.

6.3.10 References and Notes

[1] P. G. de Gennes, J. Prost, *The Physics of Liquid Crystals*, Clarendon Press, Oxford **1993**, Chapters I and II.
[2] P. Pfeuty, G. Toulouse, *Introduction to the Renormalization Group and to Critical Phenomena*, John Wiley and Sons, New York **1977**.
[3] J. Thoen, in *Phase Transitions in Liquid Crystals*, (Eds.: S. Martellucci, A. N. Chester) Plenum, New York **1992**, pp. 155–174; C. W. Garland, *ibid*, pp. 175–187; J. Thoen, *Int. J. Mod. Phys. B*, **1995**, *9*, 2157; in *Liquid Crystals in the Nineties and Beyond*, (Ed.: S. Kumar) World Scientific Publ. Co, **1995**, pp. 19–80.
[4] P. E. Cladis, W. van Saarloos, D. A. Huse, J. S. Patel, J. W. Goodby, P. L. Finn, *Phy. Rev. Lett.* **1989**, *62*, 1764; G. Dee, J. S. Langer, *Phys. Rev. Lett.* **1983**, *50*, 383.
[5] M. A. Anisimov, P. E. Cladis, E. E. Gorodetskii, D. A. Huse, V. E. Podneks, V. G. Taratuta, W. van Saarloos, V. P. Voronov, *Phys. Rev.* **1990**, *A41*, 6749, See also: P. E. Cladis, *J. Stat. Phys.* **1991**, *62*, 899.
[6] G. Grinstein, private communication, **1980**, Pfeuty and Toulouse [2].
[7] A recent review is: J. R. Dorfman, T. R. Kirkpatrick, J. V. Sengers, *Annu. Phys. Chem.* **1994**, *45*, 213.
[8] A. Schenzle, H. R. Brand, *Phys. Rev.* **1979**, *A20*, 1628.
[9] H. R. Brand, S. Kai, S. Wakabayashi, *Phys. Rev. Lett.* **1985**, *54*, 555.
[10] A recent overview of this aspect of critical phenomena is: B. Schmittmann, R. K. P. Zia, *Statistical Mechanics of Driven Diffusive Systems*, Academic Press, New York **1995**.
[11] See for example: L. Benguigui, D. Collin, P. Martinoty, *J. Phys. I (Paris)* **1996**, *6*, 1469. In this paper, the authors show that the dynamic exponent, *y*, associated with high frequency damping at a second order phase transition is frequency dependent so that simple scaling laws obtained from mode coupling calculations should not apply. However, in some smectics, where the regular part of the specific heat behavior is large compared to the fluctuation amplitude, deviations from these laws may be too small to measure.
[12] For a recent perspective on fluctuation effects in nematic liquid crystals see: Z. H. Wang, P. H. Keyes, *Phys. Rev. E*, **1996**, *54*, 5249.
[13] H. R. Brand, H. Finkelmann, *Liquid Crystal Elastomers*, Vol. III and references therein.
[14] M. Olbrich, H. R. Brand, H. Finkelmann, K. Kawasaki, *Europhys. Lett.* **1995**, *31*, 281.
[15] P. G. de Gennes, *Mol. Cryst. Liq. Cryst.* **1971**, *12*, 193. See also ref: [1]
[16] F. Hessel, H. Finkelmann, *Polym. Bull.* **1985**, *14*, 375; H. Leube, H. Finkelmann, *Makro-*

mol. Chem. **1990**, *191*, 2707, *ibid* **1991**, *192*, 1317.

[17] See for example: W. Wedler, P. Hartmann, U. Bakowsky, S. Diele, D. Demus, *J. Mater. Chem.* **1993**, *2*, 1195.

[18] L. J. Yu, A. Saupe, *Phys. Rev. Lett.* **1980**, *45*, 1000. The system is mixtures of potassium laurate, 1-decanol and D_2O. This system also exhibits a reentrant transition (cf. the Chapter in this Handbook on Reentrant Phase Transitions). The higher temperature nematic is rod-like while the lower temperature nematic is disc-like. In addition, the isotropic liquid is also re-entrant in this system.

[19] H. Brand, J. Swift, *J. Phys. Lett.* **1983**, *44*, L-333. C. Cajas, J. B. Swift, H. R. Brand, *Phys. Rev.* **1983**, *A28*, 505 (statics); *ibid* **1984**, *A30*, 1579 (dynamics).

[20] A. Saupe, P. Boonbrahm, L. J. Yu, *J. Chem. Phys.* **1984**, *81*, 2076.

[21] M. Roy, J. P. McClymer, P. H. Keyes, *Mol. Cryst. Liq. Cryst. Lett.* **1985**, *1*, 25.

[22] W. McMillan, *Phys. Rev.* **1971**, *A4*, 1238; *ibid.* **1972**, *A6*, 936; *ibid* **1974**, *A9*, 1720.; K. K. Kobayashi, *Phys. Lett.* **1970**, *A31*, 125.

[23] P. G. de Gennes, *Solid State Commun.* **1972**, *10*, 753; P. G. de Gennes, J. Prost, *The Physics of Liquid Crystals*, Clarendon, Oxford **1993**, Chapter X.

[24] C. W. Garland, G. Nounesis, *Phys. Rev. E.* **1994**, *49*, 2964 and references therein.

[25] cyano-octylbiphenyl and cyano-nonylbiphenyl respectively. The small discontinuities in 8CB and 9CB escaped the fine net of adiabatic calorimetry (J. Thoen, H. Marynissen, W. van Dael, *Phys. Rev. Lett.* **1984**, *52*, 204; *Phys. Rev.* **1982**, *A26*, 2888; H. Marynissen, J. Thoen, W. van Dael, *Mol. Cryst. Liq. Cryst.* **1985**, *124*, 195) and high resolution X-ray diffraction (B. M. Ocko, R. J. Birgeneau, J. D. Litster, *Z. Phys.* **1986**, *B62*, 487) but were caught by the front propagation test [5].

[26] B. I. Halperin, T. C. Lubensky, S. K. Ma, *Phys. Rev. Lett.* **1974**, *32*, 292 for superconductors and B. I. Halperin, T. C. Lubensky, *Solid State. Comm.* **1974**, *14*, 997 for the N–SmA transition.

[27] P. Sonntag, PhD Thesis, University of Louis Pasteur de Strasbourg I (1996) and to be published (1997).

[28] T. C. Lubensky, *J. Chim. Phys.* **1983**, *80*, 31.

[29] T. C. Lubensky, G. Grinstein, R. Pelcovits, *Phys. Rev.* **1982**, *B25*, 6022.

[30] D. R. Nelson, J. Toner, *Phys. Rev.* **1981**, *B24*, 363 and J. Toner, *Phys. Rev.* **1982**, *B26*, 462. A. J. McKane, T. C. Lubensky, *J. Phys. (Paris) Lett.* **1982**, *43*, L217.

[31] A. R. Day, A. J. McKane, T. C. Lubensky, *Phys. Rev.* **1983**, *A27*, 1461.

[32] B. R. Patton, B. S. Andereck, *Phys. Rev. Lett,* **1992**, *69*, 1556; B. S. Andereck, B. R. Patton, *Phys. Rev.* **1994**, *E49*, 1393.

[33] See for example [27].

[34] P. E. Cladis, *Phys Lett.* **1974**, *48A*, 179 and unpublished. This is a qualitative result. While it may not be possible to make elastic constant measurements for K_2 and K_3 deep enough into the N–SmA critical regime to establish quantitative certainty about critical exponents, there should be no problem for K_1 which does not diverge at this transition.

[35] Unlike para-MBBA, the ortho-MBBA isomer is not liquid crystalline. It was synthesized by Gary Taylor at Bell Labs (ca. 1970) to 'fine tune' liquid crystal electro-optic response for devive applications. While the shape of *p*-MBBA is rod-like, that of *o*-MBBA resembles a bent rod.

[36] F. Brochard, *J. Phys. (Paris)* **1973**, *34*, 411. F. Jähnig and F. Brochard, *J. Phys. (Paris,* **1974**, *35*, 301. See also: K. A. Hossain, J. Swift, J.-H. Chen and T. C. Lubensky, *Phys. Rev.* **1979**, *B19*, 432.

[37] K. Kawasaki, *Phys. Rev.* **1966**, *150*, 291; *Ann Phys.* **1970**, *61*, 1. See also review article [7].

[38] Ch. Gähwiller, *Phys. Rev. Lett.* **1972**, *28*, 1554; *Mol. Cryst. Liq. Cryst.* **1973**, *20*, 301. P. Pieranski, E. Guyon, *Phys. Rev. Lett.* **1974**, *32*, 924.

[39] See for example: P. E. Cladis in *Nematics: Mathematical and Physical Aspects*, (Eds.: J.-M. Coron, J.-M. Ghidaglia, F. Helein) Kluwer Academic Publishers, Boston **1990**; and H. R. Brand, C. Fradin, P. L. Finn, P. E. Cladis, *Phys. Lett. A* **1997**.

[40] P. Martinoty et al. (*Phys. Rev. E* **1997**) at the first-order SmC–SmF transition where nevertheless, fluctuations are observed by anomalies in ultrasonic measurements and H. Yao, T. Chan, C. W. Garland, *Phys. Rev.* **1995**, *E51*, 4584 at the first-order SmC–SmI critical point. See also [11].

[41] Type II superconductors in a magnetic field (and in the appropriate geometry) are characterized by the following properties. P. G. de Gennes, *Superconductivity of Metals and Alloys*, W. A. Benjamin, Inc. (pub.), New York, **1966**. The Meissner effect is not observed except in weak fields, $H<H_{c1}$. For $H>H_{c1}$, magnetic field lines penetrate but the penetration is not complete. de Gennes points out that this region was first experimentally discovered by Schubnikov in 1937 but is known as the vortex state from the microscopic theory derived by A. Abrikosov, *Zh. Eksp. Teor. Fiz.* **1957**, *32*, 1442 [*Soviet Physics JETP* **1957**, *5*, 1174]. For $H>H_{c2}$, no expulsion of flux is observed however superconductivity is not completely destroyed. Experimentally the transition at H_{c2} is second order. In the interval $H_{c2}<H<H_{c3}\approx1.69H_{c2}$, there is a super-

conducting surface sheath. P. B. Vigman, V. M. Filev, *Zh. Eksp. Teor. Fiz.* **1975**, *69*, 1466 [*Soviet Physics JETP*, **1975**, *42*, 747] suggested looking for the analogue of H_{c3} in liquid crystals to support de Gennes' analogy [23].

[42] T. C. Lubensky, *Physica A*, **1995**, *220*, 99.

[43] S. R. Renn, T. C. Lubensky, *Phys. Rev.* **1988**, *A38*, 2132.

[44] J. W. Goodby, M. A. Waugh, S. M. Stein, R. Pindak, J. S. Patel, *Nature*, **1988**, *337*, 449; *J. Am. Chem. Soc.* **1989**, *111*, 8119. SmC* is the chiral analogue of SmC with a macroscopic helix axis parallel to the layer normal.

[45] P. E. Cladis, A. J. Slaney, J. W. Goodby, H. R. Brand, *Phys. Rev. Lett.* **1994**, *72*, 226; *Il Nuovo Cimento*, **1994**, *16D*, 765 and unpublished. This Slaney–Goodby compound is (S)-2-chloro-4-methylpentyl 4′-(4-n-dodecyloxypropiolyloxy)-4-biphenylcarboxylate (1202Cl4M5T).

[46] On a microscopic scale, the geometry of this model may be more reminiscent of Type I intermediate states rather than a vortex state [41]. The analogue of the vortices are the screw dislocations. A analogue of the intermediate state at the N-SmA transition is a static stripe pattern; P. E. Cladis, S. Torza, *J. Appl. Phys.* **1975**, *46*, 584.

[47] C. Dasgupta, B. I. Halperin, *Phys. Rev. Lett.* **1981**, *47*, 1556.

[48] T. Chan, C. W. Garland, H. T. Nguyen, *Phys. Rev.* **1995**, *E52*, 5000; L. Navailles, C. W. Garland, H. T. Nguyen, *J. de Phys. II* **1996**, *6*, 1243.

[49] R. D. Kamien, T. C. Lubensky, *J. Phys. (Paris) I* **1994**, *3*, 2123.

[50] D. S. Fisher, M. P. A. Fisher, D. A. Huse, *Phys. Rev.* **1991**, *B43*, 130; D. A. Huse, M. P. A. Fisher, D. S. Fisher, *Nature* **1992**, *358*, 553.

[51] The time dependence [45] of the heat singularity at TGBA–N* in the Slaney–Goodby compound is suggestive of a liquid–glass transition in this material.

[52] A. C. Ribeiro, A. Dreyer, L. Oswald, J. F. Nicoud, A. Soldera, D. Guillon, Y. Galerne, *J. de Phys. II* **1994**, *4*, 407 and unpublished, V. Vill, H.-W. Tunger, D. Peters, *Liq. Cryst.* **1996**, *20*, 547. J. W. Goodby, private communication, Capri (1996).

[53] D. Guillon, A. Skoulios, *J. Phys.* **1977**, *38*, 79.

[54] P. G. de Gennes, *Mol. Cryst. Liq. Cryst.* **1973**, *21*, 49.

[55] C. R. Safinya, M. Kaplan, J. Als-Nielsen, R. J. Birgeneau, D. Davidov, J. D. Litster, D. L. Johnson, M. E. Neubert, *Phys. Rev.* **1980**, *21*, 4149. These experiments were made on the sulfur compound, 4-n-pentyl-phenylthiol-4′-n-octyloxybenzoate ($\overline{8}$S5).

[56] M. Delaye, *J. Phys. Colloq.* **1979**, *40*, C3-350, 4-nonyloxybenzoate 4′-butyloxyphenyl (9E4).

[57] B. I. Ostrovskii, A. Z. Rabinovich, A. S. Sonin, E. L. Sorkin, B. A. Strukov, S. A. Taraskin, *Ferroelectrics*, **1980**, *24*, 309.

[58] Y. Galerne, *Phys. Rev.* **1981**, *A24*, 2284.

[59] P. Keller, P. E. Cladis, P. L. Finn, H. R. Brand, *J. Phys.* **1985**, *46*, 2203.

[60] K. Ema, J. Watanabe, A. Takagi, H. Yao, *Phys. Rev.* **1995**, *E52*, 1216 and references therein.

[61] A. Fukuda, Y. Takanishi, T. Isozaki, K. Ishikawa, H. Takezoe, *J. Mater. Chem.* **1994**, *4*, 997.

[62] L. Benguigui, P. Martinoty, *Phys. Rev. Lett.* **1989**, *63*, 774.

[63] D. Collin, S. Moyses, M. E. Neubert, P. Martinoty, *Phys. Rev. Lett.* **1994**, *73*, 983.

[64] L. Benguigui, P. Martinoty, *J. de Phys. II.*, **1997**, *7*, 225. and references therein.

[65] B. S. Andereck, J. Swift, *Phys. Rev.* **1982**, *A5*, 1084.

[66] B. I. Halperin, D. R. Nelson, *Phys. Rev. Lett.* **1978**, *41*, 121; D. R. Nelson, B. I. Halperin, *Phys. Rev.* **1979**, *19*, 2457. R. J. Birgeneau, J. D. Litster, *J. Phys. Lett. (Paris)*, **1978**, *39*, L399.

[67] A. J. Leadbetter, J. C. Frost, M. A. Mazid, *J. Phys. Lett.* **1979**, *40*, 325; R. Pindak, D. E. Moncton, S. D. Davey, J. W. Goodby, *Phys. Rev. Lett.* **1981**, *46*, 1135.

[68] C. C. Huang, *Adv Phys.* **1993**, *43*, 343.

[69] G. Nounesis, C. C. Huang, J. W. Goodby, *Phys. Rev. Lett.*, **1986**, *56*, 1712. In these experiments, the temperature oscillations were at 1 Hz.

[70] hexylalkyl-4′-pentylalkoxybiphenyl-4-carboxylate.

[71] G. Nounesis, R. Geer, H. Y. Liu, C. C. Huang, J. W. Goodby, *Phys. Rev.* **1989**, *A40*, 5468.

[72] E. Gorecka, L. Chen, W. Pyzuk, A. Króczyński, S. Kumar, *Phys. Rev.* **1994**, *E50*, 2863.

[73] J. L. Gallani, P. Martinoty, D. Guillon, G. Poeti, *Phys. Rev.* **1988**, *A37* (Rap. Comm.), 3638; *Phys. Rev. E* **1997**.

[74] P. H. Keyes, *Phys. Rev. Lett.* **1987**, *59*, 83; E. P. Koistinen and P. H. Keyes, *Phys. Rev. Lett*, **1995**, *74*, 4460 and unpublished.

[75] G. Voigts, W. van Dael, *Liq. Cryst.* **1993**, *14*, 617.

[76] Z. Kutjnak, C. W. Garland, C. G. Schatz, P. J. Collings, C. J. Booth, J. W. Goodby, *Phys. Rev.* **1996**, *E53*, 4955 and to be published.

[77] T. C. Lubensky, H. Stark, *Phys. Rev.* **1996**, *E53*, 714. These authors introduce a pseudo-scalar order parameter for this critical point.

[78] P. E. Cladis, H. R. Brand, *Phys. Rev. Lett.* **1984**, *52*, 2210.

[79] F. Hardouin, M. F. Achard, H. T. Nguyen, G. Sigaud, *J. Phys. (Paris) Lett.*, **1984**, *46*, L123. An extensive review of smectic A polymorphism is contained in F. Hardouin, A. M. Levelut, M. F. Achard, G. Sigaud, *J. Chim. Phys.* **1983**, *80*, 53.

[80] J. Prost, J. Toner, *Phys. Rev.* **1987**, *A36*, 5008.

[81] L. Wu, C. W. Garland, S. Pfeiffer, *Phys. Rev.* **1992**, *A46*, 973, *ibid* **1992**, *A46*, 6761.

[82] G. Nounesis, S. Kumar, S. Pfeiffer, R. Shashidhar, C. W. Garland, *Phys. Rev. Lett.* **1994**, *73*, 565.

[83] D. L. Johnson, *J. Chim. Phys*, **1983**, *80*, 45.

[84] R. Shashidhar, B. R. Ratna, S. K. Prasad, *Phys. Rev. Lett.* **1984**, *53*, 2141.

[85] S. Somasekhara, R. Shashidhar, B. R. Ratna, *Phys. Rev.* (Rap. Comm.) **1986**, *A34*, 2561.

[86] J. B. Swift, *Phys. Rev.* **1976**, *A14*, 2274. Swift explains that the N–SmC transition is driven first order by fluctuations in a system with a characteristic length, $2\pi/k$. In such a system, fluctuations are confined to a shell around $k \neq 0$ rather than a sphere around $k = 0$. The effect is to drive the second order N–SmC transition temperature to lower temperatures and so is pre-empted by a first-order phase transition.

6.4 Re-entrant Phase Transitions in Liquid Crystals

P. E. Cladis

6.4.1 Introduction

In a re-entrant phase transition, a higher temperature thermodynamic phase with higher symmetry (and most dramatically with greater fluidity) reappears at temperatures below a stable thermodynamic phase with lower symmetry (and less fluidity). Molecules have many more internal degrees of freedom than atoms. As a result, liquid crystals have more than 30 different thermodynamic phases, many involving broken continuous symmetries. It is not surprising then that there is not a unique mechanism to account for re-entrant behavior in liquid crystals. In addition, the characterization of liquid crystal structure (i.e. microscopic structure) versus property (i.e. macroscopic expression) relations requires measurements on many different length and time scales: thermal measurements; k-space measurements (e.g. X-ray and light scattering); direct observations with microscopes, particularly the polarizing light microscope; acoustic measurements; etc. While a phase diagram in, for example, the temperature–concentration plane or temperature–electric field plane gives a macroscopic picture, that these various techniques generate, it does not distinguish between different microscopic mechanisms for re-entrant behavior. In the absence of a theoretical framework covering all known cases of re-entrance in liquid crystals, we have sorted them into three broad classes. However, as is typical of liquid crystal phenomena, and indeed of complex natural phenomena in general, the boundaries between the different classes are not sharp.

The sequence of phase transitions, with decreasing temperature, N SmA N_{re} was discovered in 1975 in cyano compounds with two benzene rings [1] (Fig. 1, [2]). Since 1975, liquid crystals have been found to

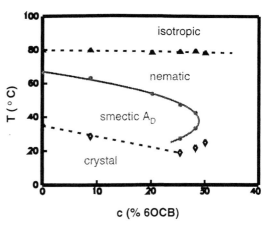

Figure 1. The N_{re} phase diagram for 8OCB/6OCB mixtures [2].

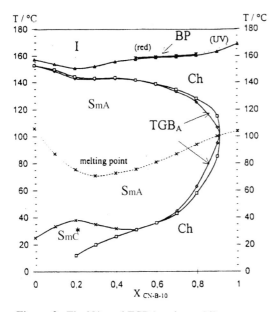

Figure 2. The N_{re}^* and TGBA$_{re}$ phases [6].

show many different examples of re-entrant behavior arising from several different mechanisms, most of which have no obvious solid-state analog. For example, the first re-entrant transition [1] is now understood to result when 'short-range effects surprisingly do the long-range ones in!' [3]. We call this category R1, re-entrance from frustration. A comprehensive review of R1 re-entrance appeared in 1988 [4] to mark the 100th anniversary of the discovery of liquid crystals. As R1 is by far the most novel example of re-entrance in liquid crystals, and robust materials exhibiting R1 re-entrance are widely available [5], it has generated the most amount of work [4]. Therefore, R1 is the transition that is currently best understood. The most recent R1 example is the phase sequence, with decreasing temperature N*, TGBA, SmA, re-entrant TGBA, N_{re}^* which is shown in mixtures involving a cyano compound with a carbohydrate link [6] (Fig. 2). These materials [6] may help develop an understanding of biologically relevant materials where more-fluid states

are known to occur at lower temperatures than less-fluid ones [7].

Re-entrance in liquid crystals can also involve temperature-driven change in short-range steric forces competing for long-range order. Molecular shape is a crucial factor in liquid crystal structure–property relations. Rod-like nematics (uniaxial positive) cannot continuously transform to disc-like nematics (uniaxial negative) without an intervening biaxial nematic phase or an isotropic liquid state in the case of discontinuous transitions. The phase sequence with decreasing temperature – isotropic liquid, uniaxial positive nematic, biaxial nematic, uniaxial negative nematic, re-entrant isotropic liquid (Fig. 3, [8]) – is an example from lyotropic liquid crystals of both a re-entrant nematic (N_{re}) and the first re-entrant isotropic liquid.

Some thermotropic liquid crystal structures are so complex both rod-like and disc-like liquid crystal phases can be stabilized at different temperatures and for different members in a homologous series [9]. For example, in double-swallow-tailed compounds with aliphatic chain lengths less than 10, 'rod-like' liquid crystal phases are observed, and in longer chains, discotic liquid crystal phases are observed [10]. We call the above re-entrant phenomenon, 're-entrance triggered by complex geometrical factors', R2. Other re-entrant phase sequences we list in R2 are the phase sequences SmA SmC$_{re}$ SmA [11] and SmC SmO$_{re}$ SmC [12] found in a symmetrical, relatively long compound [13] the homologous series of which has an alternating stability for SmO and SmC phase. As materials exhibiting R2 re-entrance tend to be in the 'exotic' limit, apart from temperature–concentration phase diagrams for given molecular structures little is known about R2 re-entrance.

The term 're-entrant' was first used to describe the reappearance of a lower temper-

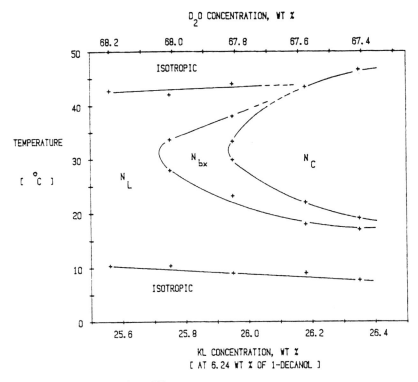

Figure 3. The first I_{re} phase [8].

ature normal metal phase in superconductors (see, for example, Cladis [4] and references therein). Re-entrance in this case, R3,
results from the competition between long-
range forces, e.g. a new order parameter
grows [14] or there is a temperature-dependent coupling between competing order parameters [15]. The phase diagram used to
discuss superconducting re-entrance is the
magnetic field–temperature plane. R3 re-
entrance has been known longer than has
re-entrance from frustration, R1. Only the
terminology is new for R3. In R3 re-entrance, fluctuations often play an important
role, leading to universal features such as
the magnificant spiral at the N_{re}–SmC–SmA
multicritical point (Fig. 4) [16]; the
N–SmA–NSmC step [17] (Fig. 5) showing
the dramatic suppression in the N–SmC
transition temperature resulting from Bra-

zovskii fluctuations at the N–SmC transition [18] compared to the N–SmA transition; and the nematic bubble when fluctuations from two SmA phases with nearly similar but different layer spacings compete
[19]. As it has been eloquently argued [20]
that re-entrance in nonpolar compounds
[21] depends sensitively on universal features of the NSmA SmC multicritical point,
we include it in R3 re-entrance.

6.4.2   R1: Re-entrance from Frustration

In the re-entrant nematic transition, a liquid
phase without translational order (nematic)
occurs at a lower temperature or higher pressure than one with one-dimensional translational order (SmA). With decreasing tem-

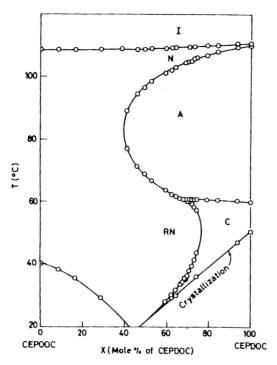

Figure 4. The N_{re}–SmA–SmC multicritical point [16].

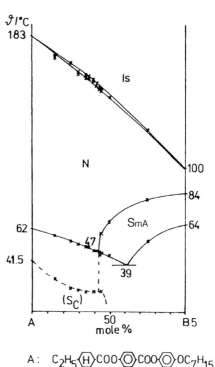

A: C_2H_5⟨H⟩COO⟨O⟩COO⟨O⟩OC_7H_{15}

B 5: C_2H_5O⟨O⟩$CH=N$⟨O⟩$CH=C-COOC_5H_{11}$
                                  CH_3

Pelzl et al., Mol. Cryst. Liq. Cryst. **168**, 147 (1989)

Figure 5. Step down to lower temperatures of the N–SmC transition line from the N–SmA transition line [17].

perature (see Fig. 1) or increasing pressure (Fig. 6), the phase sequence is N–SmA–N. The lower temperature nematic phase is called 're-entrant' [22]. This particular phase sequence was first discovered in 1975 in mixtures of cyano-Schiff base compounds (cyanooctyloxy aniline and heptyloxybenzilidene aniline) at 1 atm [1], and later in pure materials, including cyanobiphenyloctyloxy aniline (8OCB) under pressure [23]. These materials have two benzene rings.

As it was known that SmA phases of cyanobiphenyl compounds exhibit layer spacings that are larger and incommensurate with their molecular length [23], the suggestion was made [22] that the N_{re} phase in these materials was the result of dimer-type associations on a molecular level and that it could transform back to a SmA phase with layer spacing comparable to its molecular length, even when there was an intervening SmC phase (Fig. 7, [22]).

Goodby and coworkers [24] elegantly showed that re-entrance in cyano compounds resulted from a sensitive balance between dipolar and steric factors. They synthesized two benzene ring materials with an ester link that either reinforced (resulting in exclusively re-entrant nematic behavior) or opposed (resulting in the appearance of a SmC phase below the SmA) the cyanophenyl 'mesomeric relay' [24]. The conclusion is that, when dipolar forces domi-

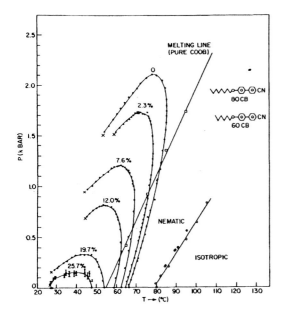

Figure 6. The N_{re} phase as a function of temperature and pressure for 8OCB/6OCB mixtures [4].

nate we have re-entrance from frustration, and when steric forces dominate the door opens to R2 and R3 re-entrance.

The incommensurate SmA phase was called SmA_d [25], where d denotes 'dimer' for the pairwise overlapped associations of aromatic cores fromed by these compounds [26], and the lowest temperature SmA phase was called SmA_I [19], where I stands for 'interdigitated', as the aliphatic chains from one layer formed of dimers are interdigitated with those of neighboring layers [22]. Because the layer spacing is about the same as the molecular length, these kinds of SmA phases are also known as SmA_1. However, stimulating the formation of monomolecular SmA phases in cyano compounds leads to the enhancement of the SmA phase at the expense of a nematic phase (Fig. 8), introducing the need to distinguish between

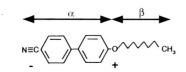

	cyano aromatic moiety: α	aliphatic chain length: β	total length: $\alpha + \beta$	model length: $\alpha + 2\beta$	Measured length
CBOOA	16.0 Å	10.4 Å	26.4 Å	36.8 Å	36.3 Å
HBAB	16.0	8.2	24.2	-	-
8OCB	13.9	9.2	23.1	32.3	30.8

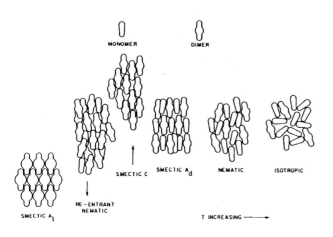

Figure 7. The re-entrant scenario for cyano compounds [22] with the model for SmA_d and SmA_I.

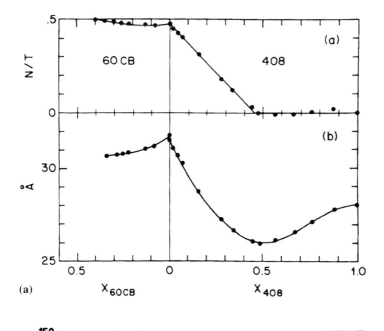

(a)

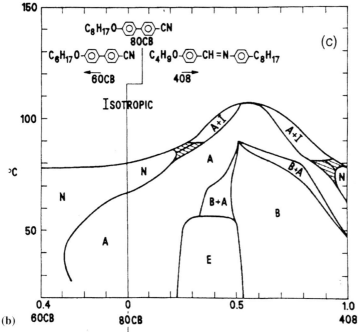

(b)

Figure 8. The number of dimer pairs (a) deduced from the observed SmA layer spacing (b) in 6OCB/8OCB and 4O8/8OCB mixtures. (c) At complete pair saturation, the number of pairs (N) compared to the total number of molecules (T) is 0.5. In a mixture of monomers, $N=0$ (c), as $N/T \to 0.5$, the SmA_d phase disappears, and when $N/T \to 0$, the nematic phase is squeezed out by the enhanced SmA_1 phase [27].

SmA_1 and SmA_I. Thus the reappearance of a SmA phase that has a layer spacing close to the molecular length may be an indication that its layer structure is more than a simple packing of single molecules (see Fig. 7, [27]).

While the N_{re} phase was also found in nitro compounds with two benzene rings [28], re-entrant phenomena in liquid crystals made a giant leapforward in 1979, when the stable N_{re}–SmA transition [29] and multiple re-entrance [30] was discovered at 1 atm

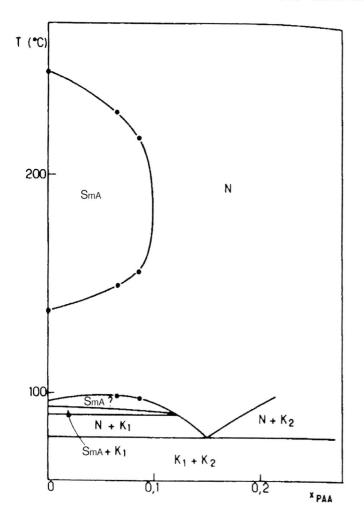

Figure 9. One of the first pure compounds to show a stable N_{re} phase (RN) with two RNA transitions [29].

by the Bordeaux group in several different cyano compounds containing three benzene rings (Fig. 9). A multiple re-entrant phase sequence with decreasing temperature is, for example, N SmA N_{re} SmA N_{re} SmA. These discoveries had a major impact on both the chemical and physical frontiers of liquid crystal research in the polymorphism of SmA and re-entrant behavior [31]. It should also be noted that the authors pointed out that the material in which they first identified a N_{re} phase [29] had previously been identified as a more ordered smectic phase [32].

One of the first temperature–concentration re-entrant phase diagrams from the Bordeaux group (Fig. 9) [29] bears an uncanny resemblance to the magnetic field–temperature phase diagram observed in a complex superconducting material at very low temperatures and very high magnetic fields (Fig. 10) [32]. This superconducting re-entrance is the closest solid-state analog to R1 re-entrance. In these systems, the internal magnetic field first shields the superconducting state from the applied magnetic field, so that the superconducting state reenters at higher fields, and then, eventually,

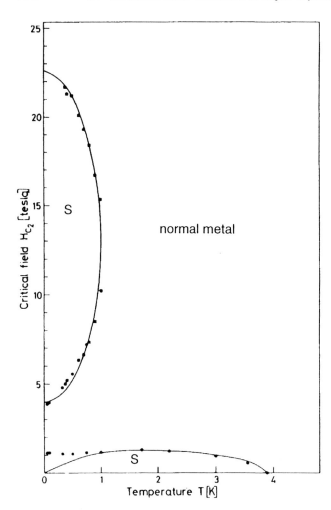

Figure 10. An example of the re-entrant phenomenon in complex superconductors [33] having a phase-diagram topology similar to that in Figure 9. At low fields the normal metal state is not re-entrant, as predicted from theory [34]. However, at extremely low temperatures and with increasing applied field, the superconducting state (S) is first restored and then destroyed again above 20 T.

when its shielding powers are exhausted, superconductivity is again destroyed [34]. However, unlike the case of magnetic superconductors [14] the normal metal state both above and below the superconducting 'nose' in Fig. 10 are similar.

An internal/external field competition [33] may be at work in the SmC^*_{re} phase observed in an applied electric field [35] when the helix structure is suppressed by an applied electric field. However, while our understanding of the re-entrant mechanism behind the behavior illustrated by Fig. 10 is relatively complete [34], nonlinear theories of smectic phase transitions [36] have not yet accounted for the observed helielec-

tric–ferroelectric–helielectric phase sequence with decreasing temperature in an applied electric field [35] or the C^* O^* C^*_{re} [13] sequence with decreasing temperature (discussed in Sec. 4).

Following the Bordeaux discovery by Hardouin et al. [29], many three-ring cyano compounds were discovered, particularly by the Halle group [37], that showed stable N_{re} phases in cyano compounds with three benzene rings. An overview of the chemical structure and re-entrant behavior of three benzene ring compounds has been given by Weissflog et al. [38].

The theoretical work by Indekeu and Berker [39] showed the new physics to emerge from R1 re-entrance. Simply put, they showed that treating molecular associations from the perspective of statistical physics, the static dimer concept [22] was consistent with an ensemble of triplets correlated on a microscopic scale. They considered a two-particle dipolar potential that had both parallel ('up–up') and antiparallel ('up–down') exchange interactions. When dipolar forces between two members of the triplet cancel, the third member experiences no force and is free to permeate from layer to layer, 'frustrating' smectic order. By allowing displacements of three dipoles relative to each other, a population of triplets is generated that has a net short-range dipole interacting with each other via short-range dipolar forces stabilizing the layered structure. The energy fluctuations associated with a triplet ensemble is evaluated using an Ising criterion. If satisfied, a layer structure can be stabilized; if not, the nematic re-enters.

The N_{re} phase shares frustration properties with a 'spin-glass'. Roughly speaking, a spin-glass results from the presence of a wide variety of competing interactions arising from random interactions in a many-body system. In a spin-glass, different regions of phase space become irretrievably separated by energy barriers. The system is called 'nonergodic', as each small piece of the system is in a different region of phase space from all others and only 'knows' its own local conditions. The Liouville theorem breaks down and the system is called 'frustrated' [40]. As the temperature is lowered, these isolated states proliferate.

In this sense, SmA_d and the N_{re} phases are frustrated. They escape being nonergodic because they are liquids in the usual sense. To underscore the uniqueness of this particular example of frustration, Berker and Indekeu [39] refer to their theory as a 'spin-gas' theory, rather than a spin-glass theory.

By combining a cyanoethyl compound with a cholesteric liquid crystal with a sufficiently high twisting power, Pelzl et al. [41] obtained re-entrant cholesteric phases. Near the N*–SmA phase transition they found the opposite color sequence resulting from the untwisting of the helix structure by SmA fluctuations from the higher temperature N*–SmA transition. They pointed out that binary mixtures exhibiting both the usual N*–SmA and the N^*_{re}–SmA transition could, in principle, be used to make temperature-sensing devices in two temperature ranges.

Vill and coworkers [6] give a summary of N^*_{re} phases, and have shown that quite different phase diagrams can result from subtle changes in molecular structure. In particular, when mixing a cyanophenyl carbohydrate based compound containing a boron link (CNB10), which shows I–BP–N* transitions with decreasing temperature, with a slightly longer compound, in which boron is replaced by carbon (CNC12) and which has a large SmA temperature range, they observed the novel phase diagram (Fig. 2) exhibiting a re-entrant TGBA phase. Contact preparations of CNC12 with (1) a relatively nonpolar smaller molecule results in a cholesteric mountain surrounded by SmA and SmC phases and with (2) a N* compound, a lower temperature N* phase ending in the middle of the SmA phase where the phase boundaries (presumably) were lost in the contact preparation [42]. In both contact preparations, the high-temperature N* phases of the two components are immiscible. The high-temperature N* phase of CNB10 is also immiscible when mixed with the same two compounds; however, in the contact mixture with the N* compound there is an almost vertical N*–SmA phase boundary rather similar to the ones observed in the presence of competing SmA and SmC phase fluctuations (see, for example,

Fig. 5), with the TGBA squeezed in between. By suitably choosing components in binary mixtures, carbohydrate based cyanophenyl compounds can exhibit features associated with both R1 and R3 re-entrance.

6.4.3   Re-entrance from Geometric Complexity

Molecular shape plays a significant role in determining the stability of liquid crystal phases. Even the number of carbon atoms in the aliphatic chain can lead to an odd–even variation in transition temperatures in a homologous series and an odd–even stability of liquid crystal phases. A dramatic example of the latter is the racemic 1-methylterephthalidene-bis(aminocinamates) (MnTAC) [13]. These compounds contain three benzene rings and are chemically identical on both sides of the center of the molecular long axis. In this homologous series, MnTAC exhibits only a SmO phase [12] when n is odd and ≥ 5, and only a SmC phase when n is even and >4. M4TAC exhibits a stable SmC_{re} transition; that is, with decreasing temperature, the phase sequence is I–SmC–SmO–SmC_{re}. The chiral analogs with $4 \leq n \leq 7$ exhibit the SmO* phase independent of the parity of n. By mixing racemic M4TAC with materials showing only either SmO* or a SmC* phase, Heppke et al. [13] obtained phase diagrams with showing SmC_{re}^{*}.

In lyotropic liquid crystals, uniaxial negative nematics (N_L, i.e. disk-like phase) were found in aqueous solutions of potassium laurate (KL), 1-decanol, and potassium chloride [43]. Within a very narrow temperature and concentration range of a heavy water solution of this ternary mixture, Yu and Saupe [8] found a novel phase diagram (see Fig. 3). In this phase diagram, N_C is uniaxial positive (rod-like). Between the

N_L and N_C phases is the first stable biaxial N phase, which Yu and Saupe [8] denoted by N_{bx}, and the first I_{re} liquid state (see Fig. 3).

Weissflogg et al. [10] give an overview of the literature on I_{re} phases in thermotropic liquid crystals. They point out that mesogens consisting of a rod-like core ending in two half-disk shaped moieties (polycatenar mesogens) have steric features between those of rod-lilke and disk-like mesogens, and can exhibit nematic, lamellar, columnar, and cubic mesophases [44]. Increasing the molecular complexity to six-ring mesogens, they found that slight variations in molecular structure can give rise to large changes in liquid crystal properties [10] and stable I_{re} phases. A fascinating example is the transition sequence with decreasing temperature found in six-ringe double-swallow-tailed compounds: I N I Cub SmC [10]. In molecular structures where conformational and corresponding entropic factors play an important role, it is possible that entropic elasticity similar to that which has been proven useful for polymers [45] (in addition to complex steric factors) should be considered.

Dowell [46] has considered the case of, for example, discotic (and lamellar) structures in which rigid aromatic moieties stack in a column (or layers), which can be stabilized by the collective dynamics of disordered alkyl chains (the 'floppy tails') surrounding the columns. The conclusion is that, as the temperature decreases, the alkyl chains become more ordered (and less dynamic), thus destabilizing the columnar stacks. Conversely, as the temperature increases the chains become more active and perhaps even entangled, leading to a more stable columnar structure, and thus there is no higher temperature N phase as is found in some discotic materials [47]. However, in other discotic materials, both the high- and

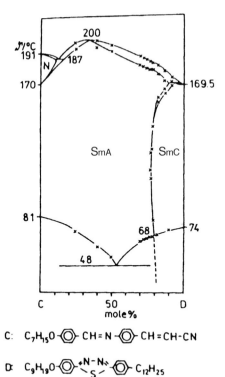

C: C₇H₁₅O —⬡— CH=N —⬡— CH=CH-CN

D: C₉H₁₉O —⬡— ⁺N-N⁺/S —⬡— C₁₂H₂₅

Figure 11. An example of the SmA–SmC–SmA$_{re}$ transition. Here the high-temperature SmA phase has a layer spacing somewhat smaller than that in a mixture of monomers, while the SmA$_{re}$ phase has a layer spacing comparable to that of a mixture of fully extended monomers [11].

low-temperature N$_{re}$ phases as well as re-entrant hexagonal columnar phases have been observed [50]. Thus, while observation of temperature-driven changes in the conformation and dynamics of the aliphatic chains help us to develop on understanding of the subtle features that suppress re-entrance in some cyano compounds [49], it seems clear that interactions of the aromatic moieties in disk-like compounds cannot be ignored, and may account for the wide variety of re-entrant behaviors observed in discotic liquid crystals.

Perhaps a somewhat simpler situation is presented by the SmA–SmC–SmA$_{re}$ transitions [11], which may be triggered by

significant temperature-dependent conformational changes in the long alkyl chains (Fig. 11). A vertical SmA–SmC phase boundary is associated the SmA–SmC–SmA$_{re}$ transition [11], where a polar compound (C in Fig. 11) with an SmA$_d$ phase is mixed with a nonpolar one (D in Fig. 11) with two relatively long aliphatic chains and only a SmC phase. X-ray studies of the layer spacings with increasing concentration of D (x_D) show three distinct regions. When $x_D < 0.5$, the layer spacing is larger than the average molecular length, indicating that in this range of concentrations the SmA phase consists of monomers and dimers. When $0.6 < x_D < 0.7$ the layer spacing corresponds to the monomer average of C and D. In the third regime, the layer spacing is smaller than the average of that of fully extended monomers as the SmC phase is approached. However, the layer spacing in the SmA$_{re}$ phase is somewhat larger than that of the high-temperature SmA phase, being approximately the same as that of the monomer average when the alkyl chains are in an all-*trans* configuration [11].

As thermotropic liquid crystal materials become more complex more sophisticated numerical computations will be needed in order to develop a corresponding hierarchy of structure–property relationships, as is currently in progress for polymers [50]. For example, the SmC to oblique columnar phase transitions [51] in six-ring, double-swallow-tailed compounds represents a worthy challenge for such computations. Furthermore, in order that the use of data on increasingly complex liquid crystal materials can be optimized in terms of their contributions to the development of new fundamental knowledge potentially relevant to polymeric and biological materials as well as new liquid crystal technologies and applications, a broader range of experimental data correlating to, for example, mechanical, magnetic, electrical, rheological, and optical properties with chemical structure is needed.

6.4.4 R3: Re-entrance from Competing Fluctuations

To describe a line of phase transitions between two SmA phases (SmA$_d$ and SmA$_I$, for example), Barois et al. [52] have used two order parameters. As the symmetries are the same, an SmA$_d$–SmA$_I$ line of first-order phase transitions should end at a critical point. However, the observation is that, as the critical point is approached and fluctuations increase on the two sides of the transition line, where the symmetry is the same but the layer spacings are different, a nematic bubble spontaneously appears embedded in SmA [19]. Mean field theories [52] cannot account for this observation.

At a second-order SmA–SmC phase transition, the symmetries are different but the layer spacing is the same. Fluctuations can drive a line of second-order SmA–SmC phase transitions to an N–SmA–SmC multicritical point (see Fig. 4) [53]. Competing N–SmA and N–SmC fluctuations pull the N phase under the SmA phase in the temperature–concentration phase plane, leading to the N$_{re}$–SmA–SmC multicritical point [16]. High-resolution studies, as a function of both concentration and pressure, resolve the fluctuation-driven N–SmA/N–SmC step (see, e.g. Fig. 5) into a universal spiral (Fig. 12) [16] around the N–SmA–SmC and the N$_{re}$–SmA–SmC multicritical points. Loosely speaking, the N–SmA transition line is dominated by N–SmA fluctuations, and the N–SmC transition line is dominated by Brazovskii fluctuations [54] that drive the N–SmC transition to lower temperatures compared to the N–SmA transition [18].

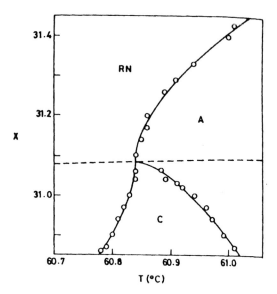

Figure 12. Detail of the N$_{re}$–SmA–SmC multicritical point showing the universal spiral topology [16] of the N–SmA–SmC multicritical point [54].

The N$_{re}$ phase has also been observed in binary mixtures of 'nonpolar' compounds [55]. Shashidhar [20] has pointed out that in these two-component systems only one component is weakly polar or nonpolar. The second component has a dipole moment that is pronounced but not as strong as that of, for example, cyanobiphenyls [21]. The Bangalore group found that practically all liquid crystals possessing a net molecular dipole moment also show near-neighbor antiparallel dipolar correlations, as evidenced by the difference in the average and isotropic values of the dielectric constant. Such correlations are absent only when the molecular structure is such that one half of the molecule is an exact mirror image of the other half. Neither of the two 'nonpolar' components showing N$_{re}$ phase behavior have this property. Nonpolar re-entrance is observed over a very small temperature range and an even narrower concentration range [21].

On further investigation, Shashidar [20] explains, a remarkable situation was found. In every case, one of the materials had an inherent SmC phase, while the other did not. As there was no observed miscibility gap, in every case there had to be a N–SmA–SMC point at very low temperatures. They also knew from their studies at high pressure [54] as well as their detailed studies on mixtures exhibiting the N–SmA–SmC and the N_{re}–SmC–SmA point (see Figs. 4 and 12) [16], that the phase diagram has the universal spiral topology where phase boundaries curve as the N–SmA–SmC point is approached. Shashidhar [20] concludes that, as re-entrance in nonpolar compounds results from the universal curvature of the phase boundaries as the N–SmA–SMC point is approached, its origin is the fluctuations associated with the N–SmA–SmC multicritical point.

6.4.5 Conclusions

In this overview of re-entrant phase transitions in liquid crystals we have identified three main causes of re-entrance: R1, re-entrance from frustration; R2, re-entrance from complex steric factors; and R3, re-entrance from competing fluctuations (e.g. 'a new order parameter grows'). Apart from (possibly) the re-entrant hexagonal transition in discotic liquid crystals, most re-entrant transitions in liquid crystals have no obvious solid-state analog. To optimize the impact of research into liquid crystal re-entrant phase transitions on fundamental knowledge and applications in general, and polymer and biological physics in particular, a braod variety of experimental data and sophisticated theories of structure–property relationships is required. So far, such a depth of investigation has been accomplished only for R1 re-entrance, as robust compounds

exhibiting R1 re-entrance are widely available. Thus, while we have learned new physics from liquid crystals by studying R1 re-entrance, the new physics to emerge from the less widely studied, more 'exotic' materials exhibiting R2 and R3 re-entrance is still to come.

6.4.6 References

[1] P. E. Cladis, *Phys. Rev. Lett.* **1975**, *35*, 48.
[2] D. Guillon, P. E. Cladis, J. B. Stamatoff, *Phys. Rev. Lett.* **1978**, *41*, 1598.
[3] N. Berker, personal communication (1996). As SmA translational order is not 'really' long range, it may not be totally irrelevant.
[4] P. E. Cladis, *Mol. Cryst. Liq. Cryst.* **1988**, *165*, 85.
[5] See e.g. P. E. Cladis, *Liq. Cryst.* (in press).
[6] V. Vill, H.-W. Tunger, *J. Chem. Soc., Chem. Commun.* **1995**, 1047. The authors found that the 11 chain homolog (1*S*,6*R*,8*R*)-8-(4'-undecoxy-pheny)-3-(4''-cyanophenyl)-2,4,7-trioxa-3-bora-bicyclo[4.4.0]decane, CNB11, exhibits TGBA$_{re}$. V. Vill, H.-W. Tunger, D. Peters, *Liq. Cryst.* **1996**, *20*, 547. Binary mixtures of CNB10 and CNC12 (Figure 2) were studied. In CNC12, boron is replaced by carbon.
[7] For example haemoglobin becomes more fluid as it is cooled towards 0°C.
[8] L. J. Yu, A. Saupe, *Phys. Rev. Lett.* **1980**, *45*, 1000.
[9] N. H. Tinh, J. Malthête, C. Destrade, *J. Phys.* **1981**, *42*, L417.
[10] W. Weissflogg, I. Letki, G. Pelzl, S. Diele, *Liq. Cryst.* **1995**, *18*, 867; *Mol. Cryst. Liq. Cryst.* **1995**, *260*, 157, and references therein.
[11] S. Diele, G. Pelzl, A. Humke, S. Wünsch, W. Schäfer, H. Zaschke, D. Demus, *Mol. Cryst. Liq. Cryst.* **1989**, *173*, 113, and references therein.
[12] The SmO and SmC phases have layered structures. In the SmC phase the molecular tilt in a layer is inclined at an angle to the layer normal and is the same in all layers. In SmO, the molecular tilt alternates in a herringbone pattern with equal and opposite tilts in neighboring layers. In the chiral case, the SmC* and SmO* phases have, in addition, a helix structure, and so they are both helielectric, although SmO* has a double-helix structure because of its herringbone structure. For a recent review of the SmO phase see: A. Fukuda, Y. Takanishi, T. Isozaki, K. Ishikawa, H. Takezoe, *J. Mater. Chem.* **1994**, *4*, 997.
[13] G. Heppke, P. Kleinberg, D. Lötzsch, *Liq. Cryst.* **1993**, *14*, 67. A. Fukuda, personal communication (1996).

[14] W. A. Fertig, D. C. Johnson, L. E. DeLong, R. W. McCallum, M. B. Maple, B. T. Matthias, *Phys. Rev. Lett.* **1977**, *38*, 987.

[15] Perhaps the earliest example of this is the temperature–concentration phase diagram where FeC alloys show a fcc phase both above and below a bcc phase. J. Billard, personal communication.

[16] S. Somasekhara, R. Shashidhar, B. R. Ratna, *Phys. Rev. A* **1986**, *34*, 2561.

[17] G. Pelzl, C. Scholz, D. Demus, H. Sackmann, *Mol. Cryst. Liq. Cryst.* **1989**, *168*, 147.

[18] J. B. Swift, *Phys. Rev. A* **1976**, *14*, 2274.

[19] P. E. Cladis, H. R. Brand, *Phys. Rev. Lett.* **1984**, *52*, 2261. F. Hardouin, M. F. Achard, N. H. Tinh, G. Sigaud, *Mol. Cryst. Liq. Cryst. Lett.* **1986**, 7.

[20] R. Shashidhar, personal communication.

[21] B. R. Ratna, R. Shashidhar, V. N. Raja, C. Nagabhusan, S. Chandrasekhar, G. Pelzl, S. Diele, I. Latif, D. Demus, *Mol. Cryst. Liq. Cryst.* **1989**, *167*, 233.

[22] P. E. Cladis, R. K. Bogardus, W. B. Daniels, G. N. Taylor, *Phys. Rev. Lett.* **1977**, *39*, 720. P. E. Cladis, R. K. Bogardus, D. Aadsen, *Phys. Rev. A* **1978**, *18*, 2292.

[23] G. W. Gray, J. E. Lydon, *Nature* **1974**, *252*, 221. J. E. Lydon, C. J. Coakley, *J. Phys. Coll.* **1975**, *36*, C1–45.

[24] P. E. Cladis, P. L. Finn, J. W. Goodby in *Liquid Crystals and Ordered Fluids*, Vol. 4 (Eds.: A. C. Griffin, J. F. Johnson), Plenum, New York, **1984**, p. 203.

[25] D. Guillon, P. E. Cladis, D. Aadsen, W. B. Daniels, *Phys. Rev. A* **1979**, *21*, 659.

[26] A. J. Leadbetter, J. L. A. Durrant, M. Rugman, *Mol. Cryst. Liq. Cryst. Lett.* **1977**, *34*, 231; unpublished data.

[27] P. E. Cladis, *Mol. Cryst. Liq. Cryst.* **1981**, *67*, 177.

[28] W. Weissflogg, N. K. Sharma, G. Pelzl, D. Demus, *Krist. Techn.* **1980**, *15*, K35. G. Pelzl, U. Böttger, S. Kallweit, D. Demus, H. Sackmann, *Cryst. Res. Technol.* **1981**, *16*, K119.

[29] Nguyen Tinh, H. Gasparoux, *Mol. Cryst. Liq. Cryst.* **1979**, *49*, 287. F. Hardouin, G. Sigaud, M. F. Achard, H. Gasparoux, *Phys. Lett.* **1979**, *71A*, 347; *Solid. State Commun.* **1979**, *30*, 265.

[30] F. Hardouin, A. M. Levelut, N. H. Tinh, G. Sigaud, *Mol. Cryst. Liq. Cryst. Lett.* **1979**, *56*, 35.

[31] For reviews of this group's work see: F. Hardouin, *Physica* **1986**, *140A*, 359. F. Hardouin, A. M. Levelut, M. F. Achard, G. Sigaud, *J. Chim. Phys.* **1983**, *80*, 53.

[32] J. C. Dubois, A. Zann, *J. Phys.* **1976**, *C3-37*, C3-35.

[33] H. W. Meul, C. Rossel, M. Decroux, O. Fischer, G. Remenyi, A. Briggs, *Phys. Rev. Lett.* **1984**, *53*, 497. In the pseudo-ternary Eu–Sn molybde-

num chalcogenides for different Eu concentrations.

[34] The authors [33] call this the Jaccarino–Peter compensation effect from V. Jaccarino and M. Peter, *Phys. Rev. Lett.* **1962**, *9*, 920.

[35] K. Kondo, H. Takezoe, A. Fukuda, E. Kuze, *Jpn. J. Appl. Phys.* **1983**, *22*, L43. H. Takezoe, K. Kondo, K. Miyasato, S. Abe, T. Tsuchiya, A. Fukuda, E. Kuze, *Ferroelectrics* **1984**, *58*, 55, and references therein.

[36] See, for example, M. Yamashita in *Solitons in Liquid Crystals* (Eds.: L. Lam, J. Prost), Springer Verlag, New York **1991**.

[37] See, for example, G. Pelzl, D. Demus, *Z. Chem.* **1952**, *21*, 152. W. Weissflogg, N. K. Sharma, G. Pelzl, D. Demus, *Krist. Techn.* **1980**, *15*, K35.

[38] W. Weissflogg, G. Pelzl, D. Demus, *Mol. Cryst. Liq. Cryst.* **1981**, *76*, 261.

[39] J. O. Indekeu, A. N. Berker, *Physics* **1986**, *140A*, 368; *Phys. Rev. A* **1986**, *33*, 1158; *J. Phys.* **1988**, *49*, 353. A. Nihat Berker, J. O. Indekeu, in *Commensurate Crystals, Liquid Crystals and Quasicrystals* (Eds.: J. F. Scott, N. A. Clark), Plenum, New York **1987**, p. 205.

[40] This term was coined by G. Toulouse after a remark by P. W. Anderson. See, for example, P. W. Anderson, in *Future Trends in Materials Science* (Ed.: J. Keller), World Scientific, New Jersey **1988**.

[41] G. Pelzl, B. Oertel, D. Demus, *Cryst. Res. Technol.* **1983**, *18*, K18. See also Vill and coworkers [6].

[42] As SmA and N* (and N) phases do not have the same symmetry, a critical point is excluded for this transition.

[43] R. C. Long, Jr., J. H. Goldstein in *Liquid Crystals and Ordered Fluids*, Vol. 2 (Eds.: J. F. Johnson, R. S. Porter), Plenum, New York **1974**, p. 147.

[44] J. Malthête, N.-H. Tinh, C. Destrade, *Liq. Cryst.* **1993**, *13*, 171.

[45] For a description of the basic aspects of entropic elasticity see P. G. de Gennes, *Scaling Concepts in Polymer Physics*, Cornell University Press, Ithaca, NY **1979**. For details of how it has been applied to polymers, see E. R. Duering, K. Kremer, G. S. Prest, *Phys. Rev. Lett.* **1991**, *67*, 3531; *J. Chem. Phys.* **1994**, *101*, 8169, and references therein. R. Everaers, K. Kremer, *Phys. Rev. E* **1996**, *53*, R37; *J. Mol. Model.* **1996**, *2*, 293.

[46] F. Dowell, *Phys. Rev. A* **1983**, *28*, 3526; **1987**, *36*, 5046.

[47] C. Destrade, H. Gasparoux, A. Babeau, N. H. Tinh, J. Malthête, *Mol. Cryst. Liq. Cryst.* **1981**, *67*, 37.

[48] N. H. Tinh, P. Foucher, C. Destrade, A. M. Levelut, J. Malthête, *Mol. Cryst. Liq. Cryst.* **1984**, *111*, 277.

[49] A. Nayeem, J. H. Freed, *J. Phys. Chem.* **1989**, *93*, 6539.

[50] See, for example, F. Müller-Plathe, *Chem. Phys. Lett.* **1996**, *252*, 419. K. Binder, C. Cicotto (Eds), *Monte Carlo and Molecular Dynamics of Condensed Matter Systems*, SIF, Bologna **1996**.

[51] W. Weissflogg, M. Rogunova, I. Letko, S. - Diele, G. Pelzl, *Liq. Cryst.* **1995**, *19*, 541.

[52] For a review of SmA polymorphism in the context of two order parameter mean field theories see, for example, P. Barois, J. Pommier, J. Prost, *Solitons in Liquid Crystals* (Eds: L. Lam, J. Prost), Springer Verlag, New York **1991**.

[53] G. Pelzl, D. Demus, *Z. Chem.* **1981**, *21 (4)*, 151. G. Pelzl, U. Böttger, D. Demus, *Cryst. Res. Technol.* **1981**, *16*, 5; *Mol. Cryst. Liq. Cryst. Lett.* **1981**, *64*, 283.

[54] R. Shashidhar, B. R. Ratna, S. Krishna Prasad, *Phys. Rev. Lett.* **1984**, *53*, 2141.

[55] G. Pelzl, S. Diele, I. Latif, D. Demus, *Cryst. Res. Technol.* **1982**, *17*, 78.

7   Defects and Textures

Y. Bouligand

7.1   Definitions, Conventions and Methods

7.1.1   Local Molecular Alignment

Ordered media are never perfect, but present deformations and even defects. In liquid crystals, the order is defined by several parameters, mainly the *local mean direction* of approximately parallel molecules, usually represented by a unit vector n chosen parallel either to the long axis if the molecule is elongated, or normal to the molecules if the molecule is discoidal. A second local variable is the *order parameter*, which corresponds to a more or less accurate alignment of molecules. The orientation of n is often chosen arbitratily ($+n$ equivalent to $-n$), since there are no polarities in the distribution of molecules, even if the chemical formula is 'arrowed', as in the classical example of 4-methyloxy-4′-*n*-butylbenzylidene aniline (MBBA) molecules. Both the parallel and the antiparallel alignment occur in equal proportions. In this case, n is called a *director*, with only the direction of the molecules being defined, with no preferred orientation.

Several conventions are used to represent directors or molecules in figures. Rod-like or disk-like molecules can be drawn as circular cylinders, either elongated or flat [1]. Another representation is given by a point for a director normal to the figure plane P, a segment of constant length for a director parallel to P, and a nail with length proportional to $\cos\alpha$ for directors lying at an angle α from P, the sharp end of the nail corresponding to the director extremity pointing towards the observer [2, 3]. The opposite orientation for nails is also adopted [4, 5].

7.1.2   Microscopic Preparations of Liquid Crystals

When a liquid crystal is introduced between two parallel glass plates (a slide and a coverslip), without any particular care, and is examined between crossed polarizers under a polarizing microscope, within a thermostated stage if necessary, the general chromatic polarization is observed and multiple patterns appear, showing that alignment is only local, with the director varying continuously over large distances. Discontinuities in the optical image suggest the presence of discontinuities in director distribution, in the form of singular points, lines and walls. Liquid crystals oppose an elastic energy to these deformations and defects. As these media are liquid, the singularities move easily until they reach an equilibrium position, this often resulting in a regular arrangement of defects and domains, called *texture*.

Descriptions of defects and textures in liquid crystals are based on concepts of *differential geometry* (see Sec. 7.4 of this Chapter). An intuitive approach is proposed in the books by Hilbert and Cohn-Vossen [6] and Coxeter [7], both of which contain chapters devoted to topology, in a style of reasoning close to that used in the study of liquid crystals. For details on polarizing microscopy see Hartshorne and Stuart [8], and for an illustrated presentation of defects in true crystals see Amelinckx [9]. Numerous micrographs and interpretative drawings can be found in early works such as those by Lehmann [10] and Friedel [11, 12] (see Chap. I of this volume). More recent books cover the essential knowledge about mesogenic molecules, phase transitions, liquid crystalline structures, and defects and their arrangement around domains, with the whole illustrated by numerous figures and micrographs, colour plates showing the main textures seen in polarizing microscopy [13, 14]. The structure and energetics of defects in liquid crystals have been reviewed by Chandrasekhar and Ranganath [15], while Kléman [16] has covered not only liquid crystals but also other ordered media and, in particular, magnetic systems.

7.1.2.1   Thermotropic Textures

Examples of textures are present in rodlets, spherulites and more or less expanded germs of a mesophase, floating within the isotropic liquid or attached to one of the two glasses, at the transition. A long and brilliant SmA rodlet or *bâtonnet* is shown in Fig. 1, contrasted against the black background due to the isotropic liquid, as observed between crossed polarizers. Smaller germs are also present and, on the left of the micrograph, the smectic A phase shows continuous shade variations and lines of discontinuity, in the form of ellipses, parabolae and hyperbolae. These conics correspond to lines of discontinuity extending either from an interface or in the bulk of the mesophase. As the liquid crystal structure is elastically deformed all around the defect, each of these lines lies at the origin of a part of the elastic energy, and when defects attach along the interface with an isotropic phase a strong minimization is obtained. Numerous defects adhere to the isotropic interface of the smectic rodlet in Fig. 1, the layers of which lie almost normal to the long axis, but are invisible in photon microscopy, their thickness being approximately equal to the molecular length. The surface defects in this bâtonnet form regular patterns with discrete rotation symmetry, and some irregularities.

Spherical germs of a twisted nematic phase (a mixture of a nematic liquid and an asymmetric compound) of different sizes and orientations, as observed under natural light are shown in Fig. 2. The molecules align according to the cholesteric model, but the helicoidal pitch is larger than in pure cholesteric phases. The existence of a periodicity creates a series of contrasted stripes. The general aspect is lamellate, with 'layers' normal to the diameter of the spherulite, but curved at the periphery, and lying perpendicular to the isotropic interface. The parallel arrangement of 'layers' can be disturbed by the coalescence of several spherulites, leading to the formation of defect lines, which often annihilate at the isotropic interface. Also due to the helical struc-

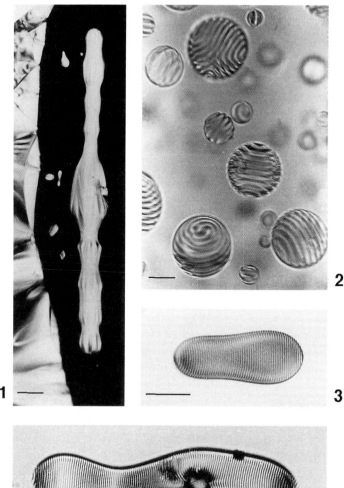

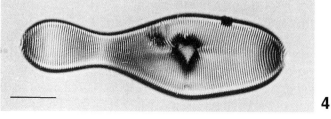

Figure 1. A smectic A phase extended on the left-hand side, and showing defect lines in the form of conics; a smectic rodlet and small germs are floating in the isotropic phase. The mesogenic product, 4-cyano-4'-n-octylbiphenyl (8CB), was added together with a small amount of Canada balsam, and the mixture was observed between crossed polarizers at room temperature. Canada balsam is an isotropic but optically active and fluid resin, extracted from the conifer *Abies balsamea*, which facilitates the production of regular smectic textures in several thermotropic liquids. Collophony, or rosin, is a stabilized pine tree resin, used to rub violin bows; Friedel used this substance not only for his violin, but also to obtain remarkable smectic textures. Scale bar: 20 µm.

Figure 2. Spherical cholesteric germs of MBBA, plus a small amount of cholesterol benzoate. There is a disclination similar to the τ^- pattern shown in Fig. 25 and a spiral decoration in another spherulite. One polarizer only. Scale bar: 20 µm.

Figure 3. Elongated germs of MBBA cholesterized by a small amount of Canada balsam as observed between parallel polarizer and analyser. Layers lie mainly perpendicular to the long axis, without defects, and orientate normally to the isotropic interface. This germ does not float within the bulk, but is slightly sandwiched between the slide and the coverslip. Scale bar: 20 µm.

Figure 4. A more developed cholesteric rodlet in the same preparation as in Fig. 3, showing a set of internal defects. Scale bar: 20 µm.

ture, spiral patterns generally appear at two opposite poles of the spherulite.

When the helical pitch is much smaller, instead of being spherical, germs grow along an axis normal to 'layers' (Fig. 3), as in smectic phases. The shape of the smectic and cholesteric germs reveals the *anisotropy of surface tension*. Molecules often tend to lie parallel to the isotropic interface in smectic and nematic phases, whereas they prefer to orientate normally to the air interface. There are strong similarities between the textures of smectic A and cholesteric phases, but the three-dimensional arrangement of cholesteric layers is often resolved in optical microscopy, whereas that of smectics is not. However, many textural patterns are similar in smectic A and cholesteric phases. Comparisons between the textures of these two phases concern mainly domains extending over distances that are very long compared to the helical pitch. For instance, when several tens of layers are present elongated cholesteric drops are observed. It should be noted that the germs shown in Figs. 3 and 4 have grown to dimensions larger than the distance from the slide to the coverslip, and are 'flattened', but a thin isotropic film is still present, separating them from direct contact with the glass plates.

7.1.2.2 Lyotropic Textures

Amphiphilic molecules have two parts, one hydrophilic and one hydrophobic, and form liquid crystals the textures of which are of great interest, since they are very close in morphology to biological materials and, in particular, cell membranes [17–19], showing a polymorphism related to that known in water–lipid systems (see Chaps. XV to XVII of Vol. 2 of this Handbook). The lamellar structure displays some usual defects and textures (Fig. 5). The bilayers are more or less separated by water (Fig. 5a, b); this

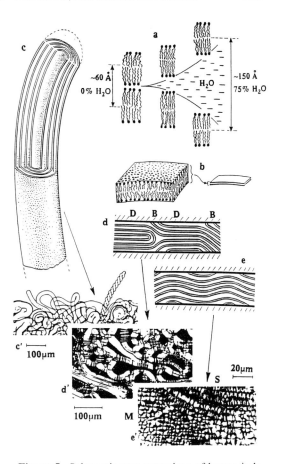

Figure 5. Schematic representations of lyotropic lamellar systems. (a) Bilayers of phospholipids (one polar end and two paraffinic chains) are more or less separated by water (after Schmitt and Palmer [18]). (b) A bilayer (sketched on a different scale) with two parallel surfaces (or two parallel lines in cross-section). (c) Myelinic finger, with cylindrically nested bilayers, the number of which is strongly reduced (possibly by a factor of 10^3), compared to the lecithin fingers in (c'). (d) The bilayers parallel to the preparation plane are said to be 'horizontal' and form a dark background (D) between crossed polarizer, whereas oblique or vertical bilayers appear bright (B), mainly if they extend at 45° from the orientation of the polarizers, as shown in (d') for lecithin. (e) Bilayers that were initially horizontal often corrugate in one or two directions to form domes and basins, each appearing as a 'Maltese cross' (M) when observed between crossed polarizers (e').

swelling is important in the fluid *myelin textures*, a very common texture in lyotropic systems, with concentrically nested layers (Fig. 5 c). Some cylindrical shapes, either straight or curved, or forming simple or double helices, are shown in Fig. 5 c′.

When a myelinic system is progressively dehydrated between the slide and the coverslip, many bilayers arrange horizontally (Fig. 5 d). These appear black between crossed polarizers, whereas those that remain vertical or oblique form brilliant stripes called 'oily streaks' (Fig. 5 d′). Further dehydration leads to the formation of air bubbles between the slide and coverslip; the bubbles extend progressively, and lateral compressions result in wrinkles that superimpose in several directions. The horizontal bilayers often transform into nested domes or nested basins, each being characterized by a 'Maltese cross', when examined between crossed polarizers (Fig. 5 e). The 'Maltese crosses' arrange in a more or less regular fashion and form polygonal mosaics, which are frequent in many liquid crystals. The oily streaks also corrugate through dehydration, giving them a cross-striated aspect in polarizing microscopy.

7.1.2.3   Liquid Monocrystals

Instead of defects and complex textures, some experimental situations allow one to obtain a very good alignment and thus are useful for comparisons with local aspects of complex preparations. Large monocrystalline domains are mainly obtained with thermotropic liquid crystals and, much less often, in lyotropic systems. Such alignments can be produced by external fields (see Sec. 9 of this Chapter) or by treating the slide and coverslip in order to create a regular *anchoring* of molecules in a uniform direction (*homeotropic* if the direction is normal to the glass, and *horizontal* if parallel

to the glass), with a preferred or *easy direction* within the interface plane (see Sec. 10 of this Chapter). In polarizing microscopy, homeotropic preparations appear uniformly dark, with more or less intense light scattering due to director fluctuations. Liquid monocrystals with horizontally aligned molecules show an extinction in four positions of the rotating stage in the polarizing microscope, each extinction being separated by 90°.

7.1.3   Images of Liquid Crystals in Polarizing Microscopy

One of the difficulties in studying liquid crystals is to attain skill in making microscopic preparations and a general knowledge about images and their interpretation. In general, defects and textures are observed within preparations at equilibrium, but heating of or pressure exerted on the coverslip produce motions and streams, which cease as a new equilibrium is progressively reached; this generally takes some minutes, but there are textures with remarkable regularities which only appear after days.

Consider a liquid crystalline slab sandwiched between a slide and a coverslip, that is sufficiently thin (of the order of micrometers) and has no strong anchoring conditions. The directors will align along practically straight lines, lying normal, oblique or parallel to the glasses, since the elastic constants prevent the occurrence of high curvatures (however, those exist in the vicinity of defects). This means that, in large domains, the length of curvature radii is notably larger than the liquid crystal thickness. Projection of the local director onto the preparation plane then corresponds to one of the two local directions of extinction between crossed polarizers, and the director is determined with the help of an additive plate, generally a quartz first-order retardation plate, a $\lambda/4$, or a compensator, as in conventional crystallography [8]. The use of quasi-monochromatic waves (e.g., sodium light) often facilitates these investigations.

When the liquid crystal forms a thick layer (20–50 µm) between the slide and the coverslip, the interpretation is more difficult. Examination under natural light or between parallel polarizers can be useful. It is remarkable that even with such thick preparations, which show an extreme complexity of textures and optics, it is still possible to obtain clear images, focused either at the upper level of the liquid crystal in contact

with the coverslip or at the bottom level, at the slide interface. This is illustrated in Fig. 6 for a cholesteric texture [20]. The picture quality is better at the coverslip level in Fig. 6 a, but both views are useful.

Some good pictures can be obtained using intermediate thickness optical sections, at horizontal levels between the slide and the coverslip. The quality of the images facilitates the preparation of stereo-pairs, in order to observe textures in three dimensions. The two examples shown in Fig. 7 were obtained simply by tilting the preparation slightly differently. The stereoviews of Fig. 7 a and b show hyperbolae branches

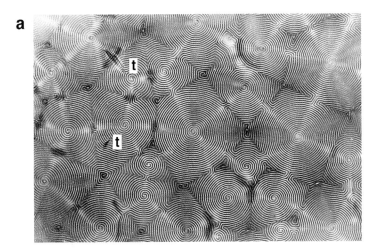

a

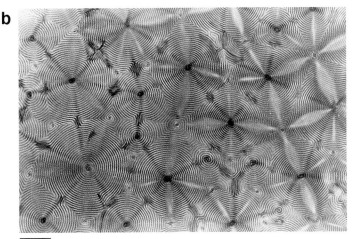

b

Figure 6. Polygonal texture in a cholesteric liquid (MBBA and Canada balsam) observed under natural light: (a) coverslip level, (b) lower glass level. t, Translation defects (there are other unmarked translation defects in the micrographs). Scale bar: 20 µm.

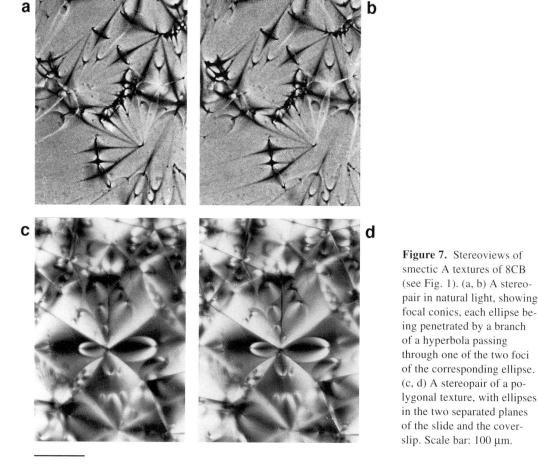

Figure 7. Stereoviews of smectic A textures of 8CB (see Fig. 1). (a, b) A stereopair in natural light, showing focal conics, each ellipse being penetrated by a branch of a hyperbola passing through one of the two foci of the corresponding ellipse. (c, d) A stereopair of a polygonal texture, with ellipses in the two separated planes of the slide and the coverslip. Scale bar: 100 μm.

passing through ellipses, a classical pair of defects in smectics, and named focal conics, that it is worth seeing in three dimensions. Figures 7c and d show a polygonal field in a smectic A phase; there are lattices of ellipses, and the slide and the coverslip levels can be distinguished. All vertices of the upper polygons superimpose vertically to central points in polygons of the bottom latice, and vice versa as in the cholesteric polygonal fields shown in Fig. 6. These textures were mainly studied by Friedel and Grandjean [11, 21].

There is another method that can be used to explore textures in three dimensions. In some favorable cases, mesogenic molecules can polymerize within the mesophase and form a resin, without any modification of the initial distribution of directors and with a texture that remains stable. The whole transformation can be observed in the hot stage in polarizing microscopy [22]. The slide and coverslip can be unstuck and the resin slab polished on its two opposing faces, to suppress the thin layers in contact with the glasses. In this way the director distributions at the glass interface and in the bulk can be compared. Semi-thin sections of the resin slab can also be prepared by ultramicrotomy, using a glass or diamond knife, and observed using polarizing microscopy.

7.1.4 Other Microscopic Methods

Techniques introduced by biologists to study cells and tissues [23, 24] have been extremely useful for the examination of liquid crystals, mainly lyotropic crystals. Microtomy methods are applied to embedded biological materials, after stabilization with small amounts of OsO_4 in water, introducing numerous cross-links into the structure. The material must be progressively dehydrated in ethyl alcohol before it is embedded in an epon resin. By using an electron-dense contrast agent (e.g. phosphotungstic acid, uranyl acetate) one can obtain excellent images, and artefacts are generally well controlled. The lamellar and middle phases of soaps or other water–lipid systems have been studied using these techniques, and excellent views of defects and textures have been obtained [25].

An alternative method is the freeze-etching technique [24, 25], which consists of producing platinum shadowed replicas of fractures created in rapidly frozen biological systems or lyotropic liquid crystals. The fracture orientation is somewhat haphazard, but occurs preferentially within the paraffinic level of the bilayers. These methods offer the possibility of viewing bilayers directly and preparing stereoviews. Beautiful pictures of liquid crystalline DNA, both cholesteric and hexagonal, have been obtained, but individual molecules are not easily resolved. However, the director distribution can be deduced from the images [26].

Microtomy and freeze-etching techniques have revealed many similarities between liquid crystals and a large series of biological materials, which show the same symmetries as nematic, smectic and cholesteric phases, but without being fluid [27–34]. The biological materials are stabilized

analogues, and it has been verified that most of these systems can be obtained in a true mesomorphic state [35, 36]. Defects and textures also are present in these biological counterparts of liquid crystals, and are associated with different shapes in tissues and organs; this aspect represents a new axis of research in biological morphogenesis [37].

7.2 Strong Deformations Resulting in Defects

7.2.1 Singular Points

Simple experimental situations lead to the formation of singular points, lines or walls. For instance, glass capillary tubes (diameter ~0.5 mm) can be treated with a surfactant to ensure radial anchoring at the inside wall, which corresponds to homeotropy [38]. When filled with a nematic liquid, the director lines are radial at the periphery and lie parallel to the capillary along its axis (Fig. 8a). The main curvatures are of the splay (mainly along the axis) and bend types. Twist is not excluded, because minimization of the curvature energy requires contributions from the three terms. As there are two possible splay orientations along the tube axis, punctual singularities occur here and there, the patterns of which are shown in meridian section in Fig. 8a [39–41].

The director lines distributed around one of these point singularities form a pattern reminiscent of the radial electric field produced by a charged particle, whereas the other singular pattern recalls the electric field about the zero point, lying in the middle of a segment the extremities of which are occupied by two identical electric charges. These singular points are very common in nematics and cholesterics. The two con-

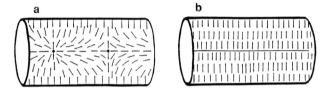

Figure 8. Textures produced by homeotropic boundary conditions in a capillary tube filled with a nematic liquid (a) or a smectic liquid (b).

figurations shown in Fig. 8a are said to be complementary, since they can fuse and disappear, or can be created by pairs (e.g. due to strong disturbances in fluid streams).

7.2.2 Singular Lines

When a nematic–smectic A transition occurs in a capillary tube (Fig. 8a), smectic layers nucleate at disclination points and a singular line forms along the axis, with cylindrical layers (as shown in Fig. 8b), but the presence of beads along the axial defect shows that the situation is less schematic [42].

Another situation occurs in a nematic liquid between two parallel glasses with horizontal anchorage, the two easy directions lying at right angles to one another [43]. The nematic liquid can 'choose' between two orientations for the twist: a right- or left-handed helix (Fig. 9). In Figs. 9a–c, a narrow domain, limited by an arc or a complete circle, is represented at the boundary between regions of left- and right-handed twist. Within this domain, the directors diverge by angles of 0–90°; this discontinuity is either attached to the coverslip (C) or the slide (S), or lies in the bulk. These narrow discontinuities, present in each figure plane, correspond to the successive sections

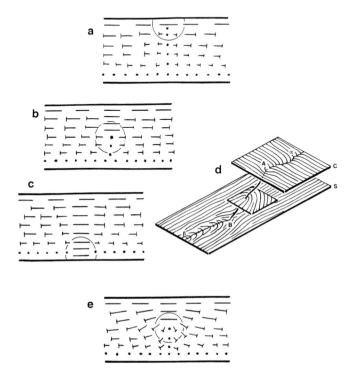

Figure 9. Defect line produced in a nematic liquid by two parallel glasses so treated that they have planar anchoring conditions, with two mutually perpendicular easy directions. The discontinuity line is either attached to the upper glass (a) or to the lower one (c), or lies in the bulk (b). (d) The three localizations of the defect thread along a path $\alpha AB\beta$, with a planar configuration in the bulk. (e) A nonplanar distribution of directors is more likely.

of a thread, attached to the coverslip along αA, penetrating the bulk from A to B, and attached to the glass slide along $B\beta$ (Fig. 9 d). The thread observed in the bulk forms a smooth curve, but when attached to glass it often follows a complex path, with a somewhat fractal aspect, as indicated in Fig. 9 d. The directors are not necessarily horizontal in this texture (Fig. 9 e).

Now, let us consider the distribution of directors along a circuit closed around such a thread of discontinuity in the bulk of the liquid. Following this circuit, it appears that the chosen orientation for the director *n* is reversed after one turn. The ribbon generated by a short segment parallel to *n*, and centred on a point M describing this circuit, is a Möbius strip and is, therefore, a non-orientable surface [44]. If the length of the circuit around the thread is progressively reduced, it appears that a discontinuous distribution of *n* is present at each point of the thread. More generally, the director field is said to form a non-orientable manifold, and physically this means that the director is not 'arrowed', there being no preferred orientation of the molecules (either parallel or antiparallel to *n*), as indicated above. This type of thread comes from a topological obstruction, a discontinuity due to the non-orientable distribution of directors at the periphery.

These two simple experimental systems show the presence in liquid crystals of two types of defect: lines and point singularities. Liquid crystals contain a large variety of lines with well-defined geometries or topologies. There are also lines that have a continuous core (for example, in the capillary tube); the axial zone corresponds to a maximum of splay and is generally considered to be a defect line, although no discontinuities apart from the singular points are present. This situation is also encountered in the third type of defect – walls.

7.2.3 Walls

Discontinuity walls do not exist in principle in usual liquid crystals, since there are no apparent limits to certain curvatures, as far as this can be observed in the vicinity of singular points and discontinuity lines. Several types of wall are usually considered:

- Twins possibly occur in liquid crystals that have a structure close to that of solid crystals, with more or less extended three-dimensional order, possibly in a smectic E phase (see Sec. 7.4.4.6).
- Defect lines arrange along narrow zones separating domains of different orientations in the liquid crystal, like grain boundaries in true crystals.
- When the anchoring conditions are sufficiently strong, at the slide and coverslip levels for instance, a reversal in the orientation of the directors can be confined in a narrow band [45]. Such zones of rapid rotation of directors are analogous to those observed in magnetic materials and, according to the relative values of the elastic constants, there are 'Bloch walls' showing mainly a twist curvature and 'Néel walls' associating bend and splay [46].

7.2.4 Interface Defects (Points and Lines)

A frequent situation in nematic phases close to the isotropic transition is the presence of a thin film of an isotropic phase separating the liquid crystal from the preparation glasses. The molecules lie either horizontally at this interface, or at an angle other than $90°$, with no preferred direction in the horizontal plane. Defects can be numerous, as shown in Figs. 10 and 11.

The two vertical threads L in Fig. 10 a and b correspond to singular lines surrounded by director lines following approximately the planar solutions of the Laplace equation $\nabla^2 \Phi = 0$, as shown by Oseen and later reviewed by Frank [54]. Φ is the azimuth of n supposed to be horizontal in a plane xy parallel to the glass plates. The solutions are of the type $= \Phi_0 + (N\varphi/2)$, where N is a positive or negative integer or zero (with $\tan \varphi = y/x$ and $\tan \Phi = n_y/n_x$), the horizontal Cartesian coordinate system xy being centred on the disclination trace. In Fig. 10 a $N = 1$, and in Fig. 10 b $N = -1$. In the recent literature, N is replaced by $s = N/2$ a multiple of $+1/2$. The fact that, in general, the three elastic constants are not equal, and that spontaneous curvatures can be present, the twist, for example, modifies the shape of the director lines and even the symmetries, but not the general aspect of these disclinations in cross-section. The expected and observed aspects between crossed polarizer in the

case $N = +1$ are pairs of dark 'brushes' attached to a central point (Fig. 10 c and d). A slight shift of the coverslip relative to the slide separates the two brushes, and a thin thread joins them, as shown in Fig. 9 d.

The preparations also show sets of four dark brushes attached to a common center, and a weak shear generally separates these patterns into pairs of dark brushes linked by a thread that is much thicker and less sharply contrasted than the threads described above. This texture was studied in resins produced by polymerized nematics, after abrasion and polishing of the upper and lower faces of the nematic analogue slab [22]. Different patterns were obtained, some of which are shown in Fig. 11 a–c, the types (a) and (c) being the most common. The corresponding aspects in polarizing microscopy are shown in Fig. 11 a'–c'. Examination using a compensator showed strong obliquities or even verticality of the directors in the core region, and continuous variations in their orientation. Splay, twist and bend are present, and there are no discontinuities along the thick threads in the bulk.

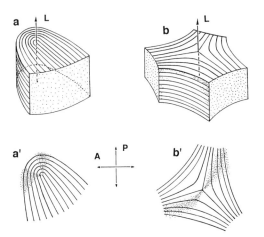

Figure 10. Two common defects in nematics, between parallel interfaces, with horizontal anchoring and no preferred orientation. (a, b) Three-dimensional views; the director lines, rather than separate segments as in Fig. 15, are represented and lie parallel or normal to the faces of the curved polyhedra. Bend and splay are concentrated in alternating sectors, the twist not being excluded from the bulk. (c, d) Corresponding aspects between crossed polarizers; the director lines are not visible under the microscope.

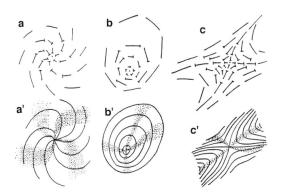

Figure 11. (a, b, c) Continuous distribution of directors within the median horizontal plane of three different types of nematic nucleus with four brushes (a', b', c'). Corresponding aspects between crossed polarizers; the curves correspond to the alignment of nails, but are not visible under the microscope.

Each of these patterns was given a name by Lehmann [10] – *halber Kernpunkt* ($s=1/2$), *halber Konvergenz Punkt* ($s=-1/2$), *ganzer Kernpunkt* ($s=1$) and *ganzer Konvergenz Punkt* ($s=-1$) – and were called *noyaux* by Friedel and Grandjean [21], or *nuclei* in English works. The assemblies of nuclei were named *plages à noyaux* [10, 11], or *nuclei textures* or *Schlieren textures* [45].

In a nematic preparation, between a horizontal slide and coverslip, the locus of vertical directors is made up of one or several lines, corresponding to intersections of surfaces $n_x(x, y, z)=0$ and $n_y(x, y, z)=0$ (within a Cartesian coordinate system x, y, z attached to the preparation, z being normal to

the preparation plane). Such lines, cut the interfaces at isolated points, where the anchoring conditions can make them singular, and their structure is represented in Fig. 12, the twist excepted [44, 47]. Associations of these interfacial singular points are shown in Fig. 13. Similarly, the locus of horizontal directors corresponds to surfaces $n_z(x, y, z)=0$, cutting the limiting interfaces along curves, also made singular by the anchoring conditions, in particular, if horizontal directors are forbidden along an isotropic interface such as the one due to the thin isotropic films that are often observed along the glasses, near the isotropic transition. This is shown in Fig. 14 for nuclei with $N=1$

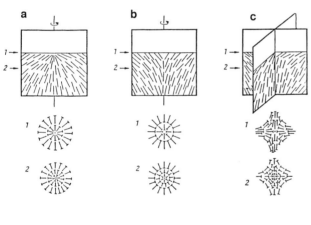

Figure 12. Singular points of the director distribution at an interface presenting a 45° anchoring angle, and the corresponding patterns in top view, at the interface (1) and just below in the bulk (2). The introduction of a twist allows one to pass continuously from the radial structure of point (a) to that of (b), with a constant revolution symmetry. Point (c) does not present this symmetry.

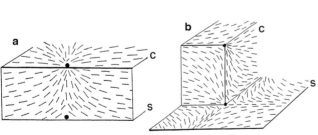

Figure 13. Vertical pairs of singular points at the slide (S) and coverslip (C) levels: (a) association of two radial points; (b) association of two non-radial points.

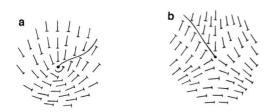

Figure 14. Distribution of directors at an interface presenting a constant anchoring angle (as in Fig. 12) and around the extremity of a thin thread, with patterns (as in Fig. 10). A discontinuity line appears unavoidably at the interface and is attached to the core of the defect. In the bulk, these lines create Néel or Bloch walls, or hybrid textures.

and $N=-1$, at the interface level, when the anchoring angle is constant and different from zero. The presence of such discontinuity lines originating from the core extremity of the nucleus is unavoidable for odd values of N.

7.2.5   The Nature of Thick and Thin Threads

Since a horizontal shift of the coverslip relative to the slide transforms a nuclei texture into a set of threads, this indicates that nuclei correspond to vertical threads with minimized length and elastic energy, this being confirmed by the generally observed stability of this texture. However, nuclei often fuse, and either disappear if their indices N are opposite, or form a new nucleus, the index N of which is the algebraic sum of the corresponding indices of the two parent nuclei [10, 11]. Similarly, threads annihilate or recombine, following the same type of laws [48, 49]. Threads were considered therefore as disclinations, the cross-sectional structure of which is that shown in Figs. 10 and 11 [16]. However, the situation is not that simple since, at any point M of a director field n (M), the director distribution within a plane P normal to n (M) generally forms patterns very similar to those shown in Fig. 11 a–c [44]. Each four-brush nucleus is continuous in the bulk and presents two singular points at its extremities, whereas a two-brush nucleus corresponds to a vertical line of discontinuity of the Möbius type. Conversely, a simple circuit closed around a thick thread joining the two singular extremities of a four-brush nucleus is not of the Möbius type.

7.3   Topological Structure of Defects

7.3.1   The Volterra Process

Defects can be defined by the geometrical operations necessary to pass from the perfect crystal to the disturbed structure. This point of view was developed in works by Volterra and Love [50, 51], and was later applied to crystals [52]. Consider an ordered material and suppose that it can be cut along surfaces and then deformed and restuck such that the two facing materials show parallel crystallographic orientations. Such operations are topological since they may change the connectivity of the material and the deformations are equivalent to symmetry operations. Such a procedure is called the 'Volterra process' and leads, in general, to a theoretical structure close to that of singular lines. This method was first introduced into the study of liquid crystals by Kléman and Friedel [2, 3].

7.3.2   The Volterra Process in Nematic, Smectic A and Cholesteric Phases

The local symmetries of nematics are all the translations, all the rotations about the director axis and all the $\pm\pi$ rotations about any axis normal to the director. The operations combining these translations and rotations also correspond to local symmetries of the nematic structure.

Topological rehandlings of nematics, after cutting, deforming and resticking, do not necessarily result in defect lines. For instance, if the two lips S_1 and S_2 of a cut surface S, limited by a line L, are separated by a translation and restuck after an eventual addition or subtraction of nematic matter, the line L would simply disappear by viscoelastic relaxation. However, if we separate the two lips S_1 and S_2 by an angle π, through rotation about an axis normal to n, and add nematic matter as indicated in Fig. 15 a–c, the material relaxes into a differently connected system, with a defect line as represented in (c) and called a disclination. Figure 15 d–f represent the situation when a piece of matter is subtracted and the two lips S_1 and S_2 are stuck after another π rotation about an axis normal to n [2, 3]. The peripheral distribution of directors corresponds nearly to that observed in many preparations, and it was concluded, more or less explicitly, that the director lines lie in parallel planes as in Fig. 10. However, all three types of deformation (splay, twist and bend) are likely to be present.

In smectic A phases the symmetries associate: all translations normal to the director and all those parallel to the director, the length of which is an integer multiple of the layer thickness; all the rotations about an axis parallel to the director; and all $\pm\pi$ rotations about axes normal to the director, either at the limit between successive layers or at the half-thickness of a layer.

Several examples of the Volterra process in smectic A phases are indicated in Fig. 16 [2, 3]. The symmetry involved can be a translation normal to the layers, in which case the defect is said to be a *dislocation*, the translation vector b or Burgers vector being either parallel to the line L as in a *screw dislocation* (Fig. 16 a and a'), or normal to L as in an *edge dislocation* (Fig. 16 b and b'). As all rotations about any axis L normal to the layers superimpose a smectic A phase onto itself, any sector centred on L can be subtracted, and one obtains, be resticking, a set of nested conic layers, which remains stable for certain boundary conditions (Fig. 16 c and c') [53]. This will be considered further in the study of focal conics (see Sec. 7.4.2). When the symmetry involved is a rotation, the defect line is called a *disclination* and is mainly characterized by the corresponding angle. The production of disclinations in smectic A phases is similar to that presented for nematics in Fig. 15. The addition of smectic layers is shown in Fig. 16 d'. The resulting viscoelastic relaxation often leads to a symmetrical disclination (Fig. 16″), L being then a three-fold axis, as in nematics (Fig. 15 c).

Disclinations are rotation defects, and are rare or absent in three-dimensional crystals, owing to their prohibitive energies, but are in general compatible with liquid crystalline structures. They were initially called 'disinclinations' [2, 3, 54], but the term was later simplified to 'disclination'.

Cholesterics are helically twisted nematics, the local symmetry of which is slightly biaxial, but close to that of nematics. If the

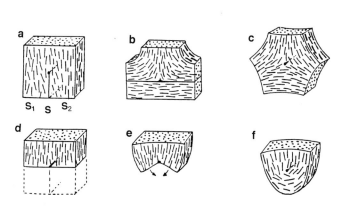

Figure 15. The Volterra process applied to a nematic liquid. A planar section limited by a line normal to the director n allows the two lips S_1 and S_2 to be separated by an angle π (b), and nematic material to be added to obtain the disclination structure (c). An initial matter subtraction creates two lips S_1 and S_2 (d), both rotated by an angle $\pi/2$ (e) and restuck to obtain another disclination (f).

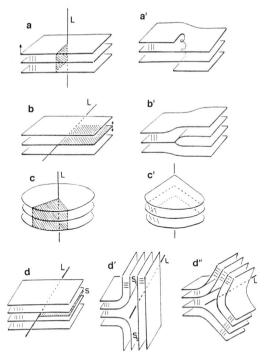

Figure 16. Creation of defects in a smectic A phase. (a, a') Creation of a right-handed screw dislocation. (b, b') Creation of an edge dislocation. (c, c') Creation of a stack of nested conic layers, as observed along focal conics. (d, d', d'') Creation of a disclination from a planar cut surface limited by a line L; a $+\pi$ separation of lips S_1 and S_2 is followed by the addition of matter and relaxation.

spontaneous twist is included in the structure definition, the symmetry group is changed and associates: all translations normal to the twist axis and all translations parallel to the twist axis that are integer multiples of the half-helical pitch; and all $+\pi$ rotations about any axis parallel either to the director or the twist axis, or normal to both [2, 3].

Examples of defect lines created in cholesterics according to the Volterra process are shown in Fig. 17. The core structure will be considered below this being replaced by a narrow cylinder. Figures 17a and b represent translation dislocations: in both cases,

the Burgers vector lies parallel to the cholesteric axis, but is either normal to the line L, as an edge dislocation, or parallel to it, as a screw dislocation, as shown in Fig. 16 for smectic A phases. No additional material is necessary to obtain the screw dislocation in the Volterra process. Figure 17c–g show the two main disclinations created (see Sec. 7.3.4).

7.3.3 A Different Version of the Volterra Process

De Gennes and Friedel [55] proposed a simple interpretation of thin threads forming loops in nematic liquids (Fig. 18a) and produced a modified version of the Volterra process by taking into account the liquid character of the mesophase. For instance, consider a perfect nematic liquid aligned along a common easy direction, defined at the surfcae of the slide and the coverslip. The medium is cut along S, a horizontal disk in the bulk, limited by a circular loop L, and two lips S_1 and S_2 are created. Each director close to S is rotated about a vertical axis by an angle $+\pi/2$ in S_1 and $-\pi/2$ in S_2, the structure being restuck along S. Such topological rehandlings are only possible in liquid ordered media, and not in true crystals. After relaxation, the director shows a $+\pi$ twist through L, between the slide and the coverslip, whereas this global twist is absent from the remainder of the preparation. The Volterra process adapted for nematic liquids is represented locally in the vicinity of L (Fig. 18b–d), in a purely planar model.

At the interface limiting a mesophase, the anchoring conditions often lead to the creation of singular lines lying either at the interface or in the bulk, when the anchoring conditions are very strong. The case of such lines at an interface in a cholesteric liquid is

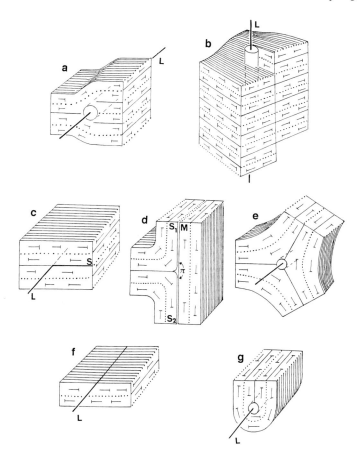

Figure 17. The Volterra process applied to cholesteric phases. The core structure is masked by a cylinder along the line L. (a, b) Edge and screw dislocation. (c–e) A section S limited by L, normal to the cholesteric axis, allows one to build either the edge dislocation (a) or a disclination (d), as in smectics (Fig. 16d and d″). (f, g) Construction of the opposite disclination. (Drawing made in collaboration with F. Livolant).

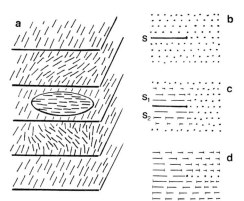

Figure 18. Planar model of a very thin thread in a nematic liquid and the modified Volterra principle used to produce it. (a) The distribution of molecules in five successive planes. (b) Directors are initially normal to the plane of the page. (c) A local twist about an angle $\pi/2$ is introduced in each lip S_1 and S_2, after a section S. (d) Relaxation occurs after resticking of the directors, which are supposed to remain horizontal.

illustrated in Fig. 19 [56]. Here, purely local rotations of directors near the interface are sufficient for the defect structure to be attained, whereas surface cuts would be necessary for defect lines to appear in the bulk. Different aspects of director distributions are shown in nematic liquids in the vicinity of a singular line at an interface, which forbids horizontal anchoring (Fig. 20). Starting from a uniform nematic alignment, a cut and a local reorientation of directors may be necessary to obtain such defect structures at an interface. (Such lines are also considered in Fig. 14).

Another example of topologically stable lines is observed in smectic C liquid crystals [57]. When layers lie horizontally between the slide and the coverslip, four-

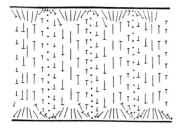

Figure 19. Distortions of a left-handed cholesteric liquid, introduced between the slide and the coverslip, with forbidden horizontal anchoring; the twist axis is horizontal between the plates.

Figure 20. Distribution of nematic directors at an interface, when horizontal anchoring is forbidden. Interface line defects appear, with three possible cross-sections (a–c).

branch nuclei resembling those shown in Fig. 11 for nematics, are frequently seen under the polarizing microscope. These arise from the oblique orientation of the molecules relative to the layers and the existence of stable arrangements defined along closed circuits within the layers (Fig. 21 a and b). These patterns are transmitted from layer to layer along nearly vertical singular lines. There are also horizontal lines of the same type, the characteristic circuits of which are shown in Fig. 21 c and d [57]. Such disclinations are produced by the Volterra process applied to a perfect smectic C liquid and, therefore, according to the following operations: an appropriate cut, a rotation of molecules within the layers of one lip, resticking and viscoelastic relaxation. Brunet and Williams [58] have described examples of such lines running parallel to layers in chiral samples of smectic C phases.

The helicoidal periodicity of smectic C* phases, due to the twisted distribution of the molecular tilt is not given by the half-pitch

as in cholesterics, but by the whole pitch p. Edge dislocations of this periodicity are easily recognizable in preparations, as they have Burgers vectors the lengths of which are a multiple of p. In principle, these defects of the helical periodicity can form without defects of the lamellar structure (Fig. 21 e). The equidistant stripes resulting from the helicoidal periodicity do not lie strictly parallel to the smectic layers. These dislocations need the presence in their core of a disclination line of the type shown in Fig. 21 d (or 21 c), as shown in Fig. 21 e.

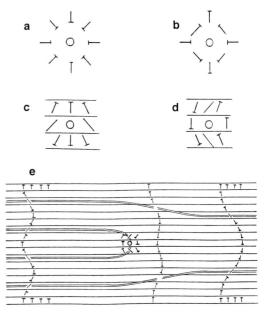

Figure 21. Schematic representation of singular lines in a smectic C phase. (a, b) The distribution of the directors in four-branch nuclei, the layers being seen from the top; open circles indicate the singularity position. (c, d) Cross-sections of singular lines parallel to layers. (e) Superimposed layers of a smectic C* phase can be devoid of defects, whereas an edge dislocation is present for the rotation periodicity (left-handed helix). The tilted molecules show a -4π rotation on the left-hand side of the diagam and only a -2π one on the right-hand side. This situation involves the presence of a line similar to (d) in the defect core. The double lines indicate the locus of directors parallel to the plane of the page.

7.3.4 Continuous and Discontinuous Defects in Cholesteric Phases

Thin threads in nematics are due to the presence of a discontinuity line, whereas thick threads correspond to the locus of vertical directors. The situation is similar in cholesterics, which are spontaneously twisted nematics, the threads in which are recognizable in large-pitch samples ($p > 10$ mm) [48, 49]. This situation is appropriate for studying the core structure of dislocations and disclinations. Let us start from a perfect cholesteric liquid (assumed to be left-handed) and isolate a virtual cylinder, the axis of which coincides with the twist axis (Fig. 22 a). The arrangement of the directors at the lateral surface of this cylinder is represented in Fig. 22 a′, a rectangle obtained by unfolding the limiting cylindrical surface, after cutting along a generator AA′ (or BB′). The positions of the directors lying within the drawing plane form a series of oblique stripes, corresponding to a double helix at the cylinder periphery, and this is also the case for positions of directors normal to this plane. Directors at $+45°$ form a quadruple helix and are represented by a series of nails. A cross-section of this cholesteric cylinder, at the level of line nm is shown in Fig. 21 a″ to recall the constant director orientation within a thin 'cholesteric layer', and it can be verified that the nail orientations along line nm in Fig. 22 a′ correspond well to the expected obliquities of the directors relative to the interface in the successive sectors of the cylinder in Fig. 22 a″.

If we now apply the Volterra process, there are two possible ways to obtain a screw dislocation, as indicated in Fig. 22 b and c, when the length of the Burgers vector is $p/2$, the half-helicoidal pitch. In a simple model it is assumed that the directors have a planar distribution, normal to the screw axis. The arrangement of directors remains that of parallel stripes in rectangles AA′BB′ transformed into parallelograms in Fig. 22 b′ and c′. However, due to the Volterra process, the double helices in Fig. 22 a and a′ are replaced by triple helices in Figures 22 b and 22 b′ and simple helices in Fig. 22 c and c′, which also are left-handed. The lateral distribution of directors along the circumference nn′ can be prolonged towards the core, the director orientation being constant along any radius. One then refinds the two classical patterns of disclinations in nematics (Fig. 22 b″ and c″), considered above in Fig. 10.

The patterns of various screw dislocations are represented in Fig. 23, either in meridian or cross-sectional views. The directors are coplanar and lie normal to the screw axis, as in Fig. 22, but have various Burgers vector lengths (multiples of $p/2$). The orientations of vectors b and p (Burgers vector and helical pitch vector) are identical if the spontaneous twist and the screw dislocation follow the same handedness. By convention, we say that the Burgers vector length b and the helical pitch p are positive when the corresponding distortions (and the spontaneous twist) are right-handed. These two lengths are negative if they both correspond to left-handed orientations. Each screw dislocation is characterized by an integer Z, which can be positive or negative, such that $b = Z p/2$; the cholesteric liquid is perfect when $Z = 0$.

The locus of the horizontal directors lying at a constant angle to the cylinder forms helices at the surface of the limiting cylinder, or a set of equidistant circles for $Z = -2$. A $180°$ arc of these helices for radial directors is represented in each of the drawings in Fig. 23, except for the case when $Z = -2$. All meridian sections of these screw dislocations are equal and superimpose through

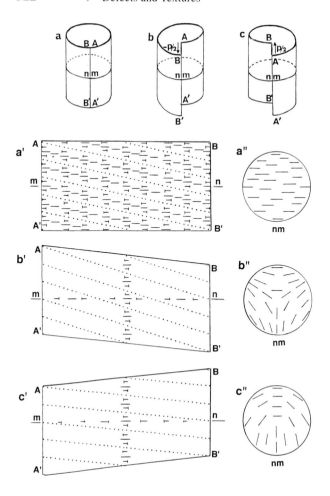

Figure 22. The distribution of directors along circular cylinders, within a left-handed cholesteric phase; the cylinder axis corresponds in (b) and (c) to the core of a screw dislocation. (a) The cholesteric order is perfect and the cylinder is cut along a generator, to be developed into a planar rectangle ABA′B′ represented in (a′), the drawing corresponding to the external view of the cylinder. Along a circle nm, directors show a definite distribution that is coherent, with a uniform alignment within the horizontal plane (a″). When the Volterra process is applied along the cylinder axis to create screw dislocations (b or c), the rectangles ABA′B′ are transformed into parallelograms (b′ or c′) and the distributions of the directors are changed along a transverse circle nm. This leads to two different patterns in the core (b″ or c″) when the molecules are assumed to remain horizontal. (Drawn in collaboration with F. Livolant.)

the corresponding helicoidal displacement. It can be verified that the pitch of this helix is $h = (Z+2)p/2$. The hatched part of the meridian section allows one to generate the whole screw dislocation by means of the corresponding helical motion. This hatched region is absent from the top figure, since the helix is replaced by a series of coaxial circles and a pure rotation suffices to generate the whole structure, with different patterns according to the section level (concentric circles, spirals, diverging radii).

The cross-sections of the various screw dislocations in cholesterics show disclination patterns similar to those encountered in nematic liquid crystals and correspond to

solutions of $\nabla^2 \Phi = 0$ in the plane [54, 59]. The core patterns corresponding to odd values of Z cannot be made continuous, since the ribbon of directors centred on a simple circuit closed about this defect line is a Möbius strip. Conversely, the disclination structures obtained for even values of Z (2, −2, −4, −6) can be made continuous, as indicated in Fig. 24 a–f, the expected meridian structure corresponding to Fig. 24 g [59–61]. The director that was normal to L in Fig. 23 is now made oblique by the presence of a component n_z which grows continuously from 0 to 1, from the periphery of the core (ρ) to its axis (ω) in Fig. 24 g. The two dotted areas correspond to zones of in-

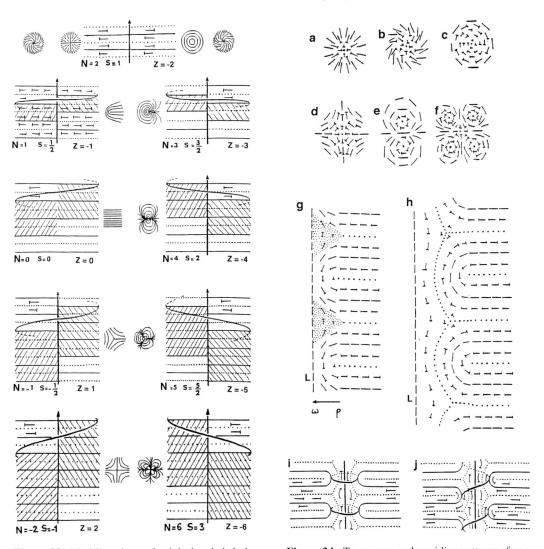

Figure 23. Meridian views of a right-handed cholesteric liquid, after the application of the Volterra process creating screw dislocations, and the corresponding patterns in cross-section. The parameters indicated below each pattern are N, $s = N/2$ and $Z = 2b/p$, the ratio of the Burgers vector to the half-helical pitch, both of which have positive length if the twist and the screw distortion are right-handed. The helical arcs indicate how one passes by helical displacement from the pattern on the left (hatched rectangle) to the corresponding one on the right (also hatched). The pitch of these helices is $h = p + b$. The cross-section patterns verify that $\Phi = \Phi_0 + N\varphi/2$, and one has $N = -Z$.

Figure 24. Transverse and meridian patterns of continuous cores in even-indexed (or non-Möbian) screw dislocations in cholesterics. (a–f) Transformations of transverse patterns for $Z = -2$ (a–c), 2 (d), −4 (e) and −6 (f), by 'escape in the third dimension'. (g) Meridian section expected from (a–d), L representing the axis of the screw line. (h) Another type of meridian section, also compatible with the production of screw dislocations along the axis L. (i, j) Two patterns deduced from (h), where the locus of the vertical directors consists of L and a set of equidistant rings with $Z = -2$ (i) or a helix with $Z = 0$ (j), the observation axis being either parallel to L or close to it.

verted twist. The meridian section in Fig. 24e is quite different: this screw dislocation was observed in cholesteric spherulites, the cholesteric layers of which are nested concentrically and, therefore, lie parallel to the external interface [62]. This defect, often called a 'radius of disclination', presents a continuous structure and has been modelled in both cross-sections and in meridian sections [63]. A planar twist zone is differentiated along the screw axis, with an orientation opposite to that of the spontaneous twist, with a pitch slightly larger than the spontaneous one. Alternative meridian structures are possible, and avoid twist inversion by the introduction of point singularities at regular intervals along the screw axis [59] (as has been supposed to occur along the disclination diameter of some cholesteric spherulites [63]). The continuous meridian structure shown in Fig. 24h also avoids twist inversions but these have been shown to occur in the two structures shown in Fig. 24i and j [49, 60].

The core structure of cholesteric disclinations was interpreted by Kléman and Friedel [2, 3]. The rotation vector considered in the Volterra process is normal to the cholesteric axis and is either parallel to the molecules or normal to them, this resulting in a core structure that is either continuous, with a longitudinal nematic alignment of directors in the core (λ disclinations), or discontinuous (τ disclinations), with a singular line of the type encountered in non-twisted nematic liquids.

Now, consider around any of these disclination lines a closed circuit C and the ribbon generated by the set of directors centred along C [44, 53]. Figure 26 shows a short straight segment of such a circuit, at a place where the director n is nearly constant, and the corresponding ribbon. One can replace a segment C_1 of this circuit by a curved segment C_1', which adds one helical turn and

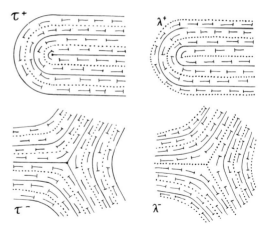

Figure 25. Local addition of a helical turn to a ribbon of directors (see text).

joins C tangentially at the two extremities of C_1. The addition of a helical turn modifies by one unit the 'linking number' of the two closed lines formed by the ribbon edges, but does not change the status of the ribbon, which is either the separation of the two faces, or their communication in the case of a Möbius strip. Note that the edges of a closed strip form either two closed curves, in a non-Möbius strip, or a single closed curve, in the case of a Möbius strip. In the study of such strips, there is also the problem of points the director of which is locally tangential to C, but in this case a slight modification of the circuit suffices to avoid such points [44].

The main result in this study is that simple circuits closed about λ disclinations give non-Möbius strips with a vector n, whereas those closed around τ disclinations are of the Möbius type. The core needs, to be continuous in order to be surrounded by non-Möbius strips. Similar ribbons built with the twist vector C centred along such circuits are of the Möbius type for the four disclinations shown in Fig. 25.

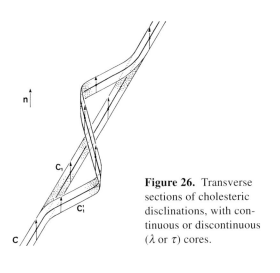

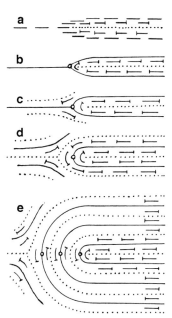

Figure 26. Transverse sections of cholesteric disclinations, with continuous or discontinuous (λ or τ) cores.

7.3.5 Core Disjunction into Pairs of Disclinations

The main variations of the core structure in edge dislocations are illustrated in Fig. 27 and 28. Consider a large-pitch cholesteric liquid introduced between two parallel glasses, with a unique direction of easy alignment lying in the plane of the page, and assume that the distance separating the two glasses is $3p/4$, intermediate between $p/2$ and p. Two types of domain are expected: those corresponding to a $+\pi$ rotation and those to a $+2\pi$ rotation from the slide to the coverslip, and these are separated by a discontinuity line analogous to that found in a very close situation, when a nematic liquid is twisted either by an angle of $+\pi/2$ or $-\pi/2$, between two parallel glasses rubbed at right angles (see Fig. 9).

Figure 28. Progressive disjunction of edge dislocations in cholesterics (left-handed twist). The open circles show the locus of vertical directors. (a) Planar model of a 'very thin thread'. (b) First step of the disjunction into a λ^- and a τ^+ dislocation. (c) Cross-section of a 'thin thread', with complete disjunction of the two disclinations λ^- and τ^+, corresponding to a Burgers vector of length $b=p/2$. (d) Edge-dislocation of the type $\lambda^-\lambda^+$, with $b=p$. (e) A $\lambda^-\lambda^+$ pair with $b=3p$.

When a large pitch cholesteric liquid is introduced within a wedge of two rubbed glasses, the 'Grandjean–Cano wedge' [64], edge dislocations form and arrange in parallel, those with large Burgers vectors lying in the thickest regions of the wedge. Planar models such as the one shown in Fig. 27 apply in the thinnest zones, whereas in thicker areas the directors present vertical components in the vicinity of the defect and this transforms the dislocation lines into pairs of disclinations (Fig. 28 a–c), with a τ^- on the left and a λ^+ on the right [48, 65]. Edge dislocations having larger Burgers vectors transform into dislocation pairs of the type $\lambda^+\lambda^-$, with a fully continuous distribution of directors (Fig. 28 d and e).

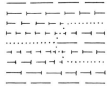

Figure 27. Planar model of a discontinuous thread separating a 2π twist zone (right) from a π one (left) (right-handed twist).

These structures are well evidenced when cholesteric layers extend normally to the preparation plane, with weak anchoring conditions, resulting in so-called 'fingerprint textures'. Disclinations are well observed in these conditions between crossed polarizers, if the pitch is large enough, and the presence of a central black zone is demonstrative of a λ core; this is not the case with a τ disclination. The $λ^+ λ^-$ associations are the most frequent. It can be verified that the set of directors centred on a circuit forming a simple loop around such disclination pairs is not a Möbius strip when the core is continuous, but is of the Möbius type when one of the two disclinations of the pair is a τ disclination. Conversely, the ribbon obtained with the cholesteric axis C along such circuits is not of the Möbius type, whatever the nature of the two disclinations forming the dislocation.

Disjunctions into disclination pairs are general in liquid crystals when the Burgers vector is large enough, and this holds for both edge and screw dislocations. This was first considered for edge dislocations in cholesterics [2, 3], but also applies to screw dislocations (see Fig. 24 i and j).

7.3.6   Optical Contrast

All the translation defects shown in Fig. 28 are easily prepared in a wedge obtained, for instance, when a coverslip lies tangentially along a generator of a cylindrical lens, both the lens and the coverslip having been treated to obtain a planar anchorage with a preferential direction. A drop of nematic liquid added together with a small amount of a twisting agent allows one to obtain a pitch of 20–30 µm. After this cholesteric mixture has been introduced between the two glasses, edge dislocations stabilize on both sides of the contact generator and form a se-

ries of parallel lines that is easily observed under natural light.

The use of a single polarizer or analyser is sufficient to vary the defect contrast considerably. As the local ellipsoid of indices is positively uniaxial and the twist is weak, the electric vector of any penetrating wave rotates as the director orientation (Mauguin's condition: the wavelength being very small compared to the cholesteric pitch) [66, 67].

If the electric vector of the incident beam is normal to the anchoring direction, the wave propagates as an ordinary ray, whereas if it is parallel the resulting extraordinary ray deviates and the image of the threads is much more shadowy and fuzzy [11, 68]. The use of a polarizer the main axis of which is perpendicular to the common anchoring direction of the two glasses gives clear-cut ordinary images of threads, much thinner than the half cholesteric pitch $p/2$, and by analysis of the whole system several types of thread can be distinguished.

Very thin threads in the vicinity of the contact generator between the cylindrical lens and the coverslip are poorly contrasted, and correspond to the planar structure shown in Fig. 28 a.

The thin threads lie in slightly thicker regions of the preparation and correspond to the model shown in Fig. 28 c. Horizontal streams decrease the contrast of these threads, which then resemble 'very thin threads', this probably being due to a horizontal alignment of directors [48, 65].

The thick threads observed at the periphery correspond to the model shown in Fig. 28 d, but often form groups of parallel lines, separated by a distance that is the half-helical pitch, as shown in Fig. 28 e.

The positions of thin and thick threads are indicated in Fig. 28 by open circles; many observations have led to the conclusion that these correspond to the locus of directors parallel to the optical axis of the microscope, i.e. normal to the preparation plane, which is usually called the vertical [48]. Most papers dealing with thick and thin threads in nematics and large pitch cholesterics generally consider that these threads represent disclination cores. This is true for thin threads, but generally not so for thick ones. The confusion arises mainly from the fact that the core of a defect line corresponds to a strong maximum of curvature and it is highly likely that vertical directors will be found in its vicinity. A good approximation is to say that thick threads represent the locus of points where $n_x=n_y=0$, and $n_z=1$, but the optics are complex, and this simply remains a first approximation.

7.3.7   Classification of Defects

As shown in every part of this section, liquid crystals are states of matter some of which are most appropriate for illustrating remarkable situations in combinatorial topology. The fluidity of mesophases is compatible with different types of curvature resulting in large series of defects, particularly disclinations, and the best tools for their definition and classification are the 'homotopy groups', the applications of which to liquid crystals are numerous [69–78], but their use is somewhat arduous; they can be understood as a mathematical generalization of the Volterra process [72]. All the 'topologically plausible' points, lines and walls and their rehandlings are predicted without exception by this theory. The Volterra process has been used to define singular lines, but not defects as points, walls and certain structures with a continuous core. The theory also considers the rehandlings of defects [48, 53]. However, it is worth remembering that among the defects that are 'topologically plausible', many of them do

not exist, since they involve considerable energies; the simplest example is that of disclinations which are topologically plausible in three-dimensional crystals – these have drawn [79] but are excluded in true crystals for energy reasons. This is also true for most liquid crystals. Geometrical and mechanical constraints are forgotten in purely topological considerations, and these are presented in the next section.

Translation and rotation vectors are still the best tools in crystallography and for defining and classifying defects in liquid crystals. We have distinguished above between edge and screw dislocations, which are translation defects characterized by a Burgers vector, either normal or parallel to the line, but these defects often lie oblique to this vector as in three-dimensional crystals, and the structure of jogs and kinks has mainly been studied in cholesterics [48]. Disclinations (or rotation dislocations) introduce a rotation vector v that is either normal to L, a *twist disclination*, encountered mainly in planar models (see Figs. 9, 18 and 27), or parallel to L, a *wedge disclination*, more frequent in smectics and cholesterics. In the latter, we deal with four vectors: the director n, the cholesteric axis C, the rotation axis v and the line axis L. Let us recall that, in λ disclinations, v is parallel to n, but normal to C. In τ disclinations, v is normal to n and C (see Fig. 25); in χ disclinations, v is normal to n, but parallel to C (see Fig. 23). Note also that these disclinations, when they run parallel to the twist axis, reproduce the structure of screw dislocations in cholesterics. The symbols λ, τ and χ are often followed by an exponent $+\pi, -\pi, +2\pi, -2\pi$, etc. corresponding to the rotation angle of the director after one 2π turn along a small circle surrounding the core line of the defect. These exponents are often reduced to +, −, 2+, 2−. Figure 25 shows a $\lambda^{+\pi}$, a $\lambda^{-\pi}$, a $\tau^{+\pi}$ and a $\tau^{-\pi}$ (often abbreviated as λ^+, λ^-, τ^+,

τ^-), whereas Fig. 23 shows a series of χ disclinations from $\chi^{+2\pi}$ to $\chi^{-6\pi}$.

The purely orientational order of nematics leads to discontinuities that are discrete structures in the form of points and thin threads. Continuous defects also exist, but some of these cannot be deformed continuously into the monocrystal and therefore should be classified as genuine defects. The presence of a spontaneous twist in a nematic liquid facilitates such situations, the main example being that of thick threads forming interlocked rings [48, 49, 53, 73].

7.4   Geometrical Constraints and Textures

7.4.1   Parallel Surfaces and Caustics in Liquid Crystals

Many liquid crystals show a lamellar aspect, due to layers of uniform thickness arranged along parallel surfaces, and this is well observed in polarizing microscopy for some cholesterics, when the layer outlines are apparent (see Fig. 6). In these two micrographs, the layers change direction abruptly along straight or curvilinear segments belonging to the hyperbola branches, since they correspond to successive intersections of nearly circular profiles centred on two different points lying in the same optical section. The layers in fact form spirals, owing to the helical structure of cholesterics, rather than circular rings, and their thickness varies due either to real changes in helical pitch, or to apparent variations when layers are oblique.

When the cholesteric pitch is less than 0.5 μm, the parallel profiles of the cholesteric layers cannot be resolved in light microscopy, but the straight or bent segments due to

abrupt changes in orientation remain well contrasted. These segments belong to hyperbola branches or ellipses, or parabolae, and form well-structured arrangements that are found not only in short pitch cholesteric but also in smectic phases [11].

Parallel contours are also evident in smectic phases viewed under a polarizing microscope. In the case of smectic C* phases, for example, these contours are due to the helical periodicity, while in smectic E and lyotropic lamellar phases they occur when defects accumulate along certain lamellae or groups of lamellae.

Normals to parallel layers form straight lines, as do the light rays with respect to waves in an isotropic medium, and they envelop two surfaces, called *focal surfaces* or *caustics* [19, 80]. However, their presence in the mesophase produces a discontinuity wall, which generally disjoins into alternate $+\pi$ and $-\pi$ disclinations [20, 81]. Caustics are often absent from liquid crystals, since layers can be poorly curved, with 'virtual caustics' located outside the liquid crystal. For instance, layers are nearly horizontal in *stepped drops* of smectics, when the conditions are homeotropic along the glass and at the air interface. They appear to interrupt along oblique 'cliffs', arranged as more or less concentric level lines, and distortions are observed in their vicinity [82]. There are also examples of parallel spherical layers, with caustics reduced to a single point. The most extreme case of degeneracy of caustics occurs when each of the two surfaces reduces to a unique singular line and not to a series of lines as indicated above.

7.4.2   Dupin's Cyclides and Focal Conics

Surfaces Σ that are envelopes of spheres centred along a curve L have their normals passing through L; conversely, all surfaces with normals converging along a curve L are envelopes of such families of spheres (Fig. 29 a). The contact curves between the spheres S and the envelope Σ are circles γ. The normals along γ converge towards A, the centre of the sphere S, and form a *revolution cone*, the axis of which is tangent to L in A. Cylindrical bilayers, which are straight, bent, or helical in myelin figures, are an example of such surfaces, which are envelopes of spheres of equal radius centred on a straight, bent or helical line L.

Each caustic surface can reduce to lines L and L′ and a simple example is presented in Fig. 29 b; the parallel surfaces are the nested and parallel tori, forming a $+2\pi$ disclination L, the whole system presenting a revolution symmetry about a vertical axis, which is itself a second defect line L′. These tori are surfaces generated by (at least) two families of circles, their meridians and parallels, which also are their lines of curvature.

The general problem of surfaces the normals of which pass through two curves L and L′ was solved by Dupin [83] and the first applications were considered by Maxwell; the general theory was developed in the 19th century [83] and in more recent articles [84].

Such surfaces Σ are envelopes of spheres in two different ways – either a family of spheres S centred on L, or a family of spheres S′ centred on L′ – and there are two spheres S and S′ tangent at any point M of Σ. The revolution cone associated with a contact circle γ of S contains L′, and vice versa. If one knows one sphere S and three spheres S′ tangent to S, the family S can be defined (and the S′ family also). The three spheres S′ cut each other at two points, O and Ω, either real or imaginary. Let O be an inversion centre with a coefficient $k = \mathbf{OM} \cdot \mathbf{Om} = O\Omega^2$, where M and m are two homologous points aligned with O. The three

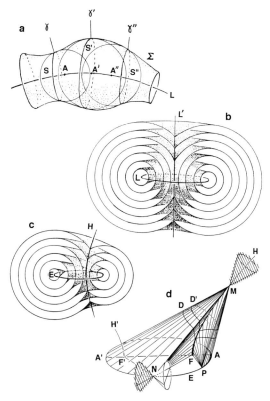

Figure 29. Cyclides are surfaces the lines of curvature of which are circles. (a) An envelope Σ of a family of spheres S, centred in A describing L, and contact circles γ. (b) Nested tori, the normals of which pass through the circle L and the axis L′, one sheet being suppressed to avoid intersection of the surfaces. (c) General aspect of Dupin's cyclides, the normals of which pass through an ellipse E and a hyperbola H. (d) A focal domain built on the arc FM of a hpyerbola and on the arc PAA′N of the ellipse E. AA′ is the major axis with the two foci FF′. H′ is the virtual hyperbola branch. D and D′ are circular intersections of Dupin's cyclides, the revolution cones being represented by their generators or director lines.

spheres S′ transform into three planes p′ and the spheres S transform into spheres s tangent to the planes p′, their envelope being a revolution cone c, and thus the inverse of the surface Σ. The tangent planes to this cone are the inverse of spheres S′, which also pass through O and Ω, and the line L′ lies in the mediator plane of OΩ. L′ is planar and belongs to all revolution cones of the normals converging on L, and is therefore a conic, an ellipse or a hyperbola or a parabola; this also holds for L. These two lines are conics, one being the locus of the vertices of the revolution cones passing through the other one. This situation is that of *focal conics*, which lie in two normal planes intersecting along their major axes, the vertices of one conic coinciding with the foci of the other one. In general, an ellipse is associated with a hyperbola of inverse eccentricity, and this leads to a set of nested toroidal surfaces, the *Dupin cyclides* (Fig. 29 c), a system that is less symmetrical than the one shown in Fig. 29 b. In this case, instead of a

revolution symmetry and a mirror plane (Fig. 29 b), there are two mirror planes at right angles to one another, which are the planes of the two focal curves.

Smectic A phases often contain pairs of focal conics (see Fig. 7 a and b) that are well contrasted in the liquid, even in the absence of polarizers. A more common situation is the presence of arcs of conics associated in pairs (Fig. 29 d). The parallel Dupin cyclides and the associated revolution cones form three mutually orthogonal systems of surfaces [7], and therefore the lines of curvature of any surface in one system are its intersections with the surfaces of the other two systems. The toroidal shape of Dupin's

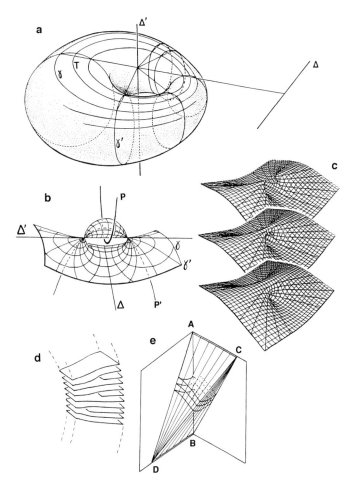

Figure 30. Dupin's cyclides and their deformations. (a) A toroidal Dupin cyclide and its circular lines of curvature γ and γ', the planes of which intersect along the axes Δ (joining O and Ω) and Δ', respectively. Two tangential planes along two symmetrical circles T intersect along Δ. (b) A parabolic Dupin's cyclide, with its two focal parabolae P and P', its circular curvature lines γ and γ', and its axes Δ and Δ'. (c) Three parallel, but separated, parabolic Dupin's cyclides. The lines represent sections by planes $x=0$, $y=0$ and $z=0$; at the top of the figure, two conical points are present, as in (b), but the spindle-shaped sheet has been removed (Rosenblatt et al. [85]). (d, e) Edge and screw dislocations 'compensate' the bend and the twist of directors in a set of equidistant surfaces normal to the director lines.

cyclide is indicated in Fig. 30 a. Conical points of Dupin's cyclides are present all along the two conics, but one of the two sheets is absent, since interpenetration of layers is impossible. The forbidden conical sheets are shown explicitly in Fig. 29 d. The case of confocal parabolae is particularly frequent in the lamellar phase of lyotropic systems [85] and the corresponding Dupin's cyclides (Fig. 30 b and c).

Micrographs of focal conics were presented in the pioneer work by Lehmann [10], and, some years later, in that by Friedel [11] who, together with Grandjean, observed the alignment of the positively uniaxial ellipsoids of indices along the straight segments joining points on the two associated conics [11]. The transition to the solid crystal obtained after cooling often shows the formation of thin crystalline lamellae lying normal to these molecular alignments between the two conics (ethylazoxycinnamate or its mixture with, for example, ethylazoxybenzoate). The paired conics remain generally recognizable in this solid texture, which is called a pseudomorphosis [11] or paramorphosis [13, 14]. By reheating the preparations one can revert to the initial liquid crystalline texture, with the focal conics located in the same places. This suggested the presence in these liquids of fluid layers forming sets of parallel Dupin's cyclides, with molecules lying normal to them [86]. Very similar textures and defects were found in soaps and other lyotropic systems, which justifies the term 'smectic' chosen by Friedel for these lamellar liquid crystals [11], the structures of which were rapidly confirmed by the first X-ray diffraction studies [87]. The persistence of various textural aspects and domains through many phase transitions is common, and several examples have been described [13, 14].

7.4.3 Slight Deformations of Dupin's Cyclides

In a perfect system of parallel Dupin cyclides, the unit vectors normal to surfaces align along straight segments joining two focal conics. A deformation resulting in a local bend modifies the layer thickness or involves edge dislocations (Fig. 30 d). Similarly, screw dislocations compensate a twist, when for instance focal conics are replaced by different curves. This is indicated in Fig. 30 e, in which the layers are assumed to extend normally to straight segments, the extremities of which belong to two rectilinear segments AC and BD instead of to two arcs of focal conics. Suppose, for example, that AB, AC and BD are parallel to the axes of a tri-rectangle trihedron. The intersections of layers with the four faces of the tetrahedron ABCD form concentric circular arcs, the concavities of which alternate, showing that layers are saddle-shaped. Moreover, successive arcs joined at their extremities do not form a closed loop, but a helical path. The geometry of this tetrahedral domain requires the presence of right-handed screw dislocations, whereas the mirror image of ABCD would lead to left-handed defects. The local density of these defects within a lamellar structure (total Burgers vector in the unit volume) is $|n \times \mathrm{curl}\, n|$ for pure edge dislocations and $|n \cdot \mathrm{curl}\, n|$ for pure screw dislocations [88].

Polygonal textures such as those shown in Fig. 6 associate not only horizontal segments (curvi- or rectilinear), but also vertical ones, joining centres to the vertices of polygons at the upper and lower faces of the mesomorphic slab, between the slide and the coverslip. The positions of these segments are shown schematically in Fig. 31 a for the cholesteric phase occupying the vertical prism AA'CC'BB'DD'. The cholesteric axes, normal to layers, join points belong-

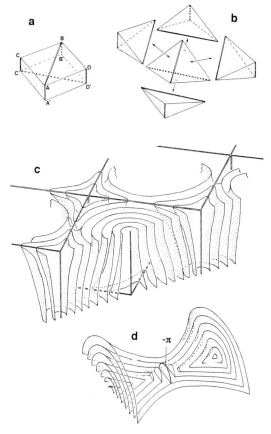

Figure 31. (a) Prismatic subunit of a polygonal field, in cholesteric and smectic A phases. Double lines represent focal segments. (b) The prismatic unit shown in (a) decomposes into five tetrahedral domains. (c) General aspect of the layers in a square polygonal field; double lines represent the array of focal segments. (d) Very acute conical points along focal segments are not only rounded, but are also rehandled by a $-\pi$ disclination. This actually introduces two conical walls, but the layers also are slightly rounded and more focal curves are involved.

ing to the two contours A'ABB' and CC'D'D indicated by a doube line. Therefore, the prism decomposes into five tetrahedra (Fig. 31 b) filled with saddle-shaped cholesteric layers. In each tetrahedron two opposite edges, called focal curves or focal segments, play the role of focal conics. These segments often are conics, but not in a focal position.

The whole texture is made of such prisms, juxtaposed in continuity. These divergences from the Dupin cyclides model can result in the presence of either translation dislocations or layer thickness variations, or in layers that are slightly oblique to segments joining the conjugated contours. The cholesteric layers extend continuously across the triangular faces joining the tetrahedra, except for the focal segments A'ABB' and CC'D'D. There are no other discontinuities across the vertical faces joining all the prisms into a texture such as the one shown in Fig. 10 a and b. Note, however, that the two radii of curvature show a discontinuity along the oblique edges of the tetrahedral domains, which is probably attenuated by slight deformations.

Polygonal textures often form square lattices in cholesteric and in smectic A phases. The global arrangement of nested parallel layers, between the horizontal slide and coverslip is shown in Fig. 31 c. The layers lie vertically at the upper and lower faces of the mesophase and are oblique in between, but they are not perfectly perpendicular to straight lines joining the conjugated segments, and thus show important thickness variations, particularly in the vicinity of focal segments. This is often corrected by a rehandling such as the one illustrated in Fig. 31 d.

7.4.4   Textures Produced by Parallel Fluid Layers

The main patterns present in these textures are shown in Fig. 5 for lyotropic lamellar phases and in Fig. 32 for thermotropic liquid crystals.

7.4.4.1   Planar Textures

Layers lie parallel to the slide and the coverslip, a more or less frequent situation in

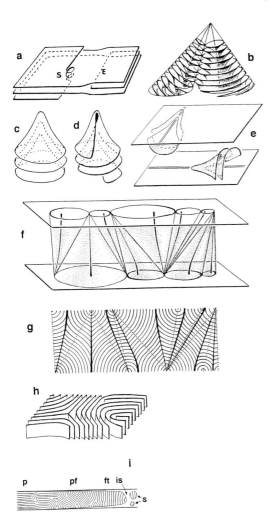

Figure 32. The main types of texture in layered liquid crystals. (a) Planar texture with edge (E) and screw (S) dislocations. (b) Two tangent focal domains, and their tangent ellipses and the converging arcs of hyperbola. (c) Nested 'hats' along a focal line. (d) Screw superimposed on a focal line. (e) Focal lines attached to the upper and lower interfaces. (f) Focal domains tangent along a series of coplanar generators; contacts between domains are parallel or antiparallel. (g) Section of (d) along the symmetry plane, showing the discontinuities in the curvature at the antiparallel contacts. (h) General patterns in fans, associating $+\pi$ and $-\pi$ disclinations and edge dislocations (and also focal curves, not shown). (i) Schematic distribution of textures in the vicinity of the isotropic phase in lamellar thermotropic liquids; ft, fan texture; is, isotropic phase of the mesogen; p, planar texture; pf, polygonal fields; s, spherulites.

smectic A and cholesteric phases, when the glasses have not been treated for molecular anchoring. These textures are produced regularly in smectic A phases, where strong homeotropic conditions prevail, while a planar anchoring with an easy horizontal direction is required for cholesterics. In general, edge and screw dislocations are present within these planar textures (Fig. 32 a) and are often seen by light microscopy of cholesteric phases of sufficiently large pitch [20], whereas these defects are detected only by electron microscopy in cryofractured, mainly lyotropic, smectic phases [89]. In some cases, edge dislocations intercalating a unique smectic layer can be visualized by means of light microscopy; an example of this has been described in smectic C phases, together with all the technical details of the optics [90]. Some edge-dislocations, however, present large Burgers vectors and disjoin into $+\pi$ and $-\pi$ disclinations; layers intercalated between these disclinations lie vertically and form brilliant stripes in polarizing microscopy or *oily streaks* (see Fig. 5 d), a common textural pattern in smectic A and cholesteric phases [11].

7.4.4.2 Focal Conics and Polygonal Textures

We have already considered the polygonal networks of cholesteric liquids, which assemble saddle-shaped layers close to Dupin's cyclides, and the focal segments of which differ from those of focal conics (see Sec. 7.4.3). Friedel and Grandjean [21] described in smectic A phases a texture much closer to Dupin's cyclides, with genuine focal conics. As for cholesterics, this texture is obtained when a thin film of the isotropic phase of the mesogenic compound separates the mesophase from the two glasses.

The mesophase itself is divided into focal domains, each one being defined by its two conjugated arcs of conics (see Fig. 29 d), which determines uniquely the corresponding family of parallel Dupin cyclides. Each domain is limited by fragments of revolution cones joining along edges, which are either generators or arcs of these conics. The fragments of Dupin cyclides filling these domains are essentially saddle-shaped (see Fig. 32 b), but are hat-shaped along the focal curves, with a strong maximum of curvature and layer thickness (see Fig. 32 c). This can be viewed in cholesterics by using light microscopy (see Fig. 6) and in lyotropic smectic phases by using electron microscopy [89]. Screw dislocations sometimes superimpose on focal curves, as shown in Fig. 32 d [54]. When focal lines run horizontally, attached to the upper and lower interfaces limiting the mesophase, the layers form half-cones, which also present a rounded apex (Fig. 32 e).

Focal domains in smectic A phases are tangent along generators of their limiting cones, the contact being either parallel or antiparallel ' in Fig. 32 f, where the major axes of ellipses are aligned, and this allows one to obtain a view of layers in the mediator plane (Fig. 32 g). The dashed lines in Fig. 32 g represent the contact generators between the tangent focal domains, either parallel and without discontinuity of the radii of curvature, or antiparallel and with a discontinuity. It is worth remembering that, as for cholesterics, these discontinuities of curvature present in the model are attenuated or smoothed out by a rapid but continuous change in the sign of the curvature. Such tangent domains are illustrated in Fig. 33, which shows a lattice of ellipses, just below the coverglass. The ellipses are separated into different sectors, which end in a polygonal field at the top left-hand side of the micrograph. The sectors are separated by lines that are particular polygon edges, and in fact each sector belongs to a polygon.

The ellipses are tangent to other ellipses and to the polygon edges. Their focal hyperbolae converge to a point on the opposite interface, which is a vertex of the conjugated polygonal lattice. This means that, within the polygon plane, the major axes of the ellipses converge to a point that is the vertical projection of a vertex common to several polygons of the conjugated lattice. When three ellipses are tangent, the free space left between them is often occupied by another ellipse at the base of a narrower focal domain, tangential to the surrounding domains, and the interstices can in turn be filled by narrower cones. It has been proposed that this iterative process persists down to a few molecular lengths [91], and therefore below the resolving power of the light microscope. This view is justified in models that keep all layers normal to the horizontal interfaces, thus limiting the polygonal field. It is also justified by energy estimates and by cryofracture images, which often show very local conical deformation involving a few bilayers only [89]. However, in thermotropic smectic phases forming polygonal networks, very small ellipses seem to be absent, their size rarely being less than the 1/20 of the polygon diameter. On the contrary, the interstices are assumed to form spherical domains, the layers being spherical and centred on the polygon vertex, where the hyperbolae of the surrounding focal domains converge [92].

7.4.4.3   Fan Textures

Layers that are mainly vertical and the presence of disclinations (see Fig. 32 h) are the two very common characteristics of fan textures in smectic and cholesteric phases. The name 'fan texture' comes from the circular contours that are often visible around disclinations in, for example, cholesterics, and from the frequent radial decorations (see

Figure 33. Polygonal field (top left) and elongated polygons, with their long edges converging at a $+\pi$ disclination. 4-Cyano-4′-n-octyl-biphenyl plus Canada balsam; crossed polarizer. Scale bar: 50 µm.

Fig. 33) seen in smectic A phases. Focal curves and translation dislocations are not excluded from fan textures. The fan patterns themselves can be absent from certain arrangements of disclinations, which however belong to the family of fan textures. This is illustrated by the rhombus in Fig. 34a and b, where two $+\pi$ and two $-\pi$ disclinations are associated with several focal curves at the top and bottom levels of the mesophase.

7.4.4.4   Texture Distribution in Lamellar Mesophases

When smectic A or cholesteric phases are in equilibrium with their isotropic phase, the mesophase is present in the preparation in the form of bâtonnets or droplets, and extends here and there over larger domains, with a definite distribution of textures. Fans lie in the vicinity of the isotropic interface, whereas planar textures are rarely in contact with it, and polygons are observed along zones separating fans from planes.

a

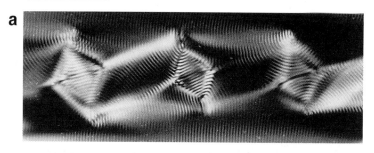

b

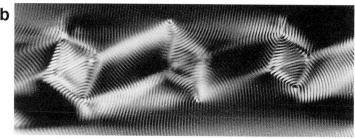

c

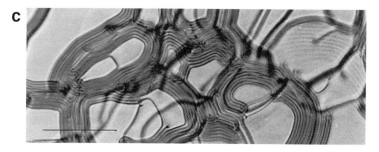

d

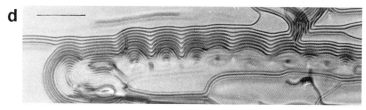

Figure 34. Cholesteric textures: (a–c) MBBA plus Canada balsam; (d) MBBA plus cholesterol benzoate. (a, b) Two views at the coverslip and slide levels of a fan texture, with lozenges associating two $+\pi$ and two $-\pi$ disclinations, linked by pairs of focal lines. (c) Planar domains separated by walls of nearly vertical layers. (d) One of the walls is associated with horizontal and vertical focal lines, appearing as intercalated black spots. (a, b) Crossed polarizers; (c, d) natural light. Scale bars: (a, b) 20 μm; (c, d) 50 μm.

As indicated above and illustrated in Fig. 32i, layers lie horizontally in planes and show some translation dislocations. They are oblique in polygonal fields and generally cross the focal lines at points where they are strongly hat-shaped, whereas elsewhere they are saddle-shaped and dislocations are present, often superimposed on the focal lines. In fans, the layers are vertical and all types of defect (disclinations, focal curves, dislocations) are present.

This texture distribution is common, but corresponds to a rather schematic model, indicating that high-energy defects, such as disclinations, are found mainly in the vicinity of the isotropic transition. Other situations are observed in thick preparations of cholesterics, for example, where planar domains can be interrupted by walls of vertical layers (Fig. 34c and d), due to edge dislocations disjoining into disclination pairs (see Fig. 5d). Despite these particular examples, the presence of disclinations remains a general character of fan textures.

The textures described above belong to the same phase and are not separated by sharp interfaces. The frontiers are fuzzy and are often the sites of hybrid, but interesting, textures. Among these, at the limit between polygons and fans, one finds occasional chevrons, the organization of which is shown in Fig. 35 [93].

Chevrons and related textures are common in smectic phases but, at the horizontal interfaces, the directors must be adjusted to the anchoring conditions [88, 93]. They are regularly produced when opposite sides of polygons are extremely elongated in a given direction [20]. This leads to arrangements as those represented in Fig. 35. When vertical anchoring of layers is preferred, ellipses or parabolae can form at the interfaces, with the corresponding focal domains, but the layer anchoring remains oblique in the interstices between domains [93]. This situation is also realized in Fig. 33 with a radiating series of extremely elongated polygons, centered around a $+\pi$ disclination.

7.4.4.5   Illusory Conics

The analyis of focal conic systems in liquid crystals is made difficult by the presence in pictures of lines, which often resemble focal conics or focal curves but actually are not. It was indicated above that conics such as polygon edges are not in a focal position, and that many lines differ from conics but play a focal role. Figure 36 shows an example of a cholesteric texture simulating the presence of a series of parabolae, but this aspect is due to complex behavior of the polarized light, associated with a frequent moiré pattern. This texture is simply interpreted in Fig. 37 on the basis of an examination of the orientations of the layer and the focal lines, which are straight or slightly curved in this texture, but not parabolic.

7.4.4.6   Walls, Pseudowalls and Broken Aspects

True walls are defined by a director discontinuity extending over a surface within a liquid crystal, but these defects are excluded from nematic, cholesteric and smectic A phases (see Sec. 7.2.3). Grain boundaries comparable to those of true crystals are associated with translation dislocations, which lie parallel within the wall or form

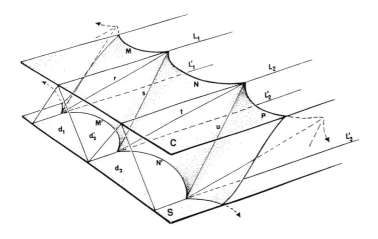

Figure 35. A model of chevron textures (the chevrons are viewed from the top) showing alternating angular and rounded contours, at both the slide and the coverslip level. Prismatic domains d and d' are filled with antiparallel conical sheets, in continuity along the generators r, s, t and u. Parabolic contours M, N and P join and form angles along the lines L at the coverslip level C, and similarly arcs M', N' and P' join and form angles along the lines L' at the slide level S.

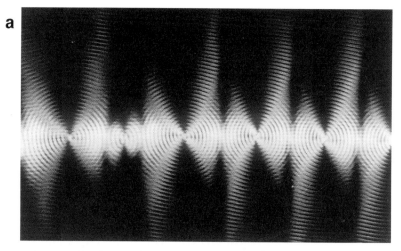

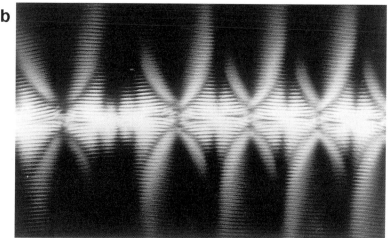

Figure 36. A cholesteric texture simulating the presence of focal parabolae. (a, b) Optical sections at the coverslip and slide levels. MBBA plus cholesterol benzoate, crossed polarizers. Scale bar: 20 µm.

diverse networks [9]. The presence of such defects is probable in smectic phases having two-dimensional order within layers, and in certain columnar phases. This also holds for smectic systems, when the hexagonal or rectangular arrangements are coherent over several lamellae (smectics B, E, G, H, etc.) the order being three-dimensional.

When smectic layers are mainly horizontal, the existence of walls often leads to *mosaic textures*, each domain extending over the entire thickness of the preparation. A *platelet texture* is observed when the do-mains of the horizontal layers are much thinner than the mesomorphic slab, and superimpose here and there, with their distinct orientations [13, 14]. When vertical layers bend concentrically about $+\pi$ disclinations, circular bands of different colour or shade appear, indicating changes in the crystallographic orientations along grain boundaries. These banded textures are common in orthorhombic smectic E phases.

A *planar wall* has been described in smectic C* phases, giving rise to a particular chevron texture. This has been shown

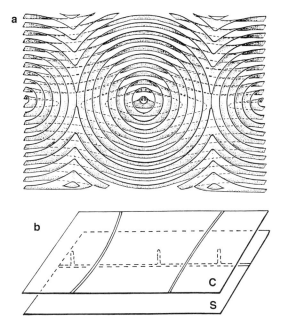

Figure 37. Model of the arrangement of layers and focal lines in the texture shown in Fig. 36. (a) Distribution of layers. (b) Vertical and horizontal focal lines at the top (coverslip C) and bottom (slide S) levels of the mesophase.

by high-resolution X-ray scattering in surface stabilized ferroelectric samples obtained by cooling from the smectic A phase [94]. At the smectic C transition, the layers progressively tilt symmetrically in two opposite directions, by an angle of about 10°, whereas molecules are tilted horizontally within layers at about 20° from the normal to the layers. The long axis of molecules show a quasi uniform orientation within this wall and the two domains, but the ferroelectric polarization changes abruptly across the wall and breaks its apparent symmetry. This type of chevrons leads to complex textures [95].

In smectic A phases, such a change in the orientation of layers along a wall involves a discontinuity of molecular alignment. If the wall is replaced by a narrow zone of strong-

ly curved layers, this involves an equal splay of molecules, and the structure is generally handled by dividing it into a set of focal curves or even disclinations. The case for smectic C phases is different, as such walls often are compatible with a uniform alignment of molecules, facilitated by a dilatation modulus of the layer thickness that is much smaller than that for smectic A phases. This leads to noticeable undulation instabilities [96]. Molecules rotate within layers and, conversely, layers show different possible orientations with respect to a molecular alignment. In chevrons such as those shown in Fig. 35, the layers are at a constant angle to the lines L and L', an angle which can be chosen to be that of the tilt [57]. The discontinuity in the layer orientation in the ferroelectric display can be smoothed by focal curves and other defects, which are weakly contrasted because the molecular orientation is not really altered by their presence.

Singular lines often attach to interfaces. Among the numerous examples given here is the case that occurs in nematic phases at the slide or coverslip level (see Fig. 9) and that of large-pitch cholesteric phases at various interfaces (see Figs. 19 and 20). Such lines are due to the presence of horizontal directors that are forbidden at a horizontal interface. The *locus of horizontal directors* in the bulk form a surface, which shows a higher mean refractive index and appears contrasted under natural light or in phase-contrast microscopy. The clear-cut interface line seems to extend into a wall within the bulk, but actually this is a mere optical illusion, which is the origin of the so-called Grandjean planes in cholesteric phases [44].

Broken aspects are classical in focal conics or fan textures of tilted smectics, and also originate from singular lines attached to an interface. This is observed in smectic C phases when disclinations like those shown in Fig. 21 a–d adhere at an interface and separate two different molecular orientations. The smectic layers lie almost vertically (i.e. perpendicular to the interface), whereas the molecules are horizontal within the layers at the interface contact. However, the molecular orientation remains continuous in the bulk. The lines limiting these interfacial domains are either parallel or normal to layers and this creates a contour composed of straight segments alternating with arcs at right angles and with some interspersed irregularities [97]. These interface textures superimpose broken patterns onto smectic focal domains and fans, not only in smectic C phases but also in more ordered types of tilted smectic phases such as SmF, SmG and SmH [13, 14].

7.4.5   Origin of Spirals in Chiral Liquid Crystals

Cholesteric layers show spiral contours (Fig. 6 a and b) around polygon centres and vertices (see Fig. 6 a and b). Analogous patterns decorate cholesteric droplets (see Fig. 2). Such spirals have also been found in thin sections of stabilized analogues of cholesterics, i.e. in biological materials assembling long biopolymers such as chitin, in various arthropod carapaces (crabs, insects) [27–33]. To explain these patterns, let us start from cholesteric layers arranged as nested toroidal surfaces (see Fig. 29 b). This situation is shown schematically along a meridian plane in Fig. 38 a; the left-handed twist and three section planes (1–3) are illustrated in Fig. 38 b–d. The hatched zone in Fig. 38 a indicates the part of the drawing to be suppressed in order to obtain a situation similar to that of layers in contact with an isotropic phase of the mesogen at level 1 (Fig. 38 b). Going up Fig. 38 a one passes continuously from a uniform alignment of directors (d) to a double-spiral pattern (c) and a pair of $\lambda^{+\pi}$ disclinations (b). These three steps are assembled into a unique pattern in Fig. 38 e, which indicates how a continuous deformation resembling a whorl produces a disclination pair and a spiral the orientation of which is linked to that of the twist.

Figure 31 d shows how the presence of a $-\pi$ disclination eliminates very acute conical shapes of nested layers. This arrangement of layers creates a rhombic and conical domain in the midpart of most of the polygon edges in Fig. 6 a. Careful examination of the layers under natural light at a level close to the coverslip generally shows a structure like the one shown in Fig. 38 f. The $\lambda^{-\pi}$ disclination forms a helical half-loop [20] due to the cholesteric twist, a situation very similar to that described in other helical cholesteric patterns (see Sec. 7.3.4).

Spirals and concentric circles are common in the focal domains of smectic C* phases [97]. They are produced by lines such as those shown in Fig. 21 c and d. These disclination lines have often been observed in preparations of vertical or slightly tilted layers. The molecules lie almost parallel to the glass plates at their contact, with uniform alignment, whereas the helical arrangement is present in the bulk. Equidistant and parallel disclinations of the type indicated shown in Fig. 21 c and d separate the chiral and non-chiral domains; these have been called 'unwinding lines' [58]. They are present at two levels: in the neighbourhood of the upper and lower plates; and adjoining the thinner regions of the preparation, where they form loops or simple helical lines [97].

When focal domains are differentiated in smectic C* phases, with ellipses attached to a glass, the molecules also lie parallel to the glass, and this prevents the formation of a helical structure, which does, however, occur at a distance in the bulk. Unwinding

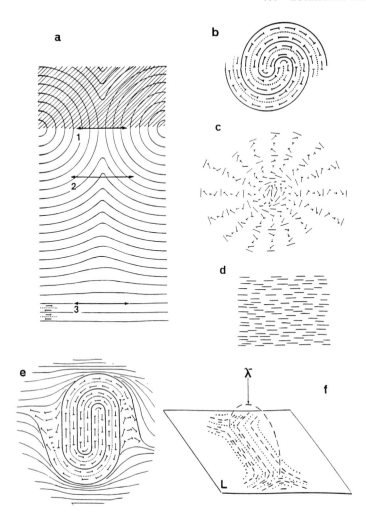

Figure 38. Continuous passage from a focal line to disclination pairs in a cholesteric liquid. (a) Meridian section of the texture corresponding to a three-dimensional representation of the structure shown in Fig. 29b; the nail convention indicates the presence of a left-handed twist. (b–d) Distribution of directors at levels 1–3 in (a), presented with the nail convention. (e) An imagined pattern to show in one picture the continuous passage from a uniform alignment at the periphery to a whorl-like texture, with two $\lambda^{+\pi}$ disclinations in the core. In (b) and (e), the separated segments corresponding to horizontal directors are represented by continuous lines. (f) Helical shape of the $\lambda^{-\pi}$ disclination in Fig. 31d and the lozenges of polygonal edges shown in Fig. 6.

lines therefore appear, and produce patterns the principle of which is closely related to that described above for cholesterics. The main differences are that in smectic C* phases, the distance between two successive lines is p, the helical pitch, whereas in cholesterics the periodicity is $p/2$, and the double spiral in cholesterics is reduced to a single one in smectic C phases. One can also observe concentric circular lines, the successive radii of which differ by p, and this is explained by the presence of a screw line superimposed on the vertical focal line, as in cholesterics (see Fig. 32d).

7.4.6   Defects and Mesophase Growth

Defects, particularly screw dislocations, play an important role in the growth of true crystals [9]. Liquid crystal germs present the defects that were described in detail by Friedel and Grandjean [21] for smectic A phases. The germs elongate perpendicular to the mean direction of the layers, the surface tension being anisotropic, and the focal domains present in the bâtonnet are arranged such that the layers lie normal to the isotropic interface. A focal line is often present along the axis of the bâtonnet, and screw dislocations can be superimposed [54], which could accelerate the transition and facilitate the elongation.

Cholesteric rodlets and spherulites having parallel layers can be devoid of inner defects, but this does not prevent their growth. However, surface points or surface lines are present. Some $-\pi$ disclinations, resulting from germ coalescence, are frequent, but disappear by confluence with the isotropic interface, as for nematic droplets.

Spherulites showing concentric layers present a disclination radius or diameter, but this structure is due to a topological constraint and does not seem to be linked to liquid crystal growth. Very rapid growth of cholesteric phases often generates screw dislocations of the two types shown in Fig. 24 i and j, and this has been filmed by Rault in p-azoxyanisol added to cholesterol benzoate [98, 99]. Slow growth does not result in the production of these defects.

Experiments have also been done at phase transitions from nematic to smectic A phases, starting from a twisted situation between two glasses with a strong planar anchoring, the two easy directions being at right angles (see Fig. 9). Thin threads are present at the junction of twisted zones (left-handed and right-handed). These transform at the smectic transition into a broken helical line, composed of a series of triplets comprising (1) a focal curve, (2) a segment of a $+\pi$ disclination and (3) a focal curve, the two focal segments (1 and 3) being conjugated [100]. Considerations such as those developed for Fig. 30 e show that numerous screw dislocations attach to the $+\pi$ line and form a radiating pattern that is visible under the light microscope, which suggests the presence of important Burgers vectors. Lyotropic lamellar systems often show undulated or helical arrangements of focal domains [101], possibly similar to those of thermotropic phases [100].

7.4.7   Defect Energies and Texture Transformations

Translation dislocations do not strongly modify the orientation of the directors or layers at long distances, as disclinations in liquid crystals do, and focal conics occupy an intermediate position (see Fig. 32 i). This indicates a gradation of energies, which will be considered below using some examples.

7.4.7.1   Disclination Points

Let us begin with simple defects, such as those that occur in capillary tubes (see Fig. 8). In the left-hand part of Fig. 8 a there is a radial point in the vicinity of which the directors are supposed to be aligned radially: $n\,(\mathrm{M}) = r/r$, with $r = \mathrm{OM}$ and r being the positive distance separating M from O, the core of the defect. Then $\mathrm{div}\,n = 2/r$; the twist and bend, whether spontaneous or not, are absent, and the density of the elastic energy is given by

$$\frac{\mathrm{d}F}{\mathrm{d}v} = \frac{2k_{11}}{r^2} \quad \text{or} \quad \mathrm{d}F = 8\pi k_{11}\,\mathrm{d}r$$

Integrating the elastic energy over a sphere centred on O, of radius R, yields

$$F = 8\pi k_{11} R$$

Such a purely radial arrangement of directors is more plausible when twist and bend are forbidden, particularly in smectic drops immersed in an isotropic fluid creating homeotropic conditions. This is realized for some spherical 4-cyano-4'-n-octylbiphenyl (8CB) droplets immersed in water or glycerol (Fig. 39 a), but for most droplets in this domain the spherical symmetry is reduced to revolution symmetry by a focal domain attached at the centre, with smectic layers normal to the interface (Fig. 39 b). This means that interfacial tension depends on the angle of directors at the interface, with

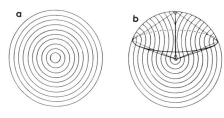

Figure 39. Spherical smectic droplets of 8CB in water or glycerol. (a) Presence of a radial singularity. (b) A focal domain associating a focal radius and a circle. The distribution of the layers is drawn in a meridian plane.

a strong minimum for molecules lying normally and a less marked one for those parallel to this interface with water or glycerol. In nematics, director lines form diverse patterns around singular points, associating splay, twist and bend in different proportions [102]. In a spherical nematic droplet floating in an isotropic medium, with the directors parallel to the interface, two singularities appear at two diametrically opposite poles, with radial arrangements similar to the one shown in Fig. 12 a. For a spherulite of radius R, with $k_{11}=k_{33}=k$, and assuming that the twist is absent, $F=5\pi kR$. This means that the volume energy due to a radial point at the centre of the spherulite is appreciably higher than that of two surface points. This confirms that defects 'prefer' to join an outer interface in order to minimize their energy, and this is reinforced by the fact that nematic molecules generally align with the isotropic interface, rather than lying normally.

7.4.7.2   Disclination Lines

As shown in Fig. 8 b, lateral homeotropy in a capillary tube filled with a smectic A phase results in pure splay that is distributed homogeneously about a disclination line, together with cylindrical symmetry. Note, however, that core rehandlings are observed [42]. If r represents the transverse compo-

nent of OM (O being an arbitrary point of the line), then

$$\operatorname{div} n = \frac{1}{r} \quad \text{and} \quad \frac{dF}{dv} = \frac{k_{11}}{2r^2}$$

$$\text{or} \quad dF = \pi L\left(\frac{k_{11}}{r}\right)dr$$

where L is a given length of the singular line. The energy cannot be integrated from 0 to R, as for the radial point, and a cut-off is required. A radius core R_c that is close to molecular dimensions is introduced; E_c is the core energy per unit length. Hence the line energy density E in the capillary or in a myelin form is

$$E = E_c + \pi k_{11} \ln\left(\frac{R}{R_c}\right)$$

The disclination patterns usually considered in energy calculations in nematics are the solutions of a Laplace equation $\nabla^2\Phi=0$ of the type

$$\Phi = \Phi_0 + \frac{N\varphi}{2} = \Phi_0 + s\varphi$$

(see Sec. 7.2.4; in energy calculations, the topological parameter $s=N/2$ is now generally used rather than N). In this case $k_{11}=k_{22}=k_{33}=k$, and

$$\operatorname{div} n = \cos(\Phi - \varphi)\frac{d\Phi}{rd\varphi}$$

$$(n \times \operatorname{curl} n)^2 = \frac{d\Phi^2}{dr^2 + r^2\,d\varphi^2}$$

from which we deduce

$$2\frac{dF}{dv} = \frac{ks^2}{r^2} \quad \text{or} \quad dF = \frac{k\pi Ls^2\,dr}{r}$$

$$E = E_c + k\pi s^2 \ln\left(\frac{R}{R_c}\right)$$

For $s=1$, we now re-find the energy of the radial disclination of a smectic liquid in a capillary tube. At the nematic transition, this disclination relaxes, with an 'escape in the third dimension' (see Fig. 8 a), the splay lying mainly along the axis and the bend at the periphery. In the absence of twist, one has: $r/R = \tan \theta/2$ (where R is the inner radius of the capillary, r is the distance from point M to the axis, and θ is the angle separating the director at point M from the axis). This means that the director lines have a meridian profile of constant shape and, to a first approximation, the textures in two different tubes of radii R_1 and R_2 differ only by a pure dilation of ratio R_2/R_1. The total deformation remains constant and the energy does not depend on the inner radius R. The energy per unit length is $E = 3 k \pi$, in disclinations $s = +1$, with a continuous core [39–41]. For $s = -1$, $E = k \pi$. The passage from a planar structure of a disclination to a continuous core suppresses the factor $\ln(R/R_c)$, and represents a non-negligible energy saving.

Vertical disclination lines normal to horizontal layers in smectic C phases also form nuclei. Polarizing microscopy shows that these nuclei have four branches, and when examined in projection onto the layer plane the observed patterns correspond to $\Phi = \Phi_0 + s \varphi$, with $s = \pm 1$. It has been demonstrated that the tilt angle of molecules with respect to the normal to layers is variable, but decreases to zero in the vicinity of the disclination core [103]. This also resembles an 'escape in the third dimension', and is mainly due to the low value of the dilatation modulus B.

Interactions of disclinations have been studied in *nematic* and *smectic C schlieren textures*, which associate nuclei interpreted as vertical disclinations. Assuming a purely horizontal distribution of directors and an 'elastic isotropy', within any horizontal plane and in the vicinity of disclination D_i, centred on O_i, we have $\Phi = \Phi_i + s_i \varphi_i$. At any point M in the texture, the director is defined as $\Phi = \Phi_0 + \Sigma s_i \varphi_i$, but this superposition principle requires that $R_c \ll r_{12} \ll r$ [104]. The energy of a disclination pair is given by

$$E = k \pi (s_1 + s_2)^2 \ln\left(\frac{R}{R_c}\right)$$
$$- 2 k \pi s_1 s_2 \ln\left(\frac{r_{12}}{2 R_c}\right)$$

The first term disappears when $s_1 = -s_2$, a common situation in schlieren textures of nematic and smectic C phases. In general, $\Sigma s_i/m \sim 0$, m being the total number of nuclei in the texture [11, 45]. The second term in the energy represents the interaction: disclinations of opposite sign ($s_1 s_2 < 0$) attract because variations in E and r_{12} occur in parallel [45], whereas they repel when they are of like signs ($s_1 s_2 > 0$).

7.4.7.3   Focal Curves

The energy of a focal domain defined by the ellipse E and the corresponding hyperbola branch, extending from one focus of E to infinity, is:

$$E = k_{11} \pi L (1 - e^2) \ln(a/R_c)$$

L being the ellipse perimeter, e its eccentricity and a its half major axis. The energy of focal domains was also calculated for pairs of parabolae [105].

It was indicated in Fig. 32 c that layers along the conics are not exactly conical but present a rounded apex, with a local thickness maximum. Assuming curvature radii very large relative to the layer thickness, θ being the angle between the normal to layer and the focal curve, whose maximum θ_∞ remains small, the energy associatings play and layer dilatation, is per unit length:

$$E = \pi k_{11} (\theta_\infty)^2 \ln(1.46 R/r)$$

with $r \sim 2\lambda/\theta_\infty$, λ being the penetration length $(k_{11}/B)^{1/2}$ [106].

7.4.7.4   Translation Dislocations

All the expressions given above for disclination and focal curve energies are similar to those for screw and edge dislocations in true crystals [9, 15], but the situation is different for translation dislocations in smectics. The presence of a screw dislocation of weak Burgers vector does not introduce any splay and does not really modify the layer thickness even at small distances [16]. The energy is reduced to a core term and therefore interaction terms are absent in the frame of the linear theory. This is not the case for edge dislocations in this approximation [106], and one has the expression:

$$E = E_c + k_{11} b^2/2\lambda R_c$$

with $\lambda^2 = k_{11}/B$, b being the Burgers vector length. Actually, edge dislocations in smectics generally split into parallel disclinations $+\pi$ and $-\pi$, distant of $b/2$ [16] and this introduces a pure splay in domains of nested cylindrical layers, whose energy is: $\pi k_{11} \ln(b/2R_c)/2$. A third term τ_c is to be added and corresponds to the core energy of the $-\pi$, what leads to a more complex expression:

$$E = k_{11} b^2/2\lambda R_c + \pi k_{11} \ln(b/2R_c)/2 + \tau_c$$

When parallel edge dislocations lie in the same plane, their interaction is negligible; but if they lie at different levels, dislocations attract or repel according to their relative positions and orientations [107].

7.4.7.5   Simulations of Defects and Textures

Computer models have been used to simulate the distribution of directors in nematic disclinations and textures, the partial differential equations being replaced by finite difference equations [108, 109]. For instance, in a square lattice, the cell i, the director of which is specified by an angle Φ_i, has four neighbouring cells i'. If the elasticity is isotropic $\nabla^2\Phi = 0$, the interaction energy is $E_i = \Sigma \sin^2(\Phi_{i'} - \Phi_i)$. Differentiating with respect to Φ_i gives the best orientation in cell i (generally a minimum of E_i, not a maximum) if the $\Phi_{i'}$ only differ by a few degrees:

$$\frac{\sin 2\Phi_i}{\cos 2\Phi_i} = \frac{\Sigma_i \sin 2\Phi_i'}{\Sigma_i \cos 2\Phi_{i'}}$$

The total energy over the model is then:

$$E = \sum_{i,i'} \sin^2(\Phi_{i'} - \Phi_i)$$

Note that a knowledge of Φ along the cells at the periphery is generally sufficient to deduce the entire pattern, since it leads to a linear system with an equal number of equations and unknowns, plus a determinant different from zero, as indicated in early work on Dirichlet's problem [110]. It is also possible to start from a given distribution of Φ, with various boundary conditions such as those known to produce disclinations, and to recalculate the different Φ_i averaged over the four neighbouring cells. This leads first to a decrease in E, the process being repeated iteratively towards a complete relaxation. One can verify that a planar $+2\pi$ disclination disjoins into a pair of $+\pi$ disclinations, due to the presence of the s^2 term in the energy: The total energy of two disclinations $s = 1/2$ represents the half-energy of a single disclination $s = 1$. Obviously, there are textures where integer indices are preferred, for many reasons: escape in the third dimension, point singularities, certain anchoring conditions, and topological constraints (such as in smectic C phases) all forbid half-integer values of the s indices.

These methods have been refined to give a constant splay/bend ratio in purely planar systems, and they allow one to reproduce the well-known aspects of disclinations in pure bend or pure splay situations. These models have also been extended to three dimensions by taking the twist into account, and one can re-find the geometries of director lines in capillary tubes or other cases of 'escapes in the third dimension'.

7.4.7.6　Defect Nucleation

Many defects originate from a non-parallel coalescence of germs (Fig. 40). In addition, mechanical constraints generate edge and screw dislocations, or another pattern called 'elementary pinch' (Fig. 41), which is also found in cholesterics and probably at the origin of focal domains in smectics. This pattern seems, therefore, to be essential in texture transformations.

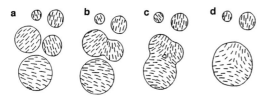

Figure 40. Schematic representation of defects created by the coalescence of nematic germs.

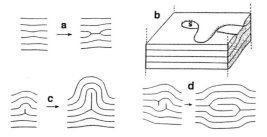

Figure 41. Creation of defects in a lamellar liquid crystal. (a) Edge dislocation due to a layer rupture. (b) Deformation of an edge dislocation into the helical structure of a screw dislocation. (c) Differentiation of an elementary pinch. (d) A pair of elementary pinches transforming into a pair of edge dislocations.

Elementary pinches differentiate from local fluctuations that are able to diverge and produce large defects, if certain constraints are present. In cholesterics, the formation of elementary pinches is continuous, since it involves the nucleation of a $\lambda^- \lambda^+$ pair [48]. These pinches are produced in the same way in nematics submitted to electric fields (see Fig. 5 a in Stieb et al. [111]). In large-pitch cholesterics, the transition from a constrained aligned state to the relaxed twisted state also involves the nucleation of these elementary pinches. The constrained state is due, for example, to strong homeotropic conditions, and is relaxed by a progressive increase in the distance separating the two horizontal plates [112], or by a transverse electric field, if the dielectric anisotropy is negatively uniaxial [113, 114].

Many textures arise from a field or a dynamics created in a previously aligned liquid crystal [115, 116]. Such evolutions are also observed in cholesteric mixtures in equilibrium with the isotropic phase, with fans along the interface, and with polygons and planar textures generally at a greater distance from the isotropic transition. Gentle heating modified the positions of the textures by means of local transformations, but hydrodynamic streams are absent, or almost so. Cholesteric layers, which are initially planar, arrange progressively into domes and basins, and the transformation is continuous, the explanation being derived from Fig. 38 a–e. This also arises from local fluctuations forming directly elementary pinches, which easily transform into spiral patterns.

7.4.7.7　Textures and Defect Associations

Defects form complex networks, but between these singular structures the liquid crystal is regularly curved. The geometry is often well defined, and one may think that a knowledge of the network would be sufficient to rebuild the complete director distribution. Actually, this is a difficult problem, the solution of which depends on its accurate formulation. For instance, two parallel disclinations λ^+ and λ^- can be associated to give either an edge dislocation (as in Fig. 28 d and e) or an elementary pinch (see Fig. 41 b), or very different textures. On the contrary, the presence of two conjugated focal curves defines unambiguously the director distribution within the corresponding domain, with the exception of the exact layer position.

Three types of defect association have been encountered in this discussion: *superposition* of defects, *concurrence* of lines at singular points and *conjugation* between two lines or two points.

Among superpositions, the first case is simply that of lines of identical nature, for instance translation dislocations that join or disjoin, with additive laws for the Burgers vectors. As indicated above, the energy of these dislocations increases as the square of the Burgers vector b^2, and in principle they split into series of elementary dislocations. However, in practice, they are seen to assemble or separate in, for example, cholesterics. This fact underlines the importance of boundary conditions and the presence of other defects in their vicinity. Another common association is that of two $+\pi$ disclinations, which add to form a 2π disclination. The reverse transformation occurs during the course of water evaporation in a myelin form which is squashed between a slide and a coverslip. The dissociation of one line into several lines often occurs at definite points, or in restricted volumes, and these singular points have their own topologies and energies that are rarely considered in theories. Focal curves can branch into pairs and ramify (see Fig. 27 in Friedel and Grandjean [21]). Trees of focal conics have even been photographed in the bulk, near an air–smec-

tic A interface, with several ramification steps, as in fractals [117]. Lines of different nature can also be superimposed. Screw dislocations often run along focal lines (see Fig. 32 d), and disclination lines can play the role of focal curves (see Fig. 33 and other examples in Bouligand [20]).

The main systems of concurrent lines are those of converging hyperbolae at the vertices of polygons in smectic phases and tangent conics (ellipses and hyperbolic edges of polygons), which are often replaced by systems of branching focal curves. Focal curves can attach at definite intervals along disclinations. All the points considered where such sets of lines converge, also have their own topologies and energies, depending on the textural environment.

The vertices of two focal conics are singular, since they correspond to a reverse in the orientation of the materialized conic sheets of Dupin's cyclides. These points often disjoin into pairs of lower energy (see Fig. 31 d and Fig. 11 in Bouligand [53]). At a transition from a smectic A to a smectic C phase disclination lines appear in focal domains, linking the two conics, and are of the type shown in Fig. 21 a and b [118]. Their origin is easily understood from Figs. 14 and 25 in Bouligand and Kléman [57].

Much more work has been concerned with systems of conjugated lines, mainly pairs of focal conics and pairs of disclination lines of opposite indices, forming typical patterns of dislocations or elementary pinches. Such associations also exist between singular points showing complementary topological properties, as occur in the capillary tube (see Figs. 8 a, 13 a and b). Note that certain associations are strictly obligatory. For example, a singular point in the bulk of a nematic liquid, or in a large-pitch cholesteric, always lies along a thick thread, since these threads correspond to the locus of vertical directors (i.e. parallel to the microscope axis), and all possible directions (vertical in particular) are represented in the immediate vicinity of such points. This thick thread is itself vertical in the case in Fig. 8 a for an observer looking from above the capillary tube, and the pattern seen under a polarizing microscope is that of nuclei (see Fig. 4.16 in Meyer [47]).

7.4.7.8   Crystals of Defects

The distribution of defects in mesophases is often regular, owing to their fluidity, and this introduces pattern repeats. For instance, square polygonal fields are frequent in smectics and cholesteric liquids. Such repeats occur on different scales – at the level of structural units or even at the molecular level. Several types of amphiphilic mesophase can be considered as 'made of defects'. In many examples the defect enters the architecture of a unit cell in a three-dimensional array and the mesophase forms a 'crystal of defects' [119]. Such a situation is found in certain cubic phases in water–lipid systems [120] and in blue phases [121]. Several blue phases have been modeled as being cubic centred lattices of disclinations in a 'cholesteric matrix'. Möbius disclinations are assumed to join in groups of 4×4 or 8×8, but in nematics or in large-pitch cholesterics such junctions between thin threads are unstable and correspond to brief steps in recombinations. An isotropic droplet or a Ginsburg decrease to zero of the order parameter probably stabilizes these junctions in blue phases.

Similarly, the recently discovered smectic A* phases [122, 123] are layered like ordinary smectic A phases, but show a regular twist and the director follows the distribution that is classical in cholesteric liquids, whereas bend and twist are forbidden in principle (curl $n = 0$). Layers are disrupted by nearly equidistant twist grain boundaries, each of which is composed of quasi-equidistant and parallel screw dislocations, normal to the helical axis, and well evidenced by freeze–fracture studies [124].

In columnar liquid crystals also, twist and splay are excluded in principle, but twist boundaries are plausible when these systems are strongly chiral [125]. The existence of such cholesteric columnar phases awaits confirmation.

7.5   References

[1] M. J. Press, A. S. Arrott, *J. Phys. Paris* **1975**, *36*, C1-177.
[2] M. Kléman, J. Friedel. *J. Phys. Paris* **1969**, *30*, C4-43.
[3] J. Friedel, M. Kléman in *Fundamental Aspects of Dislocation Theory* (Eds.: J. A. Simmons, R. de Wit, R. Bullough), National Bureau of Standards, Washington, DC, Special Publication 317, No. I, **1970**, pp. 607–636.
[4] Y. Ouchi, K. Ishikawa, H. Takezoe, A. Fukuda, K. Kondo, S. Era, A. Mukoh, *Jpn. J. Appl. Phys.* **1985**, *24* (Supp. 2), 899.
[5] N. A. Clark, T. P. Rieker, *Phys. Rev. A* **1988**, *37*, 1053.
[6] D. Hilbert, S. Cohn-Vossen, *Anschauliche Geometrie (Geometry and the Imagination)*, Chelsea, New York **1990**.
[7] H. S. M. Coxeter, *Introduction to Geometry*, 2nd ed., Wiley, New York **1989**.
[8] N. H. Hartshorne, A. Stuart, *Crystals and the Polarizing Microscope*, 4th edn, Arnold, London **1970**.
[9] S. Amelinckx, *Solid State Phys.* **1964**, Suppl. 6, 1.
[10] O. Lehmann, *Flüssige Kristalle*, Verlag W. Engelmann, Leipzig **1904**.
[11] G. Friedel, *Ann. Phys.* **1922**, *18*, 273.
[12] G. Friedel, *Colloid Chemistry*, Vol. 1, Chemical Catalog Co., **1926**, pp. 102–136.
[13] D. Demus, L. Richter, *Textures of Liquid Crystals*, 2nd edn, VEB Deutscher Verlag, Leipzig **1980**.
[14] G. W. Gray, J. W. G. Goodby, *Smectic Liquid Crystals, Textures and Structures*, Leonard Hill, Philadelphia **1984**.
[15] S. Chandrasekhar, G. S. Ranganath, *Adv. Phys.* **1986**, *35*, 507. S. Chandrasekhar, *Liquid Crystals*, Cambridge University Press, London **1977**.

[16] M. Kléman, *Points, Lines and Walls in Liquid Crystals, Magnetic Systems and Various Ordered Media*, Wiley, Chichester, NY **1983**. M. Kléman, *Rep. Prog. Phys.* **1989**, *52*, 555; *Liq. Cryst.* **1989**, *5*, 399.

[17] D. Chapman (Ed.), *The Structure of Lipids*, Methuen, London **1965**.

[18] J. Nageotte, *Morphologie des Gels Lipoïdes, Myéline, Cristaux Liquides, Vacuoles, Actualités (Sci. Ind., 431/434)*, Hermann, Paris **1936**. F. O. Schmitt, K. J. Palmer, *Cold Spring Harbor Symp. Quant. Biol.* **1940**, 94. D. M. Small, M. C. Bourgès, D. G. Dervichian, *Biophys. Biochim. Acta* **1966**, *125*, 563. F. B. Rosevear, *J. Am. Oil Chem. Soc.* **1954**, *31*, 628.

[19] Y. Bouligand in *Geometry in Condensed Matter Physics*, World Scientific, Singapore **1990**, Chap. 4; *J. Phys. Paris* **1990**, *51*, C7-35.

[20] Y. Bouligand, *J. Phys.* **1972**, *33*, 715. Y. Bouligand in *Dislocations in Solids* (Ed.: F. R. N. Nabarro), North Holland, Amsterdam **1979**, Chap. 23.

[21] G. Friedel, F. Grandjean, *Bull. Soc. Mineral.* **1910**, *33*, 409.

[22] Y. Bouligand, P. E. Cladis, L. Liébert, L. Strzelecki, *Mol. Cryst. Liq. Cryst.* **1974**, *25*, 233.

[23] W. A. Jensen, R. B. Park, *Cell Ultrastructure*, Wadworth, Belmont, CA **1967**. D. W. Fawcett, *The Cell*, 2nd edn, W. B. Saunders, Philadelphia **1981**.

[24] H. Moor, K. Mühlenthaler, *J. Cell. Biol.* **1963**, *17*, 609.

[25] R. R. Balmbra, J. S. Clunie, J. F. Goodman, *Proc. R. Soc. London, Ser. A* **1965**, *285*, 534. J. F. Goodman, J. S. Clunie in *Liquid Crystals & Plastic Crystals*, Vol. 2 (Eds.: G. Gray, P. A. Winsor), Ellis Horwood, New York **1974**, pp. 1–23.

[26] F. Livolant, *Physica* **1991**, *A176*, 117; *J. Mol. Biol.* **1991**, *218*, 165.

[27] Y. Bouligand, *Solid State Phys.* **1978**, Supp. 14, 259.

[28] Y. Bouligand in *Liquid Crystalline Order in Polymers* (Ed.: A. Blumstein), Academic Press, New York **1978**, Chap. 8.

[29] Y. Bouligand, *J. Phys. Paris* **1969**, *30*, C4-90.

[30] S. Caveney, *Proc. R. Soc. London, Ser. B* **1971**, *178*, 205.

[31] M.-M. Giraud-Guille, H. Chanzy, R. Vuong, *J. Struct. Biol.* **1990**, *103*, 232.

[32] M.-M. Giraud-Guille, *Int. Rev. Cytol.* **1996**, *166*, 59.

[33] Y. Bouligand, *Tissue Cell* **1972**, *4*, 189.

[34] Y. Bouligand, *7th Int. Congr. Electron Microscopy, Grenoble* **1970**, *3*, 105.

[35] M.-M. Giraud-Guille, *Calcif. Tissue Int.* **1988**, *42*, 167.

[36] M.-M. Giraud-Guille, *J. Mol. Biol.* **1992**, *224*, 861. L. Besseau, M.-M. Giraud-Guille, *J. Mol. Biol.* **1995**, *251*, 197.

[37] Y. Bouligand in *Physics of Defects* (Eds.: R. Ballian, M. Kléman, J.-P. Poirier), North Holland, Amsterdam **1980**, pp. 780–811.

[38] J. E. Proust, L. Ter Minassian, E. Guyon, *Solid State Commun.* **1972**, *11*, 1227.

[39] C. Williams, P. Pieranski, P. E. Cladis, *Phys. Rev. Lett.* **1972**, *29*, 90.

[40] R. B. Meyer, *Phil Mag.* **1973**, *27*, 405.

[41] A. Saupe, *Mol. Cryst. Liq. Cryst.* **1973**, *21*, 211.

[42] P. E. Cladis, *Phil. Mag.* **1974**, *29*, 641. P. E. Cladis, A. E. White, W. F. Brinkman, *J. Phys. Paris* **1979**, *40*, 325.

[43] A. M. J. Spruijt, *Solid. State Commun.* **1973**, *13*, 1919.

[44] Y. Bouligand, *J. Phys. Paris* **1974**, *35*, 215.

[45] J. Nehring, A. Saupe, *J. Chem. Soc., Faraday Trans. II* **1972**, *68*, 1.

[46] M. Kléman in *Dislocations in Solids* (Ed.: F. R. N. Nabarro), North Holland, Amsterdam **1980**, pp. 5, 245–297, 351–402. G. Ryschenkov, M. Kléman, *J. Chem. Phys.* **1976**, *64*, 413.

[47] R. B. Meyer, *Mol. Cryst. Liq. Cryst.* **1972**, *16*, 335.

[48] Y. Bouligand, *J. Phys.* **1974**, *35*, 959.

[49] Y. Bouligand in *Dislocations in Solids*, Vol. 5 (Ed.: F. R. N. Nabarro), North Holland, Amsterdam **1980**, pp. 301–347.

[50] V. Volterra, *Ann. Ecole Normale Sup.* **1907**, *24*, 401.

[51] A. E. H. Love, *A Treatise on the Mathematical Theory of Elasticity*, Dover, New York 1944.

[52] J. Friedel, *Dislocations, International Series Monographs on Solid State Physics*, No. 3, Pergamon, Oxford **1964**.

[53] Y. Bouligand in *Physics of Defects* (Eds.: R. Ballian, M. Kléman, J.-P. Poirier), North Holland, Amsterdam **1980**, pp. 668–711.

[54] F. C. Frank, *Discuss. Faraday Soc.* **1958**, *25*, 19.

[55] P. G. de Gennes, J. Friedel, *C. R. Heb. Acad. Sci., Paris B* **1965**, *268*, 257.

[56] P. E. Cladis, M. Kléman, *Mol. Cryst. Liq. Cryst.* **1972**, *16*, 1.

[57] Y. Bouligand, M. Kléman, *J. Phys. Paris* **1979**, *40*, 79.

[58] M. Brunet, C. Williams, *Ann. Phys. Paris* **1978**, *3*, 237.

[59] Y. Bouligand, M. Kléman, *J. Phys. Paris* **1970**, *31*, 1041.

[60] J. Rault, *Phil. Mag. A* **1973**, *28*, 11.

[61] M. Kléman in *Dislocations in Solids*, Vol. 5 (Ed.: F. R. N. Nabarro), North Holland, Amsterdam **1980**, pp. 245–297.

[62] C. Robinson, J. C. Ward, R. B. Beevers, *Faraday Soc. Discuss.* **1958**, *25*, 29.

[63] Y. Bouligand, F. Livolant, *J. Phys. Paris* **1984**, *45*, 1899.

[64] R. Cano, *Bull. Soc. Fr. Minér. Cristallogr.* **1967**, *90*, 333; **1968**, *91*, 20.

[65] G. Malet, J. Marignan, O. Parodi, *J. Phys.* **1978**, *9*, 863. G. Malet, J. C. Martin, *J. Phys. Paris* **1979**, *40*, 355.

[66] C. Mauguin, *Bull. Soc. Minéral.* **1911**, *34*, 6, 71.

[67] P. G. de Gennes, *The Physics of Liquid Crystals*, Oxford University Press, Oxford **1974**.

[68] F. Grandjean, *Bull. Soc. Fr. Minéral.* **1919**, *42*, 42.

[69] G. Toulouse, M. Kléman, *J. Phys. Paris* **1976**, *37*, L-149.

[70] V. Poenaru, G. Toulouse, *J. Phys. Paris* **1977**, *38*, 887.

[71] M. Kléman, L. Michel, G. Toulouse, *J. Phys. Paris* **1977**, *38*, L-195.

[72] M. Kléman, *J. Phys. Paris* **1977**, *38*, L-199.

[73] Y. Bouligand, B. Derrida, V. Poenaru, Y. Pomeau, G. Toulouse, *J. Phys. Paris* **1978**, *39*, 863.

[74] M. Kléman, L. Michel, *J. Phys. Paris* **1978**, *39*, L-29; *Phys. Rev. Lett.* **1978**, *40*, 1387.

[75] V. Poenaru in *Ill-Condensed Matter* (Eds.: R. Ballian et al.), *Les Houches XXI Summer School Proc.*, 1978, p. 265.

[76] M. D. Mermin, *Rev. Mod. Phys.* **1979**, *51*, 581.

[77] H. R. Trebin, *Adv. Phys.* **1982**, *31*, 195.

[78] M. V. Kurik, O. D. Lavrentovich, *Sov. Phys. Usp.* **1988**, *31*, 196.

[79] W. F. Harris, *Sci. Am.* **1977**, *237 (6)*, 130.

[80] M. Born, E. Wolf, *Principles of Optics*, 6th edn, Pergamon, Oxford **1980**.

[81] Y. Bouligand, *J. Phys. Paris* **1973**, *35*, 603.

[82] F. Grandjean, *Bull. Soc. Fr. Minéral.* **1916**, *39*, 164.

[83] C. Dupin, *Applications de Géométrie et de Mécanique*, Bachelier, Paris **1822**. J. C. Maxwell, *Q. J. Pure Appl. Maths.* **1867**, 34; *On the Cyclide*, Scientific Papers, Cambridge University Press, Cambridge **1890**, *2*, 144–159. C. M. Jessop, *Quartic Surfaces*, Cambridge University Press, Cambridge **1916**. G. Darboux, *Géométrie Analytique*, Gauthier-Villars, Paris **1917**. H. Bouasse, *Optique Cristalline, Polarisation Rotatoire, Etats Mésomorphes*, Delagrave, Paris **1925**.

[84] W. Bragg, *Trans. Faraday Soc.* **1933**, *29*, 1056. J. P. Meunier, J. Billard, *Mol. Cryst. Liq. Cryst.* **1969**, *7*, 421. M. Kléman, *J. Phys. Paris* **1977**, *38*, 1511.

[85] C. S. Rosenblatt, R. Pindak, N. A. Clark, R. Meyer, *J. Phys. Paris* **1977**, *38*, 1105.

[86] G. Friedel, G. Grandjean, *C. R. Acad. Sci., Paris* **1911**, *152*, 322.

[87] M. de Broglie, E. Friedel, *C. R. Acad. Sci., Paris* **1923**, *176*, 738. E. Friedel, *C. R. Acad. Sci., Paris* **1925**, *180*, 269.

[88] Y. Bouligand, *J. Phys. Paris* **1972**, *33*, 525.

[89] M. Kléman, C. Williams, C. E. Costello, T. Gulik-Krzywicki, *Phil. Mag.* **1977**, *35*, 33.

[90] S. T. Lagerwall, S. Stebler, *Les Houches Summer School Proc.* **1981**, *XXXV*, 757.

[91] R. Bidaux, N. Boccara, G. Sarma, P. G. de Gennes, O. Parodi, *J. Phys. Paris* **1973**, *34*, 661.

[92] J. P. Sethna, M. Kléman, *Phys. Rev. A* **1982**, *26*, 3037.

[93] Y. Bouligand, *J. Microsc. France* **1973**, *17*, 145.

[94] T. P. Rieker, N. A. Clark, G. S. Smith, D. S. Parmar, E. B. Sirota, C. R. Safinya, *Phys. Rev. Lett.* **1987**, *59*, 2658.

[95] N. A. Clark, T. P. Rieker, *Phys. Rev. A* **1988**, *37*, 1053.

[96] D. Johnson, A. Saupe, *Phys. Rev. A* **1977**, *15*, 2079.

[97] L. Bourdon, J. Sommeria, M. Kléman, *J. Phys. Paris* **1982**, *43*, 77.

[98] J. Rault, *Solid State Com.* **1971**, 9, 1965.

[99] J. Rault, *Phil. Mag.* **1973**, *28*, 11.

[100] C. E. Williams, *Phil. Mag.* **1975**, *32*, 313.

[101] M. B. Schneider, W. W. Webb, *J. Phys. Paris* **1984**, *45*, 273.

[102] F. R. N. Nabarro, *J. Phys. Paris* **1972**, *33*, 1089.

[103] S. T. Lagerwall, J. Stebler, *Ordering in Strongly Fluctuating Condensed Matter* (Ed.: T. Riste), Plenum, New York **1972**, p. 383.

[104] C. M. Dafermos, *Q. J. Mech. Appl. Math.* **1970**, *23*, 49.

[105] M. Kléman, *J. Phys. Paris* **1977**, *38*, 1511.

[106] P. G. de Gennes, *C. R. Hebd. Séanc. Acad. Sci., Paris* **1972**, *275B*, 549.

[107] M. Kléman, C. Williams, *J. Phys. Paris, Lettres* **1974**, *35*, L49.

[108] S. E. Bedford, T. M. Nicholson, A. H. Windle, *Liq. Cryst.* **1991**, *10*, 63. S. E. Bedford, A. H. Windle, *Liq. Cryst.* **1993**, *15*, 31.

[109] T. Kimura, D. G. Gray, *Liq. Cryst.* **1993**, *13*, 2330; *Macromolecules* **1993**, *26*, 3455.

[110] H. B. Philips, N. Wiener, *J. Math. Phys. MIT 2nd Ser.*, March 1923.

[111] A. Stieb, G. Baur, G. Meier, *J. Phys. Paris* **1975**, *36-C1*, 185. A. E. Stieb, *J. Phys. Paris* **1980**, *41*, 961.

[112] F. Lequeux, P. Oswald, J. Bechhoefer, *Phys. Rev. A* **1989**, *40*, 3974.

[113] P. Ribière, S. Pierkl, P. Oswald, *Phys. Rev. A* **1991**, *44*, 8198.

[114] F. Lequeux, M. Kléman, *C. R. Hebd. Acad. Sci., Paris* **1986**, *303*, 765; *J. Phys. Paris* **1988**, *49*, 845.

[115] F. Rondelez, J. P. Hulin, *Solid State Commun.* **1972**, *10*, 1009. F. Rondelez, H. Arnould, *C. R. Hebd. Acad. Sci., Paris B* **1971**, *273*, 549.

[116] P. Oswald, J. Behar, M. Kléman, *Phil. Mag. A* **1982**, *46*, 899.

[117] J. B. Fournier, G. Durand, *J. Phys. II France* **1991**, *1*, 845.

[118] A. Perez, M. Brunet, O. Parodi, *J. Phys. Paris* **1978**, *39*, L-353.

[119] W. Helfrich in *Physics of Defects* (Eds.: R. Ballian, M. Kléman, J.-P. Poirier), North Holland, Amsterdam **1980**, Course 12, pp. 716–755.

[120] H. Delacroix, A. Gulik, T. Gulik-Krzywicki, V. Luzzati, P. Mariani, R. Vargas, *J. Mol. Biol.* **1988**, *204*, 165; **1993**, *229*, 526.

[121] S. Meiboom, S. Sammon, D. W. Berreman, *Phys. Rev. A* **1983**, *28*, 3553.

[122] J. W. Goodby, M. A. Waugh, S. M. Stein, E. Chin, R. Pindak, J. S. Patel, *Nature* **1989**, *337*, 449.

[123] G. Strajer, R. Pindak, M. A. Waugh, J. W. Goodby, *Phys. Rev. Lett.* **1990**, *64*, 1545.

[124] K. J. Ihn, J. A. N. Zasadzinski, R. Pindak, *Science* **1992**, *258*, 275.

[125] R. D. Kamien, D. R. Nelson, *Phys. Rev. e, Statistical Physics* **1996**, *53*, 650.

8 Flow Phenomena and Viscosity

F. Schneider and H. Kneppe

A complete assessment of the literature on the hydrodynamics of liquid crystals is beyond the scope of this handbook. Therefore, only the most important fundamentals for the description of flow phenomena will be discussed. Further details can be found in the references cited at the end of each section and in the review articles and books of Porter and Johnson [1], de Gennes [2], Stephen and Straley [3], Jenkins [4], de Jeu [5], Vertogen and de Jeu [6], and de Gennes and Prost [7]. Two review articles of Leslie [8, 9] are exclusively devoted to this subject. Experimental results for uniaxial nematic liquid crystals are described in Vol. 2 A of the recently published *Handbook of Liquid Crystals*.

The hydrodynamic theory for uniaxial nematic liquid crystals was developed around 1968 by Leslie [10, 11] and Ericksen [12, 13] (Leslie–Ericksen theory, LE theory). An introduction into this theory is presented by F. M. Leslie (see Chap. III, Sec. 1 of this Volume). In 1970 Parodi [14] showed that there are only five independent coefficients among the six coefficients of the original LE theory. This LEP theory has been tested in numerous experiments and has been proved to be valid between the same limits as the Navier–Stokes theory. An alternative derivation of the stress tensor was given by Vertogen [15].

In 1971 the Harvard group [16] presented a different form of the constitutive equations which gives very compact expressions for systems with nearly uniform director orientation. The results of both theories agree. Nevertheless, the LEP presentation is preferred in most of the experimental and theoretical studies in this field.

The constitutive hydrodynamic equations for uniaxial nematic calamitic and nematic discotic liquid crystals are identical. In comparison to nematic phases the hydrodynamic theory of smectic phases and its experimental verification is by far less elaborated. Martin et al. [17] have developed a hydrodynamic theory (MPP theory) covering all smectic phases but only for small deformations of the director and the smectic layers, respectively. The theories of Schiller [18] and Leslie et al. [19, 20] for SmC-phases are direct continuations of the theory of Leslie and Ericksen for nematic phases. The Leslie theory is still valid in the case of deformations of the smectic layers and the director alignment whereas the theory of Schiller assumes undeformed layers. The discussion of smectic phases will be restricted to some flow phenomena observed in SmA, SmC, and SmC* phases.

Although the constitutive hydrodynamic equations for nematic and polymeric liquid crystals are identical the following discussion is only devoted to monomeric systems. Polymeric liquid crystals exhibit a great variety of special effects which restrict the application of standard hydrodynamic theory. Reviews on this topic are given by Wissbrun [21] and Marrucci and Greco [22].

8.1 Nematic Liquid Crystals

Most of the hydrodynamic effects observed in nematic liquid crystals can be explained by the LEP theory. Its generality is sufficient for the following discussion. According to this theory the viscous part of the stress tensor σ_{ij} for an incompressible uniaxial nematic liquid crystal is

$$\sigma_{ij} = \alpha_1 n_i n_j V_{kp} n_k n_p + \alpha_2 n_j N_i + \alpha_3 n_i N_j \\ + \alpha_4 V_{ij} + \alpha_5 n_j V_{ik} n_k + \alpha_6 n_i V_{jk} n_k \quad (1)$$

using the notation of Clark and Leslie [23] and usual summation convention. V is the symmetric part of the velocity gradient tensor

$$V_{ij} = \frac{1}{2}(v_{i,j} + v_{j,i}) \quad (2)$$

and N is the rotation of the director n relative to the fluid

$$N_i = \dot{n}_i - \frac{1}{2}(v_{i,j} - v_{j,i})n_j$$

or $\quad N = \dot{n} - \frac{1}{2}(\mathrm{curl}\, v) \times n \quad (3)$

A superposed dot denotes the material time derivative. The Leslie coefficients α_i are assumed to be independent on velocity gradient and time. They depend on temperature and pressure. Four of the coefficients are related one with another via the Parodi equation [14, 24]

$$\alpha_2 + \alpha_3 = \alpha_6 - \alpha_5 \quad (4)$$

which is a consequence of the Onsager relations. Thus, the viscous properties of an incompressible nematic liquid crystal can be described by five independent coefficients.

In the following, we assume a steady and laminar flow of the liquid crystal.

8.1.1 Shear Viscosity Coefficients η_1, η_2, η_3, and η_{12}

Figure 1 shows the system of coordinates which is used in the following discussion. The liquid crystal flows parallel to the x-axis, and the velocity gradient is parallel to the y-axis. The orientation of the director n is described by the angles Φ and Θ

$$n = (\sin\Theta\cos\Phi,\ \sin\Theta\sin\Phi,\ \cos\Theta) \quad (5)$$

Assuming that the velocity is only a function of y, the only non-vanishing components of the velocity gradient tensor are

$$V_{xy} = V_{yx} = \frac{1}{2}v_{x,y} \quad (6)$$

If the orientation of the director is fixed, for example by application of an external magnetic or electric field, Eq. (3) reduces to

$$N = \frac{1}{2}(-\sin\Theta\sin\Phi,\ \sin\Theta\cos\Phi,\ 0)v_{x,y} \quad (7)$$

If a liquid crystal is sheared between two infinite plates parallel to the xz-plane, a constant velocity gradient parallel to the y-axis results. Assuming stationary flow the stress component σ_{xy} can be calculated from

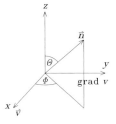

Figure 1. Director orientation with respect to the flow velocity and the velocity gradient.

Eqs. (1) and (5–7).

$$\sigma_{xy} = \Big(\alpha_1 \sin^4 \Theta \sin^2 \Phi \cos^2 \Phi$$
$$-\frac{\alpha_2}{2}\sin^2\Theta\sin^2\Phi + \frac{\alpha_3}{2}\sin^2\Theta\cos^2\Phi$$
$$+\frac{\alpha_4}{2} + \frac{\alpha_5}{2}\sin^2\Theta\sin^2\Phi$$
$$+\frac{\alpha_6}{2}\sin^2\Theta\cos^2\Phi\Big)v_{x,y} \qquad (8)$$

Collecting terms with the same dependence on the director orientation and calculation of the viscosity $\eta = \sigma_{xy}/v_{x,y}$ gives

$$\eta(\Theta,\Phi) = \frac{1}{2}(\alpha_4 + \alpha_5 - \alpha_2)\sin^2\Theta\sin^2\Phi$$
$$+\frac{1}{2}(\alpha_4 + \alpha_6 + \alpha_3)\sin^2\Theta\cos^2\Phi$$
$$+\frac{\alpha_4}{2}\cos^2\Theta$$
$$+\alpha_1\sin^4\Theta\cos^2\Phi\sin^2\Phi \qquad (9)$$

Therefore, it makes sense to introduce a new set of viscosity coefficients (see Fig. 2)

$$\eta_1 = \frac{1}{2}(\alpha_4 + \alpha_5 - \alpha_2)$$
$$\eta_2 = \frac{1}{2}(\alpha_4 + \alpha_6 + \alpha_3)$$
$$\eta_3 = \frac{1}{2}\alpha_4$$
$$\eta_{12} = \alpha_1 \qquad (10)$$

Rewriting Eq. (9) gives

$$\eta(\Theta,\Phi) = \eta_1\sin^2\Theta\sin^2\Phi + \eta_2\sin^2\Theta\cos^2\Phi$$
$$+ \eta_3\cos^2\Theta$$
$$+ \eta_{12}\sin^4\Theta\cos^2\Phi\sin^2\Phi \qquad (11)$$

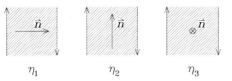

Figure 2. Definition of the shear viscosity coefficients η_1, η_2, and η_3. The outer pair of arrows symbolizes the shear flow.

Several notations for the shear viscosity coefficients are used. The notation used here stems from Helfrich [25]. A different notation was proposed by Miesowicz [26].

The shear viscosity coefficients η_1, η_2, and η_3 can separately be determined in shear flow experiments with adequate director orientations [27].

η_1: $\boldsymbol{n}\|\mathrm{grad}\,v$

η_2: $\boldsymbol{n}\|\boldsymbol{v}$

η_3: $\boldsymbol{n}\perp\boldsymbol{v}$, $\boldsymbol{n}\perp\mathrm{grad}\,v$ $\qquad (12)$

For the determination of η_{12} at least three experiments are necessary. Two for the determination of η_1 and η_2 and one experiment with a director orientation where the influence of η_{12} on the shear viscosity is maximal ($\Theta = 90°$, $\Phi = 45°$). Then, η_{12} can be calculated according to

$$\eta_{12} = 4\eta(90°, 45°) - 2(\eta_1 + \eta_2) \qquad (13)$$

Methods for the determination of the shear viscosity coefficients are described in Vol. 2 A of the recently published *Handbook of Liquid Crystals*.

8.1.2 Rotational Viscosity

Viscous torques are exerted on the director of a liquid crystal during a rotation of the director and by a shear flow with a fixed director orientation. The density Γ_i of the viscous torque is obtained by application of the Levi–Civita tensor

$$\varepsilon_{ijk} = \begin{cases} 1 & i,j,k \quad \text{cyclic} \\ -1 & i,j,k \quad \text{anticyclic} \\ 0 & \text{otherwise} \end{cases} \qquad (14)$$

on the stress tensor. Only the skew symmetric part

$$\sigma_{ij}^{ss} = \frac{\alpha_2 - \alpha_3}{2}(n_j N_i - n_i N_j)$$
$$+\frac{\alpha_5 - \alpha_6}{2}(n_j V_{ik} n_k - n_i V_{jk} n_k) \qquad (15)$$

of the stress tensor (Eq. 1) gives a contribution

$$\Gamma_i = \varepsilon_{ijk}\sigma_{kj}^{ss} = \varepsilon_{ijk}\{(\alpha_2 - \alpha_3)n_j N_k \\ + (\alpha_5 - \alpha_6)n_j V_{kp}n_p\} \qquad (16)$$

or

$$\Gamma = -\mathbf{n} \times (\gamma_1 N + \gamma_2 \mathbf{V}\mathbf{n}) \qquad (17)$$

with

$$\gamma_1 = \alpha_3 - \alpha_2; \quad \gamma_2 = \alpha_6 - \alpha_5 \qquad (18)$$

γ_1 is called rotational viscosity or rotational viscosity coefficient.

At first the viscous torque on a rotating director is calculated. If the liquid crystal is at rest the last term in Eq. (17) vanishes. A director rotation in the xy-plane gives

$$\Gamma_z = -\gamma_1 \dot{\Phi} \qquad (19)$$

Multiplying with the volume of the liquid crystal we obtain the viscous torque

$$M_z = -\gamma_1 \dot{\Phi} V \qquad (20)$$

The torque which has to be applied on the director for this rotation amounts to $-M_z$.

If a liquid crystal at rest is surrounded by a vessel the torque exerted by a rotating magnetic or electric field on the director must be transferred to the vessel.

This torque will be calculated for a completely filled vessel with quadratic cross section. The liquid crystal and the vessel are assumed to be at rest and the director is rotating in the xy-plane (see Fig. 3).

As in this case neither the director and its rotation N nor the velocity gradient tensor

V contains components depending on z, the only non-vanishing stress tensor components are σ_{xy} and σ_{yx} and in these components only terms with the coefficients α_2 and α_3 are non-vanishing. We obtain from Eqs. (3) and (1)

$$N = (-\sin\Phi, \cos\Phi, 0)\dot{\Phi} \qquad (21)$$

and

$$\sigma_{xy} = (-\alpha_2 \sin^2\Phi + \alpha_3 \cos^2\Phi)\dot{\Phi} \\ \sigma_{yx} = (\alpha_2 \cos^2\Phi - \alpha_3 \sin^2\Phi)\dot{\Phi} \qquad (22)$$

Multiplying by the areas a_y and a_x of the vessel gives the forces f_x and f_y

$$f_x = -\sigma_{xy}a_y = (\alpha_2 \sin^2\Phi - \alpha_3 \cos^2\Phi)\dot{\Phi} a_y \\ f_y = -\sigma_{yx}a_x = (-\alpha_2 \cos^2\Phi + \alpha_3 \sin^2\Phi)\dot{\Phi} a_x \qquad (23)$$

There are no forces on the base and the top plate. The total torque is, therefore,

$$M_z = (\alpha_3 - \alpha_2)\dot{\Phi} V \qquad (24)$$

which is in agreement with the torque exerted on the director (Eq. 20). In principle, Eq. (23) allows determination of the Leslie coefficients α_2 and α_3 separately. α_2 corresponds to the force on the surface if the director is perpendicular to it, α_3 corresponds to the force for the parallel case.

The magnitude of the torque does not depend on the shape of the container. It is only a function of the volume of the liquid crystal, that is the number of rotating molecules. The torque on a body with volume V suspended in a liquid crystal has the same magnitude but different sign.

The rotation of the director can be achieved by application of rotating magnetic or electric fields [5, 28]. For larger samples magnetic fields can be applied with less experimental problems. In the following we will therefore consider a liquid crystal in a rotating magnetic field with constant field strength and angular velocity. The results are identical for a rotating sample in a fixed

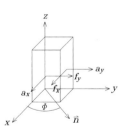

Figure 3. Surface forces on a vessel due to a rotating director.

field if the calculation is carried out with respect to the rotating sample.

The torque exerted by the magnetic field on the director amounts to

$$M = \frac{1}{2}\mu_0\chi_a H^2 V \sin 2(\Psi - \Phi) \tag{25}$$

where

$$\chi_a = \chi_\parallel - \chi_\perp \tag{26}$$

is the anisotropy of the magnetic susceptibility which is assumed to be positive and H is the magnetic field strength. Under stationary conditions and small angular velocities of the field rotation the director follows the magnetic field direction with a constant phase lag $\Psi - \Phi$, see Fig. 4. The viscous and the magnetic torque become equal

$$\gamma_1\dot{\Phi} = \frac{1}{2}\mu_0\chi_a H^2 \sin 2(\Psi - \Phi) \tag{27}$$

Increasing the angular velocity of the rotating field increases the phase lag between field and director until the maximum magnetic torque is exerted at a phase lag of $\pi/4$. The corresponding angular velocity is the critical angular velocity $\dot{\Phi}_c$

$$\dot{\Phi}_c = \frac{\mu_0\chi_a H^2}{2\gamma_1} \tag{28}$$

A further increase of the angular velocity leads to a non-stationary behaviour. The solutions of the corresponding differential equation

$$\dot{\Phi} = \dot{\Phi}_c \sin 2(\dot{\Psi}t - \Phi) \tag{29}$$

are known [29]. Their characteristic feature is a periodicity in time. Because of distur-

bances by the surrounding surfaces the predicted oscillations are usually not observable and an inhomogeneous director rotation results [30].

Measuring the torque on a sample of a nematic liquid crystal in a magnetic field rotating with an angular velocity smaller than the critical one represents a relatively simple method for the determination of the rotational viscosity coefficient. Below the critical angular velocity Eq. (24) is valid with $\dot{\Phi} = \dot{\Psi}$. Neither the phase lag $\Psi - \Phi$ nor the anisotropy of the magnetic susceptibility have to be known.

8.1.3 Flow Alignment

If a nematic liquid crystal with a fixed director orientation is sheared a torque according to Eq. (17) is exerted on the director

$$\boldsymbol{\Gamma} = \boldsymbol{n} \times \left(\frac{1}{2}\gamma_1(\text{curl}\,\boldsymbol{v}) \times \boldsymbol{n} - \gamma_2\boldsymbol{V}\boldsymbol{n}\right) \tag{30}$$

If the direction of flow and shear gradient are the same as in Fig. 1, the torque on the director becomes

$$\boldsymbol{\Gamma} = v_{x,y}\,(\alpha_3 \sin\Theta \cos\Theta \cos\Phi,$$
$$- \alpha_2 \sin\Theta \cos\Theta \sin\Phi,$$
$$\sin^2\Theta\,(\alpha_2 \sin^2\Phi - \alpha_3 \cos^2\Phi)) \tag{31}$$

The following discussion is simpler in a coordinate system which refers to the director (see Fig. 5). The Θ-axis is parallel to $z \times \boldsymbol{n}$ and the Φ-axis is parallel to the $\boldsymbol{n} \times \Theta$ vector.

Figure 4. Director alignment in a rotating magnetic field.

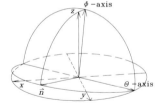

Figure 5. System of coordinates for the director orientation.

The torques with respect to this system of coordinates are

$$\Gamma_\Theta = - v_{x,y} \, (\alpha_2 + \alpha_3) \sin\Theta \cos\Theta \sin\Phi \cos\Phi \tag{32}$$

$$\Gamma_\Phi = v_{x,y} \sin\Theta \, (\alpha_2 \sin^2\Phi - \alpha_3 \cos^2\Phi) \tag{33}$$

If the director is free to rotate there will be either a rotation to a stable orientation (flow alignment, see Fig. 6) or a continuous rotation (tumbling) under the influence of the shear gradient. Which case is observed depends on the signs of α_2 and α_3. Because of thermodynamical arguments (see Sect. 8.1.9) the rotational viscosity coefficient must be positive.

$$\gamma_1 = \alpha_3 - \alpha_2 > 0 \tag{34}$$

That is, $\alpha_2 > 0$ and $\alpha_3 < 0$ is not allowed. All nematic liquid crystals investigated up to now exhibit negative α_2 values. Therefore we will discuss in the following only different signs for α_3.

Some nematic liquid crystals show positive α_3 values in the neighbourhood of a nematic/smectic transition. For positive α_3 the sign of the torque component Γ_Φ does not depend on Φ. This leads to a continuous rotation if the director is orientated in the shear plane ($\Theta = 90°$). The sign of the torque component Γ_Θ depends on Φ, that is, the director is stabilized in the shear plane for two quarters of a revolution and destabilized for the other two quarters. Because of the additional influence of surface alignment and elastic torques the real movement of the director is difficult to predict [31].

Negative α_2 and α_3 values are the most frequent combination for nematic liquid crystals. Under these circumstances no torque is exerted on the director in three orientations

(a) $\Theta = 0°$
(b) $\Theta = 90°$, $\Phi = \Phi_0$
(c) $\Theta = 90°$, $\Phi = -\Phi_0$

where

$$\tan\Phi_0 = +\sqrt{\alpha_3 / \alpha_2} \tag{35}$$

Equations (32) and (33) show that only the orientation (b) is stable against deviations from this orientation. Experiments have shown that $|\alpha_2| \gg |\alpha_3|$ is usually valid. Microscopic theories [32, 33] in which the molecules are assumed to be ellipsoids of revolution predict

$$\frac{\alpha_3}{\alpha_2} \approx \left(\frac{b}{a}\right)^2 \tag{36}$$

where a and b are the length of the ellipsoid parallel and perpendicular to the symmetry axis and a perfect parallel alignment of the molecules ($S=1$) is assumed. Thus, the flow alignment angle is usually small. For a flow aligned liquid crystal the viscosity is given by

$$\begin{aligned}
\eta_0 &= \eta_1 \sin^2\Phi_0 + \eta_2 \cos^2\Phi_0 \\
&= \eta_2 + (\eta_1 - \eta_2) \sin^2\Phi_0 \\
&= \eta_2 - (\alpha_2 + \alpha_3) \sin^2\Phi_0
\end{aligned} \tag{37}$$

neglecting a small term with the coefficient α_1. Inserting Eq. (35) yields

$$\eta_0 = \eta_2 - \alpha_3 \tag{38}$$

As α_3 is negative and usually small as compared with η_2, the viscosity under flow alignment is somewhat larger than η_2. Mostly the difference can be neglected.

Equation (38) is only valid for a uniform director orientation. In a capillary with circular cross section there are regions along the surface and the capillary axis where the orientation of the director disagrees with the

Figure 6. Flow alignment.

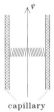

Figure 7. Flow alignment in a capillary.

capillary

flow alignment angle (see Fig. 7 and Sec. 8.1.4). By choosing proper values for radius and flow velocity and by a surface treatment the influence of these regions can be minimized.

Theoretical studies on the flow alignment can be found in the papers of Manneville [34], Carlsson [35], and Zuniga and Leslie [31].

8.1.4   Viscous Flow under the Influence of Elastic Torques

Except for the short discussion at the end of the last section, a uniform director alignment has been assumed up to now. Surface alignment and inhomogeneous fields can lead to an inhomogeneous alignment and the occurrence of elastic torques. For a complete description of hydrodynamics of nematic liquid crystals these elastic torques have to be included.

The equilibrium condition for the elastic torques corresponds to the minimum of the elastic energy under the constraint of constant director length. It is convenient to introduce a molecular field h [36]

$$h = h_S + h_T + h_B \tag{39}$$

where the three terms describe splay, twist and bend distortion of the liquid crystal.

$$h_S = k_{11} \text{grad div} n;$$
$$h_T = -k_{22}\{(n \, \text{curl} \, n)\text{curl} \, n + \text{curl}[(n \, \text{curl} \, n)n]\}$$
$$h_B = k_{33}\{(n \times \text{curl} \, n) \times \text{curl} \, n$$
$$\quad + \text{curl}[n \times (n \times \text{curl} \, n)]\} \tag{40}$$

In hydrostatic equilibrium without electric and magnetic fields the molecular field is parallel to the director and the elastic torque

$$\Gamma^e = n \times h \tag{41}$$

vanishes.

As the values of the three elastic constants are comparable and the expression for the molecular field is rather complicated, the one constant approximation

$$k_{11} = k_{22} = k_{33} = k \tag{42}$$

is often preferred and the molecular field becomes

$$h = k\nabla^2 n \tag{43}$$

Elastic terms have to be taken into account in the equations for the balance of linear and angular momentum. If the body forces are assumed to be conservative their scalar potential can be added to the pressure in the equation for the linear momentum. Nevertheless, as director rotation and shear stress are coupled both equations have to be combined for a solution of a given problem.

At low shear rates viscous torques on the director can be neglected. Then the equations for linear and angular momentum are uncoupled and the orientation of the director in the bulk can be calculated for any given surface orientation by solving the equation for angular momentum. Using the obtained director profile the equation for linear momentum can then be integrated. The apparent viscosity will be a function of the choosen surface orientation and the elastic constants.

At high shear rates the director orientation in the bulk is dominated by flow alignment. The influence of elastic torques is restricted to small boundary layers at the surfaces and regions where the velocity gradient changes sign. Therefore the apparent viscosity is close to the value for flow alignment and does not depend on the surface orientation or elastic constants.

At medium shear rates director orientation and velocity profile can only be obtained by numerical calculations. In the following the flow between parallel plates with a surface alignment parallel to the pressure gradient is calculated [37]. The orientation of the capillary is the same as in Fig. 10. The equation for linear momentum is:

$$v_{x,y} = p_{,x} \frac{y}{g(\Phi)} \qquad (44)$$

where $g(\Phi)$ is the viscosity

$$g(\Phi) = \eta_1 \sin^2 \Phi + \eta_2 \cos^2 \Phi \qquad (45)$$

The small α_1-term has been neglected. The equation for angular momentum is:

$$k\Phi'' + v_{x,y}(g(\Phi_0) - g(\Phi)) = 0 \qquad (46)$$

or with the aid of Eq. (44)

$$k\Phi'' + p_{,x}\left(\frac{g(\Phi_0)}{g(\Phi)} - 1\right) = 0 \qquad (47)$$

where Φ_0 is the flow alignment angle. This equation is solved numerically. The alignment at the surface is assumed to be $\Phi=0$. Figure 8 shows Φ as a function of y/T for $(\eta_1-\eta_2)/\eta_2=8$, $\Phi_0=0.1$, and different values of the non-dimensional parameter

$$D = \frac{T^3 p_{,x}}{k} \qquad (48)$$

where $2T$ is the distance of the plates. The different slopes of the curves for $y=0$ and T are due to the different velocity gradients in these areas.

Integration of the velocity with the aid of Eqs. (44) and (46) gives the apparent viscosity

$$\eta = \frac{g(\Phi_0)}{1 - \frac{3k\Phi'(T)}{T^2 p_{,x}}} \qquad (49)$$

$\Phi'(T)$ (that is, Φ' at the plates) is taken from the numerical solution. Figure 9 shows the viscosity ratio η_2/η as a function of $1/D = k/T^3 p_{,x}$ for $(\eta_1-\eta_2)/\eta_2=8$ and different flow alignment angles. $1/\eta$ gives an asymptotically linear dependence on $1/D$ for large D-values.

Some general predictions can be made with the aid of the scaling properties [19] of the Leslie–Ericksen equations. Neglecting the molecular inertia, the substitution

$$r^* = h\, r; \quad t^* = h^2\, t; \quad p^* = p/h^2;$$
$$\sigma_{ij}^* = \sigma_{ij}/h^2 \qquad (50)$$

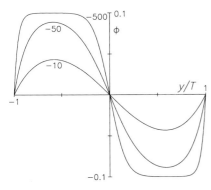

Figure 8. Alignment angle Φ as a function of y/T for different values of the non-dimensional parameter D.

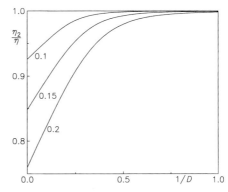

Figure 9. Viscosity ratio η_2/η as a function of $1/D=k/T^3 p_{,x}$ for different flow alignment angles.

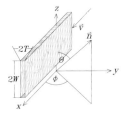

Figure 10. Transverse pressure in a capillary with rectangular cross section.

leaves the form of the differential equations unchanged. The scaling parameter h can be the gap width. Currie [37] has shown that the apparent viscosity is mainly a universal function of a non-dimensional parameter depending on the specific geometry of the flow and a parameter which depends on the surface orientation and the flow alignment angle.

Flow phenomena in the presence of elastic torques have been studied in the Couette [38, 39] and the Poiseuille geometry [40].

8.1.5   Transverse Pressure

Due to the tensor properties of the viscosity a nematic liquid crystal will generally not flow along the pressure gradient. If flow direction and director are neither parallel nor perpendicular a pressure gradient is observed transverse to the flow direction. The skew director orientation can be accomplished by application of a magnetic field to the liquid crystal flowing in a capillary. As an example the flow in a long capillary with rectangular cross section will be discussed (see Fig. 10). The width $2W$ of the capillary is assumed to be considerably larger than its thickness $2T$. Therefore the velocity gradient across the capillary width can be neglected.

The largest transverse pressure difference is observed for a director orientation in the xz-plane ($\Phi=0$). The transverse pressure can be derived from the equation for the conservation of linear momentum

$$\rho\dot{v}_i = (-p\delta_{ij} + \sigma_{ij})_{,j} \qquad (51)$$

Stationary flow and $i=z$ gives

$$-p_{,z} + \sigma_{zx,x} + \sigma_{zy,y} + \sigma_{zz,z} = 0 \qquad (52)$$

The second term vanishes because of the translational symmetry in flow direction and the last term vanishes for an infinite width of the capillary. The remaining stress tensor element amounts to

$$\sigma_{zy} = (\eta_2 - \eta_3)\sin\Theta\cos\Theta\, v_{x,y} \qquad (53)$$

and therefore

$$p_{,z} = (\eta_2 - \eta_3)\sin\Theta\cos\Theta\, v_{x,yy} \qquad (54)$$

The x-component of Eq. (51) becomes

$$-p_{,z} + \sigma_{xx,x} + \sigma_{xy,y} + \sigma_{xz,z} = 0 \qquad (55)$$

For the case studied this leads to

$$p_{,x} = (\eta_2\sin^2\Theta - \eta_3\cos^2\Theta)\, v_{x,yy} \qquad (56)$$

leading to the final result

$$\frac{p_{,z}}{p_{,x}} = \frac{(\eta_2 - \eta_3)\sin\Theta\cos\Theta}{\eta_2\sin^2\Theta + \eta_3\cos^2\Theta} \qquad (57)$$

The transverse pressure gradient passes through a maximum at approximately $\Theta=45°$. A transverse pressure for this case and an angle dependence according to Eq. (57) has been experimentally confirmed [41]. In principle this experiment can be used for the determination of viscosity coefficient ratios. Because of experimental difficulties it should only be used to demonstrate the tensor property of the viscosity of nematic liquid crystals.

If the director is orientated in the shear plane ($\Theta=90°$) a pressure gradient is also generated along the velocity gradient. The conservation of linear momentum gives for $i=y$

$$-p_{,y} + \sigma_{yx,x} + \sigma_{yy,y} + \sigma_{yz,z} = 0 \qquad (58)$$

The second term vanishes for symmetry reasons. The last term is the symmetric counterpart to the third term in Eq. (52), but vanishes under our assumptions (infinite width). Neglecting a small term with the co-efficient α_1 gives

$$p_{,y} = \alpha_6 \sin\Phi \cos\Phi \, v_{x,yy} \qquad (59)$$

and with Eq. (55)

$$p_{,x} = (\eta_1 \sin^2\Phi + \eta_2 \cos^2\Phi) \, v_{x,yy} \qquad (60)$$

the final result becomes

$$\frac{p_{,y}}{p_{,x}} = \frac{\alpha_6 \sin\Phi \cos\Phi}{\eta_1 \sin^2\Phi + \eta_2 \cos^2\Phi} \qquad (61)$$

As the thickness of the capillary is smaller than its width the pressure difference across the thickness is considerably smaller than the difference across the width.

8.1.6   Backflow

An externally applied torque on the director can only be transmitted to the surfaces of a vessel without shear, if the director rotation is homogeneous throughout the sample as assumed in Sect. 8.1.2 for the rotational viscosity. Otherwise this transmission occurs partially by shear stresses. The resulting shear flow is called backflow. As there is usually a fixed director orientation at the surfaces of the sample container, a director rotation in the bulk of the sample by application of a roating magnetic field leads to an inhomogeneous rotation of the director and to a backflow [42].

   We study a nematic layer (Fig. 11) confined between two parallel plates of infinite dimension. The director is assumed to lie in the x,y plane and the pressure is constant. The inertial terms in Eq. (51) can be neglected under usual conditions and the x compo-

nent gives

$$\sigma_{xx,x} + \sigma_{xy,y} + \sigma_{xz,z} = 0 \qquad (62)$$

The first and third term in this equation vanish due to the translational symmetry. Neglecting a term with the usually small coefficient α_1 gives

$$\frac{\partial}{\partial y}(\dot{\Phi}(\alpha_3 \cos^2\Phi - \alpha_2 \sin^2\Phi)$$
$$+ v_{x,y}(\eta_1 \sin^2\Phi + \eta_2 \cos^2\Phi)) = 0 \qquad (63)$$

and integration with respect to y leads to

$$\dot{\Phi}(\alpha_3 \cos^2\Phi - \alpha_2 \sin^2\Phi)$$
$$+ v_{x,y}(\eta_1 \sin^2\Phi + \eta_2 \cos^2\Phi) = \text{const.} \qquad (64)$$

The integration constant may still depend on time and can be determined from the boundary condition at the solid surface or in the middle of the cell. Equation (63) shows that every inhomogeneous director rotation is coupled with a velocity gradient.

   We assume a homeotropic orientation of the director at the solid surfaces and apply a magnetic field in the x,y plane with an orientation nearly parallel to the y axis. The magnetic anisotropy is assumed to be positive and the strength of the magnetic field to be small. Then the deformation will be small and one gets from Eq. (64)

$$-\alpha_2 \dot{\Phi} + \eta_1 v_{x,y} = -\alpha_2 \dot{\Phi}_{\max} \qquad (65)$$

For a common liquid crystal with negative α_2 and positive η_1 a positive angular velocity $\dot{\Phi}$ of the director will be coupled with positive shear gradients at the solid surfaces as the angular velocity vanishes there. This leads to a positive rotation of the bulk and an increase of the angular velocity of the director with respect to a fixed system of coordinates. This effect is usually described by introduction of an effective rotational viscosity coefficient γ_1^* which is smaller than γ_1.

   The exact solution for this and other cases [42, 43] shows that the effective rotation-

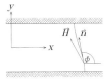

Figure 11. Director orientation under the influence of a magnetic field.

al viscosity coefficient can be as small as 25% of γ_1. Backflow effects are not observed for a planar orientation of the director and a rotation of the magnetic field in the surface plane (twist geometry).

8.1.7   Discotic Liquid Crystals

Nematic discotic liquid crystals consist of disc-like molecules with a preferred parallel orientation of their short molecular (symmetry) axes. There is no translational order. The symmetry of the liquid crystal agrees with that of usual calamitic nematic liquid crystals consisting of rod-like molecules. Accordingly, the stress tensor for calamitic and discotic nematic liquid crystals are identical. Nevertheless, there are some differences in the flow phenomena due to differences in flow alignment.

The flow alignment angle depends on the ratio of the axes of the molecules according to Eq. (36)

$$\tan^2 \Phi_0 \approx \left(\frac{b}{a} \right)^2 \tag{66}$$

The flow alignment angle for discotics should therefore be approximately 90°. A stability analysis shows that the angle above 90° is the stable one [44]. Thus for both types of nematic liquid crystals the configuration with the large dimension nearly parallel to the flow direction is the stable alignment. Figure 12 demonstrates this phenomenon.

8.1.8   Influence of Temperature and Order Parameter on the Viscosity Coefficients

The theory of Leslie and Ericksen is a macroscopic theory. Predictions for the temperature dependence of the viscosity coefficients can only be derived from microscopic theories. Because of several simplifications whose validity is not obvious the results of these microscopic theories are in some cases contradictory and at best qualitatively correct. A general feature is an exponential temperature dependence as known for isotropic liquids with comparable activation energies for all viscosity coefficients far away from the clearing point. This temperature dependence is superposed by an individual dependence on the order parameter whose influence dominates in the neighbourhood of the clearing point. Up to first order all authors predict a linear dependence of the Leslie coefficients α_2, α_3, α_5 as well as α_6 and a quadratic dependence of α_1 [12, 33, 45, 46]. The viscosity coefficient α_4 should be independent of the degree of order. The dependences of the shear viscosities on the order parameter are more complicated. Different results were obtained for the rotational viscosity. However, there are strong arguments that the rotational viscosity vanishes at the clearing point with S^2.

According to most theories the dependence on the order parameter and the gen-

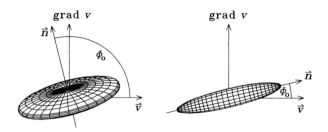

Figure 12. Flow alignment of nematic discotic and a nematic calamitic liquid crystals. The director alignment is shown schematically by the alignment of one molecule.

eral temperature dependence can be separated

$$\alpha_i = f_i[S(T)] \cdot g_i(T) \qquad (67)$$

where $g_i(T)$ is related to a relaxation process in the nematic liquid crystal. If the activation energies for the various relaxation processes are equal the ratio between two Leslie or shear viscosity coefficient should be constant far away from the clearing point which has been observed experimentally.

Kuzuu and Doi [45] obtained the following expressions for the Leslie coefficients

$$\alpha_1 = -2f \frac{p^2 - 1}{p^2 + 1} S_4 \qquad (68)$$

$$\alpha_2 = -f\left(1 + \frac{1}{\lambda}\right) S_2; \quad \alpha_3 = -f\left(1 - \frac{1}{\lambda}\right) S_2 \qquad (69)$$

$$\alpha_4 = f \frac{2}{35} \frac{p^2 - 1}{p^2 + 1}(7 - 5S_2 - 2S_4) \qquad (70)$$

$$\alpha_5 = f\left[\frac{1}{7}\frac{p^2 - 1}{p^2 + 1}(3S_2 + 4S_4) + S_2\right];$$

$$\alpha_6 = f\left[\frac{1}{7}\frac{p^2 - 1}{p^2 + 1}(3S_2 + 4S_4) - S_2\right] \qquad (71)$$

where the common factor f is

$$f = \frac{ckT}{2D_r} \frac{p^2 - 1}{p^2 + 1} \qquad (72)$$

c is the number density of molecules, $p = a/b$ the aspect ratio and D_r is the effective rotational diffusion coefficient. The parameter λ is connected to the flow alignment angle

$$\cos 2\Phi_0 = \frac{1}{\lambda} \qquad (73)$$

and can be calculated from the theory. The expression for λ contains the order parameter S_2. S_2 and S_4 are the averages over the Legendre polynomials of second and forth order of the equilibrium distribution function. S_2 corresponds to the usual order parameter. As S_4 is usually not known Ehren-

traut and Hess [33] suggest the ansatz

$$S_4 = S_2(1 - (1 - S_2)^\nu) \qquad (74)$$

with $\nu = 0.6$.

This and other mesoscopic theories assume that the molecules consist of rigid rods or ellipsoids, that a special form of the interaction potential is valid and that there are no pretransitional phenomena due to a smectic phase. As real liquid crystal molecules do not show such properties these theories only allow a very rough description of the experimental observations. A comparison between some theories has been presented by Kröger and Sellers [47].

8.1.9 Concluding Remarks

The six Leslie coefficients can not be measured directly. They can only be determined with the aid of several experimental methods which ususally lead to combinations of these coefficients. Taking into account the Parodi equation, the six coefficients can be obtained from five linear independent viscosity coefficients. Thus, the four viscosity coefficients η_1, η_2, η_3, and η_{12} and the rotational viscosity coefficient γ_1 give

$$\alpha_1 = \eta_{12}$$

$$\alpha_2 = \frac{1}{2}(\eta_2 - \eta_1 - \gamma_1)$$

$$\alpha_3 = \frac{1}{2}(\eta_2 - \eta_1 + \gamma_1)$$

$$\alpha_4 = 2\eta_3$$

$$\alpha_5 = \frac{1}{2}(3\eta_1 + \eta_2 - 4\eta_3 - \gamma_1)$$

$$\alpha_6 = \frac{1}{2}(\eta_1 + 3\eta_2 - 4\eta_3 - \gamma_1) \qquad (75)$$

Furthermore, Eqs (10), (18) and (35) lead to

$$(\eta_1 - \eta_2)\cos 2\Phi_0 = \gamma_1 \qquad (76)$$

which can be used to introduce the flow alignment angle Φ_0 into Eq. (75) instead of η_1, η_2 or γ_1. Owing to thermodynamical arguments [8] the values of the Leslie coefficients are subject to the conditions

$$\alpha_4 > 0; \quad 2\alpha_1 + 3\alpha_4 + 2\alpha_5 + 2\alpha_6 > 0;$$
$$\gamma_1 > 0; \quad 2\alpha_4 + \alpha_5 + \alpha_6 > 0; \quad (77)$$
$$4\gamma_1(2\alpha_4 + \alpha_5 + \alpha_6) > (\alpha_2 + \alpha_3 + \alpha_6 - \alpha_5)^2$$

or

$$\eta_3 > 0; \quad \eta_{12} + 2(\eta_1 + \eta_2) > \eta_3 + \gamma_1$$
$$\gamma_1 > 0; \quad \gamma_1\{2(\eta_1 + \eta_2) - \gamma_1\} > (\eta_2 - \eta_1)^2$$
$$(78)$$

The fourth condition in Eq. (77) has been omitted in Eq. (78) as it follows from the third and the fifth. Whereas η_1, η_2, η_3, and γ_1 must be positive η_{12} can have either sign.

8.2   Cholesteric Liquid Crystals

Under isothermal conditions the constitutive equations for the description of flow phenomena in nematic and cholesteric liquid crystals are identical [48]. Nevertheless, a series of novel effects are caused by the helical structure of cholesteric phases. They arise firstly because of the inhomogeneous director orientation in the undistorted helix and secondly because of the winding or unwinding of the helix due to viscous torques.

Regarding the orientation of the helical axis with respect to velocity and velocity gradient there are three main cases (see Fig. 13) which are labelled according to the indices of the shear viscosity coefficients of nematic liquid crystals. The helical axis $\boldsymbol{h}$ is assumed to be parallel to the z-axis for the following discussion. $\boldsymbol{\Phi}$ describes the angle between local director $\boldsymbol{n}$ and x-axis (see Fig. 14).

8.2.1   Helical Axis Parallel to the Shear Gradient (Case I)

The cholesteric liquid crystal is sheared between two parallel walls at $z = \pm T$. Leslie [49] and Kini [50] have studied this case with different boundary conditions. Leslie assumes that the pitch of the helix remains constant at the walls whereas Kini assumes a fixed orientation of the director at the wall with an orientation which corresponds to that of the undistorted helix. Both boundary conditions are difficult to realize in an experiment. In spite of the different orientations at the walls the results of the calculations are qualitatively the same. In the bulk the director orientation is determined by elastic and viscous torques. Transverse flow in z-direction is excluded.

The differential equations can analytically be solved for small helix distortions. If the pitch P_0 is comparable to the cell thickness $2T$ the volume velocity and the appar-

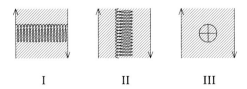

Figure 13. Orientation of the helix with respect to the shear gradient. The pair of arrows symbolizes the shear flow.

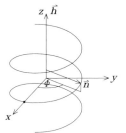

Figure 14. Orientation of the helical axis and the director with respect to the coordinate system.

ent viscosity respectively are oscillating functions of the pitch for constant cell thickness. The reason for this effect is the oscillating mean value of the director orientation. With decreasing pitch these oscillations die out and the apparent viscosity approaches the limiting value

$$\eta(P_0 = 0) = 2\eta_2\eta_3/(\eta_2 + \eta_3) \qquad (79)$$

For larger shear rates the helix will be distorted. The corresponding differential equation can only numerically be solved. The calculations of Leslie and Kini show, that for a given pitch and cell thickness the apparent viscosity starts with a plateau for small shear rates and decreases with increasing shear rates to the viscosity which is observed for flow alignment. The necessary shear rates for this transition depend on the pitch. The smaller the pitch the larger are the necessary shear rates as the resistance against a distortion increases with decreasing pitch.

The influence of static shear deformations has been studied in several papers [51, 52].

8.2.2   Helical Axis Parallel to the Flow Direction (Case II)

The flow of a cholesteric liquid crystal through a narrow capillary with circular cross section according to Fig. 15 is the most simple example for this case. The orientation at the capillary wall should be consistent with the helical structure and the pressure gradient along the capillary axis should be small. Helfrich [53] assumes that the helical structure remains undistorted and a permanent rotation of the director in the flowing liquid crystal results. The director rotation is coupled with a specific energy dissipation of

$$\frac{\dot{E}}{V} = \gamma_1\dot{\Phi}^2 = \gamma_1\left(\frac{2\pi v}{P}\right)^2 \qquad (80)$$

For small pitches and large velocities this term exceeds the energy dissipation by Poiseuille flow and the resulting flow velocity is nearly constant across the cross section (plug flow). The dissipated energy is taken from the pressure gradient along the capillary. This leads to

$$\dot{V} = \frac{R^2 P^2}{4\pi\gamma_1}p_{,z}; \quad \eta_{app} = \frac{\pi^2}{2}\frac{\gamma_1 R^2}{P^2} \qquad (81)$$

In contrast to Poiseuille flow the volume flow depends only on the square of the capillary radius. For $P \ll R$ the apparent viscosity exceeds γ_1 by far. According to Helfrich this type of flow is called permeation. Kini et al. [54] have studied the same problem using the equations of continuum theory. They could show that the velocity profile is almost flat except a small layer near the capillary wall. Neglecting the boundary layer the equations presented above were obtained.

8.2.3   Helical Axis Normal to v and grad v (Case III)

The cholesteric liquid crystal is sheared between two parallel walls. It is assumed that the walls do not influence the director orientation (weak anchoring) and that the flow velocity is parallel to the movement of

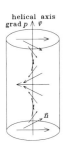

Figure 15. Flow of a cholesteric liquid crystal parallel to the helical axis.

the walls. Under these assumptions the continuum theory gives the differential equation

$$\Phi_{,\tau} = \Phi_{,\xi\xi} + k\left(1 - \frac{\cos^2\Phi}{\cos^2\Phi_0}\right) \qquad (82)$$

where

$$\tau = \frac{k_{22}}{\gamma_1 P_0^2}t; \quad \xi = \frac{z}{P_0}; \quad k = \frac{\alpha_2\, v_{x,y}\, P_0^2}{k_{22}} \qquad (83)$$

are a non-dimensional time τ, position ξ, and a non-dimensional parameter k of the system. k_{22} is the twist elastic constant and P_0 is the pitch of the undistorted helix.

For small k values the helical structure remains undistorted and its pitch constant. Then Eq. (82) allows a simple calculation of the director rotation:

$$\Phi_{,\tau} = k\left(1 - \frac{\cos^2\Phi}{\cos^2\Phi_0}\right) = k\left(1 - \frac{1}{2\cos^2\Phi_0}\right)$$
$$(84)$$

which gives

$$\Phi_{,t} = -\frac{v_{x,y}}{2} \qquad (85)$$

Thus a permanent director rotation results.

The apparent viscosity amounts to

$$\eta_{app} = \frac{1}{2}(\eta_1 + \eta_2) \qquad (86)$$

which agrees with the value for a fixed helix, as there is no rotation of the director with respect to the fluid in both cases.

For larger k values the helix becomes distorted and its pitch increases with k. Finally, the helix unwinds and a flow aligned structure appears.

8.2.4   Torque Generation Under Flow

In the case of geometry II the director of the cholesteric liquid crystal rotates during the flow through a round capillary. According to the section on the rotational viscosity this director rotation is coupled with a torque on the liquid crystal. The torque generation must be compensated by elastic deformations or additional circular flow.

For a flow of a cholesteric liquid crystal through the gap between two coaxial cylinders the torque is partly transferred to the walls of the cylinders. Fischer et al. [55] used a fixed outer cylinder and determined the torque on the inner cylinder.

Equation (20) gives for the torque on the walls

$$M = \gamma_1\dot{\Phi}V = \pm\gamma_1\frac{2\pi}{P}vV \qquad (87)$$

for a sample with volume V.

The experiments could only demonstrate the existence of the effect, as the flow in the experiment was dominated by shear flow and only a minor part (some %) resulted from permeation.

8.3   Biaxial Nematic Liquid Crystals

Shortly after the first observation of a biaxial nematic phase by Yu and Saupe [56] the hydrodynamic theory of the uniaxial nematic phase was extended to the biaxial case in several papers [57–60]. The following description is similar to that given by Saupe [59] as well as Govers and Vertogen [60], but the notation is different.

A biaxial nematic phase is characterized by two orthogonal directors a and b of length 1. Introduction of a third orthogonal vector c of length 1 simplifies the notation of the following equations. According to Saupe, Govers and Vertogen the viscous part of the stress tensor of an incompressible biaxial nematic phase is the sum of the symmetric part

$$\sigma_{ij}^S = \mu_1(a_i a_j V^{bb} + b_i b_j V^{aa})$$
$$+ \mu_2(b_i b_j V^{cc} + c_i c_j V^{bb})$$
$$+ \mu_3(c_i c_j V^{aa} + a_i a_j V^{cc})$$
$$+ \mu_4(a_i b_j + a_j b_i)V^{ab}$$
$$+ \mu_5(b_i c_j + b_j c_i)V^{bc}$$
$$+ \mu_6(c_i a_j + c_j a_i)V^{ca}$$
$$+ \mu_7(a_i b_j + a_j b_i)A^b$$
$$+ \mu_8(b_i c_j + b_j c_i)B^c$$
$$+ \mu_9(c_i a_j + c_j a_i)C^a \tag{88}$$

and the skew symmetric part

$$\sigma_{ij}^{ss} = \mu_7(a_i b_j - a_j b_i)V^{ab}$$
$$+ \mu_8(b_i c_j - b_j c_i)V^{bc}$$
$$+ \mu_9(c_i a_j - c_j a_i)V^{ca}$$
$$+ \mu_{10}(a_i b_j - a_j b_i)A^b$$
$$+ \mu_{11}(b_i c_j - b_j c_i)B^c$$
$$+ \mu_{12}(c_i a_j - c_j a_i)C^a \tag{89}$$

where

$$V_{ij} = \frac{1}{2}(v_{i,j} + v_{j,i}); \quad W_{ij} = \frac{1}{2}(v_{i,j} - v_{j,i}) \tag{90}$$

are the symmetric and the skew symmetric part of the velocity gradient tensor and

$$A_i = \dot{a}_i - W_{ij}a_j; \quad B_i = \dot{b}_i - W_{ij}b_j;$$
$$C_i = \dot{c}_i - W_{ij}c_j \tag{91}$$

are the director rotations relative to the fluid. Furthermore, the following notation is used

$$X^u = X_p u_p; \quad Y_p^u = Y_{pq}u_q; \quad Y^{uv} = Y_{pq}u_q v_p \tag{92}$$

The viscous torque is obtained according to the procedure in Sect. 8.1.2 on the rotational viscosity.

$$\Gamma_i = \varepsilon_{ijk}\sigma_{kj}^{ss} = 2\varepsilon_{ijk}(\mu_7 a_k b_j V^{ab} + \mu_8 b_k c_j V^{bc} +$$
$$\mu_9 c_k a_j V^{ca} + \mu_{10}a_k b_j A^b$$
$$+ \mu_{11}b_k c_j B^c + \mu_{12}c_k a_j C^a) \tag{93}$$

Equations (88) and (89) include the case of the stress tensor for the uniaxial nematic phase [59].

8.3.1 Shear Viscosity Coefficients

For a fixed director orientation and a flow according to Fig. 16 the velocity gradients and the director rotations relative to the fluid are

$$V_{xy} = V_{yx} = W_{xy} = -W_{yx} = \frac{1}{2}v_{x,y} \tag{94}$$

$$A_i = -W_{ij}a_j; \quad B_i = -W_{ij}b_j; \quad C_i = -W_{ij}c_j \tag{95}$$

and the viscosity $\eta = \sigma_{xy}/v_{x,y}$ becomes

$$\eta = \frac{1}{2}\{a_x^2 b_y^2(\mu_4 + 2\mu_7 + \mu_{10})$$
$$+ b_x^2 c_y^2(\mu_5 + 2\mu_8 + \mu_{11})$$
$$+ c_x^2 a_y^2(\mu_6 + 2\mu_9 + \mu_{12})$$
$$+ a_y^2 b_x^2(\mu_4 - 2\mu_7 + \mu_{10})$$
$$+ b_y^2 c_x^2(\mu_5 - 2\mu_8 + \mu_{11})$$
$$+ c_y^2 a_x^2(\mu_6 - 2\mu_9 + \mu_{12})$$
$$+ 2a_x a_y b_x b_y(2\mu_1 + \mu_4 - \mu_{10})$$
$$+ 2b_x b_y c_x c_y(2\mu_2 + \mu_5 - \mu_{11})$$
$$+ 2c_x c_y a_x a_y(2\mu_3 + \mu_6 - \mu_{12})\} \tag{96}$$

In principle, the director components can be expressed in terms of Eulerian angles. However, the resulting expression is rather lengthy and the cyclic structure of Eq. (96) is lost.

Director orientations with all directors parallel to the coordinate axes are described by the first six terms, for example, the orientation $a\|x$ and $b\|y$ corresponds to the

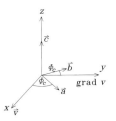

Figure 16. Director orientation in a biaxial nematic phase with respect to the flow velocity and the velocity gradient. The case $c\|z$ is shown whereas the equations in the text correspond to the general case.

viscosity coefficient $\frac{1}{2}(\mu_4 - 2\mu_7 + \mu_{10})$. The last three terms give a maximum contribution if two directors lie under $45°$ in the xy plane, i.e. they correspond to the α_1 term of uniaxial nematics. Thus, nine coefficient combinations can be determined with the aid of flow experiments.

8.3.2 Rotational Viscosity Coefficients

The last three terms in the torque Eq. (93) describe the torques due to director rotations. The three main cases are rotations around a, around b, and around c. The latter corresponds to $\dot{\Phi}_c \neq 0$ in Fig. 16. The viscous torques for these cases are

$$\Gamma_a = -2\mu_{11}\dot{\Phi}_a; \quad \Gamma_b = -2\mu_{12}\dot{\Phi}_b;$$
$$\Gamma_c = -2\mu_{10}\dot{\Phi}_c \tag{97}$$

where Φ_i describes the rotation around the i axis.

The three rotational and the nine shear viscosity coefficients according to Eq. (96) represent a complete set of coefficients.

8.3.3 Flow Alignment

It is assumed that the c director is parallel to the z axis, e.g. by application of a magnetic field. A shear flow $v_{x,y}$ exhibits a torque on the directors around the z axis. Calculation with Eq. (93) gives

$$\Gamma_c = -2\mu_{10}\dot{\Phi}_c - v_{x,y}(\mu_7 \cos 2\Phi_c + \mu_{10}) \tag{98}$$

$\Gamma_c = 0$ and $\dot{\Phi}_c = 0$ give

$$\cos 2\Phi_c = -\frac{\mu_{10}}{\mu_7} \tag{99}$$

This equation only has solutions for

$$\left|\frac{\mu_{10}}{\mu_7}\right| < 1 \tag{100}$$

Correspondingly, the flow alignment angles around all axes are

$$\cos 2\Phi_a = -\frac{\mu_{11}}{\mu_8}; \quad \cos 2\Phi_b = -\frac{\mu_{12}}{\mu_9};$$
$$\cos 2\Phi_c = -\frac{\mu_{10}}{\mu_7} \tag{101}$$

The conditions for the first two viscosity coefficient ratios correspond to Eq. (100). A discussion of the stability of the various solutions is presented in the paper of Saupe [59]. Brand and Pleiner [61] as well as Leslie [62] discuss the flow alignment without the restriction that one director is perpendicular to the shear plane.

8.4 SmC Phase

The hydrodynamic theory for SmC phases was developed by Schiller [18] and Leslie et al. [19]. Schiller as well as Leslie et al. assume a constant layer thickness, a constant tilt angle and do not include the permeation, that is, the dissipation of energy by the penetration of molecules through the layers. Furthermore, Schiller neglects the distorsion of the layers in contrast to Leslie et al. The Leslie theory is, therefore, somewhat more general. The orientation of the SmC phase is described by two directors. The director a is normal to the layers and c is parallel to the layers in the direction of the tilt. The a director, the preferred direc-

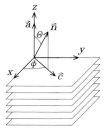

Figure 17. Definition of the directors a and c for an SmC phase.

tion of the long molecular axes, $\boldsymbol{n}$, and $\boldsymbol{c}$ lie, therefore, in a plane (Fig. 17). According to the theory of Leslie et al. [19] the viscous part of the stress tensor is the sum of the symmetric part

$$
\begin{aligned}
\sigma_{ij}^s = {} & \mu_0 V_{ij} + \mu_1 V^{aa} a_i a_j + \mu_2 (V_i^a a_j + V_j^a a_i) \\
& + \mu_3 V^{cc} c_i c_j + \mu_4 (V_i^c c_j + V_j^c c_i) \\
& + \mu_5 V^{ac}(a_i c_j + a_j c_i) + \lambda_1 (A_i a_j + A_j a_i) \\
& + \lambda_2 (C_i c_j + C_j c_i) + \lambda_3 A^c(a_i c_j + a_j c_i) \\
& + \kappa_1 (V_i^a c_j + V_j^a c_i + V_i^c a_j + V_j^c a_i) \\
& + \kappa_2 (V^{aa}(a_i c_j + a_j c_i) + 2 V^{ac} a_i a_j) \\
& + \kappa_3 (V^{cc}(a_i c_j + a_j c_i) + 2 V^{ac} c_i c_j) \\
& + \tau_1 (C_i a_j + C_j a_i) + \tau_2 (A_i c_j + A_j c_i) \\
& + 2\tau_3 A^c a_i a_j + 2\tau_4 A^c c_i c_j \qquad (102)
\end{aligned}
$$

and the skew symmetric part

$$
\begin{aligned}
\sigma_{ij}^{ss} = {} & \lambda_1 (V_j^a a_i - V_i^a a_j) + \lambda_2 (V_j^c c_i - V_i^c c_j) \\
& + \lambda_3 V^{ac}(a_i c_j - a_j c_i) + \lambda_4 (A_j a_i - A_i a_j) \\
& + \lambda_5 (C_j c_i - C_i c_j) + \lambda_6 A^c(a_i c_j - a_j c_i) \\
& + \tau_1 (V_j^a c_i - V_i^a c_j) + \tau_2 (V_j^c a_i - V_i^c a_j) \\
& + \tau_3 V^{aa}(a_i c_j - a_j c_i) \\
& + \tau_4 V^{cc}(a_i c_j - a_j c_i) \\
& + \tau_5 (A_j c_i - A_i c_j + C_j a_i - C_i a_j) \qquad (103)
\end{aligned}
$$

The notation is the same as described in Sect. 8.3. The stress tensor contains 20 independent coefficients. It fulfills the Onsager relations.

The viscous torque is obtained according to the procedure used for nematic liquid crystals

$$
\begin{aligned}
\Gamma_i = \varepsilon_{ijk}\sigma_{kj}^{ss} = {} & -2\varepsilon_{ijk}\{a_j(\lambda_1 V_k^a + \lambda_3 c_k V^{ac} \\
& + \lambda_4 A_k + \lambda_6 c_k A^c + \tau_2 V_k^c \\
& + \tau_3 c_k V^{aa} + \tau_4 c_k V^{cc} + \tau_5 C_k) \\
& + c_j(\lambda_2 V_k^c + \lambda_5 C_k + \tau_1 V_k^a \\
& + \tau_5 A_k)\} \qquad (104)
\end{aligned}
$$

8.4.1 Shear Flow with a Fixed Director Orientation

The xy plane of the system of coordinates (Fig. 18) used for the following calculation is assumed to be parallel to the smectic

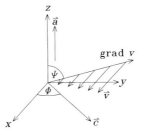

Figure 18. Definition of angles for a flow experiment in the SmC phase.

layers. The $\boldsymbol{a}$ director and the z axis are, therefore, parallel to one another. As a flow perpendicular to the layers is largely suppressed due to the permeation effect, the flow has to be parallel to the xy plane. Furthermore, it can be assumed without loss of generality for this calculation that the flow is parallel to the x axis. The $\boldsymbol{c}$ director lies in the xy plane and its direction is described by the angle Φ. The velocity gradient lies in the yz plane and is described by the angle Ψ.

Under these conditions the effective viscosity calculated from the stress tensor Eqs. (102) and (103) becomes

$$
\begin{aligned}
\eta = {} & \Big\{\tfrac{1}{2}(\mu_0 + \mu_4 - 2\lambda_2 + \lambda_5)\sin^2 \Phi \\
& + \tfrac{1}{2}(\mu_0 + \mu_4 + 2\lambda_2 + \lambda_5)\cos^2 \Phi \\
& + \mu_3 \sin^2 \Phi \cos^2 \Phi\Big\}\sin^2 \Psi \\
& + \Big\{\tfrac{1}{2}(\mu_0 + \mu_2 - 2\lambda_1 + \lambda_4)\sin^2 \Phi \\
& + \tfrac{1}{2}(\mu_0 + \mu_2 + \mu_4 + \mu_5 - 2\lambda_1 + 2\lambda_2 \\
& \quad - 2\lambda_3 + \lambda_4 + \lambda_5 + \lambda_6)\cos^2 \Phi\Big\}\cos^2 \Psi \\
& + \Big\{\kappa_1 - \tau_1 - \tau_2 + \tau_5 + 2(\tau_3 - \tau_4)\cos^2 \Phi\Big\} \\
& \cdot \sin \Phi \sin \Psi \cos \Psi \qquad (105)
\end{aligned}
$$

We define, therefore, a new set of viscosity coefficients

$$
\eta_1 = \tfrac{1}{2}(\mu_0 + \mu_4 - 2\lambda_2 + \lambda_5)
$$

$$
\eta_2 = \tfrac{1}{2}(\mu_0 + \mu_4 + 2\lambda_2 + \lambda_5)
$$

$$
\eta_3 = \mu_3
$$

$$\eta_4 = \frac{1}{2}(\mu_0 + \mu_2 - 2\lambda_1 + \lambda_4)$$

$$\eta_5 = \frac{1}{2}(\mu_0 + \mu_2 + \mu_4 + \mu_5$$
$$\quad - 2\lambda_1 + 2\lambda_2 - 2\lambda_3 + \lambda_4 + \lambda_5 + \lambda_6)$$

$$\eta_6 = \kappa_1 - \tau_1 - \tau_2 + \tau_5$$

$$\eta_7 = 2(\tau_3 - \tau_4) \tag{106}$$

and get finally

$$\eta = (\eta_1 \sin^2\Phi + \eta_2 \cos^2\Phi$$
$$\quad + \eta_3 \sin^2\Phi \cos^2\Phi) \sin^2\Psi \tag{107}$$
$$\quad + (\eta_4 \sin^2\Phi + \eta_5 \cos^2\Phi) \cos^2\Psi$$
$$\quad + (\eta_6 + \eta_7 \cos^2\Phi) \sin\Phi \sin\Psi \cos\Psi$$

The flow properties of a SmC phase with fixed director orientation and a flow parallel to the layers can, therefore, be described by seven independent viscosity coefficients. The experimental determination of these coefficients should be connected with a series of problems. If the coefficients η_4 or η_5 are determined in a capillary with a rectangular cross section with $T \ll W$ and the layer parallel to one of the plates as in Fig. 20 I. The thickness has to be constant over the whole sample with an accuracy that is not easy achieved [63,64]. There are similar problems in the measurement of the other coefficients. Minor difficulties should occur in a shear experiment with a small lateral movement of one of the plates.

8.4.2 Rotational Viscosity

Equation (104) allows to calculate the viscous torque which is exerted on the c director during a rotation around the a director with a constant tilt angle Θ. It is assumed that the liquid crystal is at rest. The only non-vanishing term in Eq. (104) is that with the coefficient λ_5.

$$\Gamma_z = -2\lambda_5 \varepsilon_{zjk} c_j C_k = -2\lambda_5 (c_x C_y - c_y C_x)$$
$$\quad = -2\lambda_5 \dot{\Phi} \tag{108}$$

The viscous dissipation requires

$$\lambda_5 > 0 \tag{109}$$

It is often interesting to compare the rotational viscosity coefficient γ_1 in the nematic phase and

$$\gamma_1^c = 2\lambda_5 \tag{110}$$

in the SmC phase. Then it has to be taken into account that $\dot{n}$ and $\dot{c}$ are different and that the torque on c is only the projection of the torque on n. A rotational viscosity coefficient suitable for a comparison with the coefficient γ_1 in the nematic phase is, therefore

$$\gamma_1^{c*} = \gamma_1^c / \sin^2\Phi \tag{111}$$

A magnetic field rotating in the xy plane leads to a rotation of the c director. The calculation is similar to that in the nematic phase. However, it has to be taken into account that the interaction with the magnetic field is different for the two phases. Principally, the magnetic susceptibility of an SmC phase is biaxial with the main susceptibilities parallel to n (χ_1), perpendicular to n in the layer plane (χ_2) and perpendicular to these two directions (χ_3). In a good approximation $\chi_2 = \chi_3$ can be assumed and the magnetic torque on the c director becomes

$$\Gamma = \frac{1}{2}\mu_0 \chi_a H^2 \sin^2\Theta \sin 2(\gamma - \Phi) \tag{112}$$

where γ is the angle between field direction and x axis and

$$\chi_a = \chi_1 - \chi_2 \tag{113}$$

In the biaxial case $\chi_1 \sin^2\Theta + \chi_3 \cos^2\Theta - \chi_2$ has to be used instead of $\chi_a \sin^2\Theta$. As in the nematic phase there is a critical velocity

$$\dot{\Phi}_c = \frac{\mu_0 \chi_a H^2 \sin^2\Theta}{4\lambda_5} \tag{114}$$

up to which the director follows the field rotation with a phase lag. λ_5 can be determined from the torque on the sample

$$M_z = 2\lambda_5 \dot{\gamma} V \tag{115}$$

8.4.3 Flow Alignment

The SmC phase is sheared between two parallel plates of infinite dimension. The distance between the plates is so large that the influence of the surface alignment at the plates can be neglected in the bulk. The orientation of the directors, the velocity and the velocity gradient is the same as already discussed for the shear flow experiment (Fig. 18). Furthermore, a rotation of the director c is now allowed. Calculation of the torque according to Eq. (104) and $\Gamma_z=0$ gives

$$2\lambda_5 \dot{\Phi} + \{(\lambda_5 + \lambda_2\cos 2\Phi)\sin\Psi \tag{116}$$
$$+ (\tau_5 - \tau_1)\sin\Phi\cos\Psi\} \, v_{,q} = 0$$

where $v_{,q}$ is the velocity gradient. For $\psi=0$ this equation gives

$$2\lambda_5 \dot{\Phi} + (\tau_5 - \tau_1)\sin\Phi \, v_{x,z} = 0 \tag{117}$$

Time independent solutions of this equation are

$$\Phi = 0; \quad \Phi = \pi \tag{118}$$

Assuming $v_{x,z}>0$, a stable flow alignment is obtained at $\Phi=0$ for $\tau_5 > \tau_1$ and at $\Phi=\pi$ for $\tau_5 < \tau_1$. For $\psi=90°$ Eq. (116) gives

$$2\lambda_5 \dot{\Phi} + (\lambda_5 + \lambda_2\cos 2\Phi)v_{x,y} = 0 \tag{119}$$

λ_5 is always positive (see Eq. 109). For $\lambda_5 > |\lambda_2|$ there is no flow alignment. For $\lambda_5 < |\lambda_2|$ a time independent solution is

$$\cos 2\Phi_0 = -\frac{\lambda_5}{\lambda_2} \tag{120}$$

which gives four solutions for Φ_0. The stable solutions are found by an expansion of Eq. (119) around the flow alignment angle ($\Phi=\Phi_0+\Delta\Phi$) for small deviations. This

leads to

$$\lambda_2 v_{x,y} \sin 2\Phi_0 < 0 \tag{121}$$

Assuming $v_{x,y}>0$ and $\lambda_2>0$ gives a stable flow alignment for $\sin 2\Phi_0<0$ and vice versa.

An extensive discussion of the flow alignment in SmC phases can be found in a paper of Carlsson et al. [65].

The viscous properties of a SmC phase with a fixed orientation of the layers can be described by nine independent coefficients. Seven coefficients can be determined with the aid of flow experiments (see Eq. 107). The investigation of the flow alignment including the relaxation time gives the three combinations λ_5, λ_2, and $\tau_5-\tau_1$. λ_2 can, however, be determined from

$$\eta_2 - \eta_1 = 2\lambda_2 \tag{122}$$

The nine coefficients can, therefore, be determined from the seven viscosity coefficients, the rotational viscosity coefficient $2\lambda_5$ and the relaxation time for the flow alignment for $\psi=0$. Instead of the two last determinations it is also possible to use two flow alignment angles at different ψ values.

A mesoscopic theory for the SmC phase has been presented by Osipov and Terentjev [66].

8.4.4 SmC* Phase

The viscous part of the stress tensor for the SmC and the ferroelectric chiral smectic (SmC*) phase agree with one another. The flow phenomena with a fixed director orientation discussed in the foregoing section can not be observed due to the inhomogeneous director orientation in the SmC* phase. However, there is a large interest in rotational movements of the director in ferroelectric displays.

In surface stabilized ferroelectric liquid crystal cells the liquid crystal is sandwiched

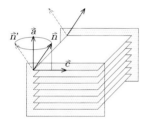

Figure 19. Director alignment in a ferroelectric liquid crystal cell.

between two glass plates which are treated to give the book shelf alignment shown in Fig. 19 with two stable director orientations at the glass plates. Due to the alignment of the director at the glass plates and due to the short distance of the glass plates the helical structure of the SmC* phase is suppressed and the orientation of the c director is nearly homogeneous within the cell. The Leslie theory assumes a rotational movement on the cone during the switching from n to n'. Other trajectories have also been discussed. Carlsson and Žekš [67] found out that the viscosities for a movement of the n director on the cone and perpendicular to it are identical. Experiments [68] have shown that the movement occurs on the cone apart from a small deviation due to the electroclinic effect.

8.5   SmA Phase

The stress tensor for the SmA phase can be obtained from the two director theory for the SmC phase by elimination of all terms containing the c director. The sum of the symmetric and the skew symmetric part becomes

$$\sigma_{ij} = \mu_0 V_{ij} + \mu_1 V_p^{aa} a_i a_j + \mu_2 (V_i^a a_j + V_j^a a_i)$$
$$+ \lambda_1 (A_i a_j + A_j a_i + V_j^a a_i - V_i^a a_j)$$
$$+ \lambda_4 (A_j a_i - A_i a_j) \qquad (123)$$

and the viscous torque

$$\Gamma_i = -2\varepsilon_{ijk} a_j (\lambda_1 V_k^a + \lambda_4 A_k) \qquad (124)$$

A rearrangement gives

$$\sigma_{ij} = \mu_1 a_i a_j V_{kp} a_k a_p + (\lambda_1 - \lambda_4) a_j A_i$$
$$+ (\lambda_1 + \lambda_4) a_i A_j + \mu_0 V_{ij} \qquad (125)$$
$$+ (\mu_2 - \lambda_1) a_j V_{ik} a_k + (\mu_2 + \lambda_1) a_i V_{jk} a_k$$

and

$$\Gamma = -2a \times (\lambda_4 A + \lambda_1 V a) \qquad (126)$$

As the symmetry of the SmA and the nematic phase agree, the structure of the stress tensors is identical and the connection between the coefficients is

$$\alpha_1 \cong \mu_1 \qquad \alpha_4 \cong \mu_0 \qquad \gamma_1 \cong 2\lambda_4$$
$$\alpha_2 \cong \lambda_1 - \lambda_4 \qquad \alpha_5 \cong \mu_2 - \lambda_1 \qquad \gamma_2 \cong 2\lambda_1$$
$$\alpha_3 \cong \lambda_1 + \lambda_4 \qquad \alpha_6 \cong \mu_2 + \lambda_1 \qquad (127)$$

Up to now the influence of the permeation effect has been neglected and the Eqs. (123–126) give only meaningful results if the flow velocity component perpendicular to the layers can be neglected.

In the other case the permeation effect, that is, an energy dissipation for a flow perpendicular to the layers, has to be taken into account. For this purpose the layer normal is assumed to be parallel to the z axis (Fig. 20) and the stress tensor becomes

$$\sigma_{ij} = \mu_1 V_{zz} \delta_{iz} \delta_{jz} + (\lambda_1 - \lambda_4) A_i \delta_{jz}$$
$$+ (\lambda_1 + \lambda_4) A_j \delta_{iz} + \mu_0 V_{ij} \qquad (128)$$
$$+ (\mu_2 - \lambda_1) V_{iz} \delta_{jz} + (\mu_2 + \lambda_1) V_{jz} \delta_{iz}$$

According to Helfrich [53], de Gennes [69], and Martin et al. [17] the permeation can be described with the additional term

$$p_z = \frac{1}{\zeta} (\dot{z} - v_z) \qquad (129)$$

in the equation of motion, $(v_z - \dot{z})$ is the flow velocity with respect to the layer structure.

In the following the flow of a SmA phase between two infinite plates will be considered. In contrast to the discussion of nematic phases these plates will be assumed to be at rest as moving plates are not always compatible with the layered structure. The three main cases are outlined in Fig. 20.

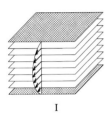

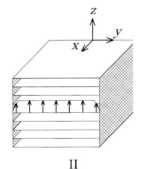

II

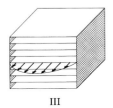

III

Figure 20. Flow of an SmA phase between two parallel plates (hatched). The layer normal is parallel to the z-axis.

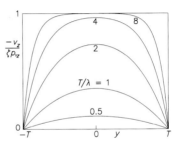

Figure 21. Velocity profile for a flow with permeation (case II of Fig. 20).

For the cases I and III parabolic velocity profiles as known for isotropic liquids result. The viscosity coefficients are

case I: $\quad \eta_1 = \frac{1}{2}(\mu_0 + \mu_2 - 2\lambda_1 + \lambda_4)$

case III: $\eta_3 = \frac{1}{2}\mu_0$ (130)

For case II the permeation has to be taken into account.

$$-p_{,z} + \eta_2 v_{z,yy} - \frac{1}{\zeta}v_z = 0 \qquad (131)$$

with

$$\eta_2 = \frac{1}{2}(\mu_0 + \mu_2 + 2\lambda_1 + \lambda_4) \qquad (132)$$

Integration gives the velocity profile

$$v_z = -\zeta p_{,z}\left(1 - \frac{\cosh y/\lambda}{\cosh T/\lambda}\right) \qquad (133)$$

where

$$\lambda = \sqrt{\eta_2 \zeta} \qquad (134)$$

is the thickness of the shear dominated part and $2T$ is the distance of the plates. Thus $T/\lambda \gg 1$ gives a plug flow and $T/\lambda \ll 1$ gives a shear dominated velocity profile (Fig. 21).

The viscosity coefficient μ_1 can only be determined in a flow experiment where the flow is neither parallel nor perpendicular to the layer normal.

The elastic behaviour of the smectic layers, the permeation effect and the viscous flow lead to a great variety of phenomena in SmA phases: effects due to the compression of the layers [70, 71], flow behind obstacles [69, 72], and instabilities [73, 74].

8.6 References

[1] R. S. Porter, J. F. Johnson in *Rheology, Theory and Applications* (Ed.: F. R. Eirich), Academic Press, New York **1967**.
[2] P. G. de Gennes, *The Physics of Liquid Crystals*, Clarendon Press, Oxford **1975**.
[3] M. J. Stephen, J. P. Straley, *Rev. Mod. Phys.* **1974**, *46*, 617–704.
[4] J. G. Jenkins, *Ann. Rev. Fluid. Mech.* **1978**, *10*, 197–219.
[5] W. H. de Jeu, *Physical Properties of Liquid Crystalline Materials*, Gordon and Breach, New York **1980**.
[6] G. Vertogen, W. H. de Jeu, *Thermotropic Liquid Crystals, Fundamentals*, Springer Series in Chemical Physics 45, Springer, Berlin **1988**.
[7] P. G. de Gennes, J. Prost, *The Physics of Liquid Crystals*, 2nd edn, Clarendon Press, Oxford **1995**.
[8] F. M. Leslie in *Advances in Liquid Crystals* (Ed.: G. H. Brown), Academic Press, New York **1979**.
[9] F. M. Leslie, *Mol. Cryst. Liq. Cryst.* **1981**, *63*, 111–128.
[10] F. M. Leslie, *Quart. J. Mech. Appl. Math.* **1966**, *19*, 357–370.
[11] F. M. Leslie, *Arch. Ratl. Mech. Anal.* **1968**, *28*, 265–283.
[12] J. L. Ericksen, *Arch. Ratl. Mech. Anal.* **1966**, *23*, 266–275.

[13] J. L. Ericksen, *Mol. Cryst. Liq. Cryst.* **1969**, *7*, 153–164.

[14] O. Parodi, *J. Physique* **1970**, *31*, 581–584.

[15] G. Vertogen, *Z. Naturforsch.* **1983**, *38a*, 1273–1275.

[16] D. Forster, T. C. Lubensky, P. C. Martin, J. Swift, P. S. Pershan, *Phys. Rev. Lett.* **1971**, *26*, 1016–1019.

[17] P. C. Martin, O. Parodi, P. S. Pershan, *Phys. Rev. A* **1972**, *6*, 2401–2420.

[18] P. Schiller, *Wiss. Z. Univ. Halle* **1985**, *34*, 61–78.

[19] F. M. Leslie, I. W. Stewart, M. Nakagawa, *Mol. Cryst. Liq. Cryst.* **1991**, *198*, 443–454.

[20] S. P. A. Gill, F. M. Leslie, *Liq. Cryst.* **1993**, *14*, 1905–1923.

[21] K. F. Wissbrun, *J. Rheol.* **1981**, *25*, 619–662.

[22] G. Marrucci, F. Greco, *Adv. Chem. Phys.* **1993**, *86*, 331–404.

[23] M. G. Clark, F. M. Leslie, *Proc. R. Soc. Lond. A* **1978**, *361*, 463–485.

[24] P. K. Currie, *Mol. Cryst. Liq. Cryst.* **1974**, *28*, 335–338.

[25] W. Helfrich, *J. Chem. Phys.* **1969**, *51*, 4092–4105.

[26] M. Miesowicz, *Bull. Acad. Polon. Sci. Lett., Ser. A, Sci. Math.* **1936**, 228–247.

[27] Ch. Gähwiller, *Phys. Lett.* **1971**, *36A*, 311–312.

[28] V. Zwetkoff, *Acta Physicochimica URSS* **1939**, *10*, 555–578.

[29] F. M. Leslie, G. R. Luckhurst, H. J. Smith, *Chem. Phys. Lett.* **1972**, *13*, 368–371.

[30] K. B. Migler, R. B. Meyer, *Physica D* **1994**, *71*, 412–420.

[31] I. Zuniga, F. M. Leslie, *Europhys. Lett.* **1989**, *9*, 689–693.

[32] W. Helfrich, *J. Chem. Phys.* **1969**, *50*, 100–106.

[33] H. Ehrentraut, S. Hess, *Phys. Rev. E* **1995**, *51*, 2203–2212.

[34] P. Manneville, *Mol. Cryst. Liq. Cryst.* **1981**, *70*, 223–250.

[35] T. Carlsson, *Mol. Cryst. Liq. Cryst.* **1984**, *104*, 307–334.

[36] P. G. de Gennes, *The Physics of Liquid Crystals*, Clarendon Press, Oxford **1975**, p. 68.

[37] P. K. Currie, *J. Physique* **1979**, *40*, 501–505.

[38] R. J. Atkin, F. M. Leslie, *Quart. J. Mech. Appl. Math.* **1970**, *23*, S3–S24.

[39] P. K. Currie, *Arch. Ratl. Mech. Anal.* **1970**, *37*, 222–242.

[40] R. J. Atkin, *Arch. Ratl. Mech. Anal.* **1970**, *38*, 224–240.

[41] P. Pieranski, E. Guyon, *Phys. Lett.* **1974**, *49A*, 3, 237–238.

[42] P. Pieranski, F. Brochard, E. Guyon, *J. Physique* **1973**, *34*, 35–48.

[43] F. Brochard, *Mol. Cryst. Liq. Cryst.* **1973**, *23*, 51–58.

[44] T. Carlsson, *J. Physique* **1983**, *44*, 909–911.

[45] N. Kuzuu, M. Doi, *J. Phys. Soc. Jpn.* **1983**, *52*, 3486–3494.

[46] M. A. Osipov, E. M. Terentjev, *Z. Naturforsch.* **1989**, *44a*, 785–792.

[47] M. Kröger, H. S. Sellers, *J. Chem. Phys.* **1995**, *103*, 807–817.

[48] F. M. Leslie, *Proc. R. Soc. A* **1968**, *307*, 359–372.

[49] F. M. Leslie, *Mol. Cryst. Liq. Cryst.* **1969**, *7*, 407–420.

[50] U. D. Kini, *J. Physique Coll. C3* **1979**, *40*, 62–66.

[51] N. Scaramuzza, R. Barberi, F. Simoni, F. Xu, G. Barbero, R. Bartolino, *Phys. Rev. A* **1985**, *32*, 1134–1143.

[52] V. G. Kamenskii, E. I. Kats, *Sov. Phys. JETP* **1988**, *66*, 1007–1012.

[53] W. Helfrich, *Phys. Rev. Lett.* **1969**, *23*, 372–374.

[54] U. D. Kini, G. S. Ranganath, S. Chandrsekhar, *Pramana* **1975**, *5*, 101–106.

[55] F. Fischer, J. Grupp, *J. Physique Lett.* **1984**, *45*, L1091–L1095.

[56] L. J. Yu, A. Saupe, *Phys. Rev. Lett.* **1980**, *45*, 1000–1003.

[57] H. Brand, H. Pleiner, *Phys. Rev. A* **1981**, *24*, 2777–2787.

[58] M. Liu, *Phys. Rev. A* **1981**, *24*, 2720–2726.

[59] A. Saupe, *J. Chem. Phys.* **1981**, *75*, 5118–5124.

[60] E. Govers, G. Vertogen, *Physica* **1985**, *133A*, 337–344.

[61] H. Brand, H. Pleiner, *J. Physique* **1982**, *43*, 853–858.

[62] F. M. Leslie, *J. Non-Newtonian Fluid Mech.* **1994**, *54*, 241–250.

[63] L. Léger, A. Martinet, *J. Physique* **1976**, *37*, C3-89–97.

[64] P. G. de Gennes, J. Prost, *The Physics of Liquid Crystals*, 2nd edn, Clarendon Press, Oxford **1995**, p. 432.

[65] T. Carlsson, F. M. Leslie, N. A. Clark, *Phys. Rev. E* **1995**, *51*, 4509–4525.

[66] M. Osipov, E. M. Terentjev, *Mol. Cryst. Liq. Cryst.* **1991**, *198*, 429–435.

[67] T. Carlsson, B. Žekš, *Liq. Cryst.* **1989**, *5*, 359–365.

[68] I. Dierking, F. Giesselmann, P. Zugenmaier, G. Pelzl, P. Schiller, *Cryst. Res. Technol.* **1992**, *27*, 727–739.

[69] P. G. de Gennes, *Phys. Fluids* **1974**, *17*, 1645–1654.

[70] R. Bartolino, G. Durand, *J. Physique* **1981**, *42*, 1445–1451.

[71] R. Bartolino, G. Durand, *J. Physique Lett.* **1983**, *44*, L79–L83.

[72] P. Oswald, *J. Physique Lett.* **1983**, *44*, L303–L309.

[73] P. Oswald, S. I. Ben-Abraham, *J. Physique* **1982**, *43*, 1193–1197.

[74] J. Marignan, O. Parodi, *J. Physique* **1983**, *44*, 263–271.

9 Behavior of Liquid Crystals in Electric and Magnetic Fields

Lev M. Blinov

9.1 Introduction

As is well known, liquid crystals are very sensitive to electric and magnetic fields [1] and it is this property which allows their application in display and other optical devices technology. The field effects may be divided into three groups. The first includes the electric- or magnetic-field-induced changes in the microscopic structure of a liquid crystal phase. This implies that the field is directed in such a way with respect to the director that, under the action of the field, the orientational state of the mesophase is unchanged (the field stabilizes the director distribution). The second group includes changes in the macroscopic structure, due to a destabilization and reorientation of the director under direct action of the electric or magnetic field. The backflow effects accompanying the distortion in the transient regime are driven by the director reorientation (primary process) and are of secondary importance. The fluid is considered to be nonconducting. The effects caused by the electric field and related to the conductivity of a liquid crystal are included in the third group (electrohydrodynamic instabilities). All these effects

are dissipative in nature and have no magnetic analogs.

Due to the strong anisotropy of liquid crystals, field effects are accompanied by dramatic changes in optical properties. The electro-optical properties of liquid crystals have been studied very actively since the beginning of the century, and many important phenomena have been discovered. The earlier theoretical and experimental results on the physical properties of liquid crystals, including certain electric field effects, have been reviewed in a variety of books [1–5], as have the details of electro-optical effects [6–9], and many results relating to the properties of liquid crystalline materials and their application in devices have been reported [10–12].

The present section is devoted to a discussion of electric field effects in various liquid crystal phases, with an emphasis on the physical aspects of the phenomena. The discussion is based on classical results, although the most important recent achievements in the field are also mentioned. As ferroelectric liquid crystals are covered in detail in other chapters in this book, they are discussed only briefly here for the sake of completeness. Electro-optical properties of polymer liquid crystals have recently been discussed in detail [13]; here, only specific features of polymer mesophases relevant to their field behavior are mentioned.

9.2 Direct Influence of an Electric or Magnetic Field on Liquid Crystal Structure

9.2.1 Shift of Phase Transition Temperatures

In general, when an external field is applied to a nonpolar or polar liquid crystal, a quadratic- or linear-in-field term, respectively, must be added to the expression for the free energy density of the medium. The quadratic-in-field energy terms describe the interaction of the electric or magnetic field with the dielectric χ_a or diamagnetic γ_a anisotropy of susceptibility:

$$g_E = \frac{1}{2} \chi_a \, E^2 \tag{1a}$$

$$g_H = \frac{1}{2} \gamma_a \, H^2 \tag{1b}$$

The linear terms describe the interaction of the fields with media, polarized or magnetized either spontaneously or by external factors other than electric field itself. In ferroelectrics and ferromagnetics the magnitude of the spontaneous polarization P_S or magnetization M_S is finite and the energy term is

$$g_E = P_S E \tag{2a}$$

$$g_H = M_S H \tag{2b}$$

As ferromagnetic mesophases are still to be discovered the linear term (Eq. (2a) or (2b)) is important only when considering ferro-

electric liquid crystals. More generally, the PE term is also used to describe other polarized systems; for example, in the discussion of the flexoelectric effects in nonpolar phases as nematics or cholesterics where the polarization P is induced by a mechanical distortion.

The field terms may stabilize or destabilize the intrinsic thermodynamic order of a mesophase, and hence increase or decrease the temperature of its transition to a less ordered phase. The shift in the transition temperature is calculated by comparing the electric (magnetic) energy with the transition enthalpy or some other competing thermodynamic quantity (e.g. elastic energy). Below we consider examles.

9.2.1.1 Second-Order Transitions

Smectic A – Smectic C Transition

In this case we must compare a gain in the electric (or magnetic) energy with a loss of the soft-mode elastic energy characterized by a change in tilt angle θ. Thus, for the electric field case, the shift in the A–C transition temperature is [7]:

$$\Delta T_{\text{C-A}} = \frac{\varepsilon_a \, (E \cdot n)^2}{8 \pi \alpha} \tag{3}$$

where n is the director, $\varepsilon_a = 1 + 4\pi\chi_a$ is the dielectric anisotropy and α is a parameter of the Landau expansion for the A–C transition:

$$g = \frac{1}{2} \alpha (T - T_{\text{C-A}}) \, \theta^2 + \frac{1}{4} b \theta^4 \tag{4}$$

and g_E is taken in the form of Eq. (1a). The sign of the shift depends on the field direction. If E is parallel to the normal h to the smectic layers and $\varepsilon_a > 0$, the field stabilizes the smectic A phase and $T_{\text{C-A}}$ decreases. An oblique field ($\varepsilon_a > 0$) induces a tilt, and thus stabilizes the smectic C phase and $T_{\text{C-A}}$ increases.

Smectic A–Smectic C* Transition

For the chiral smectic C* phase, which is ferroelectric, in addition to the quadratic term (Eq. (1a) or (1b)) the linear term proportional to polarization P must be taken into consideration. As a rule, the linear term exceeds the quadratic one and the free energy may be taken in form:

$$g = \frac{1}{2}a_0\theta^2 + \frac{1}{4}b\theta^4 + \frac{P^2}{2\chi_\perp} - CP\theta - PE \quad (5)$$

where $a_0 = \alpha(T - T_{C\text{-}A})$ describes the elasticity for $\delta\theta$ changes, P is the polarization, C is the tilt–polarization coupling constant (or piezocoefficient), and $\chi_\perp$ is the background dielectric susceptibility, which may be taken from the isotropic phase. In Eq. (5) the smectic C* is assumed to be unwound and all chiral terms important for a helical ferroelectric are omitted.

Strictly speaking, even for an infinitely small electric field the second-order A–C* transition disappears. However, the soft-mode dielectric susceptibility maximum characteristic of that transition is still observed at the "apparent" transition temper-

ature T_m. With increasing field T_m increases according to expression [14]:

$$T_m - T_{C\text{-}A} = 3\left(\frac{bC^2}{16\alpha^3}\right)^{\frac{1}{3}} E^{\frac{2}{3}} \quad (6)$$

Such a shift governed by the second term in the Landau expansion and strongly dependent on piezocoefficient C has been observed experimentally [14, 15]. The experimental points for a multicomponent ferroelectric mixture [14] and the fit using Eq. (6) is shown in Fig. 1.

9.2.1.2   Strong First-Order Transitions

Isotropic Liquid–Nematic Transition

In the case of positive dielectric anisotropy of a nematic, even a weak field makes the isotropic phase uniaxial and the N–I phase transition disappears (see Pikin [7], Chap. 4). However, the apparent N–I phase transition temperature may change with the electric or magnetic field. For $\varepsilon_a > 0$ and $E \| n$, the quadratic-in-field energy terms (Eq. (1a) and (1b)) reduce the free energy and stabilize the anisotropic phase. In the

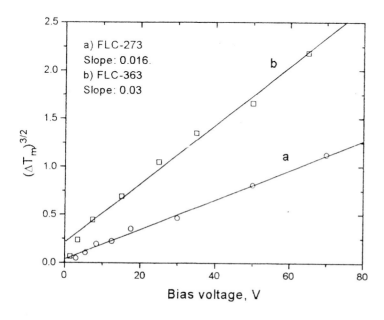

a) FLC-273
Slope: 0.016
b) FLC-363
Slope: 0.03

Figure 1. The voltage dependence of the shift in the soft-mode susceptibility maximum temperature [14]: (a) mixture 1; (b) mixture 2. Dotted lines correspond to Eq. (6). Cell thickness 10 µm.

case of an electric field, the transition temperature increases by the value [16]

$$\Delta T_{N-I} = \frac{(\varepsilon_{II} - \varepsilon_{iso})\,E^2}{8\pi \cdot \Delta H_t} \tag{7}$$

where ΔH_t is the enthalpy of the N–I phase transition, ε_{II} is the dielectric permittivity of the nematic phase parallel to the director, and ε_{iso} is the dielectric permittivity of the isotropic phase. Experimentally, for rather high positive dielectric anisotropy of the nematic phase when $\varepsilon_{II} - \varepsilon_{iso} \approx 10$, and a typical value of $\Delta H_t \approx 5 \times 10^4$ erg cm^{-3}, a shift of the order of 0.8 K is anticipated for a field strength of 10^5 V cm^{-1} (about 300 CGS units).

The shift of the first-order N–I transition is described theoretically with a Landau type expansion over two variables, orientational order parameter S, and polarization P [17]. In the case of the electric field directed along the nematic optical axis the expansion is as follows:

$$g = \frac{1}{2}\alpha(T - T_c^*)\,S^2 - \frac{1}{3}b\,S^3 + \frac{1}{4}c\,S^4$$
$$+ \frac{P^2}{2\chi_0} + \sigma P^2 S + \frac{1}{2}\kappa P^2 S^2 - PE \tag{8}$$

where χ_0 is the susceptibility of the isotropic phase, while parameters σ and κ represent the anisotropy of the susceptibility of the nematic phase. With this form of the free energy, as expected the shift of the transition is proportional to the square of the field, the coefficient α playing the role of the transition heat in Eq. (7):

$$\Delta T_c^* = \chi_0^2\,\alpha^{-1}\,(4\sigma^2\,\chi_0 - \kappa)\,E^2$$

The experimental investigation of the field-induced shift of the N–I transition was carried out using a pulse technique [17], which allowed rather strong electric fields to be applied to samples. The data [17] are shown in Fig. 2 for two nematic cyanobiphenyls, 5-CB and 6-CB, that differ in

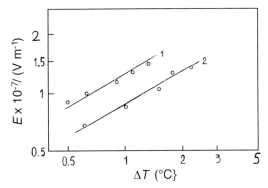

Figure 2. Change in the N–I transition temperature with the electric field in (1) 5-CB and (2) 6-CB [17]. A double logarithmic scale is used.

their latent heats of the N–I transition (1.5×10^4 erg cm^{-3} and 2.5×10^4 erg cm^{-3} for 5-CB and 6-CB, respectively). The results are consistent with both the phenomenological approach (Eq. (8)) and microscopic theory [17], operating with the number density, and the magnitude, and direction of the molecular dipoles.

At extremely high fields the isotropic phase should be indistinguishable from the nematic one, even well above the zero field transition temperature, since the uniaxial order induced by a magnetic or ac electric field in the isotropic phase will be comparable with the nematic orientational order. However, such fields are hardly accessible, even with the pulse technique. Much stronger changes in order parameter may be achieved with ferroelectric transitions (see below).

For negative dielectric anisotropy of the nematic phase, $\varepsilon_a < 0$, the external field may induce a biaxial nematic phase. With increasing field one may reach the tricritical point [18] where the first-order N–I transition becomes a second-order one.

A–C*, A–G*, C*–G*, and N–C*
Ferroelectric Transitions

For these transitions the electric-field-induced shifts of transition temperatures are linear functions of the field. The A–C* transition may be either second or first order, while the A–G* and C*–G* transitions are first order [19, 20]. Thus, to calculate the transition shift, we should compare the transition enthalpy with the linear-in-field free energy term (Eq. (2)). The Clausius–Clapeyron equation for solid ferroelectrics is well known:

$$\frac{\Delta T}{E} = \frac{\Delta P_s}{\Delta H_t} \tag{9}$$

Here, ΔP_S is the difference in spontaneous polarization between the two phases under consideration. For the first-order transition the coefficient b in the Landau expansion must be negative and a term of the sixth order in the tilt angle has to be included:

$$g = \frac{1}{2} a_0 \theta^2 + \frac{1}{4} b \theta^4 + \frac{1}{6} c \theta^6$$
$$+ \frac{P^2}{2\chi_\perp} - CP\theta - PE \tag{10}$$

The coefficient c enters the expression for the shift in the transition temperature [19, 20]:

$$\Delta T \propto \frac{4\chi_\perp}{\alpha} \left(-\frac{c}{3b} \right)^{\frac{1}{2}} E$$

Experimentally, a field strength of the order of 10^5 V cm^{-1} induces a ΔT value of the order of 1 K for a ΔH_t of about 3×10^5 erg cm^{-3} K^{-1} and $\Delta P_S \approx 300$ nC cm^{-2}. An example of the $\Delta T(E)$ experimental dependence is shown in Fig. 3.

Field dependence was also observed for the first-order N–C* transition temperature. For a field strength of about 2×10^5 V cm^{-1} the phase transition point increased by 0.6 K [21]. For a compound with a much higher spontaneous polarization in the C* phase, approximately the same shift is induced by a field one order of magnitude lower [22].

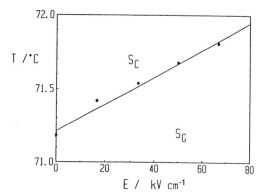

Figure 3. Electric field strength dependence for the C*–G* phase transition (compound A7, [19]).

9.2.1.3   Weak First-Order Transitions

N–A Transition

The latent heat of this transition is usually small and may even vanish if the width of the nematic temperature range is sufficiently large [23]. Thus, the transition can be either of first or second order. For the second-order transition the discontinuity in the orientational order parameter S and, hence, the dielectric (or diamagnetic) susceptibility, disappears and the field influence on both phases is the same. Thus we do not anticipate any field-induced shift in the N–A transition temperature. For the weak first-order transition there is a small discontinuity in both S and n_e dielectric (and magnetic) susceptibilities, and the shift depends on the competition between two small quantities: the difference in susceptibilities for the nematic and smectic A phases on the one hand and transition enthalpy on the other. In particular, the field may induce a change in the phase transition order, from first to the second order, as shown in Fig. 4 [24].

Smectic A–C Ferroelectric Transition*

This case has been analyzed in detail both theoretically and experimentally [25]. The

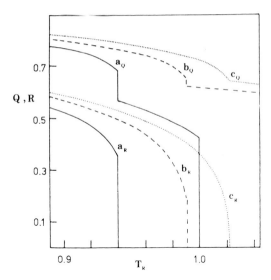

Figure 4. Field-induced changes in order parameters for a smectic A liquid crystal. Q, orientational order parameter; curves a_Q, b_Q, and c_Q show the temperature dependence of Q at various fields. R, order parameter describing the coupling of translational and orientational order, curves a_R, b_R, and c_R show the temperature dependence of R. The field increases from curves a to curves c and results in a change in the N–C transition from first to second order [24].

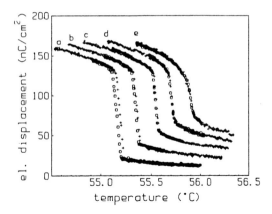

Figure 5. Temperature dependence of the electric displacement of C7 (*I*) at various field strengths. E (in kV cm^{-1}): (a) 10, (b) 20, (c) 30, (d) 40, and (e) 50. (+) Increasing and (○) decreasing temperature runs [25].

temperature dependence of the electric displacement D at various bias fields for the chiral compound C7 (**I**) is shown in Fig. 5.

$$C_2H_5O-(CH_3)C^*H-C^*H(Cl)-$$

$$-COO-\bigcirc\!\!\!-\!\!\!\bigcirc-C_7H_{15}$$

I

The change in the A–C* transition from first to second order can be clearly seen in Fig. 5. The tricritical point corresponds to the following sets of parameters: $T_C \approx 55.8°C$ (i.e. 0.8 K above the zero-field point), $E_C \approx 5 \times 10^4$ V cm^{-1} and $D_C \approx 100$ nC cm^{-2}.

A theoretical description of the experiment can be obtained using the same approach; that is, using the Landau expansion (Eq. (10)). Equation (10) has been solved numerically and the results of the calcula-

tions are shown in Fig. 6, where the molecular tilt angle, which is the order parameter of the A–C* transition, is given as a function of the external electric field for various temperatures. The parameters used to fit the experimental (Fig. 5) and theoretical (Fig. 6) curves were (in CGS units): $\alpha \approx 8.9 \times 10^5$ erg cm^{-3} K^{-1}, $b \approx -1.3 \times 10^7$ erg cm^{-3}, $c \approx 6.6 \times 10^7$ erg cm^{-3}, $T_0 = 51.4°C$, $\chi_\perp = 0.33$, $C \approx 2.9 \times 10^3$ CGS (in [25] these parameters are given in SI units).

9.2.2 Influence of the Field on Order Parameters

As the field-induced reorientation of the director occurs at rather weak fields, the geometries worth discussing here for nonpolar mesophases are: (i) the field parallel to the director and $\varepsilon_a > 0$; and (ii) the field perpendicular to the director and $\varepsilon_a < 0$. In the first case, uniaxial symmetry is conserved and the field stabilizes thermal fluctuations and increases the orientational order parameter in both the nematic and smectic A phase [26]. A qualitative picture is

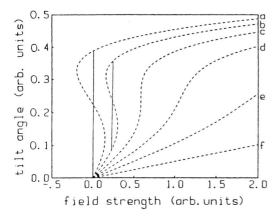

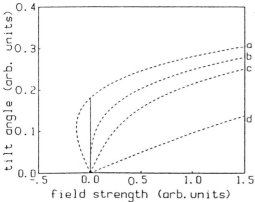

Figure 6. Theoretical tilt angle versus field strength (dashed lines): vertical solid lines indicate discontinuous transitions [25].
First-order transition, $b<0$:   (a) $T=T_c(0)$;   (b) $T_c(0)<T<T_c$;   (c) $T=T_c$;   (d) $T>T_c$;   (e,f) $T\gg T_c$. $T_c(0)$ and T_c are the zero-field first-order transition temperature and the critical point temperature, respectively. (Top)
Second-order transition, $b>0$: (a) $T<T_c$; (b) $T=T_c$; (c) $T>T_c$; (d) $T\gg T_c$. T_c denotes the second-order transition temperature. (Bottom)

shown in Fig. 4 [24]. With increasing field the orientational order parameter S increases and T_{NA} increases, as discussed above. The positional order parameter, being coupled with S also increases.

The additional optical anisotropy induced by the field is proportional to either E or $|E|$ for a weak and strong field, respectively [27, 28].

9.2.3   Changes in Symmetry

The field-induced tilt in the achiral smectic A phase (see above) results in a change in symmetry from $D_{\infty h}$ to C_{2h}. The induced tilt angle is proportional to the field squared and the factor $(T_{CA}/T-T_{C\text{-}A})^{\gamma}$, where $\gamma=1.3$ [1]. Due to the tilt, optical biaxiality can be observed in the vicinity of the transition.

9.2.3.1   Induced Biaxiality in Nematics

Another example of a symmetry change is the field-induced biaxiality in nematic liquid crystals with negative dielectric anisotropy [29]. In this case the field is applied perpendicular to the director. The latter is parallel to the z axis. The orientational order at a point r in a nematic liquid crystal is defined by the ordering matrix

$$S_{\alpha\beta}(r)=\frac{1}{2}S_0[3n_\alpha(r)\,n_\beta(r)-\delta_{\alpha\beta}]\qquad(11)$$

where S_0 is the order parameter defined with respect to a local director $n(r)$. Averaging over time and spatial fluctuations of $n(r)$ gives the ordering matrix in a principal axis system as:

$$\langle S_{\alpha\beta}\rangle\begin{bmatrix}-\dfrac{1}{2}(Q-B)&0&0\\[2mm]0&-\dfrac{1}{2}(Q+B)&0\\[2mm]0&0&Q\end{bmatrix}\qquad(12)$$

where Q and B are uniaxial and biaxial order parameters

$$Q=S_0\left[1-\frac{1}{2}\langle n_x^2+n_y^2\rangle\right]\qquad(13)$$

$$B=\frac{3}{2}S_0\langle n_x^2-n_y^2\rangle\qquad(14)$$

Now Q and B are presented in terms of the director fluctuations which are not assumed to be isotropic in the x,y plane. For $n_x^2=n_y^2$ the biaxial parameter vanishes and the uniaxial one becomes equal to the well-known parameter:

$$S = \frac{1}{2} \langle 3 \cos^2 \theta - 1 \rangle \tag{15}$$

where θ is an angle between the longitudinal axis of a molecule and the director of the nematic mesophase.

When a field is applied, for example, along the x direction, and $\varepsilon_{xx} = \varepsilon_{yy} = \varepsilon_\perp > \varepsilon_{zz} = \varepsilon_\parallel$ the n_x fluctuations are quenched to some extent and B increases. The theory [30] considers the quenching of fluctuations with various wavevectors and results in the following field dependence of the biaxial order parameter:

$$B = \frac{3 S_0 k_B T}{8 \pi K} \left(\frac{\varepsilon_a E^2}{4 \pi K} \right)^{\frac{1}{2}} \tag{16}$$

where K is an average elastic constant and k_B is the Boltzmann constant.

The field-induced biaxiality has been observed in a planar-oriented layer of compound **II**,

$$\text{C}_5\text{H}_{11}\!-\!\langle\,\text{H}\,\rangle\!-\!\langle\,\text{H}\,\rangle\!-\!\text{C}_5\text{H}_{11}$$
$$\text{CN}$$

II

with $\varepsilon_a = -5.4$, by measuring the ellipticity of laser light transmitted through a sandwich cell ($E \parallel$ light wavevector) [29].

9.2.3.2 The Kerr Effect

The Kerr effect in the isotropic phase may also be discussed in terms of the field-induced symmetry change of the medium. A birefringence is induced in an ordinary liquid if a uniform electric field is applied at right-angles to the direction of a beam passing through the cell; its value is related to the field strength E:

$$\Delta n (E) = \lambda B' E^2 \tag{17}$$

where B' is the Kerr constant and λ is the wavelength of the light.

At temperatures significantly above the transition point from the nematic to the isotropic liquid, materials forming liquid crystals behave like normal liquids. In an electric field they display the Kerr effect, with an order of magnitude equal to or less than that of nitrobenzene ($+4.1 \times 10^{-12}$ m V^{-2}). However, as the temperature decreases toward $T_{\text{N-I}}$ the Kerr constant of a nematogen changes considerably. For some substances it diverges when $T \rightarrow T_{\text{N-I}}$, being either positive or negative. In general, there is a correlation between the sign of the Kerr effect in the isotropic phase and the dielectric anisotropy of the nematic phase.

The growth of the Kerr constant is accounted for by considerable contribution of fluctuations of the orientational order parameter to dielectric properties of the isotropic phase. This contribution can be calculated within the framework of the Landau theory [31]. Field E induces the orientational order [32]

$$Q(E) = \frac{\rho \varepsilon_a E^2}{12 \pi \alpha (T - T^*)} \tag{18}$$

where ρ is the number density of molecules and α is the parameter of the Landau expansion for the N–I transition.

The field-induced order can be related to the Kerr constant

$$B' = \frac{\Delta n^0 Q(E)}{\lambda E^2} \tag{19}$$

where Δn^0 is the birefringence of the perfectly aligned liquid crystal [33]. Experiments carried out on 5-CB showed that the law $B' \propto (T - T_{\text{N-I}})^{-1}$ is fulfilled in the $T - T_{\text{N-I}}$ range from several kelvin down to 0.5 K (the Kerr constant is equal to 6×10^{-11} m V^{-2} at $T - T_{\text{N-I}} = 1$ K and $\lambda = 633$ nm). Closer to $T_{\text{N-I}}$ the law is violated and the discrepancy can be accounted for by higher order terms in the Landau expansion [33]. The microscopic theory of the Kerr effect is described by Rjumtsev et al. [17].

The Kerr effect in the pretransition temperature region has been investigated very actively [17, 34–39]. Tsvetkov et al. [39] determined the values for $B' > 0$ (phenylbenzoates) and $B' < 0$ (cyanostilbenes) close to T_{N-I}. These values exceed the absolute value of the Kerr constant for nitrobenzene by 200 and 6.5 times, respectively. Such high values of B enable the control voltage of the Kerr cell to be reduced several fold and give a decrease of an order of magnitude or more in power consumption [40].

Relaxation times τ_1 associated with order-parameter fluctuations increase when approaching T_{N-I}, but still remain short enough for light modulation (within the range 0.1–1 µS for 5-CB at $T - T_{N-I} = 5$ to 0.5 K [41]). The other time (τ_2) related to molecular relaxation is independent of temperature.

9.2.4 Specific Features of Twisted Phases and Polymers

In Section 9.2.1 of this Chapter we discussed field-induced changes in the microstructure of liquid crystals. However, field-induced unwinding of the cholesteric (macroscopic) helix (see Section 9.3.2.3 of this Chapter) shows that the transition from a twisted to a uniform nematic may also be considered as a phase transition. In the latter case the field energy term competes with a rather small elastic energy proportional to nematic-like elastic moduli and the squared wave vector of the helical superstructure Kq^2. As the pitch of the helix $p = 2\pi/q$ is large, the field threshold for the transition is very low. On the other hand, between the two extreme cases (a microstructure with a molecular characteristic dimension and a

large-pitch macrostructure) there are cases where the pitch might be smaller and the elastic moduli higher than those in cholesterics. Examples are blue phases in cholesterics and twist grain boundary (TGB) phases in smectics A and C* (TGBA and TGBC). The structures of these phases are discussed in other chapters in this book. What is essential for their field behavior is a periodic, solid-like defect structure which has its own wavevector and elasticity.

9.2.4.1 Blue Phases

There are three known thermodynamically stable blue phases: BP_I and BP_{II}, and BP_{III} (or the foggy phase). The structure of the first two is already established: BP_I is a body-centered cubic phase (symmetry group O^8 or $I4_132$) and BP_{II} is simple cubic (symmetry group O^2 or $P4_232$) [42]. The foggy phase, BP_{III}, can probably be described using one of the quasicrystal models [43].

The three-dimensional crystalline structure of blue phases with lattice periods comparable to the wavelength of visible light results in optical diffraction, which is dependent on the orientation of the light wavevector with respect to the crystalline planes. The most pronounced features of the optics of BP_I and BP_{II} are [44]: (i) selective reflection in the visible range which gives rise to the blue color of the phases (in contrast to conventional cholesterics, blue phases manifest several reflection orders); (ii) only one circular polarization, as in cholesterics, is back-scattered; (iii) blue phases are optically active and the sign of the light polarization rotation changes at the wavelengths of selective scattering maxima; and (iv) the linear birefringence is absent (i.e. blue phases are optically isotropic), but multiple scattering in perfect samples can result in a small apparent optical anisotropy. The optical properties of blue phases are completely defined by the spatially periodic tensor of the high-frequency dielectric permittivity [44]:

$$\varepsilon(\boldsymbol{r}) = \varepsilon_0 + \varepsilon_a^{ik}(x, y, z) \qquad (20)$$

where ε_0 is the average dielectric permittivity and ε_a^{ik} is its three-dimensional periodic part.

An external electric field interacts with the local dielectric anisotropy of a blue phase and contributes $\varepsilon_a E^2/4\pi$ to the energy of the liquid crystal [45]. The field distorts the cubic lattice and results in a change in the angular (or spectral) positions of Bragg's reflections. Moreover, field-induced phase transitions to novel phases have been observed [42, 46, 47]. The field can also induce birefringence parallel to the field direction, due to the optical biaxiality of the distorted cubic lattice [48].

In a weak field no phase transition occurs and a cubic lattice is distorted according to the sign of the local dielectric anisotropy of the medium. The effect is described using the electrostriction tensor γ_{iknp} [45], which relates the electric-field components to the strain tensor:

$$e_{ik} = \gamma_{iknp} E_n E_p \qquad (21)$$

The distortion of the lattice is accompanied by a change in the Fourier harmonics of the dielectric tensor (Eq. (20)) and corresponding changes in the optical properties. For example, the wavelengths of the Bragg peaks for a substance with $\varepsilon_a > 0$ may increase or decrease with increasing field depending on the field direction with respect to the crystallographic axes (Fig. 7) [49]. When the dielectric anisotropy is negative, all the signs of components of the electrostriction tensor are inverted and all the field-induced red shifts in the Bragg maxima are replaced by blue shifts, and vice versa [50]. Red and blue shifts may reach values of 26 and

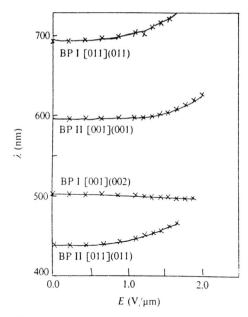

Figure 7. The wavelengths of the Bragg peaks corresponding to the planes (hkm) versus the applied field strength for BP_I (34.1 °C) and BP_{II} (34.3 °C) oriented with either a four-fold [001] or two-fold [011] axis parallel to the field [49].

35 nm, respectively, in a rather weak field of about $0.2-0.4$ V mm^{-1} and a negative ε_a [51].

With increasing external field a series of field-induced phase transitions is observed: $BP_I \rightarrow$ cholesteric, $BP_{II} \rightarrow$ cholesteric, and then cholesteric $\rightarrow$ nematic [52]. This is illustrated in Fig. 8 [42] which shows a voltage–temperature phase diagram for a mixture ($47-53$ mol%) of chiral 15-CB with 4-n-hexyloxycyanobiphenyl (60-CB). BP_I loses its stability, first transforming into the cholesteric phase, because the transition enthalpy ΔH_t is extremely small (~ 50 J mol^{-1}) for the BP_I–Ch transition. This enthalpy, normalized to a unit volume (0.2 J cm^{-3}) and compared with the difference in electrostatic energy density between the two phases $\delta\varepsilon \cdot E^2$ ($\delta\varepsilon = \varepsilon_{BP_I} - \varepsilon_{Ch} \sim 0.2$) explains the observed results. For other materials, novel phases induced by an external

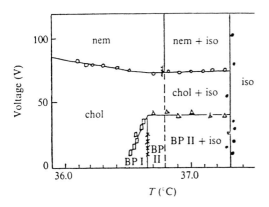

Figure 8. Voltage–temperature phase diagram for a 47–53 mol% mixture of 15-CB and 6-OCB [42].

field have been observed. Among them a tetragonal (BPX) and two hexagonal (a three-dimensional BPH^{3d}, and a two-dimensional BPH^{2d}) phases were distinguished [46, 47]. Chilaya and Petriashvili [51] observed the field-induced transition from BP$_{II}$ to the isotropic phase in a mixture with $\varepsilon_a > 0$.

When an electric field is applied to the BP$_{III}$ (foggy) phase with $\varepsilon_a > 0$, the broad selective reflection peak typical of this less ordered phase decreases in intensity and, at some threshold voltage, is replaced by a sharp peak at longer wavelength [43]. Thus, the transition to a new phase occurs. The symmetry of the new phase has not yet been established. For a system with $\varepsilon_a < 0$ [53], the increase in the field results in a considerable increase in the selective reflection peak. An explanation of such behavior and many other examples of field effects in blue phases can be found in a comprehensive review by Kitzerow [54].

9.2.4.2   Twist Grain Boundary Phases

TGB phases are observed only in chiral substances. The director is twisted in the presence of layering, the layer twist being me-

diated by the twist grain boundary walls with screw dislocations. In the TGBA phase the layering is of the smectic A type, while in TGBC* it is of the smectic C* (ferroelectric) type.

TGBA–Smectic A Transition

The electric field may push the defect walls out of the sample and induce the transition from TGBA to conventional smectic A. The field threshold of such a transition decreases dramatically on approaching the chiral nematic phase and may be of the order of 2×10^4 V cm^{-1} [55]. With increasing fields up to 1×10^5 V cm^{-1}, the apparent transition temperature decreases dramatically (by 18 K). Since in both the TGBA and SmA phases the spontaneous polarization is zero, the transition should be driven by dielectric anisotropy, which appears to be almost independent of temperature in both phases. Thus the elastic properties of the layered TGBA structure competing for the field action must play the principal role in the temperature dependence of the transition parameters.

TGBA–Smectic C Transition*

The results of an investigation of the electric-field-induced transition between these two phases is shown in Fig. 9 [56]. At any given temperature the short pitch helix of TGBA is unwound by the field and either a uniform smectic C* structure (I) or a modulated one in the form of stripes (II) or parquet (III) appears. The slope of the boundary between the TGBA and SmC* phases can be explained by the Clausius–Clapeyron equation (Eq. (9)) where, in this case, $\Delta P = P_S(C^*)$ as $P_S(TGBA) = 0$. Note that the transition temperature from TGBA to the isotropic phase is, in fact, field independent due to the much higher enthalpy of that transition.

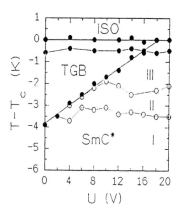

Figure 9. Voltage–temperature phase diagram obtained on heating 14P1M7 [56]. Cell thickness 2 μm.

9.2.4.3   Comment on Polymer Liquid Crystals

In addition to the effects observed in low-molecular-weight liquid crystals, an electric field causes some specific transformations in the structure of polymeric mesophases. As pointed out by Shibayev [57] the correlation length of the short-range smectic order in the nematic phase of comb-like polymers is changed considerably under the action of an electric field. The conformation and the orientational order parameter of flexible spacers separating mesogenic units from the backbone are also changed (more *gauche* isomers than *trans* isomers appear, as indicated by infrared spectroscopy [58]). The anisotropy in the orientation of polymeric backbones is induced by the electric field due to a torque exerted by the mesogenic units. In some cases the field induces smectic phases in thermodynamically stable nematic ones [59].

9.3   Distortions due to Direct Interaction of a Field with the Director

We now turn to the changes that occur in the macroscopic structure of a liquid crystal due to a destabilization and reorientation of the director under direct action of an electric or magnetic field. The external field might be coupled either to the dielectric (diamagnetic) anisotropy (magnetically or electrically driven uniform Frederiks transition and periodic pattern formation) or to the macroscopic polarization (flexoelectric effect and ferroelectric switching) of the substance. The fluid is considered to be nonconductive.

9.3.1   Nematics

9.3.1.1   Classical Frederiks Transition

The process of the field-induced reorientation of the director, called the Frederiks transition [60, 61], is usually discussed for the three typical geometries shown in Fig. 10.

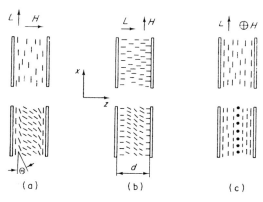

Figure 10. Three typical geometries for observing the Frederiks transition: the splay (a), bend (b), and twist (c) distortions induced by a magnetic field *H*. (Top) Initially, the director *n* is along the *L* axis, but when the field exceeds a certain threshold (bottom) distortion occurs.

The geometries correspond to the splay (with subsequent development of the bend), bend (combined with a splay) and pure twist distortions, which we may call the S-, B-, and T-distortions.

In the simplest model the director n is reoriented by the magnetic field under the action of the diamagnetic torque, which is proportional to the diamagnetic anisotropy γ_a. The corresponding contribution to the density of the nematic free energy is

$$g_H = \gamma_a (Hn)^2 \tag{22}$$

that is, the director n tends to align itself along the field $(n \parallel H)$ if $\gamma_a > 0$ and perpendicular to it $(n \perp H)$ if $\gamma_a < 0$.

The elastic torque supports the initial director orientation, fixed by the boundary conditions on the surface. In the case of strong (infinite) anchoring, the condition $n_S = n_0$ holds during the reorientation process and, as a result, a compromised director profile pertains, which satisfies the condition of the minimum free energy:

$$F_V = \int_V (g_k + g_H) \, d\tau \tag{23}$$

where g_k is the Frank energy including splay, twist and bend terms, respectively:

$$g_k = \frac{1}{2} [K_{11} (\mathrm{div}\, n)^2 + K_{22} (n \cdot \mathrm{curl}\, n)^2 + K_{33} (n \times \mathrm{curl}\, n)^2] \tag{24}$$

In the more general case of a finite director anchoring at the boundaries, the total energy of a liquid crystal is written as the sum of the bulk and the surface terms:

$$F = F_V + F_S \tag{25}$$

Les us consider the splay Frederiks transition (all the expressions are also valid for the bend effect if K_{11} and $\gamma_{\parallel}$ are replaced by K_{33} and $\gamma_{\perp}$, respectively). In the case of the uniform orientation of the director in the plane of the sample (the x,y-plane, wavevec-

tors of distortion $q_x, q_y = \infty$) the director profile $n(z)$ takes a two-dimensional form:

$$n(z) = [\cos \theta(z), \sin \theta(z)] \tag{26}$$

which can be described by elliptic functions [8, 9].

The distortion starts at a well-defined threshold field inversly proportional to the cell thickness d:

$$H_S = \frac{\pi}{d} \left(\frac{K_{11}}{\gamma_a} \right)^{\frac{1}{2}} \tag{27}$$

since, with increasing thickness, a weaker field is necessary to overcome the elastic energy $K_{11} q^2$ due to a smaller wavevector of the distortion $q = \pi/d$.

For a weak surface anchoring (surface energy $W^S < \infty$) the thickness dependence of the threshold field

$$H(W) = \frac{H_S}{d + 2b} \tag{28}$$

includes a so-called 'surface extrapolation length'

$$b = \frac{K_{11}}{W^S} \tag{29}$$

Equation (28) holds only for the zero pretilt angle of the director at the limiting boundaries. Any finite pretilt angle results in a loss of the threshold character of the effect (a continuous distortion instead of the Frederiks transition).

The dynamics of the splay and bend distortions inevitably involve the flow processes coupled with the director rotation. Such a backflow effect usually renormalizes the viscosity coefficients. Only a pure twist distortion is not accompanied by the flow. In the latter case, and for the infinite anchoring energy, the equation of motion of the director ϕ (angle variation) expresses the balance between the torques due to the elastic and viscous forces and the external field (and does not contain the fluid velocity) [62, 63]:

$$K_{22} \frac{\partial^2 \phi}{\partial z^2} + \gamma_a \boldsymbol{H}^2 \sin\phi \cdot \cos\phi = \gamma_1 \frac{\partial \phi}{\partial t} \quad (30)$$

This equation describes director rotation in the magnetic field with the inertial term $I\partial^2\phi/\partial t^2$ being disregarded, $\gamma_1 = \alpha_3 - \alpha_2$ is the rotational viscosity, α_i are Leslie's coefficients. In the limit of small ϕ angles, $\phi \ll 1$, Eq. (30) reduces to a linear form:

$$K_{22} \frac{\partial^2 \phi}{\partial z^2} + \gamma_a \boldsymbol{H}^2 \phi = \gamma_1 \frac{\partial \phi}{\partial t} \quad (31)$$

with the solution:

$$\phi = \phi_m \exp\left[1 - \exp\left(-\frac{1}{\tau_r} \right) \right] \cdot \sin \frac{\pi z}{d} \quad (32)$$

where

$$\tau_r = \frac{\gamma_1}{\gamma_a \boldsymbol{H}^2 - K_{22} \pi^2/d^2} \quad (33)$$

is the reaction or switching-on time.

The corresponding relaxation or decay time is found from Eq. (31) for $\boldsymbol{H}=0$ in a similar way:

$$\tau_d = \frac{\gamma_1}{K_{22} q^2} = \frac{\gamma_1 d^2}{K_{22} \pi^2} \quad (34)$$

Backflow effects may accompany the transient process of the director reorientation [64, 65]. The process is opposite to the flow orientation of the director known from rheological experiments. Disregarding the backflow, we can use the same equations for the splay (with K_{11}) and bend (K_{33}) small-angle distortions. The backflow effects renormalize the rotational viscosity of a nematic:

$$\gamma_1(S) = \gamma_1 - \frac{2(\alpha_3)^2}{(\alpha_3 + \alpha_4 + \alpha_6)(\alpha_3 - \alpha_2)} \quad (35a)$$

$$\gamma_1(B) = \gamma_1 - \frac{2(\alpha_2)^2}{(\alpha_4 + \alpha_5 - \alpha_2)(\alpha_3 - \alpha_2)} \quad (35b)$$

for the splay and bend distortion, respectively. The effective viscosity for the splay dis-

tortion changes by less than 1% and the bend distortion viscosity is smaller than γ_1 by 10–20%; thus the corresponding transient process including the backflow effect is faster.

Weak boundary anchoring increases both τ_r and τ_d. This is easily understood, if we remember that a finite anchoring energy W^S results in an increase in the apparent thickness of the cell according (Eq. (28)).

The classical results just discussed have been applied successfully to the determination of the elastic moduli K_{ii} and Leslie's viscosity coefficients α_i of conventional nematics. The experimental data may be found in books [3, 8–10]. The measurement of the director profile throughout the cell, including the surface layers, allows the anchoring energy to be calculated. The profile of the director distortion above the threshold of the Frederiks transition is shown in Fig. 11. The difference between curves 1 ($W^S=\infty$) and 2 ($0 < W^S < \infty$) is a measure of the anchoring energy. The simplest estimate comes from the comparison of the magnetic field coherence length at the threshold field $\boldsymbol{H}_c$ [66–68]:

$$\xi_H^c = \frac{1}{\boldsymbol{H}_c} \left(\frac{K_{ii}}{\gamma_a} \right)^{1/2} \quad (36)$$

with the sum $d+2b$ (Eq. (28)). When the applied field is much higher than the threshold of the Frederiks transition, the field coherence length becomes comparable with the surface extrapolation length itself $\xi_H \sim b$. This condition corresponds to the second threshold of the complete reorientation of a liquid crystal, including the surface layers (curve 3, Fig. 11). Thus, the second threshold field also allows b (and W^S) to be calculated.

The direction of the field-induced rotation of the director is, in principle, degenerate. The clockwise and anticlockwise directions are equally probable. Thus, domain

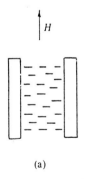

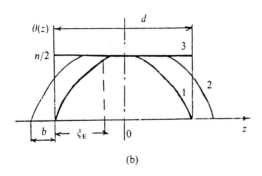

Figure 11. The geometry of the bend Frederiks transition (a) and the profile of the director deviation, induced by an external field (b).
(1) $H > H_c$, $W^S = \infty$;
(2) $H^* > H > H_c$, $0 < W^S < \infty$;
(3) $H > H^*$, $0 < W^S < \infty$.

walls appear, separating the two areas (domains) with different director orientations [69]. The wall movement influences the dynamics of the transition; the corresponding theory is presented in [70].

The Frederiks transition induced by an electric field has certain specific features not observed in the magnetic field equivalent. First, as the dielectric anisotropy is usually much higher than the magnetic one, the distortion leads to a spatial dependence of the dielectric susceptibility and a certain inhomogeneity in the electric field. Secondly, the character of the distortion may be affected by the electric conductivity and the influence of the flexoelectric effect. A discussion of these effects may be found in literature (e.g. [8, 9]).

More interesting features of the Frederiks transition are observed in the case of the simultaneous action of an electric and magnetic field. If a magnetic field is applied parallel to the nematic director $n \parallel H$ and $\gamma_a > 0$, it increases the apparent elastic coefficient by an additional quantity of $\gamma_a H^2 d^2 / \pi^2$, which results in higher values of the corresponding threshold voltages. However, this is not the only result of applying a magnetic field [71]. If a stabilizing magnetic field is applied along the homogeneously or homeotropically oriented director and a destabilizing electric field is applied perpendicular or parallel to the substrate plane,

first-order Frederiks transitions with a discontinuous jump of the director orientation at the threshold voltage may take place for sufficiently high H values.

When $E \perp H \perp n$ and both E and H are destabilizing, several distortion regions result [72]. For sufficiently high magnetic fields and small electric fields, twist distortion takes place in the direction of H. If the magnetic field is low and the electric field is strong, a splay-bend distortion occurs. In the region, where the actions of the electric and magnetic fields are comparable, a mixed splay–bend–twist deformation arises. Both second- and first-order phase transitions occur between the differently distorted regions; first-order transitions are accompanied by bistability and hysteresis phenomena [73].

The Frederiks transition in linear-chain and comb-like polymers may be treated in an analogous way as low-molecular-weight compounds, with some precautions [13], because, in general, the electric and viscoelastic properties of liquid crystal polymers are field dependent and the response of the materials to an external field is essentially nonlinear. Unfortunately, in the major part of electro-optical experiments this nonlinearity is not taken into account and results are interpreted in terms of conventional nematodynamics and constant material parameters. In addition, in only a few papers (e.g.

[74]) is a certain preliminary orientation of a polymer specified, and only then can one speak of the true Frederiks transition with a well-defined threshold voltage.

9.3.1.2 Field-Induced Periodic Structures

Under certain conditions the field-induced reorientation caused by the dielectric torque may result not in a uniform distribution of the director in the plane of a liquid crystal layer but in a spatially periodic distortion, either steady-state or transient.

Steady-State Patterns

At present, at least three types of steady-state dielectrically driven pattern are known for nematics. The electric-field-induced periodic bend distortion in the form of parallel stripes has been observed in a homeotropically oriented layer of 5-CB ($\varepsilon_a = 13$) in the presence of a stabilizing magnetic field H_z [75, 76]. The stripes with a wavevector q were parallel to the electric field E_x and stationary at low fields. It was shown that a stable periodic pattern of the director minimizes the free energy of the cell when the elastic moduli K_{11} and K_{33} are similar to each other. In these experiments the Frederiks transition is of first order, the nondeformed and deformed areas coexist at a given voltage, and the front between them may propagate along the direction y perpendicular to both fields [77].

The other stationary periodic pattern has been observed in polymer lyotropic liquid crystals [78]. The smallness of the ratio K_{22}/K_{11} of a poly-g-benzylglutamate solution results in a structure in which the period of the longitudinal domains is of the same order of magnitude as the cell thickness. Due to the high viscosity of the material ($\gamma_1 = 34\ P$) the time for domain formation exceeds 2 h. If the director was initially oriented along the x-axis (n_x) and the field (magnetic in [68]) acts along the cell normal z, the structure of the distortion may be searched for in the form

$$n_z = f(z)\cos qy; \quad n_y = g(z)\sin qy; \quad n_x = 1 \tag{37}$$

(the uniform Frederiks distortion corresponds to $q=0$).

The threshold voltage may be plotted as a function of K_{22}/K_{11} [79] (Fig. 12). The curves shown in the figure were calculated numerically for the zero anchoring energy ($W^S=0$) at both boundaries. It can be seen that for $K_{22}/K_{11} < 0.303$ the periodic splay–twist distortion is more favorable than a uniform splay.

The same result has been confirmed [80, 81] for a finite anchoring energy. The critical ratio K_{22}/K_{11} increases with increasing W^S, and reaches 0.5 for infinitely strong anchoring. The periodic distortion has been observed in a magnetic field, but an analogous case must exist in an electric field. Different geometries for studying the effect have been reviewed by Kini [82].

The steady-state modulated structure is also observed in homogeneously oriented nematic layers at frequencies of the applied

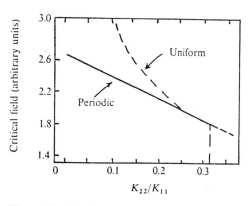

Figure 12. Calculated values of the critical field (in units of $(\pi/d)(K_{22}/\gamma_a)^{\frac{1}{2}}$) for the periodic splay–twist and uniform splay distortions [78].

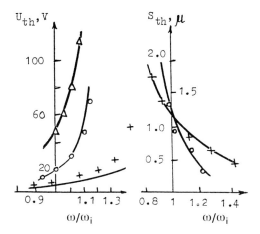

Figure 13. Frequency dependence of the threshold voltage U_{th} and domain period S_{th} for Frederiks domains near the dielectric anisotropy sign inversion frequency [83, 84]. (—) Calculated values. Experimental values: (+) $\varepsilon_0 - \varepsilon_\perp = +4.75$; (○) $\varepsilon_0 - \varepsilon_\perp = +0.35$; (△) $\varepsilon_0 - \varepsilon_\perp = +0.05$. ε_0 is $\varepsilon_\parallel$ for $\omega \to 0$.

ac voltage close to the frequency of the sign inversion ω_i of the dielectric anisotropy [83, 84]. The calculations show that the free energy is minimum when the periodic splay distortion and periodic potential distribution in the plane of a layer occur. The frequency region of the modulated Frederiks structure is very narrow near $\omega = \omega_i$. For $\omega < \omega_i$ the uniform Frederiks transition becomes more favorable, while for $\omega > \omega_i$ the threshold diverges rapidly (Fig. 13). If an initial director orientation is homeotropic the curves for the threshold voltage U_{th} and the period of the pattern at the threshold S_{th} reverse their slope with respect to the axis $\omega = \omega_i$ [84].

Transient Patterns

The formation of transient domain patterns aligned perpendicular to the initial director during the relaxation process of the magnetically driven Frederiks transition has been known since the earliest observations [85, 86]. The parallel stripes observed are, in fact, walls separating splayed regions with the clockwise and anticlockwise rotations of the director, arising due to coupling of the director orientation with a flow of liquid (the backflow effect [64] discussed above). The rate of the pattern relaxation is governed by the dynamics of the domain walls [87]. The same mechanism is responsible for the appearance of oblique structures [88].

In some cases, magnetically induced transient twist distortions have been observed in both thermotropic (MBBA [89]) and lyotropic (PBG [90]) systems. In this case, backflow effects are allowed only in a nonlinear regime, for strong distortions. The physical origin of this phenomenon could be the faster response times of modulated structures, as compared with uniform ones. When the equilibrium director distribution is approached, i.e. a relaxation process is over, the transient structures disappear. The emergence and subsequent evolution of the spatial periodicity of the transient structures have been considered theoretically [89, 90]. In addition, the pattern kinetics have been studied in detail experimentally [91] on a mixture of a polymer compound with a low-molecular-mass matrix. The polymer considerably increases the rotational viscosity of the substance and reduces the threshold for pattern formation. This indicates the possibility of recording the pattern using a video camera. A typical transient pattern is shown in Fig. 14 [91].

Some transient patterns overlapping the Frederiks transition (e.g. as observed by Buka et al. [92]) may be electrohydrodynamic in nature; these are discussed in Section 9.4 of this Chapter.

9.3.1.3   Flexo-Electric Phenomena

When liquid crystals possess electric polarization P (either spontaneous or induced by external factors), in addition to the qua-

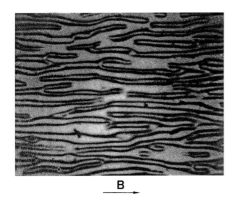

500 μm

Figure 14. Transient pattern observed in 5-CB doped with a polymer. The magnetic induction **B** is in the plane of the liquid crystal layer (thickness 100 μm) and perpendicular to the initial director (twist distortion) [91].

dratic-in-field coupling of an external electric field to the dielectrically anisotropic medium (Eq. (1)) a new, linear-in-field interaction appears (free energy density, **PE**). One of the sources of the electric polarization is an orientational distortion of the liquid crystal.

Classical Results

The macroscopic dipole moment of a unit volume (i.e. the electric polarization) may result from both the nonuniform director orientation [93]

$$P_f = e_1 n \operatorname{div} n - e_3 (n \times \operatorname{curl} n) \qquad (38)$$

and the spatially nonuniform orientational order parameter [94]

$$P_0 = r_1 (n \cdot \operatorname{grad} S) n + r_2 \operatorname{grad} S \qquad (39)$$

The flexoelectric polarization P_f depends on the curvature of the director field (div n and curl n) at constant modulus of the order parameter S. For uniform director orientation (n=constant), P_f=0. A simplified molecular picture of the phenomenon is shown in Fig. 15a, b. Equation (39) defines the so-called ordoelectric polarization P_0, which depends on the gradient of the nematic orientational order parameter S (quadrupolar in nature) and does not vanish for a uniform director distribution. For instance,

near a solid wall the nematic order parameter is a function of the z coordinate normal to the surface. The ordoelectric coefficients r_1 and r_2 are of the order of the flexoelectric coefficients e_1 and e_3, which are approximately 10^{-4} CGS. We will not discuss ordoelectricity here as it is mostly related to

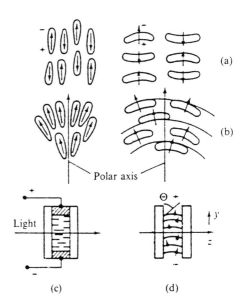

Figure 15. Flexoelectric effect: (a) structure of an undeformed nematic liquid crystal with pear- and banana-shaped molecules; (b) the same nematic liquid crystal subjected to splay and bend deformations, respectively; (c) experimental configuration; (d) distribution of the director along the z axis due to the flexoelectric effect.

the surface properties reviewed in other chapters in this book.

The flexoelectric effect results in a distortion of a dielectrically stable configuration (homeotropic nematic with $\varepsilon_a < 0$), provided that the director anchoring at the boundaries is weak (Fig. 15c). For small distortions the components of the director corresponding to the experimental geometry may be written as [95, 96]:

$$n_z = \cos \theta(z) \approx 1$$

and

$$n_y = \sin \theta(z) \approx \theta(z)$$

Then Eq. (38) reduces to

$$P_y = e_3 \frac{\partial \theta}{\partial z} \tag{40}$$

For small dielectric anisotropy ($\varepsilon_a \approx 0$) the free energy density includes only two terms: the flexoelectric and the elastic terms derived from the Frank energy (Eq. (24))

$$g = P_y E + \frac{K_{33}}{2} \cdot \frac{\partial^2 \theta}{\partial z^2} \tag{41}$$

By minimizing Eq. (41) with substitution of Eq. (40), one obtains the linear dependence of angle θ on the z coordinate:

$$\frac{\partial \theta}{\partial z} = \frac{e_3 E}{K_{33}} \tag{42a}$$

$$\theta = \frac{e_3 E}{K_{33}} z \tag{42b}$$

where $z=0$ corresponds to the center of a layer.

The resulting distortion is shown in Fig. 15d. It differs from the distortion in a Frederiks transition (see e.g. Fig. 11): the maximum angle of deviation of the director due to the flexoeffect occurs at the restricting surfaces, and the angle of deviation in the middle of the cell is zero. This is due to a particular geometry of the distortion: the

polarization under a field occurs uniformly throughout the volume and, as a result, the torque capable of rotating the director through angle θ occurs only at the surfaces. Because of this, weak anchoring of the nematic liquid crystal at the surface is a necessary requirement for the occurrence of such a flexoelectric distortion. The case of finite dielectric anisotropy and other geometries of the flexoelectric distortion have also been considered [96].

The dynamics of the flexoelectric effect have been studied [97, 98]. For a nematic layer with a geometry similar to that shown in Fig. 12c and finite dielectric anisotropy, the torque belance equation for the bulk of a sample, written in the linearized form ($\sin \theta \approx \theta$, $\cos \theta \approx 1$)

$$\gamma_1 \frac{\partial \theta}{\partial z} = K_{33} \frac{\partial^2 \theta}{\partial z^2} - \frac{\varepsilon_a E^2 \theta}{4\pi} \tag{43}$$

does not contain a destabilizing flexoelectric torque, as it is linear in the derivatives of the director. However, the flexoelectric term appears in the boundary conditions:

$$K_{33} \frac{\partial \theta}{\partial z}\bigg|_{z=\pm d/2} \pm W_i^S \theta|_{z=\pm d/2} - e_3 E = 0 \tag{44}$$

Here W_i^S corresponds to different anchoring energies at the two opposite surfaces ($i=1$ or 2), which have coordinates of $\pm d/2$ (where d is the thickness of the layer). The solution of the linear equations (Eqs. (43) and (44)) is given for various values of ε_a and W_i^S. For an ac electric field of frequency ω, two spatially periodic viscoelastic waves with wavevectors depending on the field frequency are predicted. For sufficiently high frequencies the waves decay in the vicinity of the glass plates limiting the cell. At low frequency, the waves interfere with each other in the bulk, and for $\omega=0$ the steady state distortion shown in Fig. 15d occurs.

The bulk flexoelectric response is limited by a narrow frequency region below the director relaxation frequency determined by Eq. (34). The viscoelastic wave of the director curvature decaying into the bulk of the substance has been probed experimentally by using an evanescent light wave and the modulation ellipsometry technique [99]. The wavevector of the curvature oscillations increases with increasing field frequency, $q_{\omega}=(\omega \gamma_1/2 K)^{\frac{1}{2}}$. In such a case the linear-in-field response was observed up to frequencies of the order of 100 kHz. The characteristic frequency $f_c = K/\pi \gamma_1 \lambda^2$ is determined by a crossover of two distances: the spatial period of the curvature wave $2\pi/q_{\omega}$ and the penetration length of the optical evanescent wave (a few tenths of the light wavelengths λ).

Flexoelectric Domains

The flexoelectric term δF in nematic free energy

$$\delta F_f = -\int_V [e_1 \, \boldsymbol{E}\boldsymbol{n} \, \mathrm{div}\,\boldsymbol{n} \\ - e_3 \, \boldsymbol{E}\,(\boldsymbol{n}\times\mathrm{curl}\,\boldsymbol{n})]\,\mathrm{d}\tau \qquad (45)$$

may result in a two-dimensional spatially periodic structure in the bulk. The application of an external dc electric field to a homogeneously oriented nonconductive nematic layer gives rise to a pattern of the domains parallel to the initial director orientation [100]. The period of the stripes w_{th} and the threshold voltage U_{th} of their appearance has been found [101–103] by minimizing the free energy of the nematic in an electric field, taking into account the term given by Eq. (45). A detailed experimental study of the dependence of the threshold and period of the longitudinal domains on the dielectric anisotropy and other parameters of nematic liquid crystals [104] has shown very good agreement between theory and experiment. The flexoelectric modulated structure

in nematic liquid crystals is also known as the "variable grating mode" [105–108], since for $U > U_{th}$ the period of domains w varies as $w \sim U^{-1}$ (Fig. 16).

(a)
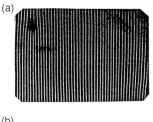

(b)
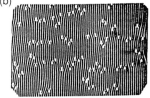

(c)

(d)
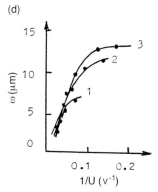

Figure 16. Flexoelectric domains in the nematic liquid crystal 4-butyl-4'-methoxyazoxybenzene [104]: (a) $U=16$ V, (b) $U=25$ V, (c) $U=50$ V ($\varepsilon_a=-0.25$, $d=11.7$ μm). (d) Variation in the domain period for different ε_a values: (1) $\varepsilon_a=-0.25$; (2) 0; (3) +0.15.

9.3.2   Twisted Nematics and Cholesterics

9.3.2.1   Twist and Supertwist Structures

Twist Effect

If the directions x and y of the planar orientation of nematic liquid crystal molecules on opposite electrodes are perpendicular to each other and the material has a positive dielectric anisotropy, then when an electric field is applied along the z axis (Fig. 17) a director reorientation occurs which is a combination of the splay, bend, and pure torsional distortions [109]. In the absence of the field (Fig. 17a), the light polarization vector follows the director and, consequently, the structure rotates the light polarization vector through an angle $\pi/2$ [110]. This specific waveguide regime (Mauguin's regime) occurs when the phase delay satisfies the condition $\Delta n\, d/\lambda \gg 1$.

As in the case of the Frederiks transition (discussed in Section 9.3.1.1 of this Chapter), the theoretical interpretation of the twist effect is based on the minimization of the free energy of the system. In this case, however, the problem is two dimensional, since both the azimuthal angle $\phi(z)$ and the tilt angle $\theta(z)$ are considered to be dependent on the z coordinate. In the case of the infinitely strong anchoring, the threshold for the distortion includes all three elastic moduli of a nematic liquid crystal [111]:

$$U_{tw} = \pi \left[\frac{\pi}{\varepsilon_a} (4\,K_{11} + K_{33} - 2\,K_{22}) \right]^{\frac{1}{2}} \quad (46)$$

When the applied voltage exceeds this threshold, the director deviates from the initial orientation so that the linear z dependence of the azimuthal angle disappears and the tilt angle becomes nonzero (Fig. 17b). The qualitative character of the functions $\phi(z)$ and $\theta(z)$ for different voltages is shown

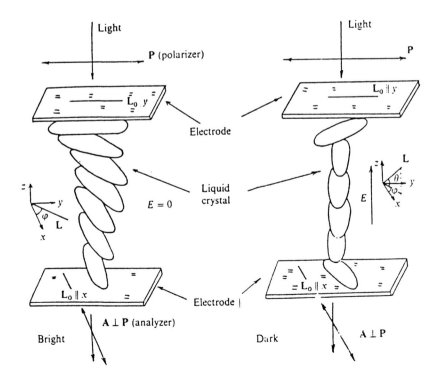

Figure 17. Twist-effect. Left: below threshold; Right above threshold

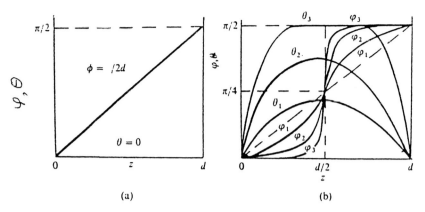

Figure 18. Dependence of director angles $\phi(z)$ and $\theta(z)$ in the twist effect for different voltages: (a) $U < U_{tw}$, (b) $U_{tw} < U_1 < U_2 < U_3$.

in Fig. 18. As the director tends to orient perpendicular to substrates the average value of $\langle \Delta n \rangle$ decreases, and for a certain voltage (the optical threshold for the twist effect) the waveguide regime vanishes.

Supertwist Effect

Supertwisted (i.e. twisted through an angle exceeding 90°) liquid crystal cells [112] are prepared from nematics doped with a small amount of an optically active material. Thus a cholesteric (or chiral nematics) with a large pitch P is created so that the helical pitch could be fitted to the boundary conditions for the directors at the substrates. For a cell with a supertwist angle ϕ_m, strong anchoring, and the zero director pretilt at the boundaries, the director distortion appears at a certain threshold voltage [113–115]:

$$U_{th}(\phi_m) = U_S$$
$$\cdot \left\{ 1 + \frac{\phi_m}{\pi} \left[\frac{\phi_m}{\pi} \left(\frac{K_{33}}{K_{11}} - \frac{2K_{22}}{K_{11}} \right) + \frac{4K_{22}}{K_{11}} \frac{d}{P} \right] \right\}^{\frac{1}{2}}$$

(47)

where $U_S = \pi(4\pi K_{11}/\varepsilon_a)^{1/2}$ is the threshold for the splay Freriks distortion in a planar nematic cell.

For a finite director pretilt at the boundaries θ_S, a sharp threshold disappears; however, it is possible to find a voltage at which the director angle at the center of the layer becomes equal to the boundary pretilt angle θ_S [116]. This voltage is very close to the optical threshold for the supertwist transition.

In supertwisted structures, bistable states are often observed [117]; for example, in the Grandjean texture with a tilted director orientation at the boundaries [118, 119]. In the absence of the tilt, the free (elastic) energy is a minimum at the thickness/pitch ratio $d/P_0 = n/2$ ($n = 1, 2, 3$, etc.), corresponding to n half-turns of the helix, located along the layer normal. If the directors at the boundaries are tilted then the corresponding free energy minima are no longer equidistant with thickness. At a certain thickness/pitch ratio, an electric field applied to a cell (the substance has $\varepsilon_a > 0$) switches the twisted structure from one state to another, differing in the number of half-turns. The relaxation of a new state might be very slow, and such a state may be considered stable. Thus, a field-induced bistability (or even multistability) effect occurs.

9.3.2.2 Instability of the Planar Cholesteric Texture

Texture Transitions

When an electric or magnetic field is applied to a liquid crystal cell, a texture transition occurs to minimize the free energy of the system. These texture changes in cholesteric liquid crystals are physically similar to the Frederiks transition in a nematic liquid crystal and result in a significant change in the optical properties of the layer. Texture transitions have been reviewed previously [8, 9] with allowance made for the sign of the dielectric or diamagnetic anisotropy, the initial texture, and the direction of the applied field. Here, we consider only the instability of the planar cholesteric texture, which has been widely discussed in recent literature.

Field Behavior of the Planar Texture

Let us discuss the case of when both the helical axis and the electric field are parallel to the normal z of a liquid crystal layer and the substance has positive dielectric anisotropy. The physical reason for the purely dielectric instability is rather well understood. When an electric field is not high enough to untwist a helix with a rather short pitch P_0, the dielectric torque tends to reorient the local directors into an arrangement similar to a homeotropically oriented nematic. In contrast, the elastic forces tend to preserve the distribution of the cholesteric layers (Fig. 19a). With not too large fields a pattern which undulates in the x and y directions appears as a compromise distribution of the director (Fig. 19b). The wavevector of this distortion along the z axis is approximately π/d. Related to the two-dimensional director pattern is a two-dimensional periodicity in the distribution of the refractive index, and hence a two-dimensional optical grating is formed. This grating is similar to the grid pattern formed due to the electrohydrodynamic instability (see Fig. 31c). With increasing voltage, the angle of deviation of the director increases, tending towards the limit $\theta = \pi/2$ for the whole layer (helix untwisting).

The threshold field for the formation of the periodic distortion has been calculated [120, 121] based on the expression for the

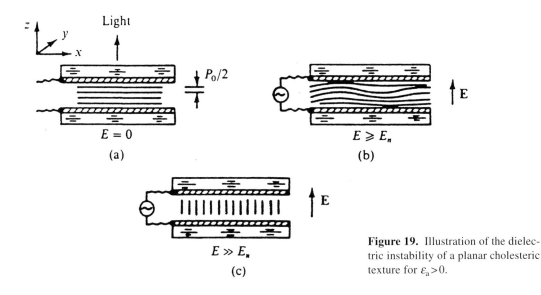

Figure 19. Illustration of the dielectric instability of a planar cholesteric texture for $\varepsilon_a > 0$.

free energy of a cholesteric liquid crystal and assuming $\varepsilon_a \ll \varepsilon$ and $d \ll P_0$:

$$g = \frac{1}{16}$$ (48)

$$\cdot \left[3 K_{33} \left(\frac{\pi}{w} \right)^2 + 8 K_{22} \left(\frac{w q_0}{d} \right)^2 - \varepsilon_a \, E^2 \right] \theta_m^2$$

In addition, it was assumed that there were only small deviations of the axis of the cholesteric helix from the normal to the glass surfaces or, equivalently, small deviations of the director from the xy planes (i.e. $\sin\theta \simeq \theta$ and $\cos\theta \simeq 1$). Futhermore, the deformation was assumed to be sinusoidal along both the x axis (the half-period equals w) and the z axis (the half-period equals d) and is described, in contrast to the nematic case, by two variables. One is the angle of deviation of the helical axis from the normal to the surfaces

$$\theta = -\theta_m \sin\left(\frac{\pi z}{d} \right) \times \sin\left(\frac{\pi x}{w} \right)$$ (49)

and the other is the difference in wavevectors for the distorted and the equilibrium helix

$$\Delta q = q - q_0 (q_0 = 2\pi / P_0)$$ (50a)

$$\Delta q = \Delta q_m \cos\left(\frac{\pi z}{d} \right) \times \sin\left(\frac{\pi x}{w} \right)$$ (50b)

The maximum compression Δq_m of the cholesteric planes and the maximum angle of deviation θ_m of the helical axis from the normal can be related to each other using purely geometric considerations:

$$\Delta q_m = \theta_m \left(\frac{q_0 w}{d} \right)$$ (51)

This value is substituted into the term with the K_{22} coefficient in Eq. (48). The numerical factors in Eq. (48) appear as a result of averaging the energy over the period of deformation [121].

The minimization of the elastic part of the free energy (Eq. (48)) (without the term

containing the field) by choosing the proper value of w gives the period of the distortion

$$w = \frac{1}{2} \left(\frac{3 K_{33}}{2 K_{22}} \right)^{\frac{1}{4}} (P_0 \, d)^{\frac{1}{2}}$$ (52)

and, using the condition $g=0$ with allowance for the field term, we find the threshold voltage for the distortion

$$U_{th} = 2 \pi^{\frac{3}{2}} (24 \, K_{22} \, K_{33})^{\frac{1}{4}} \left(\frac{d}{\varepsilon_a \, P_0} \right)^{\frac{1}{2}}$$ (53)

The threshold field is independent of the frequency up to the dispersion region ω_D of the dielectric permittivity.

The action of the field on a planar texture of a cholesteric liquid crystal for cell thicknesses comparable to the pitch and with a rigid anchoring of the directors to the surfaces does not cause two-dimensional deformations. In this case one-dimensional periodic patterns are observed [8, 9], with the orientation of the domains depending on the number of half-turns of the helix contained within the cell thickness. In fact, the domains are always perpendicular to the director in the middle of the layer.

A theoretical consideration of the case of a pitch that is comparable to the layer thickness for a purely dielectric destabilization of a planar texture in a field $E \parallel h$ has been given both numerically [122] and analytically [123, 124]. In the latter case the perturbation theory was used to search for the structure of the director field just above the threshold of the instability. Two variables, the polar angle θ and the azimuthal angle ϕ were considered, with orientation of the director at opposite walls differing by a twist angle α (pretilt angles at boundaries were also taken into account). It has been shown that two types of instability can be observed depending on the elastic moduli of the material: a total twist of the structure between two boundaries α, a ratio of the cell thickness d to cholesteric pitch P_0; and dielectric anisotropy. The first type is a homogeneous distortion analogous to the Frederiks transition in nematic liquid crystals. The second type is a periodic distortion (unidimensional in the layer plane) with a wavevector q directed at certain angles with respect to the director at boundaries.

With increasing voltage we enter one of the regimes (uniform or periodic distortion) separated by the Lifshits point where the wavevector of distortion is zero. The corresponding phase diagram has already been discussed for nematics (see Fig. 12). The periodic instability is favored when the original twist and the normalized cell thickness (d/P_0) increase and the dielectric anisotropy decreases. The principal predictions of the theory, namely, the disappearance of the periodic instability in thin cells and for small twist angles, agree with experiments performed on the Grandjean planar texture with varying d/P_0 [122].

Qualitatively similar results were obtained by Cohen and Hornreich [125]. In addition, it was claimed that the presence of a pretilt in the planar cholesteric texture decreases the tendency of the system to undergo transition to a periodically modulated (ripple) phase.

9.3.2.3 Field Untwisting of the Cholesteric Helix

Untwisting of a helix is observed only with positive dielectric or diamagnetic anisotropy and only in a field perpendicular to the helical axis h. Thus, in some cases, the director reorientation (or a texture transition) has to occur before untwisting is possible.

The Static Case

The model investigated by Meyer [126] and de Gennes [127] is illustrated in Fig. 20. Let

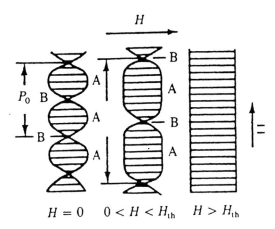

Figure 20. Schematic representation of untwisting of the cholesteric helix by a magnetic field [1].

us assume that we have a helix with pitch P_0 in the absence of a field, and that the thickness of the sample is sufficiently large (relative to P_0); thus the boundary conditions can be neglected. If $H \perp h$, then, in the initial state, in certain parts of the helix (A) the molecules are arranged favorably relative to the field, but in other parts (B) they are arranged unfavorably and tend to reorient themselves with their axes aligned along the field. As a result, the A regions will increase in size not so much due to a sharp decrease in the dimensions of the B regions (this would require too large an elastic energy $K_{22}(\partial\phi/\partial z)$), as due to a decrease in their number. Therefore, the helical pitch increases and the helix itself becomes non-ideal; that is, the sinusoidal dependence of the azimuth of the director ϕ upon the z coordinate is destroyed. When the last B region has been eliminated from the sample, the liquid crystal becomes a nematic, oriented uniformly along the field. Consequently, a field-induced cholesteric to nematic phase transition has occurred in the field.

The free energy of cholesterics in a magnetic field is

$$F = \frac{1}{2} \int \left[K_{22} \left(\frac{\partial \phi}{\partial z} - q_0 \right)^2 - \gamma_a H^2 \sin^2 \phi \right] dz \tag{54}$$

The function $\phi(z)$ should satisfy the condition for minimum F, and the corresponding Euler equation has a solution in terms of the elliptic function

$$\sin \phi(z) = \operatorname{sn} \left[\left(\frac{z}{\xi k} \right), k \right] \tag{55}$$

The values $u = z/\xi k$ and k are the argument and modulus, respectively, of the elliptic function, and

$$\xi = \frac{1}{H} \left(\frac{K_{22}}{\gamma_a} \right)^{\frac{1}{2}} \tag{56}$$

is the magnetic coherence length, see also the more general form given by Eq. (36).

The condition for the minimum of free energy (Eq. (54)) is

$$q_0 \, \xi = \frac{2 E(k)}{\pi k} \tag{57}$$

and the new pitch of the helix satisfying the condition of a minimum in F is

$$P(H) = P_0 \left(\frac{2}{\pi} \right) F(k) \, E(k) = 4 \, \xi \, k \, F(k) \tag{58}$$

Here $F(k)$ and $E(k)$ are complete elliptic integrals of the first and second kind. When $k \to 1$ the elliptic integral $F(k)$ diverges logarithmically, i.e. the helix pitch $P(H) \to \infty$. Simultaneously, when $k \to 1$, the integral $E(k) = 1$ and Eq. (57) gives the critical field for the untwisting of the helix:

$$H_c = \frac{\pi^2}{P_0} \left(\frac{K_{22}}{\gamma_a} \right)^{\frac{1}{2}} \tag{59}$$

The pitch of the helix increases gradually, beginning with infinitely small fields. Expansion of Eq. (58) gives:

$$P(H) = P_0 \left[1 + \frac{\gamma_a^2 \, P_0^4 \, E^4}{2^9 \, \pi^4 \, K_{22}^2} + \dots \right] \tag{60}$$

The increase in the pitch with increasing electric (or magnetic) field has been confirmed in numerous experiments carried out on rather thick samples [8, 9].

Field untwisting of a helix for thin cells with planar boundary conditions occurs differently from the case of an infinite cholesteric medium. For a field perpendicular to the axis of a cholesteric helix in the planar texture, a stepwise change in the pitch with increase in the field is predicted [128]. The size of the step increases with a decrease in the ratio $2 \, d/P_0$. Stepwise untwisting of the helix by an electric field perpendicular to the helix axis has been observed for the case of planar cholesteric texture with $\varepsilon_a > 0$ and with strong anchoring of the molecules to the limiting surfaces [129]. Under these conditions the relaxation of the field-induced (i.e. untwisted) state is accompanied by the formation of spatially modulated structures in the form of strips or grids.

The role of the boundary conditions has been extensively discussed in [130–133], and the influence of the flexoelectric effect has been analyzed [134].

Dynamics of the Untwisting

The dynamics of the helix unwinding is described by

$$\gamma_1 \frac{\partial \phi}{\partial t} = K_{22} \frac{\partial^2 \phi}{\partial t^2} + \gamma_a \, H^2 \sin \phi \cos \phi \tag{61}$$

The approximate solution of Eq. (61) results in the following expression for the response τ_r and decay τ_d times for the field-induced helix distortion [135]:

$$\tau_r = \frac{2 \gamma_1}{2 \, K_{22} \, q^2 \pm \gamma_a \, H^2} \tag{62}$$

and

$$\tau_{d} = \frac{\gamma_1}{K_{22} \, q^2} \qquad (63)$$

Here γ_1 and K_{22} are the rotational viscosity and twist elastic constant, q is a wavevector of the distortion mode under investigation. The plus and minus signs in Eq. (62) are related to the cases $q > q_0$ and $q < q_0$, respectively, where $q_0 = 2\pi/P_0$ is the equilibrium pitch.

In general, the dynamics of the untwisting of a helix have not been investigated sufficiently, and only the following fundamental facts have so far been established. The relaxation time is determined by the helical pitch P_0 when $d \gg P_0$, and by the thickness of the cell when $d \ll P_0$. The state of complete untwisting of the helix is metastable, and relaxation begins at defects in the structure. The response times in the field are inversely proportional to H^2.

Dynamics of Blue Phases

The field-induced distortion of the orientational state of blue phases was discussed in Section 9.2.4.3. These distortions decay rather rapidly. In the geometry of a Kerr cell (the field is perpendicular to the light wavevector) the field induces birefringence δn which results in electro-optical modulation ΔI of the transmitted light. Modulation characteristics have been studied over a wide temperature range of the blue phase with negative dielectric anisotropy $\varepsilon_a = -0.75$ [136]. The time constant for relaxation of the field-induced birefringence is well approximated by the simplest expression, Eq. (63) with $q = \pi/P_0$, where P_0 is the lattice constant and K is an averaged Frank modulus.

The dynamics of the BP_{III} (foggy) phase have some features related to the field-induced changes in the size of the domains that this phase forms.

9.3.2.4   Flexoelectric Effects

As discussed for nematics, flexoelectric effects are caused by the linear coupling of an external electric field with the flexoelectric polarization. In cholesterics, these effects are manifested in the three specific phenomena described below.

Fast Linear-in-Field Rotation of the Cholesteric Helix

This effect is observed in a geometry where the cholesteric axis h is homogeneously oriented in the plane of the cell (along x) and an electric field is applied to the electrodes of a sandwich cell along the z axis [137, 138]. In this case, the helical structure, even the ideal one, is incompatible with the planar boundary conditions, and splayed and bent regions form near the boundaries. Thus, according to Eq. (38), the flexoelectric polarization arises in those regions which can interact with the electric field. The distortion is very similar to that observed in the ferroelectric smectic C* phase (see Fig. 24) for a so-called 'deformed helix ferroelectric' effect [139].

Such a distortion results in a deviation of the optical axis in the plane of a cell. The sign and the magnitude of the deviation angle depend on the polarity and strength, respectively, of the applied field. In the field-off state the helix is undistorted and the cell behaves as a uniaxial optical plate, with the optical axis coincident with h. When the field is applied molecules leave the yz plane due to the flexoelectric deformation coming from the surface regions where the flexoelectric torque $M_f = e_f E$ is developed ($e_f = e_1 = e_3$ is assumed). Now the optical axis does not coincide with the initial orientation of the helix but forms the angle ψ with respect to it and proportional to E. The field-induced distortion of the helix is shown in Fig. 21 [137].

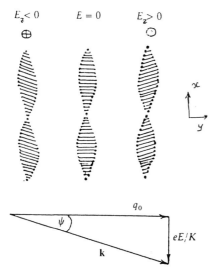

Figure 21. (Top) The pattern of the director rotation induced by an electric field applied perpendicular to the plane of the drawing [137]. (Bottom) Components of the wavevector k of the distorted helix.

In the steady-state regime, for director components parallel to the yz plane, $n_z = \cos\phi$, $n_y = \sin\phi$ (no conical distortions), the free energy density of the system in an electric field includes both the flexoelectric and dielectric terms:

$$g = \frac{1}{2} K \left(\frac{\partial\phi}{\partial y}\right)^2 + \frac{1}{2} K_{22} \left(q_0 - \frac{\partial\phi}{\partial x}\right)^2$$
$$- e_f \, E \frac{\partial\phi}{\partial y} + \frac{\varepsilon_a}{8\pi} E^2 \sin^2\phi \qquad (64)$$

where $K = K_{11} = K_{33}$.

For negligible dielectric anisotropy ($\varepsilon_a \to 0$), minimization of Eq. (64) results in $\partial\phi/\partial x = q_0$ and $\partial\phi/\partial y = e_f E/K$ (the anchoring energy is considered to be small). If the rotated axis of the helix is represented by a wavevector k then $|k|\cos\psi = q_0$ and $|k|\sin\psi = e_f E/K$ (Fig. 21) and the rotation angle is given by

$$\tan\psi = \frac{e_f E}{q_0 K} \qquad (65)$$

which is linear in E for small rotations.

For a finite value of dielectric anisotropy, with increasing field the linear response is affected by the helix unwinding [140] due to the quadratic-in-field interaction (last term in Eq. (64)). In order to increase the field-induced linear deviation of the optical axis, shot-pitch materials with small ε_a must be used [141].

Experiments show that the effect of the reorientation of the helical axis is fairly fast (in the range 10–100 μs) and the speed of the response is independent of field strength. This can be accounted for under the assumption that the pitch of the helix does not change in the field-on state, $|k| = q_0$. Then, the free energy becomes only a function of the angle ψ [137]. For small ψ, the rate of the response is determined by the dynamics of the field-free helix (Eq. (63)) with an effective viscosity coefficient instead of γ_1.

Electroclinic Response

When the helical structure of the chiral nematic phase is unwound by the influence of the limiting walls, one can observe a linear-in-field light modulation which is caused by a small molecular tilt [142]. The effect is analogous to the electroclinic effect observed in the smectic A phase as a pretransitional phenomenon in the vicinity of the A–C* transition. It is particularly strong in the vicinity of the N–A–C* multicritical point [143].

Flexoelectric Domains in Cholesterics

In Section 9.3.1.3 we discussed the longitudinal flexoelectric domains in nematics, the period of which depends on the inducing field, $w \propto E^{-1}$. When changing the sign of the field, the phases of the distortion angles $\theta(y)$ and $\phi(y)$ are changed for the opposite ones, but the direction of the wavevector of the instability (perpendicular to the initial director orientation n_0) remains unchanged. In cholesterics the mirror symmetry is broken and the same instability has specific features [144, 145].

First, the orientation of the domains depends on the number of half-turns of the helix between the cell boundaries. In the Grandjean texture the direction of the domains is different in neighboring Grandjean zones [146] for the same polarity of the field. When the polarity of the applied voltage is changed, the directions of the domain lines also change. With an oscillating field (at very low frequencies, of the order of 1 Hz), the direction of domains oscillates between two positions, characterized by an angle $\pm \alpha$ with respect to the director in the middle of the cell (in the first two Grandjean zones or in a twist nematic cell). The threshold voltage for the domain formation, and the domain period and angle α have been calculated [145].

For the zero zone with the directors at opposite boundaries parallel to each other (no twist) and zero dielectric anisotropy, both the threshold voltage and the wavevector can be calculated analytically as a function of the cholesteric pitch, despite the completely unwound state of the helix. The sign and amplitude of the angle of the field-induced domain rotation (at threshold) depends on the handedness and the pitch of the helix and the sign of the difference in the flexoelectric coefficients $e^* = e_{11} - e_{33}$:

$$\alpha_{th} = \frac{2 d K_{22}}{P_0 K_{33}} \operatorname{sign}(e^* E) \qquad (66)$$

With increasing frequency of a pulsed electric field, $\omega > \tau_d$ where τ_d is the relaxation time of the director (Eq. (34)); the deviation angle decreases and the threshold voltage diverges [146].

The other flexoelectric structure can arise in a very thin planar cholesteric layer with a certain set of material parameters. The instability occurs in the form of spiral domains [147] the handedness of which depends on the sign of the electric field. While the threshold voltage for the linear domains discussed above is almost independent of cell thickness, the threshold for spiral domains is proportional to the thickness, and the number of turns of the helix seen under a microscope in the plane of the layer increases with increasing field. Spiral domains have been observed experimentally [148].

9.3.3 Smectics and Discotics

9.3.3.1 Field Behavior of Achiral Smectics

Frederiks Transition in a Smectic A Phase

In the smectic A phase the director is always perpendicular to the plane of the smectic layers. Thus, only the splay distortion leaves the interlayer distance unchanged and only the elastic modulus K_{11} is finite, while K_{22} an K_{33} diverge when approaching the smectic A phase from the nematic one. However, the compressibility of the layered structure and the corresponding elastic modulus B should be taken into account when discussing the elastic properties of smectic phases. The free energy density for the smectic A phase subjected to the action of an external magnetic field is [1]:

$$(67)$$
$$g = \left[B \left(\frac{\partial u}{\partial z} \right)^2 + K_{11} (\operatorname{div} n)^2 - \gamma_a (n \, H)^2 \right]$$

where $u(r)$ is the displacement of the smectic layers from their equilibrium position, and the z direction coincides with the normal to the layers. Thus, it makes sense to investigate the Frederiks transition in the geometry corresponding to the splay distortion only. Let us assume a so-called 'bookshelf geometry', in which the smectic layers are perpendicular to the substrates, forming a sandwich-type cell of thickness d, and the director is oriented in the plane of the cell (e.g. along the x axis). If the magnetic anisotropy is positive ($\gamma_a > 0$) and a magnetic field is applied, the Frederiks transition should take place. The threshold for the transition coincides formally with that for a nematic (Eq. 27)); however, the amplitude of the distortion θ_m for fields exceeding the threshold is very small, being of the order of λ_S/d [149], where a smectic characteristic length

$$\lambda_S = \left(\frac{K_{11}}{B} \right)^{\frac{1}{2}} \qquad (68)$$

is of the order of few layer thicknesses. Thus, even in this, favorable case of the allowed splay distortion, the Frederiks transition is, in fact, unobservable (ghost transition). Instead, one observes a texture transition accompanied by the appearance of a number of defects.

Dielectrically Induced Texture Transitions

Texture transitions are particularly pronounced when an electric field is applied to materials having a large dielectric anisotropy. A planar texture undergoes transition to a quasihomeotropic optically transparent texture via intermediate structural defects [150,151]. The threshold voltage observed experimentally for a transition from a planar to a homeotropic texture depends on the layer thickness according to $U \propto d^{\frac{1}{2}}$ (for the Frederiks transition the threshold voltage is independent of thickness). A model that accounts for the experimental data (at least partly) has been developed by Parodi [152] who assumed the formation of transition layers between the surface and the bulk of a sample. A discrepancy between the calculated and observed periods of the texture instability may be due to a nonuniform

'chevron' structure of the original 'bookshelf' geometry. Chevron structures with smectic layers broken by a certain angle are often observed in both smectic C and smectic A layers [153].

The kinetics of the relaxation of the homeotropic field-induced texture to a focal conic one depends on the surface treatment [154–156]. A rough surface which provides a strong anchorage facilitates relaxation [156].

Frederiks Transition in a Smectic C Phase

Depending on the arrangement of the smectic layers and the director of the liquid crystal relative to the limiting surfaces, four different configurations can be identified. For each of them there are at least three alternatives for the direction of the field. Thus, 12 variations in all can be obtained and these have been studied theoretically by Rapini [149]. In most of the configurations the Frederiks transition is a ghost one, since it requires the collapse of the smectic layers. However, in three instances [157] the field only induces rotation of the director around the normal to the layers, leaving the layer structure unchanged.

Let us identify the three principal values of the dielectric permittivity as ε_i ($i = 1, 2, 3$), where ε_3 corresponds to the direction along the director, ε_1 corresponds to the direction perpendicular to the plane of the tilt, and ε_2 corresponds to the direction perpendicular to the preceding two [8, 9]. When the field is oriented along the director, a Frederiks transition is possible when $\varepsilon_2 > \varepsilon_3$. The director should rotate around the normal to smectic layers, not changing its angle Ω. The corresponding threshold field is proportional to $(\varepsilon_2 - \varepsilon_3)^{-\frac{1}{2}}$. A field perpendicular to the director and lying in the plane of the tilt induces the same distortion when $\varepsilon_2 > \varepsilon_1$. Now the threshold for reorientation of the director is proportional to $(\varepsilon_2-\varepsilon_1)^{-\frac{1}{2}}$. Finally, when the field is perpendicular to the tilt plane the same rotation takes place at threshold proportional to $(\varepsilon_1\cos^2\Omega+\varepsilon_3\sin^2\Omega-\varepsilon_2)^{-\frac{1}{2}}$.

Measurements have been made of the Frederiks transition in various smectic C liquid crystals [158–161], using the conventional sandwich cells and optically transparent electrodes. The director was oriented uniaxially by rubbing the electrodes, and the smectic layers were tilted with respect to the cell plane (a 'bookshelf' geometry). Due to positive dielectric anisotropy, an electric field applied along the cell normal (z direction) (Fig. 22) induces director rotation around the normal to the smectic layers.

The threshold field for the director deviation can be calculated neglecting the distortion of the smectic layers [158] and assuming $\varepsilon_2\approx\varepsilon_3$. The threshold depends on the angle μ between the normal and xy plane. For $\mu=0$, when the xy plane cuts half of the director cone (with an angle Ω at the apex) the threshold field is

$$E_c = \frac{\pi}{d\sin\Omega}\cdot\left(\frac{4\pi K^*}{\varepsilon_a}\right)^{\frac{1}{2}} \qquad (69)$$

where K^* is an effective elastic constant. When the cone just touches the xz plane

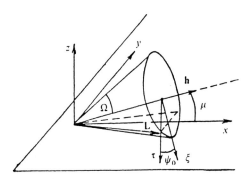

Figure 22. Geometry of the Frederiks transition for a planar-oriented smectic C liquid crystal.

($\mu=\Omega$) (Fig. 22), the threshold field increases proportionally to $\cos\Omega$. Both such cases have been studied using various substances [159–161]. For a certain set of parameters, the theory predicts a bistable behavior of the smectic C phase in a strong field. Such bistability has been observed in the form of the field-induced motion of domain walls, accompanied by a hysteresis in the process of the director reorientation. The walls appear due to the equally probable field-induced rotation of the director clockwise and anticlockwise, and separate the two areas of different director orientation.

In the absence of walls the dynamic behavior may be described, as in nematics, by adding a viscous torque to the elastic and electric ones. For the field-off state, the distortion decays with the 'nematic' time constant (see Eq. (34), where K^* should be substituted for K_{22}). The time of the response to an electric field can only be calculated for the simplest case when $\mu=0$, $\tau_E=\gamma_1/\varepsilon_a\sin^2\Omega(E^2-E_c^2)$, where E_c is defined by Eq. (69). Thus, the response of the smectic C phase is $\sin^2\Omega$ times slower than that of the nematic phase. However, in experiments the same substance often responds faster in the smectic C phase than in the nematic one [159–161]. This may be due to the smaller value of γ_1 when the motion of the director is confined by the cone surface. The same phenomenon has been observed for the ferroelectric smectic C phase [162]. The domain-wall motion makes the dynamics of switching more complicated; the field-induced wall velocity has been calculated by Schiller et al. [70].

Frederiks Transition in the Hexagonal Smectic I Phase

Frederiks transition in the hexagonal smectic I phase (and also in the smectic C phase with a weak hexagonal order [163]) has some specific features. In the smectic I phase the director $\boldsymbol{n}$ forms a fixed angle θ with respect to the smectic layer normal. The lines between the centers of gravity of neighbouring molecules within a smectic layer form a hexagonal structure and the director is oriented toward one of the six vertices of the hexagon. Thus a field-induced reorientation of the director is accompanied by a reorientation of the bond direction. The latter requires some additional energy, which must be added to the elastic term in the free energy expansion. The calculations of the threshold electric voltage for the Frederiks transition from a 'bookshelf' geometry [164] show that it may be presented in the form

$$U_I^2 = U_C^2 \left[1 + f\left(K, L, G, h, d\right) \right] \qquad (70)$$

where U_C is the 'smectic C threshold voltage' from Eq. (69); K, L, and G are elastic moduli related to the director, bond reorientation, and the coupling between the two; h is a constant related to the constraint on the director to be along one of the six preferred directions due to the hexagonal order; and d is the cell thickness. Thus, the threshold voltage for a smectic I phase, is in contrast to nematic and smectic C phases, thickness dependent.

Polar Achiral Polyphilics

Polar crystalline and mesomorphic phases are divided into two groups according to their symmetry. One group is composed of chiral molecules having point group symmetries ranging from C_1 to C_∞. Examples of this type are liquid crystalline chiral phases (smectic C*, discussed in Section 9.3.3.2 of this Chapter, F*, I*, etc.) differing in the character of the in-plane packing of their molecules within a smectic layer. These ferroelectric phases manifest reversible switching of their spontaneous polarization.

The other group of polar materials comprises mirror-symmetric (achiral) phases of point symmetry groups from C_{2v} to $C_{\infty v}$. Examples of these are many classical solid crystalline ferroelectrics, such as BaTiO, triglycinsulfate, and thiourea. Recently polar properties (probably antiferroelectric) have also been observed in a noncrystalline (mesomorphic) achiral molecular material [165,166]. The basic idea was to use the polyphilic effect to form 'building elements' of a polar phase. According to this concept, chemically different moieties of a molecule tend to segregate to form polar aggregates or lamellas which result in the formation of polar phases. The direction of the spontaneous polarization is allowed to be in the plane of the lamellas or layers (transverse ferro- or antiferroelectric) or perpendicular to the layers (longitudinal ferro- or antiferroelectric [167]). In the particular case by Tournilhac et al. [165, 166], field reversible piezoelectric and pyroelectric effects were observed in a tilted, highly disordered metastable phase [168] of a polyphilic compound.

9.3.3.2 Chiral Ferroelectrics and Antiferroelectrics

In 1975, the first ferroelectric liquid crystal (FLC), called DOBAMBC (**III**),

$$\text{C}_{10}\text{H}_{21}\text{-O} \underset{}{\text{---}} \text{N} \underset{}{\text{---}} \text{COO-CH}_2\text{-CH-C}_2\text{H}_5$$

$$\overset{\text{CH}_3}{\underset{*}{\mid}}$$

III

was synthesized [169]. Since then FLCs have been the object of intensive investigation. A contemporary version of the general phenomenological theory, including the dynamic behavior, can be found in various books [7, 170]. Many experimental results and the discussion of FLC parameters may be found in other publications [8, 9, 171, 172] and in the other chapters of this book.

A comprehensive review devoted to antiferroelectric liquid crystals has also been published [173]. Thus, for the sake of completeness, only the field effects of fundamental importance are discussed here.

Smectic C* Phase Ferroelectrics

The symmetry of the ferroelectric smectic C* phase corresponds to the polar point group C_2. When proceeding along the z coordinate perpendicular to the smectic layers, the director n and the polarization vector P directed along the C_2 axis rotate; that is, a helix of pitch h is formed.

In the presence of an external field the total free energy includes the elastic energy of director deformations F_d, the surface energy F_S, and energy F_E of the interaction of a ferroelectric phase with the field E:

$$F = F_d + F_s + F_E = \int_V g_d \, d\tau + \int_s W_s \, d\sigma$$

$$+ \int_V \left(-PE - \frac{DE}{8\pi} \right) d\tau \qquad (71)$$

In general, the elastic term includes the elasticity related to the reorientation of the director (nematic moduli K_{ii}), and to a change in the tilt angle (coefficients a, b, c, etc., of the Landau expansion; see e.g. Eq. (10)). The corresponding expression for g_d is very complex and may be found in Pikin [7]. The surface energy includes both the dispersion and the polar terms of the anchoring energy W_S. Below we consider a few rather simple cases.

The Clark–Lagerwall Effect. This effect is observed in thin surface-stabilized FLC (SSFLC) cells where the smectic layers are perpendicular to the substrates, the thickness is less than the helical pitch ($d \le h$), and the helix is unwound by the walls [174–176] (Fig. 23). The electric field of opposite polarity switches the direction of the spontaneous polarization between the UP and

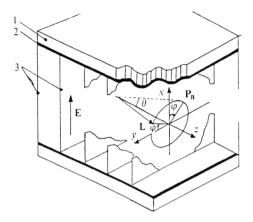

Figure 23. The Clark–Lagerwall effect. In a thin SSFLC cell the electric field of opposite polarity switches the spontaneous polarization between the UP and DOWN positions, which correspond to the LEFT and RIGHT positions of the director.

DOWN positions which correspond to the LEFT and RIGHT positions for the director that moves along the surface of a cone, the axis of which is normal to the layers and parallel to the cell substrates. In the LEFT and RIGHT positions the director remains parallel to the substrates and the SSFLC cell behaves as a uniaxial phase plate. The total angle of switching equals the double tilt angle θ.

The variation in the azimuthal director angle ϕ in the Clark–Lagerwall effect is described by the equation for the torque equilibrium, which follows from the minimization of the free energy (Eq. 71):

$$\gamma_\phi \frac{\partial \phi}{\partial t} = K \frac{\partial^2 \phi}{\partial x^2} + P_s \, E \sin\phi$$

$$+ \frac{\varepsilon_a}{4\pi} E^2 \sin\phi \cos\phi \qquad (72)$$

where the FLC is assumed to be uniaxial and γ_ϕ is the viscosity related to the motion of the director along the cone surface mentioned. When the helix is unwound by the cell walls, the first term on the left-hand side vanishes.

The polar W_p and dispersion W_d surface terms are included in the boundary conditions:

$$K \frac{\partial \phi}{\partial t} + W_p \sin \phi \pm W_d \sin 2\phi|_{x=0, d} \qquad (73)$$

For polarizations P greater than 10 nC/cm^{-2}, driving fields E less than 10 V μm^{-1}, and dielectric anisotropies $|\varepsilon_a|$ less than 1 we have

$$|\varepsilon_a| E/4\pi \ll P \qquad (74)$$

and, consequently, the third term on the right-hand side of Eq. (72) may be omitted. For $|\varepsilon_a| \approx 0$ and infinitely strong anchoring there is a threshold for the switching of a SSFLC cell [177]

$$E_c = \frac{K \pi^2}{P_s d^2} \qquad (75)$$

Here the elastic modulus K is defined in such a way that it includes θ^2 (see e.g. [7]). In the absence of backflow and at not very high voltages, the response times in the Clark–Lagerwall effect are determined by

$$\tau_\phi = \frac{\gamma_\phi}{P_s E} \qquad (76)$$

The backflow effect may decrease the response time several-fold [177]. If the inequality given by Eq. (74) is invalid, as occurs for sufficiently high fields, the response times of the Clark–Lagerwall effect increase sharply for positive ε_a values. Conversely, for negative values of ε_a, the corresponding switching times become shorter [178, 179]. For $|\varepsilon_a| E/4\pi \gg P$ the FLC switching times are approximately governed by the field squared, as in the Frederiks effect in nematic phases.

For weak anchoring conditions, two regimes of switching exist in SSFLC cells, separated by threshold field E_{th} [180]:

$$E_{th} = \frac{4W_d}{P_s d} \qquad (77)$$

For $E < E_{th}$ one observes the motion of domain walls, separating the regions of differently oriented polarization P and $-P$. The switching time is defined by the domain-wall motion. If $E > E_{th}$ (the Clark–Lagerwall regime), the switching time is determined by Eq. (76). At certain pretilt angles at the boundaries and high applied voltages, the switching may have the character of solitons or kinks in the function $\phi(z)$ spreading from one electrode to the other [181].

Deformed Helix Ferroelectric Effect. The deformed helix ferroelectric (DHF) effect observed in short-pitch FLCs [139, 182, 183] is a particular case of a more general phenomenon of the field-induced helix distortion observed in the very first investigations of FLCs [169, 184]. A theoretical explanation of the effect is given by Ostrovskii and coworkers [185–187]. The geometry of a FLC cell with the DHF effect is presented in Fig. 24. The polarizer at the first substrate makes an angle β with the helix axis and the analyzer is crossed with the polarizer. The FLC layers are perpendicular to the substrates and the layer thickness d is much larger than the helix pitch h_0. A light beam passes through an aperture $a \gg h_0$, parallel to the FLC layers, through the FLC cell.

In an electric field $\pm E$ the FLC helical structure becomes deformed, so that the corresponding dependence of the director distribution $\cos \phi$ as a function of coordinate $2\pi z/h_0$ oscillates between the two plots shown. These oscillations result in a variation in the effective refractive index (i.e. an electrically controlled birefringence appears). The effect takes place up to strengths sufficient to unwind the FLC helix

$$E_u = \frac{\pi^2 K q_0^2}{16 P_s} \qquad (78)$$

Here the elastic modulus K is defined as for Eq. (75).

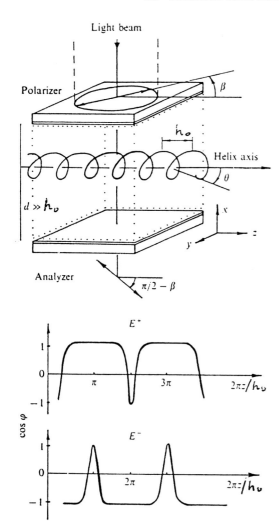

Figure 24. The deformed helix ferroelectric effect.

The characteristic response times of the effect at low fields $E/E_u \ll 1$ are independent of FLC polarization P_S and field E and are defined only by the rotational viscosity γ_ϕ and the helix pitch h_0:

$$\tau_c = \frac{\gamma_\phi}{K\, q_0^2} = \frac{\gamma_\phi\, h_0^2}{\pi^2\, K} \qquad (79)$$

For E close to the unwinding field E_u, the pitch h increases sharply to $h \gg h_0$. Consequently, the helix relaxation times τ_d to the initial state also increase: $\tau_d/\tau_c \propto h^2/h_0^2$.

This means that for $E \approx E_u$ it is possible to establish the memory state. If the helix is weakly distorted, fast and reversible switching could be obtained in the DHF mode [188].

Electroclinic Effect Near the Smectic A–Smectic C* Phase Transition. The electric field may induce a tilt of the director in the orthogonal smectic A phase near the smectic A–smectic C* phase transition (electroclinic effect [189, 191]). The electroclinic effect may be understood within framework of the Landau phase transition theory [7]. If the dielectric anisotropy is negligible, the free energy density of the smectic A phase may be expanded over the field-induced tilt angle θ:

$$g = \frac{1}{2}\, a\theta^2 + \frac{1}{4}\, b\theta^4 + \frac{P^2}{2\chi_\perp} - \mu_p P\theta - PE \quad (80)$$

where $a = \alpha(T - T_c)$, $b > 0$, $\chi_\perp$ is the background (far from the transition) contribution to the dielectric susceptibility, and μ_p is the tilt–polarization coupling constant (piezo-coefficient). Under the assumption that $b = 0$, the minimization of g with respect to θ and the polarization P results in the field-dependent tilt angle:

$$\theta = \frac{\mu_p\, \chi_\perp\, E}{\alpha\,(T - T_c)} = e_c\, E \qquad (81)$$

The coefficient e_c is known as the electroclinic coefficient. According to Eq. (81), the electroclinic coefficient becomes infinite at $T = T_c$ (in fact, it is limited by the term $b\,\theta^4$ [14]).

The dynamics of the electroclinic effect are, in fact, the same as the dynamics of the ferroelectric soft mode. The switching time may be derived from the equation for the balance of the viscous, elastic, and electric torques:

$$\gamma_\theta\, \frac{\partial \theta}{\partial t} = -\alpha\,(T - T_c)\,\theta + \mu_p\, \chi_\perp\, E \qquad (82)$$

It follows from Eq. (82) that the switching time of the electroclinic effect is independent of the electric field, and is defined only by the tilt rotational viscosity γ_θ and the elastic modulus α:

$$\tau_\theta = \frac{\gamma_\theta}{\alpha\,(T - T_c)} \qquad (83)$$

The experimental data confirm the linear increase in the tilt angle with increasing E for the electroclinic effect and independence τ_θ of E [190] (Fig. 25). The switching time τ_θ may be calculated more precisely if the fourth-order term in $b\,\theta^4$ is taken into account. In this case, a decrease in τ_θ for larger electric fields is observed [192]. At present, the electroclinic effect is the fastest of the known electro-optical effects in liquid crystals.

Domain Mode. A remarkable domain structure has recently been observed in FLCs possessing high spontaneous polarization (>100 nC cm^{-1}) [193–195]. A very stable spatially periodic optical pattern arises in layers of FLCs with their molecules oriented parallel to the substrates after a dc field treatment of the cell. The stripes are oriented perpendicular to the director orientation and their period is inversely proportional to the polarization magnitude squared, and independent of cell thickness.

The optical pattern may be observed for several months after switching off the dc field. The theoretical model is based on the consideration of a stabilizing role of the structural defects (dislocations) interacting with free charges in a FLC layer. The tilted layer structure responsible for such defects has been observed by direct X-ray investigation. The field-off domain structure has also been observed after the application of the electric field to the smectic A phase of a ferroelectric substance with very high spontaneous polarization [196]. The phenomenon is assumed to arise due to a break in the smectic layers (very similar to that just mentioned) induced by a strong electroclinic effect.

Ferroelectric Liquid Crystal Polymers. The symmetry requirements necessary for ferroelectricity in low-molecular-mass compounds are also valid for polymer mesophases. If a tilted chiral smectic phase is stable after a polymerization process, it must be ferroelectric. Following this idea the first polymeric liquid crystalline ferroelectric was synthesized by Shibayev et al. [197]. The substance is a comb-like homopolymer (*IV*)

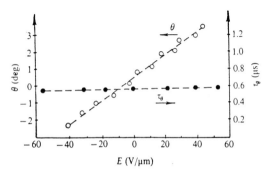

IV

with a polymethacrylate chain as a backbone and flexible spacers, COO(CH$_2$)$_{10}$–COO–, separating the phenylbenzoate mesogenic units from the backbone. Transverse (with respect to mesogenic units) dipole moments provided by several COO– groups are weakly coupled to the chiral terminal fragment, and thus the spontaneous polarization of the substance is small. Later, a variety of comb-like and main chain polymers manifesting

Figure 25. Dependence of the tilt angle θ and the response time τ_θ on the electroclinic effect in a FLC cell ($T = 25\,°C$) [190].

chiral smectic C*, F*, and other phases were synthesized (for reviews see [13, 198]). The field behavior of polymer ferroelectric liquid crystals is very similar to that typical of low-molecular-mass compounds. In general, all processes are very slow due to the high viscosity of polymer materials. An interesting feature of the polymers is their ability to form, on cooling, a glassy state with a frozen spontaneous polarization. Such solid materials have pyroelectric and piezoelectric properties.

Antiferroelectrics

A possibility of the antiferroelectric behavior of low temperature, highly ordered, chiral phases and the herringbone packing of smectic layers with an alternating tilt of the director was first discussed in connection with their very strong field-induced pyroelectric response [199]. In such a structure the direction of the in-plane polarization also alternates and the whole structure is antiferroelectric. The characteristic switching time in such phases is very long (seconds). The slowly switchable smectic X phase [200] appears to belong to the same family of antiferroelectrics. A rather fast tristable switching typical of solid antiferroelectrics has recently been reported [201–204] for chiral liquid crystalline phases of MHPOBC (**V**).

$$H_{17}C_8-O-\!\!\left\langle\!\!\bigcirc\!\!\right\rangle\!-\!\!\left\langle\!\!\bigcirc\!\!\right\rangle\!-CO_2-\!\!\left\langle\!\!\bigcirc\!\!\right\rangle\!-$$

$$-CO_2-\overset{*}{C}H(CH_3)-C_6H_{13}$$

V

A third, field-off, state, in addition to the two stable states known for the Clark–Lagerwall effect, has also been shown to be stable, and a layered structure with alternating tilt has been established by optical [205, 206] and scanning tunneling microscopy [207] techniques.

In fact, MHPOBC shows a very rich polymorphism [208]:

$$SmI_A^* - 64°C - SmC_A^* - 118.4°C$$
$$-SmC_\gamma^* - 119.2°C - SmC* - 120.9°C$$
$$-SmC_\alpha^* - 122°C - SmA - 148°C - I$$

where SmC_A^* and SmC_α^* are antiferroelectric phases, SmC* is a typical ferroelectric phase (discussed in the previous section), and SmC_γ^* is a ferrielectric phase. On a macroscopic scale the phases are twisted, with a characteristic pitch depending on temperature. Antiferroelectric phases have been observed in many other compounds, including polymers [209], and have been investigated in detail (for a review see e.g. [173]).

The application of an electric field has a considerable effect on the phase transition points in MHPOBC [210]; the phase diagram (electric field versus temperature) is shown in Fig. 26. A very narrow low-temperature ferrielectric phase is seen (FIL) between the SmC_A^* and SmC_γ^* phases mentioned earlier. In addition, even a weak field strongly influences the boundary between the SmC_α^* and SmC* phases. Further inves-

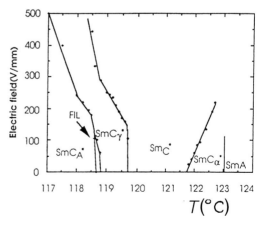

Figure 26. Electric field versus temperature phase diagram of MHPOBC [210].

tigations have shown [211] that the field-induced transition between the two phases in MHPBOC involves the formation of a variety of ferrielectric configurations. An even richer field-induced polymorphism has been observed in binary mixtures of MHPOBC and MHPOCBC [210].

A typical double hysteresis loop for electro-optical switching of the antiferroelectric SmC$^*_A$ phase is shown in Fig. 27. There are two field-induced uniform states and a zero-field third state which may be either uniform or twisted [212]. The transition from the antiferroelectric to the ferroelectric state proceeds through intermediate ferrielectric states. With increasing frequency in the range 10–100 Hz, the double hysteresis loop gradually transforms into the conventional ferroelectric loop [213]. The characteristic time corresponds to the time of the formation of domains of the antiferroelectric phase. The low-frequency switching (0.2 Hz) of the SmC$^*_\gamma$ phase reveals a triple hysteresis loop with two intermediate (+ and –) ferrielectric states and two final ferroelectric states. At higher frequency, conventional ferroelectric switching is observed.

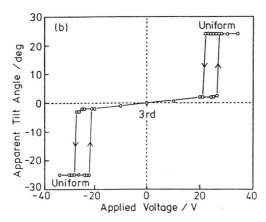

Figure 27. A double hysteresis loop typical of the electro-optical switching of the antiferroelectric phase [212].

Chiral Lyotropics and Discotics

Lyotropic Systems. Water solutions of certain amiphiphilic molecules (e.g. lipids) form lamellar lyotropic phases with a tilt of molecules to the layer normal. When the same molecules are chiral or other chiral molecules (guests) are introduced into the lamellas (host), the lamellar phase satisfies all the symmetry requirements of a ferroelectric phase (in formal analogy to SmC*). One of the best chiral additives to be introduced in lyotropic phases is cholesterol, which induces rather high spontaneous polarization in thermotropic SmC matrices [214]. In the L' phase, where matrix molecules are tilted, the chiral mixture (ethylene glycol doped with cholesterol) manifests a one order of magnitude higher piezo response than the undoped mixture. An even more striking difference between the undoped and doped samples (two orders of magnitude) is observed when a dc poling field is applied [215]. With increasing concentration of the dopant the piezo response grows. A dramatic difference between the piezo response of the undoped and the cholesterol doped samples has also been observed for water solutions [216]. Certainly the electric behavior of chiral analogs of the lyotropic phases is much more pronounced than that of their achiral counterparts and, therefore, cholesterol induces ferroelectricity in both systems studied (based on ethylene glycol and water).

Discotic Ferroelectrics. Columnar mesophases with chiral disk-shaped molecules whose normals are tilted with respect to the column axis also manifest ferroelectric switching [217]. In such phases the direction of the spontaneous polarization is perpendicular to both the column axis and to the normal to the molecular disks. The tilt of the disk forms a helix with its axis oriented along the columns. In an amphiphilic pyrene derivative, the positive and negative voltages rotate the direction of the minimum refractive index (which is perpendicular to the column axis) through a certain angle (about ±20°) clockwise or anticlockwise, respectively. The other homolog from the same series shows a field-induced change in the structure of the phase with an increase in the spontaneous polarization from 60 to 180 nC cm^{-2} [218]. The two phases are thought to have different sublattices of elliptic and polar columns. The switching times vary over a very broad range (from 1 ms to 100 s), being strongly dependent on the field (as $\tau \propto E^{-4}$) and temperature. The low-field phase is switched five times faster than its high-field counterpart. The columnar phases have some advantages over ferroelectric smectics (e.g. they are shock resistant) [218].

9.4 Electrohydrodynamic Instabilities

In this section we discuss electrohydrodynamic (EHD) instabilities, that is electric-field-induced phenomena that are caused by the flow of a liquid crystal (see also [8, 219]. The reason for the flow is electrical conductivity, which has been disregarded in previous sections. The flow may arise either independently of the anisotropic properties of substance, as in isotropic liquids (isotropic modes of the electrohydrodynamic instability), or may be driven by the conductivity anisotropy, as in liquid crystals (anisotropic modes). The threshold for EHD instabilities depends on many parameters, such as the electrical and viscoelastic properties of substance, the temperature, and the applied field frequency. Due to flow distortion of the director alignment, the instability is usually accompanied by a characteristic optical

pattern that depends on optical anisotropy of substance. This pattern formation is a special branch of physics dealing with the nonlinear response of dissipative media to external fields, and liquid crystals are well accepted model objects for the investigation of such phenomena [220]. To date, the most important results on EHD instabilities have been obtained for nematic liquid crystals (for a review see e.g. [221]).

9.4.1 Nematics

9.4.1.1 Classification of Instabilities

Electrohydrodynamic instabilities in nematics may be classified according to their threshold voltage (or field) dependence on the physical parameters of the liquid crystal (cell geometry, field frequency, etc.). The most important case for which all the typical instabilities may be observed is a planar cell with homogeneously oriented director. Instabilities in cells with homeotropic or tilted molecular orientations have some specific features, but the general mechanisms are the same. A qualitative phase diagram (the threshold voltage versus the frequency of the applied ac field) for different electrohydrodynamic instabilities in homogeneously oriented liquid crystals is shown in Fig. 28.

Among the instabilities shown there are two isotropic modes. One of these is well known from experiments on isotropic liquids and occurs at very low frequencies due to some injection processes at electrodes (injection mode). The other is also observable in ordinary liquids (such as silicon oil), but is seen particularly clearly in nematics, due to their optical anisotropy. The latter mode occurs over a wide frequency range due to ion drift to the electrodes as in electrolysis (electrolytic mode). Both isotropic

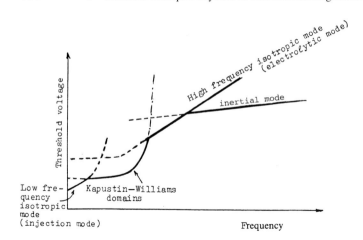

Figure 28. Threshold voltage as a function of frequency for different electrohydrodynamic instabilities in homogeneously oriented nematics.

EHD instabilities are driven by the electric-current-induced flow, even at zero conductivity anisotropy.

The other two instabilities shown in Fig. 28 may be observed only in liquid crystals (nematic, cholesteric, and smectic C). The first is the Carr–Helfrich instability, which is caused by a low-frequency electric field and occurs in the form of elongated vortices with their axis perpendicular to the original director alignment. The vortices cause a distortion of the director orientation, which is observed optically as a one-dimensional periodic pattern (Kapustin-Williams domains). The other anisotropic mode is observed only in highly conductive liquid crystals. For its interpretation the inertial term $\partial v/\partial t$ for the fluid velocity must be taken into account, which is why this mode may be called inertial mode.

The different modes may interfere with each other, even near their thresholds, and a variety of patterns may be observed. For example, 'chevrons' occur due to interference between the electrolytic and inertial modes, and a transient periodic pattern due to backflow effects accompanies the Frederiks transition. Well above the threshold, when the induced distortion is nonlinear with the applied field, many other patterns

are observed, including travelling waves or periodically pulsing modes [221].

The most important EHD instabilities are discussed below.

9.4.1.2   Isotropic Modes

Nematic liquid crystals are weak electrolytes containing a certain amount of ions. The ions may be intrinsic or created by an external electric field, which favors the dissociation of neutral molecules (chemical degradation). They may also appear as a result of electrochemical reactions at electrodes due to the injection of additional charge carriers. In some cases, the appearance of EHD instabilities in the nematic phase does not correlate with such inherent liquid crystalline properties as dielectric ε_a or conductivity σ_a anisotropy. The only physical reason for these instabilities might be a nonuniform ion distribution along the direction z parallel to electric field.

Standard Equations

To describe the field-induced motion of a liquid, a set of fundamental equations should be written which may be solved under certain assumptions. The first is the equation of motion in the Euler form:

$$\frac{\partial \boldsymbol{v}}{\partial t} + (\boldsymbol{v}\nabla)\,\boldsymbol{v} = \frac{\boldsymbol{f}}{\rho} \qquad (84)$$

where ρ is the density of the liquid, $\boldsymbol{v}$ is its velocity, t is time, and $\boldsymbol{f}$ is the force per unit volume, which includes the pressure gradient ($\operatorname{grad} p$), the gravity force $\rho \boldsymbol{g}$, the electric force $\boldsymbol{f}_E$, and forces due to viscous friction $\boldsymbol{f}_{\text{visc}}$ that depend on the fluid velocity $\boldsymbol{v}$:

$$\boldsymbol{f} = -\operatorname{grad} p + \rho\, \boldsymbol{g} + \boldsymbol{f}_E + \boldsymbol{f}_{\text{visc}} \qquad (85)$$

The second equation to be taken into consideration is the equation for the conservation of mass (or the continuity of liquid):

$$\frac{\partial \rho}{\partial t} + \operatorname{div}\rho \boldsymbol{v} = 0 \qquad (86)$$

The force exerted on the liquid by an external field is due, in the simplest case, to the space charge of density Q:

$$\boldsymbol{f}_E = Q\, \boldsymbol{E} \qquad (87)$$

In the case of a space and time dependent force $\boldsymbol{f}$, Eqs (84)–(87) have no general solution and the problem must be simplified. For example, we may ignore the gravity force and the pressure gradient and, in addition, consider the fluid to be incompressible (i.e. $\rho =$ constant). Then the friction force is written simply as [222]:

$$\boldsymbol{f}_{\text{visc}} = \eta \nabla^2 \boldsymbol{v} \qquad (88)$$

where η is the viscosity coefficient, and Eqs (84), (86), and (87) take the form of the flow continuity equation

$$\operatorname{div}\boldsymbol{v} = 0 \qquad (89)$$

and the Navier–Stokes equation

$$\frac{\partial \boldsymbol{v}}{\partial t} + (\boldsymbol{v}\nabla)\,\boldsymbol{v} = \eta \frac{\nabla^2 \boldsymbol{v}}{\rho} + \frac{\boldsymbol{f}_E}{\rho} \qquad (90)$$

As before, Eq. (90) is non-linear and has a simple solution only for low fluid velocities when the Reynolds number is small (Re $= \rho v l / \eta \ll 1$, where l is a characteristic length of an experiment). Then, the second term on the left-hand side of the Eq. (90) vanishes and we get a new form:

$$\frac{\partial \boldsymbol{v}}{\partial t} - \eta \frac{\nabla^2 \boldsymbol{v}}{\rho} - \frac{Q\boldsymbol{E}}{\rho} = 0 \qquad (91)$$

Even now there remain serious problems to be solved. In fact the charge density $Q(\boldsymbol{r}, t)$ in Eqs (87) and (91) depends not only on the external field but also on the velocity of charged fluid. Thus, Eqs (89) and (91) have to be complemented by an equation for the conservation of charge (continuity of the electric current):

$$\frac{\partial Q}{\partial t} + \operatorname{div} \boldsymbol{J} = 0 \qquad (92)$$

where $\boldsymbol{J}$ is the current density, which includes the ohmic term, the convective term due to the charge transfer by the moving liquid, and the diffusion term due to the gradient of the charge carrier density:

$$\boldsymbol{J} = \sigma\, \boldsymbol{E} + \boldsymbol{J}_{\text{conv}} + \boldsymbol{J}_{\text{diff}} \qquad (93)$$

Finally, there is a coupling between the charge and the electric field strength given by Maxwell's equation:

$$\operatorname{div}\boldsymbol{E} = \frac{4\pi\, Q}{\varepsilon} \qquad (94)$$

Equations (89)–(94) govern the behavior of an isotropic liquid in an electric field. For very weak fields the liquid is conducting but immobile. With increasing field a non-uniform spatial distribution of Q and $\boldsymbol{E}$ arises due to an ion drift to electrodes or/and charge injection, and the liquid starts moving in order to satisfy the minimum of entropy production. The corresponding critical field E_{crit} is considered to be the threshold field for the appearance of an EHD instability. Below we discuss the two particular cases mentioned earlier, namely the injection mode and the electrolytic mode.

Injection Mode

Let us consider a plane sandwich cell into which the cathode injects electrons that negatively charge neutral molecules. Due to an excess negative charge density $Q(z)$ near the cathode, a force QE is directed to the anode and tries to shift the charged layer of a liquid to the right (Fig. 29a). Since the cell is sealed and the liquid is incompressible, a circular convective flow occurs in order to reduce the internal pressure (Fig. 29b, c). This case resembles the well-known Benard problem in the thermoconvection of liquids.

The threshold of the instability may be estimated from Eqs (89)–(94). For simplicity, consider the one-dimensional case when the z component of the velocity is harmonically dependent on the transverse x coordinate:

$$v_z(x) = v_z \cos\frac{\pi}{d} x \tag{95}$$

The half-period of the cosine wave is assumed to be equal to the vortex diameter, that is to the cell thickness d. Differentiating Eq. (95) twice and substituting the result into Eq. (91) we get

$$\frac{\partial v_z}{\partial t} = \frac{v_z}{\tau_v} - \frac{QE_z}{\rho} = 0 \tag{96}$$

where

$$\tau_v = \frac{\rho\, d^2}{\eta\, \pi^2} \tag{97}$$

is a characteristic time for the relaxation of a vortex with diameter d.

In the continuity equation (Eq. (92)) we leave only the ohmic and convective terms:

$$J = \sigma E + Qv \tag{98}$$

When calculating div J it should be borne in mind that electrical conductivity results mainly in the relaxation of the space charge inhomogeneities along the x coordinate and E_z may be considered to be uniform. Thus,

$$\text{div}\, J_{\text{ohm}} = \sigma\, \frac{\partial E_x}{\partial x} = \sigma\, \frac{4\,\pi\, Q}{\varepsilon} = \frac{Q}{\tau_Q} \tag{99}$$

where

$$\tau_Q = \frac{\varepsilon}{4\pi\sigma} \tag{100}$$

is the Maxwell time for the space charge relaxation.

The convective term in the current obeys the equation:

$$\text{div}\, J_{\text{conv}} = v_z\, \frac{\partial Q}{\partial z} + Q_z\, \frac{\partial v_z}{\partial z} = \alpha \cdot v_z \tag{101}$$

where v_z is assumed to be constant and α is the gradient of the injected charge density. Therefore, the current continuity equation

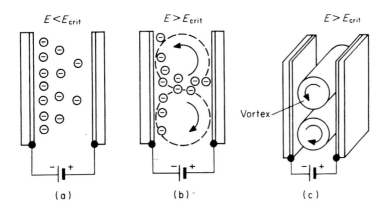

$E < E_{\text{crit}}$ $E > E_{\text{crit}}$ $E > E_{\text{crit}}$

Vortex

(a)   (b)   (c)

Figure 29. EHD instability caused by the injection mechanism. (a) Distribution of the injected charge at voltages below the threshold for convection. (b) Change in the charge distribution due to convection. (c) Cylindrical convective vortices.

(Eq. (92)) becomes

$$\frac{\partial Q}{\partial t} + \frac{Q}{\tau_Q} + \alpha \, v_z = 0 \qquad (102)$$

and we have a pair of equations (Eqs (96) and (102)) completely analogous to the set describing the thermoconvective instability [223–225].

To calculate the critical field for the injection instability it is assumed that, *precisely at the threshold*, the time dependences $v_z(t)$ and $Q(t)$ are described by an exponential growth function $\exp(st)$ with $s=0$, which means that at the threshold the velocity and charge density neither increase nor decay. This is a general principle and is called the "exchange of stability" (see e.g. [226]). Calculating the determinant for the set of linear equations (Eqs (96) and (102)) we obtain the critical field for any given injection strength α:

$$E_{\mathrm{crit}} = \frac{\rho}{\alpha \, \tau_v \, \tau_Q} = \frac{4 \, \pi^3 \, \eta \, \sigma}{\alpha \, \varepsilon \, d^2} \qquad (103)$$

Finally, assuming a linear space charge distribution along the normal to the cell, $\alpha = \partial Q / \partial z \approx Q_0 / d$, and introducing the ion mobility $\mu = \sigma / Q_0$, we arrive at the threshold voltage independent of the cell thickness in the form

$$U_{\mathrm{crit}} = 4 \, \pi^3 \, \frac{\eta \, \mu}{\varepsilon} \qquad (104)$$

This differs from the more rigorous theoretical expressions only by the use of a numerical coefficient [202, 203]. For typical parameters of a dielectric liquid ($\eta \approx 1$ P, $\mu \approx 10^{-5}$ cm^{-2} V^{-1} s^{-1} and $\varepsilon \approx 10$), $U_{\mathrm{crit}} \approx$ 1 V. However, in some cases (e.g. for conductive liquids) T may be as high as 100, or even more [227, 228]. Experimentally, in thin nematic liquid crystal cells, the critical voltage is of the order of few volts. The instability appears at a dc field or at very low frequencies (Fig. 28). The optical pattern

depends on the geometry of the experiment; in the isotropic phase the fluid motion may be observed using a strioscopic technique [229], or by following the motion of foreign particles. In nematics the areas of strong velocity gradients are easily visualized under a polarizing microscope due to the director distortion by shear. For example, in planarly oriented nematics an optical pattern in the form of beans is observed [230, 231].

Electrolytic Mode

The other important 'isotropic' instability is the so-called electrolytic mode [232, 233]. It appears in any fluid phase (isotropic, nematic, cholesteric, smectic A) within a wide frequency range in the form of small vortices near cell electrodes. The reason for the instability is a non-uniform space charge distribution (Fig. 30) due to an electrolytic process in the bulk of the substance (positive and negative ionic charges are separated during the period of the applied external ac field and move to opposite electrodes). The vortices are seen under a microscope in the form of small linear pre-chevron domains (in homogeneously oriented nematics), Maltese crosses (homeotropic orientation) and more complicated structures (cholesterics and nematics with a short-range smectic order). However, when there is no coupling between the flow and the di-

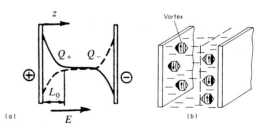

Figure 30. Electrolytic EHD mode. (a) Distribution of the density of positive (Q^+) and negative (Q^-) charges during one period of the applied ac field. (b) Vortex motion of the fluid near the electrodes.

rector (the isotropic and smectic A phases) the vortices may be observed only by using forcign floating particles. The instability has a well-defined field (not voltage) threshold that is dependent on the frequency of the applied field according to $E \propto \omega^{\frac{1}{2}}$.

To calculate the threshold it is assumed [7] that the charge density gradient along the layer thickness obeys the law:

$$\frac{\partial Q}{\partial z} = -\chi E(z) \tag{105}$$

where χ is the electrokinetic coefficient, which may be evaluated using the Maxwell equation:

$$\frac{\partial Q}{\partial z} = -\frac{\varepsilon}{4\pi} \cdot \frac{\partial^2 E}{\partial z^2} \approx -\frac{\varepsilon E}{4\pi L_D^2};$$

$$\chi = \frac{\varepsilon}{4\pi L_D^2} \tag{106}$$

Here L_D is a characteristic (Debye) length where the space charge is concentrated:

$$L_D = \left(\frac{\varepsilon D}{8\pi \sigma} \right)^{\frac{1}{2}} \tag{107}$$

where D is an ion diffusion coefficient. When the frequency ω of the applied field is increased, the ions do not have time to overcome the gap between the electrodes and concentrate in more extended regions of thickness

$$L_D(\omega) \approx L_D(0) \left[1 + \left(\frac{\varepsilon \omega}{8\pi \sigma} \right)^2 \right]^{-\frac{1}{4}} \tag{108}$$

Thus, the electrohydrodynamic instability appears in regions near the electrodes and vortices have a characteristic dimension L_D, which is smaller than the gap between electrodes.

Using the same current continuity equation (Eq. (98)) we arrive at the same set of equations (Eqs (96) and (102)) for small amplitudes of velocity v and space charge Q.

The only difference is in the characteristic relaxation time for a hydrodynamic vortex:

$$\tau_v \approx \frac{\rho}{\eta} \cdot \frac{q_x^2}{\left(q_x^2 + q_z^2 \right)^2} \tag{109}$$

Now the vortex has dimensions $w_x = \pi/q_x$ and $w_z = L_D = \pi/q_z$ along the x and z axes, respectively. The latter depends explicitly on the field frequency (see Eq. (108)).

The threshold field E_{crit} may be found using Eqs (96) and (102) and the same procedure described as earlier. For a square wave excitation with frequency f and amplitude E the different expressions are obtained in two frequency ranges: At low frequencies, $\omega \ll 4\pi \sigma/\varepsilon$,

$$E_{crit} = A \frac{(\sigma \eta)^{\frac{1}{2}}}{\varepsilon} \tag{110}$$

and at high frequencies, $\omega \gg 4\pi \sigma/\varepsilon$,

$$E_{crit} = B \left(\frac{\eta \omega}{\varepsilon} \right)^{\frac{1}{2}} \tag{111}$$

where A and B are numerical constants (given in [233, 234]). The characteristic size of the vortices at the threshold w_{crit} has also been found.

The thresholds for the formation of vortices in silicone oil, carbon tetrachloride, acetone, and in the isotropic phase of MBBA have been measured [232, 233] and it was shown that $U_{crit} \propto \omega^{\frac{1}{2}}$ and that the threshold decreases with decreasing viscosity. More detailed measurements of the dependences $U_{crit}(\omega)$ in pure and doped MBBA at various temperatures (for sandwich cells) have also been made. The threshold for the vortex motion was taken as the onset of the circulation of the solid particles in the electrode plane. The shape of the $U_{crit}(\omega)$ curves depends on the electrical conductivity. For high electrical conductivity the curves have a plateau proportional to σ in the low-frequency region and a characteristic dependence $U_{crit} \propto \omega^{\frac{1}{2}}$ independent of σ at frequencies above the critical one, which is defined by inverse time τ_Q (Eq. (100)). At the phase transition to the nematic phase, the threshold voltage of the instability has no peculiarity.

In homeotropically oriented layers the instability appears in the form of Maltese crosses (Fig. 31). The crosses are caused by small vortices (Fig. 30b) (see in [235]). In the homogeneously oriented layers, *the same* instability has a form of linear, short-pitch domains which are precursors of the

well-known *chevrons*. Thus, the high-frequency chevron mode, observed in various materials, is an isotropic mode and cannot be related to the high-frequency dielectric regime of the Carr–Helfrich instability where the threshold must diverge for zero anisotropy of conductivity (see next section). Up to now, all attempts to use the latter to interpret experimental data at high frequencies have failed. Of course, in this frequency range, at relatively high threshold fields the director of a nematic has a short relaxation time and oscillates with the field frequency, as predicted by the Carr–Helfrich–Orsay approach, but this effect is irrelevant to the onset of instability.

9.4.1.3 Anisotropic Modes

We now explicitly include in the discussion the anisotropy of the electrical conductivity σ_a of a liquid crystal. This anisotropy itself turns out to be a reason for electrohydrodynamic destabilization. First, we discuss the Carr–Helfrich mode of the instability [236, 237], which arises in a homogeneously oriented liquid crystal layer in a sandwich cell between transparent electrodes.

Carr–Helfrich Mode

When a low frequency ac electric field ($\omega \ll 1/\tau_Q$) is applied to homogeneously oriented, fairly conductive nematics, a very regular vortex motion (normal rolls) is often observed, which is accompanied by a strip domain optical pattern (Kapustin–Williams domains) [238, 239] (Fig. 31 b). The reason for this instability is the space charge accumulated in the bulk due to the anisotropy of conductivity. It appears in thin cells ($d = 10-100\ \mu m$) and has a well-defined voltage threshold that is independent of the thickness. The threshold can be easily calculated for the simplest steady-state, one-dimensional model shown in Fig. 32.

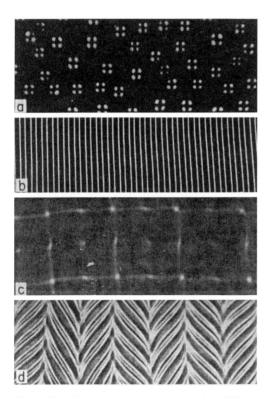

Figure 31. Optical patterns accompanying different EHD processes. (a) Electrolytic mode for the homeotropic orientation of a nematic liquid crystal. (b) Kapustin–Williams domains (KWD) in homogeneously oriented nematic. (c) Anisotropic EHD mode for the planar texture of a cholesteric. (d) A chevron structure due to interference of two instabilities (KWD and inertial mode).

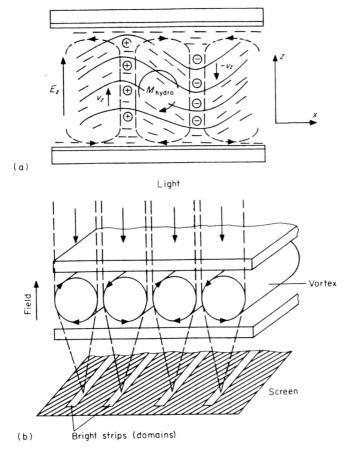

(a)

(b)   Bright strips (domains)

Figure 32. Carr–Helfrich EHD instability in nematic liquid crystals: (a) onset of the instability; (b) vortex motion of a liquid crystal and the pattern of black and white stripes in the screen plane.

The simplest model. The physical mechanism of the instability could be described as follows [236]. A homogeneously oriented nematic liquid crystal is stabilized by the elastic torque (due to surface anchoring). The dielectric torque is considered to be negligible ($\varepsilon_a = 0$). Now, let us imagine a small incident director fluctuation with a period w_x of the order of cell thickness d:

$$\theta(x) = \theta_m \cos \frac{\pi x}{w_x} \qquad (112)$$

With the field applied, this fluctuation results in a slight deviation of the electric current lines in the x direction proportional to the anisotropy of conductivity $\sigma_a = \sigma_\parallel - \sigma_\perp > 0$, and, in turn, in an accumulation of the space charge around extremum points

($\theta_m = 0$) of the director distribution (Fig. 32a). The x component of the current is

$$J_x = \sigma_\parallel E_x + \sigma_a E \theta \qquad (113)$$

where the new component of the electric field E_x is related to the space charge distribution $Q(x)$ and the director angle $\theta(x)$ by Maxwell's relationships (div $\boldsymbol{D} = 0$, div $\boldsymbol{J} = 0$):

$$\frac{\partial E_x}{\partial x} = \frac{4\pi Q(x)}{\varepsilon} \qquad (114)$$

and

$$\frac{\partial J_x}{\partial x} = \sigma_\parallel \frac{\partial E_x}{\partial x} + \sigma_a E \frac{\partial \theta}{\partial x} = 0 \qquad (115)$$

Substituting θ from Eq. (112) into Eqs (114) and (115) we have for the space charge distribution:

$$Q(x) = \frac{\sigma_a \, \varepsilon \, E \, \theta_m}{4 \pi \, \sigma_\parallel \, w_x} \sin \frac{\pi x}{w_x} \qquad (116)$$

At a certain critical voltage the destabilizing viscous torque M_V, which comes from the interaction of the space charge with an external field, becomes sufficiently large and is not compensated for by the stabilizing elastic M_k torque. The nematic liquid begins to move with a velocity v under the action of a pushing force proportional to $-Q(x) E$ in accordance with the Navier–Stokes equation (Eq. (91)) (written for the steady state regime):

$$\eta \frac{\partial^2 v_z}{\partial x^2} = -Q(x) E \qquad (117)$$

where $\eta = (\frac{1}{2})(\alpha_4 + \alpha_5 - \alpha_2)$ is a combination of Leslie's viscosity coefficients α_i.

The velocity v, in turn, supports the director fluctuation $\theta(x)$, in accordance with the torque balance equation:

$$K_{33} \frac{\partial^2 \theta}{\partial x^2} = \alpha_2 \frac{\partial v_z}{\partial x} \qquad (118)$$

Combining Eqs (116)–(118) we derive the following formula for the threshold voltage:

$$U_{crit} = E_{crit} \, d$$
$$= \left[\frac{4 \pi^3 \, K_{33} \, \sigma_\parallel \, \eta}{(-\alpha_2) \sigma_a \, \varepsilon} \right]^{\frac{1}{2}} \cdot \frac{d}{w_x} \qquad (119)$$

above which a periodic pattern of vortices forms with a period $w_x \approx d$ along the x axis. The whole process is governed by the anisotropy of conductivity. The threshold is diverged when σ_a vanishes (e.g. in the case of nematics with short-range smectic order). The threshold is proportional to the ratio η/α_2 of the two viscosities which, roughly speaking, are proportional to each other. Thus the threshold seems to be independent of viscosity and the instability may be easily observed in very viscous materials (e.g. polymer liquid crystals). The reason for this

is the compensation for two effects: (a) in very viscous media the velocity of vortex motion is lower (η coefficient); and (b) the coupling between the flow and the director (α_2 coefficient) is much stronger. A more precise expression for the threshold voltage, derived by Helfrich [236], includes also a nonvanishing value of the dielectric anisotropy. All the dependences predicted by the simplest theory have been confirmed qualitatively by experiment [8, 9]. In order to interpret the experimental data more precisely, some additional approaches have been developed.

Generalization of the Simplest Model. In the one-dimensional case, the diameter of the vortex is assumed to be equal to the cell thickness. In a more precise two-dimensional approach [240–242] this assumption is not used and the stability of the systems is studied using the variation of the wavevector $q_x = \pi/w_x$ along the x axis. The threshold voltage found is in excellent agreement with experiment [243]. Later, a three-dimensional version of the theory [244], which allows for the arbitrary anchoring energy, explained more complex domain patterns, such as oblique rolls [245–247]. Such rolls are seen for materials having a small negative dielectric anisotropy and large σ_a. If the flexoelectric term is included in the field interaction energy [248–250] the symmetry of the problem is changed; this may also account for two-dimensional vortex patterns (e.g. oblique rolls). A recent review [251] contains many interesting details concerning this problem.

In order to discuss the behavior of the instability threshold as a function of the applied field frequency, the time dependent terms $\gamma_1 (\partial \theta / \partial t)$ and $\partial Q / \partial t$ must be added to the equations for the director and charge continuity. Estimates show that the inertial term for the velocity $\partial v / \partial t$ in the Navier–Stokes equation may be disregarded even in the transient regime (except when discussing the inertial mode in strongly conducting nematics; see below). Thus we have a set of two coupled linear equations for the space charge $Q(x)$ and the curvature $\psi = \partial\theta/\partial x$ [225, 252]:

$$\frac{\partial\psi}{\partial t} + \frac{\psi}{\tau_r(E)} + \frac{QE}{\eta} = 0 \qquad (120)$$

and

$$\frac{\partial Q}{\partial t} + \frac{Q}{\tau_Q} + \sigma_{\text{eff}}\,\psi E = 0 \qquad (121)$$

This pair of equations has the same form as the set (Eqs (96) and (102)) for the injection instability. However, the fluid velocity is not included in the new set explicitly, and the destabilization is due to the coupling terms QE/η and $\sigma_{\text{eff}}\,\psi E$, where σ_{eff} is an effective constant including the anisotropy of conductivity that is responsible for the instability. The field-dependent time constant for the reaction of the director to the field was discussed earlier (Eqs (33) and (62)). Near the threshold $\tau_r \gg \tau_Q$.

The solution of set (Eqs (120) and (121)) represents a threshold voltage that is independent of frequency in the range $\omega \ll 1/\tau_Q$ (*conductance regime*) and increases critically when $\omega \to \omega_c \approx 1/\tau_Q$. Such a dependence is shown qualitatively in Fig. 28.

For frequencies above ω_c another regime is predicted by the same theory where the director *driven by flow* is oscillating clockwise and anticlockwise in the ac field. The threshold for this '*dielectric regime*' should also diverge with $\sigma_a \to 0$. In experiments, however, the modes observed at $\omega > \omega_c$ survive even at zero conductive anisotropy. Evidently, these instabilities (electrolytic and inertia modes) have a lower threshold than that predicted for the dielectric regime of the Carr–Helfrich mode.

If a substance has a small positive ε_a, the homeotropically oriented samples are stable

against pure dielectric perturbation; however, the convection may be observed due to EHD destabilization. The vortices have a short period and result in an optical pattern in the form of a short pitch grid. The period of the pattern is $(\alpha_2/\alpha_3)^{\frac{1}{2}}$ times less than d (10 times in typical nematics) [1, 8, 9, 251]. For higher ε_a the direct transition to the chaotic state may be observed [253].

Behavior Above the Threshold. At voltages higher than the threshold of the Carr–Helfrich instability, the normal rolls transform in more complex hydrodynamic patterns. One can distinguish zig-zag and fluctuating domain patterns which, in turn, are substituted by the turbulent motion of a liquid crystal accompanied by a strong (dynamic) light scattering. The corresponding 'stability diagram' [226] is shown in Fig. 33 (see the discussion of the dielectric regime above). To calculate the wavevectors and amplitude of the distortions, a set of nonlinear equations must be solved. More generally, the problem of describing the transition from a regular electrohydrodynamic vortex motion to turbulence is a part of the classical problem, concerning the transition from laminar to turbulent flow of a liquid. Some progress has been achieved

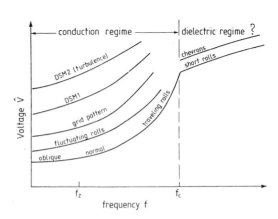

Figure 33. Voltage versus frequency phase diagram for electrohydrodynamic (convection) instabilities in the low frequency regime [226].

recently in understanding the nonlinear behavior of nematics in terms of bifurcation, phase transition, and dynamic chaos theories (for reviews see [226, 251, 254].

In the splay Frederiks transition geometry (homogeneous orientation, large dielectric anisotropy $\varepsilon_a \gg 0$) an interesting transient pattern in the form of domain stripes oriented *parallel to the initial director* has been observed [92]. The periodic pattern arises because the response of a liquid crystal placed suddenly far from equilibrium is faster to a spatially periodic director distribution than to a homogeneous orientation. The nonlinear theory of the transient patterns takes into account both the dielectric and the conductivity anisotropies [255, 256].

Dynamic Scattering of Light. As is known from the general theory of dissipative dynamic systems, after a finite number of bifurcations a system reaches dynamic chaos. This scenario is also observed in the case of EHD convective motion. With increasing voltage the velocity of the vortices increases rapidly and the periodic flow of the liquid transforms to turbulence. Turbulent motion in nematic liquid crystals results in a nonuniform distribution of the director accompanied by dynamic scattering of light [257, 258]. As the field is increased further, dynamic texturing or secondary dynamic light scattering is observed. This is related to the appearance of disclination loops or/and walls separating regions with different orientations of the director.

The light dynamic scattering effect was initially proposed for use in the manufacture of field-controllable shutters and displays, and was the starting point of modern studies of liquid crystals [259, 260].

Inertial Mode

This anisotropic mode is observed at rather high frequencies (10^3–10^6 Hz) in the case of highly conductive homogeneously oriented nematics with $\varepsilon_a < 0$. It appears in the form of long, periodic vortices of a liquid in the plane of a cell perpendicular to the initial director (Fig. 34). Such a flow results in an optical pattern of wide domains [261]. The threshold voltage of the instability diverges for vanishing conductivity anisotropy and also increases with decreasing mean conductivity, being proportional to $\sigma^{-\frac{1}{3}}$.

The physical mechanism of the instability involves the inertial term $\partial v/\partial t$ for the fluid velocity (for this reason the mode is called 'inertial'). As already mentioned, when constructing a theory for the Carr–Helfrich instability, the inertia term $\rho(\partial v_z/\partial t)$ in the Navier–Stokes equation was disregarded and two modes were predicted theoretically for different frequency ranges (the conductance and the dielectric regimes). Inclusion of the inertial term [262] led to the prediction of another high-frequency mode involving steady-state motion of the liquid and a stationary distribution of the director (as in Kapustin–Williams domains). In this case the space charge oscillates in counterphase with the external field. The approximate expression for the

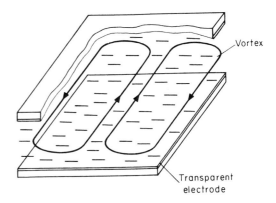

Figure 34. Convective motion of a strongly conductive nematic liquid crystal characteristic of the inertial mode.

threshold voltage of the inertial mode is

$$U_{crit} = A\omega \left(\frac{\varepsilon K}{\sigma \sigma_a} \right)^{\frac{1}{2}} \qquad (122)$$

where ε, K, and σ are average values of the dielectric, elastic, and conductivity constants. This expression agrees with most experimental observations.

The investigations of wide domains have identified the cause of the chevron (herringbone) structures. They result from interference between two modes with neighboring thresholds: the linear pre-chevron domains (deformation in the xz plane) and the wide domains (deformation in the xy plane). It has also been shown that such a herringbone structure can result from interference between the wide domains and the Kapustin–Williams domains (Fig. 31 d).

9.4.2 Cholesterics and Smectics

9.4.2.1 Cholesterics

The EHD behavior of cholesteric liquid crystals is very similar to that of nematics. When the anisotropy of the electrical conductivity is positive ($\sigma_a > 0$), the planar texture of a cholesteric liquid crystal in a field parallel to the helical axis is unstable for any sign of ε_a [263, 264]. The instability is caused by the torque induced by the electrical conductivity acting against the elastic torque of the cholesteric and, although the cause is different from the purely dielectric case (see Sec. 9.3.2.2 of this Chapter), the result obtained is the same; that is, the appearance of a two-dimensional periodic pattern for the distribution of the director.

Investigation of the EHD instability ($\varepsilon_a < 0$) in a planar Grandjean texture [265] shows that the directions of the domains alternate with a transition from one Grandjean zone to another, while the domains

always form perpendicular to the director of a cholesteric in the middle of a layer. With increasing thickness, one-dimensional deformations transform to a two-dimensional grid (Fig. 31 c).

The theory for the threshold of the instability in cells with thicknesses considerably exceeding the equilibrium pitch ($d \gg P_0$) has been considered by analogy with the case of dielectric instability [121, 266], but with allowance being made for the additional, destabilizing term in the free energy which is caused by the space charge. The frequency dependence of the threshold field for $\varepsilon_a < 0$ has been shown to be similar to that calculated for nematics. For a cholesteric liquid crystal with $\varepsilon_a > 0$ the presence of electrical conductivity is revealed by a lowering of the threshold of the instability at low frequencies.

For voltages above the threshold, as in nematics the regular pattern of periodic deformations changes to a turbulent motion of the liquid and dynamic scattering of light. However, in contrast to nematics, the cholesteric cell remains rather turbid even after removal of the voltage (dynamic scattering storage mode [267]). The residual turbidity of the cell is accounted for by the scattering of light by the focal-conic texture obtained through the influence of the flow [268]. The residual scattering texture can be erased by a high-frequency field, using a purely dielectric transition from the focal-conic to the planar texture.

If the cell walls are treated to produce a homeotropic orientation, in a cholesteric liquid crystal with $\varepsilon_a < 0$ for a particular relationship between the thickness and the helical pitch ($1 < P_0/d < 2$), a new texture (bubble domains [269]) appears under the influence of the electric field. In its initial state the liquid crystal has a homeotropic nematic structure, since the helix is untwisted by the walls. The application of a low-frequency field induces an electrohydrodynamic instability. After the removal of the field a new stable texture appears in the form of cholesteric 'bubbles' dispersed through the homeotropic nematic phase. Thus a memory is created, while erasure of the information can be achieved either by applying a high-frequency field or by mechanical displacement of glass plates.

At frequencies higher than the critical one $\omega \gg \omega_c = 4\pi\sigma/\varepsilon$ defined by inverse time τ_Q (Eq. (100)), a characteristic dependence $E_{crit} \propto \omega^{\frac{1}{2}}$ is observed for the threshold field of the convective distortion. This EHD instability has been shown [270] to be the same electrolytic mode as the one discussed above for nematic liquid crystals.

EHD instabilities have also been observed in blue phases of cholesterics [271]; the threshold voltage increases with frequency [272].

9.4.2.2   Smectics A

Anisotropic Modes

Theoretically, the Helfrich approach to anisotropic EHD instabilities in nematics may be applied to the smectic A phase, using its specific viscoelastic properties and the sign of the conductivity anisotropy. It makes sense to consider only the onset of a splay deformation in a homeotropic structure for smectic A, since $K_{33} \to \infty$. This approach is developed in Geurst and Goossens [273]. For the field directed along the normal to the smectic layers, the threshold field of an instability is given by

$$E_{crit} = q_x \left[\frac{K_{11}\varepsilon_{\perp}}{\varepsilon_{\parallel}(\sigma_{\perp}-\sigma_{\parallel})\tau_{\perp}} \cdot \frac{1+\omega^2\tau_{\perp}^2}{1-\omega^2\tau_{\perp}\tau_{\Delta}} \right]^{\frac{1}{2}} \quad (123)$$

where $\tau_{\perp} = \varepsilon_{\perp}/4\pi\sigma_{\perp}$, $\tau_{\Delta} = \varepsilon_a/4\pi(-\sigma_a)$, $\sigma_a = \sigma_{\parallel} - \sigma_{\perp}$, and q_x is the wavevector of the instability in the direction perpendicular to the

field. If it is assumed that the half-period of the instability equals the thickness of the cell, as was assumed in the Helfrich one-dimensional model for nematics, $q_x = \pi/d$, then Eq. (123) is analogous to Eq. (119), but with the condition that the viscosity coefficient η diverges in the smectic A phase so that the term including the ratio of the viscosities in the expression for the threshold field vanishes. In addition, the anisotropy of the electrical conductivity is negative in smectic A phases.

No quantitative experimental data on EHD instability in the smectic A phase are available. EHD processes are often observed in the form of the motion of a liquid, which may be detected using solid foreign particles mixed with a liquid crystal [274, 275]. In a weak field, such a motion cannot destabilize the director and the initial orientation (e.g. homeotropic) is stable. It is possible to observe the optical patterns induced by EHD processes in strongly conducting samples of smectic A phases [276–278]. An EHD instability is observed in homeotropically oriented samples of material with $\varepsilon_a > 0$ and $\sigma_a < 0$ and, in general, the situation corresponds to the model by Guerst and Goossens [273]. After switching the low-frequency field off, the optical pattern is maintained for a long period, but it can be erased by the application of a high-frequency field, which reorients the director via the dielectric torque (the controllable memory effect).

Isotropic Modes

In the pretransition temperature region above the smectic A–nematic phase transition, where the nematic phase has a certain degree of the short-range smectic A order, it is easy to detect both the motion of a liquid and the formation of domain patterns. The latter show marked differences from the domain patterns seen in pure nematic phases. For instance, in the temperature region where the ratio of the electrical conductivity $\sigma_\parallel/\sigma_\perp$ becomes less than 1, Kapustin–Williams domains are not observed with an initial planar orientation. Instead, different domains are formed at low frequencies $(\omega < \omega_c)$ of the applied field. They are arranged parallel to the initial orientation of the director and have a well-defined field strength threshold, $U_{crit} \propto d$ [275]. The threshold voltage for the Carr–Helfrich instability diverges at the temperature corresponding to the condition $\sigma_\parallel/\sigma_\perp = 1$. The disappearance of the Kapustin–Williams domains allows a new EHD mode to be seen (longitudinal domain mode). For the appearance of this new mode the sign of σ_a is irrelevant. Without doubt, the high-frequency EHD mode is caused by the isotropic mechanism of destabilization, since for $\sigma_\parallel/\sigma_\perp < 1$ the Carr–Helfrich model does not hold. Analysis shows [275] that the longitudinal domain mode is caused by the isotropic mechanism (electrolytic mode; as discussed above for nematics).

Planar oriented smectic A layers ($\sigma_a < 0$, $\varepsilon_a < 0$) also show domain instability that is EHD in nature [279]. This is probably caused by a destabilization of thin layers adjacent to the electrodes (the same isotropic mechanism). However, in this case, the conductivity anisotropy can influence the visual appearance of the effect.

9.4.2.3   Smectics C

In general, it can be anticipated that the EHD processes will result in a change in the director orientation in the smectic C phase, even when there is no change in the arrangement of the smectic layers, since the director has an extra degree of freedom – the possibility of rotation around the normal to the layers. In fact, the EHD instability is observed only in smectic C phases with a large tilt angle almost independent of temperature (C_1 phase). In C_2 smectics with a small, temperature-dependent tilt the Frederiks transition is usually observed (discussed in Section 9.3.3.1 of this Chapter). The reason for this seems to be different textures formed by the two phases [280].

The theory of the Carr–Helfrich instability in the smectic C phase has been arrived at [280] by analogy with the one presented above for nematic liquid crystals, but making allowance for the biaxiality of the former. The same set of equations for the director curvature and space charge (Eqs (120) and (121)) are valid for the C phase; however, material constants such as elastic moduli, viscosity coefficients, relaxation times for the director τ_r, and the space charge τ_Q and conductivity anisotropy σ_{eff} are more complicated. The frequency-dependent threshold voltage for the instability has been calculated for a wide range of dielectric anisotropy and the crossover of the Frederiks transition and EHD instability has been studied in detail. In particular, in the low-frequency range, both the ascending and descending types of threshold behavior with increasing frequency have been predicted.

EHD instabilities in the smectic C phase have been repeatedly observed experimentally. The formation of domains in a smectic C phase under the influence of an electric voltage was first observed in [281], then it has been studied in detail [281–283]. The results of these studies can be summarized as follows. At low frequencies, $\omega < \omega_c$, where the critical frequency ω_c is now defined taking into account the biaxiality of the smectic C phase, an EHD instability is observed (when $\sigma_a > 0$) with a threshold voltage that is independent of the cell thickness (so-called 'initial domains'). In a certain sense, this instability is similar to the

Carr–Helfrich one in nematics and may be studied quantitatively theoretically [280]. At high frequencies, an instability is observed with a characteristic frequency dependence of the threshold field $E \propto \omega^{\frac{1}{2}}$ that is independent of the layer thickness (fundamental domains). This has been regarded as an analog of the dielectric regime [282], but it can also be interpreted as the electrolytic mode [283] with some specific features. In some special cases a new domain mode is observed [284], which has been referred to the inertial (anisotropic) mode (discussed in Section 9.4.1.3 of this Chapter).

terparts. However, the inverse is not true, because liquid crystals possess strong dielectric (not magnetic) anisotropy and finite spontaneous polarization (but no spontaneous magnetization has been found in liquid crystals). Besides, a steady-state magnetic field cannot induce hydrodynamic and other dissipative processes. To a certain extent the situation may change in the near future, as mesophase have recently been synthesized that have extremely high magnetic susceptibilities [285]. The discovery of ferromagnetic mesophases cannot be excluded either.

9.5 Conclusion

In conclusion, electric field effects in liquid crystals is a well-developed branch of condensed matter physics. The field behavior of nematic liquid crystals in the bulk is well understood. To a certain extent the same is true for the cholesteric mesophase, although the discovery of bistability phenomena and field effects in blue phases opened up new fundamental problems to be solved. Ferroelectric and antiferroelectric mesophases in chiral compounds are a subject of current study. The other ferroelectric substances, such as discotic and lyotropic chiral systems and some achiral (like polyphilic) mesogenes, should attract more attention in the near future*. The same is true for a variety of polymer ferroelectric substances, including elastomers.

Until now, all the magnetic field effects have been found to have electric field coun-

---

* In fact, two new representatives of polar achiral systems have been discovered quite recently: antiferroelectric polymer-monomer mixtures [286] and ferroelectric biaxial smectic A phases composed of banana-like molecules [287].

9.6 References

[1] P. G. de Gennes, *The Physics of Liquid Crystals*, Clarendon Press, Oxford **1974**.
[2] S. Chandrasekhar, *Liquid Crystals*, Cambridge University Press, Oxford **1974**.
[3] W. H. de Jeu, *Physical Properties of Liquid Crystalline Materials*, Gordon & Breach, New York **1980**.
[4] V. A. Belyakov, A. S. Sonin, *Optika Kholestericheskikh Zhidkikh Kristallov (Optics of Cholesteric Liquid Crystals)*, Nauka, Moscow **1982** [in Russian].
[5] G. Vertogen, W. H. de Jeu, *Thermotropic Liquid Crystals, Fundamentals*, Springer-Verlag, Berlin **1988**.
[6] A. P. Kapustin, *Electroopticheskiye i akusticheskiye svoistva zhidkikh kristallov (Electro-Optical and Acoustical Properties of Liquid Crystals)*, Nauka, Moscow **1973** [in Russian].
[7] S. A. Pikin, *Structural Transformations in Liquid Crystals*, Gordon & Breach, New York **1991**.
[8] L. M. Blinov, *Electro-Optical and Magneto-Optical Properties of Liquid Crystals*, Wiley, Chichester **1983**.
[9] L. M. Blinov, V. G. Chigrinov, *Electrooptic Effects in Liquid Crystal Materials*, Springer-Verlag, New York **1993**.
[10] M. F. Grebenkin, A. V. Ivashchenko, *Zhidkokristallicheskiye Materialy (Liquid Crystalline Materials)*, Khimiya, Moscow **1989** [in Russian].
[11] M. V. Loseva, E. P. Pozhidayev, A. Z. Rabinovich, N. I. Chernova, A. V. Ivashchenko, *Itogi Nauk. Tekh. (Fiz. Khim.)* **1990**, *3*, 3.

[12] B. Bahadur, *Mol. Cryst. Liq. Cryst.* **1984**, *109*, 1.

[13] L. M. Blinov, W. Haase, *Mol. Mater.* **1993**, *2*, 145.

[14] L. M. Blinov, L. A. Beresnev, W. Haase, *Ferroelectrics*, **1996**, *181*, 187.

[15] K. Yoshino, K. Nakano, H. Taniguchi, M. Ozaki, *Jpn. J. Appl. Phys.* **1987**, *26* (Suppl. 26-2), 97.

[16] W. Helfrich, *Phys. Rev. Lett.* **1970**, *24*, 201.

[17] E. I. Rjumtsev, M. A. Osipov, T. A. Rotinyan, N. P. Yevlampieva, *Liq. Cryst.* **1995**, *18*, 87.

[18] M. J. Stephen, C. P. Fan, *Phys. Rev. Lett.* **1970**, *25*, 500.

[19] Ch. Bahr, G. Heppke, *Mol. Cryst. Liq. Cryst.* **1987**, *150B*, 313.

[20] Ch. Bahr, G. Heppke, B. Subaschus, *Liq. Cryst.* **1992**, *11*, 41.

[21] G. Andersson, K. Flatischler, L. Komitov, S. T. Lagerwall, K. Skarp, B. Stebler, *Ferroelectrics* **1991**, *113*, 361.

[22] L. M. Blinov, L. A. Beresnev, W. Haase, *Ferroelectrics*, **1996**, *181*, 211.

[23] M. A. Anisimov, *Mol. Cryst. Liq. Cryst.* **1988**, *162A*, 1.

[24] H. Hama, *J. Phys. Soc. Jpn.* **1985**, *54*, 2204.

[25] Ch. Bahr, G. Heppke, *Phys. Rev. A* **1990**, *41*, 4335.

[26] D. A. Dunmur, P. Palffy-Muhoray, *J. Phys. Chem.* **1988**, *92*, 1406.

[27] D. A. Dunmur, T. F. Waterworth, P. Palffy-Muhoray, *Mol. Cryst. Liq. Cryst.* **1985**, *124*, 73.

[28] D. A. Dunmur, T. F. Waterworth, preprint.

[29] D. A. Dunmur, K. Szumilin, T. F. Waterworth, *Mol. Cryst. Liq. Cryst.* **1987**, *149*, 385.

[30] B. Malraison, Y. Poggi, E. Guyon, *Phys. Rev. A* **1980**, *21*, 1012.

[31] N. V. Madhusudana, S. Chandrasekhar, *Liquid Crystals and Ordered Fluids*, Vol. 2 (Eds. J. F. Johnson, R. S. Porter), Plenum Press, New York **1974**, p. 657.

[32] D. Dunmur, P. Palffy-Muhoray, *J. Phys. Chem.* **1988**, *92*, 1406.

[33] D. A. Dunmur, A. E. Tomes, *Mol. Cryst. Liq. Cryst.* **1981**, *76*, 231.

[34] M. Schadt, W. Helfrich, *Mol. Cryst. Liq. Cryst.* **1972**, *17*, 355.

[35] H. J. Coles, R. R. Jennings, *Mol. Phys.* **1976**, *31*, 571.

[36] J. C. Filippini, Y. Poggi, *J. Phys. D* **1975**, *8*, 201.

[37] B. R. Ratna, M. S. Vijaya, R. Shashidhar, B. K. J. Sadashiva, *Pramana* **1975**, *1* (Suppl.), 69.

[38] R. Yamamoto, S. Ishihara, K. Moritomo, *Phys. Lett.* **1978**, *69A*, 276.

[39] V. N. Tsvetkov, E. I. Ryumtsev, A. P. Kovshik, G. I. Denis, Yu. Daugvila, *Dokl. Akad. Nauk SSSR* **1974**, *216*, 1105.

[40] J. C. Filippini, *J. Phys. D* **1975**, *8*, 201.

[41] V. N. Tsvetkov, E. I. Ryumtsev, *Mol. Cryst. Liq. Cryst.* **1986**, *133*, 125.

[42] H. Stegemeyer, Th. Bluemel, K. Hiltrop, H. Onusseit, F. Porsch, *Liq. Cryst.* **1986**, *1*, 3.

[43] D. K. Yang, P. P. Crooker, *Phys. Rev. A* **1988**, *37*, 4001.

[44] V. A. Belyakov, *Difraktsionnaya optika periodicheskikh sred slozhnoi struktury (Diffraction Optics of Periodical Media with Complex Structure)*, Nauka, Moscow **1988**, p. 255 [in Russian].

[45] V. E. Dmitrienko, *Liq. Cryst.* **1989**, *5*, 847.

[46] B. Jerome, P. Pieranski, *Liq. Cryst.* **1989**, *5*, 799.

[47] G. Heppke, B. Jerome, H. S. Kitzerow, P. Pieranski, *Liq. Cryst.* **1989**, *5*, 813.

[48] F. Porsch, H. Stegemeyer, K. Hiltrop, *Z. Naturforsch. A* **1984**, *39A*, 475.

[49] G. Heppke, B. Jerôme, H. S. Kitzerow, P. Pieranski, *J. Phys. (Paris)* **1991**, *52*, 2991.

[50] G. Heppke, B. Jerôme, H. S. Kitzerow, P. Pieranski, *J. Phys. (Paris)* **1989**, *50*, 549.

[51] G. Chilaya, G. Petriashvili, *Mol. Mater.* **1993**, *2*, 239.

[52] D. Armitage, R. J. Cox, *Mol. Cryst. Liq. Cryst.* **1980**, *64*, 41.

[53] H.-S. Kitzerow, P. P. Crooker, S. L. Kwok, G. Heppke, *J. Phys. France* **1990**, *51*, 1303.

[54] H.-S. Kitzerow, *Mol. Cryst. Liq. Cryst.* **1991**, *202*, 51.

[55] R. Shao, J. Pang, N. A. Clark, J. A. Rego, D. M. Walba, *Ferroelectrics* **1993**, *147*, 255.

[56] H.-S. Kitzerow, A. J. Slaney, J. W. Goodby, *Ferroelectrics* **1996**, *179*, 61.

[57] V. Shibayev, *Mol. Cryst. Liq. Cryst.* **1988**, *155*, 189.

[58] A. L. Smolyansky, Z. A. Roganova, R. V. Tal'rose, L. A. Kazarin, S. G. Kostromin, *Kristallografiya* **1987**, *32*, 265.

[59] H. Finkelmann, D. Naegele, H. Ringsdorf, *Macromol. Chem.* **1979**, *180*, 803.

[60] V. Frederiks, V. Zolina, *Zh. Russ. Fiz. Khim. Obshch.* **1930**, *62*, 457.

[61] V. Frederiks, V. Zolina, *Trans. Faraday Soc.* **1933**, *29*, 919.

[62] B. Brochard, P. Pieranski, E. Guyon, *Phys. Rev. Lett.* **1972**, *26*, 1681.

[63] P. Pieranski, E. Brochard, E. Guyon, *J. Phys. France* **1973**, *34*, 35.

[64] C. Z. van Doorn, *Phys. Lett. A* **1973**, *42A*, 537.

[65] M. Schiekel, K. Fahrenschon, H. Gruler, *Appl. Phys.* **1975**, *7*, 99.

[66] L. M. Blinov, E. I. Kats, A. A. Sonin, *Usp. Fiz. Nauk.* **1987**, *152*, 449.

[67] H. Yokoyama, *Mol. Cryst. Liq. Cryst.* **1988**, *165*, 265.

[68] B. Jerome, *Rep. Prog. Phys.* **1991**, *54*, 391.

[69] L. Leger, *Mol. Cryst. Liq. Cryst.* **1973**, *24*, 33.

[70] P. Schiller, G. Pelzl, D. Demus, *Liq. Cryst.* **1989**, *6*, 417.

[71] B. J. Frisken, P. Palffy-Muhoray, *Phys. Rev. A* **1989**, *40*, 6099.

[72] G. Barbero, E. Miraldi, C. Oldano, P. Taverna Valabrega, *Z. Naturforsch. A* **1988**, *43*, 547.

[73] G. Barbero, E. Miraldi, C. Oldano, *Phys. Rev. A* **1988**, *38*, 3027.

[74] V. N. Tsvetkov, L. N. Andreeva, A. P. Filippov, *Mol. Cryst. Liq. Cryst.* **1987**, *153*, 217.

[75] B. J. Frisken, P. Palffy-Muhoray, *Phys. Rev. A* **1989**, *39*, 1513.

[76] D. Allender, B. J. Frisken, P. Palffy-Muhoray, *Liq. Cryst.* **1989**, *5*, 735.

[77] P. Palffy-Muhoray, H. J. Yuan, B. J. Frisken, W. van Saarloos, *NATO ASI, Ser. B* **1990**, *225*, 313.

[78] F. Longberg, R. B. Meyer, *Phys. Rev. Lett.* **1985**, *55*, 718.

[79] V. G. Taratuta, A. J. Hurd, R. B. Meyer, *Phys. Rev. Lett.* **1985**, *55*, 246.

[80] C. Oldano, *Phys. Rev. Lett.* **1986**, *56*, 1098.

[81] W. Zimmermann, L. Kramer, *Phys. Rev. Lett.* **1986**, *56*, 2655.

[82] V. D. Kini, *Mol. Cryst. Liq. Cryst.* **1987**, *153*, 1.

[83] M. I. Barnik, L. M. Blinov, A. N. Trufanov, V. G. Chigrinov, T. V. Korkishko, *Zh. Eksp. Teor. Fiz.* **1984**, *87*, 196.

[84] V. G. Chigrinov, T. V. Korkishko, M. I. Barnik, A. N. Trufanov, *Mol. Cryst. Liq. Cryst.* **1985**, *129*, 283.

[85] E. F. Carr, *Mol. Cryst. Liq. Cryst.* **1977**, *34*, L-159.

[86] E. Guyon, R. Meyer, J. Salan, *Mol. Cryst. Liq. Cryst.* **1979**, *54*, 261.

[87] A. J. Hurd, S. Fraden, F. Longberg, R. B. Meyer, *J. Phys. France* **1985**, *46*, 905.

[88] F. Sagues, M. San Miguel, *Phys. Rev. A* **1989**, *39*, 6567.

[89] F. Longberg, S. Fraden, A. J. Hurd, R. B. Meyer, *Phys. Rev. Lett.* **1984**, *52*, 1903.

[90] A. D. Rey, M. M. Denn, *Liq. Cryst.* **1989**, *4*, 409.

[91] E. E. Pashkovsky, W. Stille, G. Strobl, *J. Phys. II, France* **1995**, *5*, 397.

[92] A. Buka, M. Juarez, L. Kramer, I. Rehberg, *Phys. Rev. A* **1989**, *40*, 7427.

[93] R. B. Meyer, *Phys. Rev. Lett.* **1969**, *22*, 319.

[94] G. Barbero, I. Dozov, I. Palierne, G. Durand, *Phys. Rev. Lett.* **1986**, *56*, 2056.

[95] D. Schmidt, M. Schadt, W. Helfrich, *Z. Naturforsch. A* **1972**, *27*, 277.

[96] A. I. Derzhanski, A. G. Petrov, M. D. Mitov, *J. Phys. (Paris)* **1978**, *39*, 273.

[97] J. Prost, P. S. Pershan, *J. Appl. Phys.* **1976**, *47*, 2298.

[98] A. I. Derzhanski, A. G. Petrov, *Acta Phys. Pol.* **1979**, *A55*, 747.

[99] L. M. Blinov, G. Durand, S. V. Yablonsky, *J. Phys. II, France* **1992**, *2*, 1287.

[100] L. K. Vistin, *Dokl. Akad. Nauk SSSR* **1970**, *194*, 1318.

[101] Yu. P. Bobylev, S. A. Pikin, *Zh. Eksp. Teor. Fiz.* **1977**, *72*, 369.

[102] Yu. P. Bobylev, V. G. Chigrinov, S. A. Pikin, *J. Phys. France* **1979**, *40*, C3-331.

[103] P. Schiller, G. Pelzl, D. Demus, *Cryst. Res. Technol.* **1990**, *25*, 111.

[104] M. I. Barnik, L. M. Blinov, A. N. Trufanov, B. A. Umansky, *Zh. Eksp. Teor. Fiz.* **1977**, *73*, 1936.

[105] Kh. Khinov, L. K. Vistin, *J. Phys. France* **1979**, *40*, 269.

[106] Kh. Khinov, *Z. Naturforsch. A* **1982**, *37*, 334.

[107] B. H. Soffer, J. D. Margerum, A. M. Lackner, D. Boswell, A. R. Tanguay, T. S. Strand, A. Sawchuk, P. Chavel, *Mol. Cryst. Liq. Cryst.* **1981**, *70*, 145.

[108] Ho-In Jeon, A. A. Sawchuk, *Appl. Opt.* **1987**, *26*, 261.

[109] M. Schadt, W. Helfrich, *Appl. Phys. Lett.* **1971**, *18*, 127.

[110] C. Mauguin, *Bull. Soc. Fr. Miner.* **1911**, *34*, 71.

[111] F. M. Leslie, *Mol. Cryst.* **1970**, *12*, 57.

[112] M. Becker, J. Nehring, T. Scheffer, *J. Appl. Phys.* **1985**, *57*, 4539.

[113] D. W. Berreman, *J. Opt. Soc. Am.* **1973**, *63*, 1374.

[114] D. W. Berreman, *Phil. Trans. R. Soc., London, Ser. A* **1983**, *309*, 203.

[115] E. P. Raynes, *Mol. Cryst. Liq. Cryst. Lett.* **1986**, *4*, 1.

[116] P. A. Breddels, H. A. van Sprang, *J. Appl. Phys.* **1985**, *58*, 2162.

[117] R. Thurston, J. Cheng, G. Boid, *IEEE Trans. Electr. Dev.* **1980**, *ED-27*, 2969.

[118] W. R. Heffner, D. W. Berreman, *J. Appl. Phys.* **1982**, *53*, 8599.

[119] D. W. Berreman, *J. Appl. Phys.* **1984**, *55*, 806.

[120] W. Helfrich, *Appl. Phys. Lett.* **1970**, *17*, 531.

[121] J. P. Hurault, *J. Chem. Phys.* **1973**, *59*, 2068.

[122] V. G. Chigrinov, V. V. Belyayev, S. V. Belyayev, M. F. Grebenkin, *Zh. Eksp. Teor. Fiz.* **1979**, *77*, 2081.

[123] P. Schiller, K. Schiller, *Liq. Cryst.* **1990**, *8*, 553.

[124] P. Schiller, *Phase Trans.* **1990**, *29*, 59.

[125] G. Cohen, R. M. Hornreich, *Phys. Rev. A* **1990**, *41*, 4402.

[126] R. Meyer, *Appl. Phys. Lett.* **1968**, *12*, 281.

[127] P. G. de Gennes, *Sol. State Commun.* **1968**, *6*, 163.

[128] R. Dreher, *Solid State Commun.* **1973**, *13*, 1571.

[129] S. V. Belyayev, L. M. Blinov, *Pis'ma Zh. Eksp. Teor. Fiz.* **1979**, *30*, 111.

[130] W. J. A. Goossens, *J. Phys. France* **1982**, *43*, 1469.

[131] J. Brokx, G. Vertogen, E. W. C. van Groesen, *Z. Naturforsch. A* **1983**, *38*, 1.

[132] A. Mochizuki, S. Kobayashi, *Mol. Cryst. Liq. Cryst.* **1993**, *225*, 89.

[133] G. Hauk, H. D. Koswig, *Cryst. Res. Technol.* **1987**, *22*, 1333.

[134] J. S. Patel, R. B. Meyer, *Phys. Rev. Lett.* **1987**, *58*, 1538.

[135] J. D. Parson, C. F. Hayes, *Phys. Rev. A* **1974**, *9*, 2652.

[136] P. R. Gerber, *Mol. Cryst. Liq. Cryst.* **1985**, *116*, 197.

[137] J. S. Patel, S. D. Lee, *J. Appl. Phys.* **1989**, *66*, 1879.

[138] S. D. Lee, J. S. Patel, R. B. Meyer, *J. Appl. Phys.* **1990**, *67*, 1293.

[139] L. A. Beresnev, L. M. Blinov, D. I. Dergachev, S. B. Kondrat'yev, *Pis'ma Zh. Eksp. Teor. Fiz.* **1987**, *46*, 328.

[140] R. Rundquist, L. Komitov, S. T. Lagerwall, *Phys. Rev. E* **1994**, *50*, 4735.

[141] R. Rundquist, M. Buivydas, L. Komitov, S. T. Lagerwall, *J. Appl. Phys.* **1994**, *76*, 7778.

[142] L. Komitov, S. T. Lagerwall, B. Stebler, G. Andersson, K. Flatischer, *Ferroelectrics* **1991**, *114*, 167.

[143] G. Legrand, N. Isaert, J. Hmine, J. M. Buisine, J. P. Parneix, N. T. Nguyen, C. Destrade, *Ferroelectrics* **1991**, *121*, 21.

[144] B. A. Umansky, L. M. Blinov, M. I. Barnik, *Pis'ma Zh. Tekh. Fiz.* **1980**, *6*, 200.

[145] B. A. Umansky, V. G. Chigrinov, L. M. Blinov, Yu. B. Pod'yachev, *Zh. Eksp. Teor. Fiz.* **1981**, *81*, 1307.

[146] G. Chigrinov, *Kristallografiya* **1983**, *28*, 825.

[147] E. M. Terent'yev, S. A. Pikin, *Kristallografiya* **1985**, *30*, 227.

[148] L. Bourdon, M. Sommeria, M. Kleman, *J. Phys. France* **1982**, *43*, 77.

[149] A. Rapini, *J. Phys. France* **1972**, *33*, 237.

[150] M. Hareng, S. Le Berre, J. J. Metzger, *Appl. Phys. Lett.* **1975**, *27*, 575.

[151] G. Pelzl, H. J. Deutscher, D. Demus, *Cryst. Res. Technol.* **1981**, *16*, 603.

[152] O. Parodi, *Solid State Commun.* **1972**, *11*, 1503.

[153] Y. Takanishi, Y. Ouchii, H. Takezoe, A. Fukuda, *Jpn. J. Appl. Phys.* **1989**, *28*, L487.

[154] H. P. Hinov, *Liq. Cryst.* **1988**, *3*, 1481.

[155] H. P. Hinov, K. Avramova, *Liq. Cryst.* **1988**, *3*, 1505.

[156] A. Jakli, A. Saupe, *Mol. Cryst. Liq. Cryst.* **1992**, *222*, 101.

[157] M. J. Stephen, J. P. Straley, *Rev. Mod. Phys.* **1974**, *46*, 617.

[158] G. Pelzl, P. Kolbe, V. Preukschas, S. Diele, D. Demus, *Mol. Cryst. Liq. Cryst.* **1979**, *53*, 167.

[159] G. Pelzl, P. Schiller, D. Demus, *Liq. Cryst.* **1987**, *2*, 131.

[160] G. Pelzl, P. Schiller, D. Demus, *Cryst. Res. Technol.* **1990**, *25*, 215.

[161] P. Schiller, G. Pelzl, D. Demus, *Liq. Cryst.* **1987**, *2*, 21.

[162] E. P. Pozhidayev, M. A. Osipov, V. G. Chigrinov, V. A. Baikalov, L. A. Beresnev, L. M. Blinov, *Zh. Eksp. Teor. Fiz.* **1988**, *94*, 125.

[163] R. Bruinsma, D. R. Nelson, *Phys. Rev. B* **1981**, *21*, 5312.

[164] P. Schiller, G. Pelzl, C. Camara, *Liq. Cryst.*, submitted.

[165] F. Tournilhac, L. M. Blinov, J. Simon, S. V. Yablonsky, *Nature* **1992**, *359*, 621.

[166] F. Tournilhac, L. M. Blinov, J. Simon, D. Subachius, S. V. Yablonsky, *Synth. Met.* **1993**, *54*, 253.

[167] J. Prost, F. Barois, *J. Chim. Phys.* **1983**, *80*, 65.

[168] L. M. Blinov, T. A. Lobko, B. I. Ostrovsky, S. N. Sulianov, F. Tournilhac, *J. Phys. II, France* **1993**, *3*, 1121.

[169] R. B. Meyer, L. Liebert, L. Strzelecki, P. Keller, *J. Phys. France Lett.* **1975**, *36*, L-69.

[170] J. W. Goodby, R. Blinc, N. A. Clark, S. T. Lagerwall, M. A. Osipov, S. A. Pikin, T. Sakurai, K. Yoshino, B. Zeks, *Ferroelectric Liquid Crystals. Principles, Properties and Applications*, Gordon & Breach, Philadelphia **1991**.

[171] L. A. Beresnev, L. M. Blinov, M. A. Osipov, S. A. Pikin, *Mol. Cryst. Liq. Cryst.* **1988**, *158A*, 1–150.

[172] T. P. Rieker, N. A. Clark, in *Phase Transitions in Liquid Crystals* (Eds S. Martellucci, A. N. Chester), Plenum Press, New York **1992**.

[173] A. Fukuda, Y. Takanishi, T. Isozaki, K. Ishikawa, H. Takezoe, *J. Mater. Chem.* **1994**, *4*, 997.

[174] N. A. Clark, S. T. Lagerwall, *Appl. Phys. Lett.* **1980**, *36*, 899.

[175] M. A. Handschy, N. A. Clark, S. T. Lagerwall, *Phys. Rev. Lett.* **1983**, *51*, 471.

[176] N. A. Clark, S. T. Lagerwall, *Ferroelectrics* **1984**, *59*, 25.

[177] T. Carlson, N. A. Clark, Z. Zou, *Liq. Cryst.* **1993**, *15*, 461.

[178] J. Z. Xue, M. A. Handschy, N. A. Clark, *Ferroelectrics* **1987**, *73*, 305.

[179] H. Orihara, K. Nakamura, Y. Ishibashi, Y. Yamada, N. Yamamoto, M. Yamanaki, *Jpn. J. Appl. Phys.* **1986**, *25*, L-839.

[180] P. Schiller, *Cryst. Res. Technol.* **1986**, *21*, 167, 301.

[181] I. Abdulhalim, G. Moddel, N. A. Clark, *J. Appl. Phys.* **1994**, *76*, 820.

[182] L. A. Beresnev, L. M. Blinov, D. I. Dergachev, *Ferroelectrics* **1988**, *85*, 173.

[183] L. A. Beresnev, V. G. Chigrinov, D. I. Dergachev, E. P. Pozhidayev, J. Fuenfschilling, M. Schadt, *Liq. Cryst.* **1989**, *5*, 1171.

[184] B. I. Ostrovskii, A. Z. Rabinovich, A. S. Sonin, B. A. Strukov, *Zh. Eksp. Teor. Fiz.* **1978**, *74*, 1748.

[185] B. I. Ostrovskii, V. G. Chigrinov, *Kristallogra-fiya* **1980**, *25*, 560.

[186] B. I. Ostrovskii, A. Z. Rabinovich, V. G. Chigrinov, in *Advances in Liquid Crystal Research and Applications* (Ed. L. Bata), Pergamon Press, Oxford **1980**, p. 469.

[187] B. I. Ostrovskii, S. A. Pikin, V. G. Chigrinov, *Zh. Eksp. Teor. Fiz.* **1979**, *77*, 1631.

[188] J. Funfschilling, M. Schadt, *J. Appl. Phys.* **1989**, *66*, 3877.

[189] S. Garoff, R. Meyer, *Phys. Rev. A* **1979**, *19*, 338.

[190] G. Andersson, I. Dahl, L. Komitov, S. T. Lagerwall, K. Sharp, B. Stebler, *J. Appl. Phys.* **1989**, *66*, 4983.

[191] L. M. Blinov, L. A. Beresnev, W. Haase, *Ferroelectrics* **1995**, *174*, 221.

[192] S. Lee, J. Patel, *Appl. Phys. Lett.* **1989**, *55*, 122.

[193] L. A. Beresnev, M. V. Loseva, N. I. Chernova, S. G. Kononov, P. V. Adomenas, E. P. Pozhidaev, *Pis'ma Zh. Tehn. Fiz.* **1990**, *51*, 457.

[194] S. A. Pikin, L. A. Beresnev, S. Hiller, M. Pfeiffer, W. Haase, *Mol. Mater.* **1993**, *3*, 1.

[195] L. A. Beresnev, E. Schumacher, S. A. Pikin, X. Fan, B. I. Ostrovski, S. Hiller, A. P. Onokhov, W. Haase, *Jpn. J. Appl. Phys.* **1995**, *34*, 2404.

[196] K. Skarp, G. Andersson, T. Hirai, A. Yoshizawa, K. Hiraoka, H. Takezoe, A. Fukuda, *Jpn. J. Appl. Phys.* **1992**, *31*, 1409.

[197] V. P. Shibayev, M. V. Kozlovsky, L. A. Beresnev, L. M. Blinov, N. A. Plate, *Polymer Bull.* **1984**, *12*, 299.

[198] M. V. Kozlovsky, L. A. Beresnev, *Phase Transitions* **1992**, *40*, 129.

[199] L. A. Beresnev, L. M. Blinov, V. A. Baikalov, E. P. Pozhidayev, G. V. Purvanetskas, A. I. Pavlyuchenko, *Mol. Cryst. Liq. Cryst.* **1982**, *89*, 327.

[200] H. R. Brand, P. E. Cladis, *J. Phys. France Lett.* **1984**, *45*, L-217.

[201] N. Hiji, A. D. L. Chandani, S. Nishiyama, Y. Ouchi, H. Takezoe, A. Fukuda, *Ferroelectrics* **1988**, *85*, 99.

[202] A. Chandani, T. Hagiwara, Y. Suzuki, Y. Ouchi, H. Takezoe, A. Fukuda, *Jpn. J. Appl. Phys.* **1988**, *27*, L-729.

[203] A. Chandani, Y. Ouchi, H. Takezoe, A. Fukuda, K. Terashima, K. Furukawa, A. Kishi, *Jpn. J. Appl. Phys.* **1989**, *28*, L-1265.

[204] A. Chandani, E. Gorecka, Y. Ouchi, H. Takezoe, A. Fukuda, *Jpn. J. Appl. Phys.* **1989**, *28*, L-1265.

[205] Y. Galerne, L. Liebert, *Phys. Rev. Lett.* **1990**, *64*, 906; **1991**, *66*, 2891.

[206] Ch. Bahr, D. Fliegner, *Phys. Rev. Lett.* **1993**, *70*, 1842.

[207] M. Hara, T. Umemoto, H. Takezoe, A. F. Garito, H. Sasabe, *Jpn. J. Appl. Phys.* **1991**, *30*, L2052.

[208] Ji Li, H. Takezoe, A. Fukuda, *Jpn. J. Appl. Phys.* **1991**, *30*, 532.

[209] K. Skarp, G. Andersson, F. Gouda, S. T. Lagerwall, H. Poths, R. Zentel, *Polym. Adv. Technol.* **1992**, *3*, 241.

[210] T. Isozaki, T. Fujikawa, H. Takezoe, A. Fukuda, T. Hagiwara, Y. Suzuki, I. Kawamura, *Jpn. J. Appl. Phys.* **1992**, *31*, L1435.

[211] K. Hiraoka, Y. Takanishi, K. Skarp, H. Takezoe, A. Fukuda, *Jpn. J. Appl. Phys.* **1991**, *30*, L1819.

[212] M. Johno, A. Chandani, J. Lee, Y. Ouchi, H. Takezoe, A. Fukuda, K. Itoh, *Proc. SID* **1990**, *31*, 129.

[213] J. Lee, A. Chandani, K. Itoh, Y. Ouchi, H. Takezoe, A. Fukuda, *Jpn. J. Appl. Phys.* **1990**, *29*, 1122.

[214] L. A. Beresnev, L. M. Blinov, E. I. Kovshev, *Dokl. Akad. Nauk SSSR* **1982**, *265*, 210.

[215] L. M. Blinov, S. A. Davidyan, A. G. Petrov, A. T. Todorov, S. V. Yablonsky, *Pis'ma Zh. Eksp. Teor. Fiz.* **1988**, *48*, 259.

[216] A. G. Petrov, A. T. Todorov, P. Bonev, L. M. Blinov, S. V. Yablonsky, D. B. Subachius, N. Tvetkova, *Ferroelectrics* **1991**, *114*, 415.

[217] H. Bock, W. Helfrich, *Liq. Cryst.* **1992**, *12*, 697.

[218] H. Bock, W. Helfrich, *Liq. Cryst.* **1994**, *14*, 345.

[219] L. M. Blinov, *Sci. Prog., Oxford* **1986**, *70*, 263.

[220] S. Kai (Ed.), *Pattern Formation in Complex Dissipative Systems and Global Dynamics*, World Scientific, Singapore **1992**.

[221] L. Kramer, W. Pesch, in *Pattern Formation in Liquid Crystals* (Eds A. Buka, L. Kramer), Springer-Verlag, New York **1995**.

[222] R. D. Feynman, R. B. Leighton, M. Sands, *The Feynman Lectures on Physics*, Vol. 2, Addison-Wesley, Reading, MA **1964**, Ch. 41.

[223] N. J. Felici, *J. Phys. France Colloq. C1* **1976**, *37*, C1–17.

[224] J. J. Felici, *Rev. Gen. Electr.* **1969**, *78*, 717.

[225] E. Guyon, P. Pieranski, *Physica* **1974**, *73*, 184.

[226] W. Zimmermann, *MRS Bull.* **1991**, *24*, 46.

[227] P. Atten, R. Moreau, *C. R. Acad. Sci.* **1970**, *270A*, 415.

[228] J. M. Schneider, P. K. Watson, *Phys. Fluids* **1970**, *13*, 1948.

[229] J. C. Filippini, J. P. Gosse, J. C. Lacroix, R. Tobazeon, *C. R. Acad. Sci.* **1969**, *69B*, 16, 736.

[230] H. Koelmans, A. M. van Boxtel, *Phys. Lett.* **1970**, *32A*, 32.

[231] H. Koelmans, A. M. van Boxtel, *Mol. Cryst. Liq. Cryst.* **1971**, *12*, 185.

[232] M. I. Barnik, L. M. Blinov, M. F. Grebenkin, A. N. Trufanov, *Mol. Cryst. Liq. Cryst.* **1976**, *37*, 47.

[233] M. I. Barnik, L. M. Blinov, S. A. Pikin, A. N. Trufanov, *Zh. Eksp. Teor. Fiz.* **1977**, *72*, 756.

[234] V. G. Chigrinov, S. A. Pikin, *Kristallografiya* **1978**, *23*, 333.

[235] A. N. Trufanov, M. I. Barnik, L. M. Blinov, V. G. Chigrinov, *Zh. Eksp. Teor. Fiz.* **1980**, *80*, 704.

[236] W. Helfrich, *J. Chem. Phys.* **1969**, *51*, 4092.

[237] E. F. Carr, *Mol. Cryst. Liq. Cryst.* **1977**, *34*, L-159.

[238] G. E. Zvereva, A. P. Kapustin in *Primeneniye ultraakustiki kissledovaniyu veshchestva (Application of Ultraacustics to Investigation of Substances)*, Moscow **1961**, Vol. 15, p. 69 [in Russian].

[239] R. Williams, *J. Chem. Phys.* **1963**, *39*, 384.

[240] S. A. Pikin, *Zh. Eksp. Teor. Fiz.* **1971**, *60*, 1185.

[241] S. A. Pikin, A. A. Shtol'berg, *Kristallografiya* **1973**, *18*, 445.

[242] P. A. Penz, G. W. Ford, *Phys. Rev. A* **1972**, *6*, 414, 1676.

[243] M. I. Barnik, L. M. Blinov, M. F. Grebenkin, S. A. Pikin, V. G. Chigrinov, *Zh. Eksp. Teor. Fiz.* **1975**, *69*, 1080.

[244] E. Bodenschatz, W. Zimmerman, L. Kramer, *J. Phys. France* **1988**, *49*, 1875.

[245] S. Kai, K. Hirakawa, *Prog. Theor. Phys.* **1978**, *64* (Suppl.), 212.

[246] M. Kohno, *Phys. Rev. A* **1989**, *40*, 6554.

[247] A. Joets, R. Ribotta, *J. Phys. France* **1986**, *47*, 595.

[248] N. V. Madhusudana, V. A. Raghunathan, K. R. Sumathy, *Pramana J. Phys.* **1987**, *28*, L311.

[249] V. A. Raghunathan, N. V. Madhusudana, *Pramana J. Phys.* **1988**, *31*, L163.

[250] W. Thom, W. Zimmerman, L. Kramer, *Liq. Cryst.* **1989**, *4*, 309.

[251] L. Kramer, W. Pesch in *Pattern Formation in Liquid Crystals* (Eds A. Buka, L. Kramer), Springer Verlag, New York **1995**.

[252] E. Dubois-Violette, P. G. de Gennes, O. Parodi, *J. Phys. France* **1971**, *32*, 305.

[253] A. Hertich, W. Decker, W. Pesch, L. Kramer, *J. Phys. France II* **1992**, *2*, 1915.

[254] I. Rehberg, B. L. Winkler, M. Torre Juarez, S. Raenat, W. Schoepf, *Festkoerperprobleme* **1989**, *29*, 35.

[255] B. L. Winkler, H. Richter, I. Rehberg, W. Zimmermann, L. Kramer, A. Buka, *Phys. Rev. A* **1991**, *43*, 1940.

[256] A. Buka, L. Kramer, *Phys. Rev. A* **1992**, *45*, 5624.

[257] V. Frederiks, V. Tsvetkov, *Dokl. Akad. Nauk SSSR* **1935**, *4*, 123.

[258] V. Frederiks, V. Tsvetkov, *Acta Physicochim. URSS* **1935**, *3*, 879.

[259] G. H. Heilmeier, L. A. Zanoni, L. A. Barton, *Proc. IEEE* **1968**, *56*, 1162.

[260] G. H. Heilmeier, L. A. Zanoni, L. A. Barton, *IEEE Trans. Elect. Dev.* **1990**, *ED-17*, 22.

[261] A. N. Trufanov, L. M. Blinov, M. I. Barnik, *Zh. Eksp. Teor. Fiz.* **1980**, *78*, 622.

[262] S. A. Pikin, V. G. Chigrinov, *Zh. Eksp. Teor. Fiz.* **1980**, *78*, 246.

[263] F. Rondelez, H. Arnould, *C. R. Acad. Sci.* **1971**, *273B*, 549.

[264] A. Arnould-Nettilard, F. Rondelez, *Mol. Cryst. Liq. Cryst.* **1974**, *24*, 11.

[265] S. V. Belyayev, L. M. Blinov, *Zh. Eksp. Teor. Fiz.* **1976**, *70*, 184.

[266] W. Helfrich, *J. Chem. Phys.* **1971**, *55*, 839.

[267] G. H. Heilmeier, J. E. Goldmacher, *Appl. Phys. Lett.* **1968**, *12*, 132.

[268] W. Haas, J. Adam, C. Dir, *Chem. Phys. Lett.* **1972**, *14*, 95.

[269] T. Akahane, T. Tako, *Jpn. J. Appl. Phys.* **1976**, *15*, 1559.

[270] S. V. Belyayev, *Zh. Eksp. Teor. Fiz.* **1978**, *75*, 663.

[271] P. L. Finn, P. E. Cladis, *Mol. Cryst. Liq. Cryst.* **1982**, *84*, 159.

[272] H. Gleeson, R. Simon, H. J. Coles, *Mol. Cryst. Liq. Cryst.* **1985**, *129*, 37.

[273] J. A. Guerst, W. J. A. Goossens, *Phys. Lett.* **1972**, *41*, 369.

[274] M. Gosciansky, *Philips Res. Rep.* **1975**, *30*, 37.

[275] L. M. Blinov, M. I. Barnik, V. T. Lasareva, A. N. Trufanov, *J. Phys. France Colloq. C3* **1979**, *40*, C3-263.

[276] V. N. Chirkov, D. F. Aliyev, A. Kh. Zeinally, *Pis'ma Zh. Eksp. Teor. Fiz.* **1977**, *3*, 1016.

[277] V. N. Chirkov, D. F. Aliyev, A. Kh. Zeinally, *Zh. Eksp. Teor. Fiz.* **1978**, *74*, 1822.

[278] D. F. Aliyev, H. F. Abbasov, *Liq. Cryst.* **1989**, *4*, 293.

[279] N. A. Tikhomirova, A. V. Ginzberg, E. A. Kirsanov, Yu. P. Bobylev, S. A. Pikin, P. V. Adomenas, *Pis'ma Zh. Eksp. Teor. Fiz.* **1976**, *24*, 301.

[280] M. P. Petrov, A. G. Petrov, G. Pelzl, *Liq. Cryst.* **1992**, *11*, 865.

[281] L. K. Vistin, A. P. Kapustin, *Kristallografiya* **1968**, *13*, 349.

[282] B. Petroff, M. Petrov, P. Simova, A. Angelov, *Ann. Phys.* **1978**, *3*, 331.

[283] D. F. Aliyev, A. Kh. Zeinally, N. A. Guseinov, *Kristallografiya* **1981**, *26*, 867.

[284] D. F. Aliyev, *Kristallografiya* **1983**, *28*, 358.

[285] Yu. Galyametdinov, M. A. Athanassopoulou, W. Haase, I. Ovchinnikov, *Koord. Khim.* **1995**, *21*, 9.

[286] E. A. Soto Bustamante, S. V. Yablonsky, B. I. Ostrovskii, L. A. Beresnev, L. M. Blinov, W. Haase, *Chem. Phys. Lett.* **1996**, *260*, 447.

[287] T. Niori, T. Sekine, J. Watanabe, T. Furukawa, H. Takezoe, *J. Mater. Chem.* **1996**, *6*, 1231.

10 Surface Alignment

Blandine Jérôme

10.1 Introduction

The phenomenon of orientation of liquid crystals by surfaces has been known nearly as long as have liquid crystals themselves [1]. The phenomenon has mainly been studied in low-molecular-weight nematic liquid crystals, both because of the simplicity of their structure and because of the use of this type of liquid crystal in displays. Most of the present chapter is therefore be dedicated to this type of liquid crystal.

When a nematic liquid crystal is placed in contact with another phase (solid or liquid), a surface bounding the liquid crystal is created. The presence of this surface induces a perturbation of the nematic order close to it (Fig. 1a). The anisotropic interactions between the molecules located right at the surface – in the surface layer – and the other phase favors certain orientations of the surface molecules. This leads to an orientational distribution of the liquid crystal molecules in the surface layer that is generally different from the bulk nematic order. The orientational order evolves from the one induced by the surface to the one in the bulk in an interfacial region of thickness ξ_i, which is of the order of the nematic coherence length. Just outside the interfacial re-

gion, the nematic director has a preferred orientation a. This macroscopic orientation of a liquid crystal by a surface is called anchoring.

The macroscopic anchoring of low-molecular-weight nematic liquid crystals is dis-

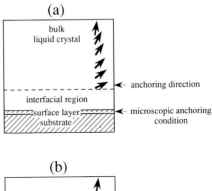

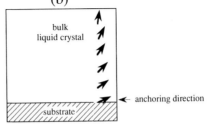

Figure 1. Schematic drawing of a liquid crystal in contact with a substrate: (a) taking into account the presence of the surface layer and interfacial region; (b) according to the macroscopic approach. The arrows indicate the director field in the bulk, represented here in a distorted configuration.

cussed in Section 10.2 and the microscopic aspects behind it in Section 10.3. The case of other liquid crystals is considered in Section 10.4. For more detailed reviews on the surface effects in liquid crystals see, for example, Sluckin and Poniewierski [2], Sonin [3], and Jérôme [4].

10.2 Macroscopic Alignment of Nematic Liquid Crystals

Until quite recently, anchoring has been described ignoring the existence of the surface layer and interfacial region, assuming that the bulk structure extends right up to the surface (Fig. 1b). The configuration of the liquid crystal is then described at each point by the director. At the surface, this director has a preferred orientation, which is the boundary condition for the field of the director in the rest of the liquid crystal. This macroscopic description works well when the director at the surface is not driven too far from its preferred orientation (by another orienting force) and when the origin of this preferred orientation is irrelevant for the problem under consideration.

10.2.1   Definitions

Both discussing anchoring in more details, it is necessary to define a terminology to describe it (for a more detailed lexicon, see Jérôme [4]).

The main concept is the anchoring induced by an interface. The energy γ_s of this interface – also called the anchoring energy – depends, among other things, on the orientation of the director at the surface. This energy has a certain number of minima obtained for orientations of the director $\boldsymbol{a}_\alpha$, which are the anchoring directions of the liquid crystal at the surface. The set of possible anchoring directions $\{\boldsymbol{a}_\alpha\}$ characterizes the anchoring induced by an interface. This anchoring can be monostable, multistable, or degenerate, depending on whether the number of elements in the set $\{\boldsymbol{a}_\alpha\}$ is one, a finite number greater than one, or infinite. It can also be planar, tilted or homeotropic, depending on whether the anchoring directions are parallel, tilted or perpendicular to the plane of the interface (Fig. 2).

In the macroscopic approach, which ignores the detailed structure of this interface, the expression of the interfacial energy γ_s has to be found following thermodynamic and symmetry considerations. γ_s is a periodic function of the azimuthal angle φ and the tilt angle θ (with respect to the surface normal) defining the orientation of the director at the surface. γ_s can thus be developed in a Fourier series [2, 10]:

$$\gamma_s(\theta,\varphi) = \sum A_{lm} Y_l^m(\theta,\varphi) \qquad (1)$$

where $Y_l^m(\theta,\varphi)$ are spherical harmonics and A_{lm} are coefficients which are non-zero only if the corresponding Y_l^m is compatible with the symmetry of the nematic phase (even l) and that of the surface.

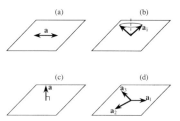

Figure 2. Some experimentally observed anchorings: (a) monostable planar (on grooved surfaces [5]); (b) degenerate tilted, also called conical (at the nematic/isotropic interface [6, 7]); (c) homeotropic (on surfactant-coated glass [8]); (d) tristable planar (on phlogopite mica [9]).

The term 'surface' means here the surface of the other phase in contact with the nematic liquid crystal. The symmetry of this surface (and of γ_s) is independent of the orientation taken by the nematic phase at the surface. In contrast, the symmetry of the interface depends on this orientation: it is the subgroup of the surface symmetry group containing the symmetry elements which leave invariant the anchoring direction effectively taken by the liquid crystal. If the other phase is a solid or liquid substrate, the surface is simply the surface of this substrate. In the case when the other phase is the gas or isotropic phase, the 'surface' is not a physical entity. However, one can still, in principle, distinguish this isotropic surface (C_∞ symmetry) from the interface with the nematic phase, the symmetry of which is C_∞, C_{2v}, and C_{1v} for homeotropic, planar, and tilted anchoring, respectively.

The interfaces at which anchoring has been mostly studied are those with crystal surfaces and treated-glass substrates. Glass treatments can be mechanical (rubbing) or chemical (deposition of a layer of surfactant, polymer, inorganic substance, etc.) or a combination of the two (for a review see Cognard [8]). The most commonly obtained anchorings are homeotropic, degenerate, and monostable planar (Fig. 2). Monostable tilted anchorings are less frequent, and multistable anchorings are seldom obtained.

The occurrence of multistable or degenerate anchorings raises the issue of the selection of anchoring directions; indeed, at each point of the surface, the director can take only one of the possible anchoring directions characterizing such anchorings. Generally speaking, the selection is made by the history of the sample, for instance cooling from the isotropic phase under a magnetic field [11, 12] or spreading [13]. In general, once the selection has been made, the anchoring direction is preserved by the sur-

face, even in the case of degenerate anchorings [11, 14]. This appears to come from the adsorption of the surface molecules onto the substrate [12]. Switching between the anchoring directions of a multistable anchoring is, however, possible [15, 16].

10.2.2 Anchoring Directions

By minimizing the interfacial energy γ_s, one can find the anchoring diagram giving the different anchorings induced by a surface depending on the values of the coefficients A_{lm} appearing in Eq. (1) (Fig. 3). These diagrams give a general description of anchoring, independent of the nature of the system. However, the limitation of this macroscopic approach lies in the fact that it is, in general, not possible to establish a relationship between the coefficients A_{lm} and the parameters governing the structure of the interface.

Anchoring diagrams show the different ways in which anchoring can change when some parameters are varied. These anchoring transitions follow the same symmetry rules as phase transitions: transitions which do not involve a change of symmetry of the interface must be first order, while when a symmetry change is involved the transition can be first or second order [10]. Anchoring transitions also appear the same way as phase transitions: nucleation and growth of domains having a new orientation for first-order transitions [19], and director fluctuations for second-order transitions [18]. These fluctuations correspond to a divergence of the susceptibility of the system for a force driving the director away from the anchoring directions: indeed, the anchoring strength (see Section 10.2.3) does go to zero at second-order anchoring transitions [20–22]. One characteristic of anchoring transitions is that this susceptibility can be

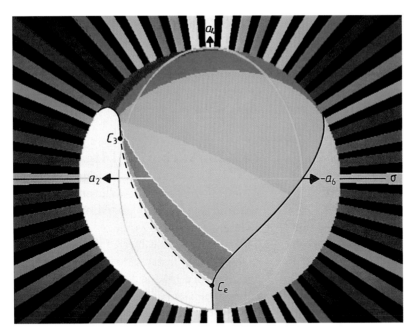

Figure 3. Anchoring diagram corresponding to a surface with C_{1v} symmetry (with mirror plane σ) inducing planar anchorings: $\gamma_s(\varphi) = a_2 \cos(2\varphi) + a_4 \cos(4\varphi) + a_6 \cos(6\varphi)$ [17, 18]. γ_s is normalized by the condition $a_2^2 + a_4^2 + a_6^2 = 1$, so that the anchoring diagram can be represented on a sphere. At each point on the sphere, the color indicates the predicted anchoring direction. The code is given by the background, the surface of which corresponds to the substrate surface. (——) First-order transition; (-----) second-order transition. C_e, Critical endpoint; C_3, tricritical point.

anisotropic (i.e. the response to an applied force can depend on the force direction) [23].

There are several examples of experimentally observed anchoring transitions (see Jérôme [4] and Pieranski and Jérôme [10], and references therein). Since anchoring transitions occurring at the surface of a liquid crystal cell induce a change of director configuration in the cell, these transitions can be used to switch a liquid crystal device between different states [24, 25].

10.2.3 Anchoring Energy

The director at an interface might depart from the anchoring directions induced by this interface under the effect of another orienting field. The way in which this happens is determined the dependence of γ_s on the orientation of the director $\mathbf{n}$ at the surface. Knowledge of this dependence is necessary in order to establish the surface boundary conditions on the director field [26]. Several methods have been developed in order to measure experimentally some characteristics of $\gamma_s(\mathbf{n})$ [27, 28]. Generally speaking, they consist in observing the change in orientation induced at a surface under the action of an external field.

Thermodynamically, γ_s can be defined by the equation [28]:

$$f = f_0 + \gamma_s\left[\mathbf{n}(z_d), z_d\right] + \int_{z_d}^{\infty} f_d\left[\mathbf{n}(z), \frac{d\mathbf{n}}{dz}(z)\right] dz \tag{2}$$

where f is the total free energy of the system per unit area, f_0 is the free energy in absence of surface and deformation in the bulk, f_d is the elastic energy density due to deformations of the director field $\boldsymbol{n}(z)$, and z_d is the position of the surface above which the nematic phase is considered to have bulk-like behavior. From this definition, it is clear that when the director field is not uniform, which is always the case in anchoring energy measurements, γ_s depends on z_d. Unfortunately, the value of z_d depends both on the technique used to measure the anchoring energy and on the structure of the interfacial region [28]. Therefore, the results of the anchoring energy measurements interpreted using this approach (as has generally been done until now) are not comparable with each other.

The dependence of the interfacial energy γ_s with the surface director was first considered by Rapini and Papoular [29] who assumed that the dependence of γ_s on the tilt angle θ was proportional to $\sin^2(\theta-\theta_s)$ where θ_s is the tilt of the anchoring direction. The coefficient of proportionality measures the ease with which the surface director can deviate from the anchoring direction and is called the anchoring strength or anchoring energy coefficient. Generally speaking, one defines two anchoring strengths, the polar and the azimuthal one, corresponding to deviations from the anchoring direction (θ_s,φ_s) at constant azimuthal angle φ and tilt angle θ, respectively:

$$W_p = \frac{1}{2}\left(\frac{\partial^2\gamma_s}{\partial\theta^2}\right)_\varphi (\theta_s,\varphi_s) \qquad (3a)$$

$$W_a = \frac{1}{2}\left(\frac{\partial^2\gamma_s}{\partial\varphi^2}\right)_\theta (\theta_s,\varphi_s) \qquad (3b)$$

All measurements of anchoring strengths give values of W_p ranging between 10^{-7} and 10^{-3} J m^{-2} ($10^{-5}-1$ erg cm^{-2}) and values of W_a one or two orders of magnitude smaller. This inequality might be intrinsic to the anchoring mechanisms, as predicted by some

theorctical calculations of $\gamma_s(\theta,\varphi)$ [30–32]. It might also come from the order-parameter gradient existing at the interface, giving rise to an order-electric polarization [33] and to a contribution to γ_s [34]: owing to symmetry considerations, this contribution is zero for W_α while it is finite for W_p. It has also been suggested that the upper limit of the measured anchoring strengths could be due to the fact that, for strong anchorings, the interfacial region would prefer to melt instead of undergoing too strong a distortion [35].

The anchoring strengths (Eq. (3)) are useful only when the director remains close to the anchoring directions. For large deviations of the surface director, the whole dependence of γ_s should be known. With few exceptions [23], measurements of γ_s are performed assuming that the θ and φ variations of γ_s are independent. Concerning the θ dependence, neither experiments [36–39] nor theoretical calculations [31, 40] (see also Jérôme [4] and references therein) have come to an agreement on whether the expression of $\gamma_s(\theta)$ should contain one or more Fourier components. Part of the discrepancies in the anchoring energy measurements could be due to bulk effects such as ferroelectricity [39] or the presence of ions [27, 41, 42].

In addition, in none of these studies was the contribution to the anchoring energy of the adsorption of liquid crystalline molecules onto the substrate [12] taken into account. In fact, γ_s contains two terms [4]:

$$\gamma_s(\boldsymbol{n}) = \gamma_s^{surf}(\boldsymbol{n}) + \gamma_s^{ad}(\boldsymbol{n};\boldsymbol{a}_{ad}) \qquad (4)$$

γ_s^{surf} is the interfacial energy in the absence of adsorption; γ_s^{ad} arises from the adsorption of molecules onto the substrate and has two opposite minima corresponding to the anchoring direction $\boldsymbol{a}_{ad}\equiv-\boldsymbol{a}_{ad}$ initially taken by the liquid crystal. The orienting field felt by the surface director therefore has uniaxial symmetry, regardless of the symmetry of γ_s^{surf}. However, γ_s^{ad} is not an intrinsic property of the surface and can be subject to changes, depending on the history of the sample [43].

From the above discussion, it is clear that the macroscopic approach is inadequate for describing situations in which the director is forced to depart from anchoring directions. A proper description of these situations should take into account the response of the interfacial region to such distortions. There have been some attempts to describe the interfacial region in a semi-macroscopic way, assuming that the nematic order is retained in the interfacial region but with a varying order parameter [22, 34, 44, 45]. These semi-macroscopic theories define a surface anchoring energy giving the energy of the surface layer as a function of its orientational state defined by the tensorial nematic order parameter. The spatial evolution of this order parameter away from the surface is calculated using the Landau–de Gennes [46] theory under the appropriate conditions corresponding to the way, in which the orientation of the director is imposed in the bulk.

The main defect in this kind of theory is the assumption that the orientational order is nematic like in the interfacial region, which is generally not true (see Section 10.3.1). Another problem is linked to the fact that the surface anchoring energy is unknown. This surface anchoring energy is difficult to measure directly and the result of any indirect measurement depends on how the response of the system to a disorientation is modeled. One way out is to use the orientational distribution of the molecules in the surface layer as a boundary condition of the nematic order [47] (see also Section 10.3.2).

10.3   Microscopic Surface Order of Nematic Liquid Crystals

10.3.1   Surface Orientational Order

An experimental technique has recently been developed to investigate the surface orientational order of liquid crystals: optical second-harmonic generation [48]. Because of its high sensitivity to surface polar ordering and its surface specificity, this method is able to probe the orientational distribution of polar monolayers of liquid crystal molecules located at the surface of a substrate, whether they are covered by a bulk liquid crystal or not.

So far, such measurements have only been performed on cyanobiphenyl molecules (n-CB and n-OCB). These molecules exhibit polar ordering (with the cyano groups pointing towards the surface) on hydrophilic substrates such as water, glass, and certain polymer films [49–51], but not on hydrophobic layers [52]. On the latter, the molecules orient with their aliphatic chains in contact with the substrate, and arrange following their natural tendency to form apolar 'dimers' (two molecules oriented head to head). In all cases where a surface dipolar ordering exists, it is lost after the first molecular layer [49].

The tilt of the hard core of cyanobiphenyl molecules is found to be always approximately the same, namely 70° from the surface normal, resulting from a balance between the dipole–dipole interaction of the polar heads and the steric interaction of the biphenyl core with the substrate [49, 53]. By means of infrared sum-frequency generation, it has also been shown that the aliphatic chains are relatively straight and point away from the surface [54].

The distribution of azimuthal orientation strongly depends on the substrate [52, 55, 56]. On glass and surfactant- and polymer-coated glass this distribution is isotropic, as would be expected from the isotropy of these surfaces. When polymer films are rubbed by translation in one direction on a piece of cloth, the polymer chains at the surface of the film are oriented along the rubbing direction [57, 58]. The liquid-crystal molecules then orient preferentially along the polymer chains (Fig. 4a). As on crystal surfaces, this anisotropic orientational distribution is the result of the direct interaction between the molecules and the substrate. Conversely, monolayers at the surface of rubbed glass, certain rubbed surfactant layers, and evaporated SiO films exhibit an isotropic azimuthal distribution [51, 52]; despite their anisotropy, these substrates have no short-range azimuthal orienting action on nematic molecules.

The azimuthal orientational distribution of surface liquid-crystal molecules can, in principle, also be obtained by imaging the molecules. However, sufficiently high res-

olution can only be obtained with scanning tunneling microscopy, which limits the use of imaging techniques to special conducting substrates, such as cleaved pyrolytic graphite [59, 60] and MoS crystals [61]. A feature common to all the images obtained is the high orientational and positional order of the molecules in the surface monolayer.

The techniques described above allow for the determination of the orientational order in the surface layer. A complete description of the liquid crystalline ordering close to a surface should also include the evolution of the orientational order in the interfacial region from the one at the surface to the one in bulk. There are, however, no experimental techniques available to measure this evolution directly. Some experimental techniques allow one to measure some integrated quantities over the whole interfacial region, which depend on the evolution of the orientational order in this region (see e.g. [62–65]). To obtain the order profile from such measurements, one needs to assume the general shape of this evolution and fit the experimental data to obtain the profile parameters. Knowledge of the surface orientational order can, however, guide the choice of model profile.

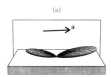

Figure 4. The three main substrate classes: (a) smooth surfaces on which surface molecules have a definite orientational distribution (represented: surface obtained on a rubbed polyimide film [52]); (b) interpenetrable surfaces of dangling chains; (c) topographies (represented: grooved surface) with a favorable (left) and unfavorable director field n. In all cases, a is the macroscopic anchoring direction.

10.3.2 Microscopic Anchoring Mechanisms at Solid Substrates

With the presently known information on the orientational order in surface layers and interfacial regions, it is possible to deduce some of the microscopic origins of the macroscopic anchoring of liquid crystals. One can distinguish three main classes of substrate, giving rise to three main classes of anchoring mechanism: smooth surfaces, interpenetrable layers, and topographies (Fig. 4). It should, however, be emphasized that the macroscopic anchoring is often due to a combination of effects, some of which are specific to certain substrates.

Smooth substrates include crystal and glass surfaces, and polymer films. The direct influence of such a substrate on the liquid crystal is limited to the molecules in the surface layer and is due to the short-range interaction of these molecules with the substrate (Fig. 4a). Once these microscopic anchoring conditions are known, the configuration of the rest of the liquid crystal (in particular the orientation of the director outside the interfacial region) can be determined from the properties of the liquid crystal regarding propagation of orientational order.

The microscopic anchoring conditions can be modeled by the energy of the surface layer as a function of its orientational order. The evolution of the orientational order away from the surface is then calculated by minimizing the sum of this surface energy and of the (orientational order dependent) bulk energy. Since the dependence of the energy of the surface layer on its orientational order is generally unknown, the orientational distribution of the surface molecules, if experimentally available (see Section 10.3.1), can be used as a boundary condition for the order in the rest of the liquid crystal [52, 55]. This surface orientational distribution does indeed seem to be strongly enforced by the surface; it is essentially independent of whether this monolayer is covered by a bulk liquid crystal or not, and whether this bulk is in the nematic or isotropic phase [52].

This type of calculation has been performed successfully, both by using a general form of the surface energy [22] and experimentally determined surface orientational distributions [66, 67], in order to explain the anchoring directions and anchoring transitions observed on different substrates. These calculations were made within the framework of the Landau–de Gennes theory [46], the liquid crystalline order at each point in space (including the interfacial region and surface layer) being described by the tensorial order parameter. The evolution of the director and the degree of molecular order in the interfacial region is coupled. This order-parameter/director coupling comes essentially from the L_2 term in the Landau–de Gennes energy; this term should therefore not be neglected (as is often the case). This implies that, since the degree of order is generally different at the surface and in the bulk, the anchoring direction is different from the mean orientation of the molecules at the surface [66–68]. This also means that the degree of surface ordering is a determinant factor in the tilt of the anchoring directions [43, 47, 55].

The Landau–de Gennes theory is, however, not applicable to all surfaces; in particular it is unable to predict the azimuthal orientation of the anchoring directions in the case of multistable anchorings, which occur on substrates with sufficiently symmetry. Indeed, reducing the information on the surface orientational distribution to a nematic tensorial order parameter suppresses the information on the symmetry of the surface. Order parameters of higher order reflecting the surface symmetry should then be included in the description of the orientational order, together with the corresponding terms in the energy of the system.

A second class of substrates corresponds to surfaces which are smooth but on which molecules with a long tail (such as surfactants) are grafted, forming a layer that the liquid crystal can penetrate (Fig. 4b). The influence of the substrate on the liquid crystal at a microscopic level can then be separated into two parts: the effect of the surface underlying the surfactant layer and the effect of the chains forming the layer. It seems that the latter always dominates; the liquid crystal adopts the orientation of the chains in the layer, which leads to a homeotropic or conical anchoring depending on the orientation of the chains [54, 69, 70].

It should be emphasized that not all substrates covered with aliphatic chains belong to this class of substrates. If the chains are closely packed, the liquid crystal molecules cannot penetrate the layer [54] and the substrate can be seen as a smooth surface belonging to the first class described above.

The third class of substrates corresponds to those having a surface of a certain topography (Fig. 4c). This topography is generally obtained by rubbing the surface with a hard material (rubbed glass or surfactant-coated glass) or by anisotropic vapor deposition (evaporated SiO films). Each point of such a surface induces a given orientational distribution of liquid crystal molecules. On all the substrates mentioned above, this local orientational distribution is isotropic [51, 52]. However, since the orientation and the height of the surface is not uniform, the microscopic anchoring conditions are not uniform over the surface, introducing distortions along the plane of the surface. The anchoring direction chosen by the liquid crystal is then the one that minimizes these distortions. For instance, grooved surfaces induce an anchoring direction parallel to the grooves. This anchoring mechanism can be modeled by minimizing the Franck–Oseen elastic free energy associated with director distortions [5].

All the anchoring mechanisms described above are based on the knowledge of the microscopic anchoring conditions. However, a complete understanding of anchoring requires knowledge of the origins of this microscopic anchoring, which should be found in the interactions between the liquid crystal molecules and the substrates. These interactions are basically unstudied.

10.3.3 The Nematic/Isotropic and Nematic/Vapor Interfaces

The case of the interface with an isotropic or vapor phase is relatively simple in that no interaction with a substrate is involved. The molecular ordering and the orientation of the nematic director at these interfaces arise mainly from the change in density and type of order taking place across the interface.

Studies of the orientational order at this type of interface concern exclusively nematic/vapor interfaces. No direct measurement of this order has been performed so far, but indirect observations have shown that the order at the free surface can be higher or lower than in the bulk [71–73]. Excess order occurs when the intermolecular interaction depends on the relative molecular positions (i.e. when the translational and rotational degrees of freedom are coupled) [74, 75]. This coupling is also essential for anchoring to occur at the surface (Fig. 5).

Since the surface breaks the bulk inversion symmetry, it creates a polarity which, in principle, should give rise to polar ordering near the surface. This ordering has been investigated in asymmetric molecules with a polar head. These molecules prefer to orient with their polar heads pointing in the direction of the more polar medium. This tends to create a polar surface layer with all the heads pointing away from the surface. However, in order to minimize the polar intermolecular interactions, the molecules tend to arrange antiparallel to each other with a partial or total overlap. This effect compensates for any surface-induced polar ordering, creating either a non-polar [76] or an antiferroelectric [49, 77, 78] surface ordering.

As far as macroscopic anchoring is concerned, it has been found experimentally that, at the nematic/isotropic interface, the director is tilted (in all the compounds so far studied, including n-CB and MBBA), with

Figure 5. Anchoring (top) and surface ordering (bottom) at the free surface of a nematic liquid crystal as a function of the strength C of the coupling between translational and rotational degrees of freedom [75]. $\perp$, Homeotropic ordering; $\parallel_{uni}$ and $\parallel_{bi}$, uniaxial and biaxial ordering parallel to the surface.

an angle θ_s from the surface normal in the range 50–80° [6,7]. At the free surface, there are more differences from one liquid crystal to the other: $\theta_s = 0°$ in n-CB and PCHn [79, 80] (because the polar head of these molecules prefers to point away from the surface), $\theta_s = 90°$ in PAA [81], and θ_s decreases to 0° as the temperature increases in MBBA and EBBA [82].

The first attempt to account for these results was made by de Gennes using the Landau–de Gennes theory [46]. This theory predicts that the tilt angle can only be 0° or 90°, depending on the elastic properties of the liquid crystals. In order to account for tilted anchorings in this framework, different effects can be added in the model, such as order electricity arising from order-parameter gradients [33].

Another series of theoretical models is based on molecular interactions, either anisotropic hard-core interactions [40, 83–85] or combining different kinds of intermolecular interactions [86, 87] (see also Jérôme [4] and references therein). The results depend greatly on the details of the models, the validity of which still needs to be checked.

10.4   Orientation of Other Liquid Crystals

The study of surface orientation of non-nematic liquid crystals is still incomplete.

Here, separate sections are dedicated to polymer liquid crystals and lyotropic liquid crystals, the latter being peculiar in that they are inhomogeneous phases.

10.4.1   Smectic and Chiral Liquid Crystals

In smectic A phases where the smectic layers are perpendicular to the molecules, the orientation of the whole structure is, in principle, fixed once the orientation of the molecules is defined by the interface. The surface orientation of achiral smectic A phases is then the same as that of the nematic phase [88, 89]. However, since splay deformations of smectic layers (director bend deformations) are forbidden and layer bend deformations (director splay deformations) require a lot of energy, smectic phases tend to adopt uniform configurations, even between two walls inducing two different orientations. In the latter case, the surface orientation of the smectic phase differs from that of the nematic phase, and depends on the layer configuration in the bulk [90, 91].

In the case of tilted smectic phases (for instance the smectic C phase), there is a degeneracy in the orientation of the director with respect to that of the layers. If a surface inducing a degenerate planar anchoring is placed perpendicular to the layers, there are two possible orientations of the director that satisfy both the surface and the liquid crystal structural constraints (Fig. 6).

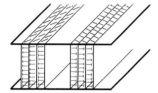

Figure 6. Two equivalent orientational states in a smectic C liquid crystal between two walls inducing a degenerate planar anchoring.

This surface bistability is at the basis of chiral smectic C surface stabilized ferroelectric liquid crystal (SSFLC) devices [92]. As their name indicates, these devices are made of thin cells in which the walls, imposing the orientation of the molecules at the surfaces, unwind the spontaneous smectic C* helix and stabilize two uniform configurations of the director in the cell. Switching between these two states can be done by applying an electric field.

The preparation of such cells requires both that the director is oriented parallel to the surface and that the layers are oriented perpendicular to the surfaces. Because the walls only have an influence on the orientation of the director, and because of the director/layer-normal degeneracy, the alignment of the layers must be achieved in the smectic A* phase, before cooling the system in the smectic C* phase [93]. Due to the positional anchoring of the layers at the surfaces [94], the location of the layers is pinned at the walls. Since the layer thickness d_A in the smectic A phase is larger than the thickness d_C in the smectic C phase, the layers must tilt in the bulk by an angle δ_C such that $d_C = d_A \cos \delta_C$ [95].

The combination of tilted orientation of the director at the surface (breaking the in-plane isotropy of the surface) and of the surface-induced up–down asymmetry [96] gives rise to new terms in the energy of the surface layer which favor splay configurations of the director [97, 98]. Such configurations have been observed at the free surface of achiral tilted smectic phases. The details of the patterns obtained depend on the type of smectic phase considered: in a smectic C phase, the director can rotate freely on a cone, while in a smectic phase presenting a bond-orientational order (smectic I, F, or L) the coupling existing between this order and the director imposes discrete rotations of the director through walls [99, 100]. The surface pattern also depends on whether or not there is a mechanism imposing the sign of the surface-induced polarization, and consequently that of the director rotation [100–103].

In a chiral compound, the liquid crystalline structure itself possesses a polarity which interacts with the surface. At the surface of a smectic C* liquid crystal, the ferroelectric polarization P points preferably either towards or away from the surface, depending on the material [104, 105]. Conversely, if the surface director of a chiral nematic liquid crystal is tilted, a polarization (dependent on the tilt angle) is created perpendicular to the tilt plane [106]. In principle, the presence of this polarization makes a contribution to the anchoring energy; this chiral contribution is, however, too small to be measured [107].

The combination of liquid crystalline chirality and surface polarization is also the origin of the so-called surface electroclinic effect in chiral smectic A phases: if the smectic layers are oriented perpendicular to the surface, the surface electric field tilts the layer normal away from the surface molecular orientation [108–110] (Fig. 7).

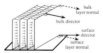

Figure 7. The surface electroclinic effect rotating the smectic layer normal away from the surface molecular orientation in smectic A* liquid crystals.

10.4.2 Polymer Liquid Crystals

The macroscopic orientation of polymer liquid crystals at surfaces can formally be described in the same way as that of low-molecular-weight liquid crystals. However, the polymer character of these materials has an influence on the surface behavior at a microscopic level. One can distinguish different aspects: loss of freedom of the mesogenic groups, change in polymer chain entropy due to the steric restriction near the surface, increase of longitudinal persistence length (along the director), and attraction/repulsion of the polymer chain at the surface.

When mesogenic groups are part of a polymer chain (in main-chain polymers) or attached by one of their ends to a main chain (in side-group polymers), they lose the ability to orient independently of each other. In particular, they might not be able to adopt the surface orientation which they would have if they were free. This effect can lead to a change in anchoring direction with respect to that of the free mesogens, as observed at the free surface of side-chain polymer liquid crystals the side-groups of which are terminated by a polar head (such as cyanobiphenyl groups) [111, 112]. As mentioned in Section 10.3.3, low-molecular-weight liquid crystalline compounds carrying a polar head orient perpendicularly to their free surface because the head prefers to point away from the surface. In a side-group polymer liquid crystal, satisfying this constraint would force the surface polymer molecules to have their main chain confined at the surface and all their side-groups pointing in the same direction. This is sterically impossible. The system solves this configurational problem by orienting the side-groups (and the surface director) parallel to the surface.

As in any polymer, the presence of a surface limits the number of configurations that the polymer chains are allowed to take in the vicinity of this surface. Since these chains and the mesogenic groups are mechanically coupled, this loss of chain entropy implies a limitation of the orientations accessible to the mesogenic groups. Moreover, the presence of a polymer chain linking the mesogenic groups increases the longitudinal persistence length along the director. These two effects can make the anchoring strength of polymer liquid crystals considerably larger than that of low-molecular-weight liquid crystals [113].

Finally, the polymer chain (the main chain in side-chain polymers or the intermesogen chain in main-chain polymers) interacts directly with the surface and can be either attracted to or expelled from the surface. For instance, side-groups terminated by a polar head and attached to a chain also carrying polar groups orient parallel to hydrophilic surfaces, instead of being tilted as the corresponding low-molecular-weight liquid-crystal molecules (see Section 10.3.1).

10.4.3 Lyotropic Liquid Crystals

The anisotropic micelles forming lyotropic liquid crystals are also oriented by surfaces. Both prolate and oblate micelles orient parallel to flat surfaces [114, 115] probably due to hard-core interactions [116]. Prolate micelles can also be azimuthally oriented by grooved surfaces, or homeotropically oriented by two-dimensional topographies [117].

However, one peculiarity of these systems is that the surfactant molecules forming the micelles are surface active and generally adsorb at the surface; this leads a modification of the surface [115]. The presence of anisotropic defects in this adsorbed layer can then induce an azimuthal anchoring of the bulk micelles. The slow reorientation of these surface defects under a magnetic field induces a slow in-plane reorientation – or gliding – of the anchoring directions [118].

10.5 References

[1] C. Mauguin, *C.R.A.S.* **1911**, *156*, 1246.
[2] T. J. Sluckin, A. Poniewierski, in *Fluid Interfacial Phenomena* (Ed. C. A. Croxton), Wiley, New York **1986**, Chap. 5, pp. 215–253.
[3] A. A. Sonin, *The Surface Physics of Liquid Crystals,* Gordon and Breach, Amsterdam **1995**.
[4] B. Jérôme, *Rep. Prog. Phys.* **1991**, *54*, 391.
[5] D. W. Berreman, *Phys. Rev. Lett.* **1972**, *28*, 1683.
[6] R. Vilanove, E. Guyon, C. Mitescu, P. Pieranski, *J. Phys. France* **1974**, *35*, 153.
[7] H. Yokoyama, S. Kobayashi, H. Kamei, *Mol. Cryst. Liq. Cryst.* **1984**, *107*, 311.
[8] J. Cognard, *Mol. Cryst. Liq. Cryst.* **1982**, *1* (Suppl.), 1.
[9] P. Pieranski, B. Jérôme, M. Gabay, *Mol. Cryst. Liq. Cryst.* **1990**, *179*, 285.
[10] P. Pieranski, B. Jérôme, *Mol. Cryst. Liq. Cryst.* **1991**, *199*, 167.
[11] J. Cheng, G. D. Boyd, *Appl. Phys. Lett.* **1979**, *35*, 444.
[12] Y. Ouchi, M. B. Feller, T. Moses, Y. R. Shen, *Phys. Rev. Lett.* **1992**, *68*, 3040.
[13] B. Jérôme, P. Pieranski, *J. Phys. France* **1988**, *49*, 1601.
[14] N. Koshida, S. Kibui, *Appl. Phys. Lett.* **1982**, *40*, 541.
[15] R. Barberi, M. Boix, G. Durand, *Appl. Phys. Lett.* **1989**, *55*, 2506.
[16] R. Barberi, G. Durand, *Liq. Cryst.* **1991**, *10*, 289.
[17] H. S. Kitzerow, B. Jérôme, P. Pieranski, *Physica A* **1991**, *174*, 163.
[18] J. Bechhoefer, J. L. Duvail, L. Masson, B. Jérôme, R. M. Hornreich, P. Pieranski, *Phys. Rev. Lett.* **1990**, *64*, 1911.
[19] P. Pieranski, B. Jérôme, *Phys. Rev. A* **1989**, *40*, 317.
[20] P. Chiarelli, S. Faetti, L. Fronzoni, *Phys. Lett.* **1984**, *101A*, 31.
[21] G. A. DiLisi, C. Rosenblatt, A. C. Griffin, U. Hari, *Liq. Cryst.* **1990**, *7*, 353.
[22] P. I. C. Teixeira, T. J. Sluckin, D. E. Sullivan, *Liq. Cryst.* **1993**, *14*, 1243.
[23] M. Nobili, G. Durand, *Europhys. Lett.* **1994**, *25*, 527.
[24] K. Ichimura, Y. Suzuki, T. Seki, A. Hosoki, K. Aoki, *Langmuir* **1988**, *4*, 1214.
[25] W. Gibbons, P. Shanon, S. T. Sun, B. Svetlin, *Nature* **1991**, *351*, 49.
[26] P. G. de Gennes, J. Prost, *The Physics of Liquid Crystals*, 2nd edn, Clarendon Press, Oxford **1993**.
[27] L. M. Blinov, A. Y. Kabayenkov, A. A. Sonin, *Liq. Cryst.* **1989**, *5*, 645.
[28] H. Yokoyama, *Mol. Cryst. Liq. Cryst.* **1988**, *165*, 265.
[29] A. Rapini, M. Papoular, *J. Phys. Coll.* **1969**, *30*, C4 54.
[30] J. Bernasconi, S. Strässler, H. R. Zeller, *Phys. Rev. A* **1980**, *22*, 276.
[31] K. Okano, N. Matsuura, S. Kobayashi, *Jpn. J. Appl. Phys.* **1982**, *21*, L109.
[32] S. Faetti, *Phys. Rev. A* **1987**, *36*, 408.
[33] G. Barbero, I. Dozov, J. F. Palierne, G. Durand, *Phys. Rev. Lett.* **1986**, *56*, 2056.
[34] H. Yokoyama, S. Kobayashi, H. Kamei, *J. Appl. Phys.* **1987**, *61*, 4501.
[35] G. Barbero, G. Durand, *Mol. Cryst. Liq. Cryst.* **1991**, *203*, 33.
[36] K. H. Yang, C. Rosenblatt, *Appl. Phys. Lett.* **1983**, *43*, 62.
[37] G. Barbero, N. V. Madhusudana, G. Durand, Z. *Naturforsch., Teil a* **1984**, *39*, 1066.
[38] H. Yokoyama, S. Kobayashi, H. Kamei, *Mol. Cryst. Liq. Cryst.* **1985**, *129*, 109.
[39] G. Barbero, G. Durand, *J. Phys. France* **1986**, *47*, 2129.
[40] J. D. Parsons, *Mol. Phys.* **1980**, *42*, 951.
[41] G. Barbero, G. Durand, *J. Phys. France* **1990**, *51*, 281.
[42] B. Valenti, M. Grillo, G. Barbero, P. Taverna Valabrega, *Europhys. Lett.* **1990**, *12*, 407.
[43] B. O. Myrvold, *Liq. Cryst.* **1995**, *18*, 287.
[44] G. Barbero, G. Durand, *J. Appl. Phys.* **1991**, *69*, 6968.
[45] M. Nobili, G. Durand, *Phys. Rev. A* **1992**, *46*, 6174.
[46] P. G. de Gennes, *Mol. Cryst. Liq. Cryst.* **1971**, *12*, 193.
[47] Y. Zhuang, L. Marrucci, Y. R. Shen, *Phys. Rev. Lett.* **1994**, *73*, 1513.
[48] Y. R. Shen, *Nature* **1989**, *337*, 519.
[49] P. Guyot-Sionnest, H. Hsiung, Y. R. Shen, *Phys. Rev. Lett.* **1986**, *57*, 2963.
[50] C. S. Mullin, P. Guyot-Sionnest, Y. R. Shen, *Phys. Rev. A* **1989**, *39*, 3745.
[51] W. Chen, M. Feller, Y. R. Shen, *Phys. Rev. Lett.* **1989**, *63*, 2665.
[52] M. B. Feller, W. Chen, Y. R. Shen, *Phys. Rev. A* **1991**, *43*, 6778.
[53] N. A. J. M. van Aerle, *Liq. Cryst.* **1994**, *17*, 585.
[54] J. Y. Huang, R. Superfine, Y. R. Shen, *Phys. Rev. A* **1990**, *42*, 3660.
[55] M. Barmentlo, R. W. J. Hollering, N. A. J. M. van Aerle, *Phys. Rev. A* **1982**, *46*, 4490.
[56] B. Jérôme, Y. R. Shen, *Phys. Rev. E* **1993**, *48*, 4556.
[57] J. M. Geary, J. M. Goodby, A. R. Kmetz, J. S. Patel, *J. Appl. Phys.* **1987**, *62*, 4100.

[58] M. F. Toney, T. P. Russel, J. A. Logan, H. Kikuchi, J. M. Sands, S. K. Kumar, *Nature* **1995**, *374*, 709.

[59] J. S. Foster, J. E. Frommer, *Nature* **1988**, *333*, 542.

[60] D. P. E. Smith, J. K. H. Hörber, G. Binnig, H. Nejoh, *Nature* **1990**, *344*, 641.

[61] M. Hara, Y. Iwakabe, K. Tochigi, H. Sasabe, A. F. Garito, A. Yamada, *Nature* **1990**, *344*, 228.

[62] K. Miyano, *Phys. Rev. Lett.* **1979**, *43*, 51.

[63] J. P. Nicholson, *J. Phys. France* **1987**, *48*, 131.

[64] W. Chen, L. J. Martinez-Miranda, H. Hsiung, Y. R. Shen, *Phys. Rev. Lett.* **1989**, *62*, 1860.

[65] G. P. Crawford, R. Stannarius, J. W. Doane, *Phys. Rev. A* **1991**, *44*, 2558.

[66] D. Johannsmann, H. Zhou, P. Sonderkaer, H. Wierenga, B. O. Myrvold, Y. R. Shen, *Phys. Rev. E* **1993**, *48*, 1889.

[67] B. Jérôme, *J. Phys.: Condens. Matter* **1994**, *6*, A269.

[68] B. Jérôme, J. O'Brien, Y. Ouchi, C. Stanners, Y. R. Shen, *Phys. Rev. Lett.* **1993**, *48*, 4556.

[69] G. Porte, *J. Phys. France* **1976**, *37*, 1245.

[70] K. Hiltrop, H. Stegemeyer, *Ber. Bunsenges. Phys. Chem.* **1981**, *85*, 582.

[71] A. W. Neumann, R. W. Springer, R. T. Bruce, *Mol. Cryst. Liq. Cryst.* **1974**, *27*, 23.

[72] D. Beaglehole, *Mol. Cryst. Liq. Cryst.* **1982**, *89*, 319.

[73] D. Langevin, *J. Phys. France* **1972**, *33*, 249.

[74] J. H. Thurtell, M. M. Telo da Gama, K. K. Gubbins, *Mol. Phys.* **1985**, *54*, 321.

[75] B. Tjipto-Margo, A. K. Sen, L. Mederos, D. E. Sullivan, *Mol. Phys.* **1989**, *67*, 601.

[76] E. F. Gramsbergen, W. H. de Jeu, *J. Phys. France* **1988**, *49*, 363.

[77] B. M. Ocko, P. S. Pershan, C. R. Safinya, L. Y. Chiang, *Phys. Rev. A* **1987**, *35*, 1868.

[78] E. F. Gramsbergen, J. Als Nielsen, W. H. de Jeu, *Phys. Rev. A* **1988**, *37*, 1335.

[79] M. G. J. Gannon, T. E. Faber, *Phil. Mag. A* **1978**, *37*, 117.

[80] S. Immerschitt, T. Kohl, W. Stille, G. Strobl, *J. Chem. Phys.* **1992**, *21*, 173.

[81] M. A. Bouchiat, D. Langevin-Cruchon, *Phys. Lett. A* **1971**, *34*, 331.

[82] S. Faetti, L. Fronzoni, *Solid State Commun.* **1978**, *25*, 1087.

[83] R. Holyst, A. Poniewierski, *Phys. Rev. A* **1988**, *38*, 1527.

[84] B. G. Moore, W. E. McMullen, *Phys. Rev. A* **1990**, *42*, 6042.

[85] M. A. Osipov, S. Hess, *J. Chem. Phys.* **1993**, *99*, 4181.

[86] J. D. Parsons, *Phys. Rev. Lett.* **1978**, *41*, 877.

[87] B. Thipto-Margo, D. E. Sullivan, *J. Chem. Phys.* **1988**, *88*, 6620.

[88] W. Urbach, M. Boix, E. Guyon, *Appl. Phys. Lett.* **1974**, *25*, 479.

[89] J. E. Proust, L. Ter-Minassian-Saraga, *J. Phys. Colloq.* **1979**, *40*, C3 490.

[90] S. J. Elston, *Liq. Cryst.* **1994**, *16*, 151.

[91] J. J. Bonvent, J. A. M. M. van Haaren, G. Cnossen, A. G. H. Verhulst, P. van der Sluis, *Liq. Cryst.* **1995**, *18*, 723.

[92] N. A. Clark, S. T. Lagerwall, *Appl. Phys. Lett.* **1980**, *36*, 899.

[93] K. Kondo, F. Kobayashi, A. Fukuda, E. Kuze, *Jpn. J. Appl. Phys.* **1981**, *20*, 1773.

[94] M. Cagnon, G. Durand, *Phys. Rev. Lett.* **1993**, *70*, 2742.

[95] M. A. Handschy, N. A. Clark, *Ferroelectrics* **1984**, *59*, 69.

[96] Y. Galerne, L. Liebert, *Phys. Rev. Lett.* **1990**, *64*, 906.

[97] R. B. Meyer, P. S. Pershan, *Solid State Commun.* **1973**, *13*, 989.

[98] J. V. Selinger, Z. G. Wang, R. F. Bruinsma, C. M. Knobler, *Phys. Rev. Lett.* **1988**, *70*, 1139.

[99] S. B. Dierker, R. Pindak, R. B. Meyer, *Phys. Rev. Lett.* **1986**, *56*, 1819.

[100] J. Maclennan, M. Seul, *Phys. Rev. Lett.* **1992**, *69*, 2082.

[101] T. J. Scheffer, H. Gruler, G. Meier, *Solid State Commun.* **1972**, *11*, 253.

[102] E. I. Demikhov, *Phys. Rev. E* **1995**, *51*, 12.

[103] J. Pang, N. A. Clark, *Phys. Rev. Lett.* **1994**, *73*, 2332.

[104] M. A. Handschy, N. A. Clark, S. T. Lagerwall, *Phys. Rev. Lett.* **1983**, *51*, 471.

[105] J. Xue, N. A. Clark, M. R. Meadows, *Appl. Phys. Lett.* **1988**, *53*, 2397.

[106] S. Tripathi, M. H. Lu, E. M. Terentjev, R. G. Petschek, C. Rosenblatt, *Phys. Rev. Lett.* **1991**, *67*, 3400.

[107] K. A. Crandall, C. Rosenblatt, R. M. Hornreich, *Liq. Cryst.* **1995**, *18*, 251.

[108] K. Nakagawa, T. Shinomiya, M. Koden, K. Tsubota, T. Kuratate, Y. Ishii, F. Funada, M. Matsuura, K. Awane, *Ferroelectrics* **1988**, *85*, 427.

[109] J. Xue, N. A. Clark, *Phys. Rev. Lett.* **1990**, *64*, 307.

[110] W. Chen, Y. Ouchi, T. Moses, Y. R. Shen, K. H. Yang, *Phys. Rev. Lett.* **1992**, *68*, 1547.

[111] S. Immerschitt, W. Stille, G. Strobl, *Macromolecules* **1992**, *25*, 3227.

[112] G. Decher, J. Reibel, M. Honig, I. G. Voigt-Martin, A. Dittrich, H. Ringsdorf, H. Poths, R. Zentel, *Ber. Bunsenges. Phys. Chem.* **1993**, *97*, 1386.

[113] E. M. Terentjev, *J. Phys. France* **1995**, *5*, 159.

[114] M. C. Holmes, N. Boden, K. Radley, *Mol. Cryst. Liq. Cryst.* **1983**, *100*, 93.

[115] U. Kaeder, K. Hiltrop, *Mol. Cryst. Liq. Cryst. Lett.* **1991**, *7*, 173.

[116] A. Poniewierski, R. Holyst, *Phys. Rev. A* **1988**, *38*, 3721.

[117] T. Yoshino, M. Suzuki, *J. Phys. Chem.* **1987**, *91*, 2009.

[118] E. A. Oliveira, A. M. Figueiredo Neto, G. Durand, *Phys. Rev. A* **1991**, *44*, 825.

11 Ultrasonic Properties

Olga A. Kapustina

Liquid crystal acoustics is an extremely diverse science insofar as the range of the problems that is covers is concerned. In recent years many experimental and applied investigations have been performed which have given rise to new branches of this science. We present here the current state of liquid crystal acoustics and discuss the most important advances in and the future potential of the field. The classical aspects of the subject are outlined briefly, with reference to associated chapters in this book.

11.1 Structural Transformation in Liquid Crystals

Since the very beginning of the research on mesomorphism, the influence of ultrasonic fields has been of great importance; only the goals of study have changed with time. The early research on which the later developments were based was done by Lehman [1], Zolina [2], Zvereva and Kapustin [3], and Fergason [4]. Discussion about the nature of the phenomena observed simulated much work in the pioneering days [5–13]. The various principles used to interpret the influence of ultrasonic fields on the mesophase were of exceptional theoretical importance. These included interpreting the effects of the field on the molecular arrangement of the liquid crystalline state on the basis of the Leslie–Erickson hydrodynamic theory or, as Kozhevnikov later demonstrated, using models adapted from nonequilibrium hydrodynamics. This variety of approaches to the problem was a result of the complex nature of the phenomena. The effect of ultrasound on the mesophase is, as a rule, associated mainly with the onset of flows induced by nonlinear phenomena. The flow process is characterized by an essentially stationary velocity distribution in the layer. Calculation and experimental determination of the flow velocity present major difficulties. Flow-induced disturbances in the molecular arrangement can be of several kinds; in particular, they can be both of the threshold or nonthreshold type and modulated or unmodulated. Modulated disturbances always show threshold behavior. Most of the data available at present refer to the nematics, the characteristic properties of which are manifested to certain degree in the smectics and in cholesterics. Ultrasonic field effects are covered in several reviews and monographes [13–19].

11.1.1   Orientation Phenomena in Nematics

All research on the changes in the macro-structure of nematics when in an ultrasonic field is done on samples 10–360 μm thick and using generally accepted optical methods. Any changes are observed either as a result of the depolarization of light transmitted through the sample or because of light scattering by the inhomogeneous structure [20]. Such experiments have shown three kinds of layer structure disturbance: homogeneous, spatially periodic, and inhomogeneous. Naturally, intermediate states between these three extreme phenomena are possible.

11.1.1.1   Homogeneous Distortion Stage

This stage involves longitudinal waves [5, 10, 21, 34] and surface acoustic waves (SAWs) [9, 35–40], with wavevectors $k = \omega/c$ and $k_R = \omega/c_R$, respectively, where c and c_R are phase velocities and ω is the angular frequency of the wave. Typical geometries of the equipment used for experiments on homeotropically aligned nematics are shown in Fig. 1. The wavevector is parallel or perpendicular to the director n and makes an angle θ with n. The measurement of ultrasound-induced birefringence offers an excellent tool for studying orientational phenomena and checking models.

The characteristic properties of the effect that can be elucidated experimentally are [5, 9, 10, 21–40]:

- Ultrasonic longitudinal and surface waves applied to a nematic sample change the birefringence properties of the fluid in the reflective and transmission modes when the ultrasound intensity J exceeds a certain minimum value [9, 10, 22, 26, 27].

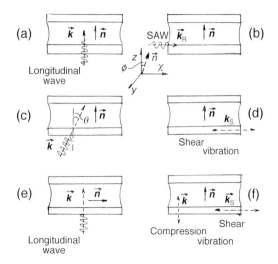

Figure 1. Equipment geometries used for studying ultrasonically induced structural transformations in liquid crystals: (a) longitudinal waves, $k \parallel n$; (b) SAW, $k_R \perp n$; (c) longitudinal waves, $k \parallel n$, $\theta \neq 0$; (d) shear vibrations, $k_S \perp n$; (e) longitudinal waves, $k \perp n$; (f) shear and compression vibrations, $k_S \perp n$, $k \parallel n$.

- The conditional clearing threshold of the ultrasound intensity J^{th} for birefringence is dependent on the layer thickness as d^{-2} [22, 25, 27].
- The frequency dependence of the threshold J^{th} is given by $J^{th} \simeq f^{-1/2}$ [17, 22, 27].
- The value of the threshold can be reduced by 'biasing' the sample with an electric field [10].
- Above the threshold: the relation between the optical transmission m of the nematic layer and the ultrasound intensity J is nonlinear [10, 22, 25, 27]; the transmission depends on the ultrasound intensity as $m \simeq \sin^2(\text{const.} J^2)$ and shows interference maxima and minima between ordinary and extraordinary light rays for monochromatic light [21, 25]; and for small values of the intensity $m \simeq J^4$ [21, 38].
- The optical transmission properties of nematic samples depend on the acoustic boundary conditions (cells with fixed

[22] or free [21] edges) and on the molecular orientation at the boundaries (planar or homeotropic configurations [27]).

- The variation in the orientation of the molecules in the ultrasonic field is observed as a system of alternating light and dark bands, the width and contrast of which depend on the ultrasound intensity. The distance between the centers of the light bands is of the order of the ultrasound wavelength [9, 12, 13, 21, 22, 27, 29, 35, 38–40]. The band configurations depend on the cell structure, the acoustic boundary conditions and the mutual orientation of the wavevector and director; these clearing patterns may be distorted by nonuniformity of the wave field inside the ultrasonic beam.

Several theories aimed at explaining the phenomena have been proposed, each of which is founded on completely different concepts. Sripaipan et al. [21] proposed a nematic layer with free ends, in which the interaction between the longitudinal oscillations (induced by the motion of the free ends of the layer in compression) and the traverse oscillations establishes steady flow of the liquid and, as a result, rotation of the molecules. However, these authors used incorrect dispersion relations and their calculations are not consistent with observed layer compression patterns. Nagai and coworkers [26, 27] hypothesized that with normal incidence of an ultrasound beam on the layer the rotation of molecules is attributable to radiation fluxes. Radiation fluxes are the steady acoustic flows caused by radiation forces in a traveling acoustic wave, the only provision being that the width of the ultrasound beam is smaller then the dimensions of the cell. In reality, radiation fluxes can only occur near the boundaries of the beam and produce a compression effect that is smaller than the one that is actually ob-

served [22, 25]. Radiation fluxes are significant when an ultrasound beam is focused so that it is obliquely incident on the layer [28]. Helfrich [33] and Chaban [34] treated molecular rotation as a threshold effect associated with the fact that the equations of motion for nematics contain nonlinear stresses proportional to the angle of molecular rotation and the particle velocity. However, the theoretical values obtained by these authors for the threshold ultrasound intensity are 2–3 orders of magnitude greater than the intensities usually observed experimentally. Dion and coworkers [41–43] hypothesized that torques created by the sound absorption anisotropy are responsible for rotating the molecules, which tend in turn in such a way as to diminish the losses in the ultrasound wave. This effect is theoretically possible, but is two or three orders of magnitude smaller than what is observed experimentally.

Kozhevnikov and coworkers [44–46] have developed the most general theory, which is based on Leslie–Ericksen hydrodynamics. These workers attributed the orientation effects to the steady inhomogeneous acoustic flows that result from the interaction between the periodic compression of the layer in the ultrasound field and the periodic motion of the liquid along the plates confining the layer. Unlike previous theories [27, 33, 34, 41–43], Kozhevnikov's model predicts the effect observed for normal incidence of the ultrasound on the layer. In the case of $k \parallel n$, according to the theory [44, 45], two physical reasons are responsible for the periodic liquid motion that occurs in the cell: vibrations of the cell plates in the layer having free ends [21, 44] and pressure gradients near the border of the beam for an infinitely wide layer [45]. In all instances, the acoustic flow velocity component along the x axis, V_x that initiates the director rotation, satisfies a well-known acoustics equation [47]. The equation for the small steady-state angle is [48]

$$\alpha_2 \frac{\partial V_x}{\partial z} = K_3 \frac{\partial^2 \psi}{\partial z^2} \tag{1}$$

Information on the velocity flows, configuration, scale, velocity distribution and the suitable distortion profiles of the director field of the layer under the action of an ultrasonic field for the qualitatively different situations presented above can be found in the literature [44, 45].

The distortion of the layer structure is usually studied by estimating the optical phase differences Δ_0 of the light that is polarized in the plane of the sandwich-cell plates, or the optical transmission m of the layer. The latter quantity can be calculated from the intensity I of the light transmitted through the layer parallel to the z axis and under crossed polarizers, the initial intensity I_0, and Δ_0 according to $m = I/I_0 \sin^2 \Delta_0/2$ [20]. It is useful to introduce the effective clearing threshold J^{th} as the ultrasound intensity (or displacement amplitude) at which $m = 0.01$ [17]. Sometimes, in order to compare theoretical prediction with experimental results it is also possible to estimate the ultrasound intensity at which $\Delta_0 = \pi$ (i.e. the depth of light modulation is 100%). Many experimental data are presented in some detail and are discussed quite throughly by Kapustin and Kapustina [17, 18]. Kozhevnikov's theory is borne out by the facts. For example, according to Gus'kov and Kozhevnikov [45], in a cell with fixed ends, $J^{th} \approx d^{-2.5} f^{-1}$. The experimental values of Jth obtained by Kapustina and Lupanov [22] and Hatakeyama and Kagawa [25] also follow this law (Fig. 2). Recent data on the qualitative behavior of Jth for some equipment geometries are summarized in Table 1 [44–46].

It has recently been shown that the flow structure depends on both the inhomogene-

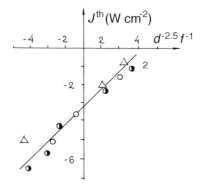

Figure 2. Effective clearing thresholds J^{th} for a 10–126 μm sample of MBBA in the ultrasound frequency range: (●) 0.4, (○) 1, (△) 3.2 MHz.

ity of the wave field in the cell and d/λ, where λ is the length of the elastic wave (see in [18]). In the ultrasound frequency range ($d/\lambda \ll 1$) the efficiency of the flow generation is high if the typical value of the wave inhomogeneity length κ is in the vicinity of λ. In this case $J^{100\%} = (4 \lambda/\kappa \delta d) J_0^{100\%}$, where $J_0^{100\%}$ is the value of $J^{100\%}$ for the wave field with $\kappa \neq \lambda$; $\delta = (\rho \omega/2 \eta)^{1/2}$ is the wave number in a viscous wave. At hypersonic frequencies ($d/\lambda \gg 1$), when the sound absorption in nematics is sufficiently high, $J^{100\%} = (4 \alpha \lambda/\pi \kappa \delta d) J_0^{100\%}$. In nematics the sound absorption coefficient α is usually about 20 dB cm^{-1}. The preceding discussion is summarized in Table 2 using the terminology of Russian physists for the comparison of the flow efficiency [18].

Information concerning the ultrasound-induced birefringence in homeotropic nematics when the k makes an angle with n (see Fig. 1 c) can be found in the literature [19, 28, 30]. The theoretical aspects of the problem cannot be detailed here and the reader is referred to the paper by Zhukovskaya et al. [46], wherein the flow mechanism of the aligning effect of an ultrasound wave or beam on a sample layer is discussed. In this case the flow is induced by nonline-

Table 1. Effective clearing thresholds.

Type of wave	θ (°)	Boundary conditions	Equipment geometry	Effective threshold	Remarks
Compression wave	0	Acoustically rigid	Cell of radius R	$J^{th} \simeq d^{-2.5} f^{-1}$	–
Compression wave	0	Acoustically soft	Infinitely wide cell; ultrasound beam of radius R	$J^{th} \simeq d^{-2.5} f^{-0.5}$	thick layer
				$J^{th} \simeq d^{-2.5} f^{-0.5}$ $\times \exp[-\delta_0(R-r)]$	thin layer
Compression wave	0	Acoustically soft	Infinitely wide cell; beam in the form of a strip	$J^{th} \simeq d^{-2.5} f^{-0.5}$	–
Compression wave	$\theta \neq 0$	Acoustically soft	Infinitely wide cell; beam of radius R	$J^{th} \simeq d^{-1.25}$ $J^{th} \simeq d^{-0.7}$	first maximum second maximum
SAW	90	Acoustically soft	–	$\xi_{0R}^{th} \simeq (fd)^{-0.75}$	two cell eigenmodes

Here value of R is the radius of a ultrasonically irradiated circular zone: an ultrasonic beam of radius R is incident on the layer of infinite width or an ultrasonic wave is incident on the layer of radius R pressed between a transparent plate and a thin wafer or elastic film; the value of r is a component of a cylindrical coordinate system (r, φ, z) with z-axis directed along the normal to the layer plane, while the boundary of the layer have the coordinates $z = \pm d/2$; the value of $\delta_0 = k/2\,\delta d$ is the coefficient of attenuation of longitudinal waves in the layer; the source of these waves in the layer is nonuniform compression near the boundaries of the ultrasonic beam or it is the nonuniformity of compression of the layer as a result of bending of the thin wafer or film at the edges. So that the value of $\delta_0(R-r)$ describes the attenuation of longitudinal waves during propagation along the width of the liquid crystal layer in the radial direction both into and outside of the ultrasonically irradiated region; the value of ξ_{0R}^{th} is the threshold of displacement amplitude of SAW.

Table 2. Values of $J^{100\%}$ versus the frequency and order of the inhomogeneity of the field.

Layer thickness (µm)	Frequency (Hz)	Order of inhomogeneity, κ	$J^{100\%}$ (mW cm^{-2})
10	10^6	λ	10
10	10^6	2λ	20
10	10^9	–	1

ar boundary forces due to the interaction between the longitudinal and viscous waves. According to this theory, the maximum of the effect corresponds to the maximum of the acoustic transparency of the cell, while the relationship between J^{th} and d depends on the angle of incidence of the ultrasound wave or beam (see Table 1). This theory correlates with the experimental results reported by Hareng and coworkers [19, 30].

There have been unanimously verified reports of steady-state distortion of a homotropically aligned nematics in a SAW field [9, 35–40]. In this case the wave field in the layer is caused by the mixing of several cell eigenmodes that have different phase velocities excited in the layer when a SAW propagates in one of the bounding plates. Thus the number of modes depends on d/λ. Earlier work concentrated more on the phenomena themselves rather than their interpretation [9, 39]. Calculated values obtained using theories based on viscoelastic effects and the parametric instability [34–37] show considerable discrepancy with experimental data. Miyano and Shen [38] have investigated flows in a nematic layer with the SAW propagation along the substrate (see Sec. 11.2 of this chapter).

11.1.1.2   Spatially Periodic Distortion Stage

Domains of different nature in homeotropically and planar oriented layers may appear under the action of longitudinal or shear waves only above a certain threshold. The directions of the domains, their width, and the threshold value depend on the type of the wave, the wave frequency, and the layer thickness.

Russian physists were the first to observe ultrasonically induced domains in non-oriented samples [5]. Later, Italian and Japanese physists investigated the appearance of domains in a homeotropic layer subjected to shear vibrations, created by the ac motion of one of the cell plates in its plane (see Fig. 1 d) [49–52].

The domain lines always aligned perpendicular to the shear direction. A theoretical attempt to describe the domains [53] was concerned with a periodic shear strain, but this model does not fit what is observed experimentally: in particular, the threshold values of the shear amplitude to not fit the experimental values in terms of magnitude, do not exhibit the experimentally observed dependence on the frequency and layer thickness, and fail to give the domain dimensions. Kozhevnikov [54] has analyzed the effect on the basis of the equations of the hydrodynamics of nematics, retaining the quadratic terms proportional to the product of the angle of rotation of the molecules and the velocity of the liquid. The model leads to the following picture of the effect. A periodic shear, in the case of a random slowly varying and periodic (along the layer) deviation of the molecules from the normal, creates eddies oscillating at the shear fre-

quency. The interaction of the eddies with the initial shear field gives rise to an average moment which increases the deflection of the molecules. The threshold amplitude and domain size are given by

$$\xi_0^{\text{th}} = \left(\frac{2K_3}{\rho}\right)^{1/2}\left(\frac{\pi}{d^2\omega}\right)F(q,p); \quad L \simeq \frac{\pi}{q_0} = 2d \tag{2}$$

where $p = \pi/d$, $q = \pi/L$, and q_0 is the wavenumber at the threshold of the effect; for information concerning the form of function $F(q,p)$ see Kozhevnikov [54]. Figure 3 shows the theoretical and experimental dependences of the threshold shear amplitude on the frequency for various values of d [49, 52, 55]. It is evident that in the range of validity of the calculations, Kozhevnikov's theory and the experimental data agree well. According to Kozhevnikov [54], the theory considers the frequencies f satisfying the inequalities $\pi K_3/2\,\gamma_1\,d^2 \ll f \ll \rho c^2/2\,\pi\eta$ and $f < \eta/2\,\pi\rho d^2$.

Russian physists [56, 57] have performed systematic investigations of the condition

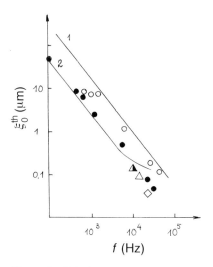

Figure 3. Calculated (lines 1 and 2) and measured (symbols) threshold displacements ξ_0^{th} versus frequency for various layer thicknesses: (1) 20, (2) 100, (○) 20, (□) 66, (●) 100, (△) 105, (▲) 200 μm.

under which domains appear in the planar layer provided that $\mathbf{k} \perp \mathbf{n}$ (see Fig. 1 e). In this situation the domain lines are orthogonal to the director $\mathbf{n}$ in the undisturbed state. The results were interpreted on the basis of the model proposed by Kozhevnikov [58] who developed a new approach for describing the structural transformations. The model is based on nonequilibrium hydrodynamics and considers both the ultrasonic relaxation processes and the anisotropy of the viscoelastic properties of nematics (the dynamic modulus of elasticity and the bulk viscosity). A schematic drawing of a cell which illustrates the model is shown in Fig. 4. The mechanism of the effect lies in the amplification of the random and nonuniform distortion of the planar structure when oscillating ultrasonically produced vortex flows interact with the director field. The anisotropic shear stresses in the layer that generate these flows are given by

$$\sigma_{xz} = \psi_0 \left(\mu_3 \, \vartheta_{zz} + U_{zz} \Delta E \right)$$

where U_{zz} and ϑ_{zz} are the compression and the rate of compression of the medium in the ultrasound field; μ_3 and ΔE are the real and imaginary parts of the modulus of elasticity (both depend on the relaxation time of the

order parameter τ_1 and the transformations of finite chains of molecules τ_2). The spatial harmonic of the form $\psi \simeq \psi_0 \sin q_x \cdot \sin pz$ is amplified most. The amplitudes of the threshold particle velocity and domain size are given by

$$\vartheta_0^{th} = \frac{1}{2\pi c} \left[\frac{K_3 \, c \, \eta \, F(S_{min})}{(\alpha_5 - \alpha_6) \Delta E} \right]^{\frac{1}{2}} d^{-1},$$

$$L = \frac{d}{S_{min}^{1/2}} \qquad (3)$$

where S_{min} is the value of the parameter $S = (q^2/p^2)$ that minimizes the function $F(S)$. The expression for this function can be found in Kozhevnikov [58]. At low and high frequencies, which are determined by limiting values of the parameter $B = \pi^2 \eta \mu_3 / \rho d^2 \Delta E$, minimum values of $F(S)$ correspond to $S_{min} = 3.7$ and 4.8. This gives for the spatial period L the values 0.52 d and $0.48 \, d$, which are in agreement with the experimental data (Fig. 5 a). In the frequency range satisfying the condition $\omega \tau_1 > 1$, the amplitude of the threshold particle velocity is virtually independent of temperature and follows the law: $\nu_0^{th} \simeq \Delta T^{-1/3}$ [58]. This agrees with the experimentally observed data (Fig. 5 b). The foregoing discussion illustrates that the new approach proposed by Russian physists is valid.

11.1.1.3   Inhomogeneous Distortion Stage

This stage involves the phenomena corresponding to higher excitation: at a definite threshold the nematic layer exhibits orientational turbulence, corresponding to observable light scattering. Kapustin and Dmitriev [5] were the first to observe this effect on a nonoriented sample in a field of longitudinal waves. Later, Kessler and Sawyer [7] found that the intensity of this scattering depends nonlinearly on the excitation level. By analogy with a similar phe-

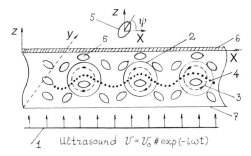

Figure 4. Diagram of a cell with domains in cross-section: (1) Ultrasound, (2) system of oscillatory vortex flow with the velocity $v_z' \simeq \cos qz$, (3) stationary flows, (4) profile of distortion of the nematic orientation, (5) nematics molecule, (6) acoustically rigid boundary, (7) acoustically soft boundary.

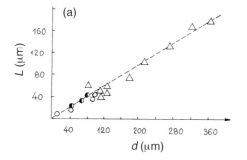

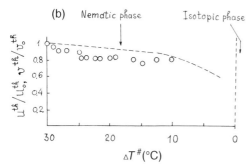

Figure 5. Ultrasonically induced orientational instability in planar nematics. (a) Spatial period of domains versus layer thickness, $f = 3.2$ MHz, $T = 30.2\,°C$. (b) Calculated and measured threshold parameters characterizing the wave field in the layer versus temperature of the nematics; the running values of Uth (tension) and v^{th} are normalized to the values of U_0^{th} and v_0^{th}, corresponding to the lower limit of the test temperature interval ($T = 22.8\,°C$), $\Delta T = T_{N-I} - T$.

nomenon in an electric field, this effect is called the acoustic dynamic scattering mode (DSM).

The characteristic properties of this effect are as follows [22, 56, 59, 60]:

- The intensity and angular distribution of the scattered light are independent of the wavelength and polarization of the incident light, and the scattering is directed mainly forward.
- The intensity and frequency of the ultrasound play an important role. Instability begins just above a threshold value J_s^{th}. Sometimes domains appear first, and then

gradually switch over to the DSM with increasing intensity of the ultrasound. The effect reaches maximum scattering (saturation) at an excitation level of about 6–8 times J_s^{th}. The threshold value is virtually independent of the substance, and hardly varies from the thickness of the layer ($J_s^{th} \approx 1/d^2$) [22].

- The initial orientation of the molecules in relation to k and to the boundary surfaces does not play a critical role, but it does have importance in technical application [22, 56, 59].
- The rise and decay times are influenced by many parameters (viscosity, sample thickness, intensity, etc.) [17, 59].

This problem has not yet been studied theoretically. However, the progress made to date in the area of experimental research clearly foreshadows rapid advances in the future. The flows generated by ultrasound create optical inhomogeneities in the layer that act as individual scattering centers. Bertolotti and coworkers [50, 60] have measured the coherence length of scattered light on the hometropic samples. The transition to the disordered state is accompanied by a drop in the coherence length to a rather small value, and the sample behaves as a thermal source containing a large number of statistically independent scattering centers. The sample thickness and observation angle do not play a critical role. Many features of the acoustic DSM have been studied with regard to technical applications [59]; this applies in particular to rise and decay times [17, 59]. Several technical studies have been concerned with contrast ratio [17, 59]. Kapustina [61] has recently reported on the acoustic DSM induced by ultrasound (1.5 MHz) in a polymer-dispersed nematic film.

11.1.2   Cholesterics in an Ultrasonic Field

The behavior of cholesterics in an ultrasonic field has generally been far less well researched than that of nematics. Only a few aspects of the problem have been examined since the systematic investigations done by Russian and Italian physists [17, 58, 62–65]. It may therefore seem to be somewhat stretching a point to draw an analogy with the phenomena in nematics, but this is a valuable way to obtain an overview of the following rather complex phenomena:

- periodic two-dimensional distortion, analogous to nematic domains (k parallel to the helical axis);
- excitation of the storage mode via turbulent flow, analogous to acoustic DSM in nematics (k parallel or perpendicular to the helical axis);
- erasing of the storage mode (k parallel to the helical axis);
- focal–conic texture transformation into a planar one (k parallel to the normal to the layer);
- bubble domains texture formation (k perpendicular or parallel to the helical axis);
- fingerprint texture transformations (k parallel to the normal to layer).

11.1.2.1   Periodic Distortion

An ultrasonic field applied parallel to the helical axis h ($k \parallel h$) can cause a square grid-like pattern deformation of the planar texture [17, 58, 62]. This can be observed in nematic/cholesteric mixtures with a helix of large pitch. According to experimental data, the spatial period of distortion follows the equation $L \simeq (p_o d)^{1/2}$ and tends to decrease slightly with frequency. The threshold particle velocity in the wave ϑ_0^{th} is practically

independent of frequency. By analogy with nematics, Kozhevnikov [62] described this phenomenon within the framework of an approach based on nonequilibrium hydrodynamics, but introduced the added assumption of an initial deformation of the layered systems in the crystal axis direction. The threshold amplitude ϑ_0^{th} and the size of the domains L in the equilibrium layers or the layers extended along the crystal axis are determined. As with nematics, two frequency bands (limiting value of $B = 38 \eta \pi^2 / \rho \omega^2 \tau d^2$) appear here also to be due to dissimilar behaviors of the threshold and domain size relative to d, p_o, f, etc. In particular, for $\omega \tau \gg 1$ ($B \ll 1$, high frequency) the value of $\vartheta_0^{th} \simeq (p_o d)^{-1/2}(1-l)^{1/2}$, whereas in the case of $\omega \tau \ll 1$ ($B \gg 1$, low frequency) the value of ϑ_0^{th} is also independent of frequency and follows the law: $\vartheta_0^{th} \simeq (d/p_o)^{1/2}$ $(1-l)^{1/2}$. Here $l = \Delta / \Delta_c$ is a relative value of the layer extension (Δ_c is the critical extension of the layered system when the domains appear under static deformation [48, 66] and Δ is the extension of the layer). According to the theory [58], the domain size L is approximately $(p_o d)^{1/2}$ over all the frequency range. The correlation between the experimental data and this theory is satisfactory only if the initial deformation of the layered system is taken into account; the relative value of the layer extension is then found to be close to $0.8 - 0.9$.

11.1.2.2   Storage Mode

The interest in this area is due more to the possibilities of technical application than theoretical considerations. Gurova, Kapustina, and Lupanov were the first to transform the uniform planar texture of a nematic/cholesteric mixture sample into a focal–conic one by means of an ultrasound field [17, 59]. The helical structure is still present in this state, having the same pitch on a microscopic level. Macroscopically, the sample is broken up into a myriad of randomly oriented domains with the size of a few micrometers, and consequently it strongly scatters visible light of all wavelengths. Such a texture is said to be stable for up to several days, during which time it gradually reverts to the planar one. The initial texture can be restored at any time by applying an ac pulse (20 kHz) [59, 64]. Russian physists have described the ultrasonically initiated storage effect [64]. Their data for the contrast ratio and for the rise and decay times have not as yet been improved upon. As mentioned above, the stored information slowly decays, but it persists in certain mixtures for several days after ultrasound excitation has been removed. Thickness, pitch and boundary conditions were found to play a crucial role in the memory properties of the planar to focal–conic texture transformations [17, 59].

11.1.2.3   Focal–Conic to Planar Texture Transition

Ultrasound waves may also create an oriented structure. This effect was observed in a sample with a focal–conic texture subjected to ultrasound (0.4 and 3.2 MHz [17, 67]). According to Hiroshima and Shimizu [67], the transmission of a sample (a nematic/cholesteric mixture) increases with increasing ultrasound intensity tending towards a value characteristic of a planar texture.

11.1.2.4   Bubble Domain Texture

Gurova and Kapustina were the first to describe ultrasonically induced bubble domains in large pitch nematic/cholesteric mixtures [17, 63]. It was found that relaxation of an ultrasound-induced perturbation at a critical layer thickness, which is com-

mensurate with the pitch of the helix, is accompanied by the formation of a bubble domain, which is a strong light scatterer. These data and the mechanism of the formation and stabilization of bubble domains have been discussed within the framework of the Akahane and Tako model [68]. This model attributes the formation and high stability of the bubble domains to the presence of defects (disclinations) in the sample, which pin the domains. The model does not take into account the interaction of the domains, and is thus valid in the case of low density domain packing. Test bubble domains have provided the first possibility for a quantitative analysis of the model, since they have a rather low packing density. The correlation of the theory with the experimental data is fully satisfactory (Fig. 6).

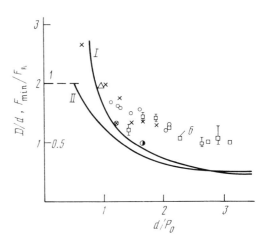

Figure 6. Comparison of Akahane and Tako's model and the experimental data. The calculated bubble domain diameter D and the free energy F_{min}, normalized to the layer thickness d and the free energy of the homeotropic structure F_h, respectively, versus the ratio $d/\bar{P}_0$ (curves I and II). Pitches (μm): (●) 3, (⊗) 4.2, (△) 10.5, (×) 48 ($f=0.2$ MHz). Frequency (MHz): (○) 0.5, (□) 3 ($P=48$ μm).

11.1.2.5   Fingerprint Texture Transition

Bertolotti et al. [65] have studied the behavior of a well-known fingerprint texture [69] formed in a sample (a nematic/cholesteric mixture) due to periodic compression in the frequency range 0.1 Hz to 130 kHz (free-edge cell). All the data obtained were analysed within the framework of Press–Arrot's model [70].

11.1.3   Smectic Phase in an Ultrasound Field

There has been no systematic study of this area because at present no potential technological applications of smectics in conjunction with ultrasound are known. Thus we have only a fragmentary picture from the few isolated investigations that have been made of smectic phases in ultrasound fields [5, 59, 71, 72].

Visual observations have shown that the optical transparency of a nonoriented sample of a smectic A phase under the action of longitudinal waves is diminished as a result of light scattering due to fluctuations in the refractive index [73]; transparency is not restored to its initial value when the disturbance is removed [5]. Italian physists have performed experiments on homeotropic smectic A layers and observed an ultrasonically induced spatially periodic structure distortion, including the formation of domains with relaxation times of up to several hours [71]. Russian physists have acknowledged the ability of smectic A phases to preserve remanent strain [72]. They investigated the behaviour of a homeotropic sample under the action of longitudinal waves. According to these data the smectic A phase exhibits an ultrasonically induced storage mode, where the remanent optical transparency barely differs from the value observed at the onset of the action of the ultrasound waves. It this case the storage effect is highly pronounced. The thickness of the sample and the method of preparation of the plate cell surfaces were found to affect both the contrast ratio and the efficiency of the transition from the transparent to the opaque structure. As in cholesterics, recovery of the initial orientation in smectic phases is realized by the application of an electric pulse (20 kHz, 200 V) [72]. All the above-described studies were concentrated more on the phenomena themselves than on their interpretation.

11.2   Wave Interactions in Nematics

All meaningful contributions in this area are relatively recent because, without the support of Vuzhva's theory [40], earlier observations have remained partly inexplicable [35, 36, 42, 74]. Since the systematic studies done by Miyano and Shen [38], only a few aspects of the problem have been examined. Three simple cases can be differentiated:

– the interaction of two nematic cell eigenmodes with different phase velocities [40];
– the interaction of the nematic cell eigenmode and the viscous wave [75];
– the interaction of the elastic and viscous waves [76].

The mathematical models proposed by Kozhevnikov and coworkers [44–46] paved the way for the development of a general wave interaction theory based on the unified concept: steady-state inhomogeneous acoustic flows. The validity of this theory [40, 75] is borne out by the following observations related to the traditional geometry. A homeotropically oriented nematic, several tenths of a micrometer thick, was placed between the substrate (1) and a glass plate (2) and observed under a microscope (Fig. 7).

In a first scenario (Fig. 7a), an interdigital transducer (3) generated the SAW at the substrate (1). Within the range of frequencies that satisfy the inequalities $d \ll \lambda$ and $d \ll \lambda_R$, the SAW excites only two nematic cell eigenmodes with phase velocities c_1 and $c_2 \simeq c/2$. The velocity components of the modes obey the following conditions: $\vartheta_{1x} \ll \vartheta_{1z}$, $\vartheta_{2z} \ll \vartheta_{2x}$ [40]. If the frequency ω is above 10^{-6} s^{-1}, the wave numbers δ and k_R are such that $\delta \gg k_R$ and $\delta d \gg 1$. According to Vuzhva's model, the velocity distri-

bution of acoustic flows resulting from the mode interactions in the xz plane and the associated director tilt angle ψ distribution are given by

$$V \sim \vartheta_{01} \vartheta_{02} \omega^{-1} \sin(k_R - k) x F(\delta, z, d),$$

$$\psi \sim \vartheta_{01} \vartheta_{02} \omega^{-1} \sin(k_R - k) x \left(\frac{d - z^2}{d} \right) \quad (4)$$

The acoustic flow patterns and the director field distortion profile are shown in Fig. 8. (For the form of the function $F(\delta, z, d)$, see Anikeev et al. [40].)

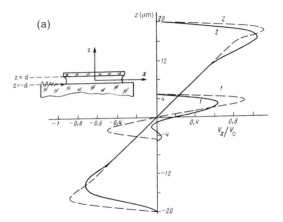

(a)

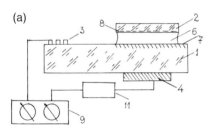

(a)

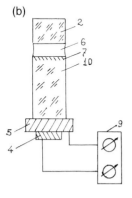

(b)

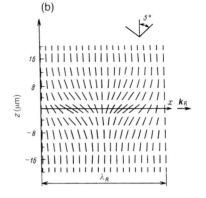

(b)

Figure 7. Equipment geometries for studying the wave and acousto-electrical interactions in nematics: (1) substrate (y cut, x oriented quartz), (2) glass plate, (3) interdigital transducer, (4) shear transducer (y cut quartz), (5) compression transducer (x cut quartz), (6) nematics, (7) mirror coating, (8) optically transparent electrode, (9) generator, (10) waveguide (substrate), (11) phase meter.

Figure 8. (a) Acoustic flow patterns for 10 and 40 μm MBBA samples (curves 1 and 2) at frequencies of (——) 6.47 and (---) 28.6 MHz for the flow velocity amplitude $V_0 = 10^{-4}$ cm·s^{-1} and $\sin(k_R - k) x = 1$. (b) Director field distortion profile in the xz plane for a 40 μm MBBA sample at a frequency of 6.47 MHz.

In a second scenario, the nematics also filled a flat capillary formed by the substrate (1) and the glass plate (2). A second transducer (4) in addition to the interdigital transducer (3) was placed under the substrate (1) (Fig. 7a). This second transducer was used to generate the viscous wave in the nematic layer. One of two nematic cell eigenmodes with a phase velocity c_2 quickly decays along the layer, and the viscous wave was able to interact only with the faster eigenmode. For information concerning the conformation, the scale and velocities of the flow resulting from the interaction the components of $\vartheta_x(\delta, \omega, z, t)$ and $\vartheta_z(\omega, z, x, t, \mathbf{k})$, and the tilt angle of director caused by this flow, the reader is referred to Anikeev et al. [75] and Bocharov [76].

It is of interest to compare the change in the optical responses of nematics due to the distortion of the director field caused by the flows in the first and second cases. The optical transmission m of the nematic layer between crossed polarizers when disturbed by a SAW with and without a viscous wave is shown in Fig. 9a. Under the action of the SAW only, $m \simeq \xi_{0R}^8$; this is in agreement with other reported results [38, 40]. With a viscous wave, $m \simeq \xi_{0R}^4$ [75]. The correlation between the calculated data and the measured values of m is evident [75]. It should be emphasized that in the case of the combined two-wave action the effective threshold is an order of magnitude lower.

Finally, in a third scenario the nematics is placed between the substrate (10) in the form of a bar and a thick glass plate (2) (Fig. 7b). One transducer, the plate (5) is bonded to the bar and creates the elastic wave of compression. The other transducer, the plate (4) is placed under the plate (5) and used to generate the viscous wave in the nematics. The data obtained using this geometry ($f = 15$ MHz) were reported by Bocharov [76]. It should be noted that only under

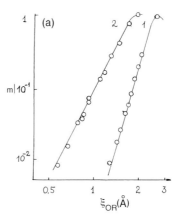

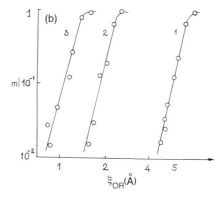

Figure 9. Optical transmission of the layer versus the amplitude of the SAW. (a) Viscous wave amplitude 0 (curve 1) and 2 Å (curve 2), $d = 40$ μm, $f = 30$ MHz, $\lambda_0 = 0.633$. (b) Optical transmission versus the amplitude of the SAW in the electric field: no voltage (curve 1), 4.5 and 4.6 V (curves 2 and 3), $d = 40$ μm, $f = 6$ MHz. (——) calculated data; (○) experimental data.

conditions of efficient viscous and elastic wave interactions is the optical response of the nematic layer subjected to the same regularities that result from the unified acoustic flow model. For example, the phase difference is $\Delta_0 \simeq \xi_{0x}^2 \xi_{0z}^2$ (where ξ_{0x} and ξ_{0z} are the viscous and elastic wave amplitude, respectively). This result corresponds well with Vuzhva's theory [75] which is based on the concept of acoustic flows. Bocharov [76] proposed a design for a new reliable acoustic holography system based on the coherent wave interaction effects considered above. Such a system, which uses the viscous wave as the reference wave, is more elaborate than earlier convertors [22, 41, 56, 77], and has the advantages of providing an order of magnitude higher sensitivity and a doubled dynamic range.

11.3   Acousto-electrical Interactions in Nematics

The interest in this area is due more to the possibilities of technical applications than to theoretical considerations. Today, three combinations of equipment geometry and the value of the dielectric anisotropy $\Delta\varepsilon$ can be differentiated:

- a crossed electric field (E directed along the z-axis; see Fig. 1) and $\Delta\varepsilon > 0$;
- a crossed electric field and $\Delta\varepsilon < 0$;
- a longitudinal electric field (E directed along the x axis) and $\Delta\varepsilon < 0$.

With the exception of the early work by Kapustin [14], who concentrated on the combined acousto-electric orientational effect in polycrystalline nematics, all basic studies of this phenomena have been done in recent years and have considered mainly homeotropically aligned nematics. The subject has been summarized in some reviews [14–18], which also outline potential applications.

Most studies have dealt with case (1) above. This equipment geometry can be used either to suppress an orientational effect [39, 78, 79] or to erase the ultrasonic images on the convertor surface [26, 59]. Nagai and Iizuka [26] have shown that under an applied electric pulse the erasure time is reduced by one or two orders of magnitude. Of the results that have reported, the influence of the anchoring energy on the acous-

to-electrical interaction mechanism is of particular interest [79].

Studies of acousto-electric interactions that fall into case (2), above, have been reported by various authors [30, 32, 75, 76, 80–84].

According to Kagawa et al. [32], the crossed electric field results in an increase in the sensitivity of nematics to ultrasound action; however, the rise and decay times are impaired [80]. Perbet et al. [30] controlled successfully all the characteristics mentioned above by switching the frequency of the electric field and by using a nematics with a frequency-dependent sign reversal of the dielectric anisotropy. For the ZLI-518 this frequency was affected in the range 200 Hz ($\Delta\varepsilon = +0.8$) to 20 kHz ($\Delta\varepsilon = -0.4$) and the critical frequency was 10 kHz. Ezhov and coworkers [80, 81] have tried to treat the acousto-electrical interaction problem for the peculiar geometry where $n \perp k$, by introducing a too simple assumption of the field's additivity, which is not supported by the results obtained. Hayes [83] was the first to interpret a mechanism of acousto-electric interactions for the case where $n \parallel k$ within the framework of the acoustic flows model. However, this concept is not completely borne out by the observed results. Several other authors have also tried to build a bridge to other models to account for the critical behavior of the threshold for the acousto-optic effect near the Frederiks transition. Akopyan et al. [74] performed an analysis on the basis of the unfaithful Helfrich's model [33] and drew mistaken conclusions. Strigazzi and Barbero [84] discussed the improvement obtained in the contrast ratio of an ultrasonic image by means of a crossed electric field. However, according to their calculations in the vicinity of the Fredericks transition, increasing the voltage applied to the electrodes gives rise to a small change of an average director tilt angle only. This is inconsistent with the reported experimental data [75, 85].

The most extensive theoretical and experimental studies of this problem have been done by Bocharov [76], who used an apparatus in which a SAW was excited at the substrate (1) (Fig. 7 a). The nematics was sandwiched between the glass plate (2) and the substrate (1) bearing transparent electrodes, to which an ac voltage ($f \approx 50$ Hz) was applied. The dependence of the layer optical transmission m on the amplitude of the SAW with and without an applied electric field is shown in Fig. 9 b. The threshold value of the Frederiks transition was 4.65 V. It is evident that in the vicinity of this transition the sensitivity of the nematics to the action of the SAW increases by more than an order of magnitude. According to Vuzhva's theory, near the threshold of the Frederiks transition ($E \rightarrow E_0$)

$$\frac{m(E)}{m(0)} = \frac{1}{\pi^2} \left(\frac{E_0 - E}{E_0} \right)^{-4} \qquad (5)$$

which is in a good agreement with experimental results [75, 85]. These data have contributed to a coherent unified physical picture of the phenomena based on the general acoustic flows model.

Case 3 has been particularly well investigated by Belova [82] with the result that the transmission of a layer in the SAW and the longitudinal electric fields can be smaller than $m = 0.01$. However, in these experiments the optical clearing pattern was found to have high spatial homogeneity along the layer, which is inconsistent with many of the other data reported [35, 38–40]. It is likely that the SAW-created distortion in the nematics is masked by the flexo-electric effect one [68].

It should be noted also that all the results so far reported indicate great promise for practical applications [18].

11.4   Ultrasound Studies of Liquid Crystals

11.4.1   Ultrasonic Spectroscopy

In the last 25 years, the investigation of the propagation of ultrasound in liquid crystals has made a considerable contribution to the study of the following aspects of these systems:

- the phase transitions between two ordered states or between a mesophase and the isotropic state (see Chap. VII, Sec. 6 of this volume);
- the relaxation phenomena characteristic of the mesophase;
- the anisotropy of the acoustic properties in the ordered phase;
- the behaviour of acoustic parameters in variable (rotating and pulsating) magnetic fields.

Modern experimental techniques used for such studies include the fixed- and variable-path pulse-position methods as well as microwave measurement techniques. In most cases the ultrasound and microwave measurements complement each other nicely.

Since 1970 the main anomalies in the propagation of ultrasonic waves in non-oriented samples of liquid crystals have been discovered: these are the phase velocity dispersion and the nonclassical ($\alpha/f^2 =$ constant) behavior of the ultrasound absorption coefficient due to the relaxation phenomenon. The anomalies are manifested in the megahertz frequency range and are most pronounced in the vicinity of phase transitions. Kapustin [14] was the first to begin extensive investigations on the anomalies, and look at all types of liquid crystals. Most of the work done since 1980 has involved the use of a dc magnetic or electric field to measure the anisotropy of the properties of ultrasound in ordered samples. It has been established that the anisotropy of the velocity and absorption depends on the angle between the direction of the wave propagation and the applied field. The behavior of these properties of ultrasound in oriented liquid crystals does not obey classical theory and depends on the nature of the relaxation process. Relaxation mechanisms have been widely discussed in the literature; the data have been summarized in some reviews [13, 15, 56] and monographs [14, 16, 18, 86–89]. The essential factors concerning the nature of relaxation processes have been derived. These results can be grouped as follows:

– processes related to the relaxation of the order parameter (Landau–Khalatnikov mechanism [73]) and the relaxation of the order parameter fluctuations [90];
– processes resulting from the rotation of the molecules about their long and short axes and the translation of molecules;
– processes due to molecular motions, when the finite chains of molecules move as a whole, or full conformational transformations.

Most of the modern 'structure' models [18] can give only a qualitative picture of the phenomena, taking into account scalar parameters (temperature, pressure, rate of change in volume) and analyzing the isotropic component of absorption, which depends on the direction of propagation of the wave. Aero [91] has proposed a new approach to the treatment of the problem. According to his model, the propagation of the ultrasound wave produces nonequilibrium changes in the long-range orientational and short-range translation order due not only to the oscillations in the themperature and pressure, but also to shearing of the medium. Thus his model takes into account three

specific mechanisms of ultrasound absorption ('critical', 'normal', and classical). The numerical estimates, obtained with this model, of the velocity and absorption of ultrasound for nonoriented and ordered nematics are consistent with the experimental data over a wide frequency range (2–500 MHz) [91].

For details concerning the behavior of the parameters of ultrasound waves in cholesterics the reader is refered to the literature [14–16, 18, 86, 92, 93]. However, the existence of an absorption gap (~1.5 dB) in cholesterics should be noted; this is observed experimentally at a wavelength that is close to the helix pitch [92, 93].

The absorption of ultrasound in smectic phases is significantly more anisotropic than that in nematics, and even the velocity has a measurable anisotropy of about 5%. Details of the behaviour of SmA, SmB, SmC and SmE phases can be found in the literature [14–16, 18, 86, 94–97]. The usual approach to the analysis of smectic phases, based on the linear theory of elasticity and hydrodynamics, results in the relationship $\alpha \simeq f^2$, which does not agree with the experimental data. In the low-frequency range the coefficients α_1, α_2, α_4 and α_5 demonstrate singularity, induced by nonlinear effects, in the form of ω^{-1}. This results in a linear frequency dependence of the ultrasound absorption. The corrections for the coefficients of elasticity B and K_1, taking into account the nonlinear fluctuation effects in smectic phases, depend on the wavevector of the smectic phase layer structure q_s: $B \simeq (\ln q_s)^{-4.5}$, $K_1 \simeq (\ln q_s)^{2.5}$ [96, 97]. In these relationships the numberical value of coefficient is very small; the vector q_s is associated with the sample dimensions and does not vanish. Thus only a small correction to the results of the linear theory of elasticity is needed.

11.4.2   Photoacoustic Spectroscopy

The revival of photoacoustic (PA) spectroscopy in the early 1970s stimulated the development of various photothermal detection techniques. The common feature of these techniques is the detection in a test sample of the oscillations in temperature that result from exposure of the sample to a modulated light beam [98]. The sensitivity of the cell depends on its radius, the length of the air column, the coefficient of thermal diffusion in the air, the frequency modulation and the size of the light beam [98]. Rosencwaig [99] was the first to demonstrate the possibility of applying the PA method to the study of the phase state of liquid crystals. The data obtained using PA spectroscopy can also be used to analyze pretransitional effects (see Sec. 6.3 of this Chapter) in liquid crystal. Martinelli et al. [100] have elucited the mutual relationship between the thermodynamic properties of liquid crystals and the main characteristics (phase and amplitude) of the PA signal. A new method for determining each thermal parameter individually has been developed on the basis of these data [100]. For more detailed information on the PA effect, see Sec. 6.2.1 of this Chapter.

11.4.3   Acoustic Emission

For a long time it was thought that acoustic emission is inherent only to solids. However, it is now known that such emission is also a characteristic of liquid crystals, which have many structural features (defects, monopoles, solitones, etc.) [73]. The nature of the emission is closely connected with the type of liquid crystal arrangement. Therefore, if a dynamic process developing in the test sample is enacted on the defect, the release of elastic energy leads to acoustic emission. One simple situation, which makes it possible to observe this phenomenon, consists of varying the temperature in the test sample which then passes through a number of phase transitions [101]. It was found that in 1 cm thick nonoriented samples of MBBA, COOB, and CBOOA acoustic activity was most highly pronounced at the nematic–isotropic transition [102]. The high sensitivity of the method allows its use as a means of monitoring the alignment stability of mesophase samples.

11.4.4   Monitoring Boundary Effects

An important parameter describing the interaction of a mesophase with a solid substrate is the anchoring energy W (see Sec. 10 of this Chapter). New methods have been described of estimating W by means of the optical response of the nematic cell to an acoustic action: pressure variations [101], shear vibration [103], and joint shear and longitudinal vibrations [104]. A qualitative relationship between the value of W, consistent with the test cell and its optical response to the shear vibration, was discovered by Belova [103], using the equipment geometry shown in Fig. 1 d. A method has been described [105] for quantitatively estimating W. The method is based on the nonlinear interactions of the oscillations of the director with the field of velocities of flow of the nematic material due to a joint effect of shear and compressive vibrations of the same frequency, but shifted in phase by $\pi/2$ [106]. These nonlinear interactions are particularly important at amplitudes and frequencies satisfying the inequality $0.1 \gamma_1 \beta \omega \xi_{0x} d/K_3 \gg 1$. In this case the relationship between W and the amplitude of shear vibration ξ_{0x} is [105]

$$W \simeq \beta \gamma_1 (\xi_{0x}^{max})^2 \frac{(\Delta n k_0 \, d)^{1/2}}{4 d \omega} \qquad (6)$$

where ξ_{0x}^{max} is the amplitude corresponding to the first maximum of the function $m = f(\xi_x)$, $\beta = \xi_{0z}/\xi_{0x}$ (see Fig. 1 f). Kapustina and Reshetov [104] have reduced Eq. (6) to a form more convenient for subsequent analysis and report the results of tests.

11.4.5   Acoustic Microscopy

There are several types of acoustic microscopes, which differ from one another mainly in the principle of transforming the acoustic image into a visible one. The lens scanning acoustic microscope is widely used [107]. The specification of a modern acoustic microscope is as follows: frequency range 50–3000 MHz, resolution 50–0.1 µm, magnification 50–5000, and depth of penetration 1 µ to 1 mm.

Acoustic microscopy is rather sensitive to inhomogeneities in micro-objects, including optically opaque ones, and can also be used to detect disturbances in adhesion, flaking, microcracks, pores, deviations from the prescribed thickness of a layer, and coatings. Possible applications of the method are investigations of the topography and morphology of smooth and textured surfaces. It may also be useful in the qualitative monitoring of multilayer structures and composite materials (e.g. polymer dispersed nematics or cholesterics) in which the components are very similar in terms of their optical characteristics but have different acoustic characteristics.

Studies of the action of hypersound on nematics have shown that a reliable acoustic microscope can be constructed from nematics and standard optical components without the need for the scanner and focuser that

are usually included. According to rough estimates for an equipment geometry like the one shown in Fig. 1 a, the acoustic-optical conversion sensitivity over the hypersonic frequency range is given by

$$J^{100\%} = \frac{4\rho c^2 k_1 k_2 K_3}{(\alpha_2 d \delta^2)} \left[\frac{105\lambda_0}{\Delta n d} \right]^{1/2} \qquad (7)$$

here k_1 and k_2 are components of the elastic wavevector $k = k_1 + i k_2$. For a 10 µm nematic layer at 10^9 Hz the sensitivity is estimated to be $J^{100\%} = 1 \text{ mW cm}^{-2}$.

11.5   References

[1] O. Lehman, *Das Kristallisationsmikroskop und die damit gemachten Entdeckungen insbesondere die flüssigen Kristalle*, Vieweg Verlag, Braunschweig **1910**.
[2] V. V. Zolina, *Trudy Lomonosovskogo Inst., AN SSSR*, **1936**, 11.
[3] G. E. Zvereva, A. P. Kapustin, *Appl. Ultraacoust. Study Matter* **1961**, *15*, 69 [in Russian].
[4] J. L. Fergason, *Sci. Am.* **1964**, *211*, 77.
[5] A. P. Kapustin, L. M. Dmitriev, *Kristallografiya* **1962**, *7*, 332.
[6] A. P. Kapustin, *Kristallografiya*, **1969**, *14*, 943; *Isv. Vysshikh Uchebnykh Zavedenii, Fiz.* **1967**, *11*, 55 [in Russian].
[7] L. W. Kessler, S. P. Sawyer, *Appl. Phys. Lett.* **1970**, *17*, 440.
[8] A. P. Kapustin, Z. Kh. Kuvatov, A. I. Trofimov, *Izv. Vysshikh Uchebnykh Zavedenii, Fiz.* **1971**, *4*, 150; *Uchenye Zapiski Ivanovskogo Pedagogicheskogo Inst.* **1972**, *96*, 64 [in Russian].
[9] L. E. Davis, J. Chambers, *Electron. Lett.* **1971**, *7*, 287.
[10] H. Mailler, L. L. Likins, T. R. Teilor, J. L. Fergason, *Appl. Phys. Lett.* **1971**, *18*, 105.
[11] M. Bertolotti, S. Martellucci, F. Scudieri, D. Sette, *Appl. Phys. Lett.* **1972**, *21*, 74.
[12] A. P. Kapustin, O. A. Kapustina, *Krist. Techn.* **1973**, *8*, 237.
[13] O. A. Kapustina, *Sov. Phys. Acoust.* **1974**, *20*, 1.
[14] A. P. Kapustin, *Electrooptical and Acoustical Properties of Liquid Crystals*, Nauka, Moscow **1973**, p. 120 [in Russian].
[15] G. G. Natal, *J. Acoust. Soc. Am.* **1978**, *65*, 1265.
[16] A. P. Kapustin, *Experimental Studies of Liquid Crystals*, Nauka, Moscow **1978**, p. 130 [in Russian].

[17] O. A. Kapustina, *Acoustooptical Phenomena in Liquid Crystals* (Ed. M. M. Labes), Gordon & Breach, London **1984**, pp. 1–164.

[18] A. P. Kapustin, O. A. Kapustina, *Acoustics of Liquid Crystals,* Nauka, Moscow **1986**, pp. 1–247 [in Russian].

[19] J. N. Perbet, M. Hareng, S. Le Berre and B. Mourrey, *Rev. Tech. Thomson CSF* **1979**, *11*, 837.

[20] M. Born, E. Wolf, *Principles of Optics,* Pergamon Press, New York **1970**.

[21] Ch. Sripaipan, Ch. F. Hayes, G. T. Fang, *Phys. Rev. A* **1977**, *15*, 1297.

[22] O. A. Kapustina, V. N. Lupanov, *Sov. Phys. Acoust.* **1977**, *23*, 218.

[23] R. Bartolino, M. Bertolotti, F. Scudieri, D. Sette, A. Sliwinski, *J. Appl. Phys.* **1975**, *46*, 1928.

[24] H. Bruchmuller, *Acoustica* **1978**, *40*, 155.

[25] T. Hatakeyama, Y. Kagawa, *J. Acoustic. Soc. Jpn.* **1976**, *32*, 92; *J. Sound Vibr.* **1976**, *46*, 551.

[26] S. Nagai, K. Iizuka, *Jpn. J. Appl. Phys.* **1974**, *13*, 189; **1978**, *17*, 723; *Mol. Cryst.* **1978**, *45*, 83–101.

[27] S. Nagai, A. Peters, S. Candau, *Rev. Phys. Appl.* **1977**, *12*, 21.

[28] S. Candau, A. Ferre, A. Peters, G. Waton, P. Pieranski, *Mol. Cryst. Liq. Cryst.* **1980**, *61*, 7.

[29] I. Lebran, S. Candau, S. Letcher, *J. Acoust. Soc. Am.* **1978**, *63*, 55; *J. Phys. (Paris)* **1979**, *40* (Suppl.), 298.

[30] I. N. Perbet, M. Hareng, S. Le Berre, *Rev. Phys. Appl.* **1979**, *14*, 569.

[31] Ch. F. Hayes, *Mol. Cryst. Liq. Cryst.* **1980**, *59*, 317.

[32] Y. Kagawa, T. Hatakeyama, Y. Tanako, *J. Sound Vibr.* **1974**, *36*, 407.

[33] W. Helfrich, *Phys. Rev. Lett.* **1972**, *29*, 1583.

[34] I. A. Chaban, *Sov. Phys., Acoust.* **1979**, *25*, 67.

[35] O. A. Kapustina, A. A. Talashev, *Sov. Phys., Acoust.* **1974**, *19*, 397.

[36] O. A. Kapustina, Yu. G. Statnikov, *Sov. Phys., JETP* **1973**, *37*, 117; *Sov. Phys., Acoust.* **1974**, *20*, 154.

[37] E. N. Kozhevnikov, I. A. Chaban, *Sov. Phys., Acoust.* **1976**, *21*, 550.

[38] K. Miyano, Y. R. Shen, *Appl. Phys. Lett.* **1976**, *28*, 473, 699; *Phys. Rev. A* **1977**, *15*, 2471.

[39] S. Sato, H. Uedo, *Jpn. J. Appl. Phys.* **1981**, *20*, 511.

[40] D. I. Anikeev, Yu. V. Bocharov, O. A. Kapustina, A. D. Vuzhva, *Sov. Phys., Acoust.* **1989**, *35*, 563.

[41] J. L. Dion, A. D. Jacob, *Appl. Phys. Lett.* **1977**, *31*, 490; *IEEE Trans. Ultrason., Ferroelec. Freq. Contr.* **1987**, *34*, 550.

[42] J. L. Dion, *CR Hebd. Sean. Acad. Sci. B* **1977**, *284*, 219; **1978**, *286*, 383; *J. Appl. Phys.* **1979**, *50*, 2965.

[43] J. L. Dion, A. LeBlanc, A. D. Jacob, *Acoust. Imaging* **1982**, *10*, 151.

[44] E. N. Kozhevnikov, *Sov. Phys., Acoust.* **1981**, *27*, 297; *Sov. Phys. JETP* **1982**, *55*, 96.

[45] N. K. Gus'kov, E. N. Kozhevnikov, *Sov. Phys., Acoust.* **1983**, *29*, 21.

[46] E. I. Zhukovskaya, E. N. Kozhevnikov, V. M. Podol'skii, *Sov. Phys. JETP* **1982**, *56*, 113.

[47] L. K. Zarembo, V. A. Krasil'nikov, *Introduction to Non-linear Acoustics*, Nauka Moscow **1966** [in Russian].

[48] S. A. Pikin, *Structural Transformation in Liquid Crystals,* Nauka, Moscow **1981**, p. 336 [in Russian].

[49] F. Scoudieri, *Appl. Phys. Lett.* **1976**, *29*, 398; *Ann. Phys.* **1978**, 311.

[50] F. Scudieri, A. Ferrari, M. Bertolotti, D. Apostol, *Opt. Commun.* **1975**, *15*, 57.

[51] F. Scudieri, M. Bertolotti, S. Melone, G. Albertini, *J. Appl. Phys.* **1976**, *47*, 3781.

[52] Y. Kagawa, T. Hatakeyama, Y. Tanaka, *J. Sound. Vibr.* **1975**, *41*, 1; *J. Acoust. Soc. Jpn.* **1975**, *31*, 81.

[53] I. A. Chaban, *Sov. Phys. Acoust.* **1978**, *24*, 145; **1985**, *31*, 77.

[54] E. N. Kozhevnikov, *Sov. Phys.*, JETP **1986**, *64*, 793.

[55] G. N. Belova, E. I. Remizova, *Sov. Phys., Acoust.* **1985**, *31*, 171.

[56] O. A. Kapustina, V. N. Lupanov, *Sov. Phys., JETP* **1976**, *44*, 1225; *Wiss. Z. Univ. Halle,* **1977**, *24*, 49.

[57] D. I. Anikeev, O. A. Kapustina, V. N. Lupanov, *Soc. Phys., JETP* **1991**, *73*, 109.

[58] E. N. Kozhevnikov, in *Proc. Sixth All-Union Conf.,* Chernigovski Pedagogicheski Inst., Chernigov, Vol. 1, **1988**, p. 121.

[59] I. N. Gurova, O. A. Kapustina, V. N. Lupanov, *Advances in Liquid Crystal Research and Applications* (Ed. L. Batal), Pergamon Press/Akademiai Kiado, Oxford/Budapest, **1980**.

[60] M. Bertolotti, F. Scudieri, E. Sturla, *J. Appl. Phys.* **1978**, *49*, 3922.

[61] O. A. Kapustina, *Sov. Phys., Acoust.* **1991**, *97*, 153.

[62] E. N. Kozhevnikov, *Sov. Phys., JETP* **1987**, *65*, 731.

[63] I. N. Gurova, O. A. Kapustina, *Liquid Crystals,* **1989**, *6*, 525; *Sov. Phys., Acoust.* **1989**, *35*, 262.

[64] O. A. Kapustina, V. N. Lupanov, G. S. Chilay, *Sov. Phys., Acoust.* **1978**, *24*, 76.

[65] M. Bertolotti, I. Sbrolli, F. Scudieri, *Advances in Liquid Crystal Research and Applications,* Pergamon Press/Akademiai Kiado, Oxford/Budapest **1988**, pp. 433–439; *J. Appl. Phys.* **1982**, *53*, 4750.

[66] W. Haas, J. Adams, *Appl. Phys. Lett.* **1974**, *25*, 535.

[67] K. Hiroshima, H. Shimizu, *Jpn. J. Appl. Phys.* **1977**, *16*, 1889.

[68] T. Akahane, T. Tako, *Mol. Cryst. Liq. Cryst.* **1977**, *38*, 251.

[69] P. E. Cladis, M. Kleman, *Mol. Cryst. Liq. Cryst.* **1972**, *16*, 1.

[70] M. J. Press, A. S. Arrot, *J. Phys. (Paris)* **1976**, *37*, 387.

[71] F. Scudieri, A. Ferrari, E. Gunduz, *J. Phys. (Paris)* **1979**, *40* (Suppl.), 90.

[72] O. A. Kapustina, V. N. Lupanov, V. M. Shoshin, *Sov. Phys. Acoust.* **1980**, *26*, 406.

[73] P. G. De Gennes, *The Physics of Liquid Crystals,* Oxford University Press, London **1974**.

[74] R. S. Akopyan, B. Ya. Zel'dovich, N. V. Tabiryan, *Sov. Phys. Acoust.* **1988**, *34*, 337.

[75] D. I. Anikeev, Yu. V. Bocharov, A. D. Vuzhva, *Zh. Tekh. Fiz.* **1988**, *58*, 1554; *Liq. Cryst.* **1989**, *6*, 593.

[76] Yu. V. Bocharov, Dissertation, Moscow **1989** [in Russian].

[77] P. Greguss, *Acoustics* **1973**, *29*, 52.

[78] T. F. North, W. G. B. Britton, R. W. B. Stephens, in *Proc. Ultrasonic Int. Conf.*, London **1975**, pp. 120–123.

[79] H. Hakemi, *J. Appl. Phys.* **1982**, *53*, 6137.

[80] S. Ezhov, S. Pasechnic, V. Balandin, *Pis'ma Zh. Tekh. Fiz.* **1984**, *10*, 482.

[81] E. V. Gevorkyan, S. Ezhov, *Trudy All-Union Correspondence Inst. Mech. Eng.* **1984**, *36*, 46 [in Russian].

[82] G. N. Belova, *Sov. Acoust.* **1988**, *34*, 13.

[83] C. F. Hayes, in *Liquid Crystals and Ordered Fluids,* New York **1978**, pp. 287–296.

[84] A. Strigazzi, G. Barbero, *Mol. Cryst. Liq. Cryst.* **1983**, *103*, 193; *Mol. Cryst. Liq. Cryst. Lett.* **1982**, *82*, 5.

[85] Yu. V. Bocharov, A. D. Vuzhva, *Pis'ma Zh. Tekh. Fiz.* **1989**, *15*, 84; **1988**, *14*, 1460.

[86] K. Miyano, J. B. Ketterson, *Phys. Rev. A* **1975**, *12*, 615.

[87] P. K. Khabibullaev, E. V. Gevorkyan, A. S. Lagunov, *Rheology of Liquid Crystals,* Allerton Press, New York, **1994**, Chs 3.2, 6.3, 6.4.

[88] H. Kelker, R. Hatz, *Handbook of Liquid Crystals,* **1980**, Ch. 3.

[89] S. Candau, S. V. Letcher, in Advances in Liquid Crystal, ed. G. H. Brown, **1978**, vol. 3, 168–235.

[90] V. G. Kamensky, E. I. Katz, *Sov. Phys., JETP* **1982**, *56*, 591.

[91] E. L. Aero, Ph. D. Dissertation, Leningrad, **1982**, p. 408 [in Russian]; *Fiz. Tverdogo Tela* **1974**, *16*, 1245.

[92] I. Muscutariu, S. Bhattacharya, J. B. Ketterson, *Phys. Rev. Lett.* **1975**, *35*, 1584.

[93] J. D. Parsons, C. F. Hayes, *Sov. State. Commun.* **1974**, *15*, 429.

[94] E. I. Katz, V. V. Lebedev, *Sov. Phys., JETP Lett.* **1983**, *37*, 709.

[95] D. Mazenko, S. Ramaswary, J. Toner, *Phys. Rev. A* **1983**, *28*, 1618.

[96] E. I. Katz, *Sov. Phys., JETP* **1982**, *56*, 791.

[97] G. Grinstein, R. A. Pelcovitz, *Phys. Rev. A* **1982**, *26*, 915.

[98] V. P. Zharov, V. S. Letokhov, Laser *Opticoacoustic Spectroscopy,* Nauka, Moscow **1984** [in Russian].

[99] A. Rosencwaig, in *Optoacoustic Spectroscopy and Detection* (Ed. Yoh-Han Pao), Academic Press, New York **1977**, Ch. 8; *Photoacoustics and Photoacoustic Spectroscopy,* Wiley, New York, **1980**.

[100] M. Martinelli, U. Zamit, F. Scudieri, J. Martellucci, J. Quartieri, *Nuov. Cim.* **1987**, *90*, 557.

[101] L. M. Blinov, S. A. Davidian, N. N. Reshetov, D. V. Subachyus, S. V. Yablonsky, *Soc. Phys., JETP* **1990**, *70*, 902.

[102] F. Scudieri, T. Papa, D. Sette, M. Bertolotti, *Ann. Phys.* **1978**, 263.

[103] G. N. Belova, *Kristallografia,* **1988**, *33*, 1320.

[104] O. A. Kapustina, V. N. Reshetov, *Sov. Phys., JETP* **1984**, *70*, 122.

[105] O. A. Kapustina, *Mol. Cryst. Liq. Cryst.* **1990**, *179*, 173.

[106] O. A. Kapustina, E. N. Kozhevnikov, G. N. Yakovenko, *Sov. Phys., JETP* **1984**, *80*, 483.

[107] B. Hadimioglu, J. S. Foster, *J. Appl. Phys.* **1984**, *56*, 1976.

12   Nonlinear Optical Properties of Liquid Crystals

P. Palffy-Muhoray

12.1   Introduction

Optical nonlinearity is manifested by changes in the optical properties of a medium due to light propagating in that medium. A large number of distinct nonlinear optical processes have been observed; these are not only of fundamental scientific interest, but are also of great importance for both existing and potential applications in technology and industry.

The field of nonlinear optics originated in the early 1960s with experimental work on optical second harmonic generation [1], and theoretical work by Bloembergen and co-workers [2] on optical wave mixing. Since that time, the field has undergone explosive growth due to the development of high intensity laser sources with short pulse duration, and the discovery of materials with large optical nonlinearities.

Liquid crystals are orientationally ordered fluids; they are 'soft' materials in the sense that their physical properties can be readily altered by even modest fields. They are particularly well suited for optical applications requiring switching. In addition to their usefulness in display technology, they are also becoming increasingly important candidate materials for applications in non-

linear optics. A number of reviews of optical nonlinearities of liquid crystals have appeared [3–7]. This section provides a brief overview of the nonlinear optical response of liquid crystals and possible areas of application.

12.2   Interaction between Electromagnetic Radiation and Liquid Crystals

Liquid crystals are anisotropic fluids with optical properties similar to those of birefringent crystals. They are often inhomogeneous and, as a consequence, the dielectric tensor is a function of position. The inhomogeneity may be a property of the phase, such as the helical structure of chiral phases, or it may be the result of deformations.

12.2.1   Maxwell's Equations

Maxwell's equations in a dielectric material with no free charges are (in SI units):

$$\gamma_s(\theta,\varphi) = \sum A_{lm} Y_l \tag{1}$$

$$\nabla \times \mathbf{H} = \frac{\partial}{\partial t}(\varepsilon_0 \mathbf{E} + \mathbf{P}) \tag{2}$$

$$\nabla \cdot (\varepsilon_0 \mathbf{E} + \mathbf{P}) = 0 \tag{3}$$

$$\nabla \cdot \mathbf{B} = 0 \tag{4}$$

where $\mathbf{E}$ and $\mathbf{D}$ are the electric field and displacement, $\mathbf{B}$ and $\mathbf{H}$ are the magnetic induction and intensity, ε_0 is the permittivity of free space, $\mathbf{P}$ is the electric polarization, and higher order multipoles have been ignored. If the magnetic permeability μ, defined by $\mathbf{B} = \mu \mathbf{H}$, is a constant, then the wave equation

$$c^2 \frac{\mu_0}{\mu} \nabla \times \nabla \times \mathbf{E} + \frac{\partial^2 \mathbf{E}}{\partial t^2} = -\frac{1}{\varepsilon_0} \frac{\partial^2 \mathbf{P}}{\partial t^2} \tag{5}$$

where μ_0 is the permeability of free space and c is the speed of light, follows. The polarization $\mathbf{P}(r,t)$ at position r at time t is a time-varying source term which is a function of $\mathbf{E}(r,t)$. The optical response of the material is thus governed by the constitutive equation relating $\mathbf{P}$ and $\mathbf{E}$.

12.2.2 Nonlinear Susceptibility and Hyperpolarizability

The polarization $\mathbf{P}$ is, in general, a nonlinear function of $\mathbf{E}$. Expanding $\mathbf{P}$ in terms of $\mathbf{E}$ gives, for the case where $\mathbf{E}$ can be written as a sum of plane waves [8] $\mathbf{E}_i = \mathbf{E}_{i0} \exp[i(k_i \cdot r - \omega t)]$:

$$\mathbf{P} = \varepsilon_0 \left\{ \chi^{(1)} \mathbf{E} + \chi^{(2)} \mathbf{E}\mathbf{E} + \chi^{(3)} \mathbf{E}\mathbf{E}\mathbf{E} + \ldots \right\} \tag{6}$$

where $\chi^{(n)}$ is the nth order susceptibility. The susceptibilities are tensors which depend on the wavevectors k_i and frequencies ω_i of the fields; explicitly [9], the contribution to the polarization from nth order effects is

$$P_\alpha^{(n)}(\omega, k) = \varepsilon_0 \chi_{\alpha\beta\gamma\ldots}$$
$$(-\omega; \omega_1, \omega_2 \ldots, -k; k_1, k_2, \ldots)$$
$$E_\beta(\omega_1, k_1) E_\gamma(\omega_2, k_2) \ldots \tag{7}$$

and momentum and energy conservation give $k = k_1 + k_2 + \ldots$ and $\omega = \omega_1 + \omega_2 + \ldots$. Equations (5) and (6) indicate how the nonlinear susceptibilities give rise to nonlinear phenomena, from harmonic generation to intensity-dependent effects. It is worth noting that, since the susceptibilities depend on frequency, in experiments involving pulsed lasers the observed polarization depends not only on the optical frequency but also on the pulse duration. In the first approximation [7], it is reduced from the steady state value by the factor of $\tau_l / \sqrt{\tau_l^2 + \tau^2}$, where τ_l is the laser pulse duration and τ is the response time of the material.

The nonlinear polarization can arise through the nonlinear polarizabilities of the constituent molecules, or through the collective response of molecules the individual polarizability of which may be linear.

The induced electric dipole moment p of individual atoms and molecules can be written in terms of the local electric field $\mathbf{F}$ and the polarizabilities, similar to Eq. (6):

$$p = \alpha \mathbf{F} + \beta \mathbf{F}\mathbf{F} + \gamma \mathbf{F}\mathbf{F}\mathbf{F} + \ldots \tag{8}$$

where α is the linear polarizability, and β and γ are the second- and third-order hyperpolarizability tensors, respectively.

Calculations of atomic and molecular hyperpolarizabilities usually proceed via time-dependent perturbation theory for the perturbed atomic states. Even for molecules of modest size, the calculation of the complete set of unperturbed wavefunctions, and exact calculation of the hyperpolarizabilities, is prohibitively difficult. Liquid crystals typically consist of organic molecules with aromatic cores, and there is considerable experimental [10] and theoretical [11, 12] evidence to indicate that the dominant contribution to the polarizabilities originates from the delocalized π-electrons in conjugated regions of these molecules. Even considering only π-electrons the calculations rapidly become computationally intensive; for this reason, hyperpolarizabilities of liquid crystal cores are usually calculated within the framework of the free-electron Hückel model [11], or using the Pariser–Parr–Pople model [13], which includes electron–electron interactions. Other approaches include self-consistent field theory [14], which includes electron–electron interactions and singly and doubly excited configurations; and simulations of molecules with known nonlinear constituents [15].

The susceptibilities are related to the polarizabilities via the macroscopic polarization $\boldsymbol{P} = \langle \rho \boldsymbol{p} \rangle$, where ρ is the number density and $\langle \rangle$ denotes the ensemble average. In general, susceptibilities are not simply proportional to the orientationally averaged hyperpolarizability of the same order. Both the orientation and density of molecules depend on the field, and local field corrections contain nonlinear contributions. Consequently, high order hyperpolarizabilities contribute to lower order susceptibilities, and low order hyperpolarizabilities contribute to higher order susceptibilities [16]. The large changes in the orientation of liquid crystals in electric fields is one example of anisotropic linear polarizability giving rise to third-order susceptibility. The relation between second-order macroscopic susceptibilities and molecular hyperpolarizabilities is discussed by Singer et al. [17].

12.3   Nonlinearities Originating in Director Reorientation

Liquid crystals possess strongly anisotropic linear susceptibilities originating in the orientational order of non-spherical molecules. Due to this anisotropy, liquid crystals experience body torques in the presence of applied fields, which may give rise to director reorientation and a change in the dielectric tensor. This mechanism is responsible for extremely large nonlinear susceptibilities in liquid crystals. For reasons of symmetry, the dielectric tensor of nonpolar nematics is a function of the square of the electric field; hence reorientational effects contribute to the intensity-dependent susceptibility $\chi^{(3)}(-\omega; \omega, -\omega, \omega)$.

12.3.1   D.C. Kerr Effect

The most widely exploited effect in display applications is the influence of low frequency electric fields on the birefringence. Here, due to a distortion of the director, elements of the dielectric tensor change as a function of the applied field. From Eq. (6), the contribution of the third-order susceptibility $\chi^{(3)}(-\omega; 0, 0, \omega)$ to the dielectric tensor is given by $\varepsilon = \varepsilon_0 \{ \boldsymbol{I} + \chi^{(1)} + \chi^{(3)} \boldsymbol{E}_{dc} \boldsymbol{E}_{dc} \}$. Although the response is non-local, it is possible to obtain a crude estimate of the average susceptibility; $\chi^{(3)} \simeq \Delta\varepsilon / (\varepsilon_0 \, E_{dc}^2)$, where $\Delta\varepsilon$ is the change in ε due to the field. If the threshold voltage for the Freedericksz transition [18] V_{th} is applied to a cell of thickness d, then

$$\chi^{(3)}(-\omega; 0, 0, \omega) \simeq \frac{d^2}{V_{th}^2}(n_e^2 - n_o^2) \qquad (9)$$

where n_o and n_e are the ordinary and extraordinary refractive indices, and, for 5CB in a 20 μm thick cell, $\chi^{(3)} \simeq 4 \times 10^{-10}$ (SI). The sample-size dependence of $\chi^{(3)}$ is an indication of the spatial nonlocality of the response. The response time for this process is just the turn-on time of the cell [18].

12.4   Optical Field-Induced Reorientation

Optical field-induced director reorientation is responsible for the largest nonlinear optical susceptibility observed in liquid crystals, the largest in any known material. Although the process is slow, the nonlinearity is about 10^9 times greater than that of CS_2. Because of its magnitude, the orientational nonlinearity of liquid crystals has been termed 'giant optical nonlinearity' (GON).

The prediction [19] that a low power optical field can induce appreciable director reorientation just above the dc field induced Freedericksz transition has been verified experimentally [20, 21] concurrently with experimental and theoretical work on optical reorientation [22–24]. Since then, it has become one of the most intensively studied nonlinear optical effects in liquid crystals [3]. The phenomenon originates from the tendency of the director to align parallel to the electric field of light due to the anisotropic molecular polarizability. The free energy density arising from the interaction of a plane electromagnetic wave and the liquid crystals is

$$F_{opt} = -\frac{1}{2}E \cdot D - \frac{1}{2}B \cdot H$$
$$= -\frac{(E \times H) \cdot k}{\omega} = -\frac{I n_p}{c} \quad (10)$$

where k is the wavevector and c/n_p is the phase velocity. For materials with $n_e > n_o$, this is minimized if the polarization is along the director, since then $n_p = n_e$ is a maximum.

Experimental results and theoretical curves are shown in Fig. 1 [24]. A 250 µm thick cell containing the liquid crystal 5CB with homeotropic alignment at the cell walls was irradiated with light from an Ar$^+$ laser propagating at an angle α to the cell normal. For normal incidence ($\alpha=0$), the optical-field-induced torque causes the director field to distort above a threshold intensity I_{th}.

The results may be simply interpreted by noting that light-induced torques are opposed by elastic torques, and the free energy density has the form

$$F \simeq \frac{1}{2}K\left(\frac{\partial\theta}{\partial z}\right)^2 - \frac{I n_o}{2c}\left(\frac{n_e^2 - n_o^2}{n_e^2}\right)\sin^2\theta \quad (11)$$

where K is the bend elastic constant and θ is the angle the director makes with

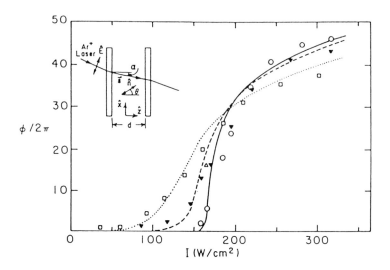

Figure 1. Experimentally observed laser-induced phase shift in 5CB.

the normal to the cell walls. If $\theta = 0$ at the walls, minimizing F gives the threshold

$$I_{th} = \frac{\pi^2 K c n_e^2}{d^2 n_o (n_e^2 - n_o^2)}.$$

It is interesting to note that there exists a material constant

$$P_0 = \frac{Kc}{n_e^2 - n_o^2}$$

with the units of power, which sets the scale for optical-field-induced effects. For 5CB, $P_0 \simeq 3$ mW. The nonlinear response is again nonlocal; the susceptibility may be estimated as in the case of the dc Kerr effect;

$$\chi^{(3)}(-\omega; \omega, -\omega, \omega) \simeq \frac{\varepsilon_0 c (n_e^2 - n_o^2)}{I_{th} n_e^2} \quad (12)$$

$$= \frac{d^2}{\pi^2 K} \frac{\varepsilon_0 n_o (n_e^2 - n_o^2)^2}{n_e^4}$$

and for 5CB in a 20 μm cell, $\chi^{(3)} \simeq 10^{-11}$ (SI). The response time again is the turn-on time of the cell:

$$\tau_{on} = \frac{\gamma d^2 I_{th}}{\pi^2 K (I - I_{th})} \quad (13)$$

$\gamma = 7 \times 10^{-2}$ Pa s and, for a 20 μm cell, for $I = 2 I_{th}$, $\tau_{on} \approx 300$ ms. For intense fields, fast response is possible. Director reorientation in 5CB due to 6 ns pulses at 1.06 m from a Q-switched Nd:YAG laser was observed [25] with response times of the order of 10–100 μs.

Anomalously low threshold intensities have been observed in nematics doped with dyes [27]. A theoretical model has been proposed [29], with the key feature that the interaction between dye and liquid crystal molecules is altered if the dye is optically excited. It has been found that the contribution of the dye to the nonlinear susceptibility may change sign as function of wavelength [29].

The optical-field-induced twist transition in planar samples has also been studied [30]. Here, because of the tendency of the polarization of the optical field to follow the principal axes of the dielectric tensor, the threshold intensity is found to be approximately four orders of magnitude greater than in the homeotropic case. In general, angular momentum from the radiation field is transferred to the liquid crystal [31], leading to precession of the director and a rich variety of nonlinear dynamic behavior [32, 33].

As a consequence of director reorientation, self-diffraction can occur in nematics [8, 34]. Laser-induced gratings can arise because two waves with ordinary and extraordinary polarizations propagate with different phase velocities. Wave mixing experiments have been performed in nematics, where two beams focused on the sample create an intensity grating. The phase grating which results from the nonlinear response of the material is then used to diffract light; the diffraction efficiency is proportional to the square of the third-order susceptibility, while the temporal profile of the diffracted beam gives information about the response time of the nonlinearity. Picosecond holographic gratings have been used to characterize the relaxation of orientation in liquid crystal films [36, 37]. Wave mixing has also been used to study optically generated hydrodynamic excitations in smectic A liquid crystals [38]. Cholesteric liquid crystals are important for laser applications. Optically induced reorientation has been studied [39]. Bistability due to light-induced pitch change of the cholesteric helix has been predicted [40] and retro-self-focusing caused by pitch dilation due to the optical field has been observed [41]. Optical-field-induced reorientation has also been studied in nonplanar geometries, such as PDLC films [42] and glass capillary arrays [43–45]. There is also evidence of photorefraction due to reorientation due to photo-induced charges [46].

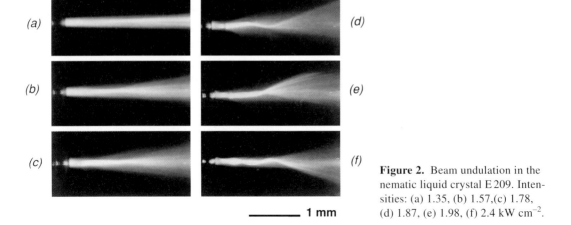

Figure 2. Beam undulation in the nematic liquid crystal E 209. Intensities: (a) 1.35, (b) 1.57,(c) 1.78, (d) 1.87, (e) 1.98, (f) 2.4 kW cm^{-2}.

Optical pattern formation in liquid crystals is of considerable interest; here, typically far-field patterns appear as result of optical field induced modulation of the director in directions transverse to the direction of propagation. Patterns have been observed in resonant cavities [47], systems with feedback mirrors [48], and in systems without feedback [49]. The propagation of a laser beam in a capillary filled with a nematic has been studied, showing self-focusing, undulation and filamentation [50]. Photographs, showing the beam profile as a function of intensity in a 1.5 mm capillary filled with the nematic mixture E209 are shown in Fig. 2.

Obtaining the full solution of Maxwell's equations together with the torque balance equations for the liquid crystal remains one of the most challenging problems in the nonlinear optics of liquid crystals [51].

12.5   Nonlinearities without Director Reorientation

Optical fields can affect the optical properties of liquid crystals without changing the orientation of the director. These nonlinearities originate from changes in the degree of orientational order, density, molecular conformation and electronic response.

12.5.1   Optical-Field-Induced Orientational Order

If a linearly polarized beam is propagating in the isotropic phase of a liquid crystal, the optical field will tend to align the molecules because of their anisotropic polarizability. Due to anisotropic interactions between the molecules, this alignment is enhanced, and the corresponding susceptibility diverges as the critical temperature associated with the phase transition is approached. The optical Kerr effect in the isotropic phase of nematics, as well as the effects of the optical field on the nematic–isotropic transition have been examined theoretically [52] by incorporating the orienting effects of the optical field into the Maier–Saupe theory [53]. Observations of optical-field-induced order in the isotropic phase of MBBA [54, 55] were shown to be in good agreement with the predictions of the Landau–de Gennes model.

Writing the free energy in terms of the scalar order parameter S gives

$$F = \frac{1}{2}a_0(T-T^*)S^2 + \ldots - \frac{In_p}{c} \quad (14)$$

Noting that, to lowest order, the refractive indices are linear in S, one obtains [7]

$$\chi^{(3)}(-\omega;\omega,-\omega,\omega) \simeq \frac{\varepsilon_0(n_e^2 - n_o^2)}{15k\rho(T-T^*)} \quad (15)$$

since $a_0 \simeq 5k\rho$ where ρ is the number density. For 5CB, this gives $\chi^{(3)} \simeq 8 \times 10^{-18}/(T-T^*)$ (in SI units). The response time is (see Eq. (13))

$$\tau = \frac{\gamma}{5k\rho(T-T^*)} \quad (16)$$

which has been verified by experiment [55, 56]. It is interesting to note that experimental measurements of $\chi^{(3)}$ with short laser pulses do not show this divergence [57] at $T = T^*$, since $\chi^{(3)}_{meas.} \simeq \chi^{(3)}\tau_l/\tau$ is independent of temperature [58].

In addition to affecting the orientation of the long axes of the liquid crystal molecules, evidence for reorientation of short axes by optical fields has been reported in smectic A materials [59].

12.5.2   Thermal Effects

Changes in temperature arising from the absorption of laser radiation can give rise to large changes in optical properties. The thermo-optic coefficients dn/dT in nematics are large; larger than in most other materials. The magnitude of the thermally induced nonlinearity can be calculated from the temperature rise due to absorption. In the adiabatic limit, $\Delta T \simeq (\alpha I \tau_l)/C$, where α is the linear absorption coefficient, τ_l is the laser pulse duration, and C is the specific heat per volume. For the nonadiabatic case, in the steady state limit, ΔT may be estimated by replacing τ_l by the thermal diffusion time

$\tau = l^2/D$, where D is the thermal diffusivity and l the relevant diffusion length. The susceptibility due to laser heating is given, since $\chi^{(3)} \simeq \Delta\varepsilon n c/I$, by

$$\chi^{(3)}(-\omega;\omega,-\omega,\omega) \simeq \frac{2n^2 c\varepsilon_0\alpha\tau}{C}\frac{dn}{dT} \quad (17)$$

in the steady state case. If $\tau_l \ll \tau$, the susceptibility is reduced by the factor τ_l/τ.

In general, both the linear absorption coefficient and the thermal diffusivity are anisotropic. Typical values for the thermal diffusivity are [60] $D_\parallel = 1.25 \times 10^{-7}$ m^2 s^{-1} and $D_\perp = 7.9 \times 10^{-8}$ m^2 s^{-1}. The absorption of light by nematic liquid crystals is typically very small in the visible region, $\alpha \approx 10$ m^{-1} for the nematic mixture E7 and for MBBA in the isotropic phase [61]. From Eq. (17), for $l = 10$ μm and $\tau \simeq 1$ ms, and estimating $C \simeq 2 \times 10^6$ J m^3 K^{-1} and $\alpha \simeq 6$ m^{-1}, for 20 ns pulses, $\chi^{(3)} \simeq 5 \times 10^{-18}$ (SI).

A wide variety of studies involving thermal effects in liquid crystals have been carried out. The dynamics of the generation of and diffraction from thermal gratings in nematic liquid crystals has been the subject of considerable interest, with good agreement between experiment and theory [62, 63].

12.5.3   Conformational Effects

In 1981, self-diffraction was observed in nematic MBBA [64] under circumstances where director reorientation could not occur. The measured nonlinear susceptibility was as large as the 'giant' nonlinearity associated with director reorientation; the mechanism proposed was photostimulated conformational changes of the liquid crystal molecules. Since the linear polarizability of the excited conformer differs from that of the molecule in the ground state, there is a direct contribution to the bulk susceptibility. The metastable conformers also act as impurities, and reduce the orientational order.

Photoinduced conformational changes in azo-dye-doped liquid crystals have been used for real-time holography [65]. The orientation of azo dyes by an optical field, the Weigert effect [66], has been used to induce orientation of liquid crystals both in the bulk and in alignment layers [67, 68].

12.5.4 Electronic Response

The fastest optical nonlinearity originates from the electronic response. To lowest order, $\chi^{(2)} = \langle \rho \beta \rangle / \varepsilon_0$ and $\chi^{(3)} = \langle \rho \gamma \rangle / \varepsilon_0$, where β and γ are molecular hyperpolarizabilities. As in the case of nonlinear susceptibilities, hyperpolarizabilities describe a variety of processes. Optical harmonic generation will be considered separately.

Contributions to the third-order susceptibility $\chi^{(3)}(-\omega; \omega, -\omega, \omega)$ from third-order hyperpolarizability have received considerable attention. Picosecond measurements have been carried out on liquid crystals in the isotropic phase [69, 70]; from the critical power for self-focusing, the nonlinear refractive index could be determined. More recently, the Z-scan method [71, 72] has been used to determined the susceptibility of a variety of liquid crystals [73]. It appears that in a variety of materials, the dominant contribution on the nanosecond timescale is excited state absorption, from a two-photon excited state [57, 74, 75]; thus this is a fifth-order process. This mechanism appears promising for optical power limiting applications [76].

Knowledge of the nonlinear susceptibilities in an aligned nematic phase allows, in principle, determination of the elements of the hyperpolarizability tensor of liquid crystal molecules if the pulse durations used in the measurements are short enough. Recent measurements of nonlinearities arising from optical-field-induced orientation processes in picosecond self-diffraction experiments [27] in the nematic and isotropic phases, suggest that pulses in the femtosecond range may be necessary to identify electronic contributions uniquely. A general review of organic materials for third-order nonlinear optics is given in Soileau et al. [77].

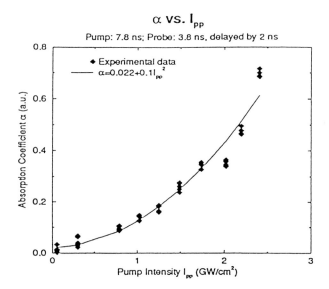

Figure 3. Intensity dependence of the absorption coefficient in the nematic liquid crystal 5CB.

12.6   Optical Harmonic Generation

Optical harmonic generation was one of the first areas of investigation in the field of nonlinear optics of liquid crystals; it has continued to receive a great deal of attention. The aim of the first studies was to observe second harmonic generation (SHG) from cholesterics; subsequent work has involved other phases, third harmonic generation (THG), and harmonic generation from surfaces as well as the bulk.

12.6.1   Bulk Second Harmonic Generation

The second-order susceptibility $\chi_{ijk}^{(2)}(-2\omega, \omega, \omega)$ is a third-rank tensor. For materials with inversion symmetry, in the electric dipole approximation $\chi_{ijk}^{(2)}=0$ for all i, j, k [78]. Nematic liquid crystals possess $D_{\infty h}$ symmetry; therefore, in the dipole approximation they are not be expected to give rise to SHG. However, SHG was observed in aligned samples of MBBA [79, 80]. It has been proposed that the origin of this is second harmonic polarization arising from the quadrupolar susceptibility and gradients of the optical field [81]. More recent work suggests that flexoelectric effects arising from thermal fluctuations [82] or, more generally, electric polarization due to spatial variation of the dielectric tensor [83] may be responsible.

Cholesteric liquid crystals are not centrosymmetric, and might be expected to give rise to SHG. They possess D_{∞} symmetry, which allows for one independent nonvanishing element of the second-order susceptibility. If the director is along the α direction, this is $\chi_{\gamma\alpha\beta}^{(2)}=-\chi_{\gamma\beta\alpha}^{(2)}$. For nonabsorbing materials Kleinman's rule holds, according to which elements of a nonlinear susceptibility tensor are invariant under any permutation of their indices [84]. As a consequence, $\chi^{(2)}$ vanishes, and SHG is disallowed. Contrary to these expectations, early experiments detected SHG in cholesteryl carbonate [85]. Subsequent work showed that the second harmonic signal originated from crystalline particles in the sample, and that the melting of these leads to disappearance of the second harmonic signal [86]. The blue phases of cholesterics may be studied by SGH [87].

If a dc electric field is present in addition to the optical field, the third-order nonlinearity can give rise to SHG. This results from contributions to the polarization of the form

$$P(2\omega) = \chi^{(3)}(-2\omega; \omega, \omega, 0)E_{\text{opt}}E_{\text{opt}}E_{\text{dc}} \tag{18}$$

Assuming that $\chi^{(3)}$ arises from the third-order hyperpolarizability, elements of $\chi^{(3)}$ give information about the degree of orientational order. Measurements of electric-field-induced second-harmonic generation (EFISH) have been carried out [88] on 5CB, showing the expected temperature dependence. In liquid crystals, E_{dc} may originate from order or flexoelectric effects, or other mechanisms resulting in electric polarization.

Ferroelectric chiral smectic C phases lack inversion symmetry, and are distinguished by spontaneous helicoidal electric polarization. The first experiment to measure SHG from ferroelectric liquid crystals was carried out [89] on unaligned samples under a dc electric field. Phase-matched SHG in ferroelectric liquid crystals has been carried out by using an electric field to unwind the helix [90]. Mechanical deformations in chiral smectic C elastomers have been shown to give rise to SHG [91]. A great deal of work has been carried out recently in studying SHG in ferroelectric liquid crystals [93], with particular emphasis on the relation between molecular structure and second-harmonic efficiency [94–96].

12.6.2   Surface Second Harmonic Generation

Second harmonic generation from surfaces has been shown to be a powerful probe of surface interactions [97], which can yield information about the structure and alignment of liquid crystal molecules on surfaces and at interfaces [98]. The role of the interface is similar to that of the dc electric field in electric-field-induced SHG, that is, to provide a symmetry-breaking field the direction of which is defined by the surface normal.

In general, liquid crystal molecules do not have the $D_{2\infty}$ symmetry of the uniaxial nematic phase. Since an interface acts as a field, its presence can provide polar order. Such surface polar ordering, confined to a single molecular layer, has been observed [99]. Surface SHG can also be used to probe the orientational distribution at the surface, and anchoring transitions [100].

12.6.3   Third Harmonic Generation

Third harmonic generation is characterized by the susceptibility $\chi^{(3)}(-3\omega; \omega, \omega, \omega)$ which is nonvanishing for all materials. Measurements of THG have been carried out on MBBA in the nematic and isotropic phase [101]. Considerable information about orientational order can be obtained from $\chi^{(3)}(-3\omega)$ [102].

In cholesterics, the situation regarding phase matching is more complex than in nematics. The normal modes are plane polarized waves in a coordinate system co-rotating with the director [87], and hence the wavevectors are functions of both the frequency and the cholesteric pitch. This pitch dependence can be used to achieve phase matching. In cholesterics, the momentum of the electromagnetic field is not necessarily conserved during harmonic generation; exchange of momentum between the field and the periodic structure of the cholesteric is possible in a 'coherent optical Umklapp process' [103].

12.7   Materials and Potential Applications

For a wide variety of proposed applications, material requirements are large and fast optical nonlinearities coupled with low losses. Low molecular weight liquid crystals have nonlinear susceptibilities which range from nine orders of magnitude greater than that of CS_2 on millisecond and slower time scales, to approximately the same as that of CS_2 on the picosecond scale. Novel materials, such as biaxial nematics and liquid crystal elastomers, offer the possibility of novel nonlinear behavior. Polymeric liquid crystals are particularly promising because of the possibility of a large nonlinear electronic response due to extended conjugated regions, as well as a fast reorientational response due to restricted reorientation of mesogenic units. They can also be used as host materials for nonlinear chromophores. Scattering losses, which can be considerable in low molecular weight materials, are reduced in polymer films.

The most widely utilized optical nonlinearity of liquid crystals is electric-field-induced director reorientation, which is used in display application. Although it is usually dealt with under the heading of 'electro-optics', it is a nonlinear optical process involving waves with frequencies 0 and ω.

Optically bistable devices are of considerable interest for information storage and processing applications. Bistable Fabry–Perot resonators [104] may be useful as dynamic optical memories as well as elements for information processing. Optically sensitive alignment layers [62], as well as a wide variety of laser addressable materials [105–107] promise the use of liquid crystals as materials for optical information storage.

Binary optical computing has been demonstrated [108] using the liquid crystal light valve made by Hughes [109].

Development of other spatial light modulators, using nematic and ferroelectric liquid crystals [110], and polymer dispersed liquid crystals [111] is in progress. Such liquid crystal light modulators can be used in binary optical computing, where high parallel processing density [112] compensates for the relatively slow device response time. A nonlinear liquid crystal film can act as an optical switching device [113] in signal processing applications.

The high optical damage threshold of liquid crystals makes them well suited for optical power limiting (OPL). This can occur for input energies as low as 0.3 μJ for 5 ns pulses at 532 nm [76]. Polymer dispersed liquid crystal films have been proposed for OPL applications both on the hybrid mode [114] and in the all optical mode using reorientational and thermal effects [115] to induce nonlinear scattering. Optic fibers with liquid crystalline cores have promise for both electro-optic and nonlinear optical applications; their potential for OPL has been investigated [26, 27, 34]. Liquid crystal can also be used as nonlinear couplers between optic fiber waveguides. Optical solitons, potentially useful in signal processing applications, have been theoretically considered in liquid crystals [116].

Ferroelectric liquid crystals are becoming increasingly promising for SHG devices and Pockels modulators [117]. The second harmonic intensity may be modulated [118]. A review of advanced liquid crystal polymers is given in Dubois [119]. Potential applications also include phase conjugation and real-time holography.

Nonlinear optical effects in liquid crystals encompass a remarkably wide range of phenomena, and offer broad potential for applications [120]. There can be little doubt that liquid crystals will play an increasingly important role in the nonlinear optical devices of the future.

12.8   References

[1] P. A. Franken, A. E. Hill, C. W. Peters, G. Weinreich, *Phys. Rev. Lett.* **1961**, *7*, 118.
[2] J. A. Armstrong, N. Bloembergen, J. Ducuing, P. S. Pershan, *Phys. Rev.* **1962**, *127*, 1918.
[3] N. V. Tabiryan, A. V. Sukhov, B. Ya. Zel'dovich, *Mol. Cryst. Liq. Cryst. Special Topics XIX* **1986**, *136*, 1.
[4] I. C. Khoo in *Progress in Optics*, Vol. 26 (Ed. E. Wolf), North Holland, New York **1988**.
[5] I. Janossy in *Perspectives in Condensed Matter Physics* (Ed. L. Miglio), Kluwer, Dordrecht **1990**.
[6] I. C. Khoo, *Liquid Crystals: Physical Properties and Nonlinear Optical Phenomena*, Wiley, New York **1995**.
[7] P. Palffy-Muhoray in *Liquid Crystals: Applications and Uses* (Ed. B. Bahadur), World Scientific, Singapore **1990**.
[8] Y. R. Shen, *The Principles of Nonlinear Optics*, Wiley, New York **1984**.
[9] D. C. Hannah, M. A. Yuratich, D. Cotter, *Nonlinear Optics of Free Atoms and Molecules*, Springer Series in Optical Sciences (Ed. D. L. MacAdam), Springer, New York **1979**.
[10] B. F. Levine, C. G. Bethea, *J. Chem. Phys.* **1975**, *63*, 2666.
[11] S. Risser, S. Klemm, D. W. Allender, M. A. Lee, *Mol. Cryst. Liq. Cryst.* **1987**, *150B*, 631.
[12] D. N. Beratan, J. N. Onuchic, J. W. Perry, *J. Chem. Phys.* **1987**, *91*, 2696.
[13] S. Risser, D. W. Allender, M. A. Lee, K. E. Schmidt, *Mol. Cryst. Liq. Cryst.* **1990**, *179*, 335.
[14] J. R. Heflin, K. Y. Wong, O. Zamani-Khamiri, A. F. Garito, *Mol. Cryst. Liq. Cryst.* **1988**, *160*, 37.

[15] R. Pachter, S. S. Patnaik, R. L. Crane, W. W. Adams, *SPIE Proc.* **1993**, *1916*, 2.

[16] G. R. Meredith in *Nonlinear Optics: Materials and Devices*, Springer Proceedings in Physics (Eds C. Flytzanis, J. L. Oudar), Springer, New York **1986**.

[17] K. D. Singer, M. G. Kuzyk, J. E. Sohn, *J. Opt. Soc. Am. B* **1987**, *4*, 968.

[18] L. M. Blinov, V. G. Chigrinov, *Electrooptic Effects in Liquid Crystal Material*, Springer, New York **1994**.

[19] R. M. Herman, R. J. Serinko, *Phys. Rev. A* **1979**, *19*, 1757.

[20] I. C. Khoo, S. L. Zhuang, *Appl. Phys. Lett.* **1980**, *37*, 3.

[21] I. C. Khoo, *Phys. Rev. A* **1981**, *23*, 2077.

[22] A. S. Zolotko, V. F. Kitaeva, N. Kroo, N. N. Sobolev, L. Chillag, *JETP Lett.* **1980**, *32*, 158.

[23] B. Ya. Zel'dovich, N. V. Tabiryan, *Sov. J. Quantum Electron.* **1990**, *10*, 440.

[24] S. D. Durbin, S. M. Arakelian, R. Y. Shen, *Opt. Lett.* **1981**, *6*, 411.

[25] P. S. Pershan, *Phys. Rev.* **1963**, *130*, 919.

[26] H. Hsiung, L. P. Shi, Y. R. Shen, *Phys. Rev. A* **1984**, *30*, 1453.

[27] I. Janossy, T. Kosa, *Opt. Lett.* **1992**, *17*, 1183.

[28] I. Janossy, *Phys. Rev. E* **1994**, *49*, 2957.

[29] T. Kosa, I. Janossy, *Opt. Lett.* **1995**, *20*, 1230.

[30] E. Santamato, G. Abbate, P. Maddalena, *Phys. Rev. A* **1987**, *36*, 2389.

[31] E. Santamato, B. Daino, R. Romagnoli, M. Settembre, Y. R. Shen, *Phys. Rev. Lett.* **1986**, *57*, 2423.

[32] E. Santamato, G. Abbate, P. Maddalena, L. Marrucci, *Phys. Rev. Lett.* **1990**, *64*, 1377.

[33] V. Carbone, G. Cipparrone, C. Versace, R. Bartolino, C. Umeton, *Mol. Cryst. Liq. Cryst.* **1994**, *251*, 167.

[34] S. M. Arakelian, Y. S. Chilingarian, *IEEE J. Quantum Electron.* **1986**, *QE-22*, 1276.

[35] G. Eyring, M. D. Fayer, *J. Chem. Phys.* **1984**, *81*, 4314.

[36] M. D. Fayer, *IEEE J. Quantum Electron.* **1986**, *QE-22*, 1437.

[37] H. J. Eichler, R. Macdonald, B. Trösken, *Mol. Cryst. Liq. Cryst.* **1993**, *231*, 1.

[38] G. F. Kventsel, B. I. Lembrikov, *Mol. Cryst. Liq. Cryst.* **1995**, *262*, 591.

[39] P. Maddalena, G. Arnone, G. Abbate, L. Marrucci, E. Santamato, *Mol. Cryst. Liq. Cryst.* **1995**, *261*, 113.

[40] H. G. Winful, *Phys. Rev. Lett.* **1982**, *49*, 1179.

[41] J.-C. Lee, S. D. Jacobs, A. Schmid, *Mol. Cryst. Liq. Cryst.* **1987**, *150b*, 617.

[42] P. Palffy-Muhoray, B. J. Frisken, J. Kelly, H. J. Yuan, *SPIE Proc.* **1989**, *1105*, 33.

[43] H. J. Yuan, H. Lin, P. Palffy-Muhoray, *Mol. Cryst. Liq. Cryst.* **1992**, *223*, 229.

[44] H. Lin, P. Palffy-Muhoray, *Opt. Lett.* **1994**, *19*, 436.

[45] I. C. Khoo, H. Li, *Appl. Phys. B* **1994**, *59*, 573.

[46] I. C. Khoo, H. Li, Y. Liang, *Opt. Lett.* **1994**, *19*, 1723.

[47] T. Vogeler, M. Kreuzer, T. Tschudi, N. V. Tabiryan, *Mol. Cryst. Liq. Cryst.* **1994**, *251*, 159.

[48] R. Macdonald, H. Danlewski, *Mol. Cryst. Liq. Cryst.* **1994**, *251*, 145.

[49] G. Hu, P. Palffy-Muhoray, N. V. Tabiryan in *Licht- und Teilchenoptik, Ann. Rep.*, Tech. Hochsch. Darmstadt **1992**.

[50] E. Braun, L. Faucheux, A. Libchaber, *Phys. Rev. A* **1993**, *48*, 611.

[51] E. Braun, L. Faucheux, A. Libchaber, D. W. McLaughlin, D. J. Muraki, M. J. Shelley, *Europhys. Lett.* **1993**, *23*, 4.

[52] J. Hanus, *Phys. Rev.* **1969**, *178*, 420.

[53] W. Maier, A. Saupe, *Z. Naturforsch., Teil A* **1958**, *13*, 564; **1959**, *14*, 882; **1960**, *15*, 287.

[54] J. Prost, J. R. Lalanne, *Phys. Rev. A* **1973**, *8*, 2090.

[55] G. K. Wong, Y. R. Shen, *Phys. Rev. Lett.* **1973**, *30*, 895.

[56] P. A. Madden, F. C. Saunders, A. M. Scott, *IEEE J. Quantum Electron.* **1986**, *QE-22*, 1287.

[57] P. Palffy-Muhoray, T. Wei, W. Zhao, *Mol. Cryst. Liq. Cryst.* **1994**, *251*, 19.

[58] C. W. Greef, *Mol. Cryst. Liq. Cryst.* **1994**, *238*, 179.

[59] J. R. Lalanne, J. Bouchert, C. Destrade, H. T. Nguyen, J. P. Marcerou, *Phys. Rev. Lett.* **1989**, *62*, 3046.

[60] W. Urbach, H. Hervet, F. Rondelez, *Mol. Cryst. Liq. Cryst.* **1978**, *46*, 209.

[61] S.-T. Wu, K.-C. Lim, *Appl. Opt.* **1987**, *26*, 1722.

[62] I. C. Khoo, R. Normandin, *IEEE J. Quantum Electron.* **1985**, *QE-21*, 329.

[63] I. C. Khoo, S. Shepard, *J. Appl. Phys.* **1983**, *54*, 5491.

[64] S. G. Odulov, Yu. A. Reznikov, M. S. Soskin, A. I. Khizhnyak, *Sov. Phys. JETP* **1982**, *55*, 854.

[65] A. G. Chen, D. J. Brady, *Opt. Lett.* **1992**, *17*, 441.

[66] F. Weigert, M. Nakashima, *Z. Phys. Chem.* **1929**, *34*, 258.

[67] W. M. Gibbons, P. J. Shannon, Shao-Tang Sun, B. J. Swetlin, *Nature* **1991**, *351*, 49.

[68] P. J. Shannon, W. M. Gibbons, S. T. Sun, *Nature* **1994**, *368*, 532.

[69] M. J. Soileau, S. Guha, W. E. Williams, E. W. Van Stryland, H. Vanherzeele, *Mol. Cryst. Liq. Cryst.* **1985**, *127*, 321.

[70] M. J. Soileau, E. W. Van Stryland, S. Guha, E. J. Sharp, G. L. Wood, J. L. W. Pohlmann, *Mol. Cryst. Liq. Cryst.* **1987**, *143*, 139.

[71] M. Sheik-bahae, A. A. Said, E. W. Van Stryland, *Opt. Lett.* **1989**, *14*, 955.

[72] M. Sheik-bahae, A. A. Said, T. H. Wei, D. J. Hagan, E. W. Van Stryland, *IEEE J. Quantum Electron.* **1990**, *26*, 760.

[73] L. Li, H. J. Yuan, G. Hu, P. Palffy-Muhoray, *Liq. Cryst.* **1994**, *16*, 703.

[74] K. J. McEwan, R. C. Hollins, *SPIE Proc.* **1994**, *2229*, 122.

[75] W. Zhao, J. H. Kim, P. Palffy-Muhoray, *SPIE Proc.* **1994**, *2229*, 131.

[76] A. Hochbaum, Y.-Y. Hsu, J. L. Fergason, *SPIE Proc.* **1994**, *2229*, 48.

[77] H. S. Nalwa, *Adv. Mater.* **1993**, *5*, 341.

[78] P. A. Franken, J. F. Ward, *Rev. Mod. Phys.* **1963**, *35*, 23.

[79] S. M. Arakelian, G. L. Grigorian, S. Ts. Nersiyan, M. A. Nshayan, Yu. S. Chilingaryan, *Sov. Phys. JETP Lett.* **1978**, *28*, 186.

[80] S. M. Arakelian, G. A. Lyakhov, Yu. Chilingaryan, *Sov. Phys. Usp.* **1980**, *23*, 245.

[81] Ou-Yang Zhong-can, Xie Yu-zhang, *Phys. Rev. A* **1985**, *32*, 1189.

[82] M. Copic, T. Ovsenik, *Europhys. Lett.* **1993**, *24*, 93.

[83] L. Li, H. J. Yuan, P. Palffy-Muhoray, *Mol. Cryst. Liq. Cryst.* **1991**, *198*, 239.

[84] P. G. Harper, G. S. Wherrett (Eds), *Nonlinear Optics*, Academic Press, London **1977**.

[85] I. Freund, P. M. Rentzepis, *Phys. Rev. Lett.* **1967**, *18*, 393.

[86] G. Durand, C. H. Lee, *Mol. Cryst.* **1968**, *5*, 171.

[87] V. A. Belyakov, *Diffraction Optics of Complex-Structured Periodic Media*, Springer, New York **1992**.

[88] S. K. Saha, G. K. Wong, *Appl. Phys. Lett.* **1978**, *34*, 423.

[89] A. N. Vtyurin, V. P. Yermakov, B. I. Ostrovsky, V. F. Shavanov, *Kristallografiya* **1981**, *26*, 546.

[90] M. I. Barnik, L. M. Blinov, N. M. Shtykov, *Sov. Phys. JETP* **1984**, *59*, 980.

[91] I. Benne, K. Semmler, H. Finkelmann, *Macromolecules* **1995**, *28*, 1854.

[92] A. Taguchi, Y. Ouchi, H. Takezoe, A. Fukuda, *Jpn. J. Appl. Phys.* **1989**, *28*, L 997.

[93] D. M. Walba, M. B. Ros, N. A. Clark, R. Shao, K. M. Johnson, M. G. Robinson, J. Y. Liu, D. Doroski, *Mol. Cryst. Liq. Cryst.* **1991**, *189*, 51.

[94] D. M. Walba, D. A. Zummach, M. D. Wand, W. N. Thurmes, K. M. Moray, K. E. Arnett, *SPIE Proc.* **1993**, *1911*, 21.

[95] M. Loddoch, G. T. Marowsky, H. Schmidt, G. Heppke, *Appl. Phys. B* **1994**, *59*, 591.

[96] M. Ozaki, M. Sakuta, K. Yoshino, B. Helgee, M. Svensson, K. Skarp, *Appl. Phys. B* **1994**, *59*, 601.

[97] Y. R. Shen, *Nature* **1989**, *337*, 519.

[98] P. Guyot-Sionnest, H. Shiung, Y. R. Shen, *Phys. Rev. Lett.* **1986**, *57*, 2963.

[99] C. S. Mullin, P. Guyot-Sionnest, Y. R. Shen, *Phys. Rev. A* **1989**, *39*, 3745.

[100] B. Jerome, *Mol. Cryst. Liq. Cryst.* **1994**, *251*, 219.

[101] K. Y. Wong, A. F. Garito, *Phys. Rev. A* **1986**, *34*, 5051.

[102] H. Matoussi, G. C. Berry, *Mol. Cryst. Liq. Cryst.* **1992**, *223*, 41.

[103] J. W. Shelton, Y. R. Shen, *Phys. Rev. A* **1972**, *5*, 1867.

[104] M. Kreuzer, H. Gottschling, R. Neubecker, T. Tschudi, *Appl. Phys. B* **1994**, *59*, 581.

[105] G. Hu, T. Kosa, P. Palffy-Muhoray in *ECLE '95 Abstracts*, Ljubljana **1995**.

[106] R. S. Akopyan, N. V. Tabiryan, T. Schudi, *Phys. Rev. E* **1994**, *49*, 3143.

[107] M. Kreuzer, T. Tschudi, R. Eidenschink, *Mol. Cryst. Liq. Cryst.* **1992**, *223*, 219.

[108] S. A. Collins, *SPIE Proc.* **1989**, *1080*.

[109] W. P. Bleha, L. T. Lipton, E. Wiener-Avenar, J. Grienberg, P. G. Reif, D. Casasent, H. B. Brown, B. V. Markevitch, *Opt. Eng.* **1978**, *17*, 371.

[110] D. Armitage, J. I. Thackara, W. D. Eades, *Liq. Cryst.* **1989**, *5*, 1389.

[111] K. Takizawa, H. Kikuchi, H. Fujikake, K. Kodama, K. Kishi, *J. Appl. Phys.* **1994**, *75*, 3158.

[112] K. M. Johnson, M. A. Handschy, L. A. Pagano-Stauffer, *Opt. Eng.* **1987**, *26*, 385.

[113] I. C. Khoo, *Appl. Phys. Lett.* **1982**, *40*, 645.

[114] J. Y. Kim, P. Palffy-Muhoray, *J. Appl. Phys.* **1989**, *66*, 362.

[115] F. Simoni, G. Cipparrone, C. Umeton, G. Arabia, G. Chidichimo, *Appl. Phys. Lett.* **1989**, *54*, 896.

[116] Y. S. Yung, L. Lam in *Solitons and Chaos in Optical Systems* (Eds.: H. C. Morris, D. Heesernan), Plenum, New York **1990**.

[117] M. Schadt, K. Schmitt, *Appl. Phys. B* **1994**, *59*, 607.

[118] I. Drevensek, T. Renato, M. Copic, *Mol. Cryst. Liq. Cryst.* **1994**, *251*, 101.

[119] J. C. Dubois, *Polym. Adv. Technol.* **1995**, *6*, 10.

[120] N. V. Tabiryan, T. Vogeler, T. Tschudi, *SPIE Proc.* **1994**, *2175*, 191.

13 Diffusion in Liquid Crystals

F. Noack[†]

13.1 Introduction

Diffusion, that is molecular mass transport by, usually, thermally activated particle motions, is a property of liquid crystals which is relatively poorly understood both experimentally and theoretically, despite countless efforts. In general this process does not strongly affect the orientational order of the mesophase, and therefore it seems not to be of primary technical importance. Furthermore, the available experimental and theoretical results are often inconsistent because of great difficulties encountered in developing and applying adequate research procedures, and so most standard textbooks [1–3] do not treat the subject systematically. Nevertheless, diffusion constants can reveal directly details about the dynamics of liquid crystalline order and thus are involved in the understanding and tailoring of other material quantities, such as anisotropic viscosities and electrical conductivities, needed to optimize the performance of liquid crystal applications. This chapter covers some basic theoretical concepts and common experimental techniques used to study diffusion in liquid crystals, and summarizes some general results and references.

13.2 Theoretical Concepts

13.2.1 The Diffusion Tensor

The diffusion sensor or diffusivity **D** of a system of n identical or specially labeled molecules is defined by two phenomenological, related linear diffusion laws which, on the one hand, describe a particle current $J_r \equiv \partial n / \partial t$ within the sample in any selected direction r, perpendicular through an area A, by the concentration gradient $\partial c / \partial r$ of the particles in the direction of J_r (Fick's first law) and, on the other hand, express the time development of the concentration $\partial c / \partial t$ by the Laplace derivative $\Delta \equiv \partial^2 c / \partial r^2$ (Fick's second law), so [4, 5]

$$\left(\frac{\partial n}{\partial t}\right) = -\mathbf{D}\, A\left(\frac{\partial c}{\partial r}\right) \tag{1a}$$

and

$$\left(\frac{\partial c}{\partial t}\right) = +\mathbf{D}\left(\frac{\partial^2 c}{\partial r^2}\right) \tag{1b}$$

If the molecules under consideration are different from the host system, one must distinguish between *solute* or *binary* diffusion and *self*-diffusion. Due to the liquid crystalline order, described by the director field $n\,(r, t)$ and the order parameter S, diffusion in mesophases is usually anisotropic. In general, **D** will be a second-rank tensor, the symmetry of which is determined by the symmetry of the liquid crystal. In an (x, y, z) axis system with the first principal axis parallel to n and the other two principal axes perpendicular to n, the diffusion tensor can be diagonalized and thus involves, at most three independent components ($D_{xx}=D_{\perp 1}$, $D_{yy}=D_{\perp 2}$, $D_{zz}=D_{\parallel}$) [4], and for uniaxial symmetry, as for common nematic (N), smectic A (SmA), or smectic B (SmB) phases, one has $D_{\perp 1}=D_{\perp 2}=D_{\perp}$. Hence as a rule, the diffusion tensor of liquid crystals has the simpler form

$$\mathbf{D}=\begin{pmatrix} D_{\perp} & 0 & 0 \\ 0 & D_{\perp} & 0 \\ 0 & 0 & D_{\parallel} \end{pmatrix} \qquad (2)$$

with two characteristic diffusion constants or coefficients of the order or 10^{-11} to 10^{-10} m^2 s^{-1}. The anisotropy ratio $\sigma=D_{\perp}/D_{\parallel}$ is a material parameter, which can be larger or smaller than 1; for nematic phases it is typically near ½. Above the transition to the isotropic liquid, where no macroscopic order persists, **D** discontinuously becomes a scalar D_{iso}. Often $\langle D \rangle \equiv (2D_{\perp}+D_{\parallel})/3$ is defined as an average diffusion constant, which is expected to be similar to D_{iso}.

The most common, idealized solution of Eq. (1a) describes the concentration profiles $c(r, t)$ obtained for an infinite medium by focalizing at the beginning ($t=0$) some labeled particles at $r=r_0$, i.e. $c(r, 0)=c_0 \delta(r-r_0)$. Using the symmetry of Eq. (2) and (x, y, z) coordinates aligned along the main tensor axes, one obtains a three-dimensional, ellipsoidal Gaussian concentration distribution [4, 5]:

$$c(r-r_0, t)=c_0 \left[(4\pi)^3 D_{\perp}D_{\perp}D_{\parallel}t^3\right]^{-\frac{1}{2}} \qquad (3)$$

$$\times \exp\left[-\frac{(x-x_0)^2}{(4D_{\perp}t)}-\frac{(y-y_0)^2}{4D_{\perp}t)}-\frac{(z-z_0)^2}{(4D_{\parallel}t)}\right]$$

which for convenience is sometimes interpreted (with $c_0=1$) as the conditional probability or propagator $P(r_0|r,t)$ to find a molecule at time t in the volume element dr at r, if it was initially at position r_0. Equation (3) is directly or indirectly the basis for the experimental techniques used to study diffusivities. In particular, the mean square displacement at time t

$$\langle (r-r_0)^2 \rangle = (4D_{\perp}+2D_{\parallel})\,t \qquad (4)$$

obtained by averaging $(r-r_0)^2$ of individual molecules with Eq. (3), provides a relatively easy and transparent means of estimating the diffusion constants. Furthermore, it should be noted that the propagator concept indicates the way to generalize Fick's laws by including additional terms [6], a behavior that has been found necessary in solid-state physics and is also discussed for liquid crystals.

13.2.2 Basic models

Assuming the validity of Fick's laws for liquid crystals, various theories [7–15] have been developed to describe the diffusion constants $D_{\parallel}$ and $D_{\perp}$, in particular their anisotropy ratio σ, in terms of the mesophase order and other adequate macroscopic and microscopic material parameters. However, in view of the unsatisfactory agreement with experimental data, so far this has only achieved limited success. By transforming the diffusion tensor from a local (cluster) to the laboratory director frame and by taking the orientational ensemble average, Blinc et al. [7] showed that the two tensor components of **D** for *thermotropic* N, S$_A$, and S$_C$ phases should be coupled to the order parameter S, the average diffusion constant $\langle D \rangle$, and the limiting values $D_{\parallel}^0$, $D_{\perp}^0$ of a perfectly ordered cluster ($S=1$) in the form

$$D_\perp = \langle D\rangle(1 - S) + SD_\perp^0 \tag{5a}$$

$$D_\parallel = \langle D\rangle(1 - S) + SD_\parallel^0 \tag{5b}$$

This result was later confirmed and refined by Chu and Moroi [8, 9], who evaluated $\langle D\rangle$ explicitly by using a power series of the momentum autocorrelation function, and also suggested geometric estimations for the D^0 values in a perfectly ordered cluster region. With such extensions, Eq. (5) finally gives in the case of nematic order [8]

$$D_\perp = \langle D\rangle\left[1 - \frac{1-\gamma}{2\gamma+1}S\right] \tag{6a}$$

$$D_\parallel = \langle D\rangle\left[1 + \frac{2(1-\gamma)}{2\gamma+1}S\right] \tag{6b}$$

where in terms of the velocity $v(t)$, diameter d, and length L of rod-like molecules, one has

$$\langle D\rangle = \tfrac{1}{3}\int\langle v(t)\,v(0)\rangle\,\mathrm{d}t \tag{6c}$$

$$\gamma = D_\perp^0 / D_\parallel^0 \simeq \frac{\pi d}{4L} \tag{6d}$$

Hence the predicted macroscopic anisotropy ratio σ is independent of $\langle D\rangle$, and for $S = 1$ is equal to $(\pi d)/(4L)$; for $S = 0.6$ with $d/L = 3$, Eq. (6) gives $\sigma = 0.5$. Relations equivalent to Eqs. (6a) and (6b) have been reported by Leadbetter et al. [10]. In smectic phases the final expression for γ becomes more complicated and lengthy, since a periodic potential barrier, hindering the diffusion process perpendicular to the smectic layers (i.e. $D_\parallel$) must be included [9–12]; this is often done by using an additional Boltzmann-factor for $D_\parallel$ in Eq. (6b), so that despite the fact that $d < L$, observed ratios σ and γ even considerably larger than unity can be explained by such a modification of the theoretical concept. Hess et al. [13] have more recently developed expressions rather different from Eqs. (5a, 5b) by an affine transformation of the isotropic diffusion law

for hard spherical particals to an affine space with aligned, nematically ordered uniaxial ellipsoids (long axis a, short axis b), which leads to

$$D_\perp = D_0\,\alpha\left[Q^{-\frac{2}{3}} + \frac{1}{3}Q^{-\frac{2}{3}}(Q^2-1)(1-S)\right] \tag{7a}$$

$$D_\parallel = D_0\,\alpha\left[Q^{-\frac{4}{3}} - \frac{2}{3}Q^{-\frac{2}{3}}(Q^2-1)(1-S)\right] \tag{7b}$$

with

$$\alpha = \left[1 + \frac{2}{3}(Q^{-2}-1)(1-S)\right]^{-\frac{1}{3}}$$
$$\times\left[1 + \frac{1}{3}(Q^2-1)(1-S)\right]^{-\frac{2}{3}} \tag{7c}$$

$D_0 = D_{\mathrm{iso}}$, and $Q = a/b$. Equations (7a) and (7b) imply that $\langle D\rangle = D_0$ for $S = 0$ and any value of Q; the main distinctions between Eqs. (6) and (7) arise from the dissimilar dependence on the geometrical parameters Q and γ.

A quite different model for nematic diffusivities, based on Oseen's hydrodynamic theory of isotropic liquids, was elaborated by Franklin [14]; it describes the diffusivity components in terms of the five Leslie viscosities α_1 to α_5, a scalar friction constant ξ, two geometric molecular quantities [14] μ and ϕ, and the order parameter S by the Franklin relations

$$D_\perp = kT\left[\frac{1}{\mu\zeta} + \right. \tag{8a}$$
$$\left. + \frac{5+S}{12\pi\mu^2\phi(-2.75\alpha_2 + 2.25\alpha_3 + 6\alpha_4 - 1.5\alpha_5)}\right]$$

$$D_\parallel = kT\left[\frac{1}{\mu\zeta} + \right. \tag{8b}$$
$$\left. + \frac{2-S}{12\pi\mu^2\phi(4\alpha_1 + 1.5\alpha_2 + 6.5\alpha_3 + 6\alpha_4 + 10\alpha_5)}\right]$$

where k is Boltzmann's constant and T is the absolute temperature. Equation (8) obviously extends Einstein's diffusivity–viscosity law [4] of isotropic liquids.

13.2.3   Model Refinements

Because of the numerous, poorly known parameters these theories and reported modifications [6, 15] or extensions for *solute* molecules [6, 12, 16, 17] and *lyotropic* systems [18] are mainly qualified to calculate the anisotropy of the diffusion constants, whereas estimations of their absolute magnitudes and temperature dependences need material data that are usually not available. Therefore it is quite common to describe experimental results by simple or, with the free-volume concept, modified heuristic Arrhenius laws [19]:

$$D_i = D_i^0 \exp\left(\frac{-E_i}{kT}\right) \qquad (9a)$$

or

$$D_i = D_i^0 \exp\left(\frac{-F_i}{T-T_g}\right)\exp\left(\frac{-E_i}{kT}\right) \qquad (9b)$$

where i is either $\perp$ or $\parallel$; as usual D_i^0 denotes the pre-exponential factor, E_i is the activation energy, F_i is the free volume parameter, T is the absolute temperature, and T_g is the glass temperature of the thermally activated diffuse mass transport. It should be noted that either Eqs. (6), (7), or (8) exactly predicts this type of behavior, but over small mesophase ranges the deviations from Eqs. (9a) and (9b) are, as a rule, minor and hence difficult to detect experimentally. However, in view of the shortcomings of the model it is interesting to note that the more recent theoretical developments have shifted to a large degree to *molecular dynamics* (MD) and *Monte Carlo* (MC) simulations [13, 15, 20–22], where the geometry of the molecules (rods, disks, ellipsoids), and the shape of the interaction potentials (Lennard–Jones, Coulomb, Gay–Berne) are more easily controlled by means of powerful computers than in the model approaches.

13.3   Experimental Techniques

Measurements of the diffusion constants of liquid crystals involve difficulties not encountered for normal liquids or solids, as in liquid crystals there is both anisotropy and viscous flow. As a consequence, data obtained with different techniques often differ greatly, and so in recent years researchers have sometimes preferred to test the quality of the models using results from molecular dynamics simulations [13, 15]. The first investigation of solute diffusion in a nematic mesophase was reported by Svedberg [23] as early as 1917, and the first detailed review on thermotropic mesophases was presented in 1982 by Krüger [24]. A review on lyotropic systems by Lindblom and Orädd [25] followed in 1994. One distinguishes between *direct* methods, where mass transport is studied directly by observing the evolution of the diffusion profile according to Eq. (3), and, for more the practicable initial conditions of Eq. (1), *indirect* methods, where such profiles govern other, usually more accessible system quantities [4, 5, 24]. In order to illustrate the main problems, these methods are briefly described below.

13.3.1   Tracer Techniques

Tracer techniques directly measure mass transport by means of putting adequate labels on some molecules, which are initially positioned with well-defined geometry on the sample surface to allow the observation of the propagating diffusion profile. In liquid crystals this has been achieved by radioactive tracers (i.e. isotope labels) or optical tracers (i.e. color labels). Suitable radioactive tracers can be realized by replacing ^1H by ^3H or ^{12}C by ^{14}C. Optical tracers (dyes, impurities, solutes) make use of molecules of similar size as the host liquid crystal (e.g. *m*-nitrophenol in *p*-azoxyanisole) [23]. Both approaches involve problems, namely the synthesis of a suitable radiotracer and the availability of a dye which really measures self-diffusion and not solute diffusion. Therefore, few results based on radioactive labels have been reported [23, 26, 27], and optical studies [28–31] are often considered not to show self-diffusion correctly. If the radiotracer or dye concentration profiles are measured as a function of time, penetration depth, and orientation for specific tracer injections, the diffusion constants can be evaluated by using the pertinent solution of Fick's laws (e.g. Eq. (3)). Several refinements have been used to broaden the limits of such procedures with regard to the profile analysis, and to take into account isotope or mass effects [24].

13.3.2 Quasielastic Neutron Scattering

Quasielastic neutron scattering (QENS) is a rather indirect method with many limitations. It makes use of the small ('quasielastic') energy shift that neutrons experience in any scattering by a moving particle, say by the diffusive translations of protons on a molecule. Mathematically, the normalized scattered neutron intensity as a function of kinetic neutron energy E (or frequency $\omega \equiv 2\pi E/h$) is related to the time Fourier transform of the dynamic pair-distribution function $G(r, t)$ of the sample material [6, 32]. Hence in Fick's approximation the linewidth increase $\Delta\omega(q)$ of the outgoing scattered beam for a selectable momentum transfer $\Delta p \equiv h q/(2\pi)$ with wavevector q, compared with the ingoing monochromatic beam linewidth [32]

$$\Delta\omega(q) = 2[D_\| q_\|^2 + D_\perp q_\perp^2] \qquad (10)$$

in principle allows one to determine the two diffusion constants. Provided the instrumental resolution is sufficient for both energy and momentum, the diffusional broadening by Eq. (10) can be separated reliably from effects due to the generally much faster molecular rotations [33], and the beam geometry does not hinder the necessary director alignment. Most early QENS measurements did not check these conditions critically, but later work gave better results [34, 35] which are almost consistent with more directly obtained data.

13.3.3 Magnetic Resonance

Magnetic resonance (NMR, ESR) with nuclear (NMR) or electronic spins (ESR) provide a broad spectrum of direct and indirect techniques for diffusion studies of pure liquid crystals and solute molecules, and thus provide the most successful means of investigating these materials. Indirect methods consider the strongly model-dependent effects of diffusion on spin-relaxation times such as T_1, $T_{1\rho}$, T_{1d}, T_{1q}, and T_2, or on lineshape parameters such as the line-width $\Delta\nu$ or the second moment M_2 of suitable spin signals [36, 37]. This implies similar difficulties as with QENS, since any molecular motion induces signal changes, and an unambiguous assignment to $D_\perp$ and $D_\|$ proves problematical. Therefore many early papers reported incorrect results, as a rule too small diffusivities. Most problems have now been overcome by special nuclear spin–echo methods, which make it possible to observe directly the magnetization decay due to diffusion by applying strong magnetic field gradients, either stationary or pulsed. This requires an analysis of Fick's diffusion term for the spin magnetization M_T transverse to the applied NMR Zeeman field B_0 according to the generalized Bloch equation [36, 38]:

$$\frac{\partial M_T}{\partial t} = -\frac{M_T}{T_2} + \mathbf{D}\frac{\partial^2 M_T}{\partial r^2} \qquad (11a)$$

where T_2 is the transverse relaxation time of the spins. In order to get the components of $\mathbf{D}$, the field gradients $(G = \partial B_0/\partial r$ must be oriented parallel and perpendicular to the director n by suitable gradient coils and, to evaluate them reliably, the diffusion-damping term must dominate the relaxation damping. At present, it is still a big problem to realize such conditions satisfactorily [24, 25, 38]. Explicitly, the diffusion decay of a spin with a gyromagnetic ratio γ is detectable in a selected direction i if

$$D_i \gtrsim \frac{10}{\gamma^2\, T_2^3\, G_i^2} \qquad (11b)$$

which means that, in view of the typically short T_2 times (<50 µs) of mesophases, very strong gradients G_i (>1 to 10 T cm^{-1}) are necessary. Such 'brute-force' methods have recently been tried on solids [39], but they have not yet been used for liquid crystals where they would perturb the director field in a rather uncontrollable way. The presently used alternatives make use of lengthening the effective T_2 by a more or less sophisticated averaging of the dipolar interactions; for example, by adding poorly ordered solute molecules [24], by exciting special solid-like echos [24], by multiple radiofrequency pulses which reduce the relaxation term [40], or by orienting the director under the magic angle (54.74°) with respect to the Zeeman field [24, 41–43]. Results obtained in this way are illustrated in the next section.

Nematic phases give rise to more difficulties than smectic phases [24]. For chiral nematic (cholesteric) systems, where $\mathbf{D}$ may depend on the pitch length p, field-gradient methods are particularly problematical because of the coupling between p and B_0 [1, 2]. Thus several groups [44, 45] have described alternative procedures based on deuteron or proton spin lineshape calculations without the need for field gradients, which allow one to measure the diffusivity along the helical axis (i.e. $D_\perp$). However, in addition to this geometric restriction, the technique is rather indirect.

13.4 Selected Results

Mainly by using optical tracer and NMR pulsed-field-gradient techniques, the diffusivities of thermotropic and lyotropic materials have been studied more or less extensively in most of the known familiar liquid crystalline mesophases [24, 25, 38], including: nematic, cholesteric, and smectic (A, B, C, C*, G, H) order; cubic, lamellar, and hexagonal phaes; rod-like, discotic, and polymer molecules; biological systems; and often also for the related isotropic liquids. A large part of the data obtained has been collected and commented on in the designated reviews. However, in view of the large deviations between different experimental methods, and the rather minor distinctions between alternative models, the value of such numerous works in the understanding of the underlying processes is generally highly unsatisfactory, since only results for isotropic phases [38] (which, due to the liquid-like long T_2 instrumental NMR requirements, can be achieved easily) are consistent or undisputed. Some general findings and problems are outlined in the final sections of this chapter.

13.4.1   The Experimental Dilemma

Figure 1 illustrates, using the results of selected studies on the familiar nematogens *p*-azoxyanisole (PAA) [23, 46–50] and methoxybenzylidenebutylaniline (MBBA) [50–54], that the diffusion constants and their temperature dependences obtained by independent research groups and techniques differ significantly, particularly for the nematic mesophase where the greatest problems occur. Even results obtained using the so-called 'direct' techniques disagree well outside the specified error limits, obviously due to methodical problems. This should be kept in mind when comparing such data with subtle details of theoretical models, and when discussing distinctions between unalike materials or phases.

13.4.2   Nematic Mesophase

The NMR pulsed-field-gradient measurements given in Fig. 1 [50] for PAA and MBBA, and similar results for countless other materials [24, 42, 43] such as *n*-alkyl- or *n*-alkoxycyanobiphenyls (*n*-CBs, *n*-OCBs [43, 55], Fig. 2) in the nematic phase always reveal, within the error limits, a simple Arrhenius-type temperature behavior of both $D_\parallel$ and $D_\perp$, with only slightly different activation energies (20–70 kJ mol^{-1}), and so one obtains nearly temperature independent anisotropy ratios of 0.5–0.7. The transition to the isotropic phase is discontinuous, but not in the way expected according to $\langle D \rangle$. Such findings are in remarkable disagreement with many results obtained by radioactive and optical tracer and NMR T_1 methods (e.g. with optical measurements of 5-CB [56])

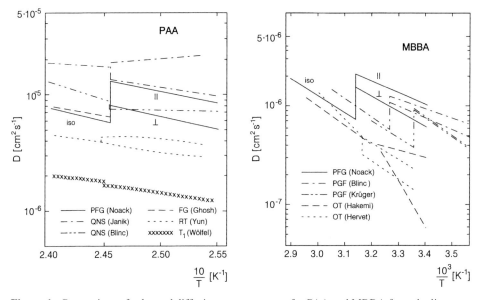

Figure 1. Comparison of selectred diffusion measurements for PAA and MBBA from the literature to illustrate the scatter of experimental data. The methods are explained in the text; data are taken from the following studies: Yun and Fredrickson [26], Noack et al. [42, 50], Blinc et al. [46, 52], Janik et al. [47], Ghosh and Tettamanti [49], Wölfel [48], Krüger and Spiesecke [51], Hakemi and Labes [53], and Hervet et al. [54]. Note that even the isotropic-to-nematic transition temperatures of the samples deviate considerably.

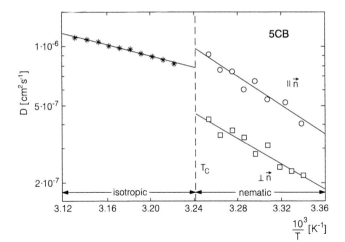

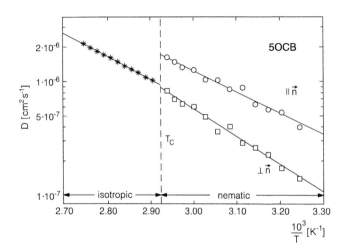

Figure 2. Arrhenius plots of pulsed-field-gradient diffusion constants [43, 55] of nematic and isotropic 5-CB and 5-OCB to illustrate the quality of the Arrhenius fits and the changes due to the slightly different side-groups of the two cyanobiphenyls. The activation energies for 5-CB are $E_{iso} = 29.6$, $E_\perp = 60.4$, and $E_\parallel = 69.7$ kJ mol^{-1}, and for 5-OCB $E_{iso} = 38.3$, $E_\perp = 46.4$, and $E_\parallel = 34.8$ kJ mol^{-1}. Neither the Arrhenius behavior nor the discontinuity at the phase transition can be described [43] satisfactorily by the available theoretical models [8, 13, 14]. Note that the major distinctions between the plots result from the different temperature ranges and clearing temperatures T_c.

and they also deviate from theoretical predictions [8, 13, 14] which would indicate a more visible influence of the non-Arrhenius variations upon the order parameter or the Leslie viscosities. In comparison with these strong inconsistencies, the distinctions between the various pulsed-field-gradient results, which must be ascribed to the different averaging procedures (by solute, multiple radiofrequency pulses, magic angle orientation, etc.), are relatively small. Some additional NMR lineshape studies indicate [45] a strong decrease in $D_\perp$ (i.e. of the transport along the helix axis) if the ordering of the phase is chiral.

13.4.3 Nematic Homologues

The literature contains very few systematic reports on the diffusion anisotropy of homologous molecules [35, 42, 43, 50] that allow a more critical analysis of the basic model parameters (diameter, length) than of chemically dissimilar systems. Figure 3 compares some results [42, 50] for members of the 4,4′-di-n-bialkoxyazoxybenzene (or PAA) series in both the nematic and the isotropic state, with the number $n' = n - 1$ of –CH$_2$ side-groups varying from 0 to 6. It can be seen that, compared at constant temperature difference ΔT relative to the individu-

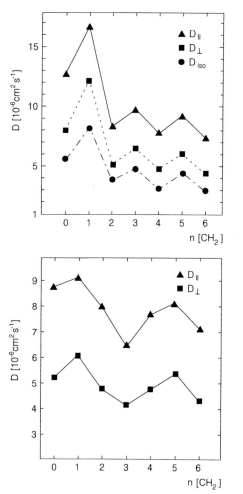

Figure 3. Diffusion constant alternation of seven homologous molecules in the n-bialkoxybenzene series ($n = 1$ to 7, $n' = 0$ to 6 –CH$_2$ groups) near the phase transition temperatures ($T = T_c \pm 0.5°C$) in the isotropic and nematic phases (top), and at constant temperature ($T = 119°C$) in the nematic phase [42, 50]. Note that the discontinuity at T_c ($D_\parallel > D_\perp > D_{iso}$) disagrees with theoretical predictions, and the alternation is opposite to that predicted by calculations using the Chu–Moroi model [8] (Eqs (6a) and (6b)).

al clearing points T_c, the diffusivity components show an odd–even alternation parallel to the clearing temperature, which changes to a $n' = 0$–3–6 alternation also exhibited by the nematic activation energies if compared at constant temperture T [50]. In the first case one has $D_\parallel > D_\perp$ and, surprisingly,

$D_\perp > D_{iso}$, independent of the alternation. Such data, which were later confirmed by QNS studies [35], and analogous results for several n-OCB homologs [43] ($n = 5$, 6, 7, 8) cannot be described by the reported theories [8, 13, 14], although the cyanobiphenyls show the predicted order $D_\parallel > D_{iso} > D_\perp$ if a more sophisticated field-cycling magic angle averaging is applied [43].

13.4.4   Smectic Mesophases

The most striking distinction from nematic diffusivities is that NMR pulse-field-gradient measurements [24, 56] of smectic mesophases show Arrhenius-type behavior of $D_\perp$ and $D_\parallel$ with strongly different activation energies, namely $E_\perp < E_\parallel$. As a consequencs, the two plots can cross [24], either within the smectic phase or outside by extrapolation, and with decreasing temperature the anisotropy ratio σ can change from <1, as for nematics, to ≫1. Values up to 10 have been reported for the SmA phase of homologous alkanoylbenzylideneaminoazobenzenes (e.g. C$_{12}$-BAA [56]), whereas in the more familiar SmA phase of terephthalylidene-bis(4-n-butylaniline (TBBA) one finds only $\sigma \approx 1$ due to the plot crossing [24]. Qualitatively, this has been explained using the Chu–Moroi concept [8, 9] assuming liquid-like diffusion within the layers ($D_\perp$) but solid-like jumps from layer to layer ($D_\parallel$). At phase transitions between different smectic order, the pre-exponential factors, activation barriers, and anisotropy ratio change discontinuously; no quantitative models are yet available.

13.4.5   Solute Diffusion

Despite some principal restrictions, it is generally easier to determine the diffusion coefficients of solute molecules [23, 24, 28–31] rather than true self-diffusion. However, the obvious distinctions have long been underestimated, and at present the two quantities cannot be related reliably. Solute diffusion constants do not follow the familiar relation of being reciprocally proportional to the square-root of the mass [5] (sometimes a strong concentration dependence is observed [57]) and theoretical estimations about the significance of the solute shape [16, 17] have not been examined sufficiently. Nevertheless, several findings are well established [24]. Due to the smaller order of the solute, the diffusion anisotropy σ in *nematic* systems turns out smaller than that of the host (i.e. near 1). On the other hand, for *smectic* phases, where the anisotropy depends primarily on the different activation potentials E_i for jumps within or perpendicular to the layers, σ becomes even larger than for self-diffusion. Values of σ up to 100 have been reported for TTE molecules (1,1,1-trichloro-2,2,2-trifluoroethane) in C_{12}-BAA [24, 56], which originate from the higher $E_\parallel/E_\perp$ ratio of the solute compared with the host. Usually, **D** values are smaller at phase transitions in the low-temperature phase [24], but the opposite jump is also observed [41].

13.4.6 Lyotropic Mesophases

Diffusivities of binary, ternary and multicomponent liquid crystalline mixtures, e.g. of soap (potassium laurate (PL), water [25, 58], and lipid (dipalmitoylphosphatidylcholine (DPPC) [25, 59] systems in lamellar, hexagonal, cubic, nematic and micellar mesophases [25, 60, 61] have been studied extensively by pulsed-field-gradient NMR [25] and optical techniques [62], partly because of their intimate relation to the structure and dynamical performance of biological membranes [18]. The main distinction from thermotropic phases is that for layered structures a noticeable diffusion occurs only *within* the layers (i.e. lateral, frequently written as D_L, but in our notation $D_\perp$), whereas it is negligibly small and difficult to detect *across* the layers [60–62] (transverse migration, for bilayers denoted by 'flip-flop'); so the mobility is essentially two dimensional, and the anisotropy ratio is so great that it is seldom specified explicit-

Table 1. Diffusion constants and anisotropy ratios of typical thermotropic and lyotropic liquid crystals considered in the text (Extensive data and references are collected in Krüger [24], Lindblom et al. [25], and Kärger et al. [38].

Material	Temp. (°C)	Phase	$D_\perp \times 10^{-11}$ m^2 s^{-1}	σ	Refs.
PAA	119	N	55	0.62	42, 50
MBBA	28	N	2.9–7.1	0.57	42, 43, 50
5-CB	30	N	2.8	0.43	43, 55
5-OCB	35	N	1.4	0.35	43, 55
TBBA	175	SmA	25	~1.0	24
	135	SmC	8.5	2.5	24
C_{12}-BAA	120	SmA	5.0	4.8	24, 56
	115	SmB	0.25	~2.5	24, 56
PL (28% H$_2$O)	80	Lamellar	24		25, 58
DPPC (25% H$_2$O)	52	Lamellar (L$_\alpha$)	0.5–0.9	>20	25, 59, 60

PAA, 4,4'-di-*n*-bialkoxy-azoxybenzene; MBBA, 4-methoxybenzylidene-4'-*n*-butylaniline; 5-CB, 4'-*n*-pentyl-4-cyanobiphenyl; 5-OCB, 4'-*n*-pentoxy-4-cyanobiphenyl; TBBA, terephthalylidene-bis(4-*n*-butylaniline); C_{12}-BAA, C_{12}-benzylideneaminoazobenzene; PL, potassium laurate; DPPC, dipolmitoylphosphatidylcholine.

ly. In the easy direction diffusion constants are of similar order of magnitude as in thermotropics, namely typically about 10^{-11} to 10^{-10} m^2 s^{-1}.

13.4.7   Selected Diffusion Constants

Table 1 lists a few pulsed-field-gradient measurements of diffusion constants for some familiar mesophases considered in the text. Extensive data have been collected by Krüger [24] and by Lindblom and Orädd [25]; further more recent references have been summarized by Kärger et al. [38].

13.5   References

[1] G. Vertogen, W. H. de Jeu, *Thermotropic Liquid Crystals, Fundamentals*, Springer Verlag, Berliln **1988**, 201.

[2] P. de Gennes, J. Prost, *The Physics of Liquid Crystals*, Clarendon Press, Oxford **1993**, 250.

[3] L. M. Blinov, V. G. Chigrinov, *Electrooptic Effects in Liquid Crystal Materials*, Springer Verlag, Berlin **1994**, 89.

[4] W. Jost, K. Hauffe, *Diffusion, Methoden der Messung und Auswertung*, Dr. Dietrich Steinkopff Verlag, Darmstadt **1972**, 1.

[5] H. J. V. Tyrrel, K. R. Harris, *Diffusion in Liquids, a Theoretical and Experimental Study*, Butterworths, London **1984**, 56.

[6] A. Ferrarini, P. L. Nordio, G. J. Moro in *The Molecular Dynamics of Liquid Crystals* (Eds. G. R. Luckhurst, C. A. Veracini), Kluwer, Dordrecht **1994**, 41.

[7] R. Blinc, M. Burgar, M. Luzar, J. Pirš, L. Zupančič, S. Žumer, *Phys. Rev. Lett.* **1974**, *33*, 1192.

[8] K. S. Chu, D. S. Moroi, *J. Phys. Colloq.* **1975**, *36*, C1-99.

[9] K. S. Chu, D. S. Moroi, *Mol. Cryst. Liq. Cryst.* **1981**, *67*, 109.

[10] A. J. Leadbetter, F. P. Temme, A. Heidemann, W. S. Howells, *Chem. Phys. Lett.* **1975**, *34*, 363.

[11] F. Volino, A. J. Dianoux, A. Heidemann, *J. Phys. Lett.* **1979**, *40*, L-583.

[12] G. Moro, P. L. Nordio, U. Segre, *Mol. Cryst. Liq. Cryst.* **1984**, *114*, 113.

[13] S. Hess, D. Frenkel, M. Allen, *Mol. Phys.* **1991**, *74*, 765.

[14] W. Franklin, *Phys. Rev., Ser. A* **1975**, *11*, 2156.

[15] S. Tang, G. T. Evans, *J. Chem. Phys.* **1993**, *98*, 7281.

[16] M. E. Moseley, A. Loewenstein, *Mol. Cryst. Liq. Cryst.* **1982**, *90*, 117.

[17] A. Kozak, D. Sokolowska, J. K. Mosciki, *Proc. ECLC*, Bovec, Slovenia **1995**, 178.

[18] R. M Clegg, W. L. C. Vaz in *Progress in Protein–Lipid Interactions* (Eds. A. Watts, J. J. H. H. M. De Pont), Elsevier, Amsterdam **1985**, 173.

[19] M. H. Cohen, D. Turnbull, *J. Chem. Phys.* **1959**, *31*, 1164.

[20] M. P. Allen, D. J. Tildesley, *Computer Simulations of Liquids*, Clarendon Press, Oxford **1989**, 182.

[21] A. Alavi, D. Frenkel, *Phys. Rev.* **1992**, *45*, R5355.

[22] G. Krömer, D. Paschek, A. Geiger, *Ber. Bunsenges. Phys. Chem.* **1993**, *97*, 1188.

[23] T. Svedberg, *Kolloidzeitschrift* **1918**, *22*, 68.

[24] G. J. Krüger, *Phys. Rep.* **1982**, *82*, 229.

[25] G. Lindblom, G. Orädd, *Prog. NMR Spectrosc.* **1994**, *26*, 483.

[26] C. K. Yun, A. G. Fredrickson, *Mol. Cryst. Liq. Cryst.* **1970**, *12*, 73.

[27] A. V. Chadwick, M. Paykary, *Mol. Phys.* **1980**, *39*, 637.

[28] F. Rondelez, *Solid State Commun.* **1974**, *14*, 815.

[29] H. Hakemi, M. M. Labes, *J. Chem. Phys.* **1975**, *63*, 3708.

[30] M. Daoud, M. Gharbia, A. Gharbi, *J. Phys. II (France)* **1994**, *4*, 989.

[31] T. Moriyama, Y. Takanishi, K. Ishikawa, H. Takezoe, A. Fukuda, *Liq. Cryst.* **1995**, *18*, 639.

[32] P. A. Egelstaff, *An Introduction to the Liquid State*, Clarendon Press, Oxford, **1992**, 217.

[33] J. A. Janik, *Acta Phys. Pol.* **1978**, *A54*, 513.

[34] J. Töpler, B. Alefeld, T. Springer, *Mol. Cryst. Liq. Cryst.* **1973**, *26*, 297.

[35] M. Bée, A. J. Dianoux, J. A. Janik, J. M. Janik, R. Podsiadly, *Liq. Cryst.* **1991**, *10*, 199.

[36] A. Abragam, *The Principles of Nuclear Magnetism*, Clarendon Press, Oxford **1961**, 265.

[37] J. H. Freed, A. Nayeem, S. B. Rananavare in *The Molecular Dynamics of Liquid Crystals* (Eds. G. R. Luckhurst, C. A. Veracini), Kluwer, Dordrecht, **1994**, 71.

[38] J. Kärger, H. Pfeifer, W. Heink, *Adv. Magn. Reson.* **1988**, *12*, 1.

[39] I. Chang, F. Fujara, B. Geil, G. Hinze, H. Sillescu, A. Tölle, *J. Non-Cryst. Sol.* **1994**, *172*, 674.

[40] R. Blinc, J. Pirš, I. Zupančič, *Phys. Rev. Lett.* **1973**, *30*, 546.

[41] S. Miyajima, A. F. McDowell, R. M. Cotts, *Chem. Phys. Lett.* **1993**, *212*, 277.

[42] G. Rollmann, Thesis, University of Stuttgart **1984**.

[43] J. O. Mager, Thesis, University of Stuttgart **1993**.

[44] N. A. P. Vaz, G. Chidichimo, Z. Yaniv, J. V. Doane, *Phys. Rev.* **1982**, *A26*, 637.

[45] R. Stannarius, H. Schmiedel, *J. Magn. Reson.* **1989**, *81*, 339.

[46] R. Blinc, V. Dimic, *Phys. Lett.* **1970**, *31A*, 531.

[47] K. Otnes, R. Pynn, J. A. Janik, J. M. Janik, *Phys. Lett.* **1972**, *38A*, 335.

[48] W. Wölfel, Thesis, University of Stuttgart **1978**.

[49] S. Ghosh, T. Tettamanti, *Chem. Phys. Lett.* **1980**, *69*, 403.

[50] F. Noack, *Mol. Cryst. Liq. Cryst.* **1974**, *113*, 247.

[51] G. Krüger, H. Spiesecke, *Z. Naturforsch. Teil a* **1973**, *28*, 964.

[52] I. Zupančič, J. Pirš, M. Luzar, R. Blinc, *Solid State Commun.* **1974**, *15*, 227.

[53] H. Hakemi, M. M. Labes, *J. Chem. Phys.* **1974**, *61*, 4020.

[54] H. Hervet, W. Urbach, F. Rondelez, *J. Chem. Phys.* **1978**, *68*, 2725.

[55] F. Noack in *Proc. AMPERE Summer Institute on Magnetic Resonance* (Eds. R. Blinc, M. Vilfan), J. Stefan Institute, Ljubliana, Slovenia **1993**, 18.

[56] M. Hara, S. Ichikawa, H. Takezoe, A. Fukuda, *Jpn. J. Appl. Phys.* **1984**, *23*, 1420.

[56a] G. J. Krüger, H. Spiesecke, R. van Steenwinkel, F. Noack, *Mol. Cryst. Liq. Cryst.* **1977**, *40*, 103.

[57] H. Hakemi, *Mol. Cryst. Liq. Cryst.* **1983**, *95*, 309.

[58] R. T. Roberts, *Nature* **1973**, *242*, 348.

[59] A. L. Kuo, C. G. Wade, *Biochemistry* **1979**, *18*, 2300.

[60] P. Ukleja, J. W. Doane in *Ordering in Two Dimensions* (Ed. S. K. Sinha), Elsevier, Amsterdam **1980**, 427.

[61] G. Lahajnar, S. Žumer, M. Vilfan, R. Blinc, L. W. Reeves, *Mol. Cryst. Liq. Cryst.* **1984**, *113*, 8592.

[62] R. Homan, H. J. Pownall, *Biochim. Biophys. Acta* **1988**, *938*, 155.

Index